Principal Events o

Modern age; end _____ First human societies; large scale extinctions o_____ species; repeated glaciation.

Appearance of humans; volcanic activity; decline of forests; grasslands spreading.

Appearance of anthropoid apes; rapid evolution of mammals. Formation of Sierra Mountains.

Appearance of most modern genera of mammals and monocotyledons; warmer climate.

Appearance of hoofed mammals and carnivores; heavy erosion of mountains; angiosperms and gymnosperms dominate.

First placental mammals; most modern angiosperm families develop.

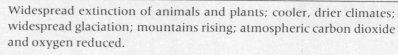

Appearance of monocots; oak and maple forests; first modern mammals; beginning of extinction of dinosaurs. Formation of Andes, Alps, Himalayas, and Rocky Mountains.

Appearance of birds and mammals; rapid evolution of dinosaurs; first flowering plants; shallow seas over much of Europe and North America.

Appearance of dinosaurs; gymnosperms dominant; extinction of seed ferns; continents rising to reveal deserts.

Widespread extinction of animals and plants; cooler, drier climates; widespread glaciation; mountains rising; atmospheric carbon dioxide and oxygen reduced.

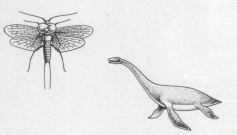

Appearance of reptiles; amphibians dominant; insects common. Gymnosperms appear; vast forests; great life abundant. Climates mild; lowlying land; extensive swamps; formation of enormous coal deposits. Many sharks and amphibians; large scale trees and seed ferns; climate warm and humid.

Appearance of seed plants; ascendance of bony fishes; first amphibians; small seas; higher, drier lands; glaciations.

Atmospheric oxygen reaches second critical level. Explosive evolution of many forms of life over the land; first land plants and animals. Great continental seas; continents increasingly dry.

Appearance of vertebrates, but invertebrates and algae dominant. Land largely submerged. Warm climates worldwide.

Atmospheric oxygen reaches first critical level. Explosive evolution of life in the oceans; first abundant marine fossils formed; trilobites dominant; appearance of most phyla of invertebrates. Lowlying lands; climates mild.

Life confined to shallow pools, fossil formation extremely rare. Volcanic activity, mountain building, erosion, and glaciation.

BIOLOGY
The World of Life

Fourth Edition

BIOLOGY
The World of Life

Fourth Edition

Robert A. Wallace
University of Florida

Scott, Foresman and Company
Glenview, Illinois   London, England

To Jayne and mornings and roses

Acknowledgments for illustrations and other copyrighted materials in this book appear on the page with the copyrighted material or in the Acknowledgments section at the back of the book, which is to be considered an extension of the copyright page.

Library of Congress Cataloging-in-Publication Data

Wallace, Robert Ardell
  Biology, the world of life.

  Includes index.
  Bibliography: p.
  1. Biology. I. Title.
QH308.2.W35   1987 574   86-24829
ISBN 0-673-16602-3

  45678—VHJ—9190898887

PREFACE

It seems like a long time since I wrote the first version of *Biology: The World of Life.* I've written other books since then, and in the intervening years, even as I wrote new editions, I've watched biology change—but never so fast and so dramatically, it seems to me, as in these past few years. Because our view of life and ways of seeing it have changed so swiftly, this edition of the book has departed more drastically from the original format than any other edition. Those who know the book from earlier editions will immediately see some of the changes, but there may well be a sharp ring of familiarity to it, too.

One thing I didn't change was my notion of how the story of biology should be told. I call it a story because that's just what it is. Furthermore, it's a good story. It's full of mystery, drama, humor, hostilities, friendships, luck, and surprises. It's a fascinating tale, rife with subplots, and one of my goals has been to let the tale be told. I'm totally at odds with those who, in the interest of sounding ''scientific,'' make biology stuffy or boring. Biology, after all, is the study of life, and life is, above all things, not boring. I am again prepared to be described as being a bit light-hearted and informal here and there. And sometimes I may have let an opinion through. However, I believe that bringing a topic to life and seeing the lighter side of the thing enhances learning. If my opinions or views should generate discussion and thoughtfulness, so much the better.

Whereas the style may not have changed much, the story and format have. The most noticeable change is probably in the general appearance of the book. (For example, it's a little bigger and almost all of the art is new, with full color throughout.) There are many more special essays on selected topics. And there are new chapters: Scientists and Their Science, Advances in Genetics, The Immune System, The Senses, and Human Populations. Several other chapters, such as reproduction, development, behavior, and those relating to the systems, have been reorganized or divided for more convenient handling.

It will also quickly become evident that the book is, in large part, about our own species. It addresses the human condition in two ways. First, it includes much more material on the structure and function of the human body than did previous editions. Second (perhaps more importantly), the latter chapters of the book emphasize the special position of the human animal among the other species of the earth. One might wonder why we consider such things as Mideast oil, South African gold, nuclear terrorism, and population planning when biology is supposed to be about how cabbages grow, where ducks live, and why leaves fall. We will indeed learn about cabbages and ducks and leaves, but we will also consider our increasingly special role in the pageant of life. If we are to keep this planet livable, we've got to learn about every facet of life, and the sooner the better. So let's begin here.

Acknowledgments

Although a book is said to be "by" its author, the author is really almost a symbol of a small army of people who helped get the book out. I am glad to have the chance here to mention some of those people who were involved with this project.

First, I would like to gratefully acknowledge the support of the General Manager of Scott, Foresman's College Division, Jim Levy. As they once said of Ed Sullivan, "No one knows how he does what he does, but he's the best there is at it." Jim's confidence in my projects is greatly appreciated.

Dick Welna, Editor-in-Chief, again lent his firm support, and he directed the project through often very delicate situations. I particularly enjoyed our lively conversations. He's one of the bright lights in this business, and I look forward to working with him again.

Jack Pritchard, Editorial Vice-President, Math/Science, was again a man for all seasons, which is a good thing because any book goes through several seasons. Jack showed his firm grasp of the publishing business by being able to effectively manage and contribute in a variety of ways as the situation changed.

Rebecca Strehlow, my editor at Scott, Foresman for the last two years, was the "hands on" person in charge for almost the entire project. She knows even the most arcane aspects of publishing, it seems, and was able to understand what was needed at each step and to "translate" from one developmental branch to another. Carrie Dierks stepped in to put the finishing touches to the developmental process.

The project editing and design people, working under the firm and experienced guidance of John Beasley, Director of Editorial Services, were responsible for coordinating an enormous amount of detail. This was effectively managed by the project editor, Marisa L. L'Heureux. This work greatly benefited from the caring and competent eye of Art Director Barbara Schneider, who often wore the concerned expression of a real "book person"—someone who knows excellence and demands it. The striking photographs you will see were uncovered by the diligent efforts of photo editor Mary Goljenboom. Advertising Manager Meredith Hellestrae constantly searched for new ways to get the word out. An author is fortunate to be able to work with someone with Meredith's attitude—she is strong-minded and receptive, simultaneously.

No book can go far without the strong support of those who actually take it to the readers. The sales and marketing division of Scott, Foresman is among the strongest in the business, thanks largely to the demanding and exacting nature of Carl Tyson. He is the first sales manager I've seen who checks the technical details of a text before lending his support to it. Marketing Manager George Duda is, well, George Duda. He is not only dedicated, but relentless. If he believes in a project, he simply cannot be swayed from it. I am grateful he believes in mine.

Any book such as this must go through countless reviews. Some were for very detailed content and some for style. The project has indeed benefited from the careful readings, comments, and suggestions of the following people.

Bonnie Amos, *Baylor University*

John E. Butler, *Humboldt State University*

William W. Byrd, *Arkansas State University*

Galen E. Clothier, *Sonoma State University*

Charles F. Denny, *University of South Carolina—Sumter Campus*

Gary E. Dolph, *Indiana University at Kokomo*

William Dunscombe, *Union County College*

William A. Emboden, *California State University, Northridge.*

Wayland L. Ezell, *St. Cloud State University*

Douglas Fratianne, *The Ohio State University*

John L. Frola, *The University of Akron*

Laurence Fulton, *American River College*

William Gnewuch, *Sacred Heart University*

Garland F. Hicks, Jr., *Valparaiso University*

Ronald D. Humphrey, *Prairie View A & M University*

Ross E. Johnson, *Washburn University of Topeka*

Charles Leavell, *Fullerton College*

William H. Leonard, *Clemson University*

Teresa, C. Minter-Procter, *Porterville College*

James Morrow, *Allan Hancock College*

Maria Nakamura-Chapin, *St. Petersburg Junior College (Clearwater Campus)*

John D. Pasto, *Middle Georgia College*

Peter Pedersen, *Cuesta College*

Richard G. Pendola, *Community College of Rhode Island*

Dennis K. Poole, *Gulf Coast Community College*

Robert C. Romans, *Bowling Green State University*

Orlando A. Schwartz, *University of Northern Iowa*

Dr. William R. Sigmund, *Slippery Rock University*

Joseph Stevenson, *St. Louis Community College at Florissant Valley*

James L. Strayer, *Washtenaw Community College*

William J. Thieman, *Ventura College*

David E. Youker, *Sauk Valley College*

It is always a pleasure to work with such able scientists as Robert Andersen, who helped us through many a sticky problem with detail, especially in some of the more innovative art (even though, as he lamented, we never got around to his specific areas of expertise).

I should add that my efforts have been greatly supported by the personal encouragement and expectations of my great friend, Jim Cook.

I would also like to thank my colleagues here at the University of Florida. It's good to be able to call on the expertise of such a helpful group of competent people.

Finally, I would like to thank my long-time coauthor Jerry Sanders. He helped just as diligently on this project as on those with his name on them—a stalwart friend and one of my favorite people.

Robert A. Wallace

CONTENTS

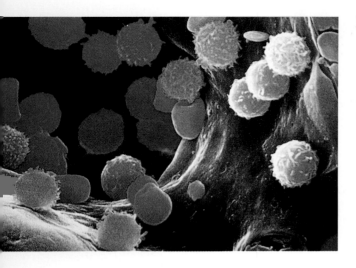

PART FIVE

SYSTEMS AND THEIR CONTROL *327*

Part One

LIFE AND THE FLOW OF ENERGY

Chapter 1

A BRIEF HISTORY AND THE ENCHANTED ISLES

Struggling and writhing, the young man lay on his back, his arms and legs flailing at the cool Argentine air. It was a most curious sight. Yet, there he was, twisting convulsively on the ground, dust and bits of grass flying around him. Who could have guessed that this lad, only months earlier, had been walking his dog down an English lane on his way to have lunch with his wealthy uncle? Furthermore, who could have known that this same young man would, years later, throw the entire Western world into an anxious turmoil? And why was he now rolling on the ground?

Curious eyes watched from the Argentine pampas. They had never seen anything like this before. Some animals darted for cover, but one large bird, overcome with curiosity, stepped closer for a better look. As the bird drew nearer, young Charles Darwin, aglow with good health and a crack shot to boot, leaped to his feet, snapped his rifle to his shoulder, and dropped the curious bird on the spot. That night the crew of the *Beagle* would dine on well-prepared rhea.

FIGURE 1.1 Charles Darwin at the age of 29. After Darwin's return, he and Emma lived in London for a while, but soon moved to Down House in a tiny community in the rolling British countryside. At right are the Cambridge botanical gardens as they appeared when Darwin was a student there.

FIGURE 1.2 Rheas are known for their vile personalities as much as for their obsessive curiosity.

This was not the most curious thing that Darwin had ever done. His shipmates, in fact, had come to expect strange things from this bright and energetic young man with his notable strength and amazing endurance. He had climbed hill after hill, just to be the first Englishman to view the magnificent beyond. He relentlessly collected all sorts of plants and animals and would go sleepless, working night and day, to retrieve some fossil before the ship again set sail. His shipmates had marveled at his strength, and their admiration increased when he once saved a group that had become stranded without supplies. He was happy and exuberant these days, but, in time, this was all to change.

But what was young Darwin doing in South America? After all, he was of the British upper class, given to riding and hunting, good food, and an occasional rousing game of blackjack. He was well-liked and not unusual in most respects, but he had not always pleased his father.

Darwin's father was a huge, commanding man. He had tried to send Charles to medical school when the boy was sixteen. But Charles couldn't stand the sight of blood and almost fainted while first witnessing surgery. Then young Darwin tried the clergy. He spent three years at Cambridge studying theology, but even then he spent much of his time wandering around the countryside, building his amateur naturalist collections. Indeed, he seemed to be more concerned with beetles than beatitudes. He was not alone in his interest, however. At that time, the English countryside was alive with amateur butterfly collectors, rock hounds, and plant fanciers whose position and wealth permitted them to indulge in such hobbies. Even so, Darwin's academic prowess had been so thoroughly unremarkable that at one point, his father had told his trifling son, ''You care for nothing but shooting, dogs, and rat catching, and you will be a disgrace to yourself and all your family.''

THE VOYAGE OF THE *BEAGLE*

It must be admitted that at this point in Charles Darwin's career there was little to suggest that he had a mind that was to be regarded as one of the most brilliant and inquisitive in history. In a short time, however, this diffident young man was to hear, through his friend the Reverend John Henslow, of an offer of free passage on a survey ship called the *Beagle* (Figure 1.3). A naturalist was needed for a voyage that was to last five years. There would be no pay, and the person chosen would have to sleep in a hammock in the cramped chart room (although he would be permitted to share the captain's table). Armed with Henslow's recommendation, Darwin eagerly applied for the position, but was nearly rejected because of the shape of his nose. Captain Fitzroy, himself only twenty-three, believed that the nose reflected the character of its bearer, and Darwin's nose just didn't show much character.

Darwin's family required some persuasion to accept this ''madcap scheme,'' which they considered scarcely suitable for a prospective clergyman. And evidently, Charles had his own trepidations about such a momentous decision. In a letter to his sister Susan he wrote:

> Fitzroy says the stormy sea is exaggerated; that if I do not choose to remain with them, I can at any time get home to England; and that if I like, I shall be left in some healthy, safe and nice country; that I shall always have assistance; that he has many books, all instruments, guns, at my service. . . . There is indeed a tide in the affairs of men, and I have experienced it. Dearest Susan, Goodbye.

FIGURE 1.3 The *Beagle*, a solid wooden ship about 25 meters long, was one of the royal navy's several vessels used to chart foreign waters and to carry explorers and scientists over the world in Britain's quest for greater power and increased trade.

Ultimately all the arrangements were made, and in 1831, the H.M.S. *Beagle* set out from Devonport with Charles Darwin, now a ship's naturalist, gazing, perhaps a bit apprehensively, at the slowly retreating shoreline of his homeland. As the heavy wooden vessel (which was about the size of a tugboat) creaked and groaned its way across the Atlantic toward South America, Darwin's worst fears were realized. Shipboard life was tougher than he expected, but worse yet, he tended to get seasick! He made the best of it, though, as the boat continued relentlessly, captained by tough, young Fitzroy, perhaps one of the world's best navigators. At long last, the *Beagle* reached South America and headed down the coast on the first leg of its voyage—sailing past the coasts of Brazil and Argentina, weaving through the terrible pounding gales of Cape Horn, and finally turning northward along the desolate coasts of Chile and Peru.

Fortunately, there were periods of respite when the *Beagle* dropped anchor to put foraging parties ashore. Darwin wasted no time getting to land. After resting a bit and regaining his land legs, he was irresistably drawn deeper into these new places—places that harbored all manner of new and fascinating things. Darwin took copious notes on everything he saw and, following his natural urges, he brought back to the ship all manner of things, much to the amusement and occasional dismay of the crew. Since the ship sometimes remained anchored for months, he ventured far inland into the wild South American terrain. Darwin, an excellent rider himself, soon developed the greatest respect for the riding skills of the rugged gauchos who often accompanied him.

To understand just what this adventure meant to Charles Darwin (and would later mean to the world), we must stop at this point to consider the state of scientific thinking at that time. What was known of the biological world? What prejudices or beliefs did Darwin have? More important, what seeds of ideas? What hunches?

THE HISTORY OF AN IDEA

New scientific ideas have appeared throughout human history. Many of these ideas have been erroneous, of course, but they often provided at least a framework for the expansion of our knowledge. Nevertheless, new ideas—and even new facts—have not always been met with enthusiasm. In the best of times, when an assumption did not fit what was known about the real world, the assumption was discarded. The notions that remained, then, were likely to be based to some extent on the available facts, and as new information came to light, our body of scientific knowledge expanded.

It would be satisfying somehow to say that scientific knowledge progressed stepwise from ancient times until the present, knowledge and understanding accumulating all the while until we reached the modern crescendo of scientific expansion, all to the good of humankind. Alas, this is not the case.

As far back as the early civilizations of Babylonia, Egypt, and Greece, scholars were trying to figure out the nature of the world in which they lived. Some of these early scientists were looking for some unifying concept, some thread of continuity. However, after the fall of these ancient civilizations, scientific curiosity steadily declined. For a period of more than ten centuries, between A.D. 200 and 1200, there were virtually no scientific advances at all. In fact, much of what had been known was forgotten. Although the Greek mathematician Eratosthenes had calculated the circumference of the earth to within fifty miles more than 200 years before the birth of Christ, in 1492, Columbus had trouble convincing anyone except a few Moorish astronomers that the world was round!

During these centuries, the writings of a few ancient scholars were preserved in dusty monasteries where they were dutifully mulled over and transcribed from one parchment to another. But instead of active investigation to verify statements or ideas, people turned to religious authority for absolute answers, seeking some sort of order in a world that was beyond their understanding. As a result, science and religion became hopelessly intertwined, so the "scientific" statements that were otherwise testable became religious doctrine and were, therefore, not to be questioned. Thus, by a twist of fate and politics, the church became, on one hand, the seat of higher learning and, on the other hand, a formidable opponent of new ideas. The Polish astronomer Copernicus (1473–1543), who voiced the heretical theory that the earth was not the center of the universe, escaped retribution by dying shortly after his work was published. Later, when the great Galileo (1564–1642) produced detailed evidence that the earth did indeed revolve around the sun, his writings were banned in Rome and he was summoned before the Inquisition and forced to recant his belief in the Copernican theory. (The Catholic Church only recently forgave Galileo.)

Nevertheless, as the world became more predictable, and people began to feel less at the mercy of magical forces and unseen beings, there was a great surge of scientific thinking. In fact, the very year Galileo died marked the birth of one of the greatest scientists of all time—Isaac Newton (1642–1727). By the age of twenty-four, Newton had already formulated the idea of universal gravitation, and in 1685, he presented a set of carefully structured laws of motion that were to revolutionize the physical sciences.

THE BEGINNINGS OF BIOLOGY

Until the eighteenth century, science was largely limited to topics dealing with the inanimate, such as mathematics, astronomy, and physics. The study of living things was largely exempt from such investigation because of the philosophical and religious influences of that time. After all, it was one thing to search for the physical laws that described salts, stones, or stars, but quite another to probe at the essence of life. Implicit in the reverence for life was the notion that since people are living things, life must surely have some special purpose, some grand design. People simply weren't ready to see themselves as just another physical phenomenon.

In addition, scientists of that day believed that all species, or kinds, of living things were created in their present form—in other words, that they had not changed during their time on earth. This was certainly the view of the Swedish botanist Carl von Linné (1707–1778) (see Figure 1.4) who devised a system of classification for all living organisms, naming them in Latin. In his fondness for Latin, he even called himself Carolus Linnaeus.

FIGURE 1.4   The Swedish naturalist, Carolus Linnaeus, the founder of the scientific method of naming living things. He apparently conceived of species as unchangeable entities, the products of divine creation. He lived and worked in the century before Darwin. Here, also, are four plant species named by Linnaeus.

(A) Arizona century plant *(Agave americana)*

(B) Ox-eye daisies *(Chrysanthemum levcanthemun)*

(C) Mountain laurel *(Kalmia latifolia)*

(D) Wild strawberries *(Fragaria vesca)*

FIGURE 1.5 Today, biologists often associate the name Lamarck with error. Lamarck, we know, *was* wrong, but we must keep in mind that, in Darwin's time, he was a major intellectual force. He was an early evolutionist and, furthermore, he believed that humans themselves had evolved. However, he also believed that the use of organs would ensure their transmission in strengthened form to the next generation. He was most ridiculed in the popular press for asserting that giraffes, by continually reaching for higher leaves, would tend to have offspring with yet longer necks. Somehow, he argued, the experiences of an individual could influence hereditary material and, hence, alter future generations.

One small departure from this idea was suggested in 1753 by Linnaeus' French contemporary, George-Louis Leclerc De Buffon (1707–1788). In 1753, Buffon proposed that in addition to those animals that had originated in the creation, there were also lesser families "conceived by Nature and produced by Time." He explained that changes of this kind were the result of imperfections in the Creator's expression of the ideal.

Interestingly, a decade later, someone else joined the doubters of the fixity of species. This turned out to be Erasmus Darwin (1731–1802), the grandfather of Charles. Erasmus was a peculiar fellow, not only a physician but an amateur naturalist who wrote about botany and zoology, often in rhyme. In his ramblings, he referred almost incidentally to certain relationships among animals that, strangely, heralded a number of important ideas that would later be embraced by Charles. Among these were the importance of competition in the formation of species, the effect of environment on changes in species, and the heritability of these changes. As you might expect, there has been a great deal of speculation on the extent to which Charles may have been influenced by Erasmus. However, the influence may not have been very great because Charles apparently never had much use for the musings of his grandfather.

Other people of that era were also beginning to toy with the notion of the changeability of species and the heritability of those changes. In France, Jean Baptiste de Lamarck (1744–1829) (Figure 1.5), a protégé of Buffon's, boldly suggested that not only had one species given rise to another, but that humans themselves had arisen from other species. A passionate classifier, Lamarck believed that every organism has its position on the "scale of nature"—with humans, of course, firmly at the top, thereby revealed as the highest form of life. Lamarck also made the important observation that the fossil animals found in older layers of rock seemed to be somewhat simpler than those in more recently deposited rock. This difference suggested to him that the older ones had gradually given rise to the more recent ones (or become "higher"). He couldn't immediately account for how such changes might have arisen, but he finally surmised that there was some "force of life" that caused an organism to generate new structures or organs to meet its biological needs. Once formed, such structures continued to develop through use, and their development in the parents was inherited by the offspring. In this way, he said, the structure became perfected in succeeding generations.

Lamarck's most famous example of such change was his theory of how the giraffe developed a long neck. He maintained that the long neck evolved as each generation of giraffes stretched their necks in an effort to reach the topmost branches of trees. He argued that this effort altered the animals' hereditary materials so that a longer neck was passed along to the offspring (see Figure 1.10). This example did not bring him great respect. In fact, it brought guffaws of derision and even today it is cited as a classic case of scientific error.

Whereas Lamarck's arguments did little to persuade his lecture audiences, he did create some lively discussions in intellectual circles. Society at large, however, held to the firm conviction that each form of life had arisen through special creation. The matter seemed settled. The English were more interested in discussing the French Revolution. They were still discussing it when Charles Darwin was born.

TIME AND THE INTELLECTUAL MILIEU

The intellectual climate of England was far more conservative than that in France when Charles Darwin was born. The English had been horrified by the brutality of the French Revolution, and any ideas held by the "French atheists" were usually either dismissed out of hand or viewed with extreme suspicion. Consequently, the church continued to hold strong sway over the sciences in England.

But as everyone knows, the English are an irreverent lot, and even a small group of Englishmen is likely to have at least *one* rebellious soul. And among the British scientific community there were undoubtedly those who simply didn't care *what* the French were doing, and who, furthermore, believed that life changed on a changing planet. It has even been suggested that this sentiment was by now so strong that it set the stage for Darwin's theory, that his idea could not fail in such an environment, that the time was right.

However, this was apparently not the case. In fact, even after Darwin had published *Origin of Species* (Figure 1.6) and the idea was catching on, he observed:

> It has sometimes been said that the success of the *Origin* proved that the subject was in the air, or that men's minds were prepared for it. I do not think that this is strictly true, for I occasionally sounded out a few naturalists, and never happened to come across a single one who seemed to doubt about the permanence of species. . . . I tried once or twice to explain to able men what I meant by Natural Selection, but signally failed.

FIGURE 1.6   Interior of the Old British Museum, Montague House, where some of the most influential scientists worked. It was into this environment that Darwin introduced his controversial ideas in the *Origin of Species*.

THE ORIGIN OF SPECIES

BY MEANS OF NATURAL SELECTION,

OR THE

PRESERVATION OF FAVOURED RACES IN THE STRUGGLE
FOR LIFE.

By CHARLES DARWIN, M.A.,

FELLOW OF THE ROYAL, GEOLOGICAL, LINNEAN, ETC., SOCIETIES;
AUTHOR OF 'JOURNAL OF RESEARCHES DURING H. M. S. BEAGLE'S VOYAGE
ROUND THE WORLD.'

LONDON:
JOHN MURRAY, ALBEMARLE STREET.
1859.

DARWIN'S THEORY OF EVOLUTION

When Charles Darwin set out on the voyage of the *Beagle,* he had no quarrel with the prevailing notion that life had originated through special creation and that species were fixed in form. Furthermore, he was aware that many scientists felt that the goal of science should be twofold: first, to discover how nature worked and, second, to use the findings to demonstrate the wisdom of the Creator. Darwin hoped to join in the parade of scientists and clergy as they marched arm in arm to produce a better world. Instead, he was to reluctantly whistle the parade to a stop.

His questions about life did not leave him alone among scientists. The physical sciences were less hampered by religious tenets, and there were signs of rumblings and stirrings in these disciplines, disturbances that would one day make it easier for some to fall into step with Darwin as he slowly marched away from the parade. Among the physical scientists who were also marching to a different drummer and developing bold new ideas about the earth was Charles Lyell (1797–1875) (Figure 1.7), himself only a few years older than Darwin. Lyell had set forth many of his new ideas in his book *Principles of Geology,* the first volume of which was published before the *Beagle* set sail. Darwin had acquired a copy and had asked to have the second volume sent to him en route. In his book, Lyell (who was to become a close friend of Darwin) rejected the religionists' view that the earth had been created in the year 4004 B.C. Lyell contended that it was much, much older. Since the earth was old, he theorized that its most dramatic features had not appeared suddenly through catastrophic upheavals, as had been thought, but were instead the result of slow, steady processes that occurred over an exceedingly long period of time. Darwin immediately seized on the implications. Lyell argued that there was *time*—time, Darwin would come to believe, in which things could change. But the idea was not yet ripe. Darwin was toying with only a seed of an idea. He had been thoroughly schooled in traditional beliefs, and he was not inclined to reject them.

While Darwin continued to divide his time aboard the *Beagle* between reading and hanging over the rail, the heavy, wooden boat creaked and groaned across the Atlantic and made its way up the west coast of South America.

To escape his constant seasickness, Darwin wasted no time getting ashore at every chance. And he wasted no time once ashore. He collected what he could and made notes about everything. For example, he made notes about the South American plant and animal life and the fossil beds he found there. Quite early on he was struck by how living things could vary so markedly from one place to the next. He collected shells from the Atlantic shore and found that they were not like those picked up on the Pacific beaches. He wrote about how birds and mammals differed from one place to the next. He noted that in some cases, species changed gradually, one type giving way to another almost unnoticeably. But in other cases, one kind of organism would suddenly disappear, another having appeared in its place.

The Galapagos Islands

When the *Beagle* finally left the shores of South America, an impatient Captain Fitzroy, concerned with completing charts, measuring harbors,

FIGURE 1.7 Charles Lyell, perhaps the greatest geologist of Darwin's time and a close friend of Charles'.

FIGURE 1.8 Darwin's finches include the peculiar woodpecker finch that uses twigs much as a woodpecker uses its beak to extract insects.

and preparing the way for British commerce, set the sails of his sturdy craft (a dog to windward) for a straight run to the Galapagos, a chain of islands lying about 580 miles off the coast of Ecuador. It was these miserable islands that history would most closely associate with an unsuspecting Darwin.

When the anchor clattered into the shallow waters of St. Stephen's harbor of the island the British called Chatham, Darwin scrambled ashore as usual, but as soon as he had looked around, he was almost ready to leave. Chatham was rough, crude, and barren, and he didn't like it. But as he began to explore the place, he encountered a very strange, and even fascinating, assortment of animals. "A little world in itself," he wrote, "with inhabitants such as are found nowhere else." There were lizards three feet long, grazing on seaweed beneath the turbulent sea—in Darwin's words, they were "imps of darkness, black as the porous rocks over which they crawl." And he saw the giant tortoises that had for years been captured by seafarers to be stacked upside down on the decks of their boats where the hardy beasts could somehow survive for months, providing fresh meat on the long voyages.

Darwin was more interested in the plants than the animals and soon spent most of his time "botanizing" over the dry and barren islands. He was struck by the strange animals of the islands, however, and collected not only the marine lizards, but also their land-bound cousins further inland as well. He also collected various kinds of birds, many of which he was convinced were undescribed species. Among these were a number of finches (Figure 1.8). They were a motley group, but definitely finches, some not unlike finches he had collected on the South American mainland.

At first, he put all the specimens from the islands into a single group for storage on the boat. Then one day while he was examining a few of the finches, he noticed that two of them taken from different islands differed in the size and shape of their bills. This struck him as odd. The importance of his observation was driven home as he was walking the four miles to the settlement of political outcasts, banished from Ecuador, who had been sent to Charles Island. His companion that day was the acting British governor of the island who informed Darwin that he was able to tell from which island any of the tortoises came. He explained that they differed, for example, in the size and shape of their carapace (shell) and in the length of their extremities. Darwin wondered if each island was somehow producing its own forms of creatures, and from that day on, he carefully separated his collections from each island. This was to prove a critical decision once he was home in England. Years later, Darwin had his finch collections examined by British specialists and it was decided that there were thirteen species, differing primarily in the size and shape of their beaks. Darwin surmised that these birds must have come originally from the South American mainland, since the volcanic islands of the Galapagos would have been formed later than the continent. But why were these birds so different from those on the mainland, and why did the assortment on each island differ so much from the one on the next? Twenty years would pass before Darwin came up with an answer.

As the voyage continued, Darwin continued not only his collecting, but also his questioning of how things came to be. His letters and observations were reaching England and had the scientific community anxiously awaiting more of his findings. In fact, he was told that upon his return, he would be invited to take his place among the British scientific establishment. The letters waiting at various ports of call encouraged him and sent him bounding into the hills of each new land, his hammer joyously ringing against the rocks.

When Darwin accepted the position as ship's naturalist he had expected to be gone about two years, but five years were to pass before his return. When the *Beagle* finally made its way back to England, Darwin was greeted enthusiastically by his family, the scientific community, and his dog. His reception among the scientists of the day was a warm one, and immediately the questions began. What had he seen? What had he brought back? A new phase of work had just begun.

The Impact of Malthus

Darwin was grateful to be back among his friends, his new colleagues, and his books. After a flurry of activity, he married his cousin, Emma Wedgewood, and retreated to the country and began to enjoy the quiet mornings when he could find time to work. In his reading, he came across an old essay by the Reverend Thomas Malthus (1766–1834) that was probably the first clear warning of the dangers of human overpopulation. In the essay, which appeared in 1798, Malthus pointed out that populations tended to increase in a geometric progression, and that if humans continued to reproduce at the same rate, they would inevitably outstrip their food supply and create a teeming world full of "misery and vice." The message may have been theological, but Darwin applied the idea to his own work and concluded that species have a high reproductive potential, but that not all individuals reproduce, because of differences in their survival abilities. The idea was that populations are kept in check in part because not all animals survive long enough to reproduce.

Malthus' paper set Darwin to thinking. He calculated that even a pair of elephants, notoriously slow breeders, could produce 19 million progeny in only 750 years. Yet it seemed that through the years, the number of elephants on the earth stayed about the same. Something was obviously interfering with their reproductive output. But was that "something" exerting an equal effect on all elephants, or did individuals differ in their ability to reproduce? Were some less successful at leaving offspring? And if this were the case, Darwin wondered what factors determined which ones were to be successful.

NATURAL SELECTION

Darwin's answer came to him in part because of his background as a country gentleman. He was familiar with the principles of **artificial selection;** he knew that through careful selection of animals for mating, breeders were able to accentuate desired characteristics in the offspring. For example, by mating only the offspring of the greatest milk producers, breeders could develop high-yield dairy cattle. And by carefully selecting only the offspring of good laying hens for breeding, they could eventu-

The Galapagos Islands

On Captain Fitzroy's charts, the Galapagos were only a group of islands to be visited as the *Beagle* left the coast of South America. Years later, his friendship with Charles Darwin would be severely strained largely because of what young Darwin would find in this forsaken place. Darwin at first thought the islands were strange and repugnant. But as he roamed about, he discovered places of indescribable beauty. He found plants and animals unlike those anywhere else. More important, though, he found creatures somewhat like those he had seen on the mainland; a finding that would one day help launch his grand theory.

All the higher Galapagos Islands (over 600 meters) show vegetational zones. The dry coastal belt turns green only during the January—May rainy season, while the higher areas, marked by fog and drizzle throughout the year, may have several distinct zones. At the lowest level are the remarkable mangroves (below), rooted in seawater and supporting a great variety of life. The volcanic origins of some of the islands are revealed by calderas, such as this one on Fernandina Island (right). A variety of life forms live in such sheltered places, many in close association with the mineral lakes found there. The arid coastal zone (bottom) is marked by cactus (particularly opuntia) and a variety of shrubs adapted to arid conditions.

The highland zone of Santa Cruz Island (above) bears the heavy, varied vegetation typical of humid, tropical areas. Transitional zones, such as this area on Marborough Island (below), harbor a mixture of adaptive types where lowland desert forms meet those from the lush mountain heights.

Galapagos tortoises (above) from different islands have distinctive appearances. Generally,
the higher the arch over the neck, the farther
the great reptile must reach for its favorite
food, the opuntia cactus. They are the islands'
only native grazer. The tortoises are found
wild in only a few places today. Most of them
have retreated to the great volcanic calderas
(right) where they may find solitude and
pools of water to cool their great bulk. Sally
Lightfoot crabs (below right) lend splashes of
brilliant color to the dull volcanic coastlines.

Among the most bizarre creatures on the islands are the marine iguanas, large lizards that graze on the seaweed just offshore (above left). Their rarer cousins, the land iguanas (above), are found further inland. The waved albatross (left), one of the world's greatest fliers with a wingspan of 2.5 meters, nests only on Hood Island in the Galapagos. The flightless cormorant (bottom) has followed the evolutionary pattern of many island forms; it has lost the ability to fly. Its feathers are wettable to facilitate diving, and after each fishing trip it must stand and dry itself.

Whereas some birds nest only at certain times of the year, the blue-footed booby (above) will rear a brood whenever conditions are good. If the fish supply should shift away from the islands, however, the blue-footed boobies do not hesitate to abandon eggs and young to follow the food. Its cousins, the red-footed booby and masked booby (above right), forage farther at sea, diving into the water at great speed and catching fish in their serrated bills. The Galapagos sea lions (below right) are quite approachable and even friendly (except for protective bulls with harems). Fur seals, on the other hand, have been relentlessly hunted and are now quite skittish.

A male great frigate bird (left) has managed to entice a white-throated female from a flock wheeling above the madly-displaying males as a group of them inflate their red throat pouches and coo enticingly. Swallowtail gulls (below) breed in great numbers on volcanic sea cliffs of the Galapagos. They are the only nocturnal gull, catching squid and fish from the darkened sea. When Ecuador claimed the islands in 1832, settlers soon arrived and brought with them a host of animals, such as goats, dogs, rats, donkeys, and pigs. Such introductions have sometimes proven disastrous to the natural life forms, many of which have become extinct.

The Galapagos hawk (above) is an unusual species, with many traits of both hawks and vultures. The Galapagos penguins (above right) are the northernmost members of the group. It is astonishing to watch the speed with which they move underwater in search of fish.

Areas of the islands are covered with ropy pahoehoe lava (below right), or with sharp and spiny lava beds that are virtually impassable. The remarkable origins of these islands are largely responsible for the startling array of life there.

FIGURE 1.9 Breeding programs based on artificial selection have produced plants and animals with extreme traits. Here, the obese mouse is two-and-a-half times heavier than normal mice. The rhino mouse (far right) first develops a normal coat, but, after two weeks gradually goes bald and wrinkled. The same methods have produced species of particular usefulness to humans.

ally produce hens that were veritable egg-laying machines. The results of such artificial selection could be seen in only a few generations (Figure 1.9).

Darwin envisioned some sort of process of **natural selection** analogous to the artificial selection imposed by the breeders. However, he suggested that in nature, the animals that would reproduce were selected by the environment. He suggested that natural selection would be far less efficient than artificial selection since individuals with only *somewhat* less-desirable characteristics might be able to produce at least *some* offspring, and thus their traits would take longer to disappear from the population. (Of course, those individuals with traits totally out of keeping with their environment would leave no offspring at all; hence, those traits would more quickly disappear from the population.) On the other side of the coin, the traits of those individuals with some reproductive *advantage* could be expected to *increase* through the generations.

Natural selection, then, came to be defined as the process through which those traits that encourage survival and reproduction are perpetuated and favored in a population as they replace those traits that are less reproductively advantageous.

Descent with Modification

Since the notion of natural selection depended on inequalities among members of a population, the source of the variation, the inequalities, had to be accounted for. Why are the individuals different? Darwin proposed that such variations appear randomly—that no driving force, no direction, and no design are necessary. He decided that if some new variation provided an advantage that increased the reproductive output of its bearer, it would spread through future populations. This meant that if the long neck of the giraffe is inherited and is helpful in acquiring food, then the giraffes with longer necks will be better nourished, and thus will be more likely to have the energy to leave offspring. Among

(A) Lamarck's Explanation.   (B) Darwin's Explanation.

FIGURE 1.10 Lamarck believed that acquired traits were inherited. For example, he suggested that giraffes, by continually stretching for higher leaves, would tend to have offspring with yet longer necks. In time, he explained, all giraffes would have long necks, as we see in the column at the left.

Darwin's theory, on the other hand, suggested that from among any population of giraffes, some are born with longer necks than others and that these would be able to reach more leaves. Better-fed animals would then most likely be able to rear their young, and thus longer necks would come to predominate in subsequent generations, as we see in the second column. We should probably keep in mind that when Darwin set sail on the *Beagle*, he himself was a Lamarckian, at least in broad principle.

FIGURE 1.11 Alfred Russel Wallace, a young adventurer and field biologist who arrived at many of Darwin's conclusions and goaded Charles into publishing, developed his own thesis on evolution by natural selection.

these offspring, he suggested, some will have longer necks than others, and these, in turn, will be more successful than their shorter-necked brothers and sisters.

Since long-necked giraffes would be favored through the generations, the result would be a general tendency for any generation to be composed of animals with longer necks than those of any preceding generation. (The differences in Lamarck's and Darwin's explanations of long necks in giraffes are shown in Figure 1.10.)

It is important to realize that Darwin developed his idea not only in the face of withering opposition, but also without hard proof. There was no experimental evidence he could offer, and to make things worse, he knew almost nothing about the field we now call genetics. If he had understood what was going on in genetics across the Channel in the garden of a monastery (see Chapter 8), he would have been able to save himself much grief. Because of such problems, he was reluctant to present his theory. He felt he should be able to explain the mechanism by which variation appeared.

Then, in 1858, something happened that goaded the thoughtful Darwin into action. Darwin received an unfinished paper from a young biologist working in Indonesia, Alfred Russel Wallace (1823–1913) (Figure 1.11). Wallace asked for Darwin's opinion of its merit. The paper was, in effect, a sketchy outline of the principles of natural selection. By this time, Darwin had already planned to present his theory of natural selection to the Linnaean Society of London, and, startled by the letter, and at the urging of his friends, Lyell, the botanist Joseph Hooker, and the scientific philosopher Thomas Huxley, he began to hasten his work.

Hooker and Lyell agreed to assist Darwin, whose young son had just died, by editing and condensing an earlier paper of his with a letter he had written the American botanist, Asa Gray, describing his developing principle of natural selection. Although Darwin feared he might be "scooped" (in science, being first often means being foremost), he asked that Wallace be permitted to present his paper first and receive credit for the idea, rather than have anyone think he had behaved in a "paltry spirit." The outcome was that the papers were read at the same meeting, with Darwin's presented first, in keeping with his much more substantial evidence. The papers were presented in July and published in August 1858. Darwin then went furiously to work, putting aside his idea of a huge monograph describing his theory. Instead, he completed an "abstract" of the idea: the *Origin of Species,* which was published in 1859. The first edition sold out on the first day.

Darwin's carefully formulated ideas were greeted with enthusiasm in some quarters, but, needless to say, the response was not universal. He was forced to defend his idea of "descent with modification" not only against scientists who demanded hard evidence, but also against the attacks of philosophers, theologians, and a general public who thought the idea was heretical.

The battle grew and was soon full blown. Darwin himself was poorly equipped for such a fight. He had fallen ill upon returning to England after his journey, and he never recovered his health. He had become a dedicated family man, spending a great deal of time with his wife and children. He continued his experiments and observations, but because of weakness, he generally could only work from mid-morning until noon

FIGURE 1.12A An aged Darwin, often in ill health, about a year before his death.

FIGURE 1.12B Darwin's study at Down House.

(see Figure 1.12). His infirmities have been diagnosed time and again by medical historians who first suggested that they were psychosomatic but who now seem to agree that Darwin may have contracted Chagas' disease by once allowing himself to be bitten by a benchuga bug.

While Darwin was developing his theory, he became plagued by anxiety and self-doubt. He and Emma lived a rather quiet and somewhat reclusive life at Down House in the country, and Darwin became even more withdrawn when the noisy debate over natural selection started. However, he had formidable defenders. Many of the best minds of the time leaped to the defense of the grand idea—brilliant, hard-nosed, and combative souls who savored the taste of intellectual battle. Among his most brilliant defenders was his long-time friend, the great debater Thomas Huxley (Figure 1.13). (When Lord Wilburforce, a clergyman, facetiously asked Huxley in a debate if he was related to the apes on his father's or his mother's side, Huxley is reported to have muttered, ''The Lord hath delivered him into my hands.'' He then won applause by saying that he would rather be related to an ape than to a man who

FIGURE 1.13 Thomas Huxley, a bright, energetic, and aggressive friend of Darwin's, who proved to be one of his greatest defenders.

refused to use his God-given powers of reason.) Of course, Darwin's defenders did not join the fray empty-handed. Here, after all, was a unifying concept, one that made sense of it all. It accounted for the observations. It was not to be rejected on any basis other than a better explanation. And there was none.

SUMMARY

Charles Darwin set out on the *Beagle* as a naturalist convinced of the fixity of species. The seed of the idea of evolution was planted while on this five-year voyage, but its growth was difficult due to the Victorian environment of nineteenth-century England. New ideas have traditionally been reluctantly received, and new ideas in science, in particular, have been subject to extraneous influences. Gradually, however, the rules of doing good science have emerged. Nonetheless, Charles Darwin's reluctant presentation of his theory of evolution by natural selection met with strong protest as well as grateful acceptance. His data were forged from years of hard fieldwork aboard the *Beagle*, and his premises were carefully thought out. Yet the theory required the support and defense of some of the most able scientists in England.

Chapter 2

SCIENTISTS AND THEIR SCIENCE

Charles Darwin was well aware that his theory of natural selection would offend a great many people, but he did not expect criticism of his methods. Yet it was his methods that drew some of the harshest attacks.

Darwin had both the fortune and misfortune to set forth his theory at a time when intellectual England had become interested in the philosophy of science, particularly in deciding how science should be done. The question was, what is the best way to gather valid information? There were two major schools of thought on the subject and, essentially, the disagreement was over how one arrives at scientific conclusions.

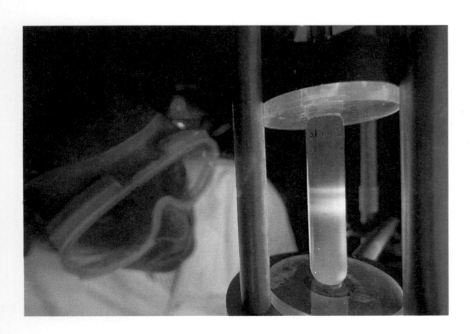

INDUCTION AND DEDUCTION

According to one school of thought, science should be done inductively. The **inductive method,** in fact, had been promoted by none other than Isaac Newton (Figure 2.1). To use the inductive method, one first gathers simple, empirical data and, from this, arrives at a generalization. Ideally, there is no goal, no premise, no hunch, and no preconception in the mind of the researcher. The inductive method works because finally, from the sheer weight of unprejudiced evidence, some general statement emerges.

But there are other ideas about how good science should proceed. The Dutchman C. Huygens, a physicist and lens grinder (who set forth the wave theory of light in opposition to Newton's particle theory), was a firm proponent of the **deductive method.** He argued that science does not always permit the absolute certainty that other fields (such as geometry) do. Huygens believed that observation or experience should suggest some general idea from which specific statements can be *derived*. The consequences, as well as the components, of the generalization can then be tested. In deductive reasoning, then, the generality is arrived at by some insight or hunch or, especially, by inductive reasoning, and the specific cases suggested by the large idea become the subject of experiments.

Almost any scientist these days feels free to get at the truth by either method, and most agree that perhaps the best science is done by employing both methods. For example, if a body of facts suggests some general principle (inductive), then specific propositions can be drawn from that principle (deductive) and tested. However, most scientists, from Darwin's day to this, probably rely more strongly on inductive evidence in developing scientific principles.

FIGURE 2.1   Isaac Newton, one of the truly important figures in the history of science. Newtonian physics was to dominate the world of the physical sciences until the twentieth century. It was Newton who, in 1687, provided final and irrefutable evidence for the Copernican doctrine of the solar system.

Science Trends and Shifting Logic

A problem in modern science is that it tends to be a bit trendy. Unfortunately, this has been true for a long time. In Darwin's era, to be "scientific" meant to operate like the physicists, with fixed laws and great precision. In fact, to be sure that he was conforming with the rules of good science, Darwin sent a draft of his paper on natural selection to one of the great physical astronomers of the time, John Hershel. He then wrote to Lyell that he was most anxious to have Hershel's opinion. It was not long in coming. Darwin later wrote Lyell:

> I have heard, by a round-about channel, that Hershel says my book is the law of higgledy-piggledy. What this exactly means I do not know, but it is evidently very contemptuous. If true, this is a great blow and discouragement.

But later, Darwin was greatly pleased when another authority on how science is done, and one of the greatest philosophers of the time, John Stuart Mill, said that Darwin's reasoning was "in the most exact accordance with the strict principles of logic."

TABLE 2.1   An Example of Inductive and Deductive Reasoning:

Inductive Reasoning

Observation: Coral atolls are usually composed of a circle of islands.

Observation: Coral atolls are formed from the deposits of living animals.

Observation: Animals without direct access to fresh seawater tend to die.

Observation: The interior of an atoll seems to be comprised of sunken coral.

Generality: Coral atolls are formed as the coral animals secrete deposits. The center, lacking nutrient-laden water, dies and sinks, leaving a ring that, in turn, breaks apart to form a circle of islands.

Deductive Reasoning

Generality: Coral atolls are formed as the coral animals secrete deposits. The center, lacking nutrient-laden water, dies and sinks, leaving a ring that, in turn, breaks apart to form a circle of islands.

Deduction: Coral atolls will have a sunken center.

Deduction: Coral animals need contact with fresh seawater.

Deduction: Seawater contains something that coral animals need.

Deduction: Coral atolls comprised of more nearly complete rings of land are probably recently formed.

But which logic? In keeping with the times, Darwin tried to adhere to the inductive method, and he said as much when he referred to a "general truth" in a letter to Thomas Huxley:

> I have got fairly sick of hostile reviews. . . . I entirely agree with you, that the difficulties of my notions are terrific, yet having seen what all the reviews have said against me, I have far more confidence in the *general truth* of the doctrine than I formerly had.

In spite of Darwin's confidence in the inductive method, he, too, occasionally lapsed into philosophical inconsistencies. He wrote in a letter to Joseph Hooker (the first person in whom he had confided regarding his developing theory) that he looked upon "a strong tendency to generalize as an entire evil," but later admitted, "I cannot resist forming one [generalization] on every subject."

In fact, some of Darwin's most insightful work was done deductively. For example, in developing his compelling theory about how coral reefs are formed he once stated, "No other work of mine was begun in so deductive a spirit as this; for the whole theory was thought out on the west coast of South America before I had seen a true coral reef." Table 2.1 illustrates how the two methods might have been used in this case.

THE SCIENTIFIC THEORY

Darwin's theory of natural selection is occasionally criticized as being "only a theory." However, many critics seem to forget that *theory* in science is a relatively lofty position. What, then, does the term mean?

Perhaps we can best describe a theory by showing how one can be developed. Suppose someone comes up with an idea—one that explains certain observed phenomena in nature. At first, it is regarded as just that, an idea. But after it has been carefully described and its premises precisely defined, it may then become a **hypothesis**—an idea that can be tested. In a sense, the hypothesis is the first part of an "if . . . then" statement. The "then" predicts the result of the hypothesis, so one can know by testing if the hypothesis is sound. A hypothesis can also stand as a provisional statement for which more data are needed. If rigorous, carefully controlled testing supports the hypothesis, more confidence will be placed in it, until it finally gains the status of a theory. The theory itself, however, may remain unproven and unprovable. A hypothesis, then, is a possible explanation to be tested, whereas a **theory** is a more-or-less verified explanation that accounts for observed phenomena in nature.

The reception a theory receives may depend on the intellectual milieu into which it is introduced. The theory of evolution, for example, caused an initial shock in the Victorian world but was assimilated by the intellectual community with surprising rapidity. The idea was accepted not only by intellectuals, but by much of the general populace as well. One possible reason for its widespread acceptance is that the deeply religious citizens of that era saw in evolution a preexisting program for change, a program authored by the deity.

The acceptance was not universal, of course. The theory had strong opposition then, even as now. There have always been those who point-

As the sophistication of science has grown over the past few centuries, so have the rules surrounding the experiment. These days, the **control** is an integral part of experimentation, largely to take the experiment out of the realm of simple observation. In a controlled experiment, the crucial factor, sometimes called a **variable,** is altered or left out. The results are compared to those generated when the variable is left in. Thus the effects of the variable itself can be determined.

Some evolutionists have argued that humans are descended from aquatic animals. One line of evidence is that we seem to have a vestigial *diving response*. To illustrate, if you submerge your face in a bowl of cold water you will not only gain the full attention of your little brother, but, in addition, your pulse rate is likely to drop. The slowed heart rate is associated with a decreased rate of metabolism. A diver with such a response could be expected to stay submerged longer.

Now, to test the reflex, the pulse of a subject could be taken normally, and then again with the face submerged. However, is it the water stimulating the receptors of the face that causes the reduced heart rate, or is the change due to something else? For example, could the bending over alone have caused the response? Controls can help answer the question in some cases. Here, a control might be to check the pulse while placing the face into an empty bowl. But how about the possibility that the lowered heart rate is due to either holding one's breath or holding one's breath while bending over? How could the experiment be altered to control for these effects?

ed to the theoretical nature of the evolutionary argument in an effort to deride the idea. This argument, however, is rather selectively applied. For example, no one has ever seen a hydrogen atom, and the behavior of hydrogen is indeed theoretical, but we can apply the theory and make water. The principles of evolution obviously strike stronger emotional chords than do the principles of chemistry.

THE SCIENTIST AS SKEPTIC

Scientific knowledge can be expanded not only through the development of hypotheses and theories or by carefully controlled experimentation (see Essay 2.1: The Controlled Experiment), but also by simple observation. For example, consider how much we learn from witnessing the birth of a volcano, from monitoring the results of an exploding star, or from simply watching wild animals (see Figure 2.2). Observations of this sort are indeed interesting, but the most reliable scientific evidence is often based on repeated testing and relentless experimentation. It is from experimentation that we generally gain our hardest evidence.

Scientists unrepentantly demand hard evidence. However, their rigor does not mean that they are eminently rational people, pristine, pure of heart, unemotional, and unfettered by personal prejudice. No scientist I know is, but, fortunately, such sterling character is not necessary. Important contributions have stemmed from hunches or personal prejudices, or from the work of someone with an ax to grind. In any case, for whatever reason those data are generated, the system works simply because most good scientists are skeptics. They are disbelievers at heart (and many love their hard-nosed image). Furthermore, because of their skepticism, any idea, theory, or experiment is certain to be attacked by someone. You can rest assured that the ideas will be examined, the

FIGURE 2.2A Much of what we know in science is due to simple observation. In some cases, there is nothing else we can do, such as when we witness the birth of a volcanic island. Yet the information is valuable. Can you see the problems with information derived from pure observation?

FIGURE 2.2B In some cases, the observer can interact with the observed. Dian Fossey told us a great deal about mountain gorillas by being able to move among them. The arrogant, tough, and dedicated Fossey was instrumental in protecting these shy beasts from poachers (who sell their hands to wealthy Europeans to be used for ash trays) until she was killed by an assailant in her mountain cabin.

theories weighed, and the experiments repeated. Alternative explanations will be suggested, and these alternatives can be met by a variety of responses from applause to hoots of derision.

Because so many scientists love to flex their minds in what they may regard as the ultimate game (albeit an important one), we often end up with good science. The system works, and we are usually left with ideas that have withstood such trials. In such an arena, simple observation has both a great disadvantage and a great advantage. The disadvantage is that a simple observation may simply be rejected, but not always by everyone and not always entirely. Pure observations are most likely to be taken seriously when the observer has an established reputation. (Unfortunately, one often hears the best stories from other sorts of observers—why don't those UFOs ever pick up a physicist?) In any case, though experimentation is fine, there are some very useful scientific ideas that have been developed simply because someone was there and paying attention.

Even with our carefully developed and well-groomed procedures for doing science, some of the most fundamental questions do not easily lend themselves to our probing. Perhaps the most basic and perplexing of these is the deceptively simple question, "What is life?"

DEFINING LIFE

It turns out that questions about the nature of life are often based as much in philosophy as in science (as we will see in Chapter 3). In fact, since the waters so quickly become muddied with disagreement, perhaps the best we can do is note a few of the most important characteristics of life. None alone can define life, but together, they form a composite that generally sets living things apart from the nonliving (Figure 2.3). With this in mind, we can say that (1) many living things *show movement*. If you're watching something and it moves, your first assumption might be that it's alive. In fact, among animals, both the hunter and the hunted often use movement as an indicator of life. (2) Living things *metabolize and grow*. Generally, living things gain mass, to some degree, over time. The building material for the new mass comes from molecules that are manufactured (as with turnips) or acquired (as with wolves) and then reorganized into new kinds of structures in the living organism. (3) Living things *reproduce*. We will see later that life is not necessarily involved with reproducing itself, but with reproducing its kinds. (4) Living things *respond*. Plants generally respond slowly, as when a potted petunia turns its leaves toward the window. A more rapid response is seen by stepping on a cat's tail. The essential point is that organisms must have ways of reacting to the changing environment of an active world. (5) Living things *evolve*. Just as life responds quickly to immediate conditions, it also changes through the generations to meet long-term shifts in condition. Such changes are most easily detected in the geological record left by fossils.

You may have noted that there are some problems in defining life. After all, rivers move, oil droplets reproduce, crystals grow, a still pond is responsive, and mountains evolve. And what about viruses? They can crystallize. Can a crystal be alive? Nonetheless, should you encounter something that has even a few of these traits, it had best be considered alive—especially if it has teeth. Obviously, this list is by no means exhaustive. What other traits could be added? And what other exceptions to these can you think of?

The Trap of Teleology

The simple word "to" has caused a great deal of confusion in many areas of science. For example, consider the phrase, "Birds migrate southward to escape winter." The statement seems harmless enough, but if interpreted literally, it implies that the birds have a goal in mind, or that they are moving under the directions of some conscious force that compels them to escape winter. Philosophers have termed such assumptions *teleology*. (*Teleos* is Greek for *end* or *goal*.) It is commonly used in reference to

FIGURE 2.3   Living things generally exhibit five characteristics that, together, set them apart from the nonliving. (A) Living things show movement, like the long, graceful strides of a mountain lion chasing its prey. (B) Living things metabolize and grow, as can be seen in the five-day growth of a daffodil from budding to blooming. (C) Living things reproduce, resulting in new generations, such as this hatching Laysan albatross. (D) Living things respond, as this cat illustrates with its ability to turn in midair to land on its feet. (E) Living things evolve, as evidenced by this fossil insect, which is a transition species representing a stage in the evolution of ants from waspy ancestors.

(A)

(C)

(B)

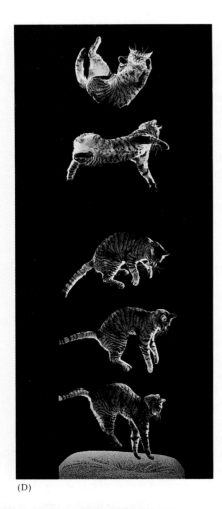

(D)

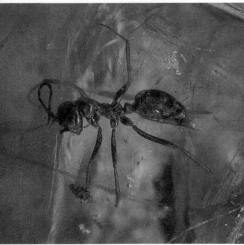

(E)

ideas that go beyond what is actually verifiable and generally implies some inner drive to complete a goal or some directing force operating above the laws of nature.

Teleological statements are rampant in the scientific literature today, partly from carelessness, but partly because scientists may feel free to communicate among themselves by using such phrasing, a kind of shorthand, when they are sure that their message will be interpreted correctly.

A biologist might say "Birds migrate southward to escape winter," (sentence 1) when he or she really means, (2) "by migrating southward, birds escape winter," or to be more precise, (3) "birds that migrate southward as winter approaches tend to escape the harsh seasonal weather of their northern summer homes." Do you see the several refinements in sentence (3) that make it more accurate than sentence (2)?

Science and Phenomenology

If you've never had the experience of looking up a word in a dictionary and coming away more puzzled than you were to begin with, try *phenomenology*. So, with apologies to my philosopher friends, let me just say that phenomenology encompasses the notion of the observer's own interpretation of an observation or body of information. Even science, then, becomes personalized, as observations are described in terms of the scientist's own experience. And, if so, biology must be the most phenomenological of all the sciences. Part of the reason is that life is so variable and changeable. Hard and fast rules are few, and we find exceptions at every turn. Each time we draw up a new rule, some beastie lurking somewhere chuckles "Not me!" Since there are so many unusual cases in the realm of life, the biologist has the obligation and opportunity to exercise choices and to make decisions based, however subtly, on a personal set of values. The biologist is thereby forced to interpret much of what he or she sees, and the interpretation is always colored by one's own tendencies, wishes, hunches, or proclivities. Just as the distinct flora of the baker's hands flavor the bread, so does the biologist interject something of himself or herself into the science, and when biases are at odds with the findings, data tend to become suspect and arguments mount. It is just such arguments that help make biology fun and keep biologists honest.

WHAT BIOLOGISTS DO FOR A LIVING

The word *biologist* may cause some to conjure up the image of a balding little man with bad posture and a squeaky right shoe, padding through aisles of dusty books on the trail of some ancient description of an extinct lizard. Others may visualize a butterfly chaser with thick glasses—the one who couldn't get a date for the prom—net poised, leaping gleefully through the bushes. Others may think of biologists as bird-watchers in sensible shoes, peering through field glasses in the cold, wet dawn in hope of catching a glimpse of the rare double-breasted seersucker. Maybe such images do fit some biologists, but there are others who search for

FIGURE 2.4 Scientists have, in recent years, become more aware of the social implications of their work, partly because of the confrontational behavior of activist groups both within and outside of the scientific community.

the mysteries of life in other places, such as in clean, well-lit laboratories amid sparkling glassware.

A few years ago it was said that a biologist is one who thinks that molecules are too small to matter, a physicist is one who thinks that molecules are too large to matter, and anyone who disagrees with both of them is a chemist. Perhaps it is true that the biologist whose scope is limited to molecules must periodically be convinced of the existence of the platypus, but no one else has been able to tell us how tiny hummingbirds are able to make it across the Gulf of Mexico. Fortunately, the sharp lines of division between such disciplines are becoming blurred and indistinct as scientists become more broadly trained and able to handle more kinds of ideas in their continuing effort to solve the "Great Puzzle."

SCIENCE AND SOCIAL RESPONSIBILITY

The question often arises, what is the social responsibility of informed scientists? Precisely what are their obligations to the rest of us, if any? Should their interests lie primarily in the continuing accumulation of more data, or are they morally obligated to help apply their findings to human problems (see Figure 2.4)? The citizen might well ask, should researchers, funded with public money, pursue their own research interests, or should they be encouraged, by law if necessary, to direct their attention to solving our immediate problems? Who should decide where the money goes? You may one day have a voice in such decisions, even if by a simple vote. Should that day come, will you know enough?

Another question is, should scientists with access to important information, "uncommon knowledge," put down the pen, shut off the computer, and step onto the soapbox? Are they obligated to tell us what they know about our problems and to suggest solutions? Or should they, as scientists, continue to generate more data and to seek yet more information? Is the information of greater value than the opinion? Furthermore, can concerned scientists assume that if the information is made available, an informed public will make the effort to understand it and to behave wisely (Figure 2.5)? (Unfortunately, even in an "enlightened" society such as ours, the general public somehow seems to remain abysmally ignorant of even the most pressing scientific questions.)

> How much radiation is safe? Can nuclear power plants explode? What is a meltdown? What sorts of meats reduce the risk of cancer? Of heart attack? Why is sunburn dangerous? What is the greatest risk of nuclear war? Should one take salt tablets if one is physically active in summer? What is CPR? What exactly is AIDS? What is nuclear winter? What does El Niño have to do with food prices? What is the Oglalla aquifer?

Some people have difficulty understanding that there is a problem. After all, breakfast this morning was warm and on time, and the traffic lights worked. The room is comfortable, and our afternoon is planned. But keep in mind as we trip lightly through these pages that in spite of feelings of complacency or security, many of these concepts are not simply academic ones that may appear on some test. Some of the things we will encounter here may have a fundamental bearing on the quality of your daily life, and, perhaps, how long you will live. You may one day be

FIGURE 2.5 The problems associated with certain scientific technologies are being vividly driven home to more Americans each year. As the public becomes more aware of such problems, there is an increasing pressure on industry to become more socially responsible. Here, crews clean up dioxin-contaminated materials at Times Beach, Missouri.

forced to make some difficult, even excruciating, decisions. Of course, you may choose to hope for the best and to ignore questions of the long run in favor of attending to immediate comforts, and who could blame you?

If one is interested in great problems these days, there is no reason to search farther than the human condition; it has the ring of a predicament. In fact, we hear that the people inhabiting this planet are, quite simply, in serious trouble. There is no need to recite the list of our problems. Why should we burden ourselves with the effects of burgeoning populations, hunger, resource maldistribution, nuclear wastes, human wastes, farming crises, dangerous air, fouled waters, dwindling mineral supplies, and international hostilities? Why should we even mention such things? After all, it's a nice day. There are things to do, people to meet, places to go . . . and perhaps an earth to save.

SUMMARY

Charles Darwin used both inductive and deductive reasoning. Inductive reasoning involves gathering data without a goal or premise and inducing the truth through unprejudiced data. Deduction involves testing some generalization suggested by some earlier observation. Any type of scientific reasoning, however, may be unfortunately influenced by trend and fashion. Science is often done by testing hypotheses and theories. A hypothesis is a carefully described and defined idea that suggests further testing. A theory is much more substantial, having survived a great deal of such testing, but is still not "proven." The study of life is rife with hypothesis and theory. The traits of life include movement, metabolism, reproduction, response, and evolution. But these traits do not define life. Biologists, in studying life, often inject much of their own personalities into their work. Their work may involve laboratory or field investigation, and they often must decide the nature of their social responsibility to the community as a whole.

Chapter 3

THE BEGINNINGS

Our best attempts at defining life seem to have a way of grinning back at us, making us feel a bit unsure of ourselves. Furrowed brows, cleaned spectacles, cleared throats, and all the other trappings of great knowledge do not diminish the simple reality that no one knows what life is. We have, though, listed some of the special qualities associated with life. With this constellation of traits in mind, then, let's see if we can utilize certain circumstantial evidence and suggest how life might have begun on the third planet from a common star.

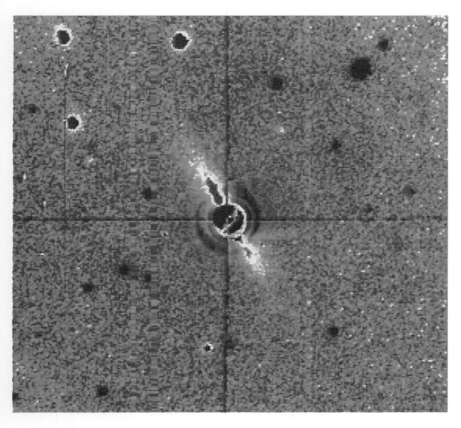

LIFE FROM NONLIFE

Questions regarding the origin of life have given rise to two major schools of thought, each with its own underlying assumptions. One is called *mechanism,* the other *vitalism.* It may seem, as we begin to define our terms, that any such arguments can only be sterile and academic, but I believe we will find that such is not the case (see Figure 3.1).

Mechanism vs. Vitalism

It will soon become apparent, if you have one ounce of poetry in your soul, that the mechanistic explanation of life is the less romantic of the two. This is because **mechanism** implies that life is the result of, and is subject to, the same laws of the universe as any other physical entity. It reduces the wondrous qualities of life to the simple interactions of mindless molecules. **Vitalism,** on the other hand, is based on the premise that living things are more than the result of molecular interaction, that life inherently possesses some undefinable and unmeasurable quality—an essence that might be called the ''life force.'' The mechanists respond that if any such force exists, it too must be physical in nature, an argument that drives the vitalists right up the wall. Since the ''life force'' of the vitalists is generally not considered to be measurable, the explanation does not present the sort of questions that can be answered by scientific investigation. As a result, we will explore the alternative, the mechanistic view. In particular, let's see if there are ways that life might have arisen based on what we know of physical and chemical principles. (We will consider such principles in more detail shortly; here we will deal only with a very general overview.) We should also remind ourselves that such ideas are highly controversial and utterly rejected in some philosophical circles. Furthermore, even the most fervent mechanists undoubtedly disagree on some major points.

One of the objections to the mechanistic view of life is based on the logical extensions of its acceptance. For example, mechanism has the perturbing tendency to cause us to reexamine some of our most sheltered, cherished, or revered traits—the ones that involve such things as

FIGURE 3.1 Can the vast array of living things on this planet, each with its special qualities and ways of making a living, be the result of the mechanics of molecular interaction? Is life simply the result of some grand equation, or does it have some unmeasurable and indefinable quality that sets it apart from nonlife?

FIGURE 3.2   Albert Szent-Gyorgi

sexuality, love, arrogance, jealousy, and patriotism, not to mention height, longevity, nesting urges, and voting tendencies. The suggestion that such things are subject in some specific way to the blind interactions of molecules is simply not acceptable to many people.

Life's Many Meanings

It seems that those who object to the mechanistic view, so far, have little to worry about. Up to this point, one cannot be very encouraged about our ever understanding the "essence of life" at the molecular level. The problem has perhaps best been expressed by one of our most capable scientists, Albert Szent-Gyorgi (Figure 3.2), who wrote:

> In my hunt for the secret of life, I started research in histology. Unsatisfied by the information that cellular morphology could give me about life, I turned to physiology. Finding physiology too complex I took up pharmacology. Still finding the situation too complicated I turned to bacteriology. But bacteria were even too complex, so I descended to the molecular level, studying chemistry and physical chemistry. After twenty years' work, I was led to conclude that to understand life we have to descend to the electronic level, and to the world of wave mechanics. But electrons are just electrons, and have no life at all. Evidently on the way I lost life; it had run out between my fingers.
>
> —Personal Reminiscences

As Szent-Gyorgi lamented, the "meaning of life" will probably never be grasped as a pure, crystalline gem of truth. It is just not likely to be that simple. Many dedicated biologists have concluded that life's many meanings, if they exist, apparently lie somewhere in the very complexity that Szent-Gyorgi tried to overcome.

THE EXPANDING UNIVERSE

As we continue our search for the theoretical beginning of life, let's take a big leap. Let's go back to the real beginning—back to the instant that the entire universe came into being. (About now you may notice just the tiniest bit of speculation about a few very minor points here and there.) The universe appeared, we are now told, some 13–20 billion years ago. And before that? Strange as it seems, we are told that there was nothing before that but a lump, that all the matter in the universe was compressed into that lump, about the size of a walnut. (Are you still with us?) And within this lump, we are told, was all the stuff of the heavens. In fact, every molecule in your body right now, at this very instant, was once in that walnut. Furthermore, the stuff that comprises your house, your shoes, and your dog was there too, as was the air you now breathe. So were the other planets and all the far-flung galaxies. It was all there (Figure 3.3).

Then the lump exploded (or began to explode, since some say it is still exploding). At the moment of the explosion, the particles that composed the lump flew in every direction, bumping, colliding, and interacting. In the course of such chance collisions, some of the particles joined. Others formed clusters. The more unstable ones immediately fell apart, but

FIGURE 3.3   One small part of the vast and mysterious realm we call the universe. At present, we are aware of life on only one small watery planet.

some had been changed by their temporary associations. Some were now able to interact in new ways with yet other kinds of particles. It may have been just such chance changes and aggregations that finally formed the great celestial array of clusters that we call our universe. Then, from amidst all this, we are told, our solar system was born. And here, on the third planet from the sun, there is hope that intelligent life may one day exist.

So, now you know how the universe came to be. Even this simplified scenario, only one of several explanations, may be a bit much, so I won't mention that astronomers also tell us that time didn't exist before the explosion, called the *Big Bang,* and that even all the empty space of the universe was also in that walnut.

Nonetheless, let's move along and assume that the earth now exists and that the celestial events that brought it about were quite impressive and worth a side trip. It is time to focus our attention on the development of the *life* that somehow managed to come into being amidst such cataclysmic events. Obviously, we will lean a great deal on intuitive or circumstantial evidence.

OUR EARLY EARTH

In order to imagine how life may have appeared here, we must visualize quite a different kind of earth from the one we know today. For example, our air is composed of about 78 percent molecular nitrogen, 21 percent molecular oxygen, 0.03 percent carbon dioxide, and traces of other gases such as argon and helium. But it seems that this was not always so. In 1936, the Russian scientist A. I. Oparin proposed that the atmosphere of the early earth contained less oxygen and much more hydrogen than it does now. The implications of this assumption are enormous. You will see why later on, but for now keep in mind that hydrogen tends to join with all sorts of elements, such as nitrogen, oxygen, and carbon. Specifically, when hydrogen joins with nitrogen, the result is ammonia. Because of this, it is assumed that the atmosphere of the earth was composed not mainly of nitrogen, but of ammonia. By the same token, there would also have been a great deal of water, formed by the union of oxygen and hydrogen. Of particular importance, carbon would have tended to join with hydrogen to form methane. Methane, as we shall see, is an organic molecule of the sort associated with living systems.

Perhaps the most difficult thing to imagine about the earth is that the earth—this earth, our home—was lifeless. There were no, nor had there ever been, ears to hear or eyes to see. There were no crusty lichens or slimy molds. Nothing lurked anywhere. The place was a dead unchronicled ball of matter covered by a very thin layer of hot swirling gases. The surface itself was hot, molten, and volcanic. In time, the surface began to cool and solidify, but the piercing shrieks and groans and deep rumblings from below startled no one. There was nothing to be frightened by at all. Heavy billowing clouds, miles thick, surrounded the darkened sphere. The murky blackness was continually split by bright, spewing gapes in the earth and thunderous lightning from above (Figure 3.4). Water vapor condensed and fell to the sterile earth, immediately exploding from the heat to be lifted skyward with a crackling hiss. It is difficult to imagine that this seething place would, in time, give rise to something called life. But, then, how do you account for your neighbors?

THE BEGINNINGS OF LIFE ON EARTH

Conjecturing about how life began is both fruitless and safe. One can never be proven right and it is impossible to be proven wrong. Nonetheless, it is a good intellectual exercise, and it can be useful in helping to understand scientific reasoning. So let's ignore all the problems associated with the question, gird our loins, and tread into an arena littered with dead guesses.

If we take the mechanistic view of life's beginnings, we must begin at the molecular level. So let's do just that and start by noting that many of the molecules associated with life are rather large and complex. The molecules that form living things today are apparently far more complex than the kind that existed when the earth was young. Thus, one can assume that the sort of life that we see today was formed by simpler molecules that somehow managed to join to form larger, more complex structures. The trouble was that those early molecules would have had almost no tendency to join; they could have coexisted quite nicely, side by side, without ever interacting at all. As time went by, however, things changed. The earth became a different kind of place and the complex molecules of the planet became more likely to interact.

According to our scenario, as the mountainous and volcanic earth continued to cool, pools of condensed water collected in the valleys. Then, as the atmosphere itself cooled, this water was lifted by evaporation, only to be returned in torrential rains of incredible proportion. The constantly falling water rushed over the earth's exposed surfaces sweeping away the salts and minerals embedded in the cooling rocks. These mineral-laden waters poured into the cracks and crevices of the parched earth, turning pattering streams into pounding rivers, and finally form-

FIGURE 3.4 The early earth; the birthplace of life (as far as we know).

ing great bodies of still waters. As the waters filled the deepest valleys and greatest canyons of the planet, the oceans were born. It was in such places that molecules of all sorts were brought together, bumping, jostling, and interacting.

Interacting? Why would they interact? Such interaction, as we will see, requires energy to make it go. It turns out (fortunately for our story) that there are a number of sources of such energy. For example, there is the sun, which produces a variety of types of radiation. Another source of energy is heat. Remember, the earth's surface was still far from placid; the searing landscape was constantly jarred with violent eruptions, and molten or hardening rock covered much of the earth's surface. Lightning continually streaked earthward from the dense, heavy clouds, jolting the molecules of the earth below. Energy, we see, was abundant.

But was all this enough? Could a mixture of water, ammonia, and simple molecular compounds, such as methane, form the complex molecules of life?

MILLER'S EXPERIMENTS

In 1953, a graduate student at the University of Chicago provided a partial answer to the question of how complex molecules could have arisen on the early earth. The student, Stanley L. Miller, built an airtight apparatus through which four gases—methane, ammonia, water vapor, and hydrogen (the gases presumably composing the atmosphere of the

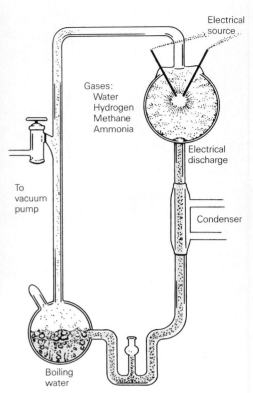

FIGURE 3.5 In 1953, Stanley L. Miller, then a graduate student, subjected a mixture of methane, ammonia, water vapor, and hydrogen to a series of electrical charges. He imagined this to be a rough duplication of conditions on the primitive earth when the "primordial soup" was subjected to bolts of lightning. The result justified his expectations. After a week, the inorganic molecules had joined to form amino acids. Here, he continues his experiments on other facets of the problem. His original apparatus is illustrated at right.

early earth)—could be circulated past electrodes (Figure 3.5). He permitted the gases to circulate together in his chamber for a week, subjecting the mixture to intermittent jolts of electricity. Then he simply analyzed the contents of the chamber. His findings were indeed startling. The experiment had produced a number of organic compounds, including amino acids, the building blocks of proteins. We will soon see some of the important ways proteins function in that complex and tightly regulated pageant called life.

Miller was already a careful scientist and he was well aware that this experiment was too important to botch. So he also ran two controls. In one case, he sterilized the gas mixture at 130° C for eighteen hours before he subjected it to the electrical power source. The yield of complex molecules was the same as in his first test. Then he ran the test exactly as before, but without the energy supplied by the electrodes. This time, there was no significant production of the critical molecules. As a result, Miller felt safe in concluding that the formation of organic molecules was not due to the presence of contaminants, but was the result of smaller molecules joining together when energized.

Miller's experiment has since been duplicated and extended many times. In similar experiments, researchers have produced yet other molecules generally associated with life. These include substances such as purines, pyrimidines, and sugars, all of which are critical to the processes of life, as we will see in Chapter 4.

THE COACERVATE DROPLET

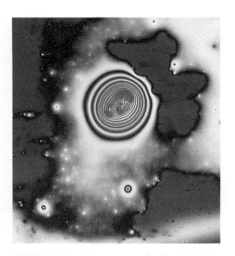

FIGURE 3.6 A coacervate droplet. Notice its organization. Such droplets have often been mistaken for living things under the microscope.

We have now traced a theoretical pathway whereby large, complex organic molecules could have been produced and maintained when the earth was young. Obviously, though, we have not even begun to answer the question about how *life* came to be. It is indeed a long step from complex molecules to the simplest forms of life. Even the simplest living thing is composed of a fantastically ordered arrangement of complex molecules. How, then, could life have arisen from some ancient hot, chemical soup?

If you expected an answer to that, you're out of luck. There are no firm answers, but we do have some ideas that bear mentioning. A. I. Oparin pointed out that *colloidal* protein molecules (see footnotes), in which the particles are suspended in a gel-like state, tend to clump together into increasingly complex masses. These may be held together by electrostatic (positive and negative) forces and thus form *coacervate droplets* (see Figure 3.6). Each such droplet consists of an inner cluster of colloidal molecules surrounded by a shell of water. The molecules of this water shell are arranged in a specific manner in relation to the colloid center. As a result, there is a clear demarcation between the colloidal protein mass and the water in which it is suspended. This separation of molecules sets the stage for the peculiar behavior of the droplet.

The Growth of Coacervates

The peculiar orientation of the water molecules around the colloidal mass gives this water shell special physical properties because, in effect, the water acts as a sort of membrane. That is, it acts as a kind of screen, allowing some molecules through while excluding others. In addition, the molecules of the colloidal mass inside tend to arrange themselves in an orderly manner. The arrangement comes about through the interaction of a number of forces in the mass, such as electrostatic charges along the molecules.

The colloid, as a result of its peculiar arrangement, has a tendency to absorb certain other molecules from the medium in which it is suspended. However, it does not absorb all molecules with equal ease. Just as the water shell only allows the passage of certain molecules, the inner colloid itself is very selective regarding which molecules it will absorb into its mass. In fact it may absorb almost all of some kinds of molecules from the medium while taking almost none of the others. As a result of this continual absorption, the droplet grows. As it grows, its colloidal mass becomes more specific, more precisely composed. Further, the increasing number of molecules comprising the colloid may allow new sorts of

---

Colloid: a substance in which dispersed particles are suspended and are so small that they do not settle out.

Gel: a colloid in a semisolid form, jelly-like in consistency.

Coacervate: an organized colloidal droplet in which the suspended particles are held together by their positive and negative electrical charges.

interactions. Some of these may result in the developing "cell" becoming even more organized.

As our developing "cell" continues its changes, the molecules at the surface of the colloid rearrange themselves into a membranelike structure just under the water layer. This new "membrane" is even more selective than the water layer. It allows even fewer kinds of molecules to pass. (See Essay 3.1 for a discussion of "membrane" formation in bodies called **proteinoids**.) Thus, with increasing selectivity in the kinds of molecules that the mass will accept, the structure tends to become more regularly organized. At this stage, the complex droplets begin to show many of the properties of living things. (Even experienced microbiologists have on occasion attempted to identify the species of such droplets.)

The Internal Structure of Coacervates

It is important to realize that a coacervate droplet provides a special environment for any molecule that it absorbs. First of all, because of the selectivity of the droplet surface, only certain types of molecules are found inside. Then, because of the compactness of the droplet, any molecules allowed in are brought into increasingly closer association. As a result, they are increasingly likely to interact.

At this point let's step back and see what we now have in our hypothetical description of cell development. What we find is a rather highly structured mass that becomes even more orderly as it grows. As a result of its orderliness and its selectivity regarding what is allowed into its interior, some molecular interactions are more likely than others. In a sense, then, the droplet "regulates," or at least influences, the types of chemical reactions that can take place inside its mass. It may have occurred to you that whereas coacervate droplets are not "alive," as we usually understand the term, they do have many of the qualities we associate with living things.

Reproducing Coacervates

Now let's take another conceptual leap and see if we can account for how such droplets were able to reproduce. This intellectual exercise is critical if we are to accept these simple droplets as ancestral to life. Again, we must rely on a good healthy imagination—held in check by circumstantial evidence. But before we get into this discussion, we should first be aware that *most* of those early coacervates did *not* reproduce. However, the earth undoubtedly harbored a wide variety of droplets. Some would have been more stable than others, and it was these that had the greatest chance of eventually reproducing.

But back to our question. How could mere coacervate droplets reproduce? Actually, there are a number of ways if one is allowed to use the term "reproduce" loosely. For example, size alone could cause reproduction. Any surviving droplets would have continued to grow as new molecules entered their masses. Finally, the mass would have become so large and unwieldy that the droplet would have fragmented as a simple

In a remarkable set of experiments by Sidney Fox and his group, a mixture of amino acids (see Chapter 4) was heated under dry conditions. The amino acids joined together to produce long chains (called polypeptides). These, in turn, were placed in hot water, which was then allowed to cool. The result was that the polypeptides joined to form small clumps called **proteinoids.** The microspheres are particularly important to the problem of life's beginnings because they suggest how membranes may have first formed around the earliest cells. The molecules of the proteinoids arranged themselves in such a way that only certain substances from the medium could pass through the outer "shells" and enter the sphere. The result was that the material inside the sphere was different, and more orderly, than the randomly mixing fluids outside. The "shell" covering these droplets was found to contain proteins and fats, the same materials composing the membranes of today's cells, as we will see later. Furthermore, as materials continued to enter the spheres, causing them to swell, they would eventually fall apart into smaller spheres that would then continue the process. This has been suggested as a very primitive form of replication in these tiny droplets that may be similar to the forerunners of life.

result of its large size, as we saw with the microspheres described in Essay 3.1. In a sense, the droplet would have broken apart under its own weight. Further, if the structure of the fragmenting droplet was highly regular and ordered, each of the resulting fragments would have been very similar to the original droplet, and thus each could then begin its own growth. This, obviously, is a very simple form of reproduction, but reproduction nonetheless.

Consider another, more complex, form of reproduction. Here, the first step would occur *within* the droplet. Furthermore, we now set the stage for the development of a very orderly and highly regulated descendancy—one that could lead toward the sorts of mechanisms by which life reproduces. This type of reproduction would be possible because of the great numbers of complex molecules that were generated in that early atmosphere.

It is not difficult to imagine that since certain molecules are attracted to other specific types, and tend to join with them, large and quite complex molecules can be produced even within the coacervate. It is conceivable that these large molecules, perhaps existing as chains, could attract other free molecules and form very specific new chains along their length. Since each molecule along the chain can join only with certain others with specific bonding tendencies, the result would be that the original molecule could direct the formation of another such molecule along its length. Should these new chains then break away from the parent "template," or "model," each would then be able to bond with other free molecules along its length, and thus direct the synthesis of new chains. Furthermore, each new chain would be likely to produce a counterpart identical to the original. (Do you see why? See Figure 3.7. A similar process will be described in detail in Chapter 7.)

No matter how the droplets reproduced, however, the point is that there are ways in which even the most complex and precise ones *could* have come to reproduce, or at least some of the molecules within their inner masses. It should be apparent that the most "successful" droplets

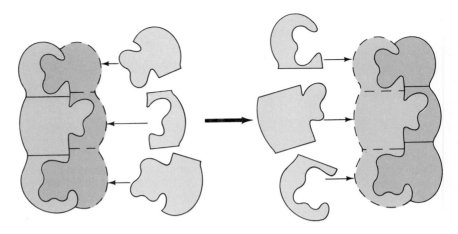

FIGURE 3.7 In this simplified diagram, the molecule at left serves as a model (or template) against which the molecule at right is formed. Conversely, should that molecule serve as a template for a new molecule, it would generate one like the original. It has been suggested that such interactions helped insure the continuance of the first complex molecules associated with life.

would be those that could produce the closest approximations of themselves since, by their very existence, they would have effectively demonstrated the stability of their molecular makeup.

Successful Droplets

According to our scenario, as the less successful droplets disappeared, the survivors could have continued to change, becoming ever more precise until they developed into the first cell-like bodies. Thus, the ratio of the more stable forms in the population of developing cells would have continued to increase. In addition, the components of the lost and decomposing droplets would have been released and made available to the increasingly efficient early ''cells.'' All the while, the more efficient types would continue to accumulate. Any existing ''cells,'' then, would have been the result of the most fortuitous of the molecular combinations in the untold number of trials over the millions of years. But finally, through such processes, according to our story, actual cells appeared. (We will see what ''actual'' cells are shortly; your intuitive definition will suffice for now.) The point at which those droplets became cells is not important, but we can be sure that there was no sudden and dramatic appearance of ''life'' on the earth. No trumpets blared; no flags waved. There were only a few peculiar droplets drifting mindlessly in the warm seas, droplets a little more organized than the rest.

You are undoubtedly aware that any description of how life appeared on earth will not be accepted by everyone. Some scientists, for example, contend that coacervate droplets probably did not give rise to complex molecules that could reproduce themselves. Instead, they say, it is more likely that complex molecules first appeared and then built something like a cell around themselves. Other people, of course, argue that there is an essential distinction between life and nonlife, that life is not merely a serendipitous accumulation of molecules, and that life—or at least human life—was created by some greater power. And then there are a host of lesser theories—including one recently suggested by a few well-known scientists who propose that life was brought here by travellers from another planet. The whole question is an academically fascinating one, but obviously it is not subject to the usual rules of scientific inquiry. So, we will now leave this intellectual exercise, having set the stage for the processes called natural selection. We will simply assume that cells appeared on the earth, and then consider the more approachable question of how they might have made their living.

Autotrophy and Heterotrophy

The molecular interactions necessary to support even the simplest forms of life are so complex that we must wonder how such interactions ever started. What triggered their development? Perhaps the impetus was partly food shortages. Certain kinds of molecules in the early environment would have been important for the processes of growth and reproduction, and as populations of life forms increased, some such "food" molecules would have fallen into short supply. In time, competing cells would have run up against "food barriers" and some would have lost the race to acquire these precious molecules. These cells would have had to develop some alternative or perish. One alternative might have been, not to find food, but to *manufacture* it. Let's see if this route might have led to increasingly complex and efficient means of competing in a newly competitive world.

Among the small changes that were continually arising by chance in early cells, some would have resulted in the cells producing various sorts of molecules, some perhaps unusual. If a cell began to produce, say, one of the rare "food molecules" (or a reasonable substitute), then the pressure to compete for that molecule would be relieved. We can call this molecule nutrient A. Let's also say that it was manufactured from a more abundant nutrient B. This would mean that cells that previously had been held in check by the scarcity of nutrient A were then able to grow and multiply—that is, until nutrient B fell into short supply. At this point, there perhaps appeared, by chance, some other slight change that enabled a cell to manufacture B from another abundant molecule, say nutrient C, and thereby to make both nutrients B and A. In time, with competition increasing, and with one molecule after another falling into short supply, long series of interactions involved in the manufacture of needed molecules would arise. Obviously, those cells best able to carry out such chains of chemical reactions would be more likely to survive.

Any such system would require an efficient energy source to drive the process. Thus, it is interesting that a chemical called ATP, which is critical in the production of energy in today's cells, is one of the molecules that appears in those laboratory experiments that simulated the conditions of the early earth.

The cells that became able to chemically manufacture their own food are now called **autotrophs** (self-feeders). As time passed, the autotrophs became quite efficient at manufacturing their food. The biochemical chains became long, complex, and refined. Today, the result is most evident in the green plants, our silent partners on the planet. As we will see, they actually use the energy of sunlight to power their intricate food-making machinery. They obviously had an advantage in the early days, and so their numbers would have multiplied. However, their food-laden bodies would not have gone unnoticed by other life forms. Other kinds of cells soon developed means of robbing the food-makers. These became the **heterotrophs** (other-feeders) (Figure 3.8).

FIGURE 3.8 Autotrophs (self-feeders) use the energy of the sun to manufacture their own food. Autotrophs come in many shapes and sizes, including (A) Volvox, (B) bur oak, and (C) green algae. Heterotrophs (other-feeders) can be divided into herbivores (which eat autotrophs), such as giraffes (D), carnivores (which eat species that eat autotrophs), such as leopards (E), and omnivores (which eat both autotrophs and heterotrophs), such as grizzly bears (F).

(A)   (B)   (C)

Today the earth abounds with heterotrophs. Some eat autotrophs (these are called **herbivores**), while some (the **carnivores**) eat the species that eat the autotrophs. Most species, however, eat both autotrophs and heterotrophs (and are called **omnivores;** *omni,* all). (Figure 3.8 shows a modern representative of each.)

Oxygen: A Bane and a Blessing

In the photosynthesis of green plants, as the energy of sunlight falls on the green pigment in the leaves, carbon dioxide and water are used to make food, and water and oxygen are released. The release of oxygen by those first photosynthesizers was a critical step in the direction of life's development. In a sense, the production of oxygen falls into the "good news-bad news" category. It's good news for us, of course, since we need oxygen, but as oxygen began to replace hydrogen as the most prevalent gas in the atmosphere, it sounded the death knell for many of the early heterotrophs. This is because oxygen is a disruptive gas, as we know from seeing the process of rusting (Figure 3.9). So, in the early days of life on the planet, many life forms were destroyed by the deadly and accumulating gas.

But we must keep in mind that there were many forms of cellular life in those days, and that life was constantly changing in all directions (with most such "hopeful experiments" doomed to failure). We can assume, however, that some of those cells would have been changed in ways that rendered them less vulnerable to the dangers of oxygen. In fact, in time, some forms would have gone one step further and begun to *utilize* the deadly gas. This ability, obviously, would have given them a great

(D)   (E)   (F)

FIGURE 3.9 Many of the seemingly most indestructible structures on earth slowly yield before the devastating power of oxygen. This neglected machinery gives mute testimony to the destructive effects of the gas.

competitive advantage over the forms that were unable to do so.

Not everyone accepts the idea that the increase in oxygen over the early earth was primarily the result of photosynthesis. Some scientists argue that most of the earth's oxygen resulted from sunlight falling on the water of the upper atmosphere and splitting it into its component parts, hydrogen and oxygen. (This notion may provide some consolation as we watch the destruction of the world's great forests and the slow poisoning of the algae-laden seas.)

A stromatolite fossil embedded in a 3.5-billion-year-old rock from Western Australia.

There is an area of Western Australia so forbidding that only the most hardened of miners and scientists go there. The miners are there for the usual reasons; the scientists go to dig for some of the oldest known rocks on earth. The rocks are interesting precisely because of their age. In 1980, a surprising announcement was made: the rocks contained fossilized remains of life that existed about 3.5 billion years ago—that is, a billion years before any other known life.

Furthermore, these rocks contained at least five different life forms. The fossils were tiny, to be sure. Most of the forms were elongated, or strandlike, and the cells lacked any sign of a nucleus, as do bacteria today. Furthermore, the organisms probably lived under a thin layer of warm water, and, since they used carbon dioxide, they were probably photosynthetic. It appeared that scientists had found one of earth's earliest pond scums.

The oldest known rocks on earth are 3.8 billion years old, and some scientists think that they may contain evidence of life also. If this is so, since the earth is estimated to be only some 4.5 billion years old, life may have appeared very suddenly after the earth's surface cooled. According to some scientists, this means that there is an increased likelihood of life appearing wherever physical conditions permit. This, in turn, suggests to some that there is a greater likelihood that life has appeared on other planets in the vast, far-flung universe.

Scientists have used data from fossils and rocks to help date the origins of the earth and life on earth. Some scientists believe that if the physical conditions permitted, life may have appeared on some other planets in our immense universe. This computer-colored strip of the surface of Mars shows some of the same elements that are found on the earth's surface, such as iron oxide (red), basalt (blue), ice and fog (turquoise), and sand (white-yellow). Far right, the surface of Mars awash in the sun's rays.

In any case, whatever its source, the rising levels of oxygen had an important effect. For one thing, as oxygen collected in the upper atmosphere, it would have formed ozone (the union of three oxygens). Ozone, we now know, acts as a shield against certain kinds of the sun's rays. Thus, the earth under the ozone shield would be protected against the terrible energy of raw sunlight. Before the ozone formed, the early cells probably took refuge five to ten meters beneath the surface of the earth's shallow seas, where the force of the ultraviolet rays was diminished. But now, with the development of a protective layer of ozone, dry land became safe to inhabit. Thus the shorelines were soon invaded by strange, primitive life forms.

We must keep in mind that the rise of the oxygen-rich atmosphere meant that life could no longer arise in the way it once had. In other words, as life developed, it destroyed the very mechanism that had produced it.

SUMMARY

No one knows how life came to be, but there are some compelling theories. The prevailing theories assume that life arose through mechanistic means, subject to physical laws. The early earth was highly energetic and chemical-laden, and life may have started in the form of coacervate droplets that had a membranelike exterior and that regulated their contents to some degree. Later they would have reproduced, first by fragmentation, then, perhaps, by chemical templates. Others suggest that once amino acids formed in the presence of heat energy, they would have joined to form complex precursors of life called proteinoids. Some early cells would have developed means to make their own food from simpler molecules and energy (autotrophy) while others would have developed ways to extract food from other life forms (heterotrophy). As oxygen increased, due to the action of autotrophs, organisms had to find ways to reduce its corrosive effects and, finally, to utilize it.

Chapter 4

THE CHEMISTRY OF LIFE

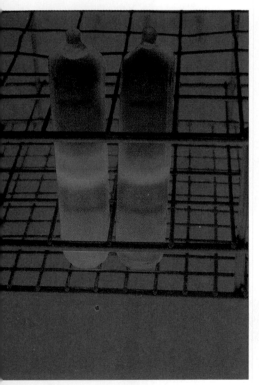

It can be argued that it really doesn't matter *how* life began; it *exists*. (We know life exists because biology is the study of life, and biology exists or you wouldn't be sitting here.) Nonetheless, as we rummaged through the various ideas that addressed the question of life's origins, you may have noticed that we quickly fell back on the behavior of molecules. These tiny particles weren't precisely defined at that time, but it should be apparent that in discussing life, we are already led to the consideration of chemistry.

However, before despondency sets in, I should hasten to point out that we will cover only a few basic points in a simple and straightforward manner and with reasonably modest goals. As you begin to feel more comfortable with the elementary concepts of chemistry, you may be surprised at how often certain familiar observations begin to fall into place and make sense. In addition, a general introduction to the chemistry of life may help you to see familiar things in a new light. This new light may reveal things you hadn't thought of, things you may find fascinating or useful.

An understanding of the basic concepts of chemistry can also broaden one's view. An established and sound scientific view may well become increasingly important as time passes (Figure 4.1). For example, can automobile emission control devices be harmful if they trap hydrocarbons and allow nitrous oxides to pass into the air? And what is the purpose of a salt-free diet? Is there any safe level of radiation? How can x-rays cause cancer? Should nuclear power plants be built? Is the risk of nuclear accidents greater than the danger of the atmospheric pollutants produced by burning coal? Why do we age? How can chemicals cause hostility? Chemistry, you see, influences our lives at a number of levels.

So here's the plan: We will begin with some basic definitions. Then we will take a look at the structure and components of life's building blocks and consider some of the ways they can interact. Later, we will

FIGURE 4.1 The importance of a basic understanding of the chemistry of life has innumerable facets. How are we affected by using the air as a chemical dump? How accurate or necessary is it to test chemicals on animals to determine how the chemicals will affect us? How do various kinds of lighting affect our body chemistry? How can research and testing into the chemistry of our brains help us control pain? Should we use male hormones to accentuate the development of "male" traits? Do mothers defend their young partly as a response to mindless chemicals circulating in their bloodstreams? How are we affected daily by the chemicals around us and within us?

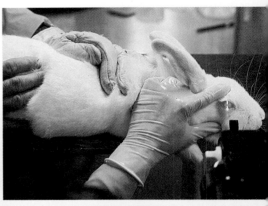

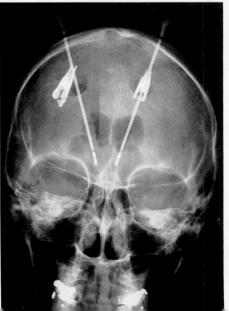

see how certain combinations of these things are important to life. Keep in mind that the principles we will consider here will serve as a foundation for many of our following discussions.

ATOMS AND ELEMENTS, MOLECULES AND COMPOUNDS

All matter, whether it exists in an alga or a distant star, is composed of elements. **Elements** are substances that cannot be divided into simpler substances by chemical means. **Atoms** are the smallest indivisible particles of these elements. Atoms *can* be divided into their component parts, but when this is done, the special qualities of that element are lost. Each element, or type of atom, has its own special chemical behavior by which it is identified. Oxygen, chlorine, and carbon are elements.

Ninety-two elements occur naturally, and about seventeen others have been synthesized in the laboratory. Each has been named and given a letter symbol. For example, the element called hydrogen is designated by the letter H. Sulfur is S. Oxygen is O. Of course, nothing is all that simple, so we find that sodium is designated by Na, from its Latin name *natrium*. The symbol standing alone usually designates one atom. For example, O refers to one atom of oxygen. The symbol 2O refers to two atoms of oxygen. But the symbol O_2 means that two oxygen atoms are joined together into one *molecule* of oxygen, the form in which oxygen usually exists in nature. **Molecules** are formed from combinations of atoms, and **compounds** are substances formed of one kind of molecule. Water, then, is a compound formed of hydrogen and oxygen atoms joined to form molecules that are designated by the symbol H_2O.

CHNOPS

CHNOPS is an elegant word that might help you to identify the six important elements that make up about 99 percent of living matter. However, if you drop the word casually at your next gathering, it is not likely that everyone will know you are referring to carbon, hydrogen, nitrogen, oxygen, phosphorus, and sulfur. If they do, you are in the company of organic chemists and should leave immediately.

The Structure of an Atom

Let's now consider what an atom is supposed to look like. We say "supposed to" because atoms cannot be seen, so all the evidence is circumstantial. (But it is impressive. People have been hanged on less compelling circumstantial evidence.) Atoms, according to theory, are composed of a dense center around which small particles spin in orbit. The center, called the **nucleus,** consists of **protons** and, usually, **neutrons.** Each proton carries one positive charge, designated by a plus sign (+).

Neutrons, as the name implies, carry no charge. The particles in orbit around the nucleus are called **electrons.** They are much smaller than protons (having about 1/1835 the mass). Each electron carries one negative charge, designated by a minus sign (−). An atom of hydrogen, the simplest element, consists of one proton and one electron, with no neutrons (Figure 4.2).

FIGURE 4.2 **FIGURE 4.2** The hydrogen atom (top) and the helium atom (bottom). Note that hydrogen has one proton (this reveals it as hydrogen) and one electron (thus it is electrically balanced). The atom below has two protons (making it helium) and two neutrons in its nucleus. The neutrons bear no charges so the atomic number of helium is two, while the atomic number of hydrogen is one. Helium is also balanced in that its two positively-charged protons in the nucleus exist with two negatively-charged electrons in orbit around the nucleus.

Electron orbit

Nucleus

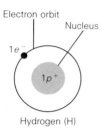

$1e^-$

$1p^+$

Hydrogen (H)

$2e^-$

$2n$
$2p^+$

Helium (He)

Figure 4.2

Helium, the next simplest element, is composed of two protons, two neutrons, and two electrons. Since an atom's **atomic number** refers to the number of protons in the nucleus, the atomic number of hydrogen is one; of helium, two. It is important to know that elements differing by even a single proton may have vastly differing chemical properties.

You may have noticed in the diagrams of hydrogen and helium that the number of positively charged protons are balanced by the negatively charged electrons, resulting in the atoms having a net neutral charge. If they have equal numbers of protons and electrons, they are said to be electrically balanced. As a final note about our examples here, it should be mentioned that the figures are not drawn to scale. In nature, the electrons are nowhere near the nucleus. In fact, if the period at the end of this sentence were a nucleus, its nearest electron would probably be across the street somewhere.

HOW ATOMS MAY VARY

Isotopes

The atoms of any given element all have the same number of protons in their nuclei. If this number were to change, the atom would, by definition, become a different element. However, the number of *neutrons* in the atoms of certain elements may vary. For example, most oxygen atoms have eight protons and eight neutrons, but some oxygen atoms may have nine, ten, or even more neutrons. Atoms that can vary in this way are called **isotopes** of an element. There are eight different neutron variants of oxygen, so oxygen has eight isotopes. Some elements have as many as twenty isotopes. Usually, all the isotopes of an element have about the same chemical properties, but in some cases, isotopes of the same element may have vastly different chemical traits.

Ions

Ordinarily, each atom of an element has the same number of negatively charged electrons as it has positively charged protons; hence, it is electrically balanced. We know that the number of protons, with their positive charges, can't change without changing the element, but an atom *can* gain or lose electrons and, when this happens, it is no longer electrically balanced. If the atom loses an electron, it is left with a net positive charge, since its protons now outnumber its electrons. If it should gain an extra electron, it then has a net negative charge. Charged particles—atoms that have either a positive or a negative charge—are called **ions.** Many ions, as we will see later, are critical to the processes of life.

ELECTRONS AND THEIR BEHAVIOR

Electrons move at almost unbelievable speeds in **orbits** around the atomic nucleus. Actually, there are a number of potential orbital paths in which an electron can move around a nucleus, each path at some specific distance from the nucleus. These potential electron paths are most easily visualized as concentric spheres, called **shells,** around the nucleus.

FIGURE 4.3 Electrons may exist in high or low energy states. An electron moving from an orbit near the nucleus to one farther out is moving from a lower to a higher energy level. "Exciting" an electron by raising it to a higher level by the input of energy is analogous to pushing a boulder up a hill. At the higher energy levels, they both increase their potential energy, which can be released if they return to their former states.

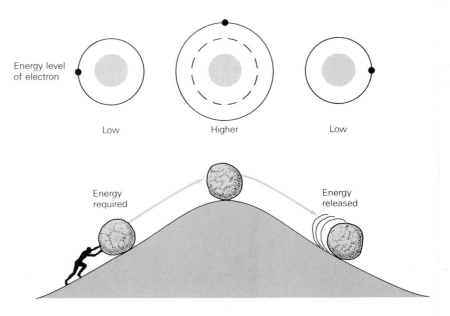

Energy level of electron

Low   Higher   Low

Energy required   Energy released

The electrons in the outer shells move much faster than those in the inner shells and, as a result, they are said to possess higher energy levels. These outer electrons, then, are in a position to best interact with other molecules.

The shells were described as *potential* electron paths because it is possible for an electron to move from one shell to another. For example, if an electron is "excited" by some external energy source, such as heat, light, or electricity, it may jump outward to a higher energy orbit. Should it later fall back into a more inner orbit, the energy that had provided its boost is released. We will shortly see how this released energy can be used in living systems. Figure 4.3 will give you an idea of some energy relationships.

Oxidation and Reduction

In some cases, an electron may be so energized that it escapes from its atom altogether. The loss of electrons in this way is called **oxidation,** because the lost electron is often recaptured by an oxygen molecule. In fact, some substances cannot lose electrons unless oxygen is available to accept them. The capture of such electrons by any element is called **reduction.** So, an atom losing an electron is oxidized, and an atom gaining an electron is reduced. This simple process of electron transfer has tremendous implications for life.

In some cases, as electrons are transferred in such reactions, they tend to travel with a proton. Together, of course, they form a hydrogen atom. So, oxidation and reduction can involve the transfer of hydrogen from one element to another. As an example, an atom of oxygen can be reduced by two hydrogen atoms, resulting in a molecule of water. Such reactions can sometimes proceed so rapidly that the effect is explosive (Figure 4.4).

FIGURE 4.4 On May 6, 1937, Germany sought to impress the world with its technological achievements, so it sent the hydrogen-filled zeppelin *Hindenburg* across the Atlantic to the United States. We were impressed, indeed: The zeppelin blew up. The hydrogen, perhaps encouraged by a spark, explosively joined with oxygen (to make water). Thirty-six people died on the approach to Lakehurst, New Jersey. Miraculously, sixty-one people survived.

FIGURE 4.5 Helium (top) and neon (bottom) are called inert atoms. They tend not to react with other elements because their charges are balanced and their electron shells are filled.

Helium (He)

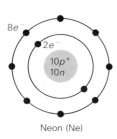

Neon (Ne)

FIGURE 4.6 Oxygen is reactive because it has only six electrons in the outer shell.

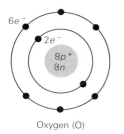

Oxygen (O)

Electrons and the Properties of the Atom

If you can remember the numbers two, eight, and eight, you will have at your fingertips the answer to very few questions. But one such question would be, what is the maximum number of electrons that can exist in the first three shells of an atom? Only a specific number of electrons can occupy any one electron shell; the first (the innermost) shell can accommodate only two, and the second and third can hold only eight each. (The fourth can hold thirty-two, and some atoms have even more shells.) However, no matter how many shells an atom has, its chemical properties are determined primarily by the outer shell. It is the one most likely to gain or lose electrons, and the atom tends to behave in such a manner as to keep this shell filled. When this outer shell is filled, the atom is said to be **inert**; that is, it is almost impossible to make it react with any other element. The outer shells of helium and neon, for example, are full. In the case of helium, with only two electrons, there is only one shell, the inner one, and it has all the electrons it can hold. Neon, as you see, has only two shells, both completely occupied (Figure 4.5). These are "inert" elements that generally do not react with other elements.

So we see, then, that atoms tend to "satisfy" themselves in two ways. First, *they tend to balance the numbers of their protons and electrons,* and second, *they tend to keep their outer shells filled.* When these conditions are met, the atoms are inert, unreactive, and not of great interest to us here. With atoms, as perhaps with people, it seems the most active and interesting ones are often the most "dissatisfied."

Why is the number of electrons in the outer shell so important? Perhaps this can best be understood by beginning with a couple of examples. Let's consider the case of oxygen (Figure 4.6). Table 4.1 tells us that oxygen has an atomic number of 8; that is, it has eight protons, and, therefore, needs eight electrons. However, since the inner shell will require two electrons, the outer shell is left with only six—two short of

TABLE 4.1 Some Elements Found in Living Matter

Element	Symbol	Atomic number	Electrons in each shell				Percentage of earth's crust
			1	2	3	4	
Hydrogen	H	1	1	0	0	0	0.10
Carbon	C	6	2	4	0	0	0.03
Nitrogen	N	7	2	5	0	0	Trace
Oxygen	O	8	2	6	0	0	46.60
Sodium	Na	11	2	8	1	0	2.90
Magnesium	Mg	12	2	8	2	0	2.10
Phosphorus	P	15	2	8	5	0	0.10
Sulfur	S	16	2	8	6	0	0.05
Chlorine	Cl	17	2	8	7	0	0.05
Potassium	K	19	2	8	8	1	2.60
Calcium	Ca	20	2	8	8	2	3.60
Iron	Fe	26	2	8	8	8	5.00

its shell requirements. So, oxygen tends to accept two electrons from other atoms.

On the other hand, note that sodium (Figure 4.7), has an atomic number of 11. This means it has two electrons in its first shell, eight in the second shell, but only one in the third shell. Since it isn't energetically feasible for a sodium atom to gain seven electrons, it tends to fulfill its outer shell requirements by losing its single outer electron. This loss means that it now only has two shells, but at least both are satisfied.

Of course, the loss of the electron (with its negative charge) ionizes the sodium. That is, it leaves it with a positive charge. Sodium ionized in this way is written Na^+. Magnesium, which has twelve electrons, is sometimes written as Mg^{++} (do you see why?). The fact that chlorine has an atomic number of 17 may give you a clue to the reason salt (NaCl) is formed so easily. Take a minute here to consider the question.

CHEMICAL BONDING

We've seen that atoms can join to form molecules, so let's take a look at how this can happen. (You already know enough to figure out that atoms can only join with certain other kinds of atoms.) By the way, it is interesting that the chemical properties of atoms may have little to do with the properties of the molecules that they form. For instance, both sodium and chlorine are deadly poisons, but together they form common table salt.

As we explore the ways in which atoms can join, you will be relieved to discover that even the most complex molecules are generally held together by only a few, and conceptually simple, forces.

FIGURE 4.7A Since sodium has eleven electrons, with ten existing in its first two shells, the third shell can harbor only one electron. Sodium cannot rectify the situation by adding seven more electrons to the third shell; the resulting charge imbalance would be too great. Instead, it tends to give up its outermost electron. Thus, sodium is highly reactive with anything that accepts electrons.

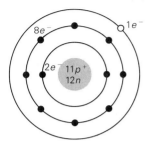

Sodium ion (Na^+)

FIGURE 4.7B Oxidation is continually taking place as electrons or hydrogen atoms are passed to oxygen. The loss of the hydrogen ions is often disruptive, being the basis for such varied processes as fire, rotting, and digestion. The process of oxidation may take place rapidly if the "energy of activation" has been provided as here, for example, by the intense heat of a match.

Ionic Bonds

First, let's reconsider how table salt is formed. We know that sodium has two electrons in its first shell, eight in its second, and only one in its third. Therefore, sodium has a tendency to give up its outer electron (at the expense of becoming ionized) in order to meet its outer-shell requirements. We also know that chlorine has two electrons in its first shell, eight in its second, and seven in its third. Hence, it can fill its outer-shell requirement rather easily by picking up a single electron. Not surprisingly, when a strong electron donor, such as sodium, encounters a strong electron acceptor, such as chlorine, there is a quick switch of an electron from one atom to the other. This satisfies the outer-shell requirements of both atoms, but as we know, it leaves the sodium, which has lost an electron, with a positive charge (Na^+). And, of course, the chlorine, with its additional electron, now has a negative charge (Cl^-).

Chemically (as well as socially, some would say), opposites attract, and the sodium and chlorine are now bearing opposite charges. As a result, they are electromagnetically attracted to each other. They draw together and join, forming the molecule NaCl, or table salt. This sort of reaction is called **ionic bonding,** and it involves the transfer of an electron from one atom to another and the subsequent union of the resulting oppositely charged ions.

FIGURE 4.8 When NaCl is placed in water, the salt molecule separates into its component sodium and chloride ions. Water has this effect on salt because of its positive and negative (polar) charge distribution.

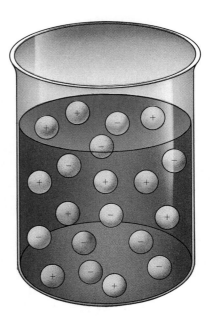

In some cases, such bonding does not occur, even when the atoms bear opposite charges. For example, when table salt is placed in water, some Na and Cl atoms dissociate into their ionic form (Figure 4.8) and they stay in this condition because of the peculiar configuration of water molecules.

One of the reasons water has such special properties is because its molecules have unevenly distributed charges. In other words, water molecules are polarized, having a positive and negative end. Because of the polarization, the positively charged end of the water molecule is attracted to the negatively charged chlorine, and the negatively charged end is attracted to the positively charged sodium. As the water molecules cluster around these charged atoms, they effectively separate the two ions. We will learn more about the peculiar qualities of water when we discuss the topic of hydrogen bonding.

Covalent Bonds

In another type of bonding, called **covalent bonding,** atoms do not exchange electrons, but *share* them. We can illustrate with oxygen again. Oxygen, as we have seen, needs two electrons to satisfy its outer shell requirements. One way it can acquire them is by sharing electrons with two hydrogen atoms, thereby forming a water molecule (H_2O). This sharing satisfies the shell requirements of both the oxygen atom and the two hydrogen atoms (Figure 4.9).

When a covalent bond is formed between two different kinds of atoms, the shared electron is usually attracted more strongly to one atom than to the other. The electron, of course, will be more powerfully drawn to the atom with the stronger positive charge. When water is formed from hydrogen and oxygen, for example, the shared electrons are pulled more strongly toward the more positively charged oxygen nucleus than they are toward the two hydrogen nuclei. it is this "imbal-

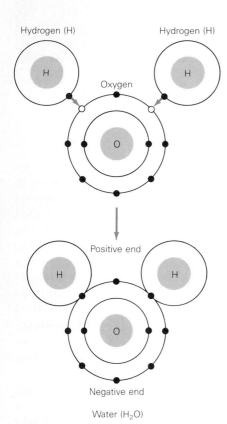

FIGURE 4.9 Covalent bonds (top) form as two hydrogen atoms move in to share their electron with an oxygen atom. The "unbalanced" structure (bottom) of a water molecule results in a stronger negative charge on one side of the molecule.

ance'' in the tendency of the electron that results in a polarized water molecule with negatively and positively charged ends.

In some cases, the electrons are about equally attracted to both elements, and, thus, the molecules are symmetrically charged, with very little polarity. But polarity is not an either-or-situation. In fact, because different kinds of molecules vary more or less in their polarity, there is no clear line of distinction between ionic and covalent bonding. Just keep in mind that in ionic bonding, an electron from one atom is completely captured by another atom, and in covalent bonding, the electron is shared by two atoms, although perhaps being more strongly attracted to one than the other.

A word about notation may be in order here. Notation involves how chemical bonds can be written. A single bond, for example, can be indicated by a single line from one atom to another. Thus, when two hydrogen atoms join, the resulting hydrogen molecule (H_2) may be written as H—H. When an oxygen atom joins with two hydrogens, bonds can be shown as H—O—H. When two oxygen atoms join together, each of them requires two electrons, so their two shared electrons form a **double bond,** O=O. Look back at Table 4.1. Do you see why nitrogen can be written as N≡N?

Hydrogen Bonds

Both covalent and ionic bonds are regarded as rather powerful forces, but there are a number of other, weaker types of bonds as well. Perhaps the most important of these is called the **hydrogen bond.** Strangely enough, hydrogen bonds are important to life precisely because of their weakness. It takes very little energy to form them, and they are easily broken.

A hydrogen bond is formed when two electronegative atoms fulfill their shell requirements by sharing a single hydrogen atom. As an example, water molecules may be held together by hydrogen bonds. The bonds are formed because, as we have seen, the electrons are more strongly attracted to the oxygen than to the hydrogens. This means that the oxygen atoms, with their electrons drawn in close about them, become somewhat electronegative. This slight displacement leaves the hydrogen atoms (the protons from which the electrons were drawn) with a slight positive charge. They therefore have a tendency to attract negatively charged particles, and the nearest such particle is likely to be the oxygen atom of the next water molecule. Thus, water molecules tend to join in a loose, but highly structured and constantly changing latticework:

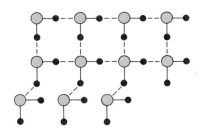

(The hydrogen bonds here are the small black dashes.)

Hydrogen bonds last for only a very brief time—about 10 raised to the negative 11th power (1/100,000,000,000) second. But, in their short existence, they bestow upon water the unusual qualities of being very fluid and, at the same time, relatively stable. The stability of water is quite easily demonstrated by considering the time it takes for it to boil. (Legend has it that if the pot is watched, the data may be unreliable.) Why does water so strongly resist being changed to a gas? It is resistant because although the heat easily ruptures the bonds between the molecules, they form new bonds with dazzling speed.

The peculiarities of hydrogen bonding also explain why it is so difficult to freeze water. Ice is crystalline; that is, it is regular and repeating in its molecular structure. But the constantly shifting molecules of water don't hold still long enough to encourage such a regular structure. The molecules are joined so weakly that they continually break and form new bonds.

Water, then, is quite stable, a trait that has given it special importance in the processes of life. But, you may ask, just why is this sort of stability so important to life?

Actually, there are a number of reasons. For humans and other terrestrial air-breathers, for example, water's stability, or resistance to change, enables us to breathe in a wide range of temperatures and in very dry environments by retarding the evaporation of moisture from our lungs. In addition, we are resistant to freezing because our bodies contain so much water. (The formation of ice crystals, of course, can rupture delicate cell membranes.) Furthermore, the unequal charge distribution of water molecules, which permits hydrogen bonding in the first place, also makes water a powerful solvent. Water can thus break down complex molecules, allowing their components to interact in new ways. In fact, all of the biochemical reactions in our bodies take place in water. Water's constantly changing structure also gives it a certain fluidity and movability that enables it to pass through our bodies' tissues and to seep deep into the earth's crust to reach the roots of the largest trees. Furthermore, because of water's powerful tendency to join, it can form "columns" and move from those roots to the highest leaves as these "columns" are drawn along as units.

THE MAGIC OF CARBON

Carbon has been described as the element most intimately associated with life, but you may have noticed that we haven't said much about it so far. Perhaps a look at Table 4.1 will show why. Carbon, you see, has six protons and six electrons, and thus it presents a special case. How might an atom like this be expected to react? After all, since two electrons will occupy its inner shell, it can only have four in its outer shell. In order to satisfy its shell requirements, then, it must either give up these four electrons or gain four more. Either of these changes, however, would throw its electric charge too far out of balance. The only way carbon can resolve this dilemma is to share its four outer electrons with other atoms through covalent bonding.

The problem becomes interesting at this point because carbon can combine with either electronegative or electropositive atoms. For exam-

FIGURE 4.10A Methane is the simplest hydrocarbon, composed of one carbon and four hydrogen atoms covalently bonded. Note that by sharing their electrons, the hydrogen atoms fill the shell requirements of the carbon.

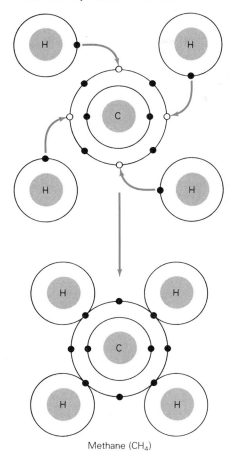

Methane (CH₄)

FIGURE 4.10B Methane is a common compound of ''swamp gas,'' which rises from the mud, lending a rather distinct odor to such places.

ple, it can share the electrons of oxygen, nitrogen, sulfur, or phosphorus atoms, or it can share electrons with hydrogen atoms. If a carbon atom simply adds four hydrogens, the result is CH₄, the swamp gas called methane (Figure 4.10).

It should be pointed out that the four hydrogens do not simply stick out at right angles, as the diagram seems to indicate. Instead, they protrude three-dimensionally in such a way as to be as far from each other as possible. Thus, the structure would be more accurately represented as:

This configuration is called a **caltrops.** In ancient times, spikes were arranged in such a way because when they were thrown in the path of an advancing army, one spike always pointed up.

In addition, because of their peculiar bonding properties, carbon atoms can interact with each other as well. In this way, carbon atoms can join together to form complex molecules, as we shall see shortly (Figure 4.11 shows some of the simpler carbon chains. It is interesting that of the some 2.2 million chemical compounds known, about 2 million contain carbon.)

FIGURE 4.11 Some of the simpler hydro-carbon chains.

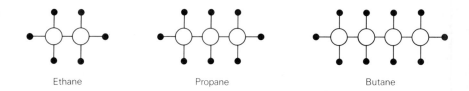

Ethane Propane Butane

The molecules produced by any combination of hydrogen and carbon atoms are called **hydrocarbons.** There are some very long hydrocarbon chains. You can see that in these kinds of molecules, the electron-shell requirements of every atom are satisfied; each carbon atom has four other atoms with which to share electrons, and each hydrogen atom is satisfied through a single covalent bond with a carbon atom.

Because of the properties of carbon, some very complex configurations can be formed. For example, carbon chains can branch. A five-carbon branch can be written as:

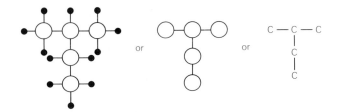

Carbon can also join ends to form rings. Five- or six-carbon rings are especially common:

The notation of such molecules can be further simplified and written as:

You should be aware that if no atom is shown where one should be expected, that atom exists and is assumed to be hydrogen. In such a case, it is assumed to be present. Other atoms attached to carbon are always noted.

Carbon tends to be associated with certain kinds of molecules. Among these are those called **functional groups.** Functional groups are molecular side groups with specific chemical responses. Among them are the **hydroxyl groups,** —OH, which combine with carbon to form alcohols,

 or — O — H

the **amino groups,** —NH2, which can form substances called amines,

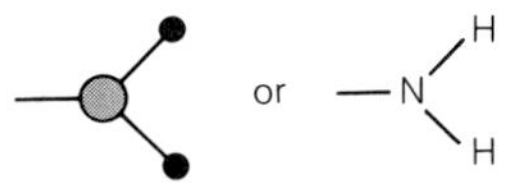

and the **carboxyl groups,** —COOH, (see Essay 4.1) which form acids:

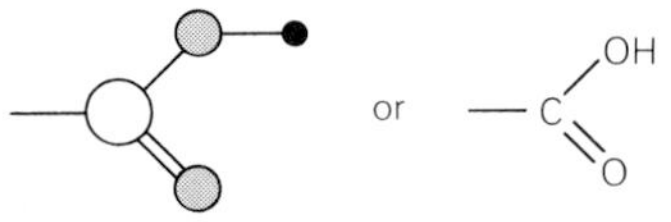

Functional groups tend to behave as a single atom. That is, they move together as a group when the larger molecule of which they were a part becomes dissociated. If they break away from the molecule and exist as unattached charged particles, they are called **free radicals.**

As we will see, the peculiar bonding of carbon not only permits the formation of large and complex molecules, but those sorts of molecules have precisely the traits that permit the complex processes of life. In fact, it can be said that because carbon is so intimately associated with the processes of living things, the chemistry of carbon is, in a sense, the chemistry of life.

CHEMICAL REACTIONS

Most of the molecules associated with life tend to join, separate, shift around, and change their nature. They interact, however, in specific ways, according to the properties with which they are endowed. Such interactions between molecules are referred to as **chemical reactions.** Let's now consider a few common chemical reactions and see why they occur. We will see that, essentially, they simply involve the breaking and forming of bonds.

To begin, consider an example in which two elements simply join. The general statement can be written:

$$A + B \rightarrow AB$$

That is, A plus B yields AB. Suppose these elements are sodium and chlorine joining to make table salt. We can simply plug in their chemical symbols and we have:

$$Na + Cl \rightarrow NaCl$$

Hydrochloric acid (HCl) is is formed in much the same way. Here, one ion each of hydrogen and chlorine forms one molecule of hydrochloric acid. This reaction, however, is a bit more complex to write since in this case, we must show the number of atoms involved. This is because the ionized hydrogen and chlorine would ordinarily be in molecular form, that is, as H_2 and Cl_2, so their union would form two molecules of hydrochloric acid. Thus, it would be written as:

$$H_2 + Cl_2 \rightarrow 2HCl$$

The hydrogen that forms water usually appears as molecular hydrogen (H_2); that is, two atoms of hydrogen are bonded to form a molecule. Of course, only one molecule of oxygen is needed to form H_2O, but oxygen also normally appears as a molecule (O_2). Here, we see, the bonds of oxygen must be broken in order for new kinds of bonds to form. Then, since each oxygen atom requires two atoms of hydrogen, we get:

$$2H_2 + O_2 \rightarrow 2H_2O$$

That is, two molecules of molecular hydrogen and one molecule of molecular oxygen form two molecules of water. Note that the same number of hydrogen and oxygen *atoms* appear on both sides of the arrow. Thus, the equation is balanced. This is an important principle.

Another important point is that some reactions can go the other way. So far, the "yields" arrow has pointed only to the right, but many reactions are reversible. For example, water can be broken down into hydrogen and oxygen:

$$2H_2 + O_2 \leftarrow 2H_2O$$

As a convenience, if the reaction is more likely to go one way than another, it is shown by two arrows, the larger one indicating the general tendency of the reaction:

$$2H_2O \rightleftharpoons 2H_2 + O_2$$

In another type of chemical reaction, the **reactants** (the molecules involved in the reaction) may simply switch partners. When molecules exchange components in this way, they form two completely different kinds of molecules. For instance, if hydrochloric acid (HCl) is combined with sodium hydroxide (NaOH), we get water (H_2O) and table salt (NaCl):

$$HCl + NaOH \rightarrow H_2O + NaCl$$

Notice, again, that the same elements, in the same quantities, appear on both sides of the reaction.

Energy of Activation

Some reactions occur very easily, but others only with great difficulty. In the easy cases, the reactions occur spontaneously if chemicals are simply mixed together. Dabbling amateur chemists sometimes blow up their labs this way. Those that proceed more reluctantly generally require an energy boost of some kind, just as a rock at the top of a hill may need a shove before it begins tumbling down. As a simple example, we know that a match can burn, but it can also just sit there in its box, doing quite nicely, thank you, without ever bursting into flame. To initiate the oxidation process, one adds energy, generally by striking the match on the back of the leg, unless the tuxedo is rented. The heat of friction provides the boost and initiates the oxidation process. Once the reaction has started, the energy is provided by the heat of the flame, and the process will continue until it reaches your fingers or until the wood is depleted. The energy boost required to initiate a chemical reaction is called the **energy of activation.** (In spontaneous reactions, normal environmental conditions provide the energy of activation.)

Acids are recognized by their sour taste. An acid is any substance that donates positive hydrogen ions (H^+) to a solution. Since H^+ is hydrogen that has lost its electron, it is simply a proton. Thus, acids can be considered proton donors. Hydrochloric acid (HCl) is a "strong" acid because it has a strong tendency to dissociate, or to release H^+ ions into its environment. It cannot be neutralized by the weak negative chloride ions, Cl^-, that are released by the same action.

Bases taste bitter and flat. They reduce the amount of H^+ in a solution. In other words, bases are proton acceptors. The ion OH^- is a powerful base because it will readily capture free H^+ ions to form H_2O. For this reason, when an acid and a base combine, they form a compound plus water. To use a familiar example:

$$NaOH + HCl \longrightarrow NaCl + H_2O$$

The acidity of a substance is measured on the pH scale. The term pH means "hydrogen power." The reference point on this scale is 7, the pH of pure water. Thus, substances with a pH value below 7 are acids and those with a pH value above 7 are bases. A substance with a pH of 7 is neutral. The scale is based on the concentration of hydrogen in one liter of water. For example, pH 3 means 10^{-3} moles of hydrogen ions per liter of water, and pH 8 means that there are 10^{-8} moles per liter. A **mole** is the number of grams of a compound equal to its molecular weight. Almost all biological processes take place in a pH environment of 6 to 8, with a few important exceptions such as digestive processes in the stomach (which occur at about pH 2). The pH level in living systems is sometimes regulated by **buffers,** acid-base pairs that serve to "soak up" small amounts of excess acid or base ions.

pH Range Showing Relative Number or H^+ and OH^- Ions in Solution

Strongest Acids	10,000,000,000,000	H^+ ions	at pH 1	10	OH^- ions	Concentrated Nitric Hydrochloric, Sulfuric Acids (stomach)
	1,000,000,000,000	H^+ ions	at pH 2	100	OH^- ions	
	100,000,000,000	H^+ ions	at pH 3	1,000	OH^- ions	
Weak Acids	10,000,000,000	H^+ ions	at pH 4	10,000	OH^- ions	
	1,000,000,000	H^+ ions	at pH 5	100,000	OH^- ions	
	100,000,000	H^+ ions	at pH 6	1,000,000	OH^- ions	Blood and Tissues of Organisms
Neutral	10,000,000	H^+ ions	at pH 7	10,000,000	OH^- ions	
	1,000,000	H^+ ions	at pH 8	100,000,000	OH^- ions	
Weak Bases	100,000	H^+ ions	at pH 9	1,000,000,000	OH^- ions	
	10,000	H^+ ions	at pH 10	10,000,000,000	OH^- ions	
	1,000	H^+ ions	at pH 11	100,000,000,000	OH^- ions	
	100	H^+ ions	at pH 12	1,000,000,000,000	OH^- ions	Concentrated Sodium Hydroxide
Strongest Bases	10	H^+ ions	at pH 13	10,000,000,000,000	OH^- ions	
	1	H^+ ion	at pH 14	100,000,000,000,000	OH^- ions	

Why, though, must a match be struck? How does the heat of friction provide the energy of activation? What does heat do? The addition of heat increases the **kinetic energy** of molecules; that is, it increases the speed of their random movements. (Molecules are in constant movement, even in solid substances.) The more active the molecules are, the greater the likelihood that they will bump into each other, and it is in just such encounters that molecular bonds are altered. The oxidation that produces flame is based on such changes in bonds.

Another way of increasing the likelihood of molecular interactions (or speeding up chemical reactions) is to increase the concentrations of the two reactants. The more molecules there are, the greater the likelihood they will interact. The same principle can be achieved by subjecting a mixture of the reactants to pressure, thereby forcing the molecules closer together. We see, then, that the chemical reactions can be encouraged or accelerated by a number of processes, such as adding heat, increasing the concentration, or increasing the pressure.

Such methods are fine for encouraging reactions in the laboratory, but how about chemical reactions in living things? Imagine the effects of heat, pressure, or varying concentrations of solutes in a delicately balanced living body. Such drastic changes would be disruptive to say the least. Obviously, living things have other ways to encourage chemical reactions; ways that are, perhaps, more compatible with the delicate mechanisms of life.

Enzymes

If you tried to impress your little brother by mixing hydrogen and oxygen to form water, his attention span would be severely taxed and he would probably wander off, because nothing would happen. However, should you coax him into watching again, add a tiny piece of platinum to the mixture. He may never watch you do anything again, because this demonstration will blow his hat off. Furthermore, after you recover your composure, you can also recover your platinum. This is because platinum in this case served as a **catalyst.** A catalyst, in the context in which we will consider it, triggers or speeds up reactions between other elements without becoming part of the product; it emerges unchanged. Catalysts do not change the type of reaction that can take place, or the direction of the reaction; they simply make it easier for the reaction to occur by lowering the necessary energy of activation. It is through this process of **catalysis**—chemical reactions initiated by catalysts—that the "cold" chemistry of life can take place.

The special types of catalysts found within living things are called **enzymes** (defined by one memorable student as "naval officers"). However, enzymes are very large molecules with very specific configurations. Technically, they are among the molecular group known as **proteins** (about which we will say more later).

The precise shape of the protein molecule is critical to the functioning of the enzyme. Its contorted configuration, produced as it loops and coils back on itself, produces a kind of groove or slot on the surface of the enzymatic mass. This slot is called the **active site** because of its role in the enzyme's chemical behavior. Lining the active site are certain functional groups, arranged very specifically. These functional groups are able to interact with certain combinations of molecules with which they come into contact and to change them in some way. The molecules they act upon are, together, referred to as **substrate.** When an enzyme alters the substrate, the resulting molecular combination is called the **product.**

Most enzymes are believed to work by placing stress on the arrangement of the substrate molecules, weakening their bonds and allowing them to interact with other molecules in new ways. In Figure 4.12, note how the inexact fit of enzyme and substrate could cause such bond changes in the substrate molecules. When the interaction is completed, the product breaks away and the enzyme, unchanged, is ready to repeat the sequence (see Figure 4.12). This whole process takes place so quickly

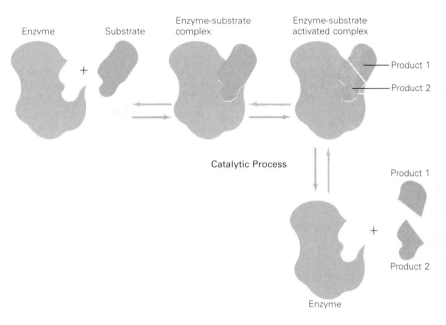

FIGURE 4.12 How an enzyme is believed to work. The enzyme, usually a complex protein, has an area that roughly fits the substrate upon which it will act. In this example, the enzyme is separating the substrate into its component parts, but in other cases, enzymes may join smaller molecules to form larger ones. It is believed that the enzyme doesn't precisely fit the substrate configuration so that when the enzyme/substrate complex is formed, the substrate is bent out of shape, or stressed, a bit. The stress exposes certain parts of the molecular substrate to other reactants in the medium, and the substrate molecule, which would otherwise have been quite stable, reacts. After the reaction, the substrate breaks away from the enzyme, leaving it unchanged and ready to initiate the same reaction with new substrates. Each type of enzyme reacts with only one kind of substrate.

that a single enzyme may go through several million such cycles each minute in a dazzling display of efficiency.

Enzymes are able to bring about several kinds of reactions. For example, they may encourage substrate molecules to *join* to other molecules. We see this as our bodies build large complex sugars from smaller subunits. Enzymatic action can also cause the substrate to be broken apart into its components. The disruption occurs when the altered substrate joins with H^+ or OH^- ions. Digestive enzymes, for example, function by such disruption. Figure 4.13 illustrates the basic enzymatic mechanism of starch formation and digestion.

FIGURE 4.13 Segments of starch and cellulose molecules. Starch is the form in which many animals store glucose units. (Notice the relationship between a link in this chain and an isolated glucose molecule.) The links (shown in shaded squares) in starch are easily broken by the digestive processes of many animals, but the links holding the units of cellulose together resist disruption by the digestive enzymes of all but a few species, such as protozoa that live in the digestive tract of termites. Since cellulose is a component of wood, we don't normally eat wood. Compare the kinds and arrangements of atoms in starch and cellulose.

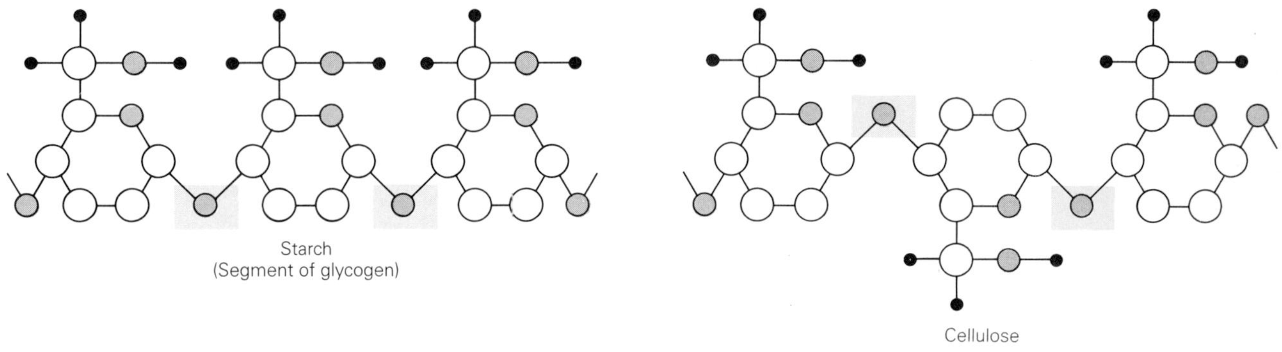

Energy Changes

As the molecules of life square dance their way through their complex pageant, they must alternately shift directions, change partners, stop, and go. But with each change, they alter their energy levels. It is important to keep in mind that chemical reactions generally result in a change in the energy level of the reactant molecules. In other words, the product of a chemical reaction will have either more or less energy than did the starting materials. If the resulting product has more energy, the process is

Starch
(Segment of glycogen)

Cellulose

said to be **endergonic** (or energy gaining); if it has less energy, it is called **exergonic** (or energy releasing).

It may have occurred to you that if a reaction results in a product with more energy than the starting materials, that energy would have come from somewhere—it would have to be *added* to the process somewhere along the line. Just as a boulder cannot roll up a hill without a push, neither can endergonic chemical reactions take place without energy somehow being expended. A general statement about endergonic reactions can be written as:

$$\text{Energy} \searrow$$
$$A + B \longrightarrow AB$$

(The addition or subtraction of some factor during a reaction is traditionally shown by a curved arrow at the step where it occurs.)

To continue our analogy, once we have gotten our boulder up the hill, it then holds considerable potential energy. Furthermore, if it should roll back down, that energy would be released along the way. The rolling down, then, is analogous to an exergonic reaction. When the product of a chemical reaction has less energy than the starting material (that extra energy being released during the reaction), the process can be summarized this way:

$$AB \longrightarrow A + B$$
$$\searrow$$
$$\text{Energy}$$

If the boulder should roll over you on its way down, the energy it releases might cause certain changes in your body. In this case, the boulder's energy would be doing *work* (causing changes). We'll see somewhat more useful ways energy can cause changes, but first let's review what sorts of molecules tend to wind their way through into this complex business we call life. We will consider only a few substances here, and they all have such familiar names that you may suspect that you already know what they are. For example, everyone knows what carbohydrates are, right? And fats? Let's see.

THE MOLECULES OF LIFE

We have considered the notion that life might one day be understood in terms of the molecules that make up living things. Whether this is so or not, we must certainly assume that the processes of life are dependent upon molecular interactions. Furthermore, in trying to retrace the development of life, we quickly became involved with the molecules of a raw and primitive earth. So although we still can't say exactly what life is and how it began, we can look at a fundamental aspect of life, its molecular composition.

Carbohydrates

The first important group of molecules we will consider that are associated with life are the "familiar" carbohydrates. (Some have discovered that it's really not easy to be very familiar with a carbohydrate.) Let's

FIGURE 4.14 Glyceraldehyde, glucose, and fructose are important simple sugars. They are called simple sugars because they are not combined with other sugars. Note that glucose and fructose may exist as straight chains or as rings. Interestingly, fructose formed from glucose in the male reproductive tract is an energy source for sperm. Glucose is an important energy source in many kinds of organisms. In humans, it is stored in the form of glycogen in a number of places, including the liver and the muscles.

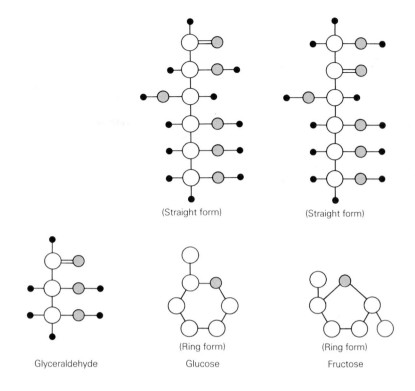

(Straight form)   (Straight form)

Glyceraldehyde   (Ring form)   (Ring form)
                         Glucose   Fructose

begin with a simple definition. **Carbohydrates** are molecules that contain carbon, hydrogen, and oxygen in a proportion of one carbon to two hydrogens to one oxygen atom (the general formula is thus CH_2O). (The simpler carbohydrates are also called **sugars,** and you may occasionally see the terms used interchangeably).

In a nutshell, carbohydrates are important because they are the principal energy source for most living things and because they serve as the basic material from which many other kinds of molecules are built. Some carbohydrates are simple, such as glyceraldehyde, glucose and fructose. Some carbohydrates, such as these latter two, may exist as either chains or rings. Other molecules of carbohydrates may be much larger, forming very long chains. The simpler carbohydrate molecules, those containing a single sugar group, are called **monosaccharides** or **simple sugars** (Figure 4.14). A carbohydrate composed of two simple sugars is called a **dissaccharide,** and more complex molecules may be called **polysaccharides.** The very large ones are simply called carbohydrates.

One way that simple carbohydrates can be joined is by **dehydration**—that is, the removal of the atoms that form water. For example, table sugar (sucrose) is formed when the atoms that comprise water are removed from glucose and fructose. We say that the dehydration permits **condensation** of the two simple sugars. Dehydration is a common chemical means of chemical reaction. On the other hand, when the components of water are *added* **(hydration)** to any molecule formed by condensation, the molecule will be broken down back into its component substances, a process called **hydrolysis.** (Figure 4.15 illustrates a simple example of condensation and hydrolysis involving sucrose.)

FIGURE 4.15 Sucrose, or table sugar, is a double sugar composed of both glucose and fructose. In the formation of sucrose, glucose loses an OH group that joins with an H from fructose to form water. The two simple sugars then join at the points at which these atoms were lost. The reaction does not proceed easily, and in living systems, an energy source must be provided. When the atoms of water are chemically added to sucrose, it breaks down into glucose and fructose, its component parts. Reactions involving loss of water are called *dehydration;* those in which water is added are called *hydration;* and they result in a chemical breaking apart known as *hydrolysis.*

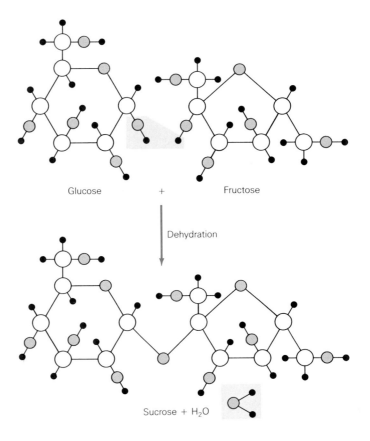

Glucose + Fructose

Dehydration

Sucrose + H_2O

Consider a vastly more complex carbohydrate also formed by dehydration, **glycogen.** Glycogen is an important carbohydrate formed by many animals. It is essentially composed of long chains of glucose units, held together, after dehydration, by certain bonding configurations called **alpha linkages,** as we saw in the starch diagram of Figure 4.13. In mammals, glycogen is the principal form in which glucose is stored in the body, chiefly in the liver. When your body, for example, is in need of glucose, glycogen is hydrolyzed (breaking the alpha linkages) and the component glucose molecules are released into your bloodstream.

Glucose units may also be joined after dehydration by a somewhat different configuration called **beta linkages,** producing **cellulose,** a relatively insoluble polysaccharide common in plants. Our digestive systems can readily break down starches into their component parts, but the beta linkages of cellulose make it resistant to the digestive enzymes of many animals. This is why people don't each much wood and rarely graze. Of course, termites eat wood and cows graze, but they have help; their digestive tracts harbor tiny organisms that can break down cellulose into its component sugars.

The hard covering, or exoskeleton, of insects is also composed of units of glucose bonded like cellulose, so that they are not easily digested by many species. However, on this point my confidence was once shaken as I followed a young man down a street in Austin, Texas, during one of the area's regular cricket invasions. He suddenly scooped a handful of the hapless insects from a wall and stuffed them into his mouth. I suppose I could have explained to him the nature of sugar bondings, but I walked into a trash can instead.

FIGURE 4.16 A fat molecule is composed of a glycerol molecule joined to fatty acids. Three molecules of water are lost in the bonding of three fatty acid molecules to a single molecule of glycerol. The fatty acids form long hydrocarbon chains the ends of which link with the glycerol molecule. The key questions that determine the nature of the fat are: how long are the chains, and are the fatty acids saturated or unsaturated (as we see in the next figure).

Fat molecules store a great deal of energy, but it is not easily retrievable. Marathon runners normally expend their available glucose after two hours of running and then must switch to fat metabolism. For some reason, the shift is considered traumatic and seems to be more easily accomplished by women than men.

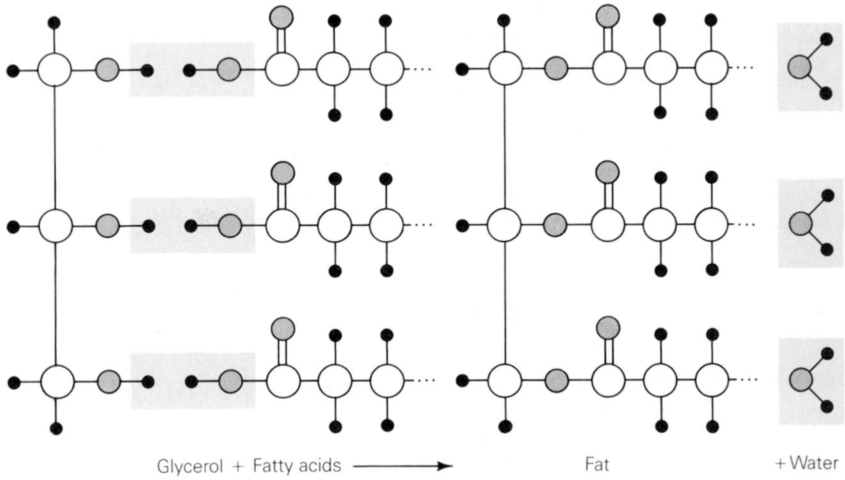

Glycerol + Fatty acids ⟶ Fat + Water

Lipids

Like carbohydrates, **lipids** are composed of carbon, hydrogen, and oxygen. Lipids, however, have far less oxygen. Also, unlike carbohydrates, lipids are insoluble in water.

There are a number of types of lipids, but **fats** are the best known. The fats are important because they store a great deal of energy. In fact, they contain twice as much energy for their weight as do carbohydrates or proteins. You would think, then, that fat people would be bursting with strength and vigor. You'll see as we go along why this is usually not the case. But first, let's take a look at the composition of fats.

Each molecule of fat consists of two types of compounds, **glycerol** and **fatty acids.** Glycerol is a chain of three carbon atoms, each with a hydroxyl (OH) group. Fats are formed by the hydroxyl groups of the carboxyl (COOH) joining with the hydrogens of the glycerol units (also called acid groups) of fatty acids after dehydration (the removal of water). The process is outlined in Figure 4.16.

Fatty acid chains can be highly variable. Most are sixteen to eighteen carbon units long, but some are shorter and some longer. Also, not only may the three fatty acid components be of different lengths, but fats may differ in the way the carbons are bonded together. For example, in some fats, the carbon chains carry all the hydrogens possible, while in others, the carbon atoms have fewer hydrogens because they are joined by a double bond (Figure 4.17). A fat with its carbon chains filled with hydrogens is said to be **saturated.** Guess what one with double bonds—that is, with room for more hydrogens—is called.

The lipids also include the **waxes,** molecules with very long and saturated fatty acid chains. These are usually solid at room temperature. And then there are the **phospholipids,** in which the third hydroxyl group of glycerol is attached to a molecule containing phosphorus. Phospholipids are important in the structure of living membranes, as we shall see.

The last lipid group we will consider is the **steroids.** The steroids are fatlike substances with a different structure than anything we have considered. Steroids are formed from four interlocking carbon rings with many types of side groups. For example, cholesterol is a steroid (Figure

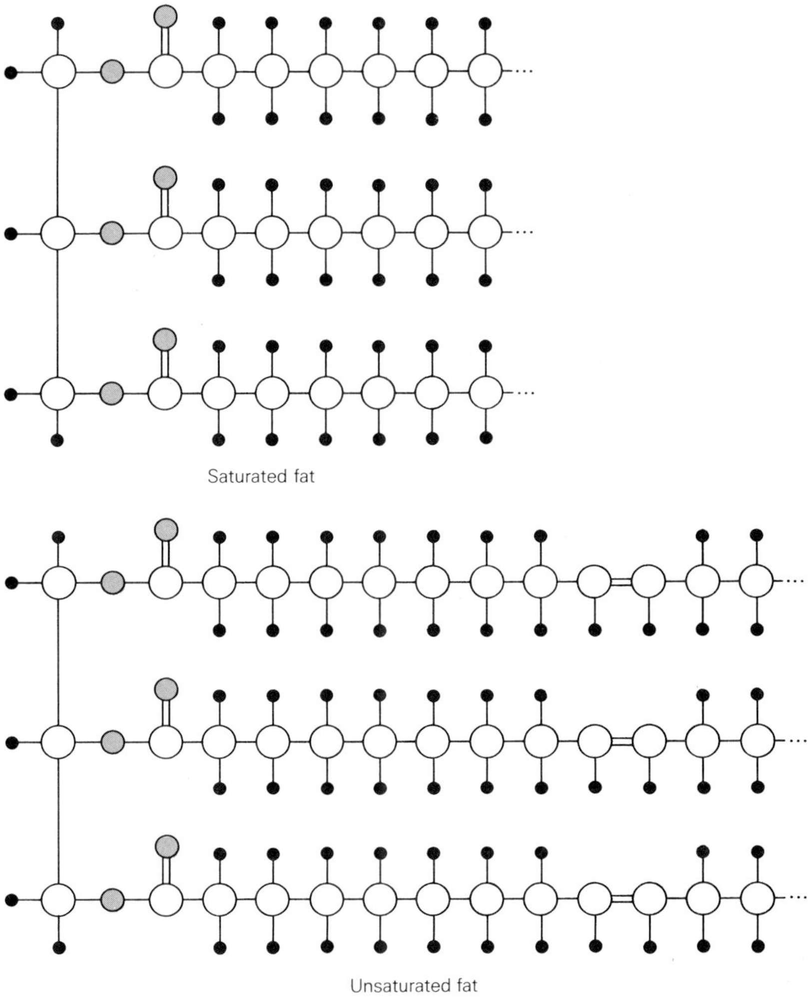

FIGURE 4.17 Unsaturated and saturated fats. In recent years, a generally overfed American populace has become increasingly concerned about the quality of all the foods available to them. For example, are our fats saturated or unsaturated? Have you heard advertisers use the term "polyunsaturated"? Househusbands or housewives may not know that unsaturated simply means that carbon atoms are linked together by covalent bonds of two pairs of shared electrons—in other words, by double bonds. Carbon atoms linked by double bonds are called unsaturated because they are able to form new bonds with other atoms. Unsaturated fats are generally oily, like olive oil, while saturated fats, like lard, are usually solid at body temperature.

Saturated fat

Unsaturated fat

4.18). Some people develop a greater interest in steroids when they learn that sex hormones are included in this group. (Certain synthetic male hormones have been abused by athletes seeking greater bulk and strength.)

Proteins

The general structure of proteins is easy to describe, but the simplicity is deceptive. Proteins are actually immense molecules with incredibly complex arrangements. Let us not be deterred, however, as we eagerly plunge ahead into new realms of knowledge.

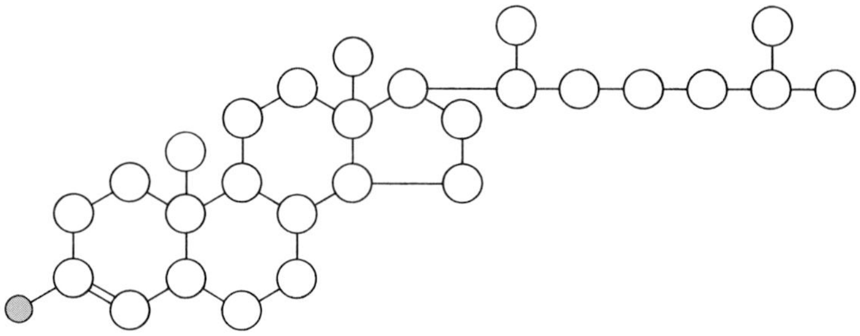

FIGURE 4.18 Cholesterol is a steroid. All steroids are composed of four basic interlocking rings, but a variety of side groups may be attached to those rings. Cholesterol is abundant in foods such as milk, butter, and eggs, and it may be synthesized in the liver as well. Cholesterol circulates in the bloodstream and, if present at high levels, it may be deposited in the aorta, the artery leaving the heart, possibly causing arteriosclerosis and heart attacks. As it circulates through the bloodstream, it may be picked up by specialized cells of the sex and adrenal glands and altered to form the steroid sex hormones. No one yet knows how these hormones work their wonders. So if steroids don't kill us outright, they at least seem to keep us in a lot of trouble.

Perhaps paradoxically, we will, of necessity, be dealing specifically in generalities. Furthermore, many of our generalities will be based on assumptions, since we know the actual structures of only a few proteins. Fortunately, however, proteins are composed of smaller building blocks that we do know something about, and we do have an idea of how these smaller units are put together.

Amino Acids

Proteins are made up of chains of smaller nitrogen-containing molecules called **amino acids.** As you may guess from the name, amino acids contain at least two kinds of functional groups, amino groups (NH_2) and acids (COOH). Both these groups, by the way, are attached to the same carbon. The rest of the molecule contains carbon and may exist in a variety of forms, including chained, branched, or looped. So, for the sake of simplicity, we'll just call the group attached to the carbon that holds the amino and the acid group, R, for radical. For a molecule to qualify as an amino acid, then, it need only have the configuration:

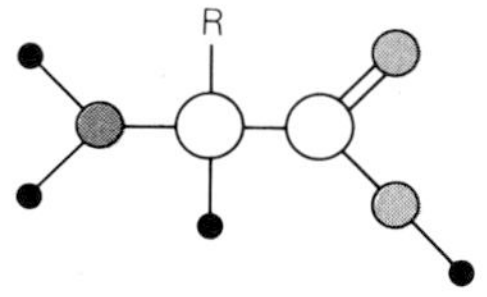

There are many amino acids, but only twenty are found in proteins. Some of these are shown in Figure 4.19. As you can see, the carbon group R determines the name of the compound.

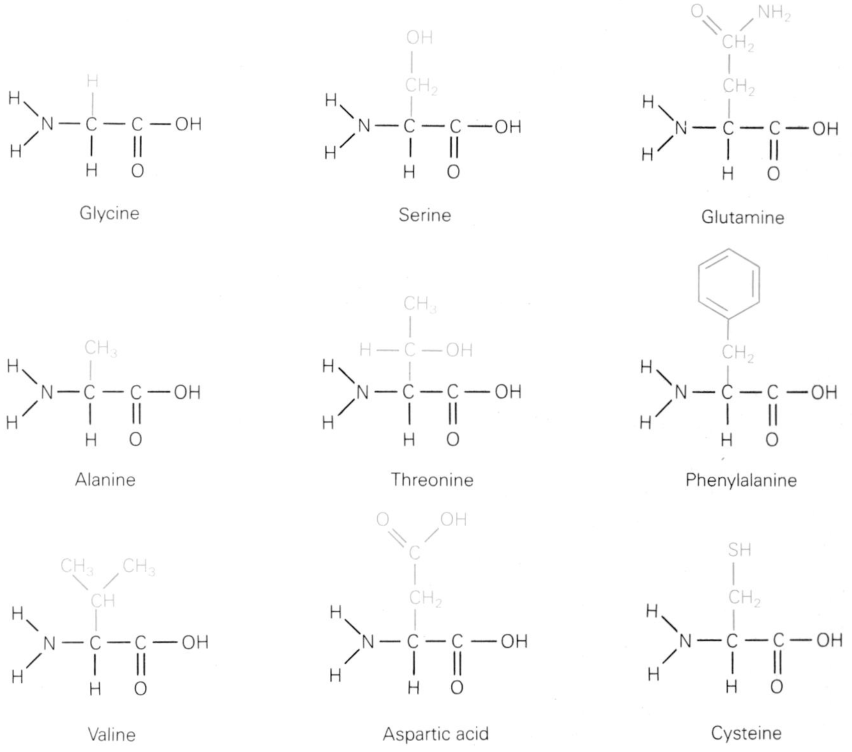

FIGURE 4.19 Some of the amino acids found in proteins. Note the variety of structures among amino acids, but also note that they all have two things in common: an amino group and an acid, or carboxyl, group. There are only twenty amino acids in all, and the body can manufacture twelve of these from other amino acids. It is essential, however, that the other eight be provided in the diet, so they are called the "essential" amino acids. It seems paradoxical that the amino acid element that is probably available from the fewest sources is nitrogen. And yet it is this same nitrogen that provides us with such problems with its disposal, as we will see in our discussion of the kidney. The importance of amino acids, of course, lies in their role in the formation of proteins.

Glycine   Serine   Glutamine

Alanine   Threonine   Phenylalanine

Valine   Aspartic acid   Cysteine

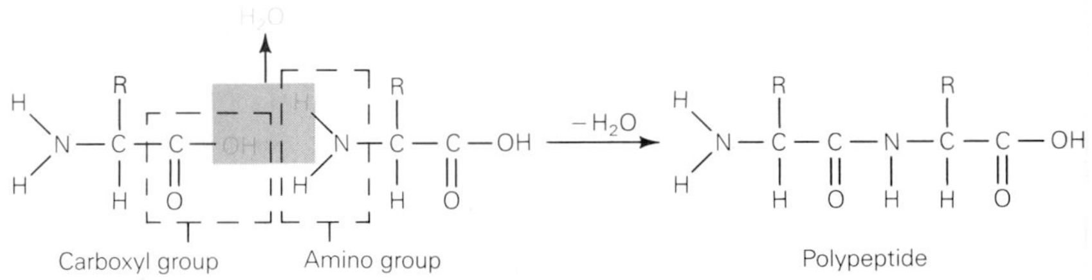

FIGURE 4.20 The carboxyl group of one amino acid can join with the amine group of the next amino acid by each contributing to the formation of water. The result is a peptide bond. Polypeptides, then, and their larger cousins, proteins, are chains of amino acids joined by peptide bonds.

Carboxyl group   Amino group   Polypeptide

Peptides

A chain of up to 300 amino acids may join together to form long **peptide chains** (or **polypeptides;** *poly,* many). Peptide chains, sometimes simply called **peptides,** are smaller than proteins, technically. However, they can join to make up a protein. The amino acids in polypeptides or proteins need not be of the same type, but they are always linked together in the same way: the OH part of the carboxyl group of one molecule joins with the H of the amine group of another by dehydration, as shown in Figure 4.20. The linkage between amino acids is called a **peptide bond.**

It is the **sequence** of amino acids in the chains that determines the biological character of the protein molecule. It has been discovered that a change in the position of even a single amino acid in a sequence of thousands can alter or destroy the activity of the protein. Imagine the incredible variety of proteins that might be possible by simply varying the number and sequence of twenty amino acids. It turns out, in fact, that although all proteins utilize only those twenty amino acids as building blocks, almost every species on earth has proteins that are peculiar to it alone.

Protein Structure

A protein molecule may be described in several ways. For example, its amino acid sequence determines its **primary structure.** (For the story of the discovery of one such sequence, see Essay 4.2.) Then, because of hydrogen bonds that form between every fourth amino acid, the protein molecule falls back on itself, forming a spiral or helical shape, somewhat like a circular staircase. These precise coils describe its **secondary structure.** Proteins that have "structural functions" in the body—that is, those that protect the body or keep it from falling apart, such as hair, nails, cartilage, bones, and tendons—all contain rather simple fibrous proteins that can be described only by their primary and secondary arrangements.

Other, more complex, proteins fold back on themselves in what is described as a **tertiary structure.** These folded and contracted proteins form dense clumps, or globules (see Figure 4.21). Enzymes, for example, are usually globular proteins with complex tertiary structures. Any such protein maintains its tertiary structure by means of hydrogen bonds, hydrophobic (water-repelling) forces, and disulfide linkages (bonds between the sulfur groups of certain amino acids along the chain). Tertiary bonds such as these are weak and can be broken easily by heat or acid, for example. Once such tertiary bonds are broken, the molecules become fibrous (part of a process called **denaturization**). Denaturization involves altering a protein in such a way that its chemical properties are

FIGURE 4.21 The tertiary folding of a hypothetical protein. *R* groups would extend outward in every direction, filling the internal areas and building a roughly globular structure. Globular proteins, then, are actually quite dense. The tertiary formation is maintained mostly by weak hydrogen bonds, by repelling interactions between different parts of the chain, and by linkages between sulfur groups. All these can be broken quite easily, for example by heat, and the former structure can never be regained. The process is called *denaturation* and, in general, it involves unfolding tightly coiled peptide chains so that they form a random configuration. Of course, denatured proteins lose their former biological activity.

In 1954, after ten years of painstaking research, Frederick Sanger and his colleagues at Cambridge University in England were able to tell us the full sequence of the amino acids in insulin, a rather small protein. This work, for which Sanger was awarded a Nobel Prize, consisted of breaking the insulin molecule into smaller pieces of protein and then analyzing the pieces. First, he subjected the entire molecule to strong acid to break it down into its component amino acids. He was then able to identify each of these amino acids by a process called chromatography. Chromatography involves filtering the product down through a column of an absorbent material for which various kinds of molecules have different affinities. Each amino acid thus adheres to the material at a different rate and so is stopped at a specific point as it moves down the column. By comparing his results with the known absorbing point of each acid, Sanger determined which amino acids were present in insulin. He was also able to learn something about their relative amounts.

To discover the sequence of the amino acids in the insulin molecules, Sanger used more specific hydrolyzing agents that break only bonds between certain molecules in the long chain. Pepsin, for example, hydrolyzes bonds only between tyrosine or phenylaline and other amino acids. To determine which end of the segment was free and which was attached, he added a chemical (dinitrofluorobenzene) that attaches to any terminal amino group and turns the compound a bright yellow.

After analyzing many such short segments, Sanger was eventually able to work out their structures, and where the terminal amino acids of these groups were identical, he could overlap them to reconstruct longer chains. For example, an overlap of the segments

$$A=C=X=N=O=F$$
$$N=O=F=G=O=I$$

yielded the longer segment

$$A=C=X=N=O=F=G=O=I.$$

Later, he was able to determine how the two component molecules of insulin were joined by bonds between sulfur atoms of cysteine molecules. Thus, after years of patiently tracking down small bits of information, one step at a time, Sanger was eventually able to put all the pieces together and tell us for the first time the structure of a relatively simple protein.

irreversibly destroyed. This is what happens when you cook the protein-rich white of an egg. The irreversibility is apparent if you have ever tried to unfry an egg.

Finally, two or more globular proteins may twine around each other to form a **quaternary structure.** Later, we will see that **hemoglobin,** the blood's oxygen-carrying molecule, has such a configuration.

Nucleotides

We should now briefly mention a group of molecules called **nucleotides.** They are not particularly complex, nor are they large, as molecules go. However, we will soon see (Chapter 6) that they have had a pivotal importance in the development of life on this planet.

FIGURE 4.22 The five nucleotides that are found in chromosomes: adenine, thymine, guanine, cytosine, and uracil.

Thymine (T)   Cytosine (C)   Uracil (U)

Pyrimidines (single ring)

Adenine (A)   Guanine (G)

Purines (double ring)

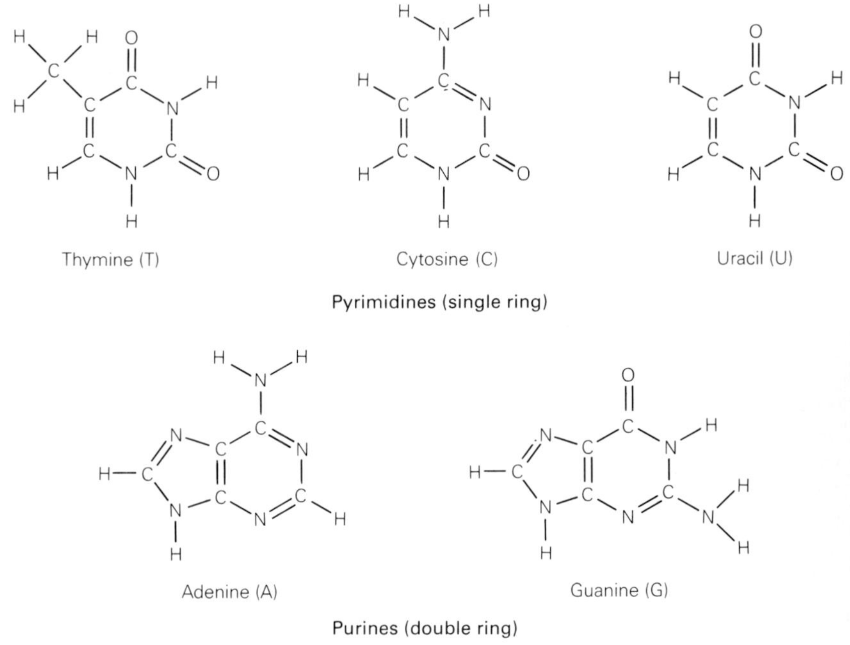

Essentially, nucleotides are composed of a sugar (called 2-deoxy-D-ribose), a phosphate group, and a nitrogen-containing base. Figure 4.22 illustrates five nucleotides that we will encounter later. Thymine, cytosine, and uracil have a single ring and are called **pyrimidines.** Adenine and guanine, with their double rings, are called **purines.** Only a few such nucleotides, we now know, provide the molecular basis for inheritance (to be discussed in more detail in Chapter 6.)

With this brief introduction to some of the important molecules of life, let's now see how these molecules interact. Their interactions, after all, fundamentally define the complex and changing nature of life.

SUMMARY

The chemicals that comprise living things are those that are found in the environment in nonliving material. Only six elements (CHNOPS) make up 99 percent of living matter. Elements are composed of atoms of a single type. A compound is comprised of a single kind of molecule. Molecules are formed from two or more kinds of atoms. Elements cannot be broken down into simpler substances by chemical means. Atoms are comprised of protons, electrons, and usually neutrons. The number of protons determines an atom's atomic number. The protons and neutrons together determine its atomic weight. Electrons orbit the nucleus along potential pathways called shells. If the outer shell is not filled, the atom can react with other atoms.

Atoms of a single element can vary in a number of ways. Elements are determined by the number of protons in their nuclei. Isotopes are atoms of a single element with different numbers of neutrons. Normally, the number of protons equals the number of neutrons. If these numbers are unequal, the atom bears a charge (+ or −) and is called an ion. Atoms tend to react with each other as they stabilize themselves by forming various kinds of bonds. Ionic bonds involve the pairing of atoms with opposite charges; covalent bonds involve atoms sharing electrons; and hydrogen bonds are formed when atoms share a single hydrogen atom. Carbon has special bonding properties because it has only four electrons in its outer shell. Because of its special properties, it can serve as the backbone for large and complex molecules.

Chemical reactions involve forming and breaking bonds. In exergonic reactions, energy is released; in endergonic reactions, it is absorbed. The reactions of life frequently require enzymes, chemicals that lower the energy of activation for specific reactions.

The molecules most associated with life are carbohydrates, fats, proteins, and nucleotides. Each has specific roles in life processes and all are carbon-based.

Chapter 5

ENERGY: THE DANCE OF LIFE

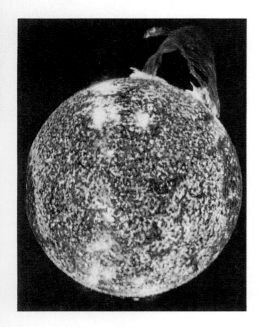

As I sit on my porch swing, the sun filters through the leaves of the forest to mottle my pages. To my left are tracks of a possum (a welcome guest) that wandered by last night. I suspect that when I was a boy growing up in Arkansas I probably ate a number of that possum's relatives. I find that this admission is best not made in some circles, but I did once mention something like that to a visitor. He was chagrined at the idea. That seemed strange to me because he was a member of the often-feared Waorani tribe of the Amazon's Conanaco River, and he and I had hunted there together and had eaten monkeys, parrots, wild pigs, and capybaras. Those were somehow okay to him, whereas the unfamiliar possum was not. (To set the record straight, possums taste better than monkeys.) We at least agreed that diets are set in complex ways, and that our food comes from a variety of sources. Some animals, then, including humans, derive part of their energy needs from the energy stored in the tissues of a variety of other kinds of animals. Their bodies contain the energy that we need. These prey animals may receive their energy from yet other animals, or they may take it from the leaves. The leaves, we will see, are able to make their food by using sunlight. Both the leaves and the animals, though, are constantly concerning themselves with their energy needs. However their energy is derived, one thing is certain: all life would abruptly end without it.

FIGURE 5.1 The energy of sunlight dances and courses its way through living systems in innumerable directions.

Energy not only stems from a variety of sources, it also exists in a variety of forms. The search for energy is a pervasive theme in the pageant of life, and while we are generally aware of the importance of energy, we often fail to understand just how many ways it reaches us and how many ways our search affects us. We will consider a few of these questions here. However, since the concept of energy involves a measure of intuitive reasoning, we must avoid the trap of false familiarity and see if we can organize our ideas a bit so that they will begin to make some sense in our search to better understand this thing called life.

Energy is defined as the ability to do work (that is, to move matter). It exists in many forms, at many levels, and it is on the move. It enables and it destroys as it ebbs and flows in countless directions throughout the universe. As it shifts about, it can be trapped and stored. Energy that is stored in this way can be released to do work at some later time, as we saw in the last chapter. There we found that molecules can be rearranged to hold more energy and then converted back to their former structure, releasing this stored energy. When this happens, the molecules revert to a lower energy level. A person holding a jar of nitroglycerine would

FIGURE 5.2 The potential energy within some chemicals can be suddenly reduced, making that energy available to do work.

probably be obsessed by a single hope: that the molecules will *not* rearrange themselves to a lower energy level. We can say, then, that a jar of nitroglycerine has **potential energy** (or **free energy**). When potential energy is released, it becomes **kinetic energy.** The jar of nitroglycerine and a rock at the top of a hill both have potential energy that can be converted to kinetic energy. With the release of kinetic energy, work can be done (Figure 5.2). (When energy is used to do work, it can be beneficial or harmful to life.)

Both potential and kinetic energy can take many different forms. For example, a car battery has potential **electrical energy.** (We might also refer to it as chemical energy.) When the electrical energy is released to turn the starter, it becomes **mechanical energy.** As the parts of the starter move, friction causes some of the initial energy from the battery to be dissipated as **heat energy.** Thus, we see that not only can energy exist in different forms, it can also be converted from one form to another.

THE LAWS OF THERMODYNAMICS

We are about to consider a very impressive topic, as you can see by the heading. However, some of the mystique may be lost when you see the simplicity of the basic laws of thermodynamics that govern the behavior of all energy, regardless of its form.

According to the **first law of thermodynamics,** energy can be neither created nor destroyed, but only changed from one form to another. This means that the energy released as our boulder topples down the hill is directly proportional to the amount of energy it took to push it up the hill in the first place. It also means that the heat energy that is "lost" as the tumbling boulder produces friction and releases heat is not actually lost, it is merely unavailable for use in that system; that is, this heat can't be used to do work.

The **second law of thermodynamics** tells us that in any energy transformation, not all the energy is transferred. Some energy will be "lost" simply because no transformations are 100 percent efficient. Thus, after each energy transformation, the available energy in a system will be slightly lower than before. Thus, as a battery turns a starter, the starter produces less energy than was available in the battery. Also, in chemical reactions, the product will contain less energy than did the reactants, unless energy is somehow added. All systems involving energy transformations, then, tend to fall to their lowest possible energy level.

Another aspect of the second law is that all processes tend toward **increased entropy.** Entropy is a measure of disorder, which means that as the available energy decreases in a system, the system "winds down" or becomes increasingly disorganized. Energy, then, must be added to a system to keep it in an organized state.

With a little imagination, you can see that these laws of energy have some interesting implications for living systems. After all, since life is highly organized, it must expend energy simply to remain organized. Therefore, living systems must have both a source of energy and some means of converting the energy from one form to another.

In the biological world, the ultimate source of virtually all energy is the sun. This energy is captured, changed, shifted about, and filtered

FIGURE 5.3A ATP is actually adenylic acid (comprised of adenine, ribose, and a phosphate) attached to two more phosphates. The last two phosphates are attached by high-energy bonds. When these bonds are broken, large amounts of energy are released, energy that is available to do work.

GOT TO GO TO WORK —
SO I'D LIKE TO CASH IN THIS GLUCOSE FOR SOME ATP

through living systems, where it is used to reduce entropy. The sun's energy generally begins its useful progression in green plants, where it is first stored in certain molecules. The various stages in life's cycles extract their share of energy from these molecules until finally all that remains are carbon dioxide and water—molecules with very little energy left. Let's now consider a few of the chemical reactions that result in your having the energy to do whatever it is you would rather be doing.

ATP: THE ENERGY CURRENCY

The standard energy currency of cells is a fairly simple molecule called **adenosine triphosphate,** or **ATP** (Figure 5.3A). ATP has been referred to as *energy currency* because it must be "spent" in order for work to be done. It is spent by breaking one of its high energy bonds, releasing inorganic phosphate (or P_i), and converting the molecule to **adenosine diphosphate,** or **ADP.** The energy that was used to form the bond that attached the phosphate is released when the bond is broken. In other words,

$$ATP \longrightarrow ADP + P_i + energy.$$

As everyone keeps trying to tell you, in order to have something to spend, work must be done, and so it is with the chemistry of life. In order to produce ATP, energy must be expended (work must be done) somewhere along the line. That energy is used to put a molecule of inorganic phosphate back onto an ADP molecule, a place it really doesn't "want" to go. It has to be forced back onto the ADP molecule by the expenditure of energy. Thus,

$$ADP + P_i + energy \longrightarrow ATP.$$

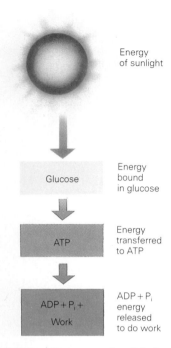

FIGURE 5.3B The energy of sunlight is used by plants to make glucose. The energy stored in glucose is then released to make ATP, which can be broken down to ADP and inorganic phosphate (P_i) with the release of energy. That energy is then available to do work.

So, ATP provides the energy necessary for life's functions, and energy is necessary for the reconstruction of ATP. Where does the energy necessary to put the phosphate on the ADP come from? Right! From the sun. What we will see now is how green plants use sunlight to form ATP, which they then use to make food (glucose), and how animals, by eating plants, or by eating plant eaters, use that stored food to form ATP of their own (Figure 5.3B).

PHOTOSYNTHESIS

Essentially, **photosynthesis** is the process by which organisms such as plants and algae utilize the energy of sunlight to make organic compounds (food). They are able to capture that energy because of their color.

Green plants get their color from pigments called **chlorophyll,** which initiate the photosynthetic mechanism by capturing the energy in light. In plants there are two kinds of chlorophyll, *a* and *b*. They, along with an orange pigment called **carotene,** comprise the pigments of most plants.

There are two distinct steps in the photosynthetic process, one that requires light (the **light-dependent reaction**) and one that doesn't (the **light-independent reaction**). The net result of this vital drama is that carbon dioxide and water, in the presence of chlorophyll and certain enzymes, are converted to glucose and oxygen. The reaction can be written:

$$6CO_2 + 12H_2O + \text{sunlight} \xrightarrow[\text{enzymes}]{\text{chlorophyll}} C_6H_{12}O_6 + 6O_2 + 6H_2O.$$

Many of the details in this process remain cloaked in mystery to this day, especially those of the light-dependent reactions. However, more is understood than you would probably care to know, so we will consider only the basic processes of photosynthesis.

The Light-dependent Reaction

Photosynthesis begins when light strikes a pigment, such as chlorophyll (we will use chlorophyll as our representative pigment), and initiates the light reaction. The pigments are found within the **chloroplasts** (see Essay 5.1).

The chloroplasts are tiny, double-membraned bodies found within the photosynthetic cells. As we will see in Chapter 6, inside the chloroplast is a jellylike matrix called the **stroma,** in which are embedded stacks of disks called the **thylakoids.** Each stack is called a **granum.** The thylakoids of different grana are joined by membranous **lamellae.** Embedded in the outer thylakoid membrane are the tiny bumps called **CF$_1$ particles,** each penetrated by a minute channel leading from inside the disk out to the stroma. (We will go over all this again in the next chapter, but for now refer to Figure 5.4.)

Photosynthesis begins when the chlorophyll within the chloroplasts is activated by the energy of light. This causes two of chlorophyll's electrons to move out into a more distant orbit; that is, to a higher energy state. From here, they can leave the chlorophyll entirely and be captured

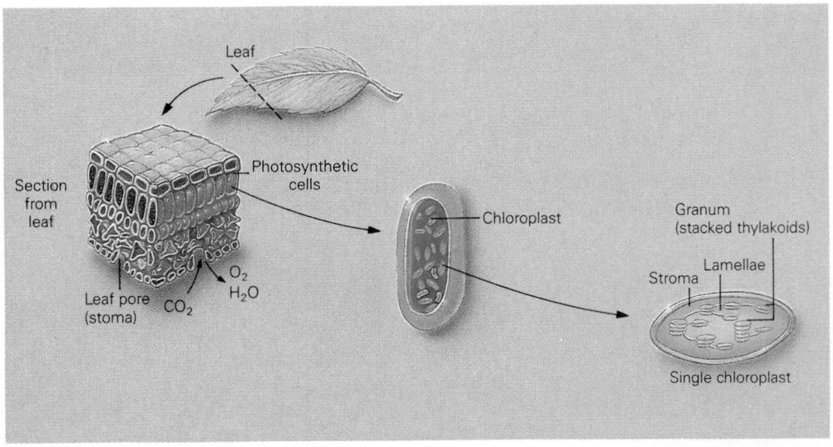

FIGURE 5.4A The location of chloroplasts within the cells of a green leaf. Tissues within the leaf contain vast numbers of photosynthetic cells, each with numerous chloroplasts. Within the chloroplasts are membranous grana, stacks of thylakoids in which the sunlight's energy first interacts with the biological realm. Surrounding the thylakoids are the clearer regions of the fluid stroma, where the final, carbohydrate-synthesizing events (light-independent reactions) occur.

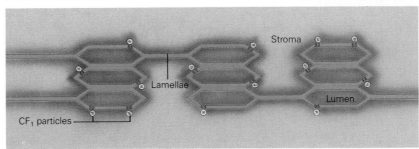

FIGURE 5.4B The position of the CF_1 particles on the thylakoid membrane. Actually, they are quite numerous and randomly scattered over the membrane surface.

FIGURE 5.5 In photosystem II, when the chlorophyll is activated by light, two of its electrons move outward to a higher energy state, from where they are captured by an electron acceptor. The lost electrons are replaced by other electrons that have been torn from water. The oxygen of the water molecule is then released as a gas.

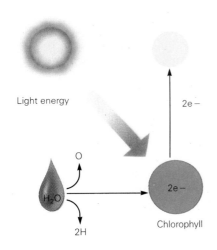

by a molecule called an **electron acceptor.** The chlorophyll replaces its lost electrons by literally tearing two of them from water molecules. When water molecules are broken apart in this way, the oxygen is released as a gas. These are the essential events of the sequence known as **photosystem II** (Figure 5.5).

From the electron acceptor in photosystem II, the electrons are passed along to the beginning of an **electron-transport chain.** The chain consists of a series of other electron-acceptor molecules, each at a lower energy level than the one preceding. This means that the electrons progressively lose their energy as they are passed "downhill" from one molecule to the next. As the electrons are passed along, some of the protons from which they were stripped follow them because of their electromagnetic attraction. As the electrons are passed to increasingly lower energy states, giving up energy at each step, that energy is used to pump some of those accompanying hydrogen ions into the interior of the thylakoids (the **lumen**). We will soon see why this is an essential step, but first let's follow the electrons through their sequence.

When the electrons have reached the lowest energy level along the chain in photosystem II, they are transferred to another pigment system called **photosystem I.** (It is designated as **I** because it is believed to have evolved first.) Once the electrons enter photosystem I, they do not languish long. Light energy again boosts them to another electron acceptor, this one at an even higher energy level than the first acceptor in system II. From this acceptor, the electrons, some accompanied by their pro-

Chlorophyll is a pigment molecule found in the chloroplasts of all green plants. The green color you see in such plants is actually caused by light rays from the center portion of the spectrum that are reflected back to your eye by the molecular structure of the chlorophyll. The colors you do not see, the reds and purples at either end of the spectrum, are the ones that are absorbed. This scheme is not coincidental. The red and violet light rays are the ones that contain the highest concentration of light energy, energy that the chlorophyll uses in the vital process of converting light energy from the sun into usable energy for all living things.

Chlorophyll is usually a mixture of two different compounds known as **chlorophyll a** and **chlorophyll b:**

$$C_{55}H_{72}O_5N_4Mg$$
Chlorophyll a
and
$$C_{55}H_{70}O_6N_4Mg$$
Chlorophyll b

Energy in plants is generally stored in the form of glucose, which is made available to animals that eat the glucose in one form or another. Glucose is first broken down by glycolysis and the product may then enter the Krebs cycle. Very little ATP is produced by these processes, but electrons are pumped across a membrane, setting up a chemiosmotic gradient. As the electrons rejoin, they power a battery of enzymes in the F_1 particles of mitochondria where a great deal of ATP is produced.

Most green plants have more chlorophyll a, although a few, including some algae, primarily utilize chlorophyll b or some related compound.

Each cell may contain as many as eighty of these small bodies that hold the chlorophyll molecules in stacks of thylakoid disks (the grana). Because of this arrangement, the largest possible reactant surface is made available for the chemical reactions involved in photosynthesis.

Chlorophyll A

tons, are passed along to a nucleotide of adenine, a complex molecule known as $NADP^+$. The NADP is thus reduced to NADPH. You will notice in Figure 5.6 that the electrons accepted by NADP are still at a high energy level.

The process leaves photosystem I short two electrons, but these are replaced by two more from photosystem II (Figure 5.6), and, as we've seen, photosystem II will recover its lost electrons from water. Because the accompanying protons (H^+) are left behind in the lumen, their concentration there builds to over a thousand times that of the stroma. At the same time, other protons are pumped into the lumen of the thyla-

FIGURE 5.6 The photosystems of photosynthesis are located in the thylakoid membrane. Protons left behind when water is disrupted and those pumped in accumulate in the thylakoid lumen. As they later pass outward through the channels in CF₁ particles, they provide the energy to make more ATP.

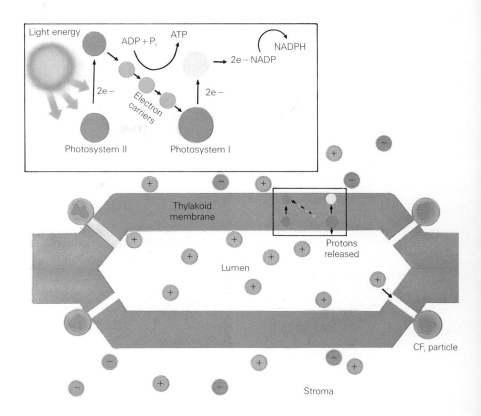

koid by the energy released from the electron transport systems. So, protons end up inside the thylakoid from two sources, from the disruption of water that occurs within the thylakoid (these are simply left there) and from protons being pumped in from the stroma.

The result of this build-up is that the hydrogen ion concentration inside the thylakoid is much greater than in the stroma surrounding the thylakoid. Because of the charges on the ions, this accumulation sets up not only a strong chemical differential, but an electromagnetic one as well. Furthermore, the two very distinct concentrations are separated only by the thylakoid membrane. This differential establishes a great reservoir of potential energy.

The differential has potential energy because there is a strong tendency for the positive charges inside the thylakoid and the negative charges outside to join and reestablish an equilibrium. Remember, though, that the outside of the thylakoid membrane that separates them is studded with tiny CF₁ particles that are penetrated by tiny channels extending into the lumen of the hydrogen-saturated thylakoid. It is through these channels that the hydrogen ions (protons) can pass outward and begin to reestablish an electrical and chemical equilibrium.

However, this is not a free ride: there is a toll for their passage. Within the CF₁ particles are a battery of phosphorylating enzymes. As the protons pass through, the energy of their movement is used to phosphorylate ADP; that is, to reattach inorganic phosphate (P$_i$) to ADP, thereby forming high-energy molecules of ATP. This, then, is how ATP is formed in photosynthesis.

The use of an electron-transport system to pump hydrogen ions across a membrane, resulting in a concentration difference that can be used to make ATP, is known as **chemiosmosis.** We will see it again when we look at the role of mitochondria in respiration. (The process was first described by a maverick scientist named Peter Mitchell who largely supports his research by revenues from his British dairy herd.)

Now we will see that the energy in the newly-formed molecules of ATP, along with the high energy molecules of NADPH, formed earlier, provide the necessary energy to juggle molecules in the light-independent stage. It is here that the nutrient molecules, such as glucose, are formed.

The Light-independent Reaction

The **light-independent reaction** derives its name because light energy is not necessary to drive this part of photosynthesis. In this case, the energy comes from ATP and high-energy electrons held by the NADPH that was produced in the light reaction. The major biochemical pathway in this reaction, sometimes called the **Calvin cycle,** is outlined in Figure 5.7.

The light-independent reaction consists of fourteen or fifteen complex steps, which, in spite of your great disappointment, we will not go into here. Just keep in mind that a carbon dioxide molecule is added to a 5-carbon sugar molecule already present in the cell. This 5-carbon sugar is **ribulose diphosphate,** or **RuDP:**

$$
\begin{array}{l}
CH_2—O—P \\
| \\
C = O \\
| \\
CHOH \\
| \\
CHOH \\
| \\
CH_2—O—P
\end{array}
$$

The additional carbon results in a very unstable 6-carbon sugar that quickly splits into two 3-carbon sugars. There follows a complex series of events in which the bonds between carbon atoms are broken and reformed, eventually producing two molecules of **phosphoglyceraldehyde,** or **PGAL.** The PGAL is the end product of photosynthesis.

PGAL may then enter into any of a number of metabolic pathways. If the plant is in need of food, PGAL can be used as is. It can also be converted to glucose or starch and then stored, or it may be converted to proteins or fats. Most of the PGAL, however, is changed back to RuDP, with an input of energy from ATP. It is then ready to enter the cycle again.

What Does a Plant Do with Its Glucose?

The photosynthetic process is indeed functional, but it requires sunlight, and sunlight is not always available. Thus, many plants store some of the glucose they produce in photosynthesis. (In a sense, perhaps, saving it for a rainy day.) Most plants do this by simply linking the glucose molecules together and storing them as starches—either in the cells that produced them or in special storage areas. The potato plant, for example, stores excess starch in swollen underground nodules that humans have been known to eat. Other plants may convert glucose to various kinds of

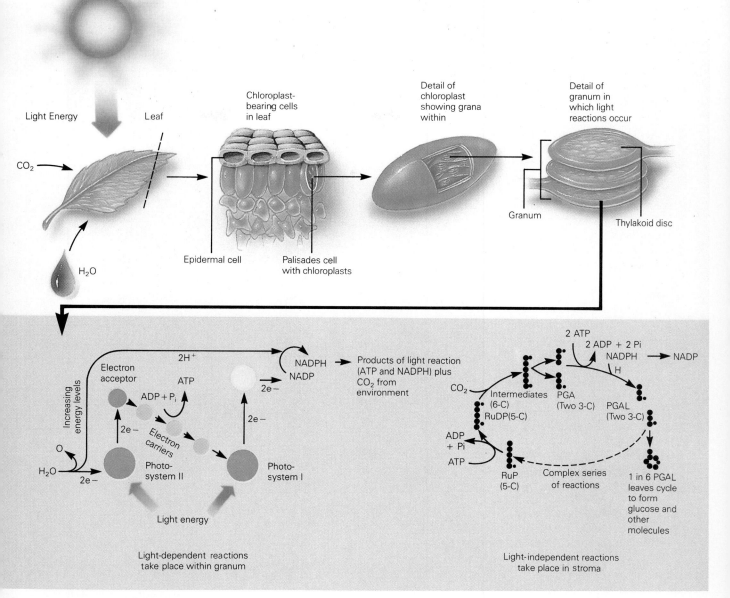

FIGURE 5.7 Photosynthesis simplified (believe it or not). It all begins when sunlight penetrates the green chloroplast-bearing cells of plants. The chlorophyll molecules in the grana, which lie embedded in the stroma of chloroplasts, absorb energy from light, enabling a pigment system called photosystem II to boost two of its own electrons (which it replaces by literally tearing two others from a water molecule) to an electron acceptor. The remaining oxygen (derived from water) is released from the plant. The electrons are passed along a chain of molecules, each at a lower energy level (down the electron-transport system) until they come to rest in a second pigment system. The protons that are left behind build to great concentrations inside the lumen. As the electrons pass along the electron-transport system, ATP is generated, and more protons are pumped into the lumen. This second system then also uses the energy of light to boost the electrons to another acceptor at yet a higher energy level. From this point, they join (reduce) NADP, forming NADPH, which, with the ATP formed during electron transport, will enter the light-independent (Calvin) cycle. As the protons cross the thylakoid membrane through channels in the CF_1 particles, the energy of their movement is used to make more ATP.

The light-independent cycle takes place outside the grana and is an extremely complex process. This is where plants will utilize CO_2 from the air to form glucose. First, RuDP is formed by adding ATP to a molecule called RuP. RuDP is the molecule that the CO_2 joins to form an unstable six-carbon intermediate that immediately splits into two three-carbon molecules. With the addition of yet more ATP and NADPH, three-carbon molecules called PGAL are formed from three-carbon PGA. PGAL is interesting in that it can form not only glucose, but fats and proteins as well. But most PGAL simply continues on, forming more RuP and continuing the cycle.

It was once suggested that all life on the planet draws its energy, in one way or another, from the sun. However, in 1977, geologists surveying the ocean bottom near the Galapagos Islands made a startling discovery. A group of excited scientists in a specially outfitted exploratory submarine found an entire community of animals drawing their energy, not from the sun, but from hot, sulfur-laden waters spewing from open vents on the seabed. A number of such vent communities have now been found. The waters there, it has been learned, are heated by the molten rock that lies beneath the earth's crust, the same phenomenon that produces volcanic action.

As shown in the figure, water seeping into cracks in the sea bottom is heated and sent upward into the frigid water of the deep ocean. Bacteria utilize the hydrogen sulfide in this water to begin an autotrophic food chain. The hydrogen sulfide is believed to provide electrons, much as water does in photosynthesis. Here, the food-making process is referred to as **chemosynthesis** (making organic compounds from the energy derived from inorganic chemical reactions). Clouds of these microscopic species set up the beginning of remarkable food chains.

Among the various kinds of animals living crowded around the vents are previously described species of giant blood-red tubeworms, nearly three meters long, agile white crabs, mussels, barnacles, and even leeches.

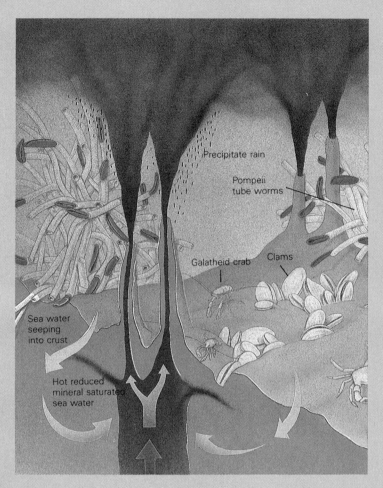

A diagram of the energy source of chemosynthetic communities beneath the sea.

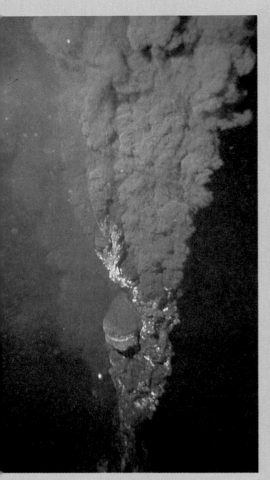

The chemosynthetic communities of the sea are host to a wide variety of life. At left, a tube worm bed in the Galapagos Rift area. In the upper right corner, galatheid crabs and mussels. These creatures draw energy not from the sun, but from the mineral-laden water that spews from the "smoker" chimneys (above). The discovery of these mysterious undersea communities was made possible through the use of small submarines such as *Alvin* (right). These submarines can carry out a variety of tasks despite the extreme pressure of the ocean depths.

sugars, such as sucrose or fructose. Sugar beets store sucrose in underground stems. Many fruits are sweet because they are laden with fructose.

Such storage is obviously a temporary measure. Eventually the plant will utilize the glucose it has made. Most of it will go to make ATP. We will see how this is done shortly, in our discussion of cellular respiration.

Let's close this part of our discussion by noting that photosynthesis is an inefficient and energy-costly process. The reason is simply because its molecular reactants, water and carbon dioxide, are in a low free-energy state—that is, they are rather stable and, thus, reluctant to participate in any chemical events. They can only be induced to react by an expenditure of some sort of energy. Fortunately for life as we know it, sunlight is plentiful and evolution has provided some organisms with the ability to use it as an energy source as part of an endless cycle of chemical events that permit the existence of most forms of life.

CELLULAR RESPIRATION: FOOD TO ENERGY

The great variety of living things on the planet plays many roles in the extemporaneous skit of life, but all life requires energy. We have seen that the most obvious storehouses of energy are the earth's green plants. We have also seen that after the energy of sunlight has been shuffled, juggled, and channeled through the photosynthetic processes, finally to end up in the bonds of glucose, it can take any of a number of routes. The plant may utilize that bond energy immediately, or it may store it in its body. It is this stored food that the heterotrophs are after, whether the heterotroph is a herbivore that eats the plant directly or a carnivore that seeks the food that was once stored in plants but now is stored in some other animal's body. Not only must heterotrophs have ways of acquiring energy, either from the plant or from each other, once they get it, they must have ways to use it. So let's see how those precious molecules are handled by the metabolic machinery of life.

The process by which cells drain high-energy molecules of their energy (thereby forming ATP) is called **cellular respiration.** The process is usually gradual and tightly controlled so that the energy is released slowly. If it were to be released all at once, there would be the crackle of exploding protoplasm followed by the cool silence of death.

Living things have developed a variety of ways to handle such high-energy molecules, essentially through four major processes: **glycolysis,** the **Krebs cycle,** the **electron-transport chain,** and **chemiosmotic phosphorylation.**

Glycolysis

The first step in the breakdown of glucose is called **glycolysis** (*glyco,* sugar; *lysis,* breakdown), a process that takes place in the cytoplasm. Glycolysis is the first step toward converting the energy within a glucose molecule to the usable energy of ATP. As we will see, though, glycolysis alone doesn't produce very much ATP.

In glycolysis, a 6-carbon glucose molecule is first destabilized by the addition of two phosphate ions in separate reactions. These ions come from two molecules of ATP, changing them to ADP. We see, then, that whereas glycolysis produces ATP, it also uses up two ATPs at the beginning of each sequence, "priming the pump" as it were. So, whereas four ATPs are produced in the process, the net gain is only two. In any case, the phosphorylated molecule quickly breaks down into two 3-carbon fragments. (The details are shown in Figure 5.8.) The two fragments are then repeatedly rearranged from one kind of molecule to another until, finally, two molecules of **pyruvate** (or **pyruvic acid**) are formed. Pyruvate is the end-product of glycolysis.

Along the way, the 3-carbon fragments are partly oxidized; that is, they give up electrons. Two electrons, with their protons, are removed from each fragment. These then reduce (or join with) two molecules

FIGURE 5.8 Glycolysis. (A) A specific enzyme is required for each step in glycolysis. Two ATPs must be invested (reactions 1 and 3), changing glucose to fructose-1, 6-diphosphate. The outcome of the first three reactions is that the fuel is made more reactive. Then the fructose-1, 6-diphosphate is split into two fragments (reaction 4). (B) In the first of two vital reactions (reaction 5), the two PGALs are oxidized as hydrogens move to two molecules of NAD (actually NAD$^+$, since this molecule bears a net positive charge until it is reduced to NADH). Then the PGALs take on another phosphate each, forming molecules called 1,3-DPG. They then each lose a phosphate to ADP, forming two molecules of ATP (reaction 6). Thus, the earlier investment of two ATPs is repaid. (C) In the final phase of glycolysis, the two 2-PGs are dehydrated (reaction 8), forming PEP, an unstable product that will react readily. Then in the last reaction (9), destabilized molecules again react with ADP, and two more ATPs are produced. This represents a net gain of two ATPs in glycolysis, along with two other energy-rich products—the two NADHs from reaction 5 and two molecules of pyruvate, the final product of glycolysis.

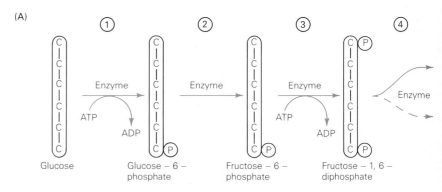

(A)

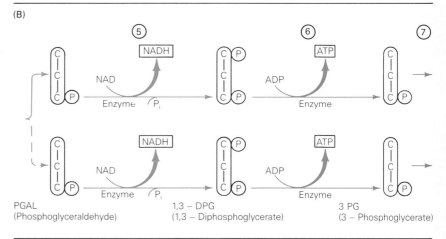

(B)

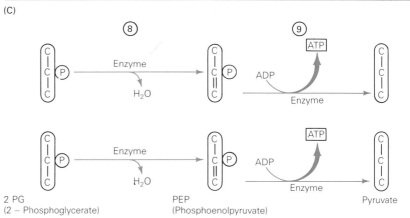

(C)

FIGURE 5.9 The anaerobic conditions used in alcoholic fermentation are maintained by the careful exclusion of oxygen. Here, the fermentation process yields rum.

called NAD (nucleotides of adenine), thereby forming two molecules of NADH. The overall reaction can be written:

$$C_6H_{12}O_6 + 2NAD + 2ATP \longrightarrow 2C_3H_3O_3 + 2NADH + 2H^+ + 4ATP.$$
(glucose) (2 pyruvic acid)

(Note the *net* number of ATP molecules formed.)

Once the pyruvic acid or pyruvate ($2C_3H_3O_3^-$) is formed, it may enter any of several different metabolic pathways. The particular route at this point may depend upon whether oxygen is present. If oxygen is present, in some species the pyruvic acid may then enter the Krebs cycle, as we shall see shortly, and produce more ATP. If oxygen is not present, the pyruvate will enter an anaerobic (without oxygen) sequence.

One such anaerobic pathway takes place in yeast and is called **alcoholic fermentation.** Here, as we see in Figure 5.9, the product is carbon dioxide and alcohol. Actually, yeasts can break down pyruvate either aerobically (with the utilization of oxygen) or anaerobically (without utilizing oxygen), but if glucose is abundant, the route is anaerobic. In yeast, the glucose is only partially broken down and forms ethanol (the ethyl alcohol present in spirits). The two molecules of NADH produced in glycolysis give up their hydrogens to help form the ethanol and then return as NAD to pick up two more hydrogens from a glucose molecule. We have known about the practical aspects of **fermentation** for a long time because early man, confronted with all the great problems of the universe, at once set about learning how to make booze.

In animals, on the other hand, anaerobic respiration produces, not alcohol, but lactic acid. It is not surprising that anaerobic animals (those that do not utilize oxygen in their energy production), such as internal parasites, are fairly sluggish. They just don't have the ATPs to be frisky. But aerobic animals—those, like us, that require oxygen for most respiration—under certain conditions may also derive energy from anaerobic processes. For example, if we are unable to get enough oxygen during prolonged exercise, our respiration may partly shift to glycolysis (aerobic and anaerobic processes can occur simultaneously). When this happens, lactic acid accumulates in the muscles, and they weaken with fatigue. If enough lactic acid accumulates, the muscles will fail to contract at all, or they may contract in an uncoordinated manner. This is why a runner who is not in condition is usually gasping for air at the end of a race and may stagger across the finish line. A conditioned runner is likely to have developed the heart and lungs along with the muscles so that sufficient oxygen is supplied to prevent the build up of lactic acid.

In mammals, lactic acid is washed from the muscles by the circulating blood and carried to the liver, where it is reconstituted to glucose. The glucose units are then linked together to form glycogen, which is stored in the liver. The glycogen can be broken down into its glucose components when glucose is needed, but the transformation is uphill—that is, it requires an expenditure of some energy. However, because the energy level of lactic acid is fairly high, the boost to glucose is a short one.

In aerobic animals, the pyruvate formed in glycolysis may take a different route. It may go through a series of reactions (the Krebs cycle) in which almost all its bond energy is extracted to make new ATPs, many more than can be formed in glycolysis. In this sequence, the energy-laden electrons of pyruvic acid are passed along to lower energy levels until they end up as mere components of humble water.

The Krebs Cycle

We have seen that in the absence of oxygen, a glucose molecule provides an aerobic organism with only two molecules of ATP. If oxygen is present, it's a whole new ball game. The pyruvate will go through a series of reactions that generate thirty-four to thirty-six more ATPs, depending on the cell type. This happens when the two pyruvate molecules formed from the glucose are shunted into the **Krebs cycle** (also called the **citric acid cycle**), named for its British discoverer, Sir Hans Krebs. Eventually, the high-energy pyruvate is broken down completely to the simple end-products, carbon dioxide and water, having released most of its energy along the way. Both the Krebs cycle and the next stage of respiration, the electron-transport chain, take place in tiny bodies, called mitochondria, within the cell (more about them later).

So if oxygen is present, the pair of 3-carbon pyruvate molecules produced by glycolysis first become oxidized—with NAD molecules again accepting the two hydrogens, becoming NADH. The pyruvic acid molecules each then lose one carbon in the form of carbon dioxide. The remaining 2-carbon fragment, **acetate,** quickly unites with a protein called **coenzyme A,** or **CoA,** to form a molecule called **acetyl-CoA:**

$$CH_3$$
$$|$$
$$O=C\diagdown_{CoA}$$

It is actually the acetyl-CoA that enters the Krebs cycle (Figure 5.10). It should be stressed that this molecule can be derived not only from glucose, but from fats and some amino acids as well (although by less direct routes).

Once the acetyl-CoA enters the Krebs cycle, the acetyl group separates from its CoA enzyme and combines with a 4-carbon molecule called **oxaloacetic acid,** or **oxaloacetate.** The result is a 6-carbon compound called **citric acid (citrate),** the common acid of lemons and limes. The citrate is then oxidized, followed by the splitting off of two carbons in the form of carbon dioxide, with the remainder reverting to oxaloacetate. This is then recycled to pick up another acetyl group from the next acetyl-CoA molecule. The cycle continues as long as acetyl-CoA is available. Details of this process are illustrated in Figure 5.10. As you can see, the stored energy in the acetyl group has been used to convert three molecules of NAD to NADH, one molecule of FAD (another electron acceptor) to $FADH_2$, and one molecule of ADP to ATP. The glucose molecule has by this time been completely oxidized. As we will see next, its products will go on to provide the cell with yet more energy after they are passed along to the electron-transport chain.

Note that the electron acceptors NAD and FAD are modified nucleotides, and each of these substances readily accepts and releases two electrons at once. Do you see a certain parsimonious efficiency in a system that makes multiple use of nucleotides in continuous oxidation and reduction processes?

FIGURE 5.10 The Krebs cycle simplified. The number of carbons is shown for each molecule. Electrons are shown being released at four reactions to reduce NAD and FAD. Keep in mind that their protons travel along (the electron and proton form hydrogen), so the reduced carriers are called NADH and $FADH_2$ (FAD carries two hydrogens). If we begin with citrate, we see the molecule losing two carbons, finally forming oxaloacetate. The six-carbon citrate is restored by the addition of the two-carbon acetyl CoA formed from glycolysis.

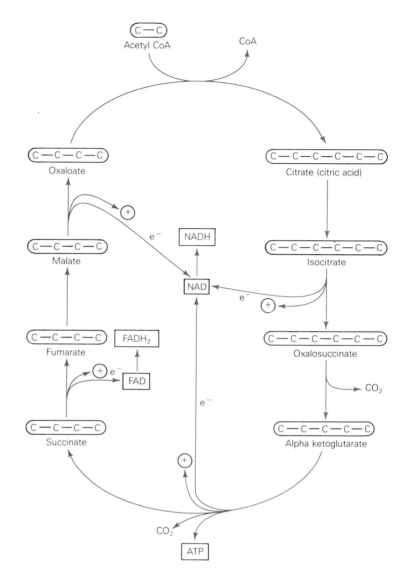

The Electron-transport Chain

In the **electron-transport chain** of respiration, electrons that had been accepted by the NAD and FAD molecules are passed along to a series of electron carriers similar to those we mentioned in photosynthesis. Here again, each acceptor is at a lower energy level than the preceding one, and again the energy held in the electrons is gradually released in a stepwise process.

The last stages here have a kind of elegant finality. At the end of the chain, the electrons join with protons to form hydrogen atoms that then combine with oxygen (acquired by breathing) to produce water.

We will next see how the energy released by the passage of electrons along the electron-transport system is used to transport hydrogen ions (protons) to the outer compartment of the mitochondria. The molecules of the electron-transport system lie within the inner membrane of the mitochondrion, thus they are well placed to shift protons to the outer compartment. Here, as we saw in photosynthesis, a chemical and elec-

The earth is constantly bathed in radiation emanating not only from the sun, but from a host of other celestial bodies. Part of the radiation that reaches us is **visible light.** Visible light, however, is only part of an **electromagnetic spectrum** that includes (in increasing order of energy) radio waves, microwaves, infrared radiation, visible light, ultraviolet radiation, x-rays, and gamma rays.

The longest waves contain the lowest levels of energy, very short waves the highest, and in between is visible light. Visible light is visible because it interacts with special pigments (light-absorbing molecules) in our eyes. It also interacts with pigments such as chlorophyll, one of the molecules that can absorb the energy of light and provide power for photosynthesis.

Physicists have described light in two ways: as **particles** (called **photons**) or as **waves.** Arguments over which concept best accounts for the behavior of light have raged for years. So let's finesse it a bit and say that light is composed of photons and moves like waves. This side-step will at least help us consider the energy of a photon in terms of its wavelength.

The various wavelengths of light are critical to life in a number of ways. Visible light, we have seen, interacts with the retinas of our eyes and provides the energy that enables green plants to grow. The longer infrared wavelengths can be reflected and dissipated as heat, and we know that warmth, of course, to some degree (no pun), is important to all life. At the other end of the spectrum, the shorter, more energetic wavelengths are usually too powerful for most forms of life to utilize since they tend to disrupt molecules, especially proteins and DNA. Ultraviolet light (UV), for example, burns our

skin and damages the retinas of our eyes. The more penetrating radiation, such as gamma rays and x-rays, is called **ionizing radiation,** because it can ionize, or break up, molecules within cells. The pieces of the disrupted molecules are very reactive and tend to enter into random and potentially harmful reactions. If such molecular fragments should disrupt the genetic material of a cell, the result may be a mutation. (We will see why shortly.) Fortunately for the life on earth, most of the more energetic and dangerous wavelengths of the sun are filtered out by ozone molecules in the stratosphere.

So, infrared warms us and ultraviolet burns us, and the wavelengths we can see lie in between. Obviously, though, there are no sharp dividing lines between visible and invisible rays. Both reds and violets become dully visible as they

enter our range of sensitivity. It is interesting that different organisms may see different parts of the electromagnetic spectrum. For instance, many insects cannot see red. Red flowers appear to them as gray. But they may see deep violet (ultraviolet) in flowers that appear white to us. And whereas insects can see ultraviolet, but not red, the reverse is true for many birds. Interestingly, many people who have had the lenses of their eyes removed because of cataracts can see ultraviolet light that is invisible to the rest of us. As we will see in Chapter 21, the reason we see colors at all is because light of different wavelengths interacts in specific ways with the color-sensitive cells in our eyes. Various animals are able to perceive different parts of the light spectrum because of physiological differences in those receptors.

FIGURE 5.11   A schematic diagram of a mitochondrion.

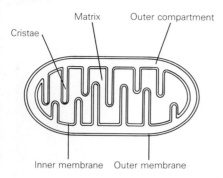

Cristae

Matrix   Outer compartment

Inner membrane   Outer membrane

FIGURE 5.12   A summary of the energy relationships within a cell. Glucose enters the cell when it undergoes glycolysis, producing two molecules of pyruvate and reducing two molecules of NAD, forming NADH. The NADH releases its hydrogen, which crosses the mitochondrial membrane into the outer compartment. The pyruvates are changed to acetyl CoA and enter the Krebs cycle. The Krebs cycle produces two molecules of CO_2 and two of ATP, but more important, it reduces NAD to NADH and FAD to $FADH_2$. These molecules then become oxidized, giving up electrons to a series of electron acceptors called the electron-transport system. As the electrons travel down this system, energy is released at each transfer, until they become spent of energy and join oxygen to form water. The energy released with their passage is used to pump protons (+) into the outer compartment where they are separated from electrons (−) in the inner compartment. The protons can rejoin the electrons only by passing through channels in the F_1 particles. The energy released by their passing is used to convert ADP to ATP.

trical differential results and, as the protons move across the membrane to reestablish an equilibrium, the energy of their movement is harnessed to phosphorylate ADP, changing it to high-energy ATP. And that, in a nutshell, is what I am about to tell you.

Chemiosmotic Phosphorylation

The mitochondrion has two membranes, an outer one and an inner one. The outer is the rather simple "covering," but the inner membrane forms a number of shelflike folds **(cristae)** that project into the mitochondrial **matrix.** The electron-transport system of the mitochondrion is embedded in the inner membrane (Figure 5.11). This membrane is also studded with stalked **F_1 particles** that appear to be similar to CF_1 particles of the chloroplast, except that their arrangement is essentially reversed: the F_1 particles project into the inner compartment from the surface of the inner membrane. The energy released in the electron-transport system is used to pump positively charged hydrogen ions into the mitochondrion's outer compartment. As they move to rejoin negatively charged hydroxyl ions (OH^-, formed by the splitting of water) in the inner compartment, they must pass through the F_1 particles where batteries of enzymes use the energy of their passage to form ATP from ADP and P_i (phosphorylation).

By way of summary, in both the chloroplast and the mitochondrion, the energy released in the electron-transport system is used to move hydrogen ions (protons). However, they are moved in opposite directions within the two bodies. In the chloroplast, they are pumped into the lumen of the thylakoid (where they join those left by the splitting of water). As they pass outward into the stroma, they must move through the CF_1 particles where they power the transformation of ADP and P_i into ATP. In respiration, on the other hand, hydrogen ions are transported by carriers across the inner membrane into the outer compartment of the mitochondrion, resulting in a concentration gradient between the outer and inner compartments. These hydrogen ions (protons) pass inward, reestablishing an equilibrium, only by passing through channels in the F_1 particles. Energy is extracted during this passage and then used to enable phosphorylating enzymes to form ATP from ADP and inorganic phosphate (P_i).

At this point, let's review the major points of respiratory reactions, beginning with glycolysis (as summarized in Figure 5.12). During glycolysis, the 6-carbon sugar glucose is split into two 3-carbon molecules. These molecules are then rearranged to pyruvic acid, and, in the process, a small quantity of ATP is produced. Yeasts can metabolize glucose either aerobically or anaerobically, but in the absence of oxygen, pyruvic acid is converted to carbon dioxide and alcohol, while in aerobic animals, in the absence of oxygen, it is converted to lactic acid. In aerobic animals, if oxygen is present, the pyruvic acid is converted to acetic acid, which combines with coenzyme A. The resulting acetyl-CoA then enters the Krebs cycle, where carbon dioxide is produced, along with some ATP, and electrons are transferred to FAD and NAD molecules. The resulting $FADH^2$ and NADH then release these electrons to the electron-transport chain, and the electrons travel down the chain to increasingly lower energy levels. The remaining electrons are finally deposited in the "elec-

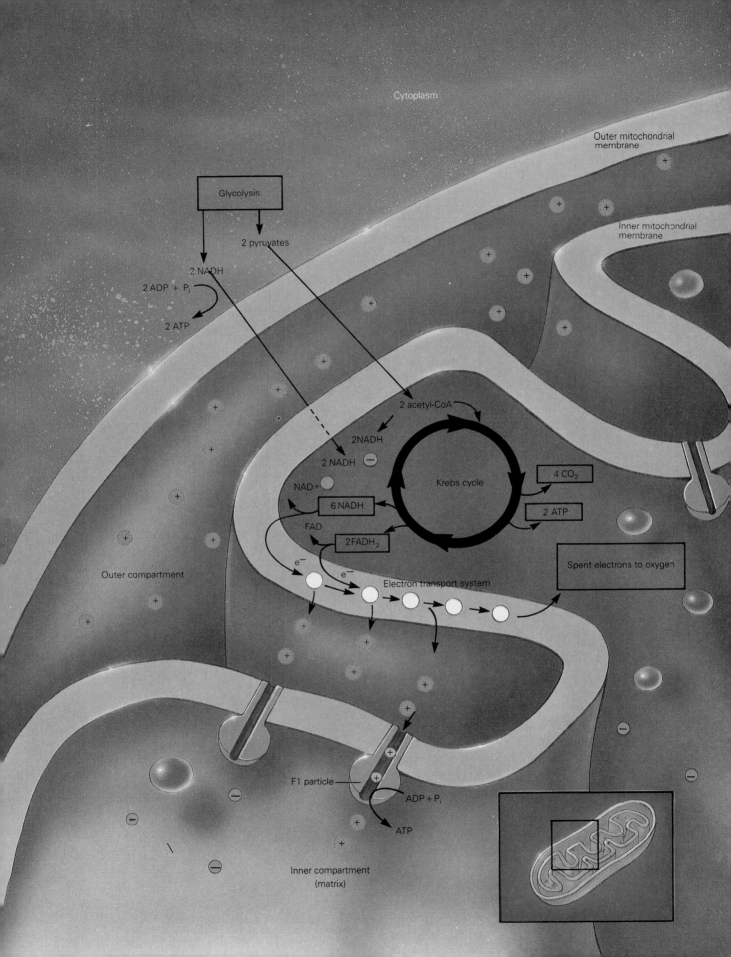

Cytoplasm

Outer mitochondrial
membrane

Inner mitochondrial
membrane

Glycolysis

2 pyruvates

2 NADH

2 ADP + P$_i$

2 ATP

2 acetyl-CoA

2NADH

2 NADH ⊖

NAD+

Krebs cycle

4 CO$_2$

6 NADH

2 ATP

FAD

2FADH$_2$

Outer compartment

e⁻

e⁻

Electron transport system

Spent electrons to oxygen

F1 particle

ADP + P$_i$

ATP

Inner compartment
(matrix)

tron dump''—that is, combined with oxygen to form water. The energy left over after these reactions appears as metabolic heat. (As a demonstration, reach over and feel the body heat of the person next to you. Explain that it's an assignment.) As the electrons are passed along to lower energy levels, energy is released that is used to pump protons out of the inner compartment of the mitochondrion, establishing a concentration gradient. As the protons move inward, attracted by the net negative charge there, they pass through enzyme-laden F_1 particles that use the energy of their passage to manufacture ATP from ADP and P_i. The process is called **chemiosmotic phosphorylation.** The ATP then powers the innumerable activities associated with life.

SUMMARY

Energy, whether potential or kinetic, is the ability to do work. Forms of energy, such as heat, chemical, electrical, or radiant, are interchangeable. Generally when energy is changed from one form to another, some is lost to the environment. Molecules can be rearranged to hold more or less energy, thereby gaining energy or releasing it.

Energy changes conform to the laws of thermodynamics. The first law states that energy can be neither created nor destroyed. Thus, the total energy in a system remains constant. The second law states that the free energy in a system (the potential energy available to do work) constantly decreases. Thus, energy must be added to a system to increase its level of free energy so that energy is available to keep it organized.

In the biological world, almost all free energy comes from the sun, but living things can capture that energy and store it in the form of high-energy bonds of certain molecules, such as ATP.

Plants use the energy in sunlight to make food from carbon dioxide and water, releasing oxygen in the process. The energy is captured by pigments in the chloroplasts and transferred through the system (primarily by shifting electrons about), including setting up a chemiosmotic gradient and passing the electrons through CF_1 particles where phosphorylating enzymes form ATP from ADP and P_i. The high-energy ATP, along with the energy of electrons held by NADPH, is then utilized to make glucose and other nutrients in the light-independent phase of photosynthesis. The energy stored in these nutrients is released in cellular respiration through glycolysis, the Krebs cycle, the electron-transport chain, and chemiosmotic phosphorylation. Glycolysis involves the anaerobic breakdown of glucose, yielding two ATPs. The product, pyruvic acid, is altered so that it can enter the Krebs cycle where more ATPs are formed. In the process, electrons are removed from the food molecules and passed along the electron-transport chain, with energy being released at each step. This energy is used to increase the chemiosmotic gradient by encouraging the accumulation of electrons within the inner compartment of the mitochondrion. As these pass outward through the F_1 particles, energy is extracted for the formation of more ATP.

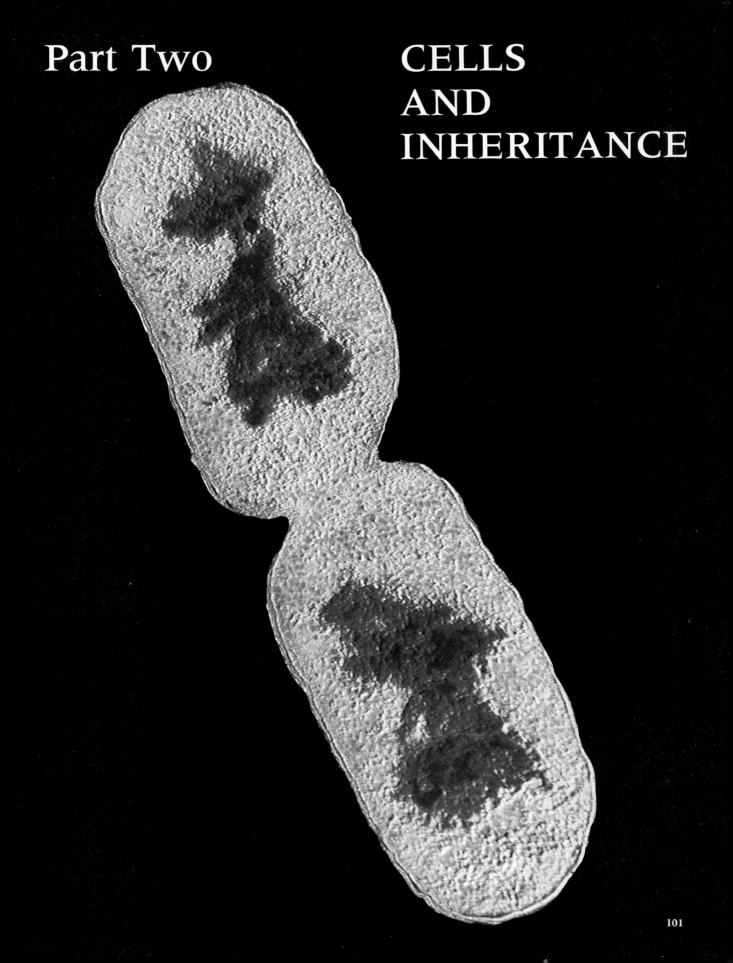

Part Two

CELLS AND INHERITANCE

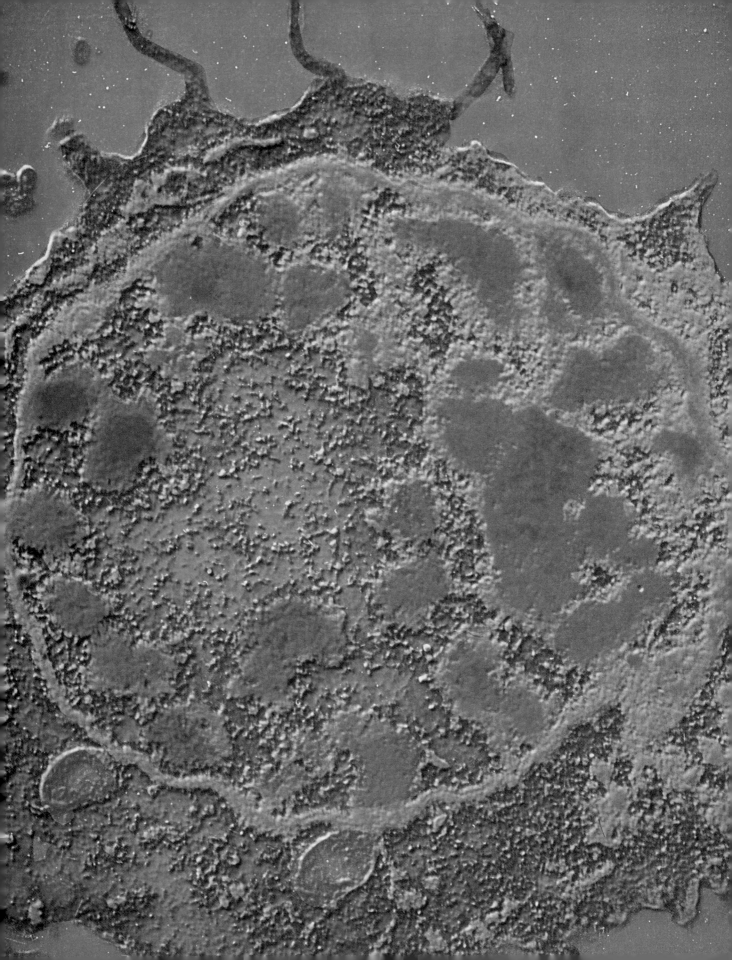

Chapter 6

THE CELL AND ITS STRUCTURE

In the mid-1600s, the prestigious Royal Society of London appointed Robert Hooke as their new curator of experiments. Hooke felt honored to have been chosen for such a position, but with the position came a problem. At the group's weekly meetings, he had to provide them with some sort of entertaining scientific demonstration. But the society was comprised of the elite scientists of that day—a rather jaded lot to be sure. They were indeed difficult to impress and almost impossible to entertain. Hooke pondered over what to show them. What would they like to see?

The scientific world was abuzz in those days with talk of lenses, bits of curved polished glass that could be used not only to restore books and letters to weakened old eyes, but to see the distant heavens more clearly. Hooke himself was interested in lenses, particularly in those that could bring into view things too small to be seen with the naked eye. In fact, he had built one of the first microscopes (Figure 6.1). With it, he managed to entertain that august body of scientists on several occasions. But as time went on, Hooke began to wonder what new uses he might make of it. What else would be interesting to see? One night while tinkering, he cut a very thin slice of cork with his penknife and placed it under his lens. Cork was interesting because people wondered how so seemingly solid a substance could float. And what do you think he saw? Nothing! That's because reflected light does not provide the illumination necessary to see objects of that nature. So he arranged to have the light pass from underneath, *through* the cork. This time Hooke did see something, and it surprised him. He wrote in his notes that the cork seemed to be composed of tiny "boxes" or "cells." He showed his cells to the society, the society was duly impressed, and a new field of science was born.

FIGURE 6.1 Hooke's primitive microscope consisted of two convex lenses at either end of a six-inch tube. The object for study was stuck on a pin attached to the base of the microscope. A flame served as the light source for the microscope.

This field, however, was a bit slow abirthing. In fact, about a century was to pass before the scientific world would begin to grasp the extreme importance of Hooke's "little boxes." In fact, it wasn't until about 1805 that the German naturalist, Lorenz Oken, formally stated what has become known as the **cell theory:** "All organic beings originate from and consist of vesicles or cells." The idea was buttressed in 1839 by two other Germans, botanist Matthias Jakob Schleiden and zoologist Theodor Schwann, who independently concluded that all living things—from oak trees to squids, from tigers to humans—are made up of cells (Figure 6.2). Another fifty years were to pass before another German, Rudolf Virchow, made the rather catchy statement, *omnia cellula e cellula*, "all cells from cells." It wasn't until about this time that people began to take the whole concept somewhat more seriously. The modern study of cells, now called **cytology** (*cyte,* cell), indeed got off to a shaky start.

Once biologists began to see the importance of understanding cells, the field grew rapidly and sporadically as one fascinating finding was jostled aside by the discovery of yet another. It was a most unsettling time. Today, though, cytologists have rather comfortably settled into their task of learning about cells. And their methods have become extremely sophisticated as they seek to answer increasingly specific questions about life's little boxes. Nonetheless, the basic definition of a cell has not changed much since cytology first began to be taken seriously. Cytologists tell us that the cell is the smallest unit that can exist as an independent organism, and that it has, at some point in its development, everything necessary for life. We should perhaps note that there are organisms that consist of only a single cell, but here we will concern ourselves primarily with multicellular species, keeping in mind that many of the principles we will discuss also apply to single-celled organisms.

FIGURE 6.2   Diversity in plant and animal cells. These cells are from multicellular organisms, and each is well adapted to a specific role.

(A) Compact bone cells

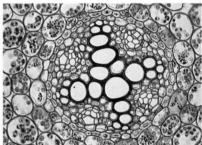

(B) Buttercup root cells

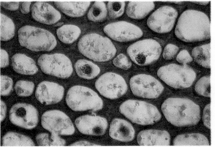

(C) Collenchyma cells in an elderberry stem

(D) Fat cells

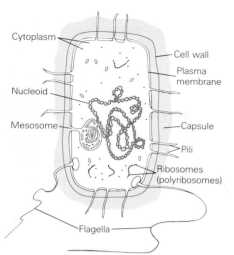

FIGURE 6.3 Compared to the typical eukaryote cell, the prokaryote cell is quite small. Further, it lacks the membrane-bound organelles of the eukaryote, including the organized nucleus. Prokaryotic DNA is naked and circular, lacking the protein complex of eukaryote chromosomes. A dense cell wall, quite different chemically from eukaryote cell walls, surrounds the membrane and is often itself surrounded by a slimy sheath. Prokaryotes may have tubelike, cytoplasmic projections known as pili. Where flagella appear, they are solid, rotating entities, anchored in the cytoplasm and cell wall, and quite unlike those of eukaryotes.

With all the attention given to cells these days it would seem that describing one should be a simple task. But this is not the case, partly because a living cell is a hotbed of activity, constantly changing with dramatic speed. But describing cells is also difficult because there are so many kinds of them.

PROKARYOTIC AND EUKARYOTIC CELLS

Before we become involved with the description and behavior of cells, we should first be clear about what kinds of cells we will be considering. We will *not* be talking about the bacteria and cyanobacteria. These are one-celled organisms that differ in several basic ways from all other cells. They are called **prokaryotic cells** (*pro,* early; Figure 6.3) to differentiate them from the **eukaryotic cells,** our primary concern here. The differences between the two types of cells are summarized in Table 6.1. The comparison will mean more to you after we wend our way through the rest of the chapter, but for now, just keep in mind that there are fundamentally two types of cells and the type that will concern us here is the eukaryotic cell.

Specialization in Eukaryotic Cells

Eukaryotic cells come in a variety of sizes and shapes, each with its own particular limits and abilities. Not only do the cells of various organisms differ, but the cells of a single organism may be quite different from each other. For example, the tiny boxlike cells beneath the skin of your hand are quite unlike the delicate threadlike nerve cells that stretch from the base of your spine down to your foot. And neither of these resemble the peculiar, pulsating cells that make up your heart.

TABLE 6.1 Differences Between Prokaryotic and Eukaryotic Cells

Structure	Prokaryotic cells	Eukaryotic cells
Nuclear membrane	Absent	Present
Membrane-bound organelles (mitochondria, Golgi bodies, etc.)	Absent	Present
Flagella	Rotating and lacking 9-2 structure	Fixed with 9-2 structure
Chlorophyll	Not in chloroplasts	In chloroplasts
Cell wall	Contains amino acids and sugars	Found in cells of higher plants
Pili	Present	Absent
Chromosomes	Nucleic acids do not form chromosomes	Includes nucleic acids and proteins

FIGURE 6.4 Human cells vary widely, permitting great specialization and division of labor within a single body.

(A) Human cardiac muscle (contractile cells)

(B) Neuromuscular junctions (neural cells)

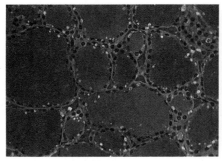

(C) Thyroid gland (secretory cells)

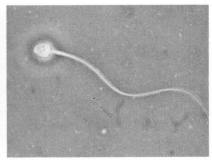

(D) Human sperm (motile cells)

One 3-inch "cell"
Surface area of membrane
54 square inches

Twenty-seven 1-inch "cells"
Surface area of membrane
162 square inches

FIGURE 6.5 The cellular structure of living things essentially breaks bodies into a number of parts. Each part has its own membrane surface and, hence, cellularity increases the total area of cell membranes in living things. Since cell membranes are living regulatory structures, the result is increased control over what enters and leaves the body's cytoplasm. Cellularity also has another advantage: that of permitting specialization as cells differentiate according to specific roles. Note that in the illustration, both "cells" have the same volume, but the subdivided one has a far greater surface area.

So the body is composed of various kinds of cells with different abilities. But cells that differ in appearance may also behave quite differently, as would be expected. To illustrate, some are highly irritable—that is, they are sensitive to environmental changes. Yet others are contractile, and others secrete fluids, while still others have long tails and can swim (Figure 6.4). This sort of variety suggests a great advantage of cellularity that might well have made the condition worth developing through evolution. That advantage is **specialization.** In other words, an organism living in a complex environment would be better equipped to meet its many varied demands if it consisted of a number of kinds of cells, each with its own special abilities. Think of the benefits of having several kinds of cells, some that could react to light, others that could distinguish particular types of pain, and yet others that could help you move toward light and away from pain. The development of cellular specialization obviously helped organisms fit more precisely into their world.

Another important advantage of an organism being comprised of many cells instead of one large cell is that the organism would have greater control over the interactions between itself and the environment. Such control would be possible because of the nature of cell membranes, the thin, living coverings of most cells. Such membranes are highly selective about what passes into, out of, and through the cells they cover. Because of this selectivity, cellularity would increase the level of control between the delicate cellular interiors and the outer environment, since with living matter broken up into cells, the relative membrane area would increase. As you can see in Figure 6.5, a 200-pound cell would have less regulatory membrane surface than a 200-pound organism composed of many cells. This is why there is no such thing as a 200-pound cell.

CYTOLOGY AND TECHNOLOGY

Cytology presents one of the clearest areas in which our knowledge is dependent almost directly on advances in technology. For example, by the early 1900s, most of the cell structure visible through the light microscope had been described. Then, with the advent of the transmitting electron microscope (TEM), a whole new dimension of the cell, its ultrastructure, was revealed. We could see things we never imagined existed. Even with the new microscope, however, we didn't know much about the cytoplasmic fluid in which tiny structures with specific roles in the life of the cell (structures called **organelles**) seemed to float. We could only assume that the organelles were suspended in some sort of "cytoplasmic matrix" or "ground substance," labels that seemed to convey information, but actually didn't. The problem was that in order to view anything with a TEM (see Essay 6.1), electrons had to pass through the substance, so it had to be sliced into ultrathin sections, and this procedure demolished any cytoplasmic detail that might have been present.

But technology saved the day again. With the high-voltage electron microscope we could see a whole new realm of cytoplasmic structure. It was no longer necessary to open the cell to see its contents. Furthermore, we could now see cell contents in three dimensions. This was indeed a remarkable step forward, but even newer technologies continued to emerge, and with each new device, we learned more. And much of what we learned turned out to be quite unexpected.

CELL COMPONENTS

Our consideration of the components of eukaryotic cells must be a bit artificial because of the complexities of such structures and the dazzling speed with which they can change. Just keep in mind that neither these descriptions nor these illustrations are likely to depict the situation for all cells, nor for any cell all the time.

It should first be made clear that the inside of a living cell is a seething, roiling mass of viscous, grainy fluid that contains tiny bodies of every description—minute structures that are constantly growing, extending, moving, multiplying, appearing, and disappearing. Such activity perhaps should not be surprising since we know that cells must maintain and repair themselves and divide and grow. We also know that the cell's sensitive cytoplasm responds to specific environmental stimuli and that as those stimuli change, so does the behavior of the cell. The vigorous activity inside living cells should serve to remind us that the goings-on within cells are incredibly complex, and that we really cannot hope to understand everything that is happening there. At least not yet.

A CLOSER LOOK AT THE CELL MEMBRANE

All cells are bounded by a delicate structure called the **cell membrane** or **plasma membrane.** The membrane is so thin that its presence was postulated on circumstantial evidence alone until the advent of the electron microscope. Modern research techniques have revealed that the membranes are composed of two layers of lipid-phosphate molecules (called phospholipids) in which are scattered various proteins. It seems, however, that the molecular arrangement of cell membranes may vary widely. For example, membrane thicknesses may be quite different from cell

The ordinary **light microscope (LM)** was once the mainstay of scientific magnification. There have been many variations on its theme, but the basic principle has remained the same. And so have the limitations of the technique—limitations imposed by the LM's low powers of resolution.

Resolution refers to the smallest distance by which two points can be separated and still be distinguished as separate points. If they are separated by a distance smaller than the resolving power of the microscope, they will be seen as a single point.

With the light microscope, the image is seen as contrasting light and dark areas as light passes through the specimen from below. The problem of resolution is therefore compounded by the inherent limitation of light diffraction, the bending of light rays as they pass through an object. The length of the light wave itself (that is, the color of the light) is a critical factor here. As the wavelength increases, so does the angle of diffraction, and its resolution is thus reduced. Under ideal conditions, the best light microscopes have a resolution of about 0.25 microns with ordinary white light. Resolution has recently been increased by the photoelectron microscope. Ultraviolet light is shined on the subject, which absorbs the light energy and reemits it as electrons that are then magnetically focused.

The resolving power of the **transmitting electron microscope (TEM),** however, is far greater yet. Thus, any object can be seen more clearly with the electron microscope, and yet other objects can be seen that are not visible at all with the light microscope.

The electron microscope, which became available in the 1930s, works on an entirely different principle. Excited electrons are drawn from a heated filament and are then directed, or focused, by magnetic fields. As the electrons pass through an object, they are deflected or absorbed to various degrees by differences in the density of the matter. They then produce an image on a coated screen or special film below the object. Electrons are easily deflected, so the specimen must be cut exceedingly thin and the operation must be carried out in a vacuum.

The wavelength of an electron beam is only about 0.005 nanometer (a nanometer is one billionth of a meter), so, theoretically, resolutions of this power are possible. A major disadvantage is that heavy-metal stains must be employed, and these are usually highly poisonous to living things. Because of this and

to cell, and the ratios of different lipids may vary from one type of cell to another. There are also indications that in some cells, the sandwich is reversed (in effect making the lipid the "bread"), and that in some cases, the proteins are globular instead of linear.

A generalized cell membrane may best be summarized in what is called the **fluid-mosaic model** in which two phospholipid layers are stabilized by specifically oriented proteins (Figure 6.6). The phospholipids are polarized so that the **hydrophilic** ("water loving") heads are closest to the protein layers, and **hydrophobic** ("water fearing") tails project inward. Globular proteins are imbedded here and there in the membrane. Some of them entirely cross the membrane and each of these may be perforated by a channel through which certain materials can pass. There may also be other kinds of molecules associated with membranes. For example, various sorts of carbohydrates may be found attached to the outer side of the membrane (specific carbohydrates identifying the cell type) (Figure 6.6). The underside of the membrane may be attached to a system of microtubules and the smaller kinds of threads and tubes, called microfilaments, that form a sort of internal support (skeleton) for the cell, called the *microtrabecular lattice* (discussed next).

Cell membranes are selectively permeable; that is, they permit certain molecules to cross them while prohibiting the passage of others. They may also actively assist some kinds of molecules in their passage, as we the necessity for extremely thin sections, there is no way living material can be examined. Thus, we can't say with certainty that what we see is not due to the method of preparation.

The **scanning electron microscope (SEM)** has far less resolving power than the standard electron microscope, but it produces the fascinating illusion of three dimensions. In SEMs, a thin beam of electrons is passed back and forth over an object, thereby scattering the electrons. They are then captured on a positively charged plate, causing it to generate a small current. The current is then amplified and fed into a cathode ray tube that projects the image caused by the reflected electrons onto a television screen.

These two micrographs of actin in a rat embryo cell illustrate the dramatic improvements in microscopes. The image to the left was produced using a fluorescent light microscope and is magnified 2,000 times. The image to the right was produced using the photoelectron microscope and is magnified 3,000 times. The photoelectron microscope offers a more detailed and clearly focused image.

will see. Thus, this membrane regulates the constituents of the cell's interior, helping to protect it from an essentially hostile environment. Not all particles entering or leaving a cell must pass through the membranous material, however; electron microscopy has revealed that the membrane surface is marked with many tiny pores through which some materials may pass. The specific roles of such pores are not well understood at present.

In a word, cell membranes are more than fluid-filled sacks. They are living, responding structures. Thus, the more membrane area an organism possesses, the greater its control over its internal environment.

FIGURE 6.6   A hypothetical model of the cell membrane. There is some argument about precisely how cell membranes are constructed. Phospholipids have water-soluble heads and water-insoluble tails. Essentially, phospholipid molecules are arranged so that their hydrophobic (water fearing), fatty acid tails are directed inward. The phospholipid layer is interrupted by globular proteins, some of which are enzymes, while others are involved in active transport or support.

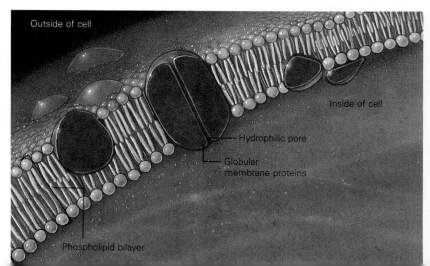

The Microtrabecular Lattice

It was recently discovered that the cell's organelles may not float freely in an amorphous cytoplasm. Instead, they are held in place by a complex bridgework called the **microtrabecular lattice** (Figure 6.7). It appears as a mazelike network of hollow fibers, extending throughout the cell, connecting and suspending the organelles in a kind of three-dimensional web. Researchers are already hard at work unlocking the secrets of this grand network and have suggested both structural and metabolic roles for it. (Others are not convinced it exists.)

There is also important evidence that the lattice holds even enzymes in place, suspended in the delicate framework. It has been suggested that precise spatial arrangement of enzymes would increase their efficiency by encouraging a specific sequence of interactions. For example, enzyme B might be held near enzyme A, so that it might more easily interact with the products of A. Enzyme C would be near B, and so forth. Such structural organization presumably would be an improvement over random enzyme movement through the cell.

Cell Walls

Plant cells have cell walls; animal cells do not. **Cell walls** are nonliving, rather inflexible, highly permeable, and strengthened by mats composed of cellulose fibers and other compounds in a tough and complex matrix (Figure 6.8). For this reason, trees are not limp.

The cell walls of plants are commercially important in a number of ways. For example, we count on cell walls to hold up our own walls as we frame our houses with wood. Also, it is for the cellulose in cell walls

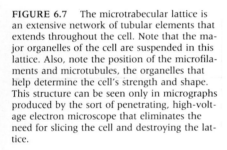

FIGURE 6.7 The microtrabecular lattice is an extensive network of tubular elements that extends throughout the cell. Note that the major organelles of the cell are suspended in this lattice. Also, note the position of the microfilaments and microtubules, the organelles that help determine the cell's strength and shape. This structure can be seen only in micrographs produced by the sort of penetrating, high-voltage electron microscope that eliminates the need for slicing the cell and destroying the lattice.

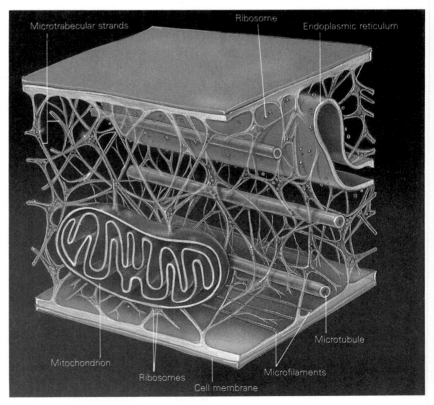

FIGURE 6.8 A micrograph of the matlike
fibers in a cell wall.

that we have leveled vast areas of our forests in response to wheedling commercials designed to increase our demand for "disposable" paper commodities. The cellulose of plant cell walls is also valuable as a major component of celluloid, rayon, cotton, and hemp. (Hemp was once provided by legally cultivating a plant called *Cannabis sativa,* later known as "killer weed" or reefer). Another component of cell walls, **lignin,** was long considered a totally useless by-product of paper manufacturing. However, researchers worked hard to find ways to alter it so that it could be sold somehow, and they cleverly managed to discover uses for it in the manufacture of synthetic rubber, vanillin, and adhesives.

Gap Junctions

In multicellular organisms, the cytoplasm of adjacent cells is often in *direct* contact. This arrangement enables one cell to quickly and effectively influence the next and encourage a rapid and precise response to stimuli. The contact is made through special membranous passages called **gap junctions** (Figure 6.9). In plants, they are known as **plasmodesmata** and appear as small holes in the cell walls through which thin strands of cytoplasm reach from cell to cell.

Mitochondria

Mitochondria are tiny structures (about the size of bacteria) that occur in almost all types of eukaryotic cells. They appear in a variety of shapes—round, elongate, and even threadlike. In some cells they seem to squirm around and move through the cytoplasm, while in other cells they appear more stationary or even immobile. They are not randomly distributed, but tend to aggregate in places where work is going on—

FIGURE 6.9 The gap junctions are essentially channels that allow the direct movement of substances between cells. Each junction is formed from a pair of "pipes" that are actually six dumbbell-shaped proteins. In the plant cell as shown here, they are called plasmodesmata.

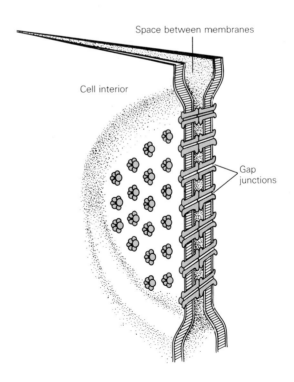

Space between membranes

Cell interior

Gap
junctions

FIGURE 6.10 A cutaway view of a mitochondrion (below). Mitochondria are double-membraned organelles involved in energy production. That energy is tied up in the energy-rich molecules of ATP. Mitochondria are about the size of some bacteria and there is some argument about how they came to be in cells. Perhaps mitochondria once "infected" cells—bacterialike. Mitochondria are concentrated in those cells and those parts of cells where the greatest work is going on.

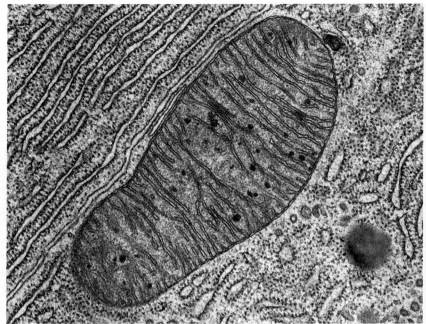

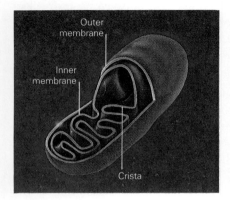

Outer membrane

Inner membrane

Crista

places where the greatest amount of energy is required. There are more mitochondria in liver cells, for instance, than there are in less active cells, such as cartilage. Not only are they more common in more active types of cells, they also accumulate in the most active part of any cell. The concentration of mitochondria at places where work is being done is not coincidental; they are the "powerhouses" of the cell, with integral roles in energy production. Specifically, they are the sites of the Krebs cycle, the electron-transport chain, and chemiosmotic phosphorylation.

A mitochondrion, as we saw in Chapter 5, is double membraned, with the inner membrane commonly folded to produce shelflike cristae that extend inward into the fluid matrix in which a number of reactions necessary for the production of energy occur (Figure 6.10). There has been some vigorous discussion over just how mitochondria and certain other organelles came to be included in cells. A major argument, supported by Lynn Margulis, is that many such organelles were once free-living organisms that long ago invaded the early cells and developed a symbiotic relationship with them (see Essay 6.2).

Golgi Bodies

In 1898, Camillo Golgi, an Italian cytologist, was experimenting with some cell-staining procedures and discovered that when he used certain stains, such as silver nitrate, peculiar "bodies" appeared in the cells. These "reticular apparatuses," as he called them, had never been noticed before, but when other workers looked for them using the same stains, they turned up in a variety of cells. However, because they could not be seen in living cells, there was a great argument over whether they really existed or were just artifacts or debris produced by the staining process itself.

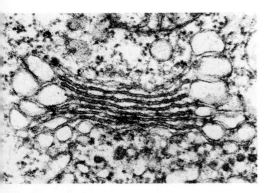

FIGURE 6.11 Golgi bodies apparently serve as "packaging centers" for the cell. They look like flattened sacs of membranes pressed against each other and seem to function in packaging the substances that are formed on the endoplasmic reticulum. Some complex molecules, such as sugars and proteins, may be assembled at the sites of Golgi bodies. Their precise role in the life of the cell is not completely understood. Whereas an animal cell may have only 10 to 20 Golgi bodies, a plant cell may contain several hundred.

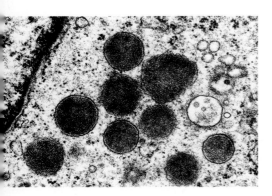

FIGURE 6.12 Lysosomes hold powerful disruptive enzymes that, if released, can destroy the cells in which they are found.

The electron microscope resolved the debate. Indeed, these strange bodies did exist, and, appropriately enough, they were named **Golgi bodies.** It was found that they had a characteristic and identifiable structure no matter what kind of cell they were found in. In every case, they appeared as a group of tiny flattened vesicles, lying roughly parallel to each other (Figure 6.11).

Even after their existence was confirmed, an argument continued over their function. What did they do for a living? In spite of a great deal of attention to the question, we're still not sure, but there is some evidence that they serve as a sort of packaging center for the cell. Enzymes and other proteins, as well as certain carbohydrates, are collected in these bodies and packaged into sacs or vesicles. In this way, they are kept apart from the rest of the cell. In some cases, the packages move to the cell membrane, and the enclosed molecules are excreted from the cell.

Lysosomes

Lysosomes are somewhat spherical and about the size of mitchondria and are, in general, distinctly unimpressive bodies (Figure 6.12). Their undramatic appearance, however, belies the rather startling role they play in the history of a cell. It is believed that lysosomes are packets of digestive enzymes that are synthesized by the cell and packaged by the Golgi bodies. The packaging is important because if these enzymes were floating free in the cell's cytoplasm, the cell itself would be digested. Christian de Duve, who first discovered the lysosomes, called them "suicide bags," and the dramatic description is not entirely unwarranted, since they can actually destroy the cell that bears them. So why would cells have ever developed such a risk to themselves? In some cases, the destruction of cells is beneficial to the organism. For example, the cells could be old and not functioning well, or they might be in a part of the body that was undergoing reduction as a part of a normal developmental process, such as in the webbing between the fingers of a developing embryo.

Lysosomes may also help dispose of unwanted mitchrondria, red blood cells, or bacteria. (Fragments of all these have been found within these organelles.) Interestingly, malfunctioning lysosomes have been associated with a number of human diseases, including cancer. Rupturing lysosomes have also been accused of contributing to the aging process.

Peroxisomes

Peroxisomes are membrane-bound bodies found in the cells of a number of organisms. They are dense structures with a striking appearance under the microscope due to a crystalline core composed of tiny tubes (Figure 6.13).

Peroxisomes contain enzymes (suggesting they may be a special class of lysosomes). Their primary enzyme is **catalase,** which helps break down hydrogen peroxide (H_2O_2) into water and oxygen. Catalase activity may be the primary way the body rids itself of hydrogen peroxide, a powerful and dangerous oxidant.

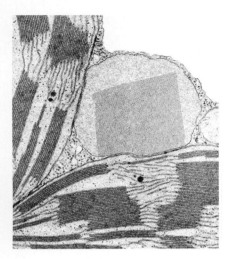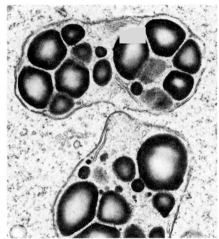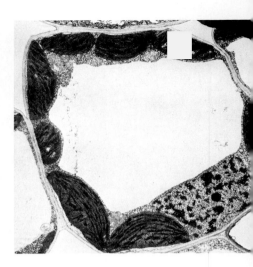

FIGURE 6.13 Peroxisomes are similar to ly-sosomes but have an unusual crystalline inner structure. They are thought to function in enzymatic oxidizing reactions—for example, by quickly oxidizing hydrogen peroxide in the cell to water and oxygen. Here a peroxisome lies wedged between two chloroplasts.

FIGURE 6.14 Plastids include colorless leucoplasts as well as green chloroplasts. They all have double membranes and their own DNA that reproduces in synchrony with that inside the nucleus.

FIGURE 6.15 Vacuoles are fluid-filled sacs; here seen as the large space occupying almost the entire interior of a cell. Some are metabolically active; others are simply storage vessels.

There are a number of kinds of peroxisomes, and they manufacture some forty enzymes besides catalase. However, no single kind of peroxisome manufactures all these enzymes and some contain no enzymes at all. The role of these latter bodies remains a mystery.

Plastids

Plastids are found in plant cells, but not in animal cells (Figure 6.14). They are interesting bodies, similar in some respects to mitochondria. For example, they are double-membraned and they have their own DNA. Strangely, their DNA may differ somewhat from that of the cells in which they reside.

There are two major types of plastids—colored and colorless. The colorless ones are called **leucoplasts.** The best known of the colored plastids are the green, chlorophyll-containing **chloroplasts.** Other chloroplasts may contain yellow or carrot-colored pigments, which, logically enough, are called **carotenoids.** The colorless plastids are usually storage bodies in which fats, starches, or proteins may be kept until needed by the cell.

Vacuoles

Vacuoles are fluid-filled sacs found in the cells of both plants and animals, as well as in microscopic organisms called protists (Figure 6.15). It is in plant cells, however, where they reach their greatest development; in fact, they may be the most conspicuous bodies in the cell. There are a number of types of vacuoles, each with a different function. Some, for example, are highly active in the cell's metabolism, while others are simply storage vessels.

In plants, the vacuoles are filled with a "cell sap" that can change volume through osmosis. In fact, it is the pressure of the swelling vacuole that forces the cell membrane against the cell wall and makes the plant tissue firm. A plant wilts when there is not enough fluid to keep its huge vacuoles filled. Besides water, the vacuole sap may contain sugars, proteins, pigments, and organic acids. It is these acids, in fact, that give oranges and lemons their tart taste.

The origin of cell organelles largely remains a mystery. How do they come to be there? How did they evolve? And from what?

One rather interesting idea was proposed by Lynn Margulis, who contends that those energy-harnessing organelles, chloroplasts and mitochondria, may be the descendants of free-living prokaryotic organisms that invaded other cells early in life's evolutionary history. This idea is known as the **symbiosis hypothesis,** and it is continually gaining support. Eventually, she suggests, the chloroplasts and mitochondria became mutually dependent on their host cells and unable to exist outside them. In time, the "living-together" arrangement evolved into a permanent marriage.

There are several lines of supporting evidence for the idea. For example, some cell organelles such as mitochondria and chloroplasts are capable of reproducing themselves, but today their reproduction is often precisely synchronized with the reproduction of the cell in which they reside.

There are also other lines of supporting evidence, including the fact that both chloroplasts and mitochondria are surrounded by a double membrane, very similar to that of prokaryotic cells from which eukaryotic cells are thought to arise. Also, both of these organelles contain their own DNA (with slight differences in code from the DNA in chromosomes). This DNA occurs in simple, circular strands, as does

the DNA of prokaryotes. Furthermore, chloroplasts have their own ribosomes, small bodies associated with energy production (discussed shortly). Interestingly, the ribosomes of these two organelles are smaller than those of the eukaryotic cell; in fact, they are about the size and shape of bacterial ribosomes (and bacteria, you recall, are prokaryotic).

The list goes on, forming a compelling body of evidence that the invasion of ancient cells by the ancestors of prokaryotes yielded many of today's cellular organelles. So far, no single alternative theory has been able to replace this encompassing scenario.

Some single-celled organisms (such as the paramecium, as we shall see) contain **contractile vacuoles** that enable them to squeeze waste material or excess water out through the cell membrane. Such vacuoles may appear and disappear in response to the organism's needs, or they may be relatively permanent structures. The latter help us distinguish one species of these tiny organisms from another.

The Endoplasmic Reticulum

The **endoplasmic reticulum (ER)** is a highly variable structure, appearing, for example, as elongate vesicles or as long, tubelike structures (Figure 6.16). Under the microscope, the ER appears as light space surrounded by membranes. Some of these membranes are studded with tiny granular particles called **ribosomes,** some are not. Those with ribosomes are called **rough ER,** and those without, **smooth ER.** (Ribosomes may also be found independently of the ER. We'll encounter these again a bit later, but for now just keep in mind that they are the sites of protein synthesis.)

Electron-microscope studies suggest that the ER is connected to the outer border of the nuclear membrane. Furthermore, there is evidence that it is also confluent with the cell membrane. The hypothesis is partly based on notions of how the nucleus may have arisen (Figure 6.17). This arrangement would result in an open channel from the nucleus to the outside of the cell. The cell might thus be able to easily transport products manufactured in the nucleus directly to the outside. But more important, with such a connection, the nucleus might be able to react quickly to changes in the cell's environment.

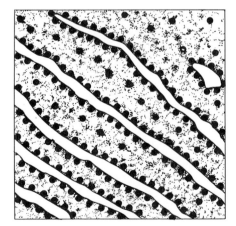

FIGURE 6.16   The endoplasmic reticulum (ER). The ER may be an extension of the cell membrane that communicates directly with the nuclear membrane. Rough ER, seen here, has ribosomes attached and is possibly involved in producing certain proteins.

FIGURE 6.17 Two opposing theories that presume to account for the nucleus within the cell. In (A), the nucleus is the primary structure and the ER and cell membrane are added secondarily as extensions of the nuclear membrane. In (B), the primal structure is the cell. The cell constituents organize themselves into a nucleus as the ER develops by inpocketings of the cell membrane.

(A)   (B)

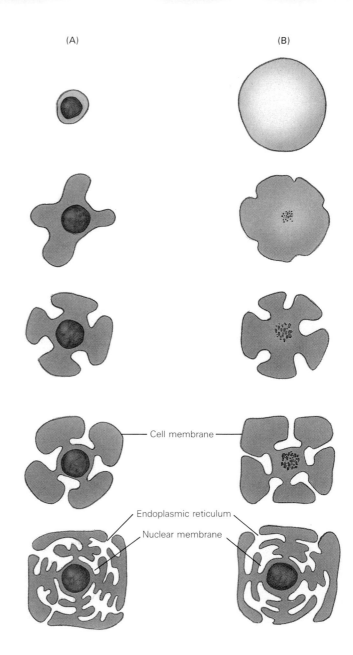

Cell membrane

Endoplasmic reticulum

Nuclear membrane

FIGURE 6.18 Centrioles. A section across the central axis of one centriole. Note the nine sets of triplet microtubules.

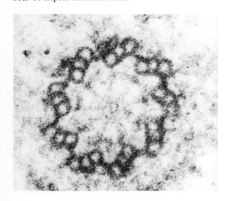

Centrioles

Centrioles are small cylindrical bodies, barely visible under a light microscope, that lie just outside the nucleus in an area of specialized cytoplasm. They are normally found in the cells of animals, algae, and some fungi; they are absent in the cells of flowering plants. As we shall see shortly, they are associated in complex ways with cell division and cell movement.

Centrioles have a characteristic structure of nine sets of microtubules running lengthwise just below their surface (Figure 6.18). Usually centrioles are paired, each lying at right angles to the other, forming a kind of "T."

FIGURE 6.19 Microtubules are composed of a tight, spiral arrangement of tiny tubulin spheres.

FIGURE 6.20 An electron microscope view of a flagellum cross-section and a cut-away diagram of a flagellum. Note the 9 + 2 arrangement of the microtubules.

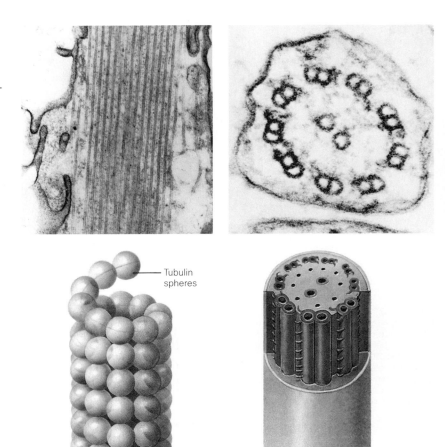

Tubulin spheres

Microtubules

Microtubules, as you might expect, look like tiny tubes (Figure 6.19). Not much is known about them, but their constituent protein, called **tubulin**, occurs in doublet spheres whose arrangement forms the tubes or cylinders. Microtubules form the core, not only of flagella, cilia, and sperm tails, but also the star-shaped "asters" and the spindle fibers that appear at mitosis.

Cilia and Flagella

Cilia and **flagella** (Figure 6.20) are hairlike extensions that project from the surfaces of certain kinds of cells. They differ only in length; cilia are about 10 to 30 microns in length, while flagella may extend to thousands of microns. Both make "beating" movements, and both may function in moving the cell along through some fluid. Cilia, in addition, move substances across the surface of a stationary cell. As examples, a sperm cell swims by beating its whiplike flagellum, and the beating cilia that line your breathing passages help sweep away airborne debris (unless you have killed them by smoking). Both cilia and flagella contain microtubules that form a 9 + 2 arrangement. Basically, the structure involves a circle of nine pairs of microtubules surrounding two single microtubules. (The arrangement in centrioles is similar, but they lack the central pair.)

FIGURE 6.21 The nuclear membrane (upper right) is pitted with "pores" that are actually biconcave indentations that allow easy movement of materials across the membrane.

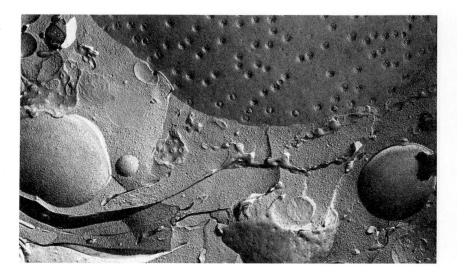

FIGURE 6.22 One of the classic experiments showing the importance of the nucleus involved cutting a living amoeba (a single-celled protist) in half, leaving one half with the entire nucleus. The deprived half died, and the nucleated half survived. Then, when the nucleus was removed from the surviving half and placed into the dying half, the latter recovered and continued a normal existence.

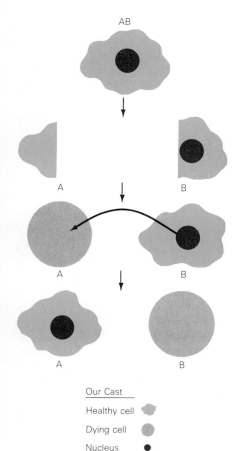

AB

A   B

A   B

A   B

Our Cast

Healthy cell

Dying cell

Nucleus

The 9 + 2 arrangement of cilia and flagella has generated much speculation about their origin. For example, some researchers believe that each cilium develops as the extension of one of a pair of centrioles. (Do you see how the structural arrangement of a cilium suggests this origin?)

The Nucleus

Perhaps the most prominent structure in most cells is the **nucleus.** It also plays a central role in the life of the cell in that it is involved in a number of critical processes, such as regulation and reproduction.

The nucleus is surrounded by a membrane that is double the width of the cell membrane, a fact that suggests two hypotheses regarding the origin of the nucleus, as you may recall. The inner and outer parts of the nuclear membrane pinch together here and there over the nuclear surface to form scattered, thin-walled areas called **nuclear pores** (Figure 6.21). Although they are not actually pores, since they do not go all the way through, they provide areas of high permeability that probably facilitate the flow of molecules between the nucleus and cytoplasm.

A number of rather straightforward experiments have illustrated the critical role of the nucleus. For example, if an amoeba is cut in half, the half with the nucleus will survive and continue to function as before. The half without the nucleus is unable to move about or capture food and soon dies. However, if the nucleus is surgically restored to this half within a few days, it recovers and is soon as good as new (Figure 6.22).

Such findings do not imply that the nucleus has absolute control over the cytoplasm. (As you might expect, real life is rarely that simple.) The cytoplasm also has certain influences on the nucleus. In addition, environmental factors may influence the nucleus. For example, in the snowshoe hare, a change in the environmental temperature prompts the nuclei of hair-producing cells to turn the animal's coat from one color to another as the seasons change (Figure 6.23).

Inside the nucleus is a fluid matrix in which a number of different types of bodies are suspended. When cells are appropriately stained, a

FIGURE 6.23 The activity of the hereditary units called genes may be influenced by the environment. We know that coat color in many animals, for example, is determined genetically. However, in the varying hare, the genetic component that dictates hair color is directly controlled by temperature. When it grows colder, the cells that had been making brown hairs change and white hairs appear. In spring, the sequence is reversed. Of course, the result is a prey animal that is difficult to see in either summer or winter.

netlike structure becomes visible within the nucleus. This material is called **chromatin** (from Greek: color) because of its affinity for the stains. At a certain period in the cell's cycle, the chromatin shortens and thickens and forms chromosomes, the structures that include sequences of genes. (As a highly educated person, you are undoubtedly aware that genes have a role in heredity. But just in case, we'll go over it later.)

Within each nucleus is a small structure called the **nucleolus.** Most nuclei contain only one nucleolus, but some kinds have more. Within the nucleolus is a large concentration of molecules called RNA, which will play an important role in another story—how protein is made.

Now that we've discussed some of the major aspects of cells, it would be nice to pull it all together and have a look at a representative cell. The problem is that there is no such thing as a "representative" cell. Cells are too diverse, and they change too rapidly to be described in a simplistic diagram. But with this small embarrassment in mind, you might review Figures 6.24 and 6.25.

HOW MOLECULES MOVE

The large multicellular organisms of the earth are faced with a number of general problems. One of these is how to move materials around through large, bulky bodies. Many of these materials must laboriously pass into and out of cells by crossing the cell membranes. There are two broad categories of such movement, **passive transport** and **active transport.**

Passive Transport

Passive transport is conceptually quite simple because it does not involve the expenditure of cellular energy. Instead, it requires only the random movement of molecules. However, as you may have suspected, in biology, even something so simple is not all *that* simple. In fact there are four kinds of passive transport: **diffusion, facilitated diffusion, osmosis,** and **bulk flow.**

FIGURE 6.24 There is no typical animal cell, but this has not stopped us from showing you one. What is depicted here is a cell as it might appear with stop-action photography. The cytoplasm is normally a seething mass of shifting composition. At no time could all these bodies be seen at once. Each body is usually discovered only under very special staining conditions. Nevertheless, all the structures you see here are believed to exist at some time in the cell's history. The shape of a cell, of course, is highly variable.

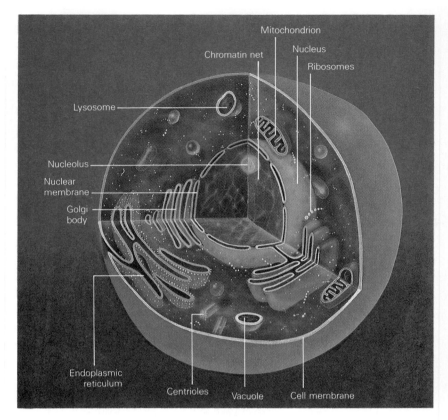

FIGURE 6.25 This plant cell is also a composite of what we are able to piece together about such cells. Some of the differences in animal and plant cells should now be apparent. Notice the heavy cell wall and the enormous vacuoles. Since plant cells are generally less active, they have fewer mitochondria. But since they are green and manufacture their own food, they have chloroplasts. Chromosomes are not yet visible in either the animal or the plant nuclei. In the conducting cells of plants, the cytoplasm disappears, leaving only the rigid cell walls to form tubelike structures through which food and water will pass.

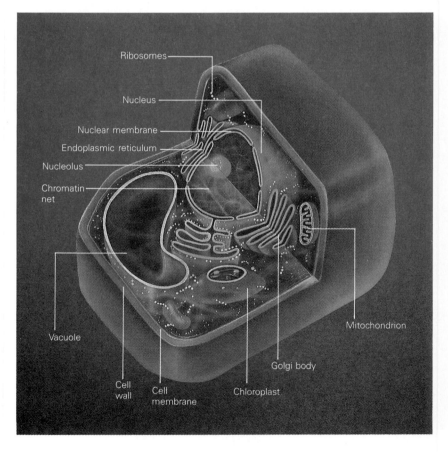

Diffusion

Physicists tell us that the molecules of any substance—gas, liquid, or solid—are constantly in motion, bumping into each other, rebounding, and changing their direction of movement. The principle would become apparent if you were to open a vial of perfume (or if you're the baser sort, try hydrogen sulfide) in a corner of a room. Soon the smell will be detectable in all parts of the room. If the solution is allowed to evaporate completely, all parts of the room will eventually smell the same (Figure 6.26). Until that happens, however, there will be a **concentration gradient,** with a greater concentration of perfume molecules nearer the vial.

The shift of molecules away from their center of concentration is due to the fact that they bump into each other with greater frequency where they are more densely packed. When they collide, of course, they rebound and change direction. However, their chances of collision are lessened when they are moving outward toward areas of lesser concentration. So there is a tendency for the *net* movement to be outward, *away* from the highest concentration gradient. *Thus, the molecules move from areas of higher concentration to areas of lower concentration*. In addition, because increased heat produces faster molecular movement, the diffusion rate is higher in a warmer system.

Molecules not only move through gaseous systems, but may also be transported by their own random motions through fluid environments, such as those in living cells. But they may not be able to move so easily from one cell to another. This is because the membranes of living cells are selective. Some substances are allowed across, some are escorted across, and some are not allowed to cross at all. In some cases, whether a molecule is allowed to pass across a membrane by simple diffusion depends on the size of the molecule in relation to the size of the pores in the membrane. Some molecules are simply too large to get through. Other molecules though, such as carbon dioxide, oxygen, and water, are able to diffuse rather freely across almost any cell membrane.

It should be stressed that in spite of its mechanical simplicity, diffusion is an important means by which small molecules move across cell membranes. Furthermore, simple diffusion works according to strict physical laws; that is, it works the same in both living and nonliving systems.

Facilitated Diffusion

Now let's consider a bit more complex sort of molecular movement. **Facilitated diffusion** is similar to simple diffusion in that the molecules are stimulated to move by environmental heat. Also, the movement is away from areas of higher concentration to areas of lower concentration. However, facilitated diffusion differs from simple diffusion in that some kinds of molecules are moved more easily than others. This is because the movement of certain molecules is enhanced by specific enzymes, called **permeases,** embedded in the cell membrane. No one yet knows how permeases work, but it is known that certain ions and compounds are quickly ushered across membranes in which these enzymes are found.

Bulk Flow

Under some conditions, the amount of water movement and the speed with which it moves cannot be accounted for by diffusion or

FIGURE 6.26   Molecules of perfume escaping from a bottle will soon permeate an entire room. At first, the molecules are concentrated near the bottle, but if evaporation is complete, in time the random movements of the molecules will disperse them evenly throughout the room. They shift from their center of concentration because they collide with greater frequency where they are more densely distributed. They are less likely to collide where they are rarer and, hence, they change direction less often as they move away from other molecules.

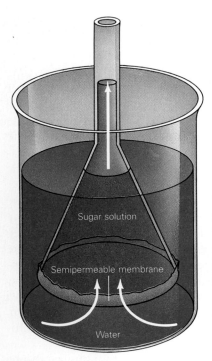

FIGURE 6.27 Osmosis will cause water molecules to move through a semipermeable membrane from a place where the water is in higher concentration to where it is lower. In the illustration, the inverted funnel is immersed in pure water, but inside the funnel, the water is diluted by sugar. Note that we're speaking of sugar diluting water (instead of the other way around) because the membrane covering the funnel will not allow the sugar molecules to pass, hence we are interested only in the movement of water molecules. The resulting movement in this case will cause the sugar solution to become more dilute, thus raising its level in the funnel.

facilitated diffusion alone. In these cases, it may be due to a process called **bulk flow,** the mass movement of fluids. Bulk flow is influenced by what is called **water potential.** Water potential is the potential ability of water to move in a mass (and thereby to do work). Actually, it is *differences* in water potential that cause bulk flow, such as exists when water potential is higher in one place than in another. Let's consider a familiar example. When water accumulates behind a dam (as in Figure 6.28), it rises higher than in the river bed below the dam. Thus, the water behind the dam has greater water potential than the water downriver. It can, indeed, do a great deal of work by turning generators or destroying villages downstream. In this case, we can say that because of gravity, the water will move from a region of greater water potential (behind the dam) to one of lesser water potential (the river bed below). Thus, water potential can be increased by *pressure;* in this case, pressure produced by gravity.

As another example of how pressure can create differences in water potential and move water, consider water being moved to an elevated holding tank by a mechanical pump. The pump is able to create a greater water potential (create a greater pressure) than the water in the elevated tank.

Water potential (and, hence, bulk flow) can also be influenced by the relative amounts of **substances in solution (solutes).** Pure water has the greatest water potential; therefore, water containing molecules or ions in solution has lower water potential. In other words, the more solute, the less water potential. The result is that the bulk flow between two solutions will be in the direction of the one that has more substances in solution. The bottom line is, bulk flow is the movement of fluids (here, water) from regions of greater water potential to regions of lesser water potential.

Osmosis

Osmosis is one of those words that has been borrowed from science and then twisted and misused beyond recognition. (Do any others come to mind?) The next time you hear the word *osmosis* in cocktail conversation, ask the speaker to define it. You will make a lifelong friend when you smirk and say, "No, osmosis is the diffusion of water across membranes, from an area of its greater concentration to one of its lesser concentration." Then say, "You see, when two solutions are separated by a selectively permeable membrane—that is, a membrane that allows only certain substances, such as water, to pass—the water will move from the solution with the greater concentration of water molecules through the membrane to the solution with the lesser concentration of water molecules." Then emphasize that the concentration of water molecules is *lower* on the side of the membrane that contains the *higher* solute concentrations (a **solute** being something dissolved in a solution). Say, further, that it doesn't really make any difference what kind of ion or molecule is in solution; it could be sugar, amino acids, or any other soluble substance; the water moves only according to the *relative* number of water molecules on either side of the membrane. The people there will greatly appreciate your discourse and you are sure to be invited back every February 30th.

FIGURE 6.28 The water behind a dam builds pressure that can be released as kinetic energy. The release can be gradual, such as by controlled spillways, or by the destruction of the structure, in which case it is released all at once, usually with devastating effects. Cells must also have ways of controlling the release of their considerable energy.

If you look around and everyone is still there, you should be prepared to answer questions, since osmosis is one of those simple principles that, at first, may seem hard to understand. For example, one might wonder, how long does osmosis go on? Theoretically, it continues until the water concentration on both sides of the membrane is equal. However, this equilibrium rarely occurs because of the fact that in living systems, molecular concentrations may rapidly change, and because other physical factors may influence the movement of the water molecules. Pressure, for example, may also influence osmotic movement. For example, in the sort of setup we see in Figure 6.27, the rising column of solute will eventually become so heavy that its pressure finally stops the net flow of water molecules. At this point, the *net* movement of water stops. (When osmotic movement apparently stops, the water actually continues to move across the membrane, but equally in both directions, so that there is no gain or loss of volume on either side.)

Active Transport

In some cases, a living cell can move molecules "uphill" across its membrane from a place of water concentration to a place of higher concentration. Such a chore is uphill because it goes against the normal direction of movement of the molecules and thus requires energy. As an example, some cells are able to maintain proper concentrations of sodium and potassium by pumping sodium out of the cell and potassium in. The mechanisms of such "pumps" are not completely understood, but it is generally assumed that some sort of carrier molecule is involved. It is visualized as a membrane protein that changes its configuration, thereby allowing the sodium and potassium to move through it (Figure 6.29). A variety of molecules can be moved by active transport in either direction, into or out of cells, but always with an expenditure of energy.

FIGURE 6.29 Cell membranes are living, responding structures that can actively move particles across themselves. In some cases, such movement must be accomplished against an osmotic gradient, and thus it requires energy (probably from ATP). Some substances move easily across the membrane on their own. These include water, oxygen, and carbon dioxide. Carbon dioxide is concentrated inside cells (where it is produced) and oxygen is concentrated outside. These, then, simply diffuse along their concentration gradients. Movement against a concentration gradient is called *active transport*. No one knows how it is accomplished, but one hypothesis is that carrier molecules pick up the molecules to be transported at one side of the membrane, diffuse across, and release them on the other side. Another hypothesis suggests that the carrier molecules change in configuration, thereby permitting specific molecules, such as sodium and potassium, to move through special channels.

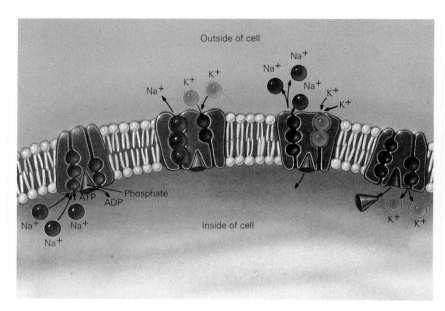

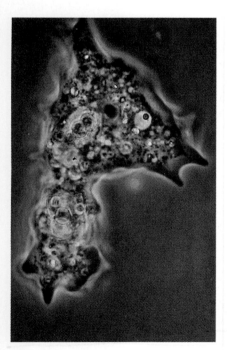

FIGURE 6.30 Food vacuoles in an amoeba. Digestive enzymes will be secreted into the vacuole. Undigested particles will be expelled through the amoebic membrane.

Endocytosis and Exocytosis

Some forms of active transport are so active that they are visible under microscopes. The best example, perhaps, is **endocytosis,** a kind of active transport that was first observed in the feeding behavior of the amoeba. When an amoeba detects a particle of food, its surrounding membrane buckles inward, forming a deepening cup around the food. The *rim* of the cup eventually draws together, and the cup pinches off from the surface, forming a membrane-bound sac, or **vacuole,** within the amoeba. The food that was outside the cell is now enclosed by the vacuole. Digestive enzymes are secreted into the vacuole and the food is broken down. The resulting nutrient molecules are passed through the vacuole membrane and out into the amoeba's inner fluids (Figure 6.30).

Endocytosis, by the way, is a general term that includes two processes. If the vacuole engulfs solid material, the process is called **phagocytosis;** if it engulfs dissolved or fluid materials, it is called **pinocytosis.**

The opposite process is equally intriguing. It is called **exocytosis,** the process by which cells *expel* materials (Figure 6.31). The material to be expelled becomes enclosed in a vacuole that eventually fuses with the cell membrane. At the point where the membrane of the vacuole joins the cell membrane, both membranes rupture and the contents of the vacuole are ejected from the cell.

We have, indeed, learned a great deal about Hooke's "little boxes" since he first passed light through that cork. Of course, many problems remain unsolved. Interestingly, some of these problems were unveiled as we came to know more about the complexities of cells. But other puzzles are almost as old as the field of cytology itself. These enduring questions have remained as monuments to life's intricacies. Even now, though, a host of dedicated researchers are determined that these monuments, too, shall fall.

FIGURE 6.31   Active transport and vacuoles. A very busy amoeba is demonstrating both endocytosis and exocytosis in all their variations. At the upper region, the amoeba is engulfing a small ciliate protozoan by phagocytosis. At the right, a channel has surrounded a solution of large molecules by pinocytosis. Eventually, both processes will create a vacuole from the cell membrane. At the lower side of the amoeba, the undigested residue from a food vacuole is being expelled by exocytosis.

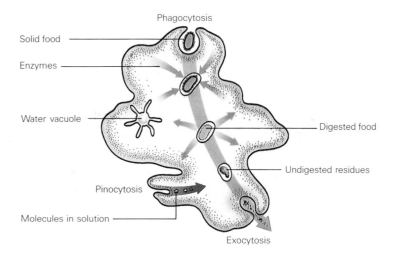

Phagocytosis

Solid food

Enzymes

Water vacuole

Digested food

Undigested residues

Pinocytosis

Molecules in solution

Exocytosis

SUMMARY

The study of cells has largely proceeded as a function of the technology available. A wide variety of cells are known, each with very special properties. Cell theory states that all living things are composed of cells. Basically, there are two kinds of cells: prokaryotic (believed to have evolved first) and eukaryotic (which includes most cells).

Cells are bounded by living membranes and contain a variety of cell organelles. Some of these organelles contain their own hereditary material and divide with the cell. Some are believed to have invaded ancestral cells and to have evolved in a mutually beneficial relationship with the cell.

An advantage of such cellularity is an increased regulatory membrane area. Multicellularity permits increased size, which then presents problems with moving fluids from place to place. Such movement is accomplished several ways: through passive transport (which includes diffusion, facilitated diffusion, osmosis, and bulk flow) and through active transport (which includes endocytosis and exocytosis).

Chapter 7

THE CELL AND ITS CYCLES

You may not be surprised to hear that when carrot cells reproduce, they form more carrot cells, and oyster cells form more oyster cells. After all, an oyster cell is very good at being an oyster cell, so when it reproduces, its best bet lies in making a very close approximation of itself. If a well-functioning oyster cell were to generate new kinds of cells, there would be no guarantee that those new cells would work—and, in fact, they probably wouldn't. So, there is an advantage in cells giving rise to new cells that are very similar to themselves. Such accurate duplication is assured by the precise mechanisms of cell reproduction.

We will now see how cells go about reproducing themselves so precisely. First, we will have a look at some cellular processes; then we will focus on what happens at the molecular level.

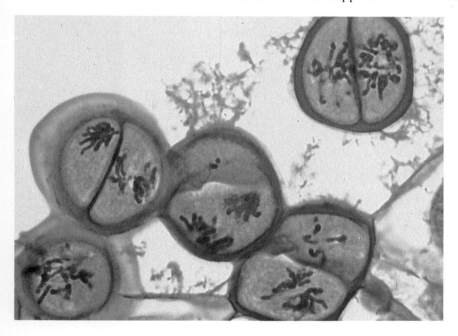

MITOSIS

One of the first questions we might ask is, Why do cells reproduce at all? That shouldn't be too hard to figure out. After all, in an active body, cells die or are worn away and must be replaced. In addition, simple growth is often based on the proliferation of cells. Many forms of life, then, demand that cells reproduce as part of maintenance and growth. So now, on to the process itself.

We should begin by noting that cells generally reproduce by dividing their constituents and then forming two new cells. The process of cell division involves **mitosis,** a mechanism by which a cell (at this stage called a **parent cell**) replicates its chromosomes. This done, the cell then divides in such a way that the chromosomal material is distributed equally to the **daughter cells.** The actual cell division is called **cytokinesis.** After mitosis and cytokinesis, each daughter cell has a full chromosome complement that is identical to that of the parent cell, and the cytoplasm of the parent cell has been distributed about equally between the daughter cells.

Interestingly, some of the cell organelles, such as the mitochondria, also reproduce themselves at cell division. Others, such as ribosomes, are synthesized anew by the daughter cells. In any case, the result is two new cells that are *about* the same size, have *about* the same number of organelles in the cytoplasm, and have *identical* chromosomal complements. Mitosis, then, is a very controlled and regulated process.

For purposes of discussion, mitosis is divided into four stages: **prophase, metaphase, anaphase,** and **telophase.** When the cell is not dividing, it is said to be at **interphase.** Interphase was once called the "resting stage," but we now know that this description is inaccurate. Interphase is a busy time, indeed, because it is then that many of the cell's components must double prior to the next mitotic division. I should add that the distinction between one phase and another is arbitrary, since there is a continuous progression from one phase to the next. Those who find comfort in labeling sometimes add further breakdowns, such as **late prophase** or **early anaphase,** so we will occasionally encounter these terms. Let's now review each of these stages in a bit more detail (Figure 7.1, pages 128–129).

At interphase, the chromosomal material of the cell may be difficult to distinguish under the microscope since it is dispersed throughout the nucleus in fine netlike strands, giving the nuclear fluid a granular appearance. At the beginning of prophase, however, the chromatin becomes more visible as it begins to gather together and thicken into long, twisted strands. Technically, it is at this point that the strands may be called **chromosomes.** During interphase, the genetic component of the cell has doubled, so now each chromosome is made up of paired strands called **chromatids,** which are joined at some point along the length by a **centromere** (Figure 7.2, page 130).

The animal cell at this stage almost doubles in size and generally becomes rounder as the cytoplasm becomes heavier and more viscous. The centromeres become very conspicuous, and the chromosomes continue to shorten and thicken as they move toward the nuclear membrane. At about this time, a peculiar **spindle** appears in the cytoplasm—a structure consisting of microtubules and appearing as thin lines that radiate across the cell from opposite poles. It is about now, too, that the nuclear membrane begins to vanish. As it disappears, the chromosomes

FIGURE 7.1 Mitosis in the endosperm of a lily. (A) Interphase: The chromatin is diffuse, spread thinly throughout the nucleus. The nucleolus is prominent. (B) Early prophase: The nuclear material begins to shorten and thicken. (C) Early prophase: The nucleolus is disappearing as the chromosomes are becoming distinct. (D) Mid prophase: The nuclear membrane breaks and microtubules appear. (E) Early metaphase: Chromosomes move toward the equatorial plane. (F) Metaphase: Chromosomes line up at the equator. (G) Early anaphase: The chromosomes begin to separate. (H) Late anaphase: the chromosomes draw nearer to the ends of the spindle. (I) Early telophase: New nuclei are becoming evident. (J) Telophase: The aggregation is complete and the two new cells are separating.

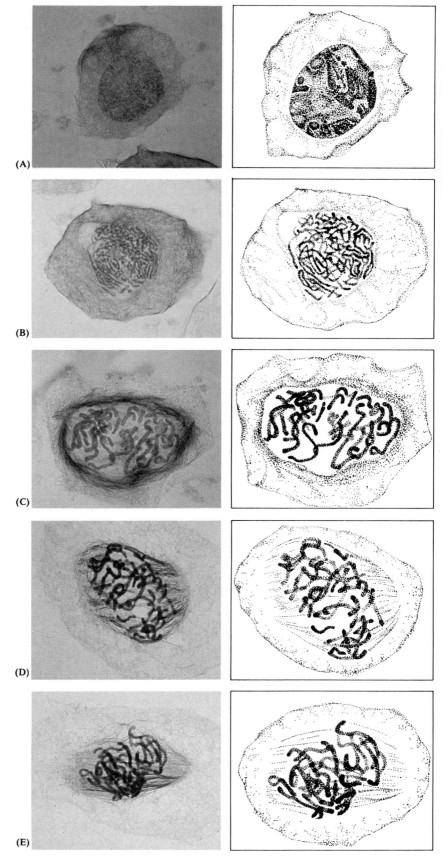

(A)

(B)

(C)

(D)

(E)

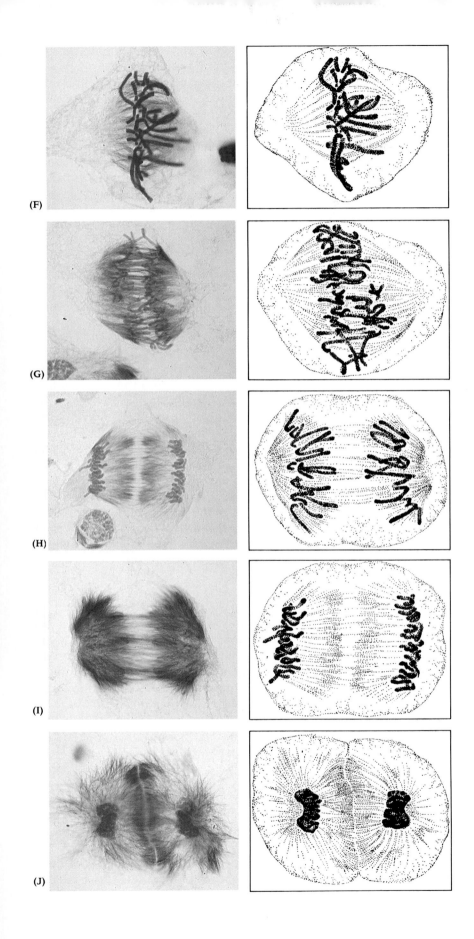

(F)

(G)

(H)

(I)

(J)

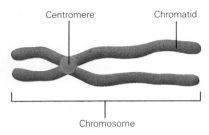

FIGURE 7.2 Chromosomes become apparent in the nuclei of cells for the first time at the beginning of prophase. The diffuse chromatin net has condensed and become twisted into strands. It is not always apparent, but chromosomes are composed of two parts called chromatids, which are joined at some point along their length to form chromosomes. When two chromatids that are joined in such a way become separated, each former chromatid is then called a chromosome. The chromatids usually lie pressed close together, but when they are distinguishable, each part extending from the centromere is called an arm. Thus, in this figure, there is one chromosome, two chromatids, and four arms.

move toward the center of the cell and line up across the spindle, marking the metaphase stage.

Anaphase begins with the division of the centromeres and the separation of the chromatids, each single strand also referred to as a chromosome. (So, chromosomes can be double stranded, and as the strands (chromatids) separate, each of these is also called a chromosome.) Subsequently, each of these single chromosomes moves along the spindle lines to opposite poles of the cell. It appears as if the thin microtubules that comprise the spindle have attached to the centromeres of the single chromosomes and are pulling each member of a pair in different directions. In any case, the chromosomes that converge at one end of the spindle are identical to those at the other end. The gathering of the identical groups of chromosomes at each spindle pole marks the beginning of the telophase stage.

The chromosomes of each group now begin to stretch out and unwind as they revert to a new interphase condition. As this happens, a nuclear membrane begins to form around each cluster of chromosomes.

In animal cells, the membrane around the old parent cell begins to pinch in between the two clumps of chromosomes **(cleavage),** finally closing together and roughly dividing the parental cytoplasm (cytokinesis). In plants, the rigid cell walls cannot pinch in. Instead, a new cell wall begins to form in the cytoplasm, roughly halfway between the divided chromosomes. In both animals and plants, each daughter cell may at first be somewhat smaller than the mother cell because it has less cytoplasm, but such differences are quickly reduced by water intake and the formation of new cell organelles. The daughter cells, by the way, may also be of unequal size since the cytoplasmic division may not be precise, but the important point is that the genetic component of each daughter cell will be identical to that of the parent cell.

Before we take a look at this process at the molecular level, let's consider another process—one that may, at first, seem similar to mitosis. Keep an eye peeled, though, because the differences are critical.

MEIOSIS

Following the steps of meiosis may answer some questions for you if you have thought much about the developmental process, especially if you are aware that eggs and sperm are simply specialized cells. One basic problem unfolds this way: at fertilization, the nuclei of an egg and a sperm are joined, and from the subsequent divisions of the fertilized egg, a new organism is formed. Because of the great precision of mitosis, each new cell of the developing individual is genetically identical to its parent cell. This means that all the cells of the developing organism have their full complement of chromosomes. Since eggs and sperm are both cells, when they combine at fertilization, why don't the resulting offspring have twice as many chromosomes as their parents?

In answering this question, we can begin by noting that the number of chromosomes is identical in virtually every cell of the body, since all the cells of any organism are genetically identical to that first fertilized egg. We say "virtually" because there are exceptions; there must be. For the union of an egg and a sperm to produce the correct number of chromosomes in the cells of the offspring, each of them must have exactly

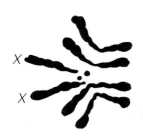

FIGURE 7.3 Pairing in chromosomes of a female *Drosophila*. Note that each chromosome in a cell has its counterpart, one having come from the mother fly, the other from the father.

half the chromosome complement of the other cells, so that the *sum* of their chromosomes results in the full complement. As it turns out, each egg and sperm indeed has exactly half the number of chromosomes characteristic of the species. Thus, when the egg and sperm are joined, the result is a primal cell with the "normal" number of chromosomes. Then, through millions of divisions of that cell (in conjunction with cell specialization, which we'll worry about later), a new individual is formed—resplendent in its own demands, hang-ups, and voting tendencies.

For virtually any species, then, eggs and sperm must have half the number of chromosomes as do other cells. But they don't start out that way. They start out with a full (double) complement of chromosomes, and somewhere along their developmental route, half the chromosomes are lost. **Meiosis,** then, is the process by which certain cells reduce their chromosome number on the way to becoming **gametes**—eggs and sperm.

The halving of chromosomal complements is possible because, as you see in Figure 7.3, chromosomes come in *pairs*. Every kind of chromosome in the nucleus has a counterpart—another chromosome very similar to itself. These matched pairs of chromosomes are called **homologues.** Chromosomes exist in homologous pairs, as you may have guessed, because the egg and sperm contribute identical *kinds* of chromosomes to the new individual. Thus, for any pair of chromosomes in the body, one can be considered the descendant of a paternal chromosome (from the father's sperm) and the other the descendant of a maternal chromosome (from the mother's egg). The result is that each parent contributes to every genetically-determined characteristic of the offspring (with special exceptions, as we shall see). So, chromosomes come in pairs—for example, twenty-three pairs in humans, nine pairs in cabbages, and forty-seven pairs in goldfish. (The numbers of chromosomes for some other species are shown in Table 7.1.) Thus, if we have an allele for some trait from one parent, we will have another allele for that trait from the other parent.

TABLE 7.1   The number of chromosomes differs for each species. Do you find it significant that the number of human chromosomes falls somewhere between that of a housefly and a sand dollar?

Chromosome numbers.[a]

Alligator	32	Garden pea	14	*Penicillium*	2
Amoeba	50	Goldfish	94	Planaria	16
Brown bat	44	Horse	64	Redwood	22
Carrot	18	House fly	12	Rhesus monkey	42
Cat	32	Human	46	Rose	14, 21, 28
Cattle	60	Lettuce	18	Sand dollar	52
Chicken	78	Magnolia	38, 76, 114	Starfish	36
Chimpanzee	48	Marijuana	20	Tobacco	48
Dog	78	Onion	16, 32	Turkey	82, 81
Fruit fly	8	Opossum	22	White ash	46, 92, 138

[a]There is no apparent significance to chromosome number as far as biologists can determine. If you feel good about your 46 (see human), check the turkey, amoeba, and cattle. What do you make of the tobacco and chimpanzee numbers? Note the variation in some plant and animal species. Plants sometimes undergo spontaneous doubling and tripling of chromosome number.

Now, let's follow the development of gametes as they begin their meiotic divisions in the **gonads** (ovaries or testes) of animals. Remember, it may seem at first that we are retracing mitosis, but this will change.

First, note that the meiotic process is divided into two parts, meiosis I and meiosis II (Figure 7.5, pages 134–135). But before meiosis begins at all—that is, during the cell's interphase—the genetic material is simply replicated. The chromatin net shortens and thickens to form distinct chromosomes, and the spindle starts to form about the time the nuclear membrane begins to disappear (all this should sound familiar so far).

The process continues as, in early prophase I, each pair of homologous chromosomes comes together. The chromosomes themselves consist of two identical chromatids (so each chromosomal group is made up of *four* chromatids at this point). As the chromosome pairs line up side by side, an important event may take place: the chromosomes may exchange parts. This exchange is called **crossing over** (Figure 7.4). The crossovers take place at various points along the chromosomes, and as chromosomal parts are exchanged, new combinations result. (We will discuss later the role of crossover in producing variation and see why such variation is so important in evolutionary processes.)

In metaphase I (as you might expect if you recall our discussion of mitosis), the chromosomes line up across the developing spindle. In meiosis, however, the chromosome *pairs* separate, with one member of each pair moving to opposite spindle poles (the movement marking anaphase I). Note that the centromeres do not divide, as they do in mitosis. Thus, the chromatids of each chromosome now travel together to the spindle poles. At telophase I, the end of the first stage of meiosis, the chromosomes begin to elongate and to again take on a netlike appearance. A nuclear membrane may now form as the cell enters interphase II. Each nucleus now contains only half the number of chromosomes of the parent cell, although each chromosome is two-stranded. Not only are there now fewer chromosomes in each cell, but they are probably qualitatively different from any of the original chromosomes as a result of crossovers.

Interphase II differs from the interphase preceding meiosis I in an important way: this time the chromosomes do not replicate themselves. Aside from this, however, the second meiotic division proceeds in the usual way. In prophase II, the chromosomes shorten and thicken, and the nuclear membrane (if it is present) begins to disappear. At metaphase II, the chromosomes line up across the equatorial plate as the spindle fibers again appear. Now, at anaphase II, in contrast to anaphase I, the centromere divides in two, each half bearing a centromere, and the single chromosomes (earlier called chromatids) move toward opposite poles. At telophase II, a nuclear membrane forms around each set of chromosomes. The cytoplasm is divided as new cell membranes appear, thus separating each group into a different cell. So where there was one cell with a full complement of chromosomes, there are now four cells, each with half the chromosomal complement (Figure 7.6, pages 136–137).

If the meiotic process has been taking place in a testicle, each of four resulting cells will undergo further changes, grow a tail, and become a sperm. However, if the meiosis was in an ovary, three of the resulting cells will form tiny, nonfunctional **polar bodies,** with only one cell sur-

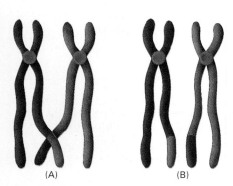

FIGURE 7.4 While homologous chromosomes are closely associated in prophase I of meiosis, they may exchange parts. Their chromatids may cross (A) and then break at the point of their crossing. Each separated part may then fuse to the other chromosome at the area of the breakage (B). New genetic combinations are created in this way.

viving to become an egg. It is not known what determines which of the four cells this will be.

There are two major take-home lessons from the story of meiosis. First, meiosis halves the chromosome number of eggs and sperm. Second, meiosis provides a means of shuffling and reorganizing the chromosomes, thus increasing variation in the offspring. This reorganization takes place in three major ways: (1) in the apparently random lineup of chromosomes at metaphase I, so that paternal and maternal chromosomes are mixed by the time a cell enters meiosis II; (2) in crossovers; and (3) in the apparently random selection of polar bodies in egg formation. You may come to have an abiding appreciation for meiosis when you realize that it is because of this process that you don't look exactly like your little brother.

THE DOUBLE HELIX

The search for the structure of the hereditary material is one of the most interesting stories in modern science. By about the middle of this century, the role of chromosomes in inheritance was well established. But a growing question was, What are they? What are they made of? Chemical analysis showed that the chromosomes themselves were composed of protein and an acid called **deoxyribonucleic acid,** or **DNA.** By 1950, it was generally accepted that the hereditary material itself was the DNA component of the chromosomes. But how was the DNA arranged in the chromosome? What did it look like? How did it duplicate itself at mitosis? And, in particular, how could it direct protein synthesis? By the middle of the century, several intensive research efforts were well under way in the United States and Britain. The prevailing notion was that there were three chains in each DNA molecule. Linus Pauling, the great American chemist, and Rosalind Franklin of England both favored this idea, but they disagreed over such details as whether the nitrogen bases stuck outward from the molecule or inward.

Then a brilliant and confident young American postdoctoral fellow ran across an ebullient and talkative English graduate student in biophysics at Cambridge University. The results of what transpired were later described by the American, James Watson, in his controversial best-seller, *The Double Helix.*

The partnership, it turned out, was a most fortunate one. The Englishman, Francis Crick, had noted that when the DNA portion of chromosomes was photographed by x-ray diffraction techniques, a peculiar image appeared—one which could only be produced by molecules arranged in a helical, or spiral, pattern. Watson's biological intuition told him that only two strands were involved. (After all, reproduction usually involved only two parents, and both chromosomes and genes generally come in pairs.) They pressed relentlessly at the question and constructed a number of unusual experimental devices. On the basis of x-ray diffraction information mainly provided by colleague Rosalind Franklin, they even built several DNA models of sticks and metal in their laboratories. Finally, they hit upon the solution and immediately published their findings (and scooped the scientific world) in a simple one-page article in the prestigious journal *Nature.*

FIGURE 7.5 Meiosis in the anther of a lily. The meiotic process is divided into two parts: meiosis I and meiosis II. Keep in mind that both the mitotic and meiotic processes are a bit different in animals because no cell wall forms; instead, the cells pinch inward to meet the new cell membrane that is being laid down between them.

Meiosis I. (A) Prophase I: The chromosomes have synapsed, forming tetrads. (B) Metaphase I: The tetrads have lined up across the spindle. (C) Anaphase I: The centromeres do not divide, but the homologous chromosomes separate and move to opposite poles. (D) Early telophase I: Chromosomes cluster at the poles. (E) Late telophase I: New nuclei are forming at each pole.

Meiosis II. (F) Early prophase II: The diffuse chromatin is beginning to condense. (G) Metaphase II: Distinct chromosomes line up across the spindle. (H) Late anaphase II: The centromeres divide. Single-stranded chromosomes move to opposite poles. (I) Telophase II: New nuclei and new cell walls are beginning to form, resulting in four haploid cells. (Photos: Carolina Biological Supply Co.)

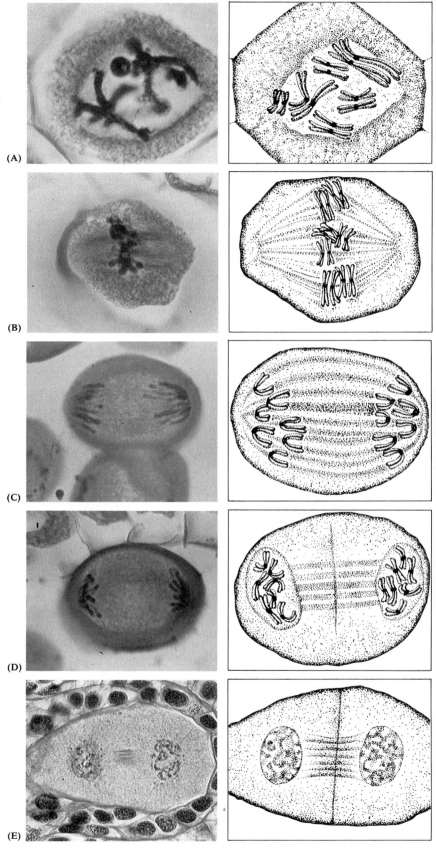

(A)

(B)

(C)

(D)

(E)

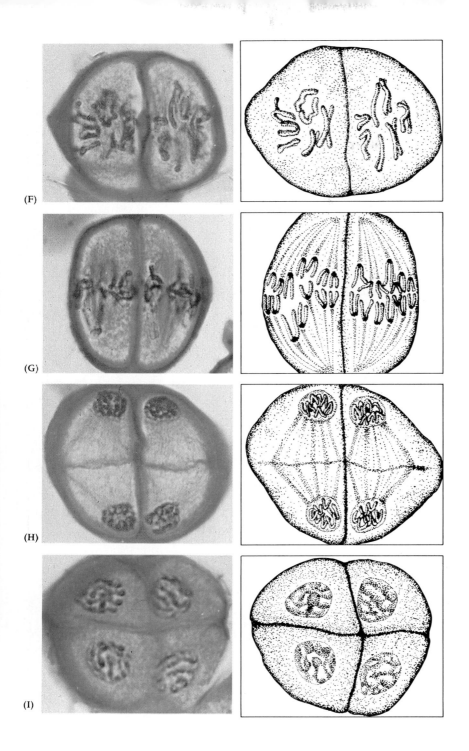

(F)

(G)

(H)

(I)

FIGURE 7.6 A graphic outline of mitosis and meiosis in animal cells. Compare the two processes carefully. For convenience, we have here an animal with *two* pairs of chromosomes. Also, to make things conceptually easier, paternal and maternal chromosomes are identified by their darkness. It doesn't matter which you want to call paternal or maternal, but it is important to notice the relative number of chromosomes from each parent that makes it into the daughter cells. Notice in particular the different behavior of the chromosomes in metaphase and anaphase of the two processes. Also keep in mind that meiosis is essentially two divisions. Finally, compare the chromosome number in mitotic telophase with that of telophase in meiosis II. Note that chromosomes are most easily counted by counting centromeres.

MITOSIS

1. INTERPHASE
Chromosomes not visible, DNA replication occurs; division preparations in progress. Centrioles replicate.

2. PROPHASE
Centrioles migrated to opposite sides; spindle forms; chromosomes become visible as they condense; nuclear membrane, nucleolus disperse.

3. METAPHASE
Chromosomes aligned vertically on cell equator. Note attachment of spindle fibers from centromere to centrioles.

4. ANAPHASE
Centromeres divide; single-stranded chromosomes move toward centriole regions.

5. TELOPHASE
Cytoplasm divides; chromosomes decondense; nuclear membrane, nucleolus reappear (reverse of prophase).

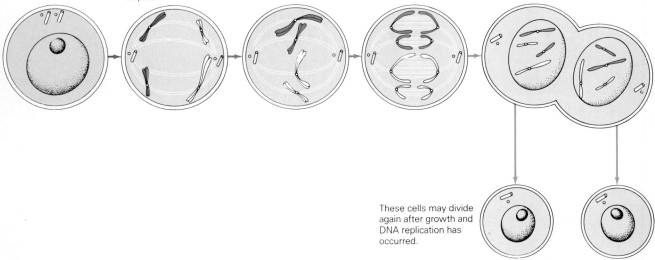

These cells may divide again after growth and DNA replication has occurred.

6. DAUGHTER CELLS
Two cells of identical genetic (DNA) quality; continuity of genetic information preserved by mitotic process.

MEIOSIS

1. INTERPHASE
Chromosomes not visible; DNA replication occurs. Centrioles replicate.

2. PROPHASE I
Mid-prophase; pairs of double-stranded chromosomes join each other; crossovers may occur shortly before metaphase.

3. METAPHASE I
Note different alignment of double-stranded chromosomal pairs at midcell and different manner of spindle attachment to centromeres.

4. ANAPHASE I
Centromeres do not divide; pairs of chromosomes (maternal and paternal) forever separated.

5. TELOPHASE I
Daughter cells not identical; chromosome number reduced by half.

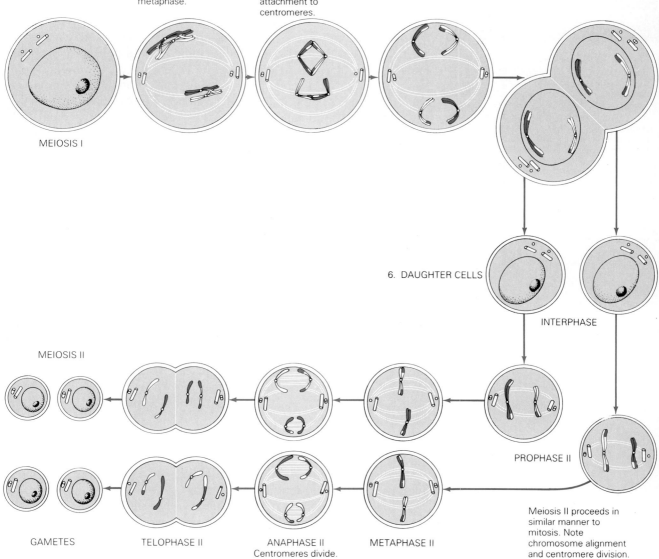

MEIOSIS I

MEIOSIS II

6. DAUGHTER CELLS

INTERPHASE

PROPHASE II

Meiosis II proceeds in similar manner to mitosis. Note chromosome alignment and centromere division.

GAMETES

TELOPHASE II

ANAPHASE II
Centromeres divide.

METAPHASE II

FIGURE 7.7 James Watson and Francis Crick, who, with a little help from their friends, won the race with the great Linus Pauling to discover the structure of DNA. They unmasked the molecule as a double helix after using such sophisticated techniques as x-ray crystallography and such mundane tools as ball and stick models. This fascinating story has been recounted in Watson's controversial book, *The Double Helix.* Watson later headed a "think tank" in New York. Crick continued to work on the puzzle and made enormously valuable contributions to the later discovery of the genetic code—the three-nucleotide sequence. Progress continues, however, and now even the revered (and very tidy) nucleotide vocabulary is under attack. It may be soon revised since it seems the story may not be as simple as has been assumed.

Watson and Crick (Figure 7.7) announced in 1953 that DNA is composed of two chains coiled around each other in the form of a double helix. (You can get the idea if you visualize a ladder twisted so that the rungs remain perpendicular to the sides.)

According to their model, the two longitudinal sides of the ladder are composed of alternating sugar and phosphate molecules. The "rungs" of the ladder are formed by four nitrogenous bases—the **purines, adenine** (A) and **guanine** (G), and the **pyrimidines, thymine** (T) and **cytosine** (C). These bases, with their sugar and phosphate components, are nucleotides (as we saw in Chapter Four).

Watson and Crick described a chain with one base per sugar-phosphate and with two bases joining together to form each rung. The bases extend inward from the ladder's sides, they said, and are held together by weak hydrogen bonds. The purines, they pointed out, could not lie adjacent to each other because they are so large they would overlap. They simply couldn't fit, and the two pyrimidines, with their single nitrogen-containing rings, were too short to reach across. The arrangement could work only if a pyrimidine joined with a purine to form the rungs (T to A and C to G) (Figure 7.8). Furthermore, according to Watson and Crick's deductions, the nucleotides along one of the strands of the double helix could follow any sequence. For example, one might "read" along a strand and find ATTCGTAACGCGT in one segment and something quite different along another segment of the same strand. Also, the strands were found to be very long, so that the necessary complexity for life could be provided by the possible variations in the sequence and the great numbers of nucleotides in any chain. In fact, if the DNA from a single cell were laid out in a straight line, it would be about five feet long! Furthermore, there are about 10 billion nucleotide parts in the 46 chromosomes of a human cell, so maybe you can (and then maybe you can't) imagine the number of possible variations in the sequences.

According to this model, if the sequence of one strand of DNA is known, the sequence of the other strand can be deduced. This is because, for example, the purine adenine (A) of one strand will always be bound to the pyrimidine thymine (T) of the other strand, and wherever cytosine (C) exists, it will be attached to guanine (G) on the other strand. With this in mind, what would be the nucleotide sequence of the complementary segment of the strand described in the preceding paragraph?

DNA Replication

As you recall from our discussion of mitosis, after the cell has divided, each chromosome replicates itself so that the chromosomal component is doubled before the next division starts. How does the chromosome, with its intertwining strands, go about it?

Actually, Watson and Crick's model of DNA structure suggested the mechanism—conceptually, a simple one. When the time comes for the chromosome to replicate itself, the weak hydrogen bonds holding the two DNA strands together break. Then the two strands simply unwind, "unzip," and open up. The strands' purines and pyrimidines are now exposed to the nuclear sap, which is loaded with a variety of molecules, among them the adenines, guanines, cytosines, and thymines. These

FIGURE 7.8 DNA forms a double helix. Note that one strand appears upside down with regard to the other. Essentially, the backbones are composed of alternating molecules of sugar and phosphate. Replication proceeds as the two strands of DNA unwind and then "unzip." Their bases are exposed to the nuclear fluid that contains a large number of free bases—thymine (T), adenine (A), cytosine (C), and guanine (G)—sugars (S), and phosphates (P). The exposed bases on the chains, then, join with their free-floating complementary bases from the nucleus. As these free bases fall into line along the strand of DNA, they become connected by a precise alignment of sugars and phosphates. Thus, each strand manufactures a strand that is identical to the one from which it had separated, and the familiar double helix is reestablished.

At left, the molecular structure of DNA is shown. At right, we see a model of DNA replication as geometric figures represent each base and the specificity of base pairing becomes obvious.

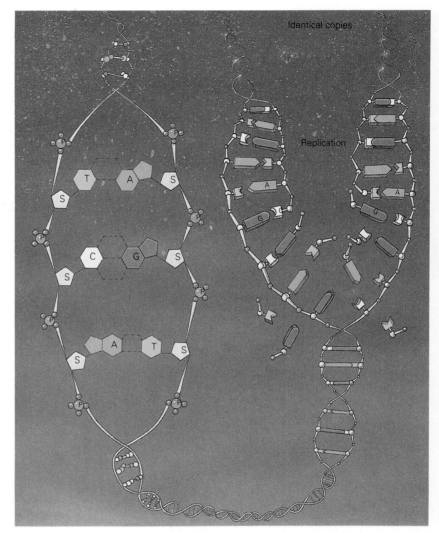

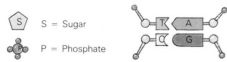

S = Sugar

P = Phosphate

free-floating molecules can then attach to the exposed bases of the open DNA strand (T to A, C to G). The original strand and the newly-forming one can then bind together by forming weak hydrogen bonds along their length. Thus, step by step, new DNA is produced, each molecule identical to the original (refer again to Figure 7.8). Furthermore, each DNA molecule will become a chromatid at the next cell division.

HOW CHROMOSOMES WORK

The extreme precision of the mitotic process suggests that the chromosomes are rather important to the cell's development and function. How, then, do they carry out their critical roles?

It will help to keep two things in mind. First, cells become *specialized* into muscle cells or nerve cells or whatever because of differences in their enzymes. Enzymes, you recall, are biocatalysts, and their role is to facilitate certain very specific biochemical events. It is through differences in the summation of such events that cells become specialized. The second thing to remember is that enzyme production is the result of chromosomal action. (An obvious question, to be answered later, is: If mitosis results in cells having the same chromosomal makeup, how do they become different from each other?) So what we are really asking here is, How do chromosomes make those peculiar proteins called enzymes?

When a chromosome begins to direct the formation of enzymes, it must first unwind, as in the replication of DNA. Now, however, only certain sections along the chromosome unwind (Figure 7.9). In order for the unwinding to occur, the hydrogen bonds between purines and pyrimidines must be broken, exposing these nucleotide bases to the nuclear fluid. As a result, free nucleotides begin to attach to the exposed bases along the DNA strands. As before, cytosine attaches to exposed guanine, guanine to exposed cytosine, and adenine to the exposed thymine. This time, however, where adenine lies exposed along a DNA molecule, it is attached, not to thymine, but to **uracil,** another nitrogen base in the cytoplasm. The nucleotides attach to each other by the usual sugar-phosphate connections, and the result is the formation of long strands, similar to the original DNA. Here, though, the DNA strands are not building complementary strands of DNA, as they do when they replicate themselves in mitosis. They are building another substance—**ribonucleic acid,** or **RNA.** RNA differs from DNA in that it

FIGURE 7.9 DNA manufactures *m*RNA in much the same way as it duplicates itself. The DNA, then, establishes the sequence of bases along the *m*RNA. However, as *m*RNA is made, thymine is replaced by uracil. As long *m*RNA molecules break away from their DNA template, they move into the cytoplasm to direct protein formation. Obviously, entire DNA strands do not unwind when forming *m*RNA. If they did, since each cell contains the same chromosome complement, they would all make the same enzymes and no specialization could occur. One of the enduring questions in biology concerns how the manufacture of *m*RNA is directed. How does one part of a chromosome come to actively make *m*RNA while other parts lie dormant, their DNA strands tightly interwound? And what causes long dormant segments to suddenly become active?

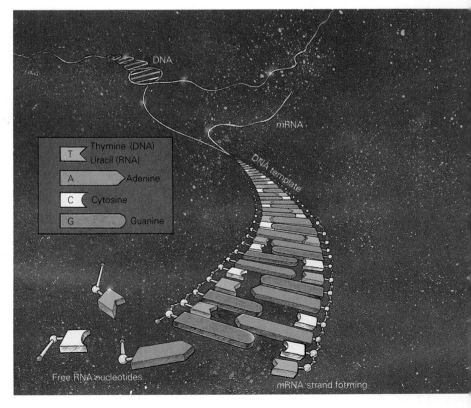

has one more oxygen atom in the sugar component of its nucleotides, and, in place of thymine, there is uracil. Also, most RNA exists as a single strand, not as a double helix.

So, as the DNA unwinds at various sections along its length, it directs the formation of RNA along these sections. This RNA then separates from the DNA strand and moves from the nucleus into the cytoplasm, each segment carrying a "message" encoded in its nucleotide sequence. This RNA is called, appropriately enough, **messenger RNA**, or **mRNA.**

The mRNA is a rather long and, as we will see, somewhat unstable molecule that contains guanine, cytosine, adenine, and uracil in sequences directed by the DNA. The sequencing of its nucleotide bases is very precise and very critical.

Before we see what mRNA does in the cytoplasm, let's learn something about how its message is carried—the mysterious **genetic code.** It may, at first, seem strange that only four nucleotide bases are sufficient to code for the thousands of proteins that make up the highly complex structures of living organisms. How, then, is the complexity achieved with only a four unit code (or "alphabet")?

Actually, you already have enough information to figure it out, but let's go through it stepwise. First, you recall that although proteins are very large and complex, they are made up of amino acids—of which there are only twenty kinds. The question, then, is how can four "letters" specify for twenty amino acids? The problem reduces to simple mathematics. If each nucleotide specified one amino acid, only four amino acids could be coded for. A two-base combination for each amino acid would provide four times as many possible combinations ($4 \times 4 = 16$), which is still not enough.

Uracil — Uridine Monophosphate
Cytosine — Cytidine Monophosphate
Thymine — Thymidine Monophosphate
Adenine — Adenosine Monophosphate (D-ribose)
Guanine — Guanosine Monophosphate

TABLE 7.2 The Genetic Code, or the "Language of Life." There are 64 Triplet Combinations. Each Triplet Is Shown with the Amino Acid or the Signal for Which It Codes.

Second letter

		U	C	A	G	
First letter	**U**	UUU } PHE UUC UUA } LEU UUG	UCU UCC } SER UCA UCG	UAU } TYR UAC UAA stop UAG stop	UGU } CYS UGC UGA stop UGG TRP	U C A G
	C	CUU CUC } LEU CUA CUG	CCU CCC } PRO CCA CCG	CAU } HIS CAC CAA } GLN CAG	CGU CGC } ARG CGA CGG	U C A G
	A	AUU AUC } ILE AUA AUG MET (start)	ACU ACC } THR ACA ACG	AAU } ASN AAC AAA } LYS AAG	AGU } SER AGC AGA } ARG AGG	U C A G
	G	GUU GUC } VAL GUA GUG	GCU GCC } ALA GCA GCG	GAU } ASP GAC GAA } GLU GAG	GGU GGC } GLY GGA GGG	U C A G

Third letter

Phenylalanine	Serine	Tyrosine	Cysteine
Leucine	Proline	Histidine	Tryptophan
Leucine	Threonine	Glutamine	Arginine
Isoleucine	Alanine	Asparagine	Serine
Methionine		Lysine	Arginine
Valine		Aspartic Acid	Glycine
		Glutamine	

So, early in the search for the genetic code it was suggested that a three-base combination was necessary to code for each amino acid. At first this was just a guess; the verifying data came later. It was reasoned, however, that with four nucleotide bases, a combination of any three would provide sixty-four possible combinations ($4 \times 4 \times 4 = 64$). But there are only twenty amino acids. So, according to this scheme, more than one nucleotide triplet can code for the same amino acid (Table 7.2).

Of the sixty-four nucleotide triplets, called **codons,** it is known that sixty-one code for specific amino acids. What about the other three, specifically, UAA, UAG, and UGA? These particular codons, interestingly enough, serve as periods in the language of life; they signal the end of a message. In addition, the codon AUG apparently doubles as an initiator, indicating the start of a message.

Now that we have some idea of the nature of the genetic code, let's see how it operates to produce the astonishing array of proteins found in living things. When the newly formed *m*RNA leaves the nucleus and enters the cytoplasm, it diffuses through the cell fluid until it encounters one of the cell's many ribosomes. It joins temporarily with a ribosome to form a ribosome-*m*RNA complex. Amino acids are brought to the complex by another RNA molecule called **transfer RNA,** or *t***RNA.** (There is also a **ribosomal RNA** in the ribosome; its role is not entirely understood.)

There are at least as many kinds of *t*RNA molecules as there are amino acids, and each kind of *t*RNA molecule can attach to only one type of amino acid. The specificity of the *t*RNA molecule is determined by the three nucleotides (forming an **anticodon**) that are found at one end. For example, if these three anticodon bases all happen to be adenine, an AAA sequence, then the *only* amino acid that the *t*RNA molecule can attach to is one called *phenylalanine.* Now, when a *t*RNA with this nucleotide arrangement with its phenylalanine encounters the *m*RNA-ribosome complex, where do you think the phenylalanine will be inserted on the ribosome-*m*RNA complex? Only at a place along the *m*RNA where there are three complementary uracils (UUU). So both the *kind* of amino acid and its *position* in the developing protein are dictated by the DNA back in the nucleus.

Figure 7.10 shows the process by which a new protein is synthesized. The ribosome attaches to one end of the *m*RNA strand and moves along its length, stopping to read off each nucleotide triplet to the *t*RNAs, which, with their amino acids, have diffused to the ribosome-*m*RNA complex. When the ribosome pauses at an *m*RNA triplet, it waits for a *t*RNA molecule with the proper complementary sequence to drift to that position. The *t*RNA briefly plugs into the *m*RNA codon and then moves away, leaving its amino acid behind. Meanwhile, the ribosome has moved along to the next codon site, where another *t*RNA falls into place, leaving its amino acid attached to the previous one. Thus, amino acids are joined one by one in the proper sequence for the formation of a specific protein. As soon as the *m*RNA is read and has directed a sequence of amino acids, it begins to break down.

Chains of amino acids formed in this way eventually form polypeptides or larger molecules of proteins. The resulting protein may be a structural material or it may be an enzyme, depending on the *m*RNA sequence that directed its formation. An enzyme formed in this way will participate in the biochemical activities of that cell and, hence, help to determine the cell's very nature (that is, answer the questions about what it looks like and what it does—perhaps the same things one might wonder about a blind date). We know that the chromosomes are identical in every cell, but how can the same DNA component result in cells of entirely different types?

We know that the fate of cells is partially directed by their enzymes. We also know that enzyme production is directed by the DNA of the chromosomes as it determines the sequence of nucleotides along *m*RNA strands which then move into the cytoplasm where they direct protein formation. Since there are many kinds of proteins within any cell, we can assume that many kinds of *m*RNA molecules may be formed along a single DNA chain. This happens because the *m*RNA molecules are formed along only short sections of the long DNA molecule. Obviously, it wouldn't do for a DNA molecule to continually transcribe *m*RNA

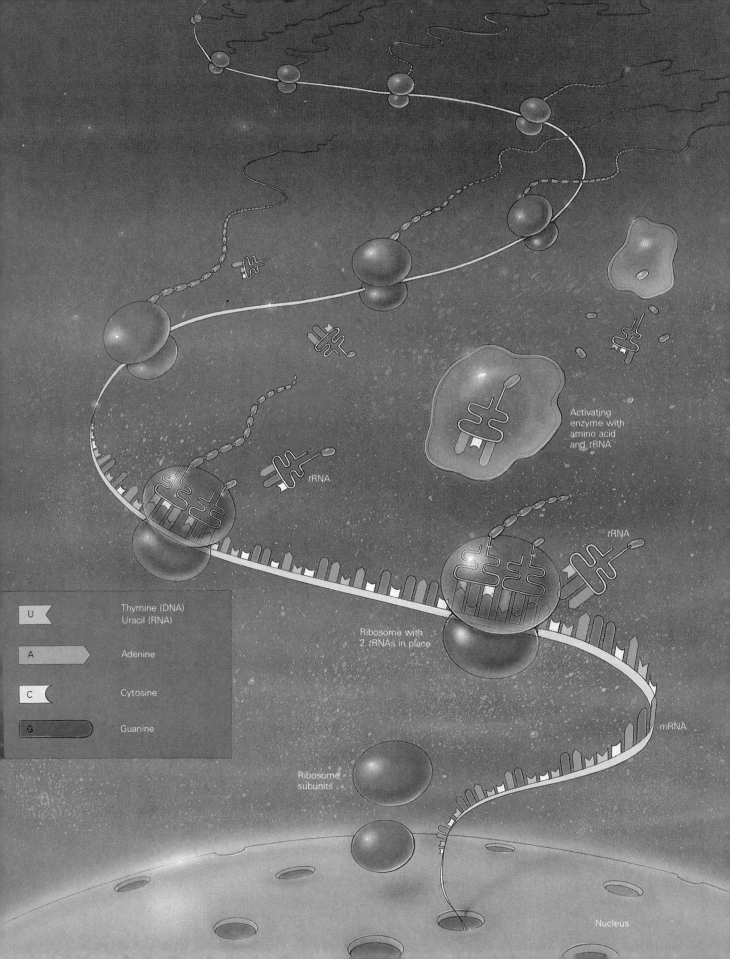

U	Thymine (DNA) Uracil (RNA)
A	Adenine
C	Cytosine
G	Guanine

Activating enzyme with amino acid and *t*RNA

*t*RNA

*t*RNA

Ribosome with 2 *t*RNAs in place

Ribosome subunits

*m*RNA

Nucleus

FIGURE 7.10 Protein synthesis. Messenger RNA is assembled along parts of separated strands of DNA. The rather large *m*RNA molecules then move through the nuclear membrane into the cytoplasm. In the cytoplasm, they are joined by ribosomes that "read" the base sequence of the molecule. A sequence of three bases (a triplet) is read at once, and each triplet codes for a particular amino acid. As a triplet is read, a smaller RNA molecule, transfer RNA, or *t*RNA, moves to the site with an amino acid in tow.

The *t*RNA has an anticodon sequence that fits the triplet's codon. When the appropriate *t*RNA locks onto the codon, it releases its amino acid to a developing chain of amino acids and then unlocks and moves away. Any *t*RNA molecules can carry only one kind of amino acid. After the *m*RNA has directed a number of amino acid sequences, it breaks apart. Long sequences of amino acids form proteins, of course, and some of these are enzymes. Enzymes determine the nature and behavior of the cell.

along its entire length since this would result in only one kind of *m*RNA, and every protein would be the same. So we see that parts of chromosomes are shut down while other parts are unwound and busily forming *m*RNA.

Cells, then, take on their specific characters as DNA directs the synthesis of different enzymes. This explanation of cell differences, as we will see, simply places the question of control one step further back. We are now faced with the question of what determines which section of a chromosome will be active. At this point, the answers become rather varied and complex, but one line of evidence suggests that the cell's location may be critical. The idea is that the immediate environment of a cell may have an effect on its chromosomes.

Continuing Problems in Cell Biology

Complex flowcharts, stroked chins, and remarkable electron microscope photographs of cell interiors may lead some people to believe that the "great questions" have been answered, and all we need to do now is to tie up a few loose ends. Nothing could be further from the truth. In fact, we have barely touched the hem of the garment (an old Arkansas phrase) in many areas of cell biology.

In spite of what may have seemed great assurance in our foregoing description, we actually know very little about such matters as mitochondrial replication, spindle formation, cell-environment interaction, cell specialization, cell growth, chromosome structure, or the permeability of membranes. In fact, we know precious few of the simplest enzymatic pathways. It seems that the intense research efforts in this area are currently overmatched by some very fundamental questions that, so far, have simply refused to yield. But new minds, vigorous and well-trained, are continuing to focus on the questions—often from odd angles—even as new technologies are developed. We can, therefore, expect a continuing stream of revelations as these people go about their work, confident that few secrets are held forever.

SUMMARY

The workings of cells are extremely precise and regulated. The regularity and coordination is revealed as they divide. A parent cell gives rise to two daughter cells that have roughly the same cytoplasmic contents and precisely the same genetic material. The genetic division is called mitosis. In preparation for mitosis, the cell completes DNA replication and begins to synthesize tubulin, which serves as the building material for the microtubules of the mitotic spindles.

Mitosis proceeds through four phases: prophase, metaphase, anaphase, and telophase. In early prophase, paired centrioles near the nucleus separate and migrate to opposite sides of the nucleus. At the spindle poles, the centrioles organize the microtubules to form asters, continuous spindle fibers, and centromeric spindle fibers. In late prophase, opposing centromeric spindle fibers attach to centromeres; the chromosomes move to form a metaphase plate across the middle of the spindle. In metaphase, individual chromatids can be easily seen.

It somehow seems highly unlikely to the young, but under the best of circumstances, they will grow old—old and ugly, some say. The changes will be almost imperceptible, normally just a slow accumulation of small changes, a wrinkle here, a backache there. Perhaps the best barometer of the change is meeting an old friend after several years. The compliments on your appearance will be effusive and directly proportional to your state of deterioration.

What brings about these changes? Why do our bodies let us down? What's happening anyway? Several things are happening. Primarily, though, cells are becoming less efficient. They haltingly carry out their tasks of removing wastes, destroying poisons, repairing genes, and manufacturing proteins. As cells weaken, so do the bodies they serve.

There are a number of factors that may contribute to the problems of aging cells. There is an accumulation of free radicals (atoms with only a single electron in their outer shells). They may satisfy their shell requirements by taking electrons that were in the process of meeting some cellular need, such as generating ATP. Certain pigments may also accumulate and pose a threat. Lipofusion, a brown pigment produced by the metabolism of fat, may choke cells and impair their delicate functions. Even glucose can be a problem. For example, in time, glucose can damage proteins, including collagen, one of the primary supportive tissues of the body. Glucose can even clog the DNA helix and halt the orderly formation of proteins.

The causes of such changes remain a mystery, but there are two major theoretical approaches to the problem: (1) the time bomb theory and (2) the wear-and-tear theory. The time bomb theory is based on the idea that we are genetically programmed to fade out. The mechanisms include a genetic clock that causes healthy cell lines to die out after about 50 doublings (cancer cells do not). A hormone clock has also been suggested. It could operate, for example, if the pituitary gland began to secrete a "killer hormone" after puberty. (Aging has been retarded in some animals by removing the pituitary and dosing them with hormones artificially.) There may also be a group of clocks, all running at different rates, but interdependent, so that when one slows (perhaps due to some environmental change or signal) the others slow as well. (*Environmental*, in this sense, implies anything that is not internally based.)

The wear-and-tear theory suggests a genetic mechanism in which random mutations can be expected to eventually alter a few important genes that control an array of other genes or that manufacture certain vital proteins. The mutation rate, of course, can be increased by environmental abuses. The level of free radicals can also build up due to

When anaphase begins, the centromeres divide, releasing the two chromatids, which are now called daughter chromosomes. The daughter chromosomes, each with its own centromere, travel to opposite poles.

Telophase begins when the daughter chromosomes reach their poles. The chromosomes uncoil, the mitotic apparatus disassembles, and the nuclear membrane forms around each group of daughter chromosomes. Telophase marks the time of actual cell division, or cytokinesis.

Cytokinesis in animal cells is referred to as cleavage. In plants, cytokinesis takes place when a new cell wall forms between telophase nuclei. Mitosis in animals occurs repeatedly during early growth and slows as the organism matures. In plants that grow continuously, mitosis occurs at a fairly constant rate.

The hereditary material is DNA, which is comprised of two strands in the form of a double helix. The strands are connected by hydrogen bonds between purines (adenine or guanine) and pyrimidines (thymine or cytosine). These, with sugar and phosphates, are called nucleotides. A series of three nucleotides codes for one amino acid in developing a protein chain.

environmental abuses (and can also be reduced environmentally, such as by ingesting vitamins E and A). Finally, cells may give way to the continual assault from a number of sources, such as the constant (and increasing) barrage of radiation and chemicals that can cause an accumulation of small mutations that slowly destroy us.

The rate at which we age, no matter what the causes, seems to be largely genetic. Some people remain vigorous and alert into their eighties, while others show the effects of time much earlier and walk a degenerative and painful path to death. It has been suggested that the best way to live long is to choose an elderly set of grandparents.

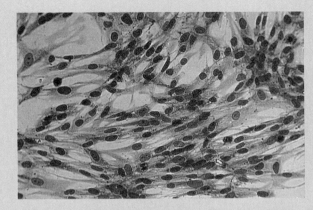
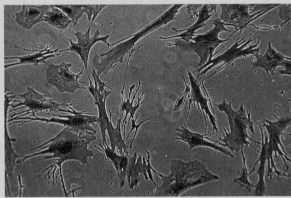

The effect of aging on our cells is vividly illustrated by the difference between these healthy young skin cells (top) and the aged skin cells (bottom).

Proteins (including enzymes) are made as a strand of DNA inside the nucleus opens up to expose its nucleic acid to the cytoplasm and complementary strand of nucleotide attaches. Segments of this RNA (called messenger RNA) move into the cytoplasm where transfer RNAs match the triplet codons in the proper sequence as the triplets are read by a ribosome. The *t*RNAs are attached to amino acids. These are released when a *t*RNA triplet matches an *m*RNA triplet. In this way, long sequences of amino acids are joined in the proper sequence to form proteins.

Chapter 8

INHERITANCE: FROM MENDEL TO MOLECULES

Charles Darwin was one of the best geneticists of his time. Charles Darwin was a terrible geneticist. That tells us something about the state of genetics in the Victorian world. But interestingly, one of the very few people in Darwin's time who knew more about genetics than Darwin did was one of the best geneticists who ever lived. In fact, some of his findings would have answered some of Darwin's most vexing questions.

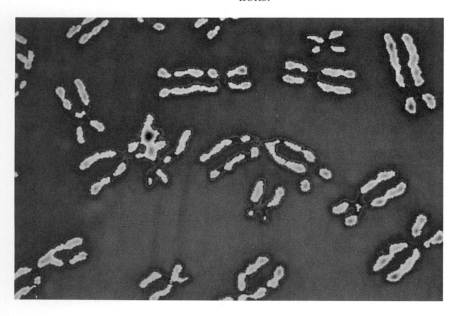

FIGURE 8.1 Gregor Johann Mendel mathematically designed his meticulous experiments on pea plants and gradually uncovered basic genetic principles. His work went largely unrecognized, even by Darwin, who desperately needed just those principles to help explain his own work.

MENDELIAN GENETICS

The answers to many of the questions about heredity that plagued Darwin were published by a contemporary, a monk named Gregor Johann Mendel (1822–1884) who was a member of an Augustinian order in Brunn, Moravia (now Czechoslovakia) (Figure 8.1).

How close did Darwin come to learning about Mendel's findings? The journal in which Mendel's work was published was found in Darwin's library. However, whereas the adjacent article is heavily marked with Darwin's scrawl, Mendel's paper was not touched. Apparently, Darwin saw the paper but was unable to grasp its implications. Keep in mind that Mendel was probably the first mathematical biologist and that Darwin was not a good mathematician. While in school, he once wrote that he was "mired firmly in the mud of mathematics" and that there he would remain.

A Kindly Old Abbot Puttering in His Garden

The Abbot Gregor Mendel was a rather remarkable man in his own right, although he is usually depicted as a kindly old man of the cloth who, while puttering around in his monastery garden, somehow stumbled upon important laws governing genetic transmission. Alas, our penchant for building myths to embellish the memory of notable people has led us astray again. The fact is that early in his life, Mendel began training himself in the sciences, and he became a rather competent naturalist. To support himself during those years, he worked as a substitute high-school science teacher. The professors at the school, noting his unusual abilities, suggested that he take the rigorous qualifying examination and become a regular member of the faculty. Mendel took the test and did surprisingly well, but he failed to qualify. The conditions for teaching high school were rigorous indeed.

However, the monastic order to which Mendel belonged was confident of his abilities and, in 1851, sent him to the University of Vienna for two years of concentrated study in science and mathematics. When Mendel returned to the monastery, he decided to put aside his teaching duties for a time in order to begin his plant hybridization studies in earnest. He was noted for his intelligence and vigor, and he applied these qualities to the study of plant breeding. He worked hard, developed new varieties of fruits and vegetables, kept abreast of the latest developments in his field, and became active in community affairs.

In 1865, while Darwin was still puzzling over the enigma of heredity, wondering how traits are actually transmitted, Mendel presented a single paper at a meeting of the Brunn Natural Science Society. His talk was politely applauded. Then the group burst into a lively discussion of the hot topic of that day—the idea of natural selection. Mendel's paper was published in the society's proceedings the following year. However, no one had a clue as to what Mendel was going on about, and the work for which he was to become so famous was met with resounding silence. But why? What had he said? Why was it so hard to understand?

Actually, Mendel said three things:

1. When parents differ in one characteristic, their offspring will be **hybrids** of that particular characteristic; however, the offspring

may bear the traits of one parent instead of showing a blend of the traits of both parents. This means that the trait of one parent must somehow be dominant over the trait of the other parent. He called this phenomenon the **principle of dominance.**

2. When a hybrid reproduces, its gametes (egg or sperm) will be of two types. Half will carry the dominant trait provided by one parent, and half will carry the recessive trait provided by the other parent. This he called the **principle of segregation.**

3. When parents differ in two or more characteristics, the occurrence of any characteristic in the next generation will be independent of the occurrence of any other characteristic. That is, the tendency toward tallness in plants is not associated with any specific seed color. The two are inherited independently. Thus, any combination of the parental components may appear in the offspring. This is called the **principle of independent assortment.**

At this point, you are probably as mystified as was Mendel's audience, but let us continue. Perhaps the best way to clarify things is to see exactly how Mendel arrived at his conclusions.

The Principle of Dominance

To begin with, Mendel based his information on a carefully planned series of experiments and, more important, on a statistical analysis of the results. The use of mathematics to describe biological phenomena was a new concept. Clearly, Mendel's two years at the university had not been wasted.

The care with which Mendel planned his projects is reflected in his selection of the common garden pea as his experimental subject. There were several advantages in this choice. Pea plants were readily available, fairly easy to grow, and Mendel had already developed some thirty-four pure strains. These strains differed from each other in very obvious ways, so there would be little difficulty in classifying the results of a given experiment. Mendel chose to study seven different pairs of characteristics:

1. Seed form—round or wrinkled
2. Color of seed contents—yellow or green
3. Color of seed coat—white or gray
4. Color of unripe seed pods—green or yellow
5. Shape of ripe seed pods—inflated or constricted between seeds
6. Length of stem—short (9 to 18 inches) or long (6 to 7 feet)
7. Position of flowers—axial (along the stem) or terminal (at the end of the stem).

Mendel's approach, a novel one at that time, was to cross two true-breeding strains that differed in only one characteristic, such as seed color. Peas ordinarily self-fertilize, so for this cross it was necessary to transfer the pollen by hand (Figure 8.2). Mendel called this original parent generation P_1 and designated their first-generation offspring the F_1 (first filial) generation. When the F_1 plants were allowed to self-pollinate, so that they crossed with each other at random, the offspring resulting from this cross were called the F_2 generation, and so on.

FIGURE 8.2 The petals of the garden pea usually ensure self-pollination; that is, the plant fertilizes itself. For his cross-pollination studies, Mendel had to open the young flower and remove the pollen (containing the male "sperm"). The pollen was then transferred to the female part (stigma) of another flower.

TABLE 8.1 Mendel's Pea Plant Experiment

Dominant form	No. in F_2 generation	Recessive form	No. in F_2 generation	Total examined	Ratio
Round seeds	5,474	Wrinkled seeds	1,850	7,324	2.96:1
Yellow seeds	6,022	Green seeds	2,001	8,023	3.01:1
Gray seed coats	705	White seed coats	224	929	3.15:1
Green pods	428	Yellow pods	152	580	2.82:1
Inflated pods	882	Constricted pods	299	1,181	2.95:1
Long stems	787	Short stems	277	1,064	2.84:1
Axial flowers	651	Terminal flowers	207	858	3.14:1

Now, when Mendel crossed his original P_1 plants, he found that the characteristics of the two parents didn't blend, as prevailing theory said they should. When plants with yellow seeds were crossed with plants that had green seeds, their F_1 offspring did not have yellow-green seeds. Instead, all of them had yellow seeds. Mendel termed the trait that appeared in the F_1 generation the **dominant trait,** but he was now left with a vexing question. What had happened to the trait that had disappeared in this cross, the **recessive trait?** After all, it had been passed along through countless generations so it couldn't have just disappeared.

The Principle of Segregation

In the next stage of his experiments, Mendel allowed the F_1 plants to randomly self-pollinate, and, lo and behold, the missing recessive trait reappeared in their F_2 offspring! Also, the ratio of F_2 plants with recessive traits to those with dominant traits was fairly constant, regardless of the particular characteristics involved. The exact results for each of the seven pairs of characteristics Mendel used in his tests are shown in Table 8.1.

Finally, in the third year of the experiments, Mendel allowed the F_2 plants to self-pollinate. He found that all those with recessive traits produced only recessive F_3 offspring. However, the F_2 plants that showed a dominant trait produced both types of F_3 offspring. One-third of them produced *only* dominant offspring. The other two-thirds produced both dominant and recessive offspring, but they produced three times as many offspring with the dominant trait as they did with the recessive trait. In other words, all the F_2 recessives and one-third of the F_2 dominants bred true. The other two-thirds of these dominants produced mixed offspring—but in the same 3:1 ratio of dominant to recessive as in the plants their F_1 parents had produced.

Let's consider a specific example of a cross between two pure lines— plants that breed true when they self-pollinate and are identical in all

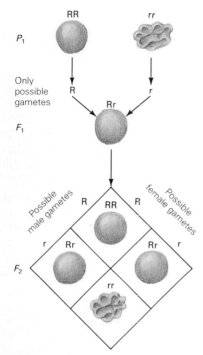

P_1

RR   rr

Only
possible
gametes   R   r

F_1 Rr

Possible
male gametes   Possible
female gametes

F_2

R	R
RR	Rr
Rr	rr

FIGURE 8.3 Arriving at the expected ratio in the F_2 generation of plants that differ in only one character (here, seed shape). The parent (P_1) generation is a cross between a pure-breeding plant with round seeds (**RR**) and a pure-breeding plant with wrinkled seeds (**rr**). Obviously, a plant that had received an **R** gene for round seeds from each parent could only pass along **R** genes to its progeny, and the same for a plant receiving only **r** genes from its parents. The F_1, then, must be **Rr** since one gene for seed shape must come from each parent. If two individuals from the F_1 are crossed, the expected ratio of their offspring can be calculated by use of a Punnett square. Remember, though, that these are only *expected* results, and this ratio would probably be approached only if a great number of crosses were made.

characteristics except one. In this case, suppose one strain always produces round seeds and the other always produces wrinkled seeds. We have already seen that the round form is dominant, so we'll designate that form as **R** and the recessive wrinkled form as **r**. Since both plants are true-breeding, we can assume that in each plant, the genetic components (from their own parents) are identical for their respective traits. That is, in one plant, the components are **RR,** and in the other, they are **rr.** Such plants are said to be **homozygous** for that particular characteristic. As a result, all the gametes they produce will be the same.

Remember that the plants are identical in all characteristics except one, so we only need to diagram the differing characteristic. Since half the gene component in the F_1 generation comes from each parent, the genetic composition, or **genotype,** of any F_1 plant would have to be **Rr.** Hence, these plants would be **heterozygous** for the characteristic in question. Now, when these plants produce gametes ("eggs and sperm"), each gamete will carry either an **R** component or an **r** component, and the combination that occurs when two gametes come together at fertilization will determine the genetic characteristic of the resulting F_2 plant.

The results we should expect in the F_2 generation can be diagrammed in a form called a **Punnett square.** As you can see in Figure 8.3, the physical characteristics, or **phenotypes,** of the F_2 seeds occur in a 3:1 ratio. However, a moment (or several moments) of reflection will show that the F_3 ratio Mendel obtained in the third year of his experiments could have been gotten only if one-fourth of the F_2 plants were homozygous dominant for the characteristic, one-half were heterozygous, and one-fourth were homozygous recessive. So the genotype ratio of this F_2 generation is described as 1:2:1.

These results led Mendel to conclude, finally, that every heritable characteristic is determined by two components, one from each parent. (Before you start feeling smug about already knowing this, remember that it was Mendel who told you.) Thus, if two homozygous parents were represented by **RR** and **rr** (as for seed form), their offspring would have to be heterozygous **Rr,** since one component would have come from each parent; and when this **Rr** individual produced gametes, half would contain an **R** component and the other half would contain the **r** component. A cross between two **Rr** individuals could then be expected to yield offspring with the genotypes **RR, rR, Rr,** and **rr,** or a genotype ratio of 1:2:1. What would be the genotypes and the phenotypes of the offspring if an **Rr** crossed with a homozygous recessive **rr** instead?

Mendel's experiment with monohybrid crosses enabled him to show that whatever it is that is actually passed along (now called a **gene** or an **allele**) is passed along in pairs. One allele is contributed by each parent. However, only one of these two components can actually be expressed. Hence, an individual can only show the effects of an allele passed along by one parent, the other allele being completely dormant. The principles of dominance and segregation implied by these results ruled out (in Mendel's mind, at least) the earlier notion that heritable characteristics were always blended in the offspring.

The Principle of Independent Assortment

In this next set of experiments, Mendel crossed **dihybrids,** true-breeding plants that differ in two characteristics. In one such experiment, plants with round **(R),** yellow **(Y)** seeds were crossed with a strain that had wrinkled **(r),** green **(y)** seeds. We saw in Table 8.1 that round and yellow are both dominant, so we shouldn't be astounded to learn that all the F_1 plants had round, yellow seeds. Since one P_1 parent was **RR YY** and the other was **rr yy,** all the F_1 plants would have to be **Rr Yy** or heterozygous for both characteristics. When these plants were allowed to self-pollinate, we would expect a random and independent assortment of the various characteristics to produce the F_2 generation shown in Figure 8.4.

It was apparent to Mendel from the results of this dihybrid cross that the segregation of genes for one characteristic was not affected by the segregation of genes for the other characteristic. Thus, he was able to deduce that genetic combinations follow the principle of independent assortment. It should be pointed out here that if you were to run a single experiment to test Mendel's results, you might end up with all wrinkled, green seeds in the F_2 generation—just as when you toss a coin ten times, it *might* come up tails each time. You could expect the results shown in Figure 8.4 only if you ran enough tests for the law of averages to apply.

FIGURE 8.4 In this case, we have the expected F_2 ratio from a dihybrid cross (a cross between plants that differ in only two characters—here, seed shape and seed color). One pure-breeding parent in the P_1 generation bore both dominant traits; the other, both recessive traits. Obviously, they could pass along only dominant and recessive genes, respectively. This means the F_1 is necessarily heterozygous for both traits. At meiosis, however, the traits assort independently. This means that when the F_1 is crossed with itself, we must account for all possible combinations in order to determine the F_2 ratio. Only one possible combination is labeled. If you were to work this out, you would get nine genetic combinations. Calculate the genotypes and phenotypes and then check page 154 to see how wrong you were. Again, keep in mind that we're only dealing with probabilities. This means that all the F_2 *could* be **rryy** (double recessives), but the odds are very much against it.

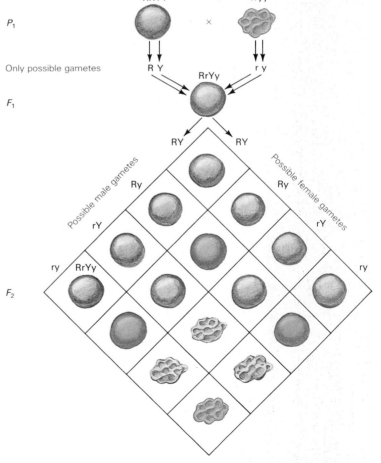

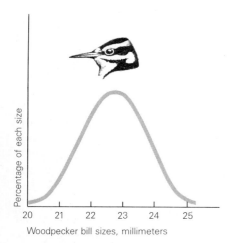

FIGURE 8.5 The bills of populations of woodpeckers vary. Some bills are long, some short, but the majority are of an intermediate length. In the population shown in this figure, there are few birds with bills 21 millimeters long and virtually none with bills of 20 millimeters. Nor do any have bills longer than 25 millimeters. Practically every woodpecker you see, though, has a bill within the 22 to 24 millimeter range. Why do you suppose this is? When a population shows a continuous distribution like this, instead of distinct classes, one can assume that the trait is controlled by several genes. Of course, a bird genetically disposed to have a long bill might not achieve that status if it has been subjected to a poor diet.

(Graph y-axis: Percentage of each size; x-axis: 20, 21, 22, 23, 24, 25; Woodpecker bill sizes, millimeters)

Answer to Figure 8.4.

Your should get:

Genotype		Phenotype
1	**RRYY**	9 round yellow
2	**RRYy**	3 round green
1	**RRyy**	3 wrinkled yellow
2	**RrYY**	1 wrinkled green
4	**RrYy**	
2	**Rryy**	
1	**rrYY**	
2	**rrYy**	
1	**rryy**	

Before we leave the subject of Mendelian genetics, you should know that there are the inevitable exceptions to all the rules we've mentioned. To illustrate, not all characteristics are expressed in dominant-recessive form. In some cases, a characteristic can appear in an offspring in an intermediate form between the characteristics of each parent. For example, four-o'clocks may have either red or white flowers. A cross between the two produces pink-flowered plants. A situation in which one characteristic is not dominant over the other is called **partial dominance.** As before, one component, or allele, of the gene is contributed by each parent, but in this case, neither allele gives way to the other. Instead, they both exert an influence on the physical characteristics determined by that gene. This means, of course, that the pink-flowered plant is heterozygous for flower color. Could a cross between two such plants produce red- or white-flowered plants again? In what ratios? What kind of gametes would these plants produce?

There are also other ways in which intermediate effects arise. For example, people do not come in only two sizes, large or small. They come in a wide range of heights, weights, and shapes, just as they differ widely in skin color, bone size, nose shape, and any number of other physical characteristics. Such variation can be accounted for if we assume that certain traits are coded for by more than one gene.

Suppose some characteristic such as the length of bill in a woodpecker were coded for by four separate genes, with "longness" indicated by a capital letter and "shortness" indicated by a lower-case letter. Then a bird with the genotype **AA BB CC DD** would be expected to have the longest bill, provided environmental factors such as diet were conducive. A bird with the genotype **Aa BB Cc DD** would have a somewhat shorter bill, and one with the genotype **Aa bb cC dd** would have a still shorter bill. Now, since there is a bill size that is "best" for a woodpecker, we would expect to find more birds with bills in the intermediate range than at either size extreme. Thus, if the bill length of a population of woodpeckers were plotted on a graph, we might expect something approaching a "normal," or bell-shaped, curve (Figure 8.5).

CLASSICAL GENETICS

It is unfortunate that those nineteenth-century biologists who did read Mendel's reports were unimpressed with his findings. It is also unfortunate that no one saw enough significance in them to bother repeating his laborious experiments. Early in the twentieth century, however, some seventeen years after Mendel died in obscurity, his work was rediscovered by three researchers who independently arrived at essentially the same conclusions. They were each elated and anxious to be the first to tell the world their findings. By now, though, enough was known for such information to be appreciated, and, in fact, it was met with great applause. The applause rang a bit hollow over the mossy grave of old Mendel, though, the man who had beaten them all.

Mendel's rediscovery set the stage for extensive research in what is now considered **classical genetics.** With Mendel's ideas as a foundation, the chromosomal theory of inheritance was soon established.

By the beginning of this century, it was assumed that the chromosomes of any organism were identical, and that meiosis was simply a

Leonardo da Vinci was the illegitimate son of Piero, a notary from the town of Vinci, and a peasant girl named Caterina. The third wife of Leonardo's father later bore a son named Bartolommeo who idolized Leonardo although he was forty-five years younger. After the death of the legendary Leonardo, Bartolommeo attempted an amazing experiment. He studied every detail of his father's relationship with Caterina. Then Bartolommeo, himself a notary by family tradition, returned to Vinci and found another peasant wench who seemed similar to Caterina, according to all Bartolommeo knew. He married her and she bore him a son whom they called Piero. Strangely, the child actually looked like Leonardo and was brought up with encouragement to follow in the great man's footsteps. The boy became an accomplished artist and was becoming a talented sculptor when he died, thus ending the experiment.

A red chalk self-portrait done by da Vinci circa 1512.

way of dividing them up so that the gametes would receive the same number. It was also believed that during meiosis, the maternal and paternal chromosomes separated and that each set traveled together at anaphase, so that any gamete would contain chromosomes from one parent or the other, but not both. In the early 1900s, however, it was found that the chromosomes are not identical, but come in a variety of sizes and shapes, and that they also come in pairs—two of each type of chromosome. Chromosomes that comprise a pair are called **homologues.** In addition, it was found that meiosis divides chromosomes not only quantitatively but also qualitatively, so that each gamete receives both the same *number* and *kind* of chromosomes.

A young graduate student at Columbia University, Walter S. Sutton, began to try to relate Mendel's findings to what was known about meiosis. His results were surprising. For example, he found that when chromosome pairs line up at meiosis, there was no way to determine which way a paternal or maternal chromosome would go at anaphase. For example, some maternal chromosomes moved to one pole and some to the other, but there was no way to determine their direction. At that time, it was not known that chromosomes were the carriers of genes. The independent assortment of chromosomes, however, confirmed a hunch Sutton had that hereditary units are associated with chromosomes. Also, since there were far more hereditary characteristics than there were chromosomes, he reasoned that each chromosome must be responsible for many genetic traits. (See Table 8.2 and Figure 8.6 for examples of human traits known to be determined by genes, and Essay 8.1 for a description of an early attempt to breed for certain traits.)

In 1910, after Sutton's report, an innocuous little insect entered the world of genetics. Thomas Hunt Morgan, also at Columbia, began a program of breeding experiments with the fruit fly (or vinegar fly), *Dro-*

FIGURE 8.6 Polydactyly, the condition of having more than five digits on the hands or feet, is a dominant trait in humans. In the top photo, a man exhibits six fingers on each hand. In the lower photo, two girls and their brother show variation in the condition. The girl at left has six toes on both feet, her sister has six toes on one foot, and her brother has the normal complement of five toes on each foot.

TABLE 8.2 Some Human Traits That Are Inherited

	Dominant	Recessive
Hair, skin, etc.	dark hair	blond hair
	nonred hair	red hair
	curly hair	straight hair
	abundant body hair	little body hair
	early baldness (dominant in males)	normal
	white forelock	self-color
	normal	absence of sweat glands
Eyes and facial features	brown	blue or gray
	hazel or green	blue or gray
	congenital cataract	normal
	normal	nearsightedness
	farsightedness	normal
	astigmatism	normal
	free ear lobes	attached ear lobes
	broad lips	thin lips
	large eyes	small eyes
	long eyelashes	short eyelashes
	high, narrow bridge of nose	low, broad bridge
Skeleton and muscles	dwarfism (achondroplasia)	normal
	normal	midget
	polydactyly (more than five digits)	normal
	syndactyly (webbed fingers or toes)	normal
	brachydactyly (short digits)	normal
	normal	progressive muscular atrophy
Some conditions affecting systems	hereditary edema (Milroy's disease)	normal
	hypertension	normal
	normal	hemophilia (sex-linked)
	normal	sickle-cell anemia
	normal	congenital deafness
	migraine	normal

Source: Adapted from C. Villee, *Biology,* 4th ed. (Philadelphia: Saunders, 1962), and Evelyn Morholt, Paul F. Brandwein, and Alexander Joseph, *A Sourcebook for the Biological Sciences* (New York: Harcourt, Brace & World, 1966).

sophila melanogaster (Figure 8.7). These little flies are ideal subjects for genetic studies. They are easy to maintain in the laboratory; they mature in twelve days and reproduce prolifically; and they have a variety of features that can be readily identified. They also have giant chromosomes in their salivary glands. Specific areas, or loci, of the chromosomes are readily distinguishable. In addition, they have only four pairs

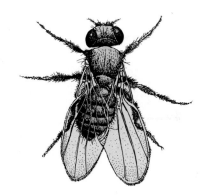

FIGURE 8.7 The fruit fly, *Drosophila melanogaster*.

of chromosomes. The insects have little economic importance, and they don't seem to bother humans much when the two species come in contact. In fact, it has been said that God must have invented *Drosophila melanogaster* just for Thomas Hunt Morgan.

Sex Determination

One of the first things Morgan did was attempt to cause genetic changes, or **mutations,** in his flies. He subjected them to cold, heat, x-rays, chemicals, and radioactive matter. At first, he found nothing, but then he noticed among a group of normal red-eyed flies that there was a single male with strange *white* eyes. Morgan was convinced that this was a mutation (a subject we will take up later). He carefully nurtured his little white-eyed specimen and crossed it with several of its red-eyed sisters. In these crosses, he found that the F_1 generation was all red-eyed, just as you would expect if white eyes were recessive. Simple enough so far—but then the story became a bit more complicated.

When the F_1 flies were allowed to interbreed, the result was 3,470 red-eyed and 782 white-eyed flies. This is not the simple 3:1 ratio you might have expected according to Mendelian principles. Also, there was another troubling factor. All the 782 white-eyed flies were males. The red-eyed flies, however, were of both sexes—2,459 females and 1,011 males. At this point, an obvious answer was suggested: only males are white-eyed. Alas, when the original white-eyed male was crossed with its F_1 red-eyed daughter, the result was 129 red-eyed females, 132 red-eyed males, 86 white-eyed males—and 88 white-eyed females! If you have a penchant for puzzles, can spare a few minutes, and think you are smart, try to figure out how this could have happened. As a clue, keep in mind that sex is determined by chromosomes. The rest of us will trudge along.

Morgan surmised that although chromosomes may segregate independently, the genetic determiners (or genes) may not. He then surmised that the sex-determining factor and the eye-color factor are somehow linked together. Hence, as one goes at anaphase, so must go the other.

By this time, Morgan knew something about the chromosomes that determine the gender of an offspring. He knew that of the four pairs of chromosomes in *Drosophila,* only one pair was responsible for gender. He even knew which ones were the sex chromosomes and which were the **autosomes,** the chromosomes responsible for other characteristics. In females, the sex chromosomes are a rod-shaped pair of chromosomes, for some reason called *X* chromosome. In males, the sex chromosomes consist of only one *X* chromosome and a J-shaped chromosome called a *Y* chromosome (Figure 8.8). The *X* chromosomes are usually larger and contain far more genes than the *Y* chromosome. This difference in size will be important to the relative genetic contributions of the two kinds of chromosomes, as we will see.

In the production of sperm, the *X* and *Y* chromosomes line up together at meiosis, so that half the gametes will contain an *X* and half will contain a *Y*. Since females have only *X* chromosomes, every egg will contain an *X*. This, then, sets the basis for gender determination in offspring. If a *Y*-bearing sperm reaches the egg first and fertilizes it, the offspring will be *XY,* and thus male. If an *X*-bearing sperm fertilizes the

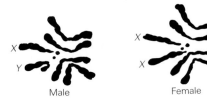

Male   Female

FIGURE 8.8 The chromosomes of male and female *Drosophila*. Note that each chromosome of one sex is represented in the genotype of the other, except for the twisted *Y* chromosome of the male. Because there are few chromosomes and they are rather easily identifiable, *Drosophila* were chosen for intensive research.

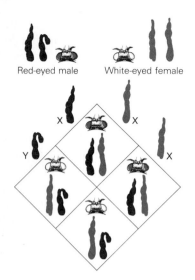

Red-eyed male   White-eyed female

FIGURE 8.9 In *Drosophila*, the gene for white eyes (a recessive gene) exists on the *X* chromosome. In females, the gene can be masked by a dominant gene for red eyes on the other *X* chromosome. But males carry no such dominant gene on their *Y* chromosome, so if a male's *X* should bear the gene for white eyes, it will be expressed.

egg, the offspring will be *XX*, a female. (This is how gender is determined in many animals including mammals, but in other organisms it may be determined differently.)

Now, regarding the question of eye color in *Drosophila*, Morgan showed that the genetic allele for eye color is located only on the *X* chromosome, and white eye color is recessive. Therefore, females will have white eyes only if both eye-color alleles are for white eyes. In males, however, when a white-eye factor turns up on the *X* chromosome, there is no dominant counterpart on the *Y* chromosome, so the white eye color will be expressed (Figure 8.9).

Sex Linkage in Humans

There have been several interesting applications of the findings regarding gender inheritance in the past few years. For example, the cells in women have tiny specks in their nuclei called **Barr bodies.** Also, white blood cells in women show a small "drumstick" attached to the nucleus (Figure 8.10). Cells from normal men lack both the Barr body and the drumstick. There was a stir at the Olympic games not long ago when a European woman who showed remarkable strength in the field events was found to be genetically abnormal, lacking the Barr bodies and drumsticks in some of her cell nuclei. Her strength was believed to be due to the "male" characteristics of certain cells. For certain athletic competitions these days, sex tests are mandatory.

Red-green color blindness in humans is caused by a recessive allele on the *X* chromosome. Thus, a woman who is heterozygous for this condition will show no symptoms of color blindness. Among her children, however, she can expect half her sons to be color-blind and half her daughters to be carriers of color blindness (as she is). In her sons, the recessive gene on the *X* chromosome is not overridden because the tiny *Y* chromosome bears no allele for color discrimination. Figure 8.11 shows the expected occurrences of color blindness in this woman's grandchildren, the F_2 generation.

There are a number of other phenomena associated with sex-linked genes that have great impact on people's lives. For example, one type of muscular dystrophy is caused by a recessive gene on an *X* chromosome. The effects of this abnormality are so pronounced, however, that males

FIGURE 8.10 Electron micrographs of the Barr body (left) and drumstick (right) from the cells of a woman. Drumsticks are found in white blood cells. The sex of an individual can be identified according to these telltale structures, normally lacking in males.

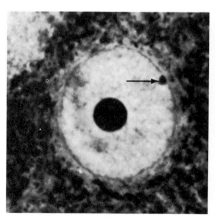
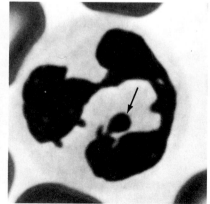

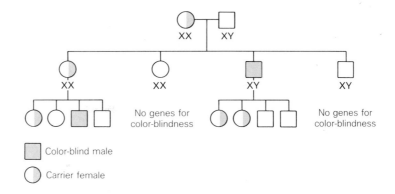

FIGURE 8.11 How color blindness is inherited. The original mating is between a carrier female and a normal male. Spouses (not shown) are presumed to be free of the color-blindness gene. The squares represent males, the circles females, and the condition (heterozygous or homozygous) is represented by color (which may seem odd, but it's convenient).

carrying the gene usually die in their teens before reproducing. As a result, females rarely have the disease, since the probability of receiving the allele from both parents is very low. However, they can carry a single gene for the condition, so that when they reproduce, they may pass it along to their offspring. Another such disorder is hemophilia, which is also a recessive gene carried on an *X* chromosome (see Essay 8.2).

In other instances, there may be an imbalance of *X* and *Y* chromosomes. For example, certain females may have cells with no Barr bodies or drumsticks. In such cases, there is only one *X* chromosome instead of two. The physical characteristics that accompany this condition, resulting in what is called Turner's syndrome, are female genitalia, but underdeveloped breasts and tiny ovaries—and for some reason, short stature. The presence of an extra *X* chromosome in males *(XXY)* also results in certain anomalies called Kleinfelter's syndrome. Such males usually have sparse body hair, some breast development, and cells that show Barr bodies and drumsticks.

A study relating to sex chromosomes that began in 1968 raised some critical social issues. The goal was to identify *XYY* males as infants and then to study their behavior. The reason was that *XYY* males are not only unusually tall, but they are twenty times more likely to be placed in a criminal institution than are genetically normal males. The study was to determine whether *XYY* males are genetically prone to antisocial behavior. The announcement caused a furor in some circles, however. It was argued that identifying such males would result in their being treated differently, and that this might, itself, result in abnormal behavior. Other people, perhaps for political or philosophical reasons, also objected to studying the role of genetics on human behavior. In any case, the program was stopped in 1975.

Gene Linkage and Crossover

Recall from our discussion of Mendel's work that chromosomes assort independently. Thus, the fate of genes on one chromosome of a pair would have to be independent of those on the homologous chromosome. If two genes are on the *same* chromosome, however, it is likely that they will be passed together to the gamete. The occurrence of two or more hereditary units on the same chromosome is called **gene linkage.**

In *Drosophila,* for example, the recessive alleles for a black body **(b)** and a curved wing **(c)** are carried on the same chromosome (Figure 8.12). The dominant alleles are for a gray body **(B)** and a normal wing

FIGURE 8.12 Curved wings in *Drosophila.* The curved wing is one of the obvious traits, due to a single gene mutation, that makes this small insect so easy for geneticists to work with. The curved wing trait is recessive, as is black body color, the normal color being gray. It was discovered later that the genes for wing shape and body color lie on the same chromosome and thus do not follow the principle of independent assortment.

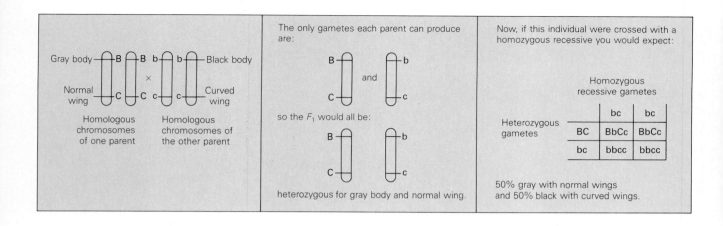

		Homozygous recessive gametes	
		bc	bc
Heterozygous gametes	BC	BbCc	BbCc
	bc	bbcc	bbcc

50% gray with normal wings and 50% black with curved wings.

The only gametes each parent can produce are:

B and b

C c

so the F₁ would all be:

B b

C c

heterozygous for gray body and normal wing.

Now, if this individual were crossed with a homozygous recessive you would expect:

Gray body — B B b b — Black body

Normal wing — C C c c — Curved wing

Homologous chromosomes of one parent   Homologous chromosomes of the other parent

FIGURE 8.13 When genes are linked by existing on the same chromosome, you would expect these results from the cross shown above. However, the linkage may be disrupted by crossovers.

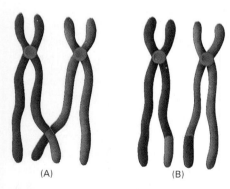

(A)   (B)

FIGURE 8.14 In meiosis, homologous chromosomes come to lie side by side in early prophase. During this time, their arms may cross, break at the point of crossing, and the parts rejoin in such a way that the chromosomes below the cross are exchanged.

(C). So if we diagram a cross between a homozygous normal fly, **BB CC,** and a homozygous recessive fly, **bb cc,** we will get the results shown in Figure 8.13. The only gametes the gray, normal-winged fly can produce are **BC,** and the only gametes the black, curved-wing fly can produce are **bc.** Hence, their F₁ offspring would all be heterozygous for both characteristics—**Bb Cc**—and would, of course, show gray bodies and normal wings.

Now, if this **Bb Cc** fly were crossed with a homozygous recessive, **bb cc,** you would expect half of their offspring to be gray with normal wings and half to be black with curved wings (the reason for this is shown in the Punnett square). This is not what happens, however. In all likelihood, this cross would produce about 37 percent gray, normal-winged flies, about 37 percent black, curved-wing flies, 13 percent *black,* normal-winged flies, and 13 percent gray, *curved*-wing flies. This kind of assortment, it turns out, can best be accounted for if two conditions are met: (1) the genes for both body color and wing shape would have to be on the same chromosome, with their alternative alleles on homologous chromosomes, and (2) the chromosomes would have to be able somehow to exchange genes.

You can probably believe condition 1 easily enough on the basis of our previous discussions, so let's consider how we might get condition 2. Recall that in meiosis, the homologous chromosomes line up during metaphase I, and as anaphase begins, the homologues of each pair move toward opposite poles. While they are lined up side by side in prophase, however, the homologues may stick together at various places and then exchange parts of their chromatids. The opposite chromatids of the homologue cross each other and then break off at their point of contact. Each broken part then fuses with the opposite portion of its homologue (Figure 8.14).

You might wonder how the chromatids manage to break at *exactly* the same place on each one, so that one chromosome does not end up with more genes than the other chromosome. Geneticists investigating crossovers are asking the same question—but they have no answer yet. We do know, however, the basic mechanism by which crossovers occur, and we also know that some crossovers take place at virtually every meiosis.

Because it was once the practice of ruling European monarchs to consolidate their empires through marriage alliances, hemophilia came to be transmitted throughout the royal families. **Hemophilia** is a sex-linked recessive condition in which the blood does not clot properly. Any small injury can result in severe bleeding and, if the bleeding cannot be stopped, in death. Hence, it has been called the ''bleeder's disease.''

Hemophilia has been traced back as far as Queen Victoria, who was born in 1819. One of her sons, Leopold, Duke of Albany, died of the disease at the age of thirty-one. Apparently, at least two of Victoria's daughters were carriers, since several of their descendants were hemophilic. Hemophilia played an important historical role in Russia during the reign of Nikolas II, the last Czar. The Czarevich, Alexis, was hemophilic, and his mother, the Czarina, was convinced that the only one who could save her son's

life was the monk Rasputin— known as the ''mad monk.'' Through this hold over the reigning family, Rasputin became the real power behind the disintegrating throne.

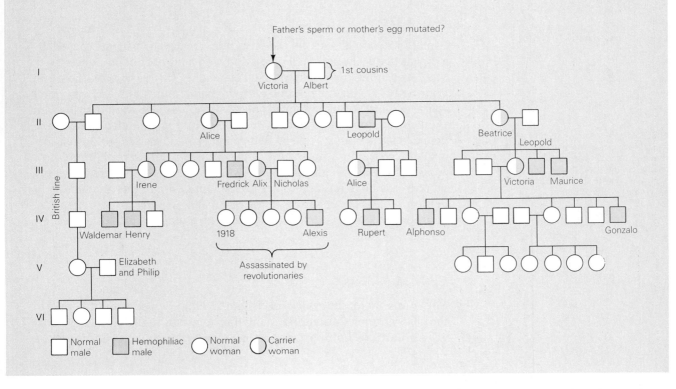

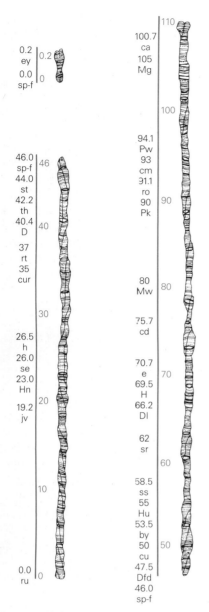

FIGURE 8.15 A map of chromosomes from the salivary gland of *Drosophila*. The units are shown in the numbered column at right. The relative positions of specific genes (such as one called jv) are shown by the numbers above the gene designation.

CHROMOSOME MAPPING

If genes are located on chromosomes, and if specific alleles are precisely exchanged through crossovers, then the genes for certain characteristics must lie at specific points along each chromosome. But how do we know which genes lie where? Finding the answer involves "mapping" the chromosome—a technique that is largely based on information from observations of crossovers.

Thomas Morgan and A. H. Sturtevant, one of the brilliant young graduate students attracted to Morgan's laboratory in those days, hypothesized that if genes were arranged linearly along the chromosome, then those lying closer together would become separated by crossovers less often than those lying farther apart. Genes lying closer together would thus have a greater *probability* of being passed along as a unit. Stated another way, the percentage of crossover is proportional to the distance between two genes on a chromosome—so *percent crossover* is the number of crossovers between two genes per 100 prophase opportunities.

As an example, suppose two characteristics, which we can call *A* and *B*, show 26 percent crossover. We can assign 26 crossover units to the distance between the two genes. Then, if some characteristic *C* turns out in breeding experiments to have 9 percent crossover with *B* and 17 percent crossover with *A*, it would be located between *A* and *B* at a point 9 units from *B* and 17 units from *A*. After the information from many such experiments has been compiled, a chromosome map can be constructed that indicates the position along the chromosome of the genes that code for certain characteristics. Figure 8.15 shows chromosome maps developed by this technique for chromosomes in the salivary gland of *Drosophila melanogaster*.

MUTATIONS

Mutations, or changes in the hereditary material, can be caused by x-rays, chemicals, ultraviolet radiation, and a number of other agents. In fact, even now, as you sit peacefully reading this fascinating account, your body is being bombarded with tiny subatomic particles from outer space—a form of radiation. If one of these particles should strike you right in the gonad and alter a nucleotide in a developing gamete, the genetic instructions carried by that gamete could be changed. Recall that a small change in a nucleotide sequence can result in an entirely new codon. If an altered gamete happens to enter the reproductive process, whatever change has taken place will be passed along to the next generation, unless the change is lethal or otherwise prevents reproduction, or unless it is reversed through a second mutation, a highly unlikely event.

Gene Mutations

Mutations can occur at two levels, the level of the gene and the level of the chromosome. Mutations involving alterations in the genes are called **gene mutations** (or **point mutations**). Gene mutations occur during the

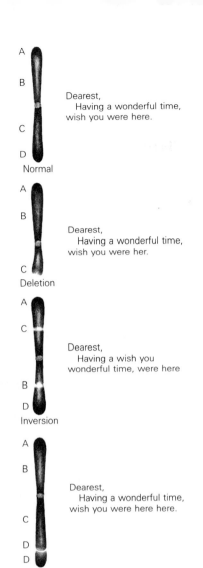

A
B

Dearest,
    Having a wonderful time,
wish you were here.

C

D
Normal

A
B

Dearest,
    Having a wonderful time,
wish you were her.

C
Deletion

A
C

Dearest,
    Having a wish you
wonderful time, were here

B
D
Inversion

A
B

Dearest,
    Having a wonderful time,
wish you were here here.

C
D
D
Duplication
or addition

FIGURE 8.16 Chromosomal mutations and their epistolary counterparts, showing the normal message with deletions, inversions, and duplications (or additions) to indicate how such changes can garble the message.

duplication of DNA. They may involve the omission of nucleotide units, the repetition of nucleotide units, or mistakes in bonding between these units. They may also involve the substitution of one nucleotide for another—either the substitution of one purine or pyrimidine for another or the substitution of a purine for a pyrimide or a pyrimidine for a purine.

Not all such changes result in a change in the phenotype, the observed trait. For example, the sequence GAC codes for the insertion of aspartic acid into a protein, and a change from GAC to GAU causes no alteration in the code. Mutations can also be masked in some cases by other mutations that compensate for the effect of the first change.

The evolutionarily important mutations, of course, are those with some observable result. Often, in fact, these small changes in genes can have enormous effects. For example, the addition or deletion of a single nucleotide could change the entire amino acid product. This is because the codons are read in groups of three nucleotides. For example, ACG—GCC—GGA codes for threonine, alanine, and glycine. If another G should be inserted after the first one, the sequence would read instead ACG—GGC—CGG—A, and the amino acids that would be inserted into the protein would be threonine, glycine, and arginine. Thus, this one added nucleotide could change every amino acid in the sequence from that point on.

Chromosome Mutations

The second level of mutations involves changes in the structure of chromosomes. These, logically enough, are called **chromosome mutations.** Such changes may occur when the chromosomes are moving about or undergoing changes—for example, during meiosis. At such times, a part of a chromosome may be broken off and lost (a **deletion**), or it may break off and reattach at its other end (an **inversion**), or a segment may be duplicated, resulting in a doubling in the number of certain genes (a **duplication**) (Figure 8.16). The most serious of these mishaps is a deletion, since the absence of a large number of genes is likely to result in death.

In some cases, mutations may have little effect on the phenotypes of the individual. For example, a slight difference in hair color as the result of a mutation might scarcely be discernible. Some mutations may also be highly beneficial in some environments, but not in others. A change that produces blue eyes makes them more sensitive to light, but this sensitivity ceases to be an advantage in environments subject to glaring sun. Mutations are not necessarily bad when viewed in the context of life's big picture. We will see later how increased variation in a population can be important to natural selection, but it should be clear that random changes are not likely to be helpful to a population that has become finely attuned to its habitat through countless generations. Suppose you raised the hood on a sports car, and, knowing nothing about motors, you made an adjustment. It may be that the change you made was just the one needed to make the car run even better, but the odds are a lot greater that you screwed something up. The odds with random mutations are about the same.

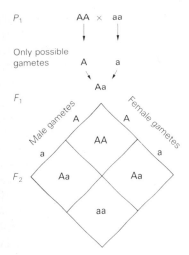

FIGURE 8.17 Here, one of the parents bears the dominant trait and one the recessive, but in the F_1, only the dominant phenotype exists, and in the F_2, the recessive is outnumbered 3 to 1. So it would seem that the recessive is disappearing from the population. However, an examination of the genotype shows that the ratio of dominant to recessive genes is precisely the same in each generation. The ratio will remain constant through the generations if mating is random, unless it is disrupted by some exterior influence such as immigration, selection, or drift. The principle here formed the basis of what came to be known as the Hardy-Weinberg principle.

POPULATION GENETICS

At one time, a genetic question making the rounds was: If brown eyes are dominant over blue eyes in humans, why doesn't everyone have brown eyes by now?

We now know the genetics of eye color to be more complicated than the question suggests, but even so, the answer is not a simple one. In fact, at the beginning of the century, many geneticists raised this point to argue against Mendel's principles. Actually, however, Mendel's laws held up quite nicely under this attack.

We can appreciate the argument by considering how genes can change in populations. We have already been using the term *population,* and you may have taken it to mean a group of individuals. You were not wrong, but at this point, we can give it a more precise meaning. In biology, a **population** designates a group of interbreeding or potentially interbreeding individuals. With this in mind, we can consider the ratios of different alleles in a population for a specific characteristic, such as eye color.

First, imagine a population of only two individuals in which one sex is homozygous for dominant trait **A** and the other sex is homozygous for recessive trait **a,** as shown in Figure 8.17. Now, we know that all their F_1 offspring will be heterozygous for that characteristic. The Punnett square then shows us that in the F_2 generation, three out of four individuals will show the dominant trait, and only one will show the recessive trait. So it might appear that we are on the way to eliminating the recessive gene from the population. However, if we now plot the F_3, F_4, F_5, and so on, generations, we will find that the proportion of dominant and recessive genes in the population has not changed at all. In fact, look again at the F_1 and F_2 generations. There is a 1:1 ratio of the two alleles even at these stages.

Of course, a population of two is unrealistic, but the example illustrates a principle that also operates in larger populations. We can show by means of Punnett squares that the frequency of alleles for any characteristic will remain unchanged in a population through any number of generations—unless this frequency is altered by some outside influence.

The principle came to light during a discussion at Cambridge University in 1908. R. C. Punnett, the young Mendelian geneticist, was having lunch with his older friend, G. H. Hardy. Punnett said he had heard an argument critical of the Mendelian approach that he couldn't answer. According to the argument, if the gene for short fingers were dominant, and the gene for long fingers were recessive, then short fingers ought to become more common each generation. Within a few generations, the critics said, no one in Britain should have long fingers.

Punnett disagreed with the conclusion, but he couldn't give a good reason for his argument. Hardy said he thought the answer was simple enough and jotted down a few equations on a napkin. He showed the amazed Punnett that given any particular frequency of genes for normal or short fingers in a population, the relative number of people with short or long fingers ought to stay the same as long as the population was not subject to natural selection or other outside influences that could lead to changes in gene frequency. (**Gene frequency** refers to the ratio of genes

FIGURE 8.18 The English mathematician G. H. Hardy. Hardy and the German physician W. Weinberg, working independently, discovered in 1908 what has come to be called the Hardy-Weinberg principle. The principle states that gene frequencies in a population remain stable over long periods of time unless the frequencies are shifted through exterior circumstances. The principle explains the continuance of recessive traits in a population, a problem that plagued many geneticists at the turn of the century.

in a population.) Punnett talked a reluctant Hardy into publishing the idea somewhere other than on a napkin (Hardy thought the notion was too trivial to be published). Others were developing the same notion, including a German physician named Wilhelm Weinberg, so the idea came to be known as the Hardy-Weinberg principle. (The principle was actually developed even earlier by an American named W. E. Castle, but for convenience and convention, we will sacrifice patriotism and propriety.) The principle is stated today as: *In the absence of forces that change gene ratios in populations, when random mating is permitted the frequencies of each allele (as found in the second generation) will tend to remain constant through the following generations.* We will consider forces that change gene ratios in populations later, but for now let's examine the basis for the principle.

First of all, if we calculate the gene frequencies in the F_2 generation from Figure 8.17, we see in the F_2 generation that one-fourth of the population is **AA,** one-half is **Aa,** and one-fourth is **aa.** So, to determine what fraction of the F_3 generation will be offspring, say, of **AA** and **Aa** crosses, we simply multiply ¼ by ½ and get ⅛. In the third generation, then, assuming random mating, we can expect one-eighth of the population to have the genotypes resulting from this cross. What part of this generation will be the offspring of **AA** and **AA** crosses? Of **Aa** and **aa** crosses?

Now, there are four possible types of crosses that could produce **AA** individuals in the F_3 generation. You can work out for yourself what they are or conserve your energy and read the answer on page 166. (For any of these four mating possibilities, the boxes there can be filled in accordingly.) Each of these combinations can be expected to produce one-eighth of the F_3 generation, so what part of the population will be heterozygous, or **Aa**? By counting the frequency of this type of gamete among the possible combinations, we see that ⅛ + ⅛ + ⅛ + ⅛ = ½. And what was the fraction of **Aa** individuals in the F_2 population? One-half. If you work out the results for all possible combinations for the F_3 generation, or for any combination in any generation after the F_3 generation, you will see that the ratio of genetic components in a population remains stable. This is why we continue to have blue eyes in our population, even though they are a recessive trait.

Considering the genetic stability suggested by the Hardy-Weinberg principle, how can gene ratios in populations change? After all, evolution is essentially just changes in gene ratios. If there were some great advantage in having brown eyes that gave brown-eyed people a better opportunity to survive and reproduce, then the ratio of brown eyes to blue eyes could shift. It could also change if only people with eyes of a certain color were considered sexually attractive, or if mutations were more likely to occur in one direction than another—say, from **a** to **A** rather than from **A** to **a.** If the population were small, shifts in gene frequencies could occur rapidly, since in a small group only a few accidental deaths could result in a great change in the overall ratios. Also, gene frequencies could change as a result of the flow between populations through immigration and emigration. Any of these things can occur in human populations, so the Hardy-Weinberg model is to some extent an artificial one. However, it helps to explain the relative constancy we see in populations around us.

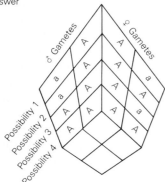

Now for those burning with a fierce love of mathematics, let us consider the mathematical statement of the Hardy-Weinberg principle. The principle is stated as

$$(p + q)^2 = 1,$$

which can be written in expanded form as

$$(p + q) \times (p + q) = 1$$

or

$$p^2 + 2pq + q^2 = 1,$$

where

p = frequency of the dominant allele (**A** in our example)
q = frequency of the recessive allele (**a** in our example)
p^2 = frequency of the homozygous dominant genotype (**AA**)
$2pq$ = frequency of the heterozygous genotype (**Aa**)
q^2 = frequency of the homozygous recessive genotype (**aa**)
1 = a population in genetic equilibrium (one in which the above conditions are met).

Such mathematics has very specific and important implications for us. For example, if we know the prevalence of the trait of albinism (the absence of pigment) in the population, we can predict, within limits, the probability that a couple will have an albino baby. Here's how this would work. Normal skin and eye pigment in humans is dominant over the albino condition **a.** The genotype **aa** occurs in about 1 in every 20,000 people. According to the Hardy-Weinberg equation, this frequency would be given by q^2, so the frequency of genotype **aa** is

$$q^2 = 1/20,000,$$

and the frequency of a single allele for this trait is, thus,

$$q = \sqrt{1/20,000} = 1/141.$$

The frequency of the dominant allele, **A,** would then be

$$p = 1 - q$$
$$\text{or } p = 1 - 1/141 = 140/141.$$

The heterozygous condition, **Aa,** would, therefore, occur in the population with a frequency of

$$2pq = 2 \times 140/141 \times 1/140$$
$$= 1/70 \text{ or } 1.4 \text{ percent.}$$

Since 1.4 percent of 20,000 is 280, this means that only 280 people in every 20,000 will be carrying a recessive allele for albinism. Hence, in the absence of a family history of this characteristic in either parent, the chance that any couple will have an albino child is very slim.

In contrast, among American blacks a disease called sickle-cell anemia (see Essay 8.3), produced by the recessive condition **ss,** appears in about 1 person in 400. Using the Hardy-Weinberg equation, we can see that the heterozygous carrier condition (**Ss**) can be expected in almost 1 person in every 10 in this population. With this kind of information, governments and health organizations can establish priorities in dealing with health problems. Whereas the bureaucratic or political mind might view 1 in 400 as too low a ratio to warrant urgent attention, 1 in 10 is a ratio that certainly cannot be ignored. However, decisions about what steps to take in such a situation are exceedingly difficult. For example,

Sickle-cell anemia is a recessive condition characterized by fragile red blood cells that collapse into a sickle shape. These cells are unable to carry oxygen and may clog the blood vessels, so that persons who have the disease suffer painful symptoms and usually die at an early age. The cause of sickle-cell anemia has been traced to a mutation in a single gene, an alteration of one nucleotide in a triplet. The result is that in one of the four complex polypeptides that comprise hemoglobin, a single amino acid is changed—and so is the capability of the entire molecule.

In persons homozygous for this condition, two of the four polypeptides are affected. These people do not normally survive long. In heterozygous persons, only one polypeptide is affected, and the red blood cells are able to function under normal oxygen requirements. Under conditions in which the body demands additional oxygen, as during strenuous exercise or at high altitudes, these blood cells are likely to collapse. Nevertheless, the heterozygous condition also carries a compensatory advantage in certain environments. Persons who are heterozygous for sickle-cell anemia are more resistant to malaria than those with normal hemoglobin. Thus sickle-cell anemia is maintained in malarial areas from Africa to India.

Sickled cells showing the ragged and collapsed configuration that is associated with their inability to carry oxygen.

should the available funds be spent on efforts to overcome only the symptoms of the disease, or should controls be exercised over marriage—the only way we now know to reduce the occurrence of the disease? Is there some possibility of controlling the problem at a level somewhere between these extremes? In any case, the number of individuals that we now know must be carriers for this condition serves as an impetus for our continuing research.

SUMMARY

Gregor Mendel derived three basic patterns of genetics by studying crosses of pure-breeding pea plants. He described the principle of dominance, the principle of segregation, and the principle of independent assortment. The principle of sex determination in *Drosophila* was discovered by T. H. Morgan. Two genes on the same chromosome are said to be linked, and some linked genes are on the chromosomes that determine sex, thus some traits appear in sex-biased ratios. The linkages can be disrupted by crossing over, when chromosomes exchange genetic material.

Mutations are changes in the hereditary material. Most are harmful. Point mutations are changes in single genes, usually as a nucleotide is altered. The addition or subtraction of a single nucleotide can change the rest of the sequence. Chromosome mutations are changes in the chromosome that may involve many genes. They may include deletions, inversions, duplications, and additions.

Population genetics involves changes in the gene frequencies within groups. The Hardy-Weinberg principle states that populations do not evolve unless the premises of the principle are broken, such as by natural (or artificial) selection, mutation, immigration, emigration, and the effects of small populations.

Chapter 9

ADVANCES
IN GENETICS

The field of genetics is presently undergoing sweeping changes largely because of a simple introduction: Madison Avenue has met the microbe. Already we find that the most conservative of financiers are interested in the mating habits of a lowly bacterium that is found in the bowels of everyone (including their employees). How did this all come to be? How did a generally harmless germ become the focus of such unexpected attention?

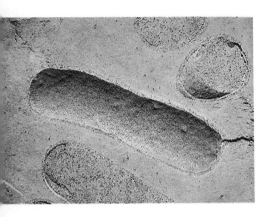

FIGURE 9.1 *Escherichia coli*, a generally harmless bacteria of the intestine, has become the focus of intense research interest in genetic engineering.

A VERY COMMON BACTERIUM

The bacterium at the center of this recent surge of interest is called *Escherichia coli* (usually simply referred to as *E. coli*) (Figure 9.1). It lives in the human intestine, where it normally does no harm. In fact, it may help make certain vitamins available to us from the waste matter in the large intestine. And, while we're discussing such delicate matters, you might like to know that a great deal of fecal matter is actually composed of this bacterium. (Toss that out the next time there's a lull in the dinner conversation.)

E. coli first gained the interest of geneticists in the 1940s. Geneticists have traditionally tended to choose only a few subjects for investigation and to compile incredibly detailed bodies of information about only a few processes. Recall that Mendel introduced the garden pea to geneticists, and not long afterwards, the hardy little fruit fly, *Drosophila*, came under their relentless probing. As researchers came to realize that genetic studies are best carried out on very simple systems, the common bacterium *E. coli* seemed a good choice as the next subject for intensive analysis.

E. coli was considered a good choice because it is genetically simple, with little internal structure. The cell itself is easily broken apart, enabling researchers to get at its cellular machinery. As described in Essay 9.1, it is readily infected by viruses, a trait that lends it to modern research methods. And finally, *E. coli* reproduces by dividing in half about every twenty minutes and, thus, experiments can progress quickly. Such rapid growth of populations means that even rare mutations will turn up in significant and measurable numbers.

BACTERIAL SEX

Because of the great body of information now accumulated about *E. coli*, and because of the bacterium's availability, it has been enlisted as the primary subject of the newest, most promising, and, to some, the most disturbing area of genetic research. The field is generally referred to as **recombinant DNA** research, and although the results have become rather commonly headlined, few people have more than a vague idea of how it works and why it has caused such a stir in scientific circles. As we consider recombinant techniques here, you can decide for yourself how comfortable you are with such research becoming increasingly routine. We will begin with a brief exposé of bacterial sex.

Many kinds of bacteria, including *E. coli*, normally divide through a procedure quite similar to mitosis. That is, they reproduce their genetic component and then pinch in two, each half receiving identical genes. The term *recombinant*, however, refers to the process of genetic recombination, and in many species, such as ours, genes are usually shuffled about and recombined through processes that involve sex. So you should not be surprised to learn that *E. coli* also has ways of recombining its genes and that the process is somewhat akin to sex.

According to those who have seen bacterial sex, *E. coli* doesn't really have the hang of it (but then again, who does?). However, in about one case in a million, two bacteria will undergo something called **conjuga-**

Viruses are remarkable small organisms, about 1/20,000 the size of bacteria, that reproduce entirely inside cells. Viruses are unique among living things—if, in fact, they are alive. Outside of a living cell they are completely inert. Under some conditions they may crystallize and appear as lifeless as ordinary table salt.

Viruses are all parasites. They can't be grown on a medium in a petri dish because they must invade living cells and use the metabolic machinery of their hosts to complete their life cycles. As soon as a virus enters a living cell, it begins to take over the metabolic machinery of that cell. It reorganizes the cell's processes so that the cell begins to engage in producing more viruses instead of continuing its normal activities. Finally, when the cell ruptures, new viruses are released to take over yet other cells.

Single or double strands of DNA or RNA are usually found inside the virus. The strands are coated with a protein layer that protects its nucleic acid from the cell's normal

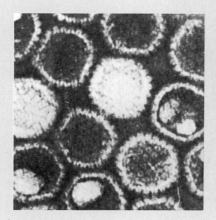

Herpes simplex virus

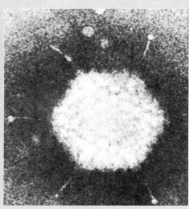

Adenovirus (common cold)

enzymes. This protein layer also gives the virus a special affinity for certain kinds of host cells. Some viruses, for example, have an affinity for bacteria. When such a bacteriophage, or bacteria-eater, accidentally bumps into a proper host cell, it attaches by its tail through covalent bonding. An enzyme of the virus digests a hole in the cell wall, and the virus then contracts the "head" and releases its nucleic acid contents into the bacterium.

To date, no viruses have been directly implicated in human cancer, but some viruses have definitely been shown to cause cancer in other animals. It is likely that some human cancers are caused by viruses, but the evidence so far is indirect. For example, virus particles have been found in the milk of women from families with a high incidence of breast cancer—particles that are identical to the virus that causes breast cancer in mice.

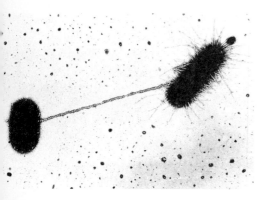

FIGURE 9.2   Conjugating bacteria. One bacterium forms a cytoplasmic tube that penetrates a second bacterium. Genetic material then moves through this tube.

tion (Figure 9.2). In conjugation, one bacterium forms a long, slender tube that penetrates the body of an adjoining bacterium. Then, as we will see, a type of genetic material moves from the bacterium that produced the bridge through the bridge into the body of the second bacterium. In this way, genes from one organism can join the genes of another organism, and as the genes are shuffled about, recombination can take place. We will consider all this in more detail shortly. For now, just keep in mind that bacteria can exchange genetic material at conjugation.

Now the story gets even stranger. In a nutshell, the chromosomes of *E. coli* have been found to be circular, and maleness has turned out to be a kind of disease. This obviously deserves a closer look.

It is now known that populations of *E. coli* are composed of two "mating types" that were (a bit hastily) labelled "male" and "female." Each type will conjugate only with the other type. The label "male" was given to the donors of genetic material during conjugation, and the label "female," to the recipients. In conjugation, once the tubelike bridge between the two mating types has formed, certain of the "male's" chromosomal rings break, and the linear results slowly begin to move across the tube from the male into the female.

FIGURE 9.3 *E. coli* with a circular, single-stranded chromosome and smaller, double-stranded plasmids.

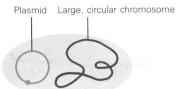

Plasmid   Large, circular chromosome

E. coli

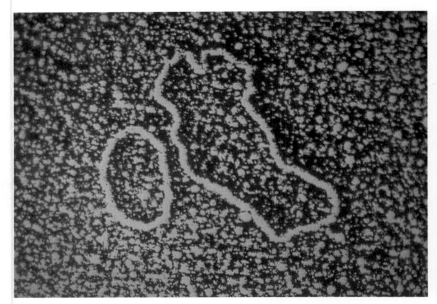

Contagious Maleness

But what's this about maleness being a kind of disease? (Some people have expected as much for years.) Experiments have shown that when male and female strains are mixed, the males remain males, but about a third of the females also become male. On the surface it seems that maleness is, indeed, somehow contagious.

The reason for this contagion was a great puzzle until it was learned that maleness is due to the presence of a tiny genetic parasite called a **plasmid** (Figure 9.3). Actually, a plasmid is a small circle of DNA found inside bacteria. Some kinds of plasmids contain only a few genes, and their activity is intensely directed toward their own reproduction. In *E. coli*, the plasmid not only contains the genes that confer "maleness," but it also directs the bacterium to form a conjugation tube. As the tube forms, the plasmid reproduces, forming linear copies of itself. These copies pass through the bridge into the female. The female is, in this way, changed to a male and rendered impermeable to penetration by any other male. (This, then, accounted for the growing number of males in the mating group.) Scientists by now knew enough to discard the old "male" and "female" labels. The males simply had the fertility factor (the F plasmid) and the females didn't. So the males were considered F+ and the females F−.

Let's now review the conjugation process with our new terminology in mind. Conjugation begins with an F+ cell developing a conjugation tube and penetrating a receptive F− cell. The DNA of the plasmid then begins to replicate, remaining circular as the copied strand, produced in linear form, breaks away and passes through the tube and into the F− cell (Figure 9.4). Once the entire sequence of DNA is inside the recipient, the strand becomes circular, and a new male (F+) is formed. You are undoubtedly grateful at finally learning about the private lives of some bacteria, but there's more. For example, there is the question of why all this is important.

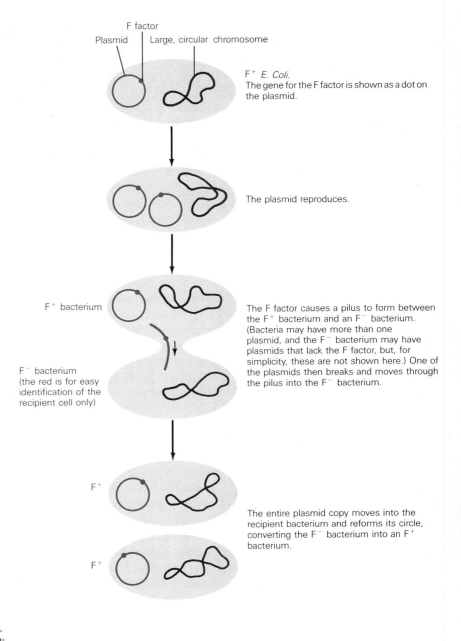

FIGURE 9.4 Conjugating *E. coli* with the newly formed F plasmid moving into F− bacterium.

F factor

Plasmid   Large, circular chromosome

F⁺ *E. Coli.*
The gene for the F factor is shown as a dot on the plasmid.

The plasmid reproduces.

F⁺ bacterium

F⁻ bacterium
(the red is for easy identification of the recipient cell only)

The F factor causes a pilus to form between the F⁺ bacterium and an F⁻ bacterium. (Bacteria may have more than one plasmid, and the F⁻ bacterium may have plasmids that lack the F factor, but, for simplicity, these are not shown here.) One of the plasmids then breaks and moves through the pilus into the F⁻ bacterium.

F⁺

The entire plasmid copy moves into the recipient bacterium and reforms its circle, converting the F⁻ bacterium into an F⁺ bacterium.

F⁺

FIGURE 9.5 Some bacteria exhibit a form of sexual behavior. Those containing F plasmids build cytoplasmic bridges to those that do not. They send copies of their F plasmid through the bridge into the recipient. If the F plasmid has first become inserted into the larger chromosome, it pulls that chromosome, or a part of it, along with it. The host chromosome may then recombine through crossing over with parts of the new chromosome.

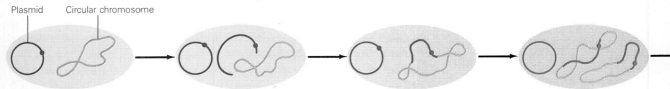

Plasmid   Circular chromosome

(A) *E. coli* with large, circular chromosome and plasmid. The F factor gene is shown as a dot on the plasmid.

(B) The plasmid reproduces and one copy breaks.

(C) The copy becomes incorporated into the larger chromosome.

(D) The larger chromosome reproduces and the inserted plasmid breaks near the F factor gene.

Why All This Is Important

Researchers have discovered that, on rare occasions, a plasmid can break and become inserted into the bacterium's own larger circular chromosome. In this way, then, bacteria can change drastically after the input of new genetic material. Furthermore, as the host DNA divides in the normal process of reproduction, the plasmid DNA divides right along with it. In this way, the plasmid DNA and the traits it codes for are passed to all the host's cloned descendants. A **clone** is a group of genetically identical organisms derived from a single individual by asexual reproduction. (Remember how quickly descendants can be produced by cells that can divide every twenty minutes.) In any case, now the story takes an even stranger twist.

Although the DNA of the plasmid becomes incorporated into the DNA of the host, the plasmid DNA retains its ability to break along its length and then to pass linearly through the conjugation tube into a recipient cell. The result is that after the DNA with the plasmid fragment duplicates itself, the plasmid material may break at a point near the gene that codes for the F+ factor, causing the whole DNA strand to become linear. Then, when the plasmid begins its migration through the conjugation tube, it drags the new cell DNA right along with it. This, then, is the stage at which the bacterial version of sex actually takes place.

Because the bacterial DNA is so long, the process is slow (the entire process taking about a hundred minutes). However, the conjugation tube itself usually can't last this long and breaks before the whole DNA molecule makes it through. Because the gene sequence coding for the F factor is among the last to go, it often doesn't make it and the recipient remains F−. The recipient, though, now has new genetic material that can combine, through crossing over, with its own chromosome. This is the point at which the genetic recombination actually takes place. The process is described in Figure 9.5. Understanding these processes has enabled us (1) to account for some of the rapid changes in bacterial populations, but more importantly it has enabled us (2) to put bacteria to work for us. All we have to do (an understatement) is to insert a desirable human gene into the plasmid. As the bacteria reproduce, they make our product right along with theirs (Figure 9.6). We can now make substances cheaply and abundantly that were once available in only trace amounts, and by controlling the genetic machinery of these

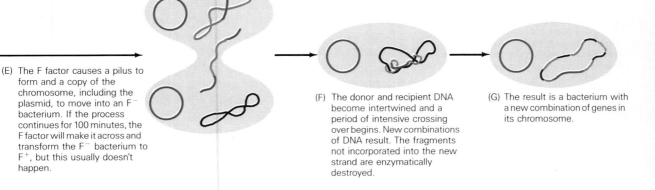

(E) The F factor causes a pilus to form and a copy of the chromosome, including the plasmid, to move into an F− bacterium. If the process continues for 100 minutes, the F factor will make it across and transform the F− bacterium to F+, but this usually doesn't happen.

(F) The donor and recipient DNA become intertwined and a period of intensive crossing over begins. New combinations of DNA result. The fragments not incorporated into the new strand are enzymatically destroyed.

(G) The result is a bacterium with a new combination of genes in its chromosome.

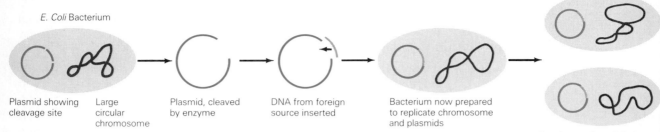

E. Coli Bacterium

Plasmid showing cleavage site

Large circular chromosome

Plasmid, cleaved by enzyme

DNA from foreign source inserted

Bacterium now prepared to replicate chromosome and plasmids

Daughter cells and their descendents will carry copies of the inserted DNA

FIGURE 9.6 Using bacteria to make human hormones. The plasmids are first separated out from ruptured bacteria and subjected to a disruptive enzyme that breaks the circle at specific points. These opened loops are then mixed with a sequence of DNA from another source, say the gene sequence that makes some human hormone. The foreign sequence is inserted at the broken ends, reforming the circular plasmid, but now with new human genes inserted. As the bacterial genes direct the formation of bacterial substances, the human genes direct the formation of human hormones. The bacteria continue to reproduce in great numbers until finally they are chemically ruptured and quantities of human hormone can be retrieved.

lowly bacteria, we can manipulate living things in ways that no one could have imagined only a few years ago. Let's look at some of the ways in which scientists have been using such findings. Maybe we can get some idea of what all the hollering has been about.

IS IT TAMPERING WITH GENES OR GENETIC ENGINEERING?

Genetic engineering, as the name implies, involves manipulating genes to achieve some particular goal. Already you can see what sorts of objections might arise. People of all stripes tend to suffer great moral indignation of one sort or another at the mere mention of the idea. But it doesn't take much imagination to see what new achievements might be on the horizon.

Perhaps the greatest threat of recombinant techniques, some would say, lies in its very promise. The possibilities of such genetic manipulation seem limitless. For example, we can mix the genes of anything, say, for example, an ostrich and a German shepherd. This may only bring to mind images of tall dogs, but what would happen if we inserted cancer-causing genes into the familiar *E. coli* that is so well adapted to living in our intestines? What if the gene that makes botulism toxin, one of the deadliest poisons known, were inserted into the DNA of friendly *E. coli* and then released into some human population? One might ask, "But who would do such a terrible thing?" Perhaps the same folks who brought us napalm and nerve gas.

Another, less cynical, concern is that well-intended scientists could mishandle some deadly variant and allow it to escape from the laboratory. Some variants have been weakened to prevent such an occurrence (see Essay 9.2), but we should remember that even after smallpox was "eradicated" from the earth, there were two minor epidemics in Europe caused by cultured experimental viruses that had escaped from a lab. One person died of a disease that technically didn't exist.

Such things, of course, are the stuff of nightmares, and when gene splicing became a real possibility (see Essay 9.3), there was an immediate furor. Many of the scientists involved met together in Asilomar, California, in February 1975 and, in some rather heated sessions, forged a set of operating guidelines to keep such potential catastrophes from ever becoming a reality. These guidelines were self-imposed and have

At one of the conferences on the risks of recombinant studies, geneticist Roy Curtiss III volunteered to produce a weakened mutant—one that could not survive outside the laboratory.

Given the go-ahead, he first produced an *E. coli* with a defective gene that prevented it from manufacturing its protective coat. The material that the gene normally made had to be provided artificially. But microbes mutate on their own, so some were soon back to producing the normal gene. Curtiss then deleted another gene necessary for coat production. But he was outfoxed by the crafty germs; they reproduced anyway. Dennis Pereira, a graduate student working with Curtiss, found that they were manufacturing a sticky substance called colanic acid, which acted as a kind of protective coat. So Curtiss and Pereira produced a microbe that couldn't make colanic acid. Finally they had a germ that depended on scientists for its livelihood. As an unexpected bonus, this new bug was sensitive to ultraviolet light. Ordinary sunlight would kill it.

One problem remained. Even dying *E. coli* can conjugate with normal *E. coli*, so an escaped germ might be able to pass its dangerous gene to a healthy colleague. Curtiss, however, altered the gene of the dependent bacteria that makes thymine. Thymine, therefore, had to be supplied, and without it, how could DNA be made? Perhaps now the bug was helpless enough.

This fermenter is a biochemical cauldron for growing gene-spliced microorganisms.

evolved into quite firm rules at this point. But not everyone was satisfied. Some communities even banned such research in their areas. Then something happened. No one is quite sure why the change came about, but much of the opposition to such research was quietly abandoned. It all happened at about the time recombinant techniques showed promise of great commercial feasibility.

How Great a Promise?

There were people who feared the results of gene splicing, but no one could deny that the promise was great and that the successes of the technique were beginning to mount. The beneficial aspects of recombinant DNA research were first proposed in the late 1970s, when a number of laboratories announced preliminary results suggesting that gene splicing could be a practical solution to a great many human problems. As these various small laboratories, often staffed by maverick graduate students, announced their findings, Wall Street reacted with great verve and fervor, and millionaires were made overnight. The hooks and ladders of financial gain tempted a number of dedicated scientists to escape the ivory towers of academia.

Progress in recombinant techniques was rapid, and some began to speak of an end to all genetic diseases and, perhaps, the eradication of cancer. It was suggested that certain kinds of cattle could be produced that would require less feed and be more disease-resistant. Perhaps high-

New techniques based on plasmid transfer now make it possible to determine the exact nucleotide sequence of any isolated section of DNA or RNA. Already, the entire base sequences of DNA in bacteriophages, plasmids, polio viruses, and human mitochondria have been determined. They produce pages and pages filled with A, T, C, and G. (There are exactly 5315 in the case of the first bacteriophage sequence to be discovered.) Your own DNA sequence of some 6.4 billion nucleotide pairs would fill a thousand volumes the size of this one. And, incredible as it sounds, the task is already well under way. Tom Maniatis, a Harvard molecular biologist, has chopped human DNA into fragments a few tens of thousands of nucleotides long and inserted them into the DNA of bacteriophages. His laboratory is virtually a DNA library of three million separate clones of cultured human DNA.

And now we have the gene machine. This is a device with Madison Avenue overtones, but it is

available for only about $30,000, and it does just what it says. An operator can create any DNA sequence simply by typing it out. The computerized machine does the rest. For example, one might know a protein amino acid sequence but be unable to isolate the corresponding mRNA. If the investigator knows the sequence of even a short segment of the gene, it is typed into the gene machine, and a probe locates the rest of the gene from a DNA library consisting of cloned chromosome segments of the species in question. If all one has is a tiny amount of protein, a highly sensitive protein-sequencing accessory is available. Samples of protein placed in the automatic sequenator

yield plants could be produced that would require less fertilization. Perhaps potatoes could be made resistant to frost.

And, in fact, some of the promise has already been fulfilled. For example, we have produced frost-resistant potatoes. Also, with recombinant techniques, scientists have manufactured human growth hormone so that thousands of growth hormone-deficient people can now attain normal height. In the past, growth hormone had to be extracted from the pituitaries of human cadavers, a tedious and extremely expensive process requiring the glands of 50 cadavers for a single dose. A few thousand seriously undersized adolescents have already been treated to some extent (as well as one man who grew to normal height after having been only four feet tall until the age of 35). Very soon, it seems, no one will have to be any shorter than he or she wants to be. Does that seem to you like a blessing or like cavalier tampering with nature?

As another example of recombinant DNA success, engineered E. coli now produces sizeable quantities of the antigen against hoof-and-mouth disease, making possible a relatively inexpensive vaccine against that costly disease. That bit of genetic engineering alone might help save the economy of Mexico.

are dismantled chemically, one amino acid at a time from one end of the polypeptide. These fragments are identified automatically and the sequence is printed out.

So, once a probable nucleotide sequence is determined, it is typed into the gene machine, which then goes through the steps of that DNA's synthesis. In about a day's time, a small quantity of the desired DNA is produced. Sections are cloned in a bacterial plasmid, which is then taken up by a receptive bacterium. Standard biological gene cloning methods then take over. Potentially, then, any enzyme the experimenter can imagine or any other kind of protein or gene can be made to order.

In radionucleotide sequencing (left), scientists can label each protein in a human gene to look for congenital disease. With a gene machine (right), scientists can create any DNA sequence simply by typing it out—the machine does the rest.

There are some other notable advances. Inexpensive human insulin has been produced by such techniques, as well as interferon, a promising antiviral substance. However, it seems that the news media sometimes jump the gun by oversimplifying the problems with such research and by unnecessarily or prematurely raising the expectations of the public.

Today, there are a great many other studies underway. Some may well offer new solutions to some vexing human problems. For example, you might be relieved to learn that we have now spliced rat growth hormone genes into mouse cells and produced giant mice. We've never had any decent giant mice before.

SUMMARY

New promise in genetic techniques is held by recombinant DNA techniques, or gene splicing, generally using the harmless intestinal bacterium *E. coli*. The single chromosome of *E. coli* is circular and often exists with smaller rings called plasmids. Some bacteria have F plasmids and are called F$^+$, those lacking it are called F$^-$. The F$^+$ plasmid causes a bacterium to build a conjugation tube to an F⁻ bacterium. The F plasmid may then move through the tube, linearly, into the F⁻ bacterium, making it F⁺. If the plasmid of the F⁺ bacterium becomes incorporated into the circular chromosome, when it breaks it may drag parts of the chromosome with it into the F bacterium. The chromosome part may then undergo crossing over with the circular chromosome of the recipient F⁻ and, in this way, the F⁻ bacterium gains new genetic combinations.

If other sorts of genes, such as human genes, can be incorporated into the bacterial chromosome when the bacterium divides and its descendants grow in number, the alien gene is duplicated along with the bacterial genes. Great quantities of the alien gene can then be recovered. Some believe this ability has great promise; others consider it a great threat.

Part Three

THE
DIVERSITY
OF CHANGING
LIFE

Chapter 10

THE PROCESSES
OF EVOLUTION

The theory of evolution has been called the cornerstone of modern biology. If it is in fact one of the more important ideas in modern history, it is a bit peculiar that the concept is so little understood in what is undoubtedly one of the best-educated and most sophisticated societies of all time. Why is the idea so comfortably accepted and routinely utilized by some, while driving others right up the wall? Why is any such ''cornerstone'' the object of so many critics' chisels? Probably for two reasons—reasons at quite different intellectual levels. One is based on what some people perceive as the theory being at odds with religious dogma. Matters of faith, however, lie outside the realm of scientific inquiry, and arguments aimed by either side clearly and inevitably miss their targets. The second reason for some of society's uneasiness with the concept of evolution is based on simple misunderstanding of its principles.

So let's see if we can clear up some of these misunderstandings while building a foundation that will help to clarify much of what will follow. To keep from further muddying the waters, we will cover only some of the major and more general concepts on which the theory is based.

FIGURE 10.1 The golden-fronted wood-pecker (top) looks remarkably similar to the red-bellied woodpecker (bottom). The range of the golden-fronted extends through dry areas from Honduras northward to central Texas, where it rather abruptly stops and that of the red-bellied begins. The red-bellied's range extends to the eastern coast, where it lives primarily in wooded habitats. The birds overlap in a narrow range in Texas, but in spite of strong similarities in appearance, vocalization, and behavior, they do not interbreed. Though they do not recognize each other as sexual partners, each does see the other as a competitor, and they mutually exclude each other from their territories as if they were of the same species. Obviously, they use different cues in the recognition of mates and competitors.

THE QUESTION OF SPECIES

We can begin with a simple question: What is a species? It may seem like a simple question, but it's not. Biologists find it quite difficult to answer, and, in fact, it is said that there is no record of any two biologists ever agreeing on an answer. If they seem to agree, it's generally because each one has stifled what he or she knows about this exception or that. We shall not be daunted by lack of consensus; we shall admit a certain ignorance and plunge right in.

The question of species seems easy because almost all of us can tell a dog from a fern (the leash slips off the fern), and we know an oyster from a horse (stirrups drag on the oyster). These, then, are clearly different species. However, in a small, narrow area in Texas, we find a vexing situation in the form of the red-bellied woodpecker and the golden-fronted woodpecker. At first sight, they are difficult to tell apart (Figure 10.1). Their voices, feeding habits, and nest sites are also very similar. Yet, they are quite different species and obviously have no trouble telling each other apart. They do not attempt to interbreed, and no hybrids have been seen. However, far more dissimilar species than these (at least to the human eye) are known to interbreed.

As an example of interbreeding between distinct types, we find wolves and hounds may interbreed, producing wolf-dog hybrids of these two different ''species.'' Are they different species after all? Most biologists use interbreeding as a criterion for a species. Since wolves and dogs interbreed, then, even while differing greatly in appearance and behavior, some scientists treat them as a single species. Others argue that the *ability* to interbreed implies the *opportunity* to interbreed and that dogs and wolves are sufficiently isolated that interbreeding between them is rare enough that they can be considered different species. However, when they do interbreed, the offspring are healthy and *able to breed among themselves* (Figure 10.2). This, too, is a common criterion for animal species (botanists may disagree). Horses and donkeys can produce mules, but the mules are sterile, so the parents are considered to be of different species. How about the grackles of Texas and Puerto Rico? Are these noisy black birds of the same species? They are identical in virtually all respects except for the smaller size of the Puerto Rican birds. They obviously don't interbreed because they are geographically distant, but some researchers believe they could and would interbreed if they ever met. There are also species that cover a large area, changing gradually across the range. The subgroups interbreed all along the range, but individuals at either end of the range are so different as to make breeding impossible. Are those individuals of the same species?

These, then, are some of the problems with defining species. So, we will accept the definition of the renowned zoologist Ernst Mayer: ''A species is a group of actually or potentially interbreeding populations that is reproductively isolated from other such groups.''

Reproductive isolation refers to the inability of one group to successfully interbreed with another. There are a number of ways reproductive isolation can be achieved, but the most obvious one is due to simple geographic barriers. If the groups can't reach each other, they can't interbreed. Also, should the groups be in contact, their mating behavior may be incompatible. For example, they may not recognize each other's signals. Then there can be the physical problems associated with mis-

FIGURE 10.2 Pawnee, a cross between a wolf and an Alaskan husky. This is a magnificent specimen, being not only vigorous and intelligent, but extremely amiable as well. People who rear animals of wild ancestry are often surprised at the independent spirit of the animals. Pawnee, for example, is quick to know your wishes and then, unlike a German Shepherd, makes his own decision about whether to comply.

matched reproductive organs. For example, some insects have very complex "lock and key" genitals, and the organs of different species simply do not fit. If mating between different species should occur, reproductive isolation may be achieved by genetic mechanisms, such as the inability of dissimilar chromosomes to pair up at meiotic prophase I. Can you think of ways that populations might be benefited by individuals not being able to successfully mate with similar species? In any case, we will assume that species exist in spite of our difficulty with forming a precise definition. We will now consider some of the ways different species might arise.

SPECIATION

Allopatric Speciation

Speciation is the formation of new species, and, today, most researchers agree that new species generally arise when subgroups become isolated from the parent group. For example, it is believed that the Abert and Kaibab squirrels of the southern and northern rims of the Grand Canyon were once the same species, but diverged when they became separated by the formation of the great gulch (Figure 10.3). The idea of geographic separation is fundamental to most notions of how animal species have formed. Technically, the process is referred to as **allopatric speciation** (*allopatric,* other land). Speciation, it is generally believed, does not usually occur (especially among animals) within a group in which individuals are free to interbreed, because any genetic differences that arose would be swamped by, and incorporated into, the existing population.

Allopatric populations are those that occupy different places. If they overlap, they are said to be allopatric only in areas of nonoverlap. Different populations that occupy the same area are said to be **sympatric** (together land), and those abutting, **parapatric** (next land) (Figure 10.4). Speciation is believed to occur primarily, perhaps exclusively, in allopatric (or perhaps parapatric) animal populations in which individ-

FIGURE 10.3 The Abert squirrel (left) and Kaibab squirrel (right) are two distinct species that live on opposite sides of the Grand Canyon. It's believed that they were once one population that was divided as the great chasm developed. The two populations followed their own paths of evolution and now are quite different and normally unable to interbreed.

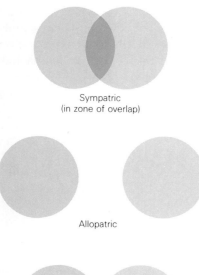

Sympatric
(in zone of overlap)

Allopatric

Parapatric

FIGURE 10.4   Populations tend to influence each other according to how greatly they overlap. Sympatric species must minimize competition by subdividing a shared habit. Allopatric populations normally do not interact. Parapatric populations may occur as populations vigorously interact and exclude each other, or they may be a result of habitat difference and not due to interaction at all. (The coastline provides an example of such habitat differences.)

FIGURE 10.5   Divergent evolution occurs when species break up into subgroups that then become increasingly different from each other. Convergent evolution involves the development of similar traits in dissimilar species of different stocks.

uals from different localities do not intermingle and have no opportunity to interbreed. So, speciation is believed to occur predominantly in allopatric populations when a single population is somehow split or a group branches off. In time, a subpopulation continues to evolve. The two groups accumulate enough genetic differences that should they somehow rejoin, they would be reproductively isolated and interbreeding would be impossible. At the point where they can no longer interbreed, we say that speciation has occurred (see Essay 10.1).

Divergent and Convergent Evolution

In determining just how and when speciation has occurred, it is important to realize that similarities may not be a reflection of evolutionary relationships. Just as different environments cause populations to diverge, a process called **divergent evolution,** sometimes similar environments can cause species to become more similar through a process called **convergent evolution.** Convergent evolution can cause a great deal of confusion for those trying to work out evolutionary histories, since physical similarity does not necessarily indicate relationship (Figure 10.5).

Sympatric Speciation in Plants

While allopatric speciation is the rule among animals, plants may form species in other ways. They may produce species through **sympatric speciation**—the formation of two species from one within the same area. In fact, sympatric speciation among plants is rather common.

Speciation Through Hybridization

Among many plants, the flowering plants in particular, new species can arise by the hybridization of existing species. This discovery was surprising, since animal hybrids are usually infertile because the chromosomes

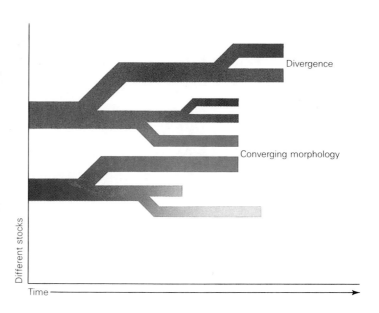

Divergence

Converging morphology

Different stocks

Time

In 1912, Alfred Wegener published a paper that was triggered by the common observation of the good fit between South America's east coast and Africa's west coast. Could these great continents ever have been joined? Wegener coordinated this jigsaw-puzzle analysis with other geological and climatological data and proposed the theory of **continental drift.** He suggested that about 200 million years ago, all of the earth's continents were joined together into one enormous land mass, which he called Pangaea. In the ensuing millennia, according to Wegener's idea, Pangaea broke apart, and the fragments began to drift northward (by today's compass orientation) to their present location.

Wegener's idea received rough treatment in his lifetime. His geologist contemporaries attacked his naiveté as well as his supporting data, and his theory was neglected until about 1960. At that time, a new generation of geologists revived the idea and subjected it to new scrutiny based on recent findings.

The most useful data have been based on magnetism in ancient lava flows. When a lava flow cools, metallic elements in the lava are oriented in a way that provides permanent evidence of the direction of the earth's magnetic field at the time, recording for future geologists both its north-south orientation and its latitude. From such maps, it is possible to determine the ancient positions of today's continents. We now believe that not only has continental drift occurred, as Wegener hypothesized, but that it continues to occur today.

Geologists have long maintained that the earth's surface is a restless crust, constantly changing, sinking, and rising because of incredible, unrelenting forces beneath it. These constant changes are now known to involve large, distinct segments of the crust known as **plates.** At certain edges of these masses, immense ridges are being thrust up, while other edges sink lower. Where plates are heaved together, the buckling at the edges has produced vast mountain ranges. When such ridges appear in the ocean floor, water is displaced and the oceans expand. (Astoundingly precise satellite studies reveal that the Atlantic Ocean grows 5 cm wider each year.)

An understanding of continental drift (or **plate tectonics**) is vital to the study of the distribution of life on the planet today. It helps to explain the presence of tropical fossils in Antarctica, for example, and the unusual animal life in Australia and South America. Continental drift provided just the sort of separation of subpopulations that would permit widespread speciation that would form the basis for widely diverging groups of primitive organisms.

As the composite maps indicate, the disruption of Pangaea began some 230 million years ago in the Paleozoic era. By the Mesozoic era, the Eurasian land mass (called Laurasia) had moved away to form the northernmost continent. Gondwanaland, the mass that included India and the southern continents, had just begun to divide. Finally, during the late Mesozoic era, after South America and Africa were well divided, what was to be the last continental separation began, with Australia and Antarctica drifting apart. Both the North and South Atlantic Oceans would continue to widen considerably up to the Cenozoic era, a trend that is continuing today. So we see that although the bumper sticker "Reunite Gondwanaland" has a third-world, trendy ring to it, it's an unlikely proposition.

Lower Triassic period
225 million years ago

Lower Cretaceous period
90 million years ago

Paleocene epoch
65 million years ago
END OF MESOZOIC ERA

of the parents are so different that they cannot pair up in prophase I of meiosis. As you undoubtedly recall, this pairing up is essential for the crossing-over processes of homologous chromosomes in the formation of gametes.

Why don't hybridizing plants have the same problem? Primarily because plants with very different appearances may be very similar genetically—at least enough to allow normal or nearly normal meiosis. Where such genetically similar species overlap, there may be extensive **hybrid swarms**—essentially, large hybrid populations. Surprisingly, such hybridization doesn't seem to result in the breakdown (through merging) of either parent species. This may be because hybrid swarms find their own niche, interacting with the environment differently than do the parent groups, thereby becoming truly a new and distinct species.

Speciation Through Polyploidy

Plants can also hybridize successfully in a quite dramatic way, one in which whole sets of chromosomes become doubled (or even doubled again). The general process is called **polyploidy** (*poly*, many). This happens spontaneously from time to time in the mitotic divisions of a growing plant. Chromosomes normally reproduce in preparation for cell division; for some reason, however, the cell itself occasionally fails to divide while it continues through its normal sequences as if nothing were unusual. These cells may then proceed with normal mitosis. The result of such doubling is that the daughter cells will end up with four complete sets of chromosomes. A cell produced in this way is called a **tetraploid** cell (*tetra*, four), and it may go on to form tetraploid tissue and even tetraploid flowers.

So hybridization and polyploidy create instant, sympatric species of plants, ready to be tested by the forces of natural selection. This versatility helps explain how flowering plants arose abruptly in evolution and how they quickly spread out over the landscape to help create the incredible diversity of plant species that so beautifully accent our world.

VARIATION IN POPULATIONS

As you watch your little brother out of the corner of your eye, you may wonder how you could have the same parents. After all, you're a sophisticate, and he can't even be trusted not to eat what he finds on his shoe. Furthermore, if you come from a large family, you may have noticed that none of the kids are very much alike. They seem to vary, not only in appearance, but in such traits as attitudes, friendliness, and even agility. If this sort of variation can be found among the children of the same parents, it is not surprising that even in an enormous population, such as exists on the streets of New York at any hour of the day (Figure 10.6), no two people will look alike—at least not much. There is, indeed, a great deal of variation among humans.

There is also great variation in other species. We can usually tell dogs apart, even without resorting to the crass and overly personal means they employ among themselves. Of course, there are species that, from our view, don't seem to vary much. They all seem to be rather alike. For example, standing on the seashore, it is difficult to tell one gull from

FIGURE 10.6 Chicagoans rushing to clothing sales exhibit a wide range of variation in their appearances, tastes, and behaviors.

another. We know, however, that they can recognize individuals among themselves. (Perhaps we all look alike to them.) The point is, variation among individuals is a common theme among the populations that inhabit the earth, although the differences may not always be apparent to us.

Since variation is so common, one might ask if it has some fundamental advantage. Indeed it does—if not for the individuals themselves, then for the population in general. The avantage of a variable population is this: In a variable population, there are likely to be some individuals with traits that will ensure their success under a wide range of conditions. Put another way, no matter how the environment should change (within limits), in a variable population there are likely to be some individuals that will fare well. Those that do, by surviving and reproducing, will cause subsequent generations to follow along new paths that are more in keeping with the new conditions. Of course, if the environment should shift again, there would be some individuals capable of surviving those conditions, too.

Now let's take a look at some of the factors that may influence variation and see just how variation itself can play an important role in certain sorts of changes through time, the sorts of changes that mark the progress of evolution.

HOW NATURAL SELECTION OPERATES ON VARIATION

We have seen that variation can arise in a population through many means. We saw earlier that Charles Darwin recognized the importance of variation as he uncovered its role in the processes of evolution. Variation, in a sense (and as he implied), provides the "raw material" upon which natural selection can act. Some of the members of a population will possess those traits that enhance reproductive efforts. Those traits (especially those borne by genes that code for them) will, therefore, be carried into future generations. Furthermore, as they are carried along, they will replace less successful alternative traits, and so they will come to increase, proportionately, in later generations.

Darwin referred to such selection as "survival of the fittest." This is sort of an unfortunate term, because many people have taken it to refer to the continued success of the best specimens. Actually, though, in the cool arithmetic of evolution, "fittest" refers solely to reproductive success. Evolutionarily, then, we can expect the greatest success from the best reproducers. Thus, each generation will be composed primarily of the offspring of the best reproducers from previous generations, and we can expect those traits that led to reproductive success to increase in frequency from one generation to the next.

How Natural Selection Can Mold Variation

Variation in a population can be described graphically, as we see in Figure 10.7A. Let's say this curve describes height in a group of people. We see, then, that most individuals in the population are of intermediate height, and that there are increasingly fewer taller or shorter people as they diverge from that intermediate height. Any trait distributed this way in a population produces the familiar bell-shaped curve.

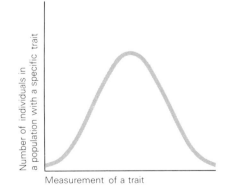

FIGURE 10.7A Normal distribution, a statistical term that is graphically depicted as a bell-shaped curve.

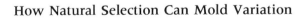

FIGURE 10.7B   If the small vertical lines signify some optimal trait, say a certain height, when selection is strong (heavier arrows), most individuals will be close to that height. When selection is weak (light arrows), that is, if it is of no great disadvantage to be of a particular height, then height in the population will vary from the optimum, producing a wider curve.

FIGURE 10.7C   Here, Y marks the optimum value for some trait, such as height. Those taller or shorter would therefore not do as well. Their numbers would be reduced by ecological pressures (shown by X and X^1). If it should become better to be taller, then Y, the optimum, would shift. Individuals at X would suffer, while those at X^1 would thrive. The result would be directional selection, producing a taller population until the population stabilized again around the new optimum.

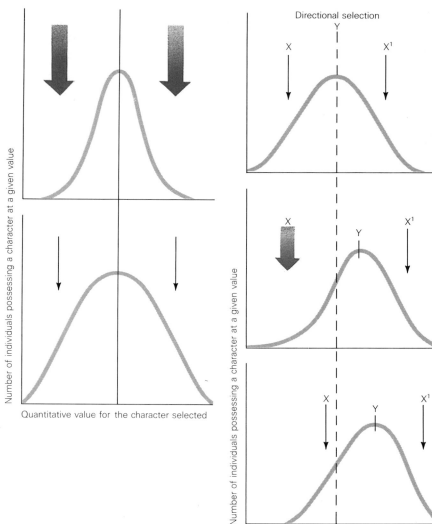

The shape of any such curve, of course, can change. For example, it can be taller and thinner, it can widen, it can be skewed to one side or the other, or it can be "bumpy." Let's examine the implications of such shapes.

First, if being of intermediate height is especially important, then those people who are taller or shorter will suffer some increased disadvantage. Because of such **high selection pressures** (strong selection against departure from the optimum), in time, there will be fewer of them, and the curve will become narrower, as we see in Figure 10.7B. If height is not so important, shorter and taller people will tend to survive and reproduce more easily, and because of **low selection pressures** (weak selection against departure from the optimum), such a population will produce a wider, more relaxed curve.

So, when the environment is stable over a long period of time, as is assumed in these cases, the population tends to cluster around a single type, with divergent forms being less successful to some degree. The result is called **stabilizing selection.** In stabilizing selection, then, the

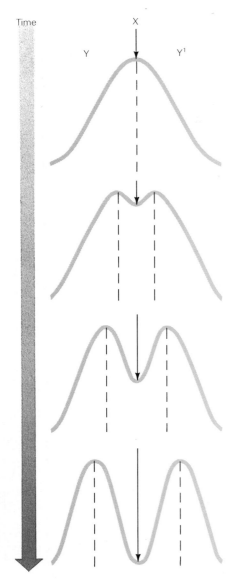

Time

X

Y   Y¹

FIGURE 10.7D Disruptive selection can produce a bimodal (two-humped) curve. Bimodal curves are produced in populations in which it is advantageous to be at one of the two extremes of a curve since the intermediate condition is selected against.

population is clustered around a mean, or average, condition that is assumed to be optimal for the prevailing conditions. Yet, conditions can change. Suppose, as we've just seen, taller becomes better, and increased height yields a new kind of advantage (remember the giraffes). The population then shifts toward the high end of the height scale, as in Figure 10.7C. Such a process is labelled **directional selection** (changes due to populations tracking a new optimum).

In some cases, there may be selection against the mean condition, resulting in an accumulation of individuals at the extremes. This is called **disruptive selection.** For example, consider a population of marine skates—large, flat fish that escape predators either by lying flat on the sea bottom covered with sand or by rapidly fleeing. Suppose most of the skates try concealment for a time and then begin to flee when a shark draws near. Others, though, perhaps bolder or more lethargic, do not move as the shark swims over. Yet others are very skittish and flee at the first sign of danger. In this population, then, we might expect selection against the average behavior as the sharks caught those that tried to flee while the sharks were near. If the others tended to escape more frequently, the population would tend to be composed increasingly of both lethargic and skittish fish (''hiders'' and ''runners''). If escape tendencies were plotted on a graph, the population would produce a **bimodal** (two-humped) **curve,** as we see in Figure 10.7D. (Such a bimodal curve can also be a result of sexual dimorphism as is described in Essay 10.2.)

VARIATION BY SINGLE-GENE AND POLYGENIC INHERITANCE

There are a number of traits in humans that seem to be due to the operation of a single allele. These are either-or conditions produced by what is called a **single-gene effect**. Figure 10.8 illustrates some such conditions. Thus, some variation is produced by the interactions of rather simple genetic systems.

Other kinds of variation, though, may be produced in more complex ways. For example, polygenic inheritance is the control of a single trait by multiple genes, as we saw in Chapter 8. Height in humans is influenced this way. Because of polygenic inheritance for height, people are not either tall or short. Instead, there is a gradual variation that enables us to generate a smooth, gradual, and continuous bell-shaped curve as each gene adds its own small influence. It can thus be assumed that if there are a number of genes that contribute to tallness, while others produce shortness, taller individuals are likely to possess a disproportionate number of the ''tall'' genes. On the other hand, those people with a preponderance of ''short'' genes will be short. By the same token, this is why parents of average height can produce exceptionally tall or short offspring. Let's say there are only seven genes that influence height (T for tall and S for short). If the parents are **TSTTTSS** × **STTTSST**, they might yield offspring that are **TTTTTTS** (very tall), **SSSTSSS** (very short), **TSTSTST** (intermediate), or any number of other possible combinations and heights.

We saw in the case of disruptive selection that there may be more than one optimum phenotype. In our example, selection for evasive techniques produced two types, hiders and runners. Thus the population

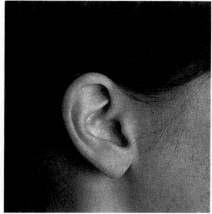
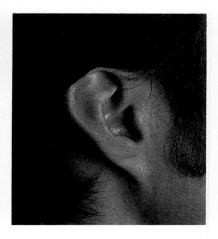

FIGURE 10.8 Only a few traits in humans are known to be controlled by single genes. Two such traits are the ability to roll one's tongue and unattached earlobes (versus attached at right).

of skates tended to be **dimorphic** (*di*, two; *morph*, form). Other populations may produce more than two types and are called **polymorphic** ("many forms"). We find a fascinating example in the brittle stars, spindly starfish, or sea stars that may have any of several distinctly different appearances. This seems to confuse predators by not allowing them to build a single image of the sea stars that would aid the predators in their search (Figure 10.9).

We see, then, that variation in populations can be of several types: continuous (slight differences, usually maintained by several genes affecting a single trait), dimorphic (two types, such as tall and short), and polymorphic (several distinct types).

We know that variation is important to the processes of evolution, so we might now consider the ways that variation can be maintained or even increased in a population.

How the Environment Can Influence Variation

We have noted some of the ways in which genetic mechanisms can increase variation in a population. Here, let's consider a few of the ways in which the environment can interact with populations in such a way as to further increase that variation. In our first example, we might imagine what would happen if the environment were to have the tendency to shift wildly or unpredictably. Obviously, populations would have difficulty "tracking" it—that is, changing to conform to the requirements set by existing conditions. For example, if taller animals do best under certain environmental conditions, the population will become taller with each generation as it approaches that optimum. But if the environment should suddenly change and begin to favor short individuals, the frequency of short individuals would begin to increase. If, before the shift to the optimum was complete, the environment were to suddenly change, favoring tallness again, the populations would be unable to stabilize around a mean. Each time the environment changed, the population would begin to shift. If the optimum continually changed, never allowing the population to stabilize, the result would be a population with high variation.

FIGURE 10.9 Brittle stars vary in their appearance. Thus, it is difficult to find one and then generalize about its appearance and use that information to find another one.

High variation can also result if the environment is "patchy"—that is, comprised of markedly different kinds of places. Because a patchy environment presents a number of kinds of habitats, subpopulations (parts of the larger group) may adapt to a number of environments simultaneously. Individuals within each patch might be expected to be rather similar as they conform to the requirements set by local conditions, but the overall population might well be quite variable. As long as members of these different subpopulations continue to interbreed, however, they will tend to "dilute" each other's adaptations. Thus, no group would be able to fully adapt to its kind of patch. One might envision a group of permanent ponds dotted across the landscape, some deeper or cooler or with grassier bottoms than the others. The frogs that inhabited those ponds might develop different traits, each adapting to the particular features of its own watery home. Gene transfer would keep speciation from occurring as the frogs occasionally visited other ponds and mated with the individuals there.

Another environmental influence or variation is geographic **clines.** Clines are produced by gradual changes across the habitat. In such cases, the geographic features of the habitat vary slightly as one moves from place to place. Geographic clines can, in turn, produce gradual changes in the organisms that inhabit these areas. For example, the eastern United States changes somewhat gradually from North to South. Some animals, such as mice and frogs, range from, say, Vermont to Louisiana, and over this range they vary gradually from one place to the next (Figure 10.10). We noted earlier that individuals at the extremes of such a range may be so different as to be unable to successfully interbreed.

These, then, are some of the ways that the environment can influence variation. We might also keep in mind that they operate with genetic influences on variation, such as meiosis and crossing-over, processes that at a different level encourage variation among individuals in populations. We can wrap up this discussion now with a well-documented example that illustrates just how natural selection can operate on variation to produce changes in gene frequency (which, after all, is what evolution is all about).

FIGURE 10.10 Species that cover vast ranges encounter a variety of environmental influences. Thus, they may vary from place to place. Where the environment varies gradually, the population may change gradually from one place to the next, so that individuals at the extremes of the range may be quite dissimilar, as we see here in these leopard frogs, which range from northern states to southern states.

ESSAY 10.2   SEXUAL DIMORPHISM

You may have noticed that boys look different from girls and that it's easy to tell a rooster from a hen. In some deep-sea fish, the sexes are quite dissimilar; the male may even be an appendage fused to a knob on the head of a female. On the other hand, it may be difficult to tell a male mouse from a female, and the sexes of seagulls are almost identical. **Sexual dimorphism**—or the different appearance of the sexes—varies widely among animals.

There are interesting evolutionary implications of sexual dimorphism. How did it start? With respect to vertebrates, one idea is that as they came under increasingly severe competition for commodities such as food, the males increased in size and strength in order to compete more successfully. The females were restrained from such changes by the demands of gestation (which placed a premium on a certain body size) and the fact that they were more closely associated with the young (which made it important that they be physically and behaviorally less conspicuous). Males that were best able to gain commodities might then have been disposed to share those with females for reproductive purposes. This meant that females might then be more likely to choose the more vigorous males as mates.

This tendency would have driven males toward increasingly garish appearances and bold, aggressive, and conspicuous behavior in order to display their vigor and to intimidate other males in disputes over commodities. Exceptions exist, of course, but generally today, males of most species, especially among birds and mammals, are larger, stronger, and more aggressive than females and are dominant over them in social interactions. It provides some entertainment, and a lot of argument, to try to determine how this general condition could have arisen.

An interesting relationship between sexual dimorphism and domestic duties exists among some species. Consider an example from birds. The sexes of song sparrows look very much alike. The males have no conspicuous qualities that immediately serve to release reproductive behavior in females. Thus, courtship in this species may be a rather extended process as pair-bonding (mating) is established. Once a pair has formed, both sexes enter into nest building, feeding, and defense of the young. The male may only mate once in a season, but he helps to maximize the number of young that reach adulthood carrying his genes. He is rather inconspicuous, so whereas he doesn't turn on females very easily, he also doesn't attract predators to the nest.

The peacock, on the other hand, is raucous and garish. He displays madly and frequently and is successful indeed. Once having seduced an awed peahen, he doesn't stay to help with the mundane chores of child rearing, but instead disappears into the sunset looking for new conquests. The peahen he has just left may be an inferior bird, not likely to be able to rear young successfully. He could have discovered this by a more careful process of mate selection. But no matter, perhaps the next conquest will be a better bird. Alas, just as he attracts females, he also attracts tigers, and his flurry of sexual activity may end in a flurry of feathers. Instead of carefully rearing one brood each season for several seasons, we see, he maximizes his reproductive success through another means.

Think of other species of birds and mammals. Which evolutionary route have they taken? What about humans?

FIGURE 10.11 Two forms of *Biston betularia*, the peppered moth, on a lichen-covered background. The dark form became predominant as soot blackened the lichen, exposing the lighter form to predatory birds. With the advent of pollution-control devices, the background is becoming lighter and the light form is returning.

INDUSTRIAL MELANISM

In the 1700s, the British countryside was as green as its poets said. The trees were numerous, and many were covered with a light, gray-green lichen. Furthermore, if one looked carefully at the lichen coverings, it was sometimes possible to discern a light-colored moth sitting concealed upon the bark. One might also occasionally spot a black moth perched rather conspicuously on the light-colored lichen. The black moths were rather rare, largely because they were easily spotted by hungry birds and quickly removed from the population.

But "progress" will have its way, and England's entrance into the industrial revolution was marked by the construction of huge, coal-burning factories. These factories belched forth their billowing, black smoke, and soon the English countryside was covered with a cloak of soot. As the trees darkened, the more common white moth became increasingly easier to see, while the darker moth became less visible. Thus, birds began to eat more of the light moths and their "light genes" began to disappear from the population. Finally, by the late 1840s, 98 percent of the moths were dark (Figure 10.11). The transition, then, from the light to the dark form was virtually completed in only fifty years. Interestingly, the use of pollution-control devices in British manufacturing has produced cleaner air, lighter trees, and a resurgence of the lighter moth.

SMALL POPULATIONS

So far we've been discussing large, randomly mating populations that have been molded by the powerful effects of natural selection. In these groups, as adaptive traits are selected, others are relentlessly winnowed out as the populations track the environment by directional selection. But what about small populations? Do they change in the same ways? Or are they under different sorts of pressures? In other words, do they shift in response to different kinds of factors?

Genetic Drift

Small populations, indeed, may be molded by different factors than those operating in larger populations. For example, in small populations, **genetic drift** may be important. Genetic drift is a change in the frequency of genes in a population due to simple chance. Changes of this sort are *not* produced by natural selection. Genetic drift is important in small populations because they would, by definition, have a smaller gene pool (genetic reservoir). This means that any random appearance or disappearance of a gene would have a relatively large impact on gene frequencies. In large populations, any random change would have little impact since it would be swamped by the sheer numbers of other genes in the population.

Founder Principle

One of the influences operating on small populations is called the **founder principle.** Occasionally, a few organisms will be separated from the main population and establish their own breeding group. For example, a few birds might be blown from the mainland to some remote island where they become the founders of a new population, or a storm might wrench a large tree loose, allowing it to float out to sea. Such a raft could bear small creatures and eventually deposit them on some new land, perhaps an island. It is in such small and isolated groups that we see the founder principle in operation.

Any small group is unlikely to bear all the genes found in the main population, nor could the genes of the small group be expected to appear in the same frequencies as those of the larger group. Certain genes might not be represented at all in the small groups, and other genes might be disproportionately common. As the organisms breed and begin propagating their own genetic ratios, it is soon apparent that they have departed from the evolutionary course of the parent population. Furthermore, if the founders must respond to different selective pressures in their new habitat, they may diverge even more rapidly from the parent population through natural selection.

To illustrate, in a cross between two animals, both with **Aa** and **Aa** genotypes, we would expect the classical 1 : 2 : 1 genotypic ratio in their offspring. However, if we were to randomly choose four of those offspring, we might end up with an **aa, AA, AA,** and **AA.** We might then use these to begin a new population. But let's say the **aa** individual had a bad day and died. The **a** allele would now be gone from the population and its alternative, the **A** allele, would have become *fixed* (reached 100 percent). This, indeed, would be an example of *very* rapid evolution (change in gene frequency in a population), and not a very likely occurrence, but it illustrates the special role of small populations in evolutionary processes.

It is because of the potential importance of seemingly insignificant changes that small populations can change, or evolve, so rapidly and, by the same token (their resulting lack of genetic flexibility), why they may be expected to become extinct quite easily.

There is another factor that places small populations at high risk of extinction. As a population grows smaller, genes that interfere with reproductive processes tend to accumulate. Such interference, for example, is given as a reason for the extinction of the American heath hen in 1930. And, even now, the wisent is at dangerously low levels and may not be able to recover. (What is a wisent?*)

SMALL STEPS OR GREAT LEAPS?

A cornerstone of Darwin's explanation of evolution as "descent with modification" was the notion that evolution is essentially the gradual accumulation of small changes over a long period of time. (The time available for life to have evolved is shown in Table 10.1 and Figure 10.12.) Such changes, he believed, can account for the sweeping changes that result in adaptation and speciation.

Recently, however, Niles Eldredge and Stephen Jay Gould, of the American Museum of Natural History, suggested that the fossil record

*A European buffalo.

Table 10.1   Geological Timetable

Era and Duration	Period	Epoch	Years Before Present (in millions)	Principal Advents of the Era
Cenozoic 65 Million Years	Quaternary	Recent	(11 thousand)	Age of man; end of last ice age; warmer climate. First human societies; large scale extinctions of plant and animal species; repeated glaciation.
		Pleistocene	1	
	Tertiary	Pliocene	11	Appearance of man; volcanic activity; decline of forests; grasslands spreading.
		Miocene	25	Appearance of anthropoid apes; rapid evolution of mammals. Formation of Sierra Mountains.
		Oligocene	36	Appearance of most modern genera of mammals and monocotyledons; warmer climate.
		Eocene	54	Appearance of hoofed mammals and carnivores; heavy erosion of mountains.
		Paleocene	65	First placental mammals.
Mesozoic 160 Million Years	Cretaceous		135	Appearance of monocots; oak and maple forests; first modern mammals; beginning of extinction of dinosaurs. Formation of Andes, Alps, Himalayas, and Rocky Mountains.
	Jurassic		181	Appearance of birds and mammals; rapid evolution of dinosaurs; first flowering plants; shallow seas over much of Europe and North America.
	Triassic		220	Appearance of dinosaurs; gymnosperms dominant; extinction of seed ferns; continents rising to reveal deserts.
Paleozoic 360 Million Years	Permian		280	Widespread extinction of animals and plants; cooler, drier climates; widespread glaciation; mountains rising; atmospheric carbon dioxide and oxygen reduced.
	Pennsylvanian		310	Appearance of reptiles; amphibians dominant; insects common. Gymnosperms appear; vast forests; great life abundant. Climates mild; lowlying land; extensive swamps; formation of enormous coal deposits.
	Mississippian		355	
	Devonian		405	Many sharks and amphibians; large scale trees and seed ferns; climate warm and humid.
	Silurian		425	Appearance of seed plants; ascendance of bony fishes; first amphibians; small seas; higher, drier lands; glaciations.
	Ordovician		500	Atmospheric oxygen reaches second critical level. Explosive evolution of many forms of life over the land; first land plants and animals. Great continental seas; continents increasingly dry.
	Cambrian		600	Appearance of vertebrates, but invertebrates and algae dominant. Land largely submerged. Warm climates worldwide.
				Atmospheric oxygen reaches first critical level. Explosive evolution of life in the oceans; first abundant marine fossils formed; trilobites dominant; appearance of most phyla of invertebrates. Lowlying lands; climates mild.
Pre-Cambrian 3500 Million Years			2700	Life confined to shallow pools, fossil formation extremely rare. Volcanic activity, mountain building, erosion, and glaciation. Photosynthetic life.

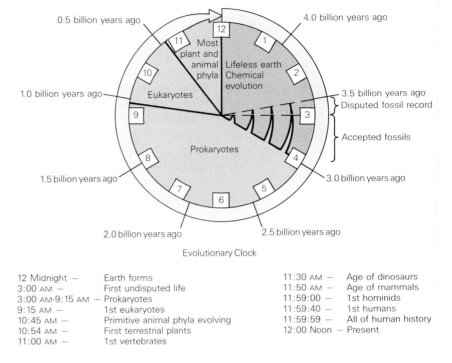

FIGURE 10.12 An earth calendar. Here, the earth's history is shrunk into a period of 12 hours—from midnight to noon. The events are in chronological order and the spacing indicates lapsed time.

Evolutionary Clock

12 Midnight —	Earth forms	11:30 AM —	Age of dinosaurs
3:00 AM —	First undisputed life	11:50 AM —	Age of mammals
3:00 AM-9:15 AM —	Prokaryotes	11:59:00 —	1st hominids
9:15 AM —	1st eukaryotes	11:59:40 —	1st humans
10:45 AM —	Primitive animal phyla evolving	11:59:59 —	All of human history
10:54 AM —	First terrestrial plants	12:00 Noon —	Present
11:00 AM —	1st vertebrates		

does not show a gradual transition of one life form to another. Rather, they have reported that one life form persists in the record for long periods of time and then suddenly is replaced by a different life form apparently descended from the older one. They say that the story of evolution is marked by long periods of stabilization in which life forms changed little, occasionally punctuated by episodes of rapid change.

They argue that the stable periods often last about three million years, before the onset of a sudden change. The new version appears quickly, leaving little evidence of intermediate forms. Eldredge and Gould labelled evolution that proceeds by such fits and starts **punctuated equilibrium.** Their notion is not universally accepted, but it has generated a great deal of lively discussion.

DEATH STARS AND DINOSAURS: THE GREAT EXTINCTIONS

It may be quite difficult to account for the beginnings of new species, but it is often easy to see how some species died out. We have built dams and knowingly doomed small pockets of isolated species. We have hunted other animals to extinction, as we did the dodo, the carrier pigeon, and the last common ancestor of the horse and zebra. However, other species have passed into extinction for reasons that continue to puzzle us. In particular, it is difficult to account for the massive, large-scale extinctions in which many species passed from the earth at once. For example, why did so many species die out with the great dinosaurs at the end of the Cretaceous?

One hypothesis advanced to account for the die-off at the end of the Cretaceous is based on data suggesting that the temperature of the earth dropped drastically about that time. The dinosaurs and the others, some say, simply died of the chilling effects of hypothermia. However, others have argued that the cooling of the earth unbalanced the sex ratios of many species. They note that the sexes of many kinds of animals, such as alligators, amphibians, and some fish are temperature dependent. That is, eggs raised in environments below a certain temperature will give rise to animals of one sex, and above that temperature, to the other sex. The argument is that as the earth cooled below a certain critical point, all the hatchlings of some species would have been of one sex and doomed to roam the earth without ever knowing the joys of parenthood. How awful.

Another explanation of the great extinctions of the Cretaceous also involves a cooling episode, but this one accounts for its origins. According to the **Alvarez hypothesis,** the earth was struck by a great asteroid that raised a cloud of dust that blocked the sun for many months and effectively stopped photosynthesis. Any such disruption of the food chain would have led to the demise of a great many species. The evidence here is circumstantial but solid. The best line of evidence is the discovery that a rare element called iridium was deposited in a fine layer over the earth at about the end of the Cretaceous. Iridium is uncommon in the earth's crust, but quite common in asteroids.

Other researchers have suggested that the earth was struck by some heavenly body that was not an asteroid, but perhaps three or four comets. (Ready for this?) The comets, they say, were from the great belt of one hundred million or so comets that lazily circle the sun far beyond the reaches of the solar system. Occasionally, though, about every 26 to 30 million years, great numbers of the comets are jerked toward our sun by the passage of a companion star to the sun. The companion star has not been located or identified, but it has been named: *Nemesis.* Journalists, with their flair for high drama, have taken to calling it the Death Star.

Any such companion star, if it exists, could be any of the hefty little "black stars," dense bodies with a gravitational pull so strong that not even light can escape. It has been suggested that the star has an extremely elliptical orbit that takes it far into the celestial realm—until its next deadly loop through the belt of comets. Geologists have found that the earth has indeed been peppered every 30 million years or so by celestial objects big enough to form craters, and that the cycles roughly match the great extinctions that paleontologists tell us have taken place on our planet. Again, there is great disagreement about the existence of Nemesis and, in fact, the periodic extinctions.

Finally, there are those who say that about every 20 to 30 million years, the earth passes through severe galactic storms of dark clouds and gas as the sun takes us through the plane of the Milky Way, and that these storms are responsible for the great surges of death on the planet.

However, on a brighter note, the sun is now passing through a clear zone, the pristine aftermath of an exploded star. So all should be clear sailing for a while . . . or so they say.

In summary, we have seen that, as Darwin suggested, variation in natural populations results in some organisms being better reproducers than others and that their kinds of genes tend to increase in populations. The best reproducers, of course, will tend to be those most in harmony with the environment. Thus, populations tend to track their environment through adaptation. Evolution, then, as we understand it today, is simply a function of basic arithmetic. Those genes that promote successful reproduction will increase in frequency.

Life on earth is subject to countless pressures as it continues striving for its very existence. It must constantly react to the nature of its situation (or its predicament) and it must change. Put simply, it must change in order to take advantage of its part of the world. The world is a variable and changeable place, and different life forms have evolved that are uniquely able to utilize one aspect of the earth or another in countless ways. In the next chapter, we will begin to explore this vast array of life and see just what the processes of natural selection have wrought.

SUMMARY

The theory of evolution has spawned a great deal of argument, but it remains the cornerstone of modern biology. Essentially it involves descent with modification as gene frequencies in populations change through time, often because of natural selection. It involves not only the formation of new species and higher-level groups, such as genera, but changes within a species in time and space as well. Species are populations of potentially interbreeding individuals that are reproductively isolated from other such groups, but the definition has many arguable facets. Animal speciation is believed to generally occur as allopatric populations diverge. Sympatric plants may speciate through hybridization and polyploidery.

Variation is the raw material upon which natural selection operates. High variation usually means selection forces are weak, and vice versa. Variation is maintained in populations through a number of means. Different sorts of selection pressures can produce stabilizing selection (as the optimum remains static), directional selection (as the optimum changes), and bimodal selection (when there is more than one optimum).

Some variation is due to single-gene effects, but most is due to polygenic inheritance. The environment can increase variation in a population if the environment shifts or if it is patchy. A classic example of the effects of natural selection in modern times is the change in the color of a moth as the British countryside changed color due to increasing and then decreasing levels of soot.

Small populations change more rapidly than larger ones, and much of the change may be due to simple chance, not selective forces. Darwinian evolution has assumed that evolution proceeds by the accumulation of small changes, but new evidence suggests it may proceed in great fits and starts (punctuated equilibrium). A number of causes have been suggested for the perceived periodic extinctions on earth, including asteroids, death stars, comets, and galactic clouds.

Chapter 11

DIVERSITY I: THE MONERA, PROTISTA, AND FUNGI

At this point, we must begin to tread softly. Although it is well-known that biologists are eminently agreeable people, when it comes to how living things are to be classified, they can be a bit testy. In fact, it may be difficult to find any two biologists who can agree on how any living things should be classified. They may even argue over what an *animal* is. That may seem strange because, of course, anyone can recognize an animal. If something has two eyes and large teeth and is after us, we are not going to be easily convinced that it's a plant. We're going to scream that "an animal" is chasing us. If we're lucky, our rescuer won't be a biologist, who would immediately demand, "Quick! Is is a pinniped? Good grief, what *order* is it in?"

Of course, there are many ways to distinguish an animal or a plant or a fungus. A toadstool is a fungus. The little elves that sit under it are animals. So where does the disagreement arise? The disagreement over classification is essentially of two sorts. Experts may disagree over just how closely species are related, and they may disagree over whether an organism is, say, a plant or an animal at all.

The latter is the more fundamental question. As an example, there is great disagreement over just what group sponges are in. And then, just because an organism bears chlorophyll is it a plant? Maybe not. We will shortly consider the perplexing *Euglena* which has chlorophyll and behaves like an animal. And what can we say of an organism that can turn into a crystal, as viruses do? What about tiny organisms composed of only a single cell that creep around and devour other tiny creatures, such as the amoeba. Are they animals? If they aren't plants, aren't they animals? It depends on who you're talking to. There are those who believe that there are only two kingdoms of living things: plants and animals. So, to them, these tiny creatures might be called animals because they certainly don't seem to be plants. To those who believe that living things should be divided into, say, five kingdoms, then, the amoeba might be removed from the animals and placed within a group of tiny one-celled species, like itself, called Protista (see Table 11.1).

FIVE KINGDOMS AND SPLITTERS AND LUMPERS

We will adhere to the five-kingdom approach in our discussion because it helps to simplify things and because many evolutionists tell us that it makes sense (Essay 11.1 describes the various biological classifications). They say that there are five major groups that are rather closely related and that are different from the other groups in important ways (Figure 11.1). (However, a newly discovered group of ancient monerans, the **archaebacteria** (*archae,* old), are clearly not closely related to other bacteria, called the **eubacteria** (*eu,* true), and some would put them in their own kingdom. However, we will take the simpler approach and lump them together until the jury returns on that one.)

FIGURE 11.1 The two- and five-kingdom schemes of living things.

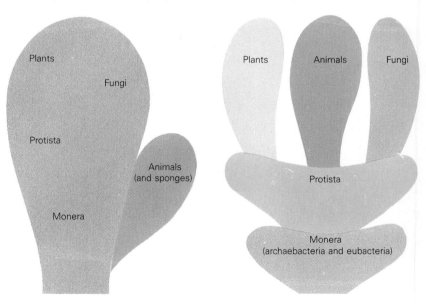

TABLE 11.1   Characteristics of the Five Kingdoms

Characteristics	Monera	Protista	Kingdom Fungi	Plantae	Animalia
1. Nuclear membrane?	No	Yes	Yes	Yes	Yes
2. Mitochondrial bodies? (present for energy production in the cells)	No	Yes	Yes	Yes	Yes
3. Ability to photosynthesize?	Some do	Some do	No	Yes	No
4. Motility? (ability to move)	Some have it, some do not	Some have it, some do not	Primarily nonmotile	Primarily nonmotile	Yes
5. Form?	One-celled	One-celled	Molds, multicellular and yeasts, single-celled	Multicellular, some algae	Multicellular

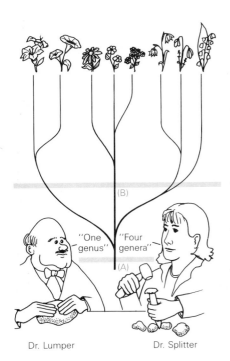

FIGURE 11.2 Both Dr. Lumper and Dr. Splitter agree that there are eight species of plants being classified. Dr. Lumper, however, places all eight in one genus, believing that the species all descended from a single species that flourished at time (A). Dr. Splitter disagrees and calls for four genera. She concludes that the eight species descended from four distinct groups that already had been established at time (B). By the looks of things, both may be right, depending on where one draws the line—literally.

The task of separating the various species into their proper groups is not easy. One problem is that there are simply so many species. Whereas there are only about 1,500 species of bacteria, there are about 8,500 species of birds. The carrot family can boast of about 3,000 species, and there are about 15,000 species of wild orchids. The beetles weigh in with about 300,000 species (about 30,000 of which live in North America).

It is a great pleasure to walk the countryside with a first-rate naturalist. These people are storehouses of information regarding the flora and fauna of their regions. They seem to know every living thing you come across. Occasionally, though, they can't name something. They may thoughtfully turn it over and over and mumble something about what it *might* be. This, in essence, is the same problem confronting **taxonomists,** those people interested in naming and classifying living things. Not only are there too many kinds of things to be handled easily, but taxonomists take different approaches to naming things.

There are two main philosophical camps among taxonomists, the "splitters" and the "lumpers." The splitters tend to place species with similar, but distinct, traits into different groups. The lumpers tend to ignore slight differences and to place similar species into the same group. It is because of such fundamental differences that we have so many disagreements regarding biological relationships (Figure 11.2).

We will follow the road most taken, then, and try to avoid the controversies. We will begin by describing the largest groups, the kingdoms, and then discuss various representatives at lower levels in the taxonomic scheme (see Essay 11.1: Classification).

The five kingdoms as generally accepted by biologists are:

Monera: The monera are prokaryotic cells (having a primitive nucleus). They include the archaebacteria, the true bacteria, and the cyanobacteria (once called blue-green bacteria or blue-green algae). The rest of the life on the planet is eukaryotic (that is, with a true nucleus).

Protista: These include some one-celled algae and the protozoa. Most single-celled eukaryotes are in this group. (See Table 11.2 for the major distinctions between prokaryotes and eukaryotes.)

TABLE 11.2 Differences between Prokaryotes and Eukaryotes

	PROKARYOTES		EUKARYOTES
	Archaebacteria	*Eubacteria*	
Cell wall	Variety of substances, often proteinaceous	Peptidoglycans	Cellulose, chitin; absent in animals and some protists
Cell membrane lipids	Modified branched fatty acids	Straight-chain fatty acids	Straight-chain fatty acids
Nuclear membrane	Absent	Absent	Present
DNA	Naked, circular	Naked, circular	Associated with chromosomal protein
Membrane-bound organelles	Absent	Absent (except for mesosome and thylakoid in some)	Present
Ribosomes	30S, 50S subunits; structural similarity to eukaryotes	30S, 50S subunits; unlike archaebacteria and eukaryotes	40S, 60S subunits (30S, 50S in chloroplasts and mitochondria)
Flagella	?	Tubular, rotating (protein flagellin)	Microtubular undulating (protein tubulin)
Photosynthetic pigments	Bacteriorhodopsin	Bacteriochlorophyll (a)	Chlorophyll a, b, c
Cell division	Fission	Fission	Mitosis and meiosis

Fungi: These include the phycomycetes, ascomycetes, and basidiomycetes.

Plants: These include the red, brown, and green algae, the mosses, and the vascular plants (such as ferns, club mosses, horsetails, and seed plants).

Animals: These include a host of familiar species from sponges to mammals.

THE KINGDOM MONERA AND A CONTINUING SIMPLICITY

The monerans are structurally the simplest of all living things. They have apparently changed little since they first appeared on the earth. In fact, forms resembling today's monerans can be found in the oldest known fossil-bearing rocks. Their simplicity is reflected in the fact that they lack nuclear membranes, membrane-bound organelles, and flagellae with microtubules having the characteristic 9 + 2 structure. Because they are simpler, some people believe that they are ancestral to the eukaryotes (the other four kingdoms).

Archaebacteria

The archaebacteria, the oldest known monerans, are rather mysterious life forms, and as we learn more about them, the more intriguing they become. They seem to be very primitive, somewhat rare, and closely related to the earliest bacteria. Some are obligate anaerobes—that is, they cannot survive in the presence of oxygen. However, they can sur-

Living things are divided, for convenience, into increasingly smaller groups. One such division separates living things into kingdoms. Thus, an organism in the animal kingdom could be expected to have more in common with another organism in its own kingdom than one from the plant kingdom. Among animals, certain ones share a number of traits in common, and these may be placed together into a phylum. Thus, all animals that have rodlike structures called notochords at some period in their development are placed in one phylum, chordata. As finer and finer divisions are made among plants and animals, one finally arrives at a group of organisms that have so many things in common that they normally interbreed. These organisms are said to be of the same species. The taxonomic breakdown is:

Classifications
Kingdom
  Phylum (zoology),
  Division (botany)
    Class
      Order
        Family
          Genus
            Species

Memory Aid
King
  Philip
    Came
      Over
        From
          Greece
            Singing

Members of two genera, then, won't share enough things in common to be able to interbreed, but they are still all within the same family. Members of one family have more things in common with other members of the order than with families from another order.

vive in a number of unusual places that do not harbor oxygen, such as hot springs, ammonia-rich habitats (like cattle feedlots), and salt marshes. Many are also common in a number of other lovely places, such as large intestines and swamp bottoms.

Eubacteria

Eubacteria, the true bacteria, are extremely small (most are only a few micrometers in length), and they exist in vast numbers virtually everywhere on earth. They range from the upper atmosphere to hot springs, polar ice caps, raw petroleum, the deepest reaches of the sea, and animal guts. Some surfaces are free of the creatures since most of them dislike acid, high temperatures, and dryness. Nonetheless, it is believed that their mass outweighs that of all plants and animals combined.

Most bacteria require oxygen, but a few can live without it (including *Clostridium botulinum*, which manufactures one of the deadliest poisons known (Essay 11.2)). Oxygen actually inhibits the growth of some bacteria such as *Clostridium tetani*, which lives deep in soil or in deep puncture wounds where there is little oxygen. Thus, deep wounds from an object lying on the soil can bring about deadly lockjaw or tetanus.

Some bacteria can synthesize their food, but others cannot (Figure 11.3 A, B, and C). Some take their nutrients from living things, such as the sorts that invade our bodies and make us sick. Others take their nutrients from material that is no longer living. These are the decomposers that often cause us such worry and expense. Any food not protected against bacteria will begin to give way before their digestive enzymes. Decomposition does have its beneficial effects. Without it, the

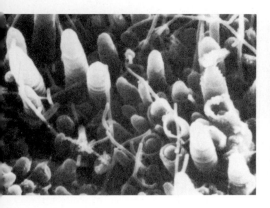 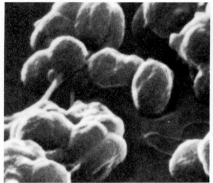 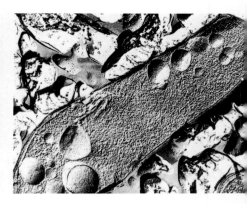

FIGURE 11.3 Bacteria are a highly varied group with very distinct life styles. At left, chemosynthetic bacteria from the surface of a mussel at a deep water volcanic vent. They derive their energy not from the sun, but from the chemical energy of the molten core of the earth. Center, *Neisseria gonorrhoeae,* the organism of gonorrhea. At right, *Pseudomonas aeruginosa* bacterium, a denitrifier that releases nitrogen held in compounds as nitrogen gas.

FIGURE 11.4 The three basic shapes of bacteria: the rounded coccus, the rodlike bacillus, and the corkscrew-shaped spirillum. (Photo far right: Reprinted with permission of the present publishers, Jones and Bartlett Publishers, Inc., from Shih and Kessel: *Living Images,* Science Books International, 1982, page 3.)

earth would be littered with corpses, a situation that would clearly interfere with dancing and marching. Also, decomposition means that the essential mineral nutrients locked in the corpses of dead organisms are available to be recycled by new generations of organisms.

Bacteria generally exist in one of three shapes, the rodlike **bacillum,** the spherical **coccus,** and the long, twisted **spirillum** (Figure 11.4).

Most bacteria produce asexually (that is, without sex) after the circular chromosome duplicates itself (Chapter 9). As the two loops of DNA separate, the cell pinches in two, one complete strand of DNA going to each daughter cell. The process is far more simple and less precise than is the mitosis of eukaryotes. The division can be accomplished in twenty minutes, which is one reason infections can spread so fast. As we saw earlier, however, some bacteria can mate. A bridge of cytoplasm is formed, and DNA passes from one to the other.

Do you like the smell of freshly turned earth? It is not the earth that you smell, but a special group of bacteria called **Actinomycetes** that live in the soil. The Actinomycetes are of particular interest because they are the source of many antibiotics (Essay 11.3), such as streptomycin, chloramphenicol, and the tetracyclines. However, the same group of bacteria produces leprosy (perhaps the least contagious and most slowly developing of all diseases since the organisms divide only once every twelve days).

One of the problems in controlling bacterial infection is that bacteria have some rather effective defense mechanisms. For example, some bac-

 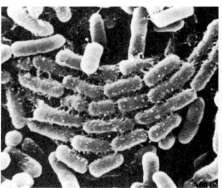 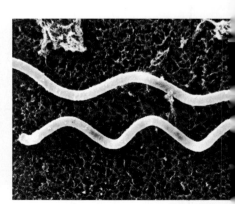

The toxin produced by the bacterium *Clostridium botulinum* is among the most deadly poisons known. Only 60 billionths of a gram will kill a human. The endospores (shrunken, dehydrated, and highly resistant forms of the organism) are found as a common inhabitant of the soil around the world (so we are literally surrounded by an incredibly dangerous life form). Fortunately, however, the spores grow only under very specific conditions, basically in an anaerobic and nutrient-rich medium. Canned food that has not been properly sterilized is an ideal home for the microorganism. If canned food has been contaminated, however, it will produce a telltale foul smell and gas (which may cause cans to swell). This is why it can be suicidal to taste suspicious food. The symptoms of botulism begin about three days after eating poisoned food and include vomiting, constipation, and paralysis of the eyes and throat as breathing becomes more difficult. The poison functions by interfering with the release of acetylcholine, a neurotransmitter. Recovery is possible if antitoxins are administered within the first 24 hours.

teria respond to a hostile environment by forming **endospores,** a kind of living kernel made from material inside the bacterial body. After an endospore is produced, the living material outside it is simply sloughed away. The resistant endospore may enable the organism to survive until the return of better times. Some endospores can survive in boiling water for several hours.

Cyanobacteria

The cyanobacteria are prokaryotes, but they are not like bacteria in the usual sense of the word. The groups differ in several important respects. Like other photosynthetic bacteria, the cyanobacteria (or blue-greens, as they are sometimes called) lack chloroplasts, their light-capturing pigments being scattered over highly folded inner membranes known as **lamellae** (Figure 11.5). However, their photosynthetic mechanism differs from that of other bacteria. Like plants, they derive their required electrons by breaking apart water molecules. (Bacteria usually get their electrons from other sources, such as hydrogen sulfide.) The blue-greens share an important ability with some bacteria: the ability to "fix" nitrogen. The earth's living systems indeed depend strongly on the ability of these two groups to incorporate atmospheric nitrogen into the molecules of life (Figure 11.6). In fact, it is in this group that we see the first "division of labor," considered to be an evolutionary advancement. It is manifest by different cell types assuming different roles—one type specializing in photosynthesis, the other in nitrogen fixation.

Not surprisingly, most species of blue-greens contain blue-green, light-absorbing pigments (green chlorophyll and blue **phycocyanin**).

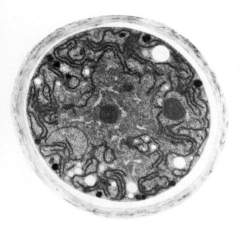

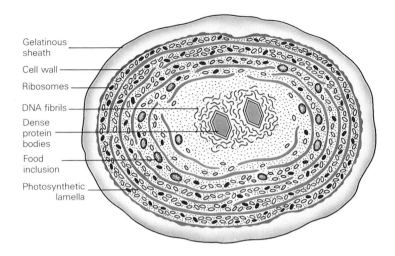

Gelatinous
sheath

Cell wall

Ribosomes

DNA fibrils

Dense
protein
bodies

Food
inclusion

Photosynthetic
lamella

FIGURE 11.5 Cyanobacteria. The blue-greens possess chlorophyll and are capable of photosynthesis. However, they are unusual among photosynthesizers in that they, like other prokaryotes, lack well-defined nuclei. Blue-greens can be found in common places like roadside ditches and in exotic places like hot springs. They contain a number of odd-shaped structures whose function is not known.

Others, though, are reddish in color, due to a red pigment, **phycoery-thrin.** (As an aside, the Red Sea is named for its occasional eruptions of red cyanobacteria.) Blue-greens may exist in long chains, but each cell is physically separated from the others without the cytoplasmic bridges that commonly link plant cells. Whereas blue-greens are abundant in the soil, they do best in water. In fact, they may become so abundant in drinking water that it not only develops a bad taste but can make one quite ill.

Cyanobacteria are a rather hardy lot. They can survive quite nicely in places where most other species cannot exist, such as in hot springs, salty lakes, and glaciers and on parched desert rocks. One reason for their success is that they require only sunlight, carbon dioxide, water, and a few minerals. If conditions are unfavorable, some species form thick, tough walls or a jellylike protective coat. Under favorable conditions, they reproduce by simply dividing—and they do it rapidly. It is suspected that they have ways of mixing their genes, but they have never been caught in the act of mating.

FIGURE 11.6 Nitrogen-fixing bacteria live in the roots of some plants and are able to incorporate the nitrogen found in the air into a water-soluble molecule that the plant can use. The plant, in turn, provides the bacteria with a suitable environment in which to live. This is one of the clearer examples in nature of a mutualistic symbiotic relationship in which two species live in intimate association to the good of both.

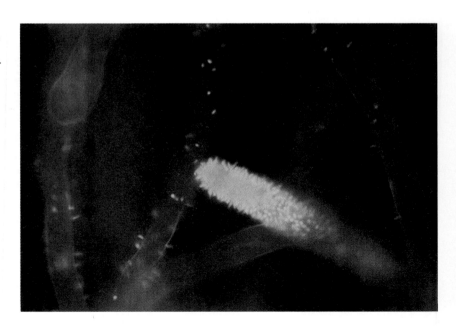

For centuries, people have been looking for ways to heal themselves from bacterial infection or to keep from falling to bacterial attack in the first place. One of the first efforts involved simple cleanliness (an unusual idea among many of our forebears). This led to efforts to perform surgery under antiseptic conditions. Then it was found that various chemicals could impede bacterial growth. The problem here was that the treatment was often as unpleasant or dangerous as the infection.

The real revolution in bacterial control, however, was spawned by those first attempts, centuries ago, to apply molds to wounds in order to control infection. By the nineteenth century, Louis Pasteur and other scientists of the time were well aware of the ability of certain microorganisms to kill other microorganisms. However, the importance of a new group of antibacterial agents, called **antibiotics,** was not really established until 1927, when Alexander Fleming isolated the active, bacteria-killing agent from a mold called **penicillium.**

Antibiotics are, in most cases, naturally-occurring chemicals produced by microorganisms, chemicals that kill or inhibit the growth of other microbes. However, scientists have now learned to augment the production of the drug by encouraging the growth of the microorganisms that produce it, and in some cases, they are able to produce antibiotics entirely artificially by chemical means.

Almost all natural antibiotics are produced by three groups of microorganisms: actinomycetes (the branching chains so common in soil), the rod-shaped bacilli, and molds. Some of the most potent antibiotics have been found in some rather peculiar places. As examples, bacitracin was isolated from bacteria found on the skinned knee of a little girl named Tracy; cephalosporin was derived from a mold taken from the sea near a sewage outlet, and streptomycin was isolated from bacteria found in the throat of a chicken. Given this, where is the next logical place to look for a new antibiotic?

Researchers plotting the molecular structure of an antibiotic. Such research has led to great advances in the development of more effective antibiotics.

THE KINGDOM PROTISTA AND ACCELERATING EVOLUTION

We now move to the eukaryotes, quite a different matter, as we shall see. The simplest of these (and the first to have evolved) form the kingdom **Protista.** The protistans are a complex and diverse group of organisms that are placed together simply because they are all single-celled eukaryotes. Their predecessors on the planet were prokaryotic autotrophs. These were rather successful creatures, probably able to rapidly multiply when conditions were good and to become dormant during hard times. About a billion years ago, eukaryotic forms appeared that gave up autotrophy and began to extract their nutrients from other organisms. These new kinds of organisms were bound by complex and efficient membranes, and they developed the ability to exchange genetic material in a kind of sexual union. They evolved rapidly in a few rather distinct directions and eventually gave rise to today's dazzling variety of plants, animals, fungi, and the small organisms called Protista.

The protistans are, today, quite a varied group. Some prowl the earth like voracious little animals, while others lie quietly and make their own

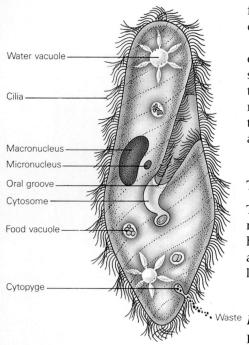

Water vacuole

Cilia

Macronucleus
Micronucleus

Oral groove

Cytosome

Food vacuole

Cytopyge

Waste

FIGURE 11.7   The paramecium is an active, spinning ciliate that inhabits pond water. Food is swept into the oral groove by cilia and is deposited into the cytostome. The cytostome enlarges and pinches off with the food inside and moves into the cytoplasm. After the food is digested, the waste is expelled through the membrane at a specific place called the cytopyge. Water balance is maintained by star-shaped water vacuoles that swell with excess water and then forcibly expel it through the membrane.

food in the manner of plants. Yet others suck up the molecules of the dead, as do today's fungi.

The protistans possess a membrane-bound nucleus and other typical eukaryotic structures. Most live alone as single-celled organisms, but some form rather specialized colonies. Because of this diversity within the group, it is assumed that many of the protistans are not very closely related. In fact, their closest common ancestor probably existed among the first life-forms. Nonetheless, because of important similarities, they are, for convenience, placed in the same kingdom.

The Animal-like Protists

The single-celled, animal-like organisms called **protozoans** ("first animals") exist all around us in a variety of habitats. Most of the protozoa, however, go unnoticed because they are microscopic, ranging from about 5 to 100 microns. However some are up to several millimeters long—large enough to threaten small insects.

Phylum Ciliophora

If you should find yourself with a microscope and unable to resist examining the water of a scummy pond, you would likely find great numbers of tiny protozoans. Some of these would be covered with tiny, hairlike cilia. These are members of the phylum **ciliophora.** The most familiar of these are the **Paramecia** (Figure 11.7). Paramecia (singular, Paramecium) are recognizable by their slipper shape, which is maintained by their outer thickened membrane, the **pellicle.** The pellicle, while holding its form, is flexible enough to enable the paramecium to bend around objects as it furiously swims through the water propelled by the wavelike actions of the **cilia** covering its body. Behind its rounded anterior (or front) end lies a deep **oral groove** into which food is swept by other cilia. The food is then forced through a mouthlike pore **(cytostome)** at the end of the groove and into a bulbous opening, which will break away and move into the cytoplasm as a **food vacuole.** Digestion takes place as enzymes enter the vacuole, the products passing out into the cytoplasm. Undigested food passes out where an opening appears at a fixed place in the pellicle. The opening is called the **cytopyge.** Excess water entering the cell is excreted by a star-shaped **contractile vacuole.** In addition, each organism contains two nuclei of unequal sizes.

Paramecia are relentless hunters, not only sweeping tiny organisms such as bacteria into their oral groove, but ensnaring larger prey by explosively discharging long, barbed shafts with trailing threads, called **trichocysts,** from beneath the pellicle. The trichocysts are also used in defense.

Paramecia normally reproduce by simply replicating their genetic material and then dividing. They may also **conjugate.** During conjugation, a kind of sexual reproduction, two paramecia join at their oral grooves and exchange genetic material. Conjugation begins when the larger of the two nuclei, called the **macronucleus,** disintegrates while the smaller **micronucleus** divides twice, halving their chromosome number and then exchanging some of the haploid products with the mating partner.

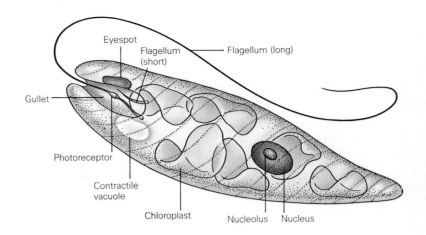

FIGURE 11.8 *Euglena* have presented an enduring puzzle to taxonomists. They behave like animals, thrashing their way through the water with a whiplike flagellum. On the other hand, they possess chlorophyll and are photosynthetic. Although they can manufacture carbohydrates, they must extract vitamins and minerals from the environment. Such flagellated mastigophora often seem to thrash about rather aimlessly, unlike the ciliates that give the impression of rushing busily on some dim errand. Flagellates are found in the bodies of animals, in the gut, in soil, in lakes and streams, and even in the sea. Some researchers believe that the earliest eukaryotes were flagellates.

Labels on figure: Eyespot · Flagellum (short) · Flagellum (long) · Gullet · Photoreceptor · Contractile vacuole · Chloroplast · Nucleolus · Nucleus

Phylum Mastigophora

If for some reason you were to continue to examine your scummy drop of water, you might find another kind of organism, one not covered by cilia but possessing a single whiplike **flagellum.** This organism is a member of the phylum **Mastigophora.** One of the best-known of the mastigophora is *Euglena* (Figure 11.8) with its long appendage (and a shorter one) protruding from the **gullet** at its rounded anterior end, leading to a saclike reservoir. The gullet in this group probably serves only as an area of attachment for the flagellum, because *Euglena* has never been seen eating. In fact, it doesn't have to eat because it contains chloroplasts and can manufacture its own food through photosynthesis. Near the gullet lies a pulsating contractile vacuole, continually expelling excess water. Nearby is another peculiar structure, a light-sensitive **photoreceptor** shielded on one side so that it can only be stimulated by light coming from a certain direction. In this way, the organism can determine the direction of that light and move toward it, enabling the chloroplasts to make food.

Interestingly, *Euglena* swims *toward* the moderate levels of light that penetrate its waters, but avoids the destructive direct rays of the sun. However, even with abundant light, *Euglena* cannot make all its food; it must absorb some minerals from the environment, much as a fungus does. In the absence of light, it may take all its food from the environment. In fact, if it is kept in the dark for long periods, the chloroplasts may disappear.

Under adverse circumstances, the *Euglena* can simply lose water, reduce its internal (metabolic) activity, and become rounded and dormant. This **cyst** may persist until the return of better times.

The *Euglena* reproduces by dividing after mitosis. It can also dehydrate, form cysts, and divide at this stage. It has never been seen engaging in any sexual activity whatsoever.

Phylum Sarcodina

That same pond water is very likely to contain a shapeless but rather famous member of the phylum **Sarcodina,** the **amoeba** (Figure 11.9). This little beast can creep about on the bottoms of still ponds by sending out **pseudopods,** extensions of the main body, and then flowing into these outstretched arms. It also uses the pseudopods to engulf food materials, which then become contained in food vacuoles. Some rela-

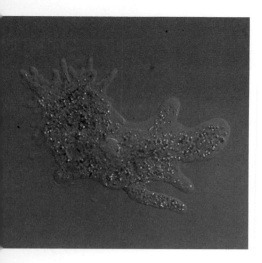

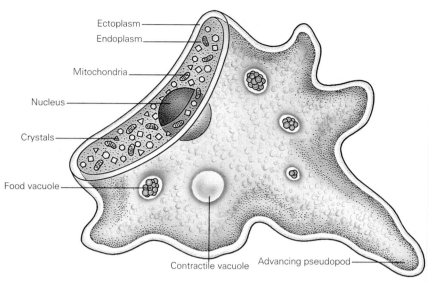

Ectoplasm
Endoplasm
Mitochondria
Nucleus
Crystals
Food vacuole
Contractile vacuole
Advancing pseudopod

FIGURE 11.9 The Sarcodina include the amoebas, small organisms that move and feed by extending a part of their body and then flowing into that extension. They are stimulated to surround any small object in which they detect salt or protein. In order to engulf food or to expel waste or excess water, their cell membranes must instantly rupture and rejoin through processes not entirely understood. Interestingly, they use the same contractile proteins that humans do, actin and myosin, in what is yet further evidence of the great kinship of life.

tives of the amoeba have not only pseudopods but flagellae as well, while others secrete a shell or cover themselves with sand grains.

Sarcodines employ rather simple means of regulating water and waste in their bodies. Amoebas, for example, take in oxygen and excrete metabolic wastes, such as ammonia, through diffusion. Excess water that flows in through the thin membrane by osmosis collects in one area and becomes encircled by a membrane, forming a **contractile vacuole.** Later, the vacuole draws near the cell membrane, then contracts and squirts out the water by temporarily rupturing the membrane.

The cytoplasm of the amoeba is divided into a clear, thin, outer layer, the **ectoplasm,** and a granular inner **endoplasm** that contains numerous crystals and mitochondria. In reproduction, mitotic division of the nucleus is followed by the cytoplasm simply pinching in two.

Interestingly, amoeboid movement relies on the same contractile proteins as do many other species, including vertebrates. Such findings are considered testimony for the common heritage of living things.

Phylum Sporozoa

The phylum **sporozoa** is a diverse group of tiny organisms. Some of them are so different from each other that they are probably more closely related to other phyla of protistans than they are to other sporozoans. Thus, it is difficult to place the group in any evolutionary line. They are placed together because they are all parasitic, living within animals. In fact, because of their reliance on other species, many of them have lost the ability to survive independently. One group has been critically important to the human species—the genus **Plasmodium,** which is the causative agent of malaria (Figure 11.10).

The Plantlike Protists

The other major group of the protists are the algae. They are the photosynthetic plantlike protists. They are extremely widespread, living in all aquatic habitats and a few terrestrial ones. We will briefly consider the two major phyla.

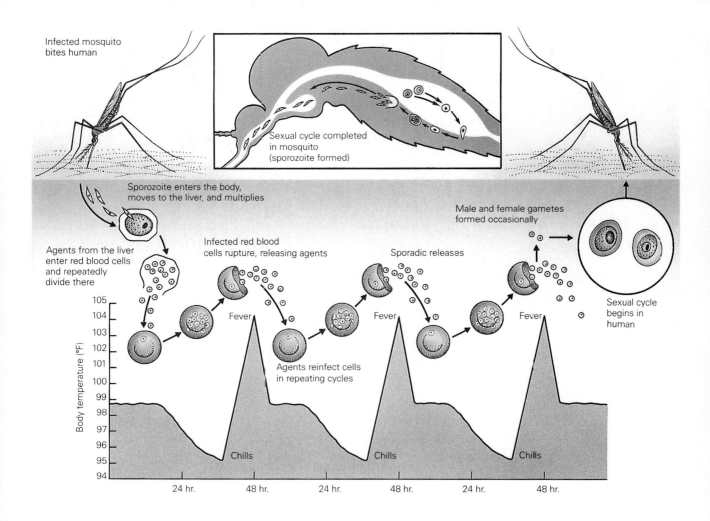

Infected mosquito bites human

Sexual cycle completed in mosquito (sporozoite formed)

Sporozoite enters the body, moves to the liver, and multiplies

Agents from the liver enter red blood cells and repeatedly divide there

Infected red blood cells rupture, releasing agents

Male and female gametes formed occasionally

Sporadic releases

Fever

Agents reinfect cells in repeating cycles

Fever

Fever

Sexual cycle begins in human

Chills

Chills

Chills

Body temperature (°F)

105
104
103
102
101
100
99
98
97
96
95
94

24 hr.   48 hr.   24 hr.   48 hr.   24 hr.   48 hr.

FIGURE 11.10   Malaria is caused by a protistan, called a *Plasmodium* (right), that develops a form called a sporozoite within the mosquito. When it is injected into the human body, it enters the liver and multiplies repeatedly there, releasing the rapidly increasing agents into the bloodstream, where they enter red blood cells. As they multiply within the cells, they are sporadically released into the blood, thereby infecting new cells. These releases mark times of high fever, followed by chills. Some of these released agents occasionally produce male and female forms. When a mosquito draws blood from an infected person, it may consume these male and female gametes, which join in the mosquito's body to form new sporozoites.

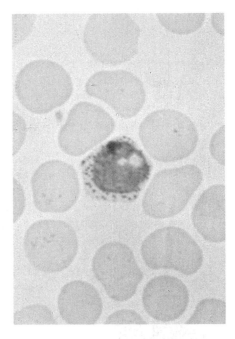

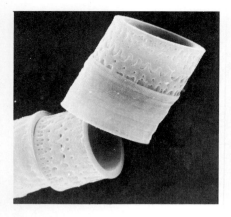
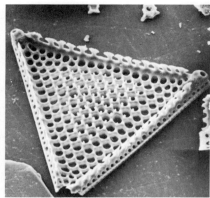

FIGURE 11.11 Diatoms are unicellular or colonial algae that live both in fresh and salt water. They possess two shells, or valves, that in some species fit together like the halves of a pillbox. The valves are covered with a variety of beads, striations, pores, and ribs that lend an exquisite beauty to these delicate organisms. They may reproduce sexually, but usually they simply undergo cell division. As you can see, they vary greatly in structure.

Phylum Chrysophyta

The **Chrysophyta** ("golden plants") comprise the yellow-green and golden brown algae and are named for their yellow carotenoid pigments, although they also possess chlorophyll *a* and *b*. This group includes the beautiful and important diatoms (Figure 11.11). Their cell walls are comprised largely of silicon, and their glassy corpses have slowly rained down upon lake and ocean bottoms for millions of years, producing very thick layers of **diatomaceous earth.**

Diatoms are very common, and they can store fat, so they comprise a very basic and critical link in aquatic food chains. It is the fat of diatoms, by the way, that imparts that beloved fishy taste to fish. This is why some predatory fish that are not part of the diatom chain, such as trout, do not have that taste.

Phylum Pyrrophyta

The **Pyrrophyta** (fire-colored plants) are microscopic, photosynthetic algae with two tinsellike flagella. They, too, are an important basic link in the sea's food chain (Figure 11.12). One group has enchanted sailors for as long as they have gone down to the sea. These are the **dinoflagellates.** When they are exposed to air, as in the wake of a ship or a churning oar, dinoflagellates will emit a bright flash of light. One of the most fascinating sights I've seen was the trail of sparkling lights trailing my skiff in the waters of the Indian Ocean as I rowed back to the ship anchored offshore one June night.

Another dinoflagellate is not so charming. *Gonyaulax* reproduces wildly at times, turning the sea a deep red and producing the dreaded "red tide." Its sheer numbers may seriously deplete the water of its oxygen and thereby suffocate other species. *Gonyaulax* also produces a powerful nerve poison that can kill vertebrates. These poisons accumulate in the bodies of the animals that eat the dinoflagellates and are ultimately consumed by vertebrates somewhere along the food chain. Thus, not only are fish killed by the thousands, but the lives of humans who eat tainted clams, for example, are threatened as well.

As we leave the world of the tiny monerans and protistans, we should remind ourselves of two things. One, their development placed other living things on their successful evolutionary trails, and, two, the monerans and protistans are still with us, a testament to the resilience of their seemingly frail bodies. The protistans set the stage for many of the groups

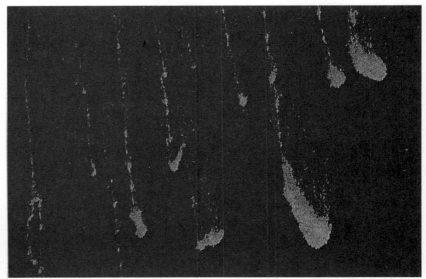

FIGURE 11.12 The red tides are produced by dinoflagellates called Pyrrophyta (below). They have a membrane-enclosed nucleus, but their chromosomes are reminiscent of those of the prokaryotes, as is their means of replicating DNA. That is, the two DNA molecules that form just before the cell divides each attach to a different part of the nuclear membrane and each then goes with that part when the nucleus pinches in two. These populations can multiply explosively at times. During such "blooms," the sea itself may become red and filled with the poison of the dinoflagellates (right). One species found mainly along the coast of the Pacific Northwest produces a poison so powerful that one millionth of a gram can kill a mouse.

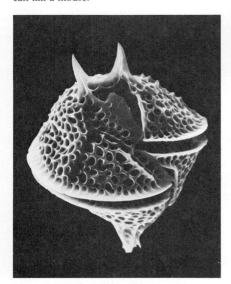

that follow by developing a nucleated cell with a complex and efficient genetic apparatus and by developing systems of handling energy by (1) creating energy-laden molecules (for example, through photosynthesis) and (2) developing (or acquiring) inclusion bodies (like mitochondria) that can utilize that energy efficiently. (Do you see the similarities between bacteria and mitochondria? Why is that similarity so important to theorists? See Chapter 6.)

THE KINGDOM FUNGI AND VARIED LIFESTYLES

A peaceful walk in a damp forest may bring you upon a most enchanting sight: a ring of delicate mushrooms, their little caps pushing up through the woodland floor (Figure 11.13). The beautiful little caps may be of various colors and different shapes, and some have bold markings. Mushrooms can be beautiful indeed. We must keep in mind that a mushroom is a **fungus.** A fungus? Athlete's foot is a fungus. So what is it we put on steaks?

We do indeed decorate our steaks with fungi, but, one would hope, a different variety than we find on our feet. (We rarely find truffles there.) Fortunately, the fungi are also a variable group, and some have certain remarkable characteristics. For example, the cell walls of some species are not made of cellulose but of chitin, the protective covering of insects. Also, most of them, including mushrooms, live saprophytically, drawing nutrients from the bodies of dead organisms. They do this by secreting enzymes into the organism and absorbing the products of digestion. A few, including the culprit on our feet, are parasites of living things.

Fungi were once placed with the true plants and are still largely considered in the domain of botanists, but they are quite different from plants in a number of ways. For example, they grow by producing long thin filaments, or **hyphae,** that join together to produce a spongy, cottony mass, the **mycelium.** The hyphae may grow outward through the soil as long filaments. Under the proper conditions, they may form small buttons that give rise to the familiar basidiocarp. (You should be aware that in this group, as in the plants, phyla are called *divisions.*)

FIGURE 11.13 Mushrooms and toadstools are charming, if occasionally deadly, organisms. Most are saprophytic; that is, they secrete enzymes into their environment that break down any dead material they encounter. The products are then absorbed into the organism. A few mushrooms and toadstools are parasitic and break down the tissue of living things. Sometimes mushrooms seem to spring forth in your yard almost overnight. This usually happens during a wet period and is only the final stage of a long developmental period. The part you see is the basidiocarp, and it springs from an immature "button." These buttons lie along the spreading mycelium beneath the surface of the soil. The ring of mushrooms one sees may simply mark the margin of that mycelium. A mycelium is a mass of hyphae, and hyphae are the thread-like structures that compose the body of a fungus.

Divisions Oomycota and Zygomycota

Both the divisions zygomycota and oomycota are quite diverse. The zygomycetes all have cell walls composed primarily of cellulose. As is characteristic of fungi, their bodies are composed of masses of tubelike hyphae (singular, hypha). Together, the hyphae comprise the mycelium. The filamentous hyphae of the zygomycetes are peculiar in that they are not composed of individual cells separated by cell walls. Instead, each hypha is a continuous strand with no cross walls. Thus, the cytoplasm can freely move, carrying multiple nuclei and other suspended organelles. The oomycetes (Figure 11.14) have been important both economically and politically.

So-called bread molds are found in more than one division, but the black bread mold *Rhizopus nigricans* is a zygomycete. The cottony growth that appears on bread is a loose mass of hyphae that arises from the germination of asexual spores. Stalks arise from this mass and at their ends they produce tiny balls called **sporangia**. These contain the asexual spores. The sporangia are supported by short, rootlike hyphae called **rhizoids** (Figure 11.15). The sexual phase of reproduction involves the tips of two different kinds of hyphae (called *plus* and *minus*) joining and forming cross walls not far behind where they touch. The cross walls isolate two groups of nuclei contained in what are now called the **gametangia**. The walls where the hyphae touch will break down, enabling the isolated nuclei to fuse, forming a **zygospore**. The zygospore then enters a period of dormancy when all but one of the diploid nuclei disappear. The surviving nucleus undergoes meiosis, producing four

FIGURE 11.14 In the oomycete that causes potato blight, an airborne spore lands on the leaf of a potato plant and germinates. As it grows, long tubelike hyphae penetrate the mesophyll cells of the host's leaves. From these, sporangia arise, each sporangium producing many spores. The potato plant dies, but not before the fungus completes this asexual, spore-forming cycle. The sporangia will rupture, and the airborne spores will drift away, each to give rise to a new organism. The Irish potato famine of 1843–1847 was due to this organism and resulted in a great wave of immigration of Irish citizens to America.

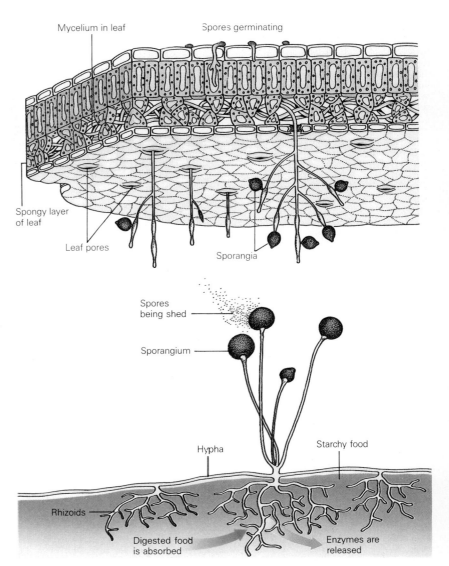

Mycelium in leaf
Spores germinating
Spongy layer of leaf
Leaf pores
Sporangia

Spores being shed
Sporangium
Hypha
Starchy food
Rhizoids
Digested food is absorbed
Enzymes are released

FIGURE 11.15 Bread mold, *Rhizopus nigricans*, will form on almost any dampened bread. The tubular hyphae form first and grow through the bread as rhizoids. Some will give rise to black knobs called sporangia, each of which may contain over 50,000 spores. They will later be released from the ruptured sporangia and be carried on currents of air.

haploid gametes. Usually, three of these disappear as the survivor begins a series of divisions that will produce the next sporangium (Figure 11.16).

Division Ascomycota

The ascomycetes are a very diverse group. Many species have a unique saclike structure called an **ascus** (Figure 11.17). The ascus is a reproductive structure that produces spores and, in many species, is the enlarged end of the hypha that lines the inside of a fleshy cup. When the spores are shed and drift away, each will form a new organism. The best-known of the ascomycetes are the yeasts, notably baker's or brewer's yeast. Other yeasts grow naturally on fruits such as grapes and aid in the winemaking process. It might be said, then, that the yeasts are the most domesticated of all the fungi, although not highly regarded as pets.

FIGURE 11.16   Sexual reproduction occurs in bread molds by a form of conjugation. The hyphae are of two types, plus and minus. If the two contact each other, bridges will join them near their tips. The nuclei at the ends of the converging bridges will form the gametes. When the bridges join, a zygote is formed. When it germinates, new hyphae are formed.

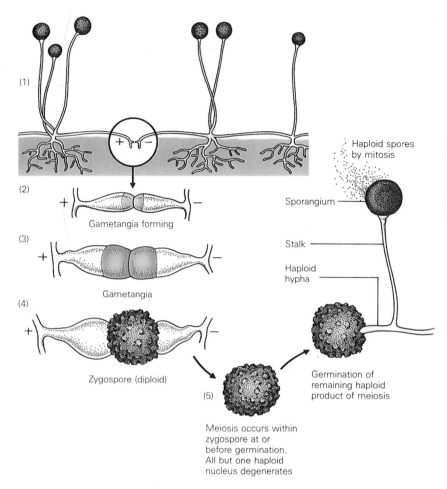

(1)

(2)
Gametangia forming

(3)
Gametangia

(4)
Zygospore (diploid)

(5)
Meiosis occurs within zygospore at or before germination. All but one haploid nucleus degenerates

Haploid spores by mitosis

Sporangium

Stalk

Haploid hypha

Germination of remaining haploid product of meiosis

FIGURE 11.17   (A) Yeasts are among the simplest of the fungi. Each cell is oval or spherical with either a cell membrane or a very thin cell wall. Yeasts can produce rapidly by budding, which begins as a protrusion from the mother cell. As the bud enlarges, the nucleus divides and one daughter nucleus then moves into the emerging bud. The two may remain attached and divide again and again, forming long chains.

(B) The ascomycetes also include the sac fungi, morels, truffles, the agents of Dutch elm disease, and a host of other plant ailments. In many species, a fleshy cup is lined with specialized hyphae (the ascus) that contain spores called ascospores. These spores are the result of the union of two different types of nuclei that then undergo meiotic division through a complex series of steps.

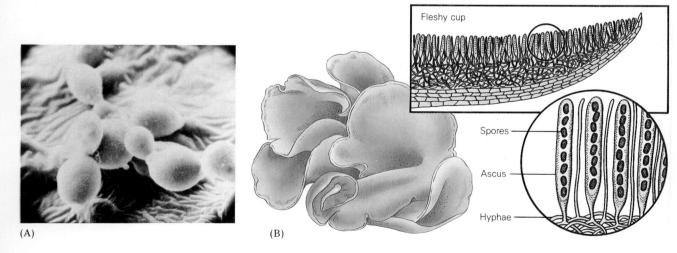

Fleshy cup

Spores

Ascus

Hyphae

(A)

(B)

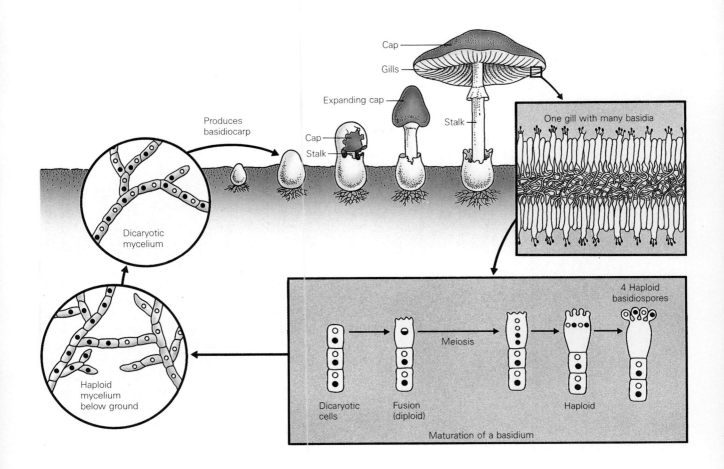

Cap

Gills

Expanding cap

Cap

Stalk

Stalk

Produces
basidiocarp

One gill with many basidia

Dicaryotic
mycelium

Haploid
mycelium
below ground

4 Haploid
basidiospores

Meiosis

Dicaryotic
cells

Fusion
(diploid)

Haploid

Maturation of a basidium

FIGURE 11.18 Life cycle of the common edible mushroom, a true fungus. Spores are dispersed by wind. If they land in a suitable place, they will germinate into primary mycelia, which are comprised of tubelike hyphae. If two hyphae of appropriate types come together, they will join by extending their walls toward each other. These extensions contain nuclei, which join at fertilization. A secondary mycelium now spreads through the soil, and some may eventually form the visible mushrooms. In sexual fusion, the diploid nucleus undergoes meiosis to form four haploid nuclei. Each nucleus is expelled (with some cytoplasm) through an extension at the tip of the basidium, a clublike structure (for which the club fungi are named). A tough wall forms around it and it becomes a spore.

Another group of the ascomycetes includes several common molds, particularly *Aspergillis, Penicillium,* and *Neurospora*—also called the brown, green, and pink molds, respectively. They all occur on bread, and *Penicillium,* of course, is the mold from which penicillin was derived. A related mold is responsible for the flavor of Camembert and Roquefort cheeses (people allergic to penicillin can react violently to these cheeses). Some of the ascomycetes are not exactly compatible with human endeavor. For example, the powdery mildews infect not only lilacs and roses, but the cereal grains as well.

The ascomycetes have both sexual and asexual phases. The most common type of asexual spore is called a **conidium** and is formed by budding at the tip of a specialized hypha called a **conidiophore.** The same mycelium that produces the conidiophore may later give rise to the spores of the ascus. These spores, however, are not formed from a simple budding but by the fusion of nuclei from two different hyphae and their subsequent mitotic and meiotic divisions.

Division Basidiomycota

The basidiomycetes, the group that includes the mushrooms (Figure 11.18), produce the most spectacular fruiting bodies of all the fungi. The puffballs may be over half a meter wide (Figure 11.19), and we also find the giant bracket fungi that grows on damp logs (Figure 11.20). The basidiomycetes are not only spectacular, but some are very dangerous—

FIGURE 11.19 Puffballs are among the club fungi, perhaps the most advanced of the fungi. They are filled with spores that on a dry day may fill the air around the puffball (especially if the puffball is prodded by a foot). They are in the same class as mushrooms and toadstools, with such elegant species as smuts, rusts, and stinkhorns.

FIGURE 11.20 The club fungi comprise the class Basidiomycetes. The basidium is covered with tiny projections that produce haploid spores. Most basidiomycetes are saprophytic, but a few, such as the smuts and rusts, are important parasites of plants.

the toadstools, for instance. The delicately hued *Amanita verna* is called the destroying angel—and with good cause since it annually kills not only many scavenging dogs (daschunds are particularly susceptible), but many inexpert human mushroom hunters as well. Wheat rust and smut fungi that attack cereal grains are also among the basidiomycetes and are so devastating to crops that they have been known to influence international relations as certain food supplies fall scarce. Not long ago, a Soviet leader threatened to "bury" us, but his country had first better conquer wheat rust, since the Soviets are among the greatest importers of our grain.

The walls of these fungi are chitinous, and the hyphae are completely compartmentalized, forming separate cells. As in the ascomycetes, however, the cell walls are perforated, permitting cytoplasm to move from one cell to the next. The spores produced by the union of different types of nuclei, which then undergo meiosis, are called **basidiospores.** (They correspond to the ascospores of the ascomycetes.) The basidiospores, however, are not enclosed, but are found on the outside of a specialized hypha called the basidium. The basidia are located on the **gills,** the papery divisions on the underside of the mushroom cap, thus you have to be pretty short to see them. When the basidiospores are released and land on a suitable surface, they begin dividing and form new mycelia.

There are a number of other fungi, but this should give you some idea of the diversity within the group. The species that do not easily fall into classification are lumped together into the orphanage called Fungi Imperfecti, but the imperfection lies with our knowledge and not with the organisms.

The fungi have played an important role in the colonization of the earth by plants and animals. By processes of decomposition, they have helped recycle the nutrients that are so critical to the species with which the fungi share the earth.

SUMMARY

Many people use the five-kingdom approach to classifying life—the monera, protists, fungi, plants, and animals. The prokaryotes may be divided into archaebacteria (the ancient form) and eubacteria. These,

Let's consider the lichens separately because stretching our system of classification in this case is asking a bit much, even for the lumpers. One reason is that lichens are comprised of two different kinds of organisms, a fungus (usually an ascomycete or a member of the Fungi Imperfecti) and an "alga" (usually a moneran or one of the true green algae described later). The relationship is believed to be mutually beneficial in that the fungus, with its probing filaments, can cling tightly to rocks, and the algae, through photosynthesis, manufactures food molecules, some of which are used by the fungus.

Lichens grow extremely slowly, and although they may be an early and critical phase of the soil-building process as they attach to barren rocks, they have attracted relatively little research attention. In fact, only recently have scientists been able to grow the alga and fungus separately and then join them to form a lichen.

Lichens are found clinging to all sorts of exposed surfaces including bare rock, soil, and tree bark. Vast areas of the arctic and subarctic regions are covered by the famed "reindeer moss" of the lichen Cladonia.

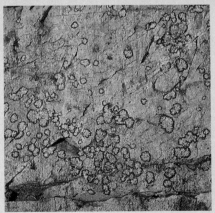

Washington lichen (above) on pine bark, British soldier lichen growing in moss (above right), Foliose lichen (center), and Washington lichen on rock (bottom).

with the cyanobacteria, comprise the monerans. The protistans are eukaryotic and may be animallike or plantlike. The animallike protists include the phyla Ciliophora, Mastigophora, Sarcodina, and Sporozoa. The plantlike phyla include the algae Chrysophyta and Pyrrophyta. The kingdom fungi are saprobes with divisions (phyla) including Zygomycota, Ascomycota, and Basidiomycota.

Chapter 12

DIVERSITY II: THE PLANTS

We humans seem to innately understand that we are firmly bound to the world of plants. We are attracted to them, and we are somehow likely to feel better when we are in their midst. We take long walks among them when we can get out, and we bring them to us when we can't. I even recall a conflicted New Yorker who covered his lawn with asphalt and then painted it green. Apartment dwellers in concrete cities often carefully tend their beloved window plants in a rather poignant homage to our bond with that silent world.

Interestingly, the kind of plant that attracts us may depend on where we were raised. I grew up in the south, and I love the dark forests. I even live in one. Californians may feel claustrophobic in such places and crave the sight of distant vistas and low-lying plants. They may be solaced by a Joshua tree silhouetted against a desert sky. Others living in the north may roam hardwood forests and mark the seasons by the color of the leaves. No matter what our preferences, though, we find plants compelling and comforting, a feeling that may stem from our long association and profound interdependence with them.

But what exactly are plants? If you asked an old man watering an African violet what a plant is, he is not likely to say, "According to the five-kingdom scheme, it is a eukaryotic photosynthetic organism." If he did, his answer would acknowledge the existence not only of the more common green plants, but also those few species that are not photosynthetic. His answer would exclude the blue-greens, which, after all, are little more than specialized bacteria. He would rule out fungi because they are not photosynthetic. He might also mutter something about the problems of classification and then say that a more precise definition is: *Plants are usually photosynthetic, have cell walls, store carbohydrates in the form of starch, are usually multicellular, and possess a peculiar life cycle involving alternating generations.*

ALTERNATION OF GENERATIONS

There are a great many species of algae and plants with both sexual and asexual phases. Often the phases alternate in what is called an **alternation of generations** in what is essentially a continuing alternating sequence of haploid and diploid forms (Figure 12.1). For example, in some plants, spores grow into organisms that do not release more spores, but rather gametes ("eggs" and "sperm"). When a male and female gamete unite, the fertilized egg will grow into an adult individual that looks entirely different from the one that had grown from the spores. When it matures, this organism will then produce spores. Gamete-producing generations are called **gametophytic,** and spore-producing generations are labelled **sporophytic.**

The spores are produced by meiosis; hence, they have only half the complement of chromosomes characteristic of the other phase and grow into individuals of generation A, which then release gametes. Of course, both the gametes have only half the "usual" number of chromosomes. When these join, forming generation B, the sporophyte is produced, restoring the full chromosomal complement. Then generation B releases spores, initiating generation A, the gametophyte, again. Nearly all plants, as well as some animals, go through an alternation of generations. However, one generation may be much more prominent than the other, depending on the species.

FIGURE 12.1   Some plants show a pronounced alternation of generations. The sporophyte generation begins when the eggs and sperm are joined in fertilization. The mature sporophyte then produces spores (male or female) whose subsequent growth comprises the gametophyte generation. The gametophytes then produce gametes ("eggs" or "sperm"), which join to form a new sporophyte.

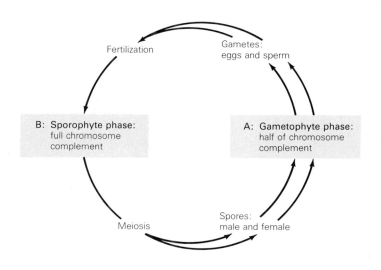

FIGURE 12.2 Red algae appeared about 600 million years ago. There are now about 4,000 species. They may exist as single cells or filaments, or they may form leaflike or stemlike structures. Red algae range in color from bright red to deep purple. There are a few fresh-water forms, but most are marine. Basically, they are comprised of branched filaments that may be embedded in a gelatinous matrix. Since many forms deposit calcium salts in their cell walls, we have a fairly good fossil record of their evolution.

The Development of Alternating Generations

Alternating generations may have evolved in response to regularly recurring stressful periods in the organism's life history. It has been suggested that the mixing of genes through sexual reproduction might be highly advantageous in such dire periods. The development of the asexual phase, it is thought, might be expected under more optimal, stable, or sheltered conditions.

The relationship between stress and reproduction is an interesting one. For example, in some species, sexual reproduction increases in periods of stress. In earlier days, steel girders were placed around surviving trees in burned-over areas. The idea was that as the tree grew and came to exert pressure against the girder, the stress would cause the tree to produce more seeds. Interestingly, many species of animals, after being subjected to some sort of harsh conditions that reduces their numbers, tend to undergo a rapid surge of reproductive activity (which, one could argue, might be a response to either stress or low numbers). Stress has even been suggested as the stimulus of those baby booms that usually follow great wars (but the effect of so many randy returning servicemen cannot be discounted). In any case, it has been suggested that the development of the sexual phase in the alternation of generations may have been due to periodic stressful conditions and that the plants developed the ability to switch to the metabolically less expensive asexual phase as the harsh conditions were regularly followed by more benign, less demanding periods.

As we begin our survey of plants, the first few divisions (phyla) we will look at are loosely called **algae.** The term once referred to aquatic plants, but has largely lost any precise scientific meaning. We will find that the red and brown algae are quite different from the rest of the plants. We will also see that although the green algae are similar to the protista in many respects, biochemically they are similar to other green plants.

DIVISION RHODOPHYTA

Rhodophyta ("red plants") are the red algae (Figure 12.2). Some are red because in addition to chlorophyll *a* and *b*, they contain a red pigment, **phycoerythrin.** Others, however, may range in color from green to red, purple or greenish-black. They are similar in many respects to the prokaryotic cyanobacteria, and some suggest that they may have evolved from them. In any case, their lineage is ancient; they closely resemble forms that lived about 600 million years ago in the warm, shallow waters of the planet.

Most red algae live in deep tropical seas. Not coincidentally, red pigment tends to trap shorter wavelengths of light (those toward the blue end of the spectrum), and it is these shorter wavelengths that penetrate deepest into the ocean's depths. The red pigments capture the energy of light and transfer it to the chlorophyll where the photosynthetic process begins. They can be single-celled and filamentous, or they may be large with what is apparently a more usual plantlike body. Even the larger species though, are composed of mats of filaments, much in the manner

FIGURE 12.3 *Fucus*, or rockweed, is a brown alga found in cold, temperate waters. It is commonly seen by walking along rocky areas when the tide is out. The body plan is essentially a repeatedly branching one, and the flattened forked thalli are held aloft by air bladders when the tide is in. *Fucus* can reproduce by fragmentation (when a part breaks away and begins to form a new plant elsewhere) or by sexual reproduction. The eggs and sperm are produced on bulbous branches of the thallus. The life cycle of *Fucus* is similar to that of the vascular plants in that there is a dominant diploid sporophyte and a reduced gametophyte generation.

of fungi. The red algae have incredibly complex and diverse reproductive lives that, in spite of your pleading for more information, are a bit beyond the scope of our discussion here.

DIVISION PHAEOPHYTA AND UNDERSEA FORESTS

The **phaeophytes** ("dusky plants") are the brown algae. If you have spent much time prowling beaches, you have undoubtedly encountered these plants (Figure 12.3). They are the shiny brown plants with stem-like and leaflike structures. However, their "stems" and "leaves" are really not stems and leaves at all. The stems lack conducting tissues, and the leaves have no veins. They possess chlorophyll *a* and *c* and a brown pigment called **fucoxanthin.** Unlike flowering plants, their cells contain centrioles, as do animal cells.

The brown algae show alternation of generations. In the simpler, smaller species, the gametophyte and sporophyte may be very similar. In the larger species, the sporophyte generation is dominant. The brown algae are among the most conspicuous of marine plants. After all, the common rockweed, *Fucus,* is in this group, as are the giant kelps along the Pacific coast. (These are among the world's largest plants, reaching 100 meters in length.) The blades or **thalli** (singular, **thallus**) are anchored to rocks by **stalks** that are attached to **holdfasts.** The thalli are lifted toward the energizing light at the surface by air-filled bladders. The great "forests" often tempt divers who risk becoming entangled in the dense growth. The famous Sargasso Sea is a thick matlike raft, comprised of Sargassum, a brown alga. This remarkable body covers over two million square miles of the Atlantic Ocean and is comprised largely of plants that have grown in shallower waters, broken off, and floated to sea.

DIVISION CHLOROPHYTA: ANCESTORS OF LAND PLANTS

The **chlorophytes** are the green algae (Figure 12.4). They are mostly fresh-water organisms, but a few species live in the sea. They are believed to be directly in the evolutionary line that gave rise to the land

FIGURE 12.4 The green algae are tiny plants, some of which float aimlessly in the seas, while others live in fresh water ponds and cover them with scum. Green algae are important parts of some food chains, and early in their history they may have given rise to land plants. They are very similar to the higher plants in their structure and biochemistry. Their green color is due to chlorophyll *a* and *b*, which is contained in chloroplasts. The plant itself ranges from unicellular flagellates, to filaments and sheets, to far more complex structures. *Acetabularia* (left) is very unusual, seemingly off on its own evolutionary pathway. It is comprised of a single wineglass-shaped cell. *Codium* (right) lacks cell walls that would separate its many nuclei. Its branching structure gave rise to its romantic common name, "dead man's fingers."

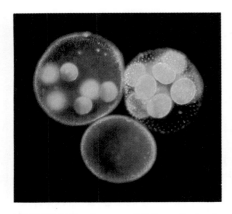

FIGURE 12.5 A group of *Volvox* colonies. In a *Volvox* colony, a few cells begin to become distinguished from the others early in their development. They will grow noticeably larger than the rest and will become reproductive cells. During asexual reproduction, each of these will divide and begin to form a daughter colony, even while it is inside the parent colony. When the divisions are complete, the new colony will depart through the wall of the parent colony. The parent colony then dies. Under certain conditions, the reproductive cells can also form eggs and sperm instead of new colonies.

plants. They share three important traits with mosses, ferns, and seed plants: (1) they store carbohydrates, (2) the cell walls are comprised mainly of cellulose, and (3) they contain chlorophyll *a* and *b*. In spite of such common traits, green algae are extremely diverse. Many exist as single-cells, while others are filamentous or even bladelike, and some form aggregates or specialized colonies (Figure 12.5). Their fragile beauty has captured the imagination of botanists for centuries. It is almost paradoxical what enchanting sights one can see by taking a closer look at the green scum from a still pond. The green algae, by the way, may be important for more than aesthetic reasons since a great deal of research attention has been focused on them as a potential source of food from the sea. Let's now leave the algae and move on to the next most complex and organized group of plants, the mosses. We will describe them here and reserve the discussion of their life cycle for Chapter 14.

DIVISION BRYOPHYTA AND INVADING THE LAND

The **bryophytes** include the mosses and their close relatives. They are an ancient lineage and are among the oldest forms of land plant. They are widely diverse and grow in a variety of places. You probably can conjure up an image of a moss, but what about a hornwort? A liverwort? (See Figure 12.6.) Mosses, as you know from popular folklore, always grow on the north sides of trees, except when they don't. It is true that in the northern hemisphere, they are more *likely* to be found on the north side of trees, but that's because the midday sun shines from the south. Mosses grow best where it's shady, cool, and wet. (One reason they need the moisture is that the motile sperm of mosses must swim to the egg.) Mosses also grow on rocks and bare soil. Some species even grow in the cracks of concrete sidewalks in cities, almost totally ignored by city dwellers.

In a sense, mosses are midway between green algae and the "higher" vascular plants that we will consider next. Their "roots" are really tiny cellular threads (**rhizoids**) that serve only as anchors; they do not absorb water. The "leaves" are flattened outgrowths, just a few cells thick.

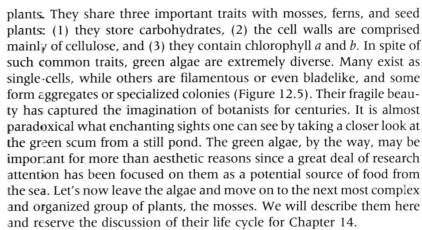

FIGURE 12.6 Hornworts (above) and liverworts (right) are bryophytes, ancient but inconspicuous species. Humans may live haughtily among such ignominious species without ever recognizing them, but when we die, they may grow ignominiously on our graves. They have never fully adapted to the terrestrial world. Their aquatic heritage is seen in their flagellated sperm, which must swim to the egg cells, and in their lack of vascular (transporting) tissues. Since they lack supportive xylem, none grow very tall.

FIGURE 12.7 Mosses are rather common bryophytes. They lack vascular systems and even roots. They survive so well partly because of their small size. (Can you see why?) The moss gametophyte consists of a central upright strand of cells with leaflike growths where most photosynthesis occurs. These outgrowths are not true leaves, nor is their "stem" a true stem since both lack conducting tissue. The plant is attached to the soil by threadlike rhizoids. Rhizoids also lack conducting tissue. The brown stalks are the sporophytes bearing swollen sporangia at their tips.

These curl up when the environment becomes dry, thereby reducing water loss. Mosses lack the conducting vessels of the vascular plants, and they often grow crowded together, supporting each other and trapping water between their tiny bodies (Figure 12.7). It is not believed that mosses gave rise to vascular plants because we have found fossil vascular plants that are older than any fossil moss.

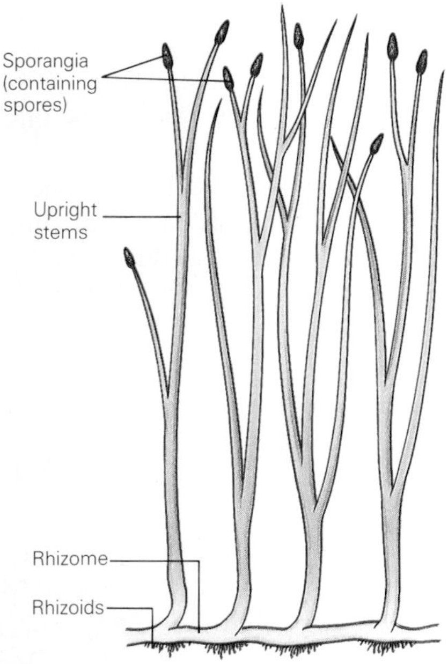

DIVISION TRACHEOPHYTA, ADAPTING TO LAND

Now we reach the **tracheophytes**, the vascular plants. Vascular implies the presence of vessels that transport food, minerals, and water throughout the plant. When we think of green plants, we usually have in mind the tracheophytes. The tracheophytes appeared some 400 million years ago, and within only 50 million years or so, they had, through divergence, formed the major evolutionary lines of land plants. Many have become extinct, but others have persisted to grace our landscape today.

The tracheophytes include the ferns, the gymnosperms (including the conifers—cone-bearers such as pine, spruce, and fir), and the flowering plants. Tracheophytes, then, usually are large land plants, and thus they have had to solve the problems of large size and terrestrial life. This means they had to be able to support themselves as they reached up toward the sunlight. They needed a way to stand anchored in the soil. They had to develop ways to withstand the deadly drying air and to conserve their water. They needed a system to conduct water upward from the ground and food downward from the leaves. They needed a way to exchange vital gases with the atmosphere, and they needed a way to achieve fertilization in the dry air. Thus, we find plants with strong fibers, conducting vessels, roots or rootlike structures, openings in the leaves or stems, waterproof and airproof coverings, and tough, resistant pollen.

Among the earliest evidence of tracheophytes are fossils of **Rhynia** plants found in fossil beds in Scotland (Figure 12.8). These ancient

Sporangia (containing spores)

Upright stems

Rhizome

Rhizoids

FIGURE 12.8 Ancient 400-million-year-old fossils have enabled us to reconstruct an early organism called *Rhynia*. We have learned that it had both underground and erect stems and that the stem tips sometimes bore sporangia. We also know that it had neither roots nor leaves. Photosynthesis was carried on in its stems, and water and minerals entered through hairlike rhizoids on the underground stem. They were gone long before our first shuffling ancestors appeared.

FIGURE 12.9 The horsetails (left) and club moss (right)—not real mosses—are ancient species with true vascular systems. Their ancestors were towering figures on the earth's landscape, but today they are represented by a few species that are likely to be found in areas abandoned by humans, such as vacant lots. These are both in the sporophyte generation.

Devonian plants (see Table 11.1) were preserved remarkably well, and we can clearly see their waxy covering (**cuticle**), openings (**stomata**) in the stem through which gases could pass, rootlike structures called **rhizomes** (that were really horizontal stems), and tiny, hairlike **rhizoids** (similar to root hairs of modern plants) through which water could pass into the plant. They had no leaves and no roots. Their most advanced feature seemed to be a tendency to form a two-pronged fork, but they did have a vascular tissue. Probably none of these gave rise to more advanced plants. There are a number of species of other primitive vascular plants still living (Figure 12.9).

Class Filicinae

The **Filicinae** are the ferns and fernlike plants of the division Tracheophyta and the subphylum **Pteropsida** (Figure 12.10). In this class we see true leaves for the first time. (**Leaves**, we can now say, are *flattened photosynthetic structures that grow from a stem and have true vascular tissue.*) Fern leaves meet these qualifications, although they are traditionally called **fronds**.

The most familiar ferns are small plants that live in the moist soils of woods and swamps. Their delicate beauty enhances a walk in the dark woods. especially if we find them while they are still pale "fiddleheads," uncurling shyly from the damp earth. Other ferns are not so small and may grow to 30 meters as they wind their way up tall trees. Still others may stand alone as **tree ferns**, rising to 20 meters in height (Figure 12.11). Ferns lack the strength of woody plants, however, so they usually grow in protected ravines or under the forest's canopy of leaves, a tendency that only heightens their aura of subtle drama.

Class Gymnospermae

The **gymnosperms** are the plants with "naked" seeds—**conifers** such as pine, spruce, fir, cedar, and their relatives. They also include cycads,

FIGURE 12.10 Ferns and fernlike plants are among the oldest vascular plants. Fossil ferns, found in deposits from the Middle Devonian, grow from the Arctic to the tropics.

FIGURE 12.11 Tree ferns are impressive plants that seem to be a vestige of some bygone era of the earth. Ferns have true stems, roots, and leaves. The fronds or "leaves" are usually divided into a number of smaller parts. These represent the sporophyte stage. Spores are produced in clusters of sporangia under the fronds. Such spores travel easily and are often among the first colonizers of new areas, such as oceanic islands formed by volcanic activity.

gingkos, and the peculiar gnetophytes. The seeds are considered naked because they are not surrounded by a fleshy **ovary**. (The difference between gymnosperms and the "flowering plants" is described in more detail in Chapter 14.) Instead, the seeds are protected (and well protected at that) by **cones**. Some pines, such as the lodgepole pine, have such tightly closed cones that only fire can open them and release the seeds.

The gymnosperms are among the most beautiful and impressive of all plants. After all, the redwoods are gymnosperms (Figure 12.12). The tallest living thing in the world is a redwood—over 100 meters tall—that was saved from lumbering interests. (A taller tree was discovered by lumberers and quickly cut down and sawed into boards before environmentalists could organize to prevent it.) The tree with the greatest bulk is another gymnosperm, a giant sequoia that reaches 80 meters in height and has a girth of 20 meters at its base. It is about 4,000 years old, but it is a youngster compared to a bristlecone pine, almost 5,000 years old, that was found in the mountains of eastern Nevada. Other gymnosperms include the hemlocks of Canada, the cypress of the southern swamps, the wiry ephedra of the deserts (Figure 12.13).

The gymnosperms first appeared during the Paleozoic era, but they didn't reach their heyday of diversity until the Mesozoic, a time when

FIGURE 12.12 The redwoods are among the most beautiful and overwhelming treasures of the plant kingdom. (However, many people see them in terms of board feet.) The trees are fire- and insect-resistant and once covered vast areas of the West Coast. They have now been decimated by public demand for resistant tomato stakes and lawn furniture that does not require paint. It would be nice to say that we have at last come to our senses, but the great trees continue to fall daily, except perhaps on Sundays.

FIGURE 12.13 The cypress family is distributed worldwide. Most are trees (right), but a few are shrublike. In all species, the leaves are small and scalelike like those of the primitive *Ephedra* (below) from whose ancestors they may be descended. Some species of cypress are called cedars, but true cedars grow in the Himalayas, Asia Minor, and, of course, Lebanon.

FIGURE 12.14   At first glance, you might mistake a cycad for a small palm tree, but they are not closely related at all. Cycads are unmasked as gymnosperms in that they bear cones. Their leaves are covered by a waxy cutin that retards water loss, and their spines probably help protect them from foraging animals. They are very slow to reproduce, and only a few species remain, mostly limited to the warmer temperate zones and to semiarid regions of the tropics.

FIGURE 12.15   The gingko, *Gingko biloba*, is represented by only one species. It was common in the Mesozoic and many species thrived then, but today it owes its survival to Eastern religionists who, for thousands of years, have planted them around temples. About 25 million years ago, gingkos similar to today's abounded in the great temperate forests of Asia, North America, and Europe. Mysteriously, they vanished from Europe and North America between 10 and 15 million years ago. The plant is highly resistant to insects, disease, and air pollution. Thus, it is planted as a "street tree" throughout the world.

great cycads shaded the dinosaurs (Figure 12.14). There were several species of gingkos (Figure 12.15) in those days, but only one hardy species remains today.

Not only are the conifers (the cone-bearing gymnosperms) abundant today, but they dominated the Mesozoic landscape as well. Such long-term success undoubtedly lies in their remarkable adaptations in fertilization, seed protection, and seed dispersal, traits that we will consider in more detail later.

Class Angiospermae

The **angiosperms** are among the most spectacular of all living things; these are the flowering plants. They are also among the most evolutionarily successful forms of life, and they include more than what we think of as flowers. For example, violets are flowering plants, but so are grasses and palms (Figure 12.16).

We don't know how the flowering plants evolved or where they came from because the first flowers apparently didn't fossilize. We do know, however, that they reached a dizzying crescendo of success in a very brief period. Fossil angiosperms appear rather suddenly and in great numbers in deposits from the Cretaceous period of about 130 million years ago. Their sudden emergence may have had something to do with a climatic change, since the worldwide climate about this time became dramatically cooler and drier. Angiosperms in those days didn't have to worry much about being eaten by a grazing dinosaur, however, because the period mysteriously marked the extinction of these great beasts as well as many conifers and cycads.

No brief introduction to the world of plants could reveal how tightly our lives are bound to theirs. In fact, the bond between animals and plants is so strong that probably it cannot yet be entirely understood. The interplay between these life forms has had a profound effect on the evolution of both, and they remain inextricably linked in countless ways today. It may be more accurate to try to understand, emotionally as well as intellectually, that we are simply different but interdependent parts of the incredible and wonderful phenomenon called life.

FIGURE 12.16 The angiosperms are a diverse group. In spite of the different appearances of these species, they are all flowering plants. There are about a quarter million species, and they have virtually dominated the earth's surface for the last 70 million years. The sporophyte stage is the most conspicuous, and it is here that we find those modified leaves called flowers. At right, wooly britches fiddleneck. Center (from left to right), jack-in-the-pulpit, milk thistle, and round-lobed hepatica. Bottom, wood lily and Canadian white violet.

SUMMARY

Much of the diversity of plants is related to their alternating generations when haploid forms (the gametophytic) may vary greatly from the diploid forms (the sporophytic). The divisions may also vary markedly in appearance and evolutionary direction. Division Rhodophyta contain chlorophyll a and b and phycoerythrin. They are highly diverse. Division Phaeophyta are common beach plants that may form dense undersea forests. Division Chlorophyta share important traits with tracheophytes. They are largely freshwater organisms. Division Bryophyta, the mosses and close relatives, are among the oldest land plants, but are probably not forerunners of tracheophytes. The tracheophytes are the vascular plants (those with conducting vessels). They include ferns, gymnosperms, and angiosperms. The gymnosperms have naked seeds; the angiosperm seeds are enclosed by ovaries.

Chapter 13

DIVERSITY III:
ANIMALS

Here's an easy question: What is an animal? But perhaps by now you won't be taken in by any such "simple" question in biology. And right you are. In fact (and perhaps not surprisingly), there is even strong disagreement among biologists about such questions. For example, many biologists might be willing to say, "An animal eats and has a nervous system." But then we might hear a disgruntled objection from the back of the room: "What about sponges?" It's true, sponges don't have digestive or nervous systems, so perhaps they should be placed in another kingdom, but who wants to deal with another kingdom? So, sponges uncomfortably reside with elk and grizzly bears in the animal kingdom. Thus, again we find ourselves stretching things a bit, compromising and making exceptions. Let's admit that we have an imperfect system, forge ahead, and see just where animals fit in our five-kingdom scheme, paying particular attention to the evolutionary processes that got them there.

THE ANIMAL KINGDOM

The first animals appeared between one and two billion years ago. We don't know much about them since the earliest ones were small and soft-bodied and didn't fossilize very well. About all they left of themselves were a few burrows and tracks for us to ponder, and, as a final insult, a few droppings. But by the early Cambrian, hard-bodied **metazoan** (multicellular) species had appeared, and we have abundant fossils from that period, although we know almost nothing about the evolutionary processes that produced them.

We will use a simplified scheme to describe the animal kingdom, considering the major phyla, a few subphyla, and the major classes within each group (Figure 13.1). (You may want to refer to Appendix A.)

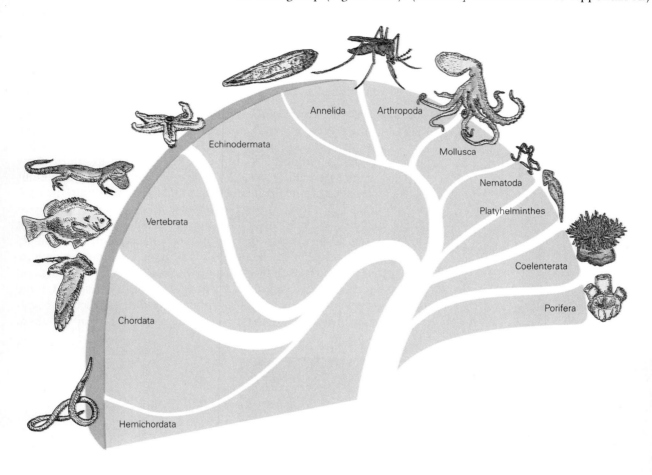

FIGURE 13.1 An evolutionary tree showing the variety of organisms that have sprung from a common stock. All the groups exist at present and, hence, are still evolving. In this scheme, general intelligence increases, in an imprecise sense, as one moves from right to left across the branches. Certainly the brainier creatures are at left, but who can say a roundworm is smarter than a flatworm? And it can be argued that the octopus and the acorn worm (representing the hemichordates) are particularly misplaced. The tree is shown this way partly for anatomical reasons. For example, the single nerve cord of the roundworm more closely approximates the nerve cord of vertebrates than does the double nerve cord of the flatworm. Do you think such a diagram makes too many assumptions?

Phylum Porifera: An Evolutionary Dead End

The Porifera, or sponges, are of three classes: calcareous (chalky) sponges, glass sponges, and bath sponges (Figure 13.2). Essentially, their bodies are like vases with holes in the sides. Water enters these holes and is expelled out the top, having been relieved of some of its oxygen and food particles. The food particles are trapped by flagellated **choanocytes,** or **collar cells,** that may pass them to **amoebocytes,** cells that can creep through the body wall, distributing the nutrients as they go. Sponges are extremely primitive animals, as demonstrated by the fact that their bodies can be shredded and passed through a sieve, and the cells will crawl back together and reassemble themselves into a functional animal. Sponges are considered to be evolutionary dead ends in that it is unlikely that their ancestors gave rise to other lines. So you needn't worry about one being in your family tree. In any case, they are successful enough, having survived on the planet for about a billion years.

Phylum Coelenterata and the Development of Radial Symmetry

The **coelenterates** are the jellyfish, polyps, sea anemones, and corals. They have radial body symmetry (described in Figure 13.3), stinging cells (**nematocysts**), tentacles, a very primitive nerve net (the first organisms we've seen with true nerves), and a saclike gut. Essentially, their bodies are comprised of two layers, an outer **epidermis** one cell thick and an inner **gastrodermis** that lines the saclike gut, **the gastrovascular cavity** (Figure 13.4). A jellylike substance called **mesoglea** separates the two layers.

Some coelenterates have two distinct body types, and in the course of their existence, they may change from one phase to another. Those in

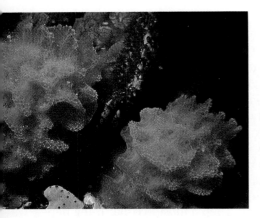

FIGURE 13.2 The glass sponges are delicate animals that are different in so many critical respects from other animals that their placement among the kingdoms has been heavily debated. However, they have been placed among the porifera for now.

FIGURE 13.3 In radial symmetry, the body is essentially spherical, disk-shaped, or cylindrical. Any plane drawn through the center produces roughly equal left and right halves. Bilateral symmetry, on the other hand, means that the body can be equally divided by one plane only. This division produces mirror-image left and right halves that contain similar structures. Perfect examples of radial or bilateral symmetry are not always found. Can you think of structures in your own body that are asymmetric—that is, neither paired nor composed of mirror-image right and left halves?

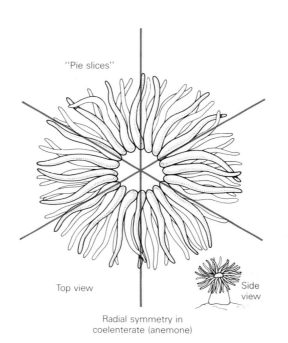

"Pie slices"

Top view

Side view

Radial symmetry in coelenterate (anemone)

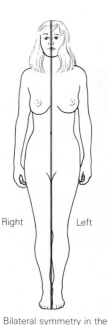

Right   Left

Bilateral symmetry in the chordate (human)

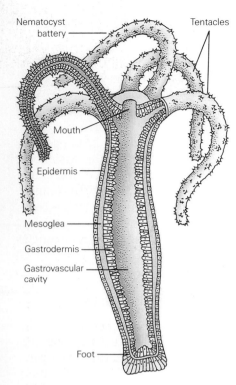

Nematocyst battery

Tentacles

Mouth

Epidermis

Mesoglea

Gastrodermis

Gastrovascular cavity

Foot

FIGURE 13.4   The *Hydra* is a coelenterate with only two cell layers and a radial body plan. The outer cell layer is called the epidermis; the inner is the gastrodermis. Between these lie an undifferentiated substance called mesoglea.

which the sedentary **polyp** (vaselike) stage predominates are in the class **Hydrozoa** and can be represented by the *Hydra* (which, incidentally, has *only* a polyp stage). In the other form, the **medusa** (belllike) stage predominates. This is the free-swimming stage, and the class is called **Scyphozoa.** It includes the jellyfish (Figure 13.5). In the class **Anthozoa** we find the sea anemones and corals (Figures 13.6 and 13.7). Corals secrete hard, limy structures and come in many sizes, shapes, and colors.

Coelenterates have a most interesting life cycle that suggests the next crucial step in evolution—a change in body symmetry. The suggestion is based on the very different appearances of the adult and larval coelenterate. The larva is called a **planula,** and it resembles a flattened mass of cells covered by cilia. It crawls about on the bottom of its watery habitat until the time comes for its next stage. Then a most peculiar thing happens. It stands on one end and its body slowly begins to change. The free end becomes cuplike as tentacles develop around the rim of the forming cup. It is soon apparent that the little flattened planula is developing into a polyp. This phase is very brief in some groups, but very extended in others (Figure 13.8). The polyp stage begins to form layers of tentacled cups. The outer layers break away (a process called ''budding'') to become male and female medusae. When their eggs and sperm join, a new planula is produced.

The evolutionary significance of the little planula is suggested by the fact that it looks a great deal like the next group we will discuss, the bilateral flatworms. Evolutionists have wondered what would happen if certain mutations caused the planula to continue in its wanderings, never to stand on end, never to become a ''vase''? What if such an ''eternal'' planula should develop the ability to reproduce? Is this how the flatworms began?

FIGURE 13.5   Jellyfish can be among the most compelling sights of the sea. Their delicate bodies pulsate, ghostlike, through the depths as they trail their deadly tentacles.

FIGURE 13.6   The anemone has a large, fleshy body that is easily distinguished from the small polyps of hydrozoans. Many anemones are colorful, resembling flowers, but their beautiful tentacles may spell death for small animals that touch them.

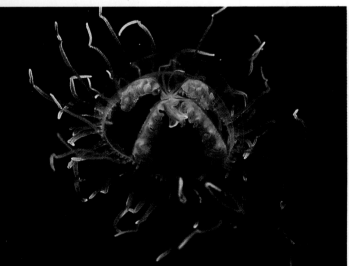

FIGURE 13.7 A coral reef is a beautiful and exciting place. Charles Darwin first noted that coral atolls begin as reefs surrounding islands. As the islands sink, the reef grows so as to keep itself in the area penetrated by sun's rays. As the animals on the inner areas are deprived of fresh seawater washing across them, they tend to die, their limey walls crumbling, leaving an expanding ring of living coral, as seen here. The ring may then break apart leaving a circle of islands. Corals are polyp forms, similar to the anemones, but generally smaller. They can be very diverse, as we see below.

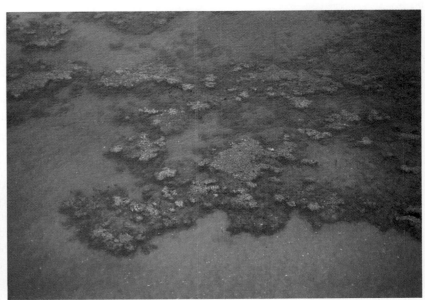

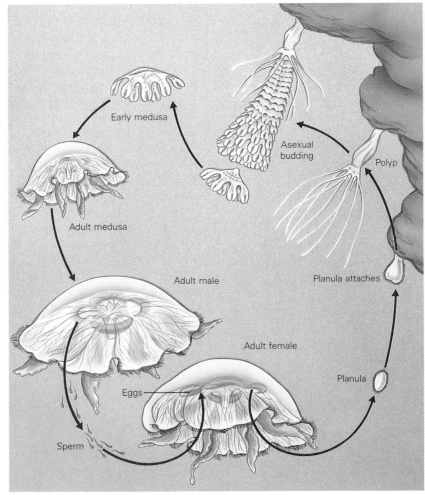

FIGURE 13.8 The planula is the larval stage in the life cycle of the coelenterates. The planula usually has an outer layer of ciliated cells surrounding a rather compact mass of cells. It has a very simple nervous system, with most neurons aggregated at the anterior end. It lacks a mouth and must depend on whatever food it has stored. It attaches to some substrate and forms the polyp stage. Medusas are formed as cuplike buds break away from the polyp stage.

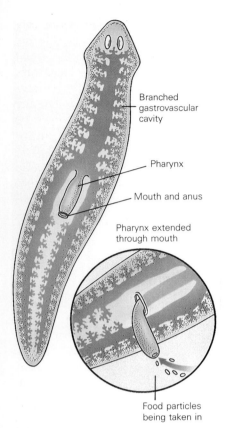

Branched gastrovascular cavity

Pharynx

Mouth and anus

Pharynx extended through mouth

Food particles being taken in

FIGURE 13.9 The planarian flatworm, with its extensive branched gastrovascular cavity, is able to evert its pharynx to ingest food.

Phylum Platyhelminthes and the Appearance of Bilateral Symmetry

The **platyhelminths** are the flatworms. These are the first group we've encountered with bilateral symmetry as adults. The evolution of such a body plan would be important because it is associated with movement in a single direction; that is, with animals that tend to move forward. With forward movement would have come an accumulation of sensory mechanisms in the leading or "head" end. This could have triggered the development of a primitive brain. Such senses would have helped in discovering food and in avoiding danger in the environment into which the animal was moving. Since food or danger would be as likely to be found on one side as the other, the nervous system of the body would tend to develop equally on both sides; hence, a wide variety of creatures have appeared with two sides and a head end.

The flatworm nervous system boasts two large nerves running parallel down the length of the animal and linked by cross bridges. The muscular system is also well developed, the first we've seen. Turbellarians live in both fresh and salt water, and their muscles, nerves, gonads, and digestive tract are specialized enough to contain true organs. This is the first group we've encountered that has three specific embryonic tissues (or "germ layers") as discussed in Chapter 16.

The best known flatworms are the **planaria,** free-swimming species of the class **Turbellaria.** Planaria are common experimental animals because they can regenerate after being mutilated and because they can learn simple tasks, such as turning a certain direction in response to a stimulus. The planarian is unlike some flatworms in having a saclike gut and a protusible pharynx that can be everted to encompass food outside the animal and draw it back into its extensive gut (Figure 13.9).

The class **Trematoda** includes a grisly group of parasites that infect the various organs of many animals—including humans (Figure 13.10). They attach to these organs by suckers. The internal systems of trematodes are well developed and include the reproductive organs of both sexes. Their life cycles may be extremely complex, involving a number of hosts, as we will see in Chapter 26.

The class **Cestoda** includes the **tapeworms.** They are textbook examples of evolution moving toward simplification rather than complexity. With each simplification, they foreclosed new evolutionary options, so their ancestral line probably gave rise to no other animal today.

The tapeworm head, or **scolex,** anchors the worm in the gut by a group of suckers and hooks (Figure 13.11). The segments, or **proglottids,** are little more than increasingly swollen sacs of eggs that break off and pass out with the host's feces. Tapeworms lack a mouth and digestive tract, but these are hardly necessary since they live in a sea of digested food; they can simply absorb nutrients through their body walls. Tapeworms grow quite long and can rob the host of enough nutrients to cause extreme weakness. A common human tapeworm can grow to be over 5 meters long, but the fish tapeworm (which can also infect humans) can reach 20 meters in length!

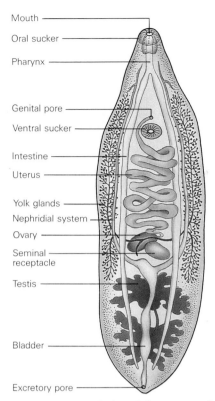

Mouth
Oral sucker
Pharynx

Genital pore
Ventral sucker

Intestine
Uterus

Yolk glands
Nephridial system
Ovary
Seminal receptacle

Testis

Bladder

Excretory pore

FIGURE 13.10 The liver fluke, a trematode, is a parasitic flatworm with rather well-developed systems.

FIGURE 13.11 The head, or scolex, of a tapeworm is crowned with a circle of hooks by which it imbeds in the intestinal wall of its host. It then begins to form a long chain of segments called proglottids, each a veritable sack of eggs. At the end of the chain, the proglottids break off and release their contents. Each egg must then pass through a complex chain of events that are largely at the mercy of sheer chance. Most perish. (Photo reprinted with permission of the present publisher, Jones and Bartlett Publishers, Inc., from Shih and Kessel: *Living Images,* Science Books International, 1982, page 84.)

FIGURE 13.12 The *Trichina,* or pork roundworm, can be transmitted to humans by eating pork that has not been properly cooked. Most roundworms, however, are harmless and live on decaying matter. It's a good thing most are harmless since only 10,000 species have been named, and it is suspected that there may be 500,000 more.

Unlike the tapeworms, the flatworms, as a group, were evolutionarily important due to their remarkable success in giving rise to new species. In fact, long ago, an ancestral form diverged into two lines that would produce quite distinct groups of animals. Actually, the two great lines formed from a seemingly small difference. In one group, the first embryonic opening to form became the mouth; in the other, it became the anus. In today's **protostomes,** the first embryonic opening becomes the mouth; the anus forms later. This group includes the annelids, arthropods, and mollusks (See Figure 13.1). In the **deuterostomes** of today, the first opening forms the anus, the mouth appearing later. This group includes the echinoderms and the vertebrates. It is because of embryological evidence such as this that we place the humble echinoderms, such as the starfish, closer to ourselves than we do the highly intelligent mollusks or the very complex social arthropods.

Phylum Nematoda and a Tubelike Body Plan

The **nematodes,** or **roundworms,** are among the most startling of all living things in many ways. You may wonder how a roundworm could be startling—until you learn that you may eat thousands each day. Roundworms live practically everywhere, and those who study them are often given to obsessive cleanliness, washing their food meticulously. Roundworms are distinctive in that they are the first group we've considered with a tubular gut—one in which food goes in one end and the waste out the other (decidedly more civilized than a saclike gut).

Most roundworms live harmlessly on decaying matter, but the best known roundworms are human parasites, such as **Trichina**—the worms that infect pork and then are passed to humans who eat incompletely cooked meat (Figure 13.12). Other notorious roundworms can cause monstrous deformities such as elephantiasis (resulting in blocked lymphatic ducts that causes grotesque swelling of the legs). Another species, about a full meter long, winds its way through human tissues, pressing and squeezing its way through and between the cells of the body and sometimes causing great pain. Treatment has traditionally involved pulling this worm out through the skin by slowly winding it around a stick. Interestingly, although the roundworms have been remarkably success-

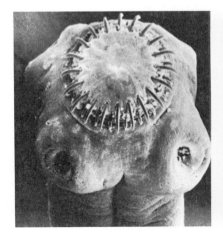
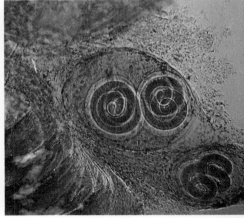

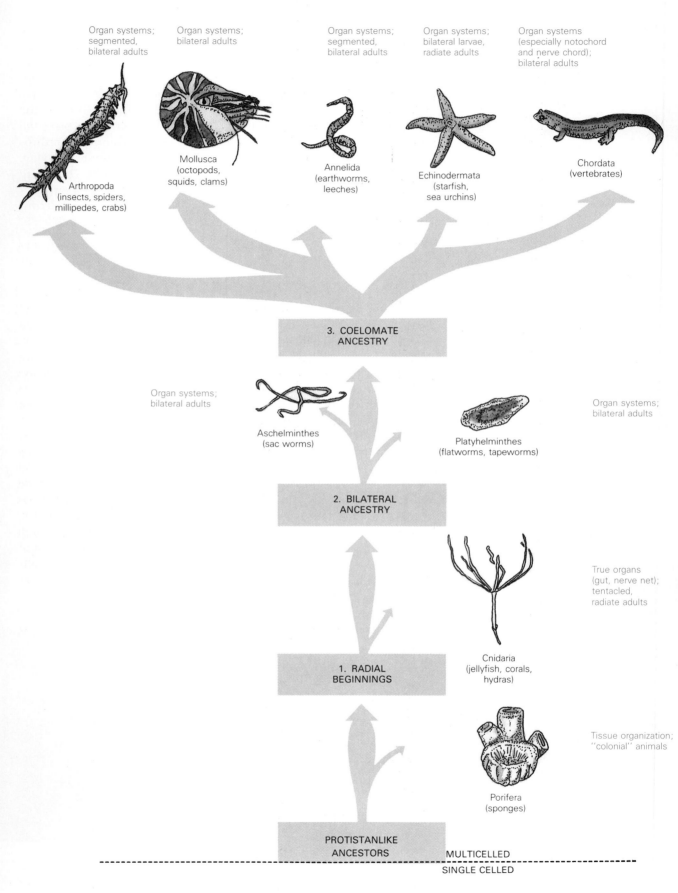

Organ systems;
segmented,
bilateral adults

Organ systems;
bilateral adults

Organ systems;
segmented,
bilateral adults

Organ systems;
bilateral larvae,
radiate adults

Organ systems
(especially notochord
and nerve chord);
bilateral adults

Arthropoda
(insects, spiders,
millipedes, crabs)

Mollusca
(octopods,
squids, clams)

Annelida
(earthworms,
leeches)

Echinodermata
(starfish,
sea urchins)

Chordata
(vertebrates)

**3. COELOMATE
ANCESTRY**

Organ systems;
bilateral adults

Aschelminthes
(sac worms)

Platyhelminthes
(flatworms, tapeworms)

Organ systems;
bilateral adults

**2. BILATERAL
ANCESTRY**

True organs
(gut, nerve net);
tentacled,
radiate adults

Cnidaria
(jellyfish, corals,
hydras)

**1. RADIAL
BEGINNINGS**

Tissue organization;
"colonial" animals

Porifera
(sponges)

**PROTISTANLIKE
ANCESTORS**

MULTICELLED
- -
SINGLE CELLED

FIGURE 13.13 An evolutionary tree showing hypothetical descendancy. The earliest animals are believed to have been protistlike. From these arose the radial creatures, then the bilateral. Coeloms appeared later. Note the strong division of the coelomates. The two groups form differently as embryos, as we will see in Chapter 16.

ful in their own right, they haven't diverged much over the course of their evolution. That is, they haven't given rise to important new groups of animals, as did the early flatworms.

Phylum Mollusca and the Advent of the Coelom

The **mollusks** are in the evolutionary line with earthworms and insects, since they are all protostomes. They did not give rise to the other protostomes, but they shared a common ancestor. In addition, they give us our first look at the **coelom.** The coelom is the mesoderm-lined area between the gut and the body wall that gives the internal organs a certain freedom of movement. Figure 13.13 shows the major evolutionary developments in animals. Notice the appearance of **tissues** (organized groups of cells), **organs** (organized and interacting groups of tissues, such as form the heart), and **organ systems** (organized and interacting groups of organs, such as form the circulatory system).

Mollusks are named for their soft bodies, although some are encased in hard shells and others have stiff shells inside their bodies (parakeets like to sharpen their bills on cuttlebones, the inner skeletons of cuttlefish). There are about 100,000 species of mollusks, and they first appeared about 500 million years ago.

All mollusks have a mass of muscle on one side of the body (as we see in the **foot** of a snail), a fold of tissue over the body (called a **mantle**), and some have a rasplike, filing tongue (called a **radula**). They have well-developed systems (Figure 13.14). The various species may differ greatly in size. Some mollusks are almost microscopic, while one clam, *Tridacna,* may be a meter wide and weigh several hundred pounds (Figure 13.15). And let's not forget the giant squids of the North Pacific that terrorize mainly people in movie theaters (Figure 13.16). In spite of the size of such giants, mollusks probably have more reason to be afraid of

FIGURE 13.14 The internal structure of the clam. Clams, like other bivalve (two-shelled) mollusks, are filter feeders that strain food particles from water flowing across their gills. Note the relationship of the various systems and compare them to what you know about vertebrates. Do you see anything unusual about the relationship of the heart and gut? How do you think the food gets from the gills to the mouth? What part do clam-eaters eat?

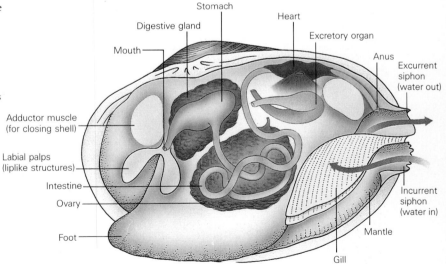

Stomach
Digestive gland
Heart
Mouth
Excretory organ
Anus
Excurrent siphon (water out)
Adductor muscle (for closing shell)
Labial palps (liplike structures)
Intestine
Ovary
Foot
Incurrent siphon (water in)
Mantle
Gill

FIGURE 13.15 The *Tridacna,* or giant clam, is a most beautiful and awesome creature when seen underwater. The mantle is a lovely, undulating blue. Divers have died by reaching a hand inside, only to have the massive shell close protectively.

FIGURE 13.16 Little is known about giant squids, but they grow to great lengths and are known to be predators.

us than we do of them. After all, we lustily devour all sorts of oysters, clams, mussels, snails, squids, octopuses, and periwinkles. (These smaller species may also serve as temporary hosts for parasites that infect humans.)

Phylum Annelida and Repeating Segments

With the **annelids,** we come to the strongly segmented worms, the most familiar being the earthworms. These are in the class **oligochaeta** ("few hairs") (Figure 13.17). Earthworms have hairlike **setae** that help them to crawl through the earth and to brace against the tugging robin. There are some earthworms, though, that a robin might hesitate to tangle with, such as the four-meter giants that rumble and gurgle their way through the Australian soil (Figure 13.18).

The repeating units of segmented animals were an important evolutionary advancement (because of the specializations they made possible, as we will see in the next group). The annelids are the first group we've seen with a circulatory system. Thus, food from the gut does not have to diffuse through the coelomic fluids. Instead it can be carried by blood vessels, a much more efficient arrangement. Such an arrangement also

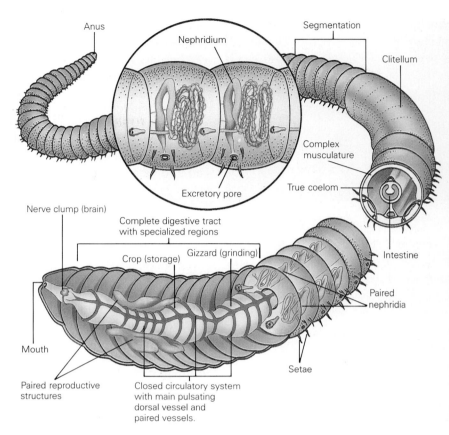

FIGURE 13.17 The body plan of an earthworm. Each segment boasts a pair of nephridia. The tubelike nephridium passes through the body septum. The beating cilia set the fluid of the body in motion so that it flows past blood vessels lying tight against the nephridium that shares that segment. Metabolic wastes from the blood thus enter the tube and are excreted, while water and other essential products are absorbed from the body fluids back into the capillaries.

FIGURE 13.18 A giant earthworm. They may grow to 4 meters in length. As they push through the earth, they often make deep gurglings and rumblings that give away their presence.

set the stage for the development of larger bodies in which oxygen and nutrients could be distributed and waste removed rapidly by special systems.

Typical of many invertebrates, the annelids have a ventral bilobed nerve cord that enlarges in each segment. A well-developed nervous system would be important in an animal with repeating segments. Note the paired vessels that extend from the pulsating ("pumping") dorsal vessel to surround the anterior part of the digestive tract.

Earthworms reproduce in an interesting fashion. Although each worm contains both testes and ovaries, they do not self-fertilize. Instead, they exchange sperm by lying head-to-tail, as each injects its sperm into the special **seminal receptacle** of its friend. Later the ring, or **clitellum** (the band that you've seen around earthworms), secretes a layer of slime that begins to move along the worm. As it passes the egg-producing cells, the eggs move into it and are joined by sperm when the ring passes the pores of the seminal receptacles. The earthworm finally shrugs off its slimy ring, and the ends immediately seal. The "cocoon" then lies in the soil while the zygotes develop inside its protective coat.

Another class, called **Polychaeta,** is interesting in that the segments of these worms are more specialized than those of the earthworm. The polychaetes have "legs," actually **parapods,** that are used in both swim-

ming and respiration, and they have heads with eyes, jaws, and tentacles (Figure 13.19). And they can bite.

The class **Hirudinea** includes the leeches, another group of beauties. Leeches, of course, suck your blood. They do this by attaching with a suckerlike head harboring three sharp teeth. After they attach, they inject an anticoagulant to keep the blood from clotting and then they suck until they are bloated. They were once used to reduce blood pressure and to take color out of the black eyes of the doctors who prescribed them. (In fact, the word "leech" originally referred to the physician, but came to be attached to the animal as well.)

FIGURE 13.19 Polychaetes are segmented worms with rather marked specialization. They have well-developed heads and they have appendages on their segments. Whereas oligochaetes, such as earthworms, have only a few bristles per segment, polychaetes have many bristles. Polychaetes live almost entirely in salt water and eat a variety of foods, from algae and detritus to small animals.

Phylum Arthropoda and the Appearance of Specialized Segments

Some have called this period in history the age of man, but it might just as well be referred to as the age of **Arthropoda.** These creatures are abundant, diverse, and successful, indeed. The arthropods are the "joint-legged" creatures, and they include millipedes, centipedes, crayfish, and, most important, insects. They not only have jointed legs but a chitinous exoskeleton, as well.

The arthropods are evolutionarily important because they are the first group to strongly capitalize on the segmented body plan. They do this by allowing various segments to specialize for different roles. Thus, we find segments with antennae, or claws, or legs, each performing different tasks. In this way, the animals can behave more flexibly and meet a greater variety of environmental challenges. Such specialized segments require a relatively complex nervous system to coordinate the activities of different body parts, and this is what we find here. There are two subphyla, the **Chelicerata** ("claw horn ones") and the **Mandibulata** ("jawed ones").

The chelicerata have claw-like fangs as their first appendages, no antennae, and the head and thorax are fused and distinct from the abdomen. Most species have four pairs of legs, and they are represented by the horseshoe crabs (class **Merostomata**) and the spiders and scorpions (class **Arachnida**) (Figure 13.20).

FIGURE 13.20 Tiny spiders have been known to strike terror in the hearts of the bravest of us. It is probably no great consolation to know that only about thirty species are dangerous to humans. The spiders are derived from an ancient branch of the arthropods, called the Chelicerata. Their hardened body makes them resistant to drying, and the body segments are largely fused, leaving two main sections, the cephalothorax (roughly comparable to the head and thorax of insects) and the abdomen. Like insects, they have tracheal tubes for breathing, and the respiratory surface is increased by an area of flaplike tissue called a book lung. Blood flows into the heart through the ostia. After mating (when the female may capture and devour her suitor), the sperm are stored in a receptacle until eggs are released from the ovary. The chelicera are the pincerlike first appendages.

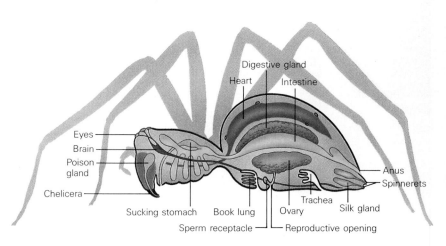

FIGURE 13.21 The anatomy of a male lobster. Notice the small heart. Keep in mind that the heart is not used to disperse oxygen. The enlarged anterior part of the stomach contains grinders, an unusual arrangement. How has a simple segmented body become modified and specialized in this group? Males in this group "rape" females by forcibly turning them over, inserting their specialized first abdominal appendage in special grooves, and filling the receptacle of the female with sperm. Fertilization will not take place until later, when she releases eggs. She will then tend the developing zygotes as they become entwined in the specialized legs called swimmerets.

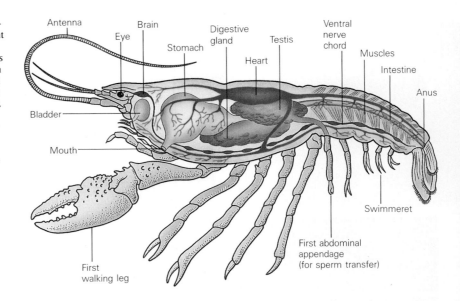

The first appendages of the Mandibulata are modified into antennae, and they have jaws that work from side to side. There are more species in the Mandibulata than in any other group of any phylum. The class **Crustacea** include the shrimps, crabs, lobsters, barnacles, and countless smaller creatures, most of which live in the water. The segmentation of the arthropods may not be obvious, partly because much of the body is covered with a continuous, smooth chitinous shell (Figure 13.21). The classes **Chilopoda** and **Diplopoda** include the centipedes and millipedes, respectively. They are not economically important, except for tropical hotels losing an occasional guest when the one crawls up the wall. The centipedes may have a mildly painful sting. Most species scavenge plant food, although a few will go after bedbugs and roaches. The centipedes have one pair of legs per segment, while the harmless millipedes have two (Figure 13.22).

Perhaps the most diverse and impressive class on earth is the **Insecta** (Figure 13.23). You probably won't be startled to learn that these are the insects. The planet is literally crawling with bugs, and so far about

FIGURE 13.22 Centipedes (left) have only one pair of legs on each segment. The head bears one pair of antennae and two pairs of mouthparts. Some species are quick and relentless hunters, incapacitating their prey by a sting from the modified front pair of appendages. Centipedes add new segments as they grow. Millipedes (right) have two pairs of legs per segment and are harmless herbivores.

FIGURE 13.23 Insects have certain standardized structures that qualify them for the title, such as head, thorax, and abdomen, but they show extreme variation on a theme as we see in the spotted lady beetle, scarlet and green leafhopper, sugar maple borer, and Ichneumon wasp. Insects, you will note, have antennae as well as feeding appendages. They comprise the most diverse class of animals on earth.

900,000 species have been described, with more being added to the list almost daily. Perhaps you are one of those people who reflexively squashes any such small creature. If so, you may have slaughtered a member of some beautiful and unknown species "because it was a bug."

The astute observer may have noticed that not all insects are harmless. If you haven't noticed, go lie down on a grassy dune; the bite of some ants can be a religious experience. (In the Amazon, the sting of the Konga ant is like someone gripping your skin with pliers and twisting hard for two hours.) Also, insects pester and parasitize our livestock, eat our crops, walk on our food when we don't know where they've been, carry parasites and diseases, and threaten us in countless other ways. At the same time, however, they provide us with honey and silk and help pollinate our crops. In the generalized insect body in Figure 13.24, we see a head attached to a thick **thorax,** followed by a segmented **abdomen.**

Phylum Echinodermata and an Evolutionary Puzzle

The **echinoderms** ("spiny skin") share the deuterostome line with our own group, the chordates. There are five living classes of echinoderms: The **Crinoidea** (primitive stalked sea lilies), the **Asteroidea** (starfish), the **Ophiuroidea** (brittle stars), the **Echiniodea** (sea urchins and sand dollars), and the **Holothuroidea** (those great, limp sacs called sea cucumbers).

The evolution of the echinoderms is interesting because they apparently once developed a longitudinal and bilateral body plan and then reverted back to the more primitive radial plan, at least in the adult stage. Their bilateral heritage is today evident only in their free-swimming, ciliated larval stage. Perhaps some ancestral form was forced, by competition or predation, back to the two-dimensional sea floor where the ability to move equally well in any direction was advantageous and the radial plan thus became superimposed on a bilateral one.

Whereas echinoderms seem to wear their skeleton on the outside, like the insects, their protective skeletal elements are actually embedded just under the skin. They also have an unusual feature called a **water vascular system,** which is a kind of hydraulic pump that extends the soft, pouchlike tube feet, with their terminal suckers (Figure 13.25). Echino-

FIGURE 13.24 The insect body plan (represented by the grasshopper). Note the breathing system. Air enters the body through the spiracles and diffuses through finely branching tubules, dead-ending in the tissues, as we will see in Chapter 19. How does this system differ from that of mammals? What aspects of the insect body are similar to those of mammals? How could such similarities have evolved? What are the major, general differences in insects and mammals?

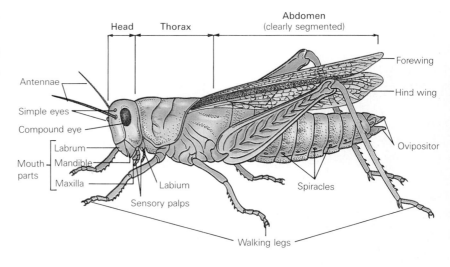

FIGURE 13.25 The peculiar water vascular system of the starfish. Water enters and leaves through the sieve plate. By shifting the water, the tube feet can be extended. When the tube feet are pressed against something solid, such as an oyster shell, strong suction can be developed as water is moved from the foot. The starfish has no head end, and any arm can lead. What are the advantages and disadvantages to such a system? The skin may seem spiny, but the calcium-laden spines are actually under the skin.

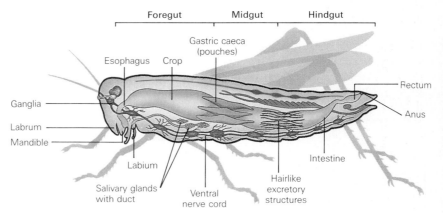

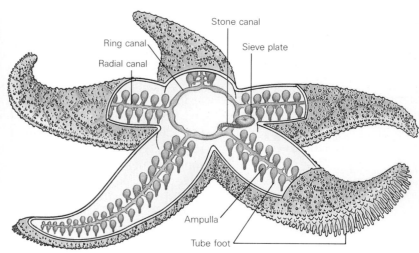

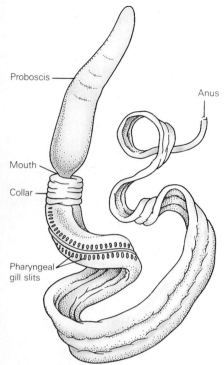

derms are sluggish creatures with poorly developed nervous systems, but they are efficient and relentless predators. For example, a starfish attacking an oyster wraps its arms around its prey and pulls tenaciously until the shell opens just a bit. Then it everts its stomach, squeezing it between the shell halves, and digests the oyster in its own shell, drawing the products into its own body.

Phylum Hemichordata and a Half-Step Along

Aside from a few biologists, not many people can muster much interest in the **hemichordates** ("half chordate"). Most are marine worms of the most uninspiring sort, and no one is likely to ask a hemichordate researcher how the day went. So why are biologists interested? Because the hemichordates occupy an important position in the evolutionary development of the vertebrates (Figure 13.26). Specifically, their embryonic development and general structure indicate that they are distantly related to the backboned animals. The relationship is revealed by the presence of pharyngeal slits and a hollow, dorsal nerve cord. They also have a stiff rod that acts as a supporting structure similar to the notochord of the chordates that we will see next.

Phylum Chordata and the Advent of Freeways

The **chordates** are named for possessing, at least at some stage, a **notochord,** or rigid, cellular rod covered with two layers of strengthening fibers. The chordates are quite a varied group, ranging from slimy little sea creatures to Las Vegas gamblers, from chipping sparrows to blue whales, from dingos to dugongs. No one is quite certain how they are related, but one hypothetical evolutionary sequence makes at least some sense (Figure 13.27).

All chordates have three basic features in common. At some time in their lives they all have (1) a notochord, (2) a dorsal, hollow nerve cord that enlarges at one end to form the brain, and (3) pharyngeal slits (or gill slits), which are openings in the throat region that lead from the alimentary canal to the outside. Pharyngeal slits are used for respiration in fishes, but in air breathers, they disappear early in embryonic development. Chordates also generally share a few less distinctive traits. For example, there is almost always a tail that extends beyond the anus at some stage. (Many children are born each year with rudimentary tails that do not fit the expectations of some parents and are quickly removed.) They are also bilaterally symmetrical, with distinct heads, some segmentation, three embryonic germ layers (Chapter 15), and a well-developed coelom.

The three chordate subphyla include the **Urochordata,** the **Cephalochordata,** and the **Vertebrata,** the group that includes us. Taking them one at a time, we notice that the Urochordates don't look like our close relatives (at least not mine). Actually, they look more like sea squirts, and that's good because that's what they are (Figure 13.28). In fact, they can only be identified as chordates when they are larvae. Once they change from their free-swimming stage, they lose their similarity to other chordates. (To illustrate how different the adults are from other chordates, they produce cellulose in their body walls.) Their strange larval

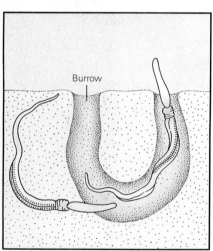

FIGURE 13.26   The hemichordate body plan. The acorn worms are rather large worms with a proboscis at their anterior end, a collar, and a long body. The mouth is on the underside between the proboscis and the collar. The acorn worm burrows in mudflats, using its proboscis in the manner of an earthworm.

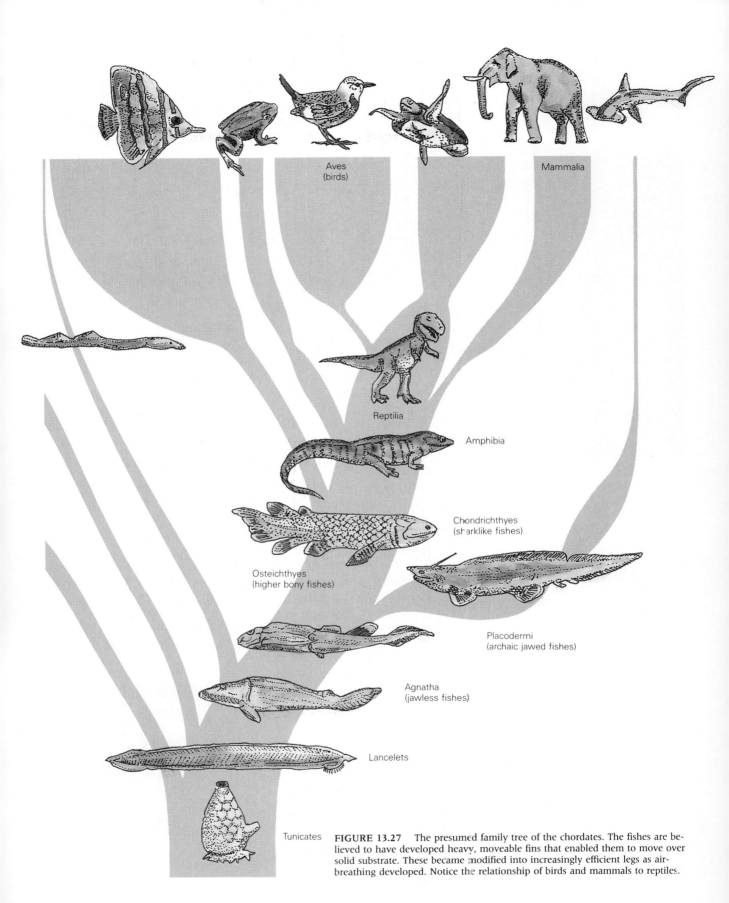

Aves
(birds)

Mammalia

Reptilia

Amphibia

Chondrichthyes
(sharklike fishes)

Osteichthyes
(higher bony fishes)

Placodermi
(archaic jawed fishes)

Agnatha
(jawless fishes)

Lancelets

Tunicates

FIGURE 13.27 The presumed family tree of the chordates. The fishes are believed to have developed heavy, moveable fins that enabled them to move over solid substrate. These became modified into increasingly efficient legs as air-breathing developed. Notice the relationship of birds and mammals to reptiles.

FIGURE 13.28 Sea squirts or tunicates. As larvae (below), they resemble tadpoles. At this stage, they are elongate and bilaterally symmetrical. The tail is stiffened by a rodlike notochord to which muscles are attached. Later they will settle down and attach, by their heads, to the ocean floor, where they will lose their brain, notochord, and tail, and develop a tougher outer coating (right).

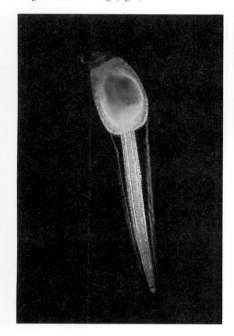

history has led some scientists to speculate that the vertebrates arose from a line of urochordates in which the larval stages never matured physically, but became able to reproduce. It is suggested that this line could have given rise to animals like the **lancelets,** which would have been ancestral to the Agnatha (jawless fishes). Evolutionists have little trouble guessing at the evolutionary route from there to vertebrates.

The lancelets, or *Amphioxus,* are the best known cephalochordates (Figure 13.29), which some say are descended from the hemichordates. They spend much of their lives burrowed in the sandy bottoms of warm seas. Their entire bodies, 5 to 7 centimeters long, are thus protected, as only their mouths are visible. The adults look a bit like larval sea squirts, but they are adult chordates with segmentation of muscles along their length.

The vertebrates are the best known and most varied of the chordates. (The general vertebrate body plan is shown in Figure 13.30.) They are distinct in that they have dorsal, hollow nerve cords protected by a series of bones called **vertebrae,** which together make up the **vertebral column** or **backbone.** This flexible backbone also serves as an attachment for a number of powerful muscles. In addition, the vertebrates have a complex skin (some with hair, glands, hoofs, feathers, scales, or fur, albeit not all on one animal). The heart is located ventrally and pumps

FIGURE 13.29 The *Amphioxus,* or lancelet. Some zoologists believe that it is descended from larval tunicates that failed to metamorphose while becoming sexually mature. Adult lancelets are fishlike, and they retain their notochord. They burrow into sandy bottoms, leaving only their mouth exposed. Food-laden water is drawn in and nutrient particles are swept along by cilia into the gut.

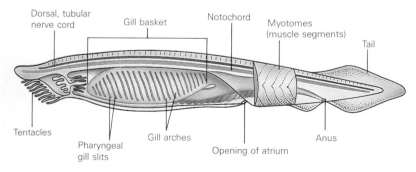

Dorsal, tubular nerve cord · Gill basket · Notochord · Myotomes (muscle segments) · Tail · Tentacles · Pharyngeal gill slits · Gill arches · Opening of atrium · Anus

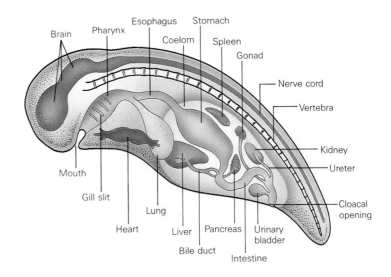

FIGURE 13.30 The general body plan of vertebrates, including humans, although humans are usually taller. All vertebrates have vertebrae (as one would hope), and they also have a dorsal nerve cord with a brain at one end. Other features shown here may or may not be present or may be present only at some time in development. These include gill slits and the notochord. Most vertebrates have some segmentation, bilateral symmetry, a head and tail, a coelom, and some way to get about.

red, iron-containing blood, and kidneys remove the waste from that blood. There is an endocrine regulatory system (Chapter 22), and the sexes are usually separate.

Class Agnatha

The class **Agnatha** ("jawless") is marked by round mouths without jaws, and it includes the lamprey and hagfish. These fish are unusual in that they retain parts of their notochord through adulthood. Biology may introduce you to some ugly animals now and then, and these two are good examples (Figure 13.31). Their behavior also may not gain them great respect. The lampreys, in particular, are not admired because of their habit of attaching by their sucker mouths to the bodies of live or dead fish and sucking out their juices. Hagfish are scavengers and **hermaphroditic** (with both male and female parts), but they produce eggs and sperm at different periods in their lives.

From the jawless fishes, we go to the jawed fishes, a group that includes some of the most fearsome creatures on earth. In large part, they are feared precisely because they have developed a jaw, a mouthpart with a hinge. With this hinge, a great number of fish have taken to biting.

FIGURE 13.31 Two beauties. The hagfish (below) and the lamprey (below right). The notochord is replaced by vertebrae. These fish have no jaws and therefore must feed by sucking the juices out of other animals. Jaws are believed to have developed from the support bars of the most anterior gill slit in some distant ancestor of these creatures. Think of how different the world would be if the simple ability to bite had never evolved.

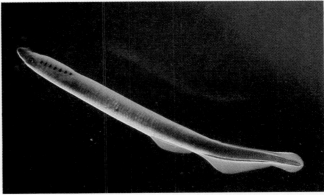

FIGURE 13.32 The shark is one of the most fabled, feared, and least understood large animals on earth. The skeleton is cartilaginous. Sharks are reported to have well-developed jaws and teeth.

Once biting became possible, animals would have begun to use their new talent in increasingly efficient ways. Should they bite pieces out of leaves? Should they bite other animals? Would it be better to take a bite out of an animal and run for it, or to stay and try to eat the whole animal? What if it, too, had teeth? The ability to bite was undoubtedly an important development in predator-prey relationships (Chapter 25), and it forced certain evolutionary directions on the bearers of jaws. For example, some biting species would have had to develop great strength, as well as visual acuity and perhaps speed or stealth. Of course, new digestive abilities would be required with any change in diet. The development of jaws was a critical change, indeed, and one that triggered any number of other modifications.

Class Chondrichthyes

The class **Chondrichthyes** ("cartilage fish") includes the sharks, rays, and skates (Figure 13.32). They, of course, have jaws, and sharks have been known to bite. As their name implies, their skeletons are comprised of cartilage rather than bone. A great deal of attention is paid to sharks in the popular press, but the reports are not always accurate. It is generally safe to say that people who make blanket statements about shark attack behavior don't know much about sharks. (I generally ignore such statements, but when I spot a shark while diving, it all comes back and I am prepared to believe anything.)

Some female sharks lay eggs, but others nurture their offspring within their bodies until the young can be born alive. The reputation of sharks is not helped by the knowledge that in some species, the first embryonic shark to develop will turn and devour its brothers and sisters while still in the uterus.

Class Osteichthyes

The class **Osteichthyes** ("bone fish") includes the bony fish (Figure 13.33). There are about 18,000 living species, and they can be found from shallow ponds and rippling brooks to the darkest ocean depths (Figure 13.34). Bony fish have evolved a number of remarkable adaptations to survive in their unpredictable world. Some use their swim bladders as lungs and can survive for months encased in mud, and some catfish can walk clumsily from one pond to the next by using their fins as

FIGURE 13.33 The bony fish. How could the fins have given rise to legs? Do you see a relationship between swim bladders and lungs? What evidence is there that their ancestors could have invaded dry land? For such evidence, do we look to living or extinct species, or both?

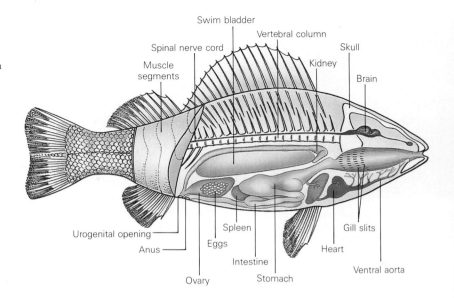

Swim bladder
Spinal nerve cord
Vertebral column
Skull
Muscle segments
Kidney
Brain
Urogenital opening
Spleen
Gill slits
Anus
Eggs
Heart
Ovary
Intestine
Stomach
Ventral aorta

limbs. The biggest bony fish are probably the swordfish, which grow to 4 meters long.

In 1939, a fishing boat plying the waters between Madagascar and Africa hauled up a great, ugly fish that no one recognized. It turned out to be a **coelacanth,** or *Latimeria,* a member of a species that had long been believed to be extinct. The "living fossil" was particularly interesting because some biologists believe that its heavy lobed fins may be of the sort that once gave rise to legs (Figure 13.35). The search for the beasts intensified and scores of these *Latimeria* have been caught and subjected to investigation.

Class Amphibia

The class **Amphibia** ("double life") includes aquatic animals that have successfully invaded the land without completely relinquishing their watery heritage. One might ask why they would give up a lifestyle, to which they were well adapted, for even a part-time rigorous existence

FIGURE 13.34 Some fish are extremely bizarre. Just by looking at this specimen, what can you deduce about its life style? Does it live in lighted waters? What does it eat? What might you guess its social life is like? Does it swim in schools? Alone? Is it a prey animal? As clues, note that there are three fish here. The males are attached as appendages of the female. The enormous jaws indicate that the fish is prepared to eat just about anything it comes across.

FIGURE 13.35 The famed coelacanth. In 1939, one of these beasts was hauled from the waters off southeast Africa, reminding us once again that some fossils yet live. Fishermen were warned in a series of posters and bulletins to keep an eye peeled for them and not to throw them back or eat them. The fishermen ate some of them anyway.

FIGURE 13.36 The amphibians have had to make many compromises as they sought to bridge the gap between water and land. The marsupial frog does not trust its eggs to the tender mercies of aquatic predators and keeps them safe in moist pouches on its back.

FIGURE 13.37 Four reptile representatives: (A) the wood turtle, (B) the California mountain kingsnake, (C) the bluntnose leopard lizard, and (D) the crocodile. The crocodilians have the most specialized heart, with the most efficient separation of oxygenated and deoxygenated blood.

on land. Perhaps because the land had been exploited less, competitors and predators were fewer, and oxygen was more readily available.

But the invasion of land also had its dangers. There was the risk of drying out and being subjected to drastic temperature fluctuations. Without the buoyancy of water, a land dweller also needed strong support and a sturdier frame, both metabolically expensive structures. So, in their search for the best of both worlds, amphibians retained their dependence on water, especially in reproduction, and they never developed skins that were impermeable to water and able to hold moisture in. But they did develop inpocketing lungs (instead of outpocketing gills) that saved water. And they developed a strong skeleton and sturdy, muscular legs. Even now, the most advanced amphibia, such as frogs, toads, and salamanders, return to lay eggs in the water, after having spent their early lives there (Figure 13.36).

Class Reptilia

The class **Reptilia** ("crawlers") includes the snakes, turtles, lizards, and crocodilians. They probably rank somewhere between spiders and bats in popularity. Snakes, in particular, often meet their fate by encounters with humans. (We are very hard on bugs and snakes.) We might remember, however, that the ancestors of those snakes once ruled the earth while our ancestors were scurrying around under leaves.

Reptiles probably evolved from an amphibian ancestor about 300 million years ago, and they, in turn, apparently gave rise to birds and mammals (Figure 13.37). Reptiles are generally considered to be "cold blooded," or *ectothermic*. That is, they are not very efficient at regulating

(A)

(B)

(C)

(D)

(A) (B) (C)

(D) (E) (F)

FIGURE 13.38 Birds have a wide range of appearances that correspond with their diverse life styles. Most can fly, and others run or swim. They are extremely specialized creatures in other ways as well. Shown here are the: (A) roseate spoonbill, (B) brown pelican, (C) American goldfinch, (D) anhinga, (E) shoebill, and (F) Andean condor.

their internal temperatures and thus they are rather strongly affected by the environment. They must therefore largely regulate their temperatures by behavioral mechanisms, such as seeking shade or basking in the sun. (Fish, amphibians, and reptiles are ectothermic, while birds and mammals are *endothermic*, as we will see in Chapter 17.) Reptiles also developed a tough outer covering that can hold water in, and **amniotic eggs** (Chapter 15) which are water-filled and protected by a leathery sheath. Thus, they can lay their eggs on dry land.

Class Aves

Birds comprise the class **Aves** and are easy to identify: they have feathers (Figure 13.38). They also have scaly legs and feet, unmasking their reptilian heritage. There are about 8,600 species of birds, most of which can fly. Some species have secondarily become grounded; that is, they are the descendants of birds that could fly, but for some reason have lost the ability. Flightless birds include the ill-tempered rhea and the powerful ostrich (ostriches have been known to disembowel predators with a single kick from a huge, clawed foot). The much smaller flightless kiwi rummages through leaves, hunting insects by a remarkable sense of smell. And then there are the penguins, now suffering reduced numbers in some areas since Antarctic vacations became more popular. Nonetheless, even the flightless birds still possess many of the specializations for flight.

The structural specializations for flight are quite pronounced, but can vary depending on the role of flight in the bird's natural history. Some birds, such as quail, spend little time in the air, but must rise quickly from the ground when threatened. At the other extreme, some species, such as frigate birds and some swifts, may rarely land, spending almost their entire adult lives in the air.

So we see that avian anatomy is geared not just for flight, but for different kinds of flight. Nonetheless, a heritage of any flight has placed special demands on the body of all birds (Figure 13.39). For example, they all have a streamlined shape and feathers that can be tightly compressed against their bodies. They also have light, hollow bones that may be filled with long saclike extensions of the lungs. Many species have remarkably developed flight muscles, and some have sacrificed the development of one gonad, thereby lightening their weight. These are all rather marked changes from the reptilian forms that gave rise to birds.

FIGURE 13.39 Modifications for flight are seen in nearly every aspect of the bird's anatomy and physiology, from its streamlined form to its elevated metabolic rate. In spite of the demands placed upon it, the skeleton is extremely light. The frigate bird, for instance, has a wingspan of just over 7 feet, yet its skeleton weighs just 4 ounces. In general, the slender, hollow bones of birds have a deceivingly delicate appearance; in fact, however, they are strong and flexible, containing numerous triangular bracings within. Flight feathers, which can weigh more than the skeleton, owe their extreme strength and flexibility to numerous vanes. These have an interlocking arrangement of hooklike barbules.

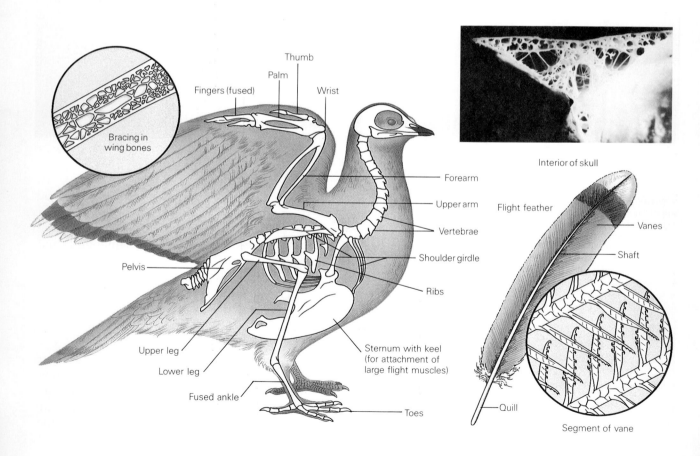

Class Mammalia

Another descendant of reptiles are the mammals. The class **Mammalia** is comprised of hairy milk-givers (such as ourselves). Mammals emerged along with the birds to share dominance as the earth grew cooler during the Cenozoic. It is probably not coincidental that both these groups are able to regulate their internal temperatures (through processes we will see in Chapter 18). The subclass **Prototheria** includes the egg-laying **monotremes,** such as the duck-billed platypus and the spiny anteater (Figure 13.40). Prototherian females lack breasts and secrete milk directly into the belly fur where it is licked up by the infants.

Another subclass, the **Metatheria,** is comprised of the **marsupials—** such as opossums, kangaroos, wombats, and koala bears (Figure 13.41). The young are generally born in a relatively undeveloped state, and most species carry their offspring in protective pouches.

As an aside, no one knew precisely how kangaroos were born until one day when a zookeeper with not much to do was standing around watching a kangaroos's vagina. To his surprise, he saw a tiny sluglike creature emerge and, using its forelegs in a kind of Australian crawl, begin to wend its way through the fur to its mother's pouch. It was an embryonic kangaroo, clearly not well developed. Subsequent research revealed that such tiny embryos attach to a nipple in the pouch and become fused to it for a time. Even while they nurse, the mother may become pregnant again, but her body can hormonally stop the development of the second embryo until the pouch is free.

The third subclass, the **Eutheria,** is composed of the **true,** or **placental, mammals.** Perhaps because we are in this group, more is known about the placentals than the other mammals. The group is rather diverse, so let's review only a few of the orders found within our subclass.

The most primitive order of Eutheria is the **Insectivora** ("insect eaters"). These include moles and shrews—small but voracious animals that live on or under the ground and are constantly and frenetically

FIGURE 13.40 The spiny anteater is a monotreme (a mammal that lays eggs and, when the eggs hatch, feeds its young with milk that oozes from the mammary glands). It has a remarkably sensitive nose and can detect insects that could have sworn they were well hidden.

FIGURE 13.41 Marsupials are rather primitive, pouched mammals. They are rather diverse, as we see here with the cus-cus, pretty face wallaby, and possum.

FIGURE 13.42 Some mammals are not very well known at all, such as this star-nosed mole. Many kinds of mammals live among us, however, and escape our attention by being furtive or living on a different schedule. Flying squirrels, for example, are very common in many areas, but since they are nocturnal, few people have seen them.

FIGURE 13.43 The capybara is a water-loving animal and the world's largest rodent, sometimes reaching 100 pounds in weight. When chased, they may jump into a river and swim away underwater, a most disconcerting habit to the Indians hunting them.

foraging for morsels such as grubs, insects, or worms (Figure 13.42). The **Rodentia** ("gnawers") are the rats, mice, beavers, and other chiselers (Figure 13.43). The **Carnivora** ("meat eaters") include cats, bears, dogs, otters, and jackals (Figure 13.44). The vertebrate carnivores all have elongated canine teeth that enable them to catch game and tear its flesh, but their hunting patterns can be quite diverse. For example, they may hunt singularly or in groups, by stealth or by chase. (Actually, there are probably few true carnivores other than some sharks and flies. Almost all other species include some vegetable matter in their diets.)

The **Artiodactyla** ("even toes") include the pigs, camels, sheep, goats, deer, cattle, hippopotamuses, and giraffes. They walk about on the tips of two toes (you may have noticed that two is an even number). Many artiodactyls can "chew their cud" by regurgitating plant food that they chew a second time to help break down tough cell walls, enabling more of the plant's nutrients to be extracted. The **Perissodactyla** ("odd toes") are the mammals that walk on a single, modified digit. These include the horses, tapirs, rhinoceroses, and zebras (Figure 13.45). The **Chiroptera**

FIGURE 13.44 Some animals are specialized in seeking out and eating the flesh of other animals. These are the carnivores. Since their prey may be quick and sensitive, or even dangerous, many carnivores have developed high intelligence and complex social systems. These lions have killed a wildebeest.

("hand wing") include the bats. These animals may not please your aesthetic sensibilities, but they are for the most part unfairly maligned. Of course, vampires do exist, and they will draw your blood, but bats can be quite beautiful and graceful, such as the fruit-eating flying foxes (with a wingspan of over a meter). In general, most bats, in fact, are helpful insect eaters (Figure 13.46). The **Lagomorpha** ("rabbit shape") include rabbits, hares, and pikas (Figure 13.47). The **Proboscidea** ("front feeder") are the elephants, the group with a nose that defies imagination, and long tusks that are actually modified incisors (Figure 13.48). Elephants have very complex social systems and interact with their environment in very intricate ways that we are only now beginning to understand. Finally, there are the **Primata** ("first"). The primates include the monkeys, chimpanzees, gorillas, orangutans, baboons, humans, and even tree shrews, lemurs, and tarsiers (Figure 13.49).

There are obviously a great many things to be learned about our own kingdom, but time is growing short because our world and its denizens are changing at a dazzling and disconcerting rate. Many animals do not

FIGURE 13.46 Bats are extremely specialized mammals that do not entirely arouse our admiration, perhaps because one species is specialized for gently fanning sleeping humans while they make a painless cut in the skin and then lap the blood. Most bats, however, are harmless or helpful. Many specialize in catching and eating insects. This epauleted bat takes nectar from flowers (here, a baobab flower).

FIGURE 13.47 The pika is often seen by backpackers high in the mountains. They are rabbitlike creatures that survive the harsh elements by taking shelter in burrows among the rocks. They cut plants in the summer months and lay them out to dry before piling them together in huge private stashes underground.

FIGURE 13.48 Elephants are intelligent and highly social animals. They may occasionally fight viciously, and tuskers have been known to fatally wound an opponent. They may also be very helpful to each other. A sick or injured elephant may be helped along by two comrades on either side. Unfortunately, elephant country is being claimed by growing populations of humans, and in some cases, hunters have killed entire small herds illegally, for the sport. Their bodies, tusks and all, have been found in great heaps, left to rot. One can now lease land adjoining the great reserves for the privilege of killing an elephant that inadvertently crosses into the leased land, again for sport.

FIGURE 13.49 Primates are fascinating creatures because of their high intelligence and physical dexterity. If they often remind us of humans, it may be because we are included in this group. We are only now discovering just how much we have in common with some of the other primate species. For example, chimpanzees may cooperate to hunt game and they may practice something akin to war between groups. Females tend their young carefully until they are able to go out on their own.

even live in the kind of world that they were born into. (We certainly don't.) Our studies and descriptions, then, must account for a past that will never exist again and a very near future quite unlike anything existing now.

SUMMARY

The first animals appeared between one and two billion years ago and have since diverged into a number of very distinct phyla, although some phyla are arbitrarily derived. The Porifera are an evolutionary dead-end, probably not strongly related to the other animals. The Coelenterata have a radial body plan and two major layers. Some have two distinct phases. With the Platyhelminthes came a longitudinal plan with bilateral symmetry and a primitive "brain." The Cestoda do not continue the evolutionary advancement just described, but degenerate toward simplification. The Nematoda developed a one-way gut rather than the more primitive saclike gut. The Mollusca are the first group with a coelom, a mesodermally-lined body cavity. With the Annelida come marked segmentation. In the Arthropoda, the segments become specialized for different tasks. The Echinodermata revert toward the radial body plan as adults, but their embryonic development (bilaterality and deuterostomy) reveals them to be relatively close evolutionarily to the Chordata. The phylum Chordata (with a notochord, pharyngeal slits, and a dorsal, hollow nerve cord) includes the subphyla Urochordata, Cephalochordata, and Vertebrata. The Vertebrata include three fishes: the jawless classes, Agnatha, the cartilaginous Chondrichthyes, and the bony fishes, the Osteichthyes. The class Amphibia make a partial transition to land. The Reptilia completed the transition and gave rise to the Aves and the Mammalia.

PART FOUR

REPRODUCTION AND DEVELOPMENT

Chapter 14

PLANT REPRODUCTION AND DEVELOPMENT

Reproduction is one of the most prevalent themes in the drama of life. Humans are keenly interested in it (or at least in the tawdry pageant that accompanies it), and while it is incessantly discussed, it remains one of the most regulated of conversational topics. The tension eases a bit, however, if the topic is *plant* reproduction. After all, what could be unseemly (or perhaps interesting) about plant reproduction? A book called *Sex in the Garden* would be perfectly acceptable in any circle once it was understood that the topic was plants.

However, perhaps plant reproduction is unfairly slighted when we bequeath our prejudices on this topic or that. The various processes involved in begetting plants can be quite interesting indeed. Furthermore, they can illustrate some fundamental aspects of reproduction in a simple form that can then be applied to more elaborate rituals in other species. So, keep an eye out for emerging general principles as we go, and perhaps we can come to some better understanding of just what's going on out there in the garden, and other places as well.

SEXUAL REPRODUCTION

Disadvantages of Sexual Reproduction

We must keep in mind that evolution rewards *only* reproductive output. (Reread that sentence.) The genes of individuals who do not reproduce maximally will be replaced by the genes of those who do. There are no values to be attached to reproduction; it is neither good nor bad. There is only the relentless arithmetic of natural selection. Thus, arithmetically, if one wished to maximize one's genes in the next generation, it would be best to produce genetic **replicas** of one's self. Yet sexual organisms dilute their genes by mixing them with those of a partner. The question of how sex arose has produced a great deal of debate in scientific circles.

Sexual reproduction demands the production of enormous numbers of gametes, although only a few will ever enter into fertilization—a great waste. Also, for some species there is a real chance that the individuals might be so sparsely distributed that two individuals, or their gametes, might not be able to find each other. For example, a plant that releases pollen into the air must not be too far from its nearest neighbor or its effort will be wasted. In point of fact, the probability of breeding failure for members of some species is very high. Obviously, if you can reproduce all by yourself, you not only insure that your offspring will bear *all* of your genes, but you eliminate the risks associated with sexual reproduction (Figure 14.1).

Advantages of Sexual Reproduction

Why, then, have sex? One advantage of sexual reproduction should be apparent if you recall that meiotic processes yield gametes that carry different kinds of genes, and normally there is no way to predict which of these gametes will eventually enter into fertilization. As a result, any individual's offspring are likely to be highly variable. This variation, we have seen, can be critical to evolutionary processes in a changing or unpredictable world.

You recall that under such circumstances, with a genetically variable population there is an increased likelihood of survival for some members of the group. Sexual reproduction also allows the union of superior new genes. Suppose two individuals bearing two newly-mutated superior genes should mate. Those offspring bearing both mutations would be at a marked advantage indeed. Another advantage of sex is that with diploidy, recessive genes can essentially lie dormant in the population. (After all, if they aren't expressed, they can't be subjected to natural selection.) These genes, then, create a reservoir of potential variation should the alternative, dominant gene suddenly become unfavorable.

Any new "good" gene appearing in an asexual population can quickly become established and cause rapid evolution as it quickly increases in following generations. This can be disadvantageous. Suppose an asexual population of pond dwellers is subjected to a period of hot, dry weather. A mutation that allows the population to survive this condition could quickly spread through the gene pool. If, however, things changed and cool, wet weather set in, the entire population might be lost. In sexually reproducing populations, the diploid condition (with masked recessive genes) and the high variation in such groups may help protect against such devastation. Even with such pronounced theoretical advantages to

FIGURE 14.1 There are those who say sexual reproduction has some marked disadvantages.

the evolution of sex, there are those who continue to effectively argue that these advantages simply do not offset the disadvantages.

REPRODUCTION AND LIFE CYCLES IN VARIOUS PLANTS

Green Algae

In most green algae, the haploid gametophyte generation is the dominant or most conspicuous phase. The plant undergoes a period of marked growth during this time (Figure 14.2). The diploid sporophyte stage is usually represented by only a single cell that seems to be waiting out its rigorous season. Whereas the sporophyte is rather unspectacular, its development may have been of enormous evolutionary importance. It is believed to have evolved from a long line of algae that reproduced solely by asexual means. Its appearance in the distant past, then, as algae were beginning to give rise to land plants, would have set the evolution of the elaborate and conspicuous multicellular sporophytes that dominate our landscape today. As the sporophyte gained importance in the rigorous terrestrial realm, it could have become increasingly large and complex until it came to house and protect the gametophytes themselves, as we see in many land plants today. (Remember that most green algae exist as single cells of two different mating types. These could then join in fertilization to form the sporophyte stage.)

In more complex green algae, gametes of the two mating types are clearly distinct in their overall appearances. In these species, the larger (the "egg") waits in a protective body for its smaller, motile partner (the

FIGURE 14.2 Life cycle of *Ulothrix*, a green alga. During the extremely abrupt sporophyte generation, the zygospore (the result of gametes joining in fertilization) undergoes meiosis and becomes filled with flagellated spores called zoospores. These spores form filaments that, depending on the phase of the plant, produce either gametangia (gamete-forming structures) or sporangia (spore-forming structures). The gametangia produce flagellated isogametes (gametes of the same appearance; *iso*, same) that can join in fertilization with gametes from other strains to imitate the sexual phase of the alga. This union results in a new zygospore. The sporangium, on the other hand, releases different sorts of spores that simply settle to begin the growth of new filaments in a form of asexual reproduction.

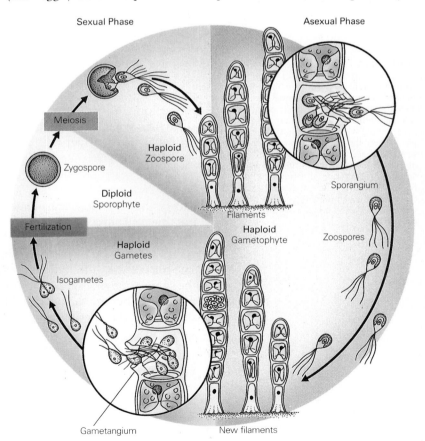

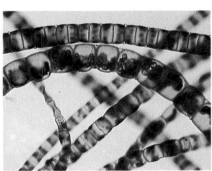

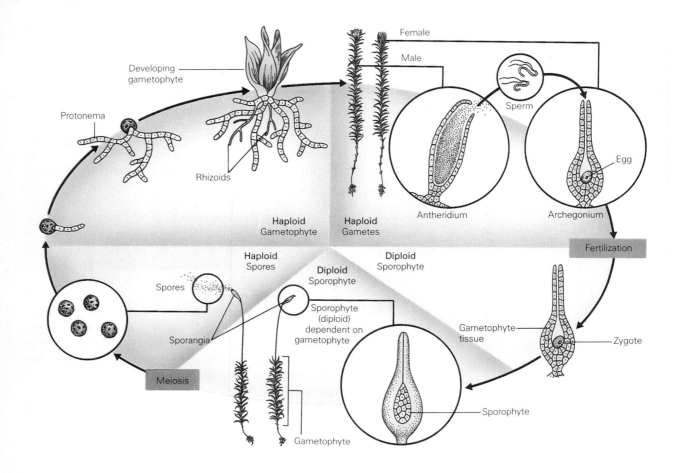

FIGURE 14.3 The moss life cycle. The haploid generation dominates. In fact, the diploid sporophyte grows out of the gametophytic tissue and depends on it for nourishment. The diploid sporophyte is essentially a stalked capsule that forms spores. These are released and borne by the wind to some suitable, moist place where they germinate and produce the slender thread, the protonema. This develops into the familiar carpet we think of as moss. Some tissues produce antheridia that form flagellated sperm, others produce archegonia that enclose a stationary egg. The union of the sperm and egg initiate the new sporophyte generation.

''sperm''). Since the egg doesn't have to move about, it can grow large and save its energy to nurture the zygote. Specialized structures could then be expected to develop that would make fertilization of the egg and protection of the resulting zygote even more efficient. This trend sets the stage for a great many important evolutionary developments, as we will see when we consider the flowering plants.

Mosses

Mosses have made a reasonably successful transition to terrestrial life, but they still have some of the same requirements for reproduction as did the ancient green algae from which they descended. Their flagellated sperm must swim to the egg, which is held in a vaselike **archegonium.** The resulting zygote undergoes meiosis and produces spores that enter the gametophytic stage as seen in Figure 14.3. The gametophytes produce either the male **antheridium** or the female archegonium. From these arise the sperm and eggs that will initiate the next sporophyte stage.

Ferns

The reproductive cycles of the tracheophytes (plants with conducting vessels) are quite varied. One of the most ancient of these cycles is represented by the ferns (Figure 14.4).

FIGURE 14.4 Many North American woodlands are enhanced by the presence of delicate ferns. In the spring, one may find them only blushing with a hint of the light green that will be their color, as they stand protectively curled, tentatively rising from the forest floor. Later they will radiate outward, full-grown, giving the forest a greenhouse atmosphere. Perhaps you have seen those peculiar brown dots on the "leaf" undersides. These are the spore-producing organs that will give rise to the next kind of generation. Ferns are particularly abundant on sodden forest floors, such as are found in rain forests, and are often seen growing from the damp, rotting logs characteristic of such areas.

FIGURE 14.5 Life cycle of the fern. In ferns, the sporophyte generation is highly dominant. After fertilization, the zygote begins to grow from the gametophytic archegonium. It soon produces its own roots and leaflike fronds as the old gametophyte withers and dies. Underneath the fronds form the dark spots called sori, within which are the spore-producing sporangia. The spores are released and, once settled, form flattened structures called prothallia. These form either sperm-producing antheridia or archegonia, in which develop the eggs. Fertilization initiates the next gametophyte stage.

In the ferns and some of their close relatives, such as horsetails and lycopods, we see a pronounced sporophyte stage adapted to life on land. However, the fern gametophyte stage is much more unprotected and vulnerable than that of the seed-bearing plants. Fern gametophytes usually lack conducting tissue, so they must remain very small since no cells can exist far from the surface that comes into contact with water. Water is necessary for their reproduction since the motile sperm from the male gametophytes must actually swim to the female gametophyte. In a sense, then, although fern sporophytes are terrestrial, the gametophyte stage is still essentially aquatic (Figure 14.5).

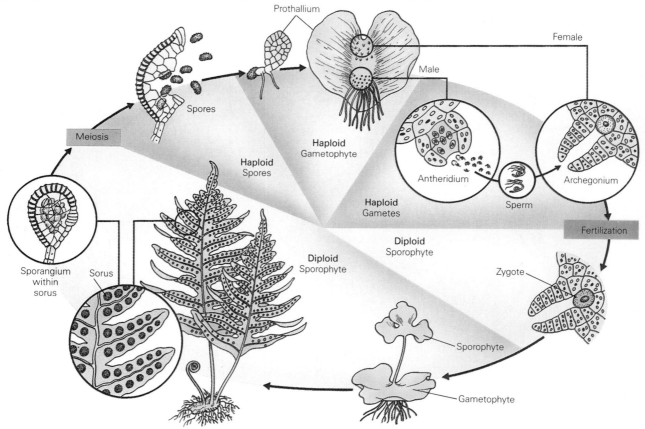

Conifers

Conifers are gymnosperms that, you will probably not be surprised to learn, bear cones. For example, pine trees bear pine cones, if all is going well. But precisely what is a pine cone? (Are you just a bit embarrassed because you really do not know what a pine cone is?) Actually, the question is not as simple as it seems. To begin with, there are two kinds of cones, male **(staminate)** and female **(pistillate)** (Figure 14.6). Early in their development, certain cells deep within the cones (the "mother cells": **microspore mother cells** in male cones, **megaspore mother cells** in female cones) undergo meiosis and produce spores. In the male cone, each spore will become a pollen grain, comprised of only three cells. These delicate pollen grains will eventually become enclosed in a thick, protective covering with thin, winglike extensions. When the scales of the male cone lift, the tiny pollen is released to be carried by the wind, perhaps eventually to settle on female cones. The evolution of tough, resistant pollen was an important step in completing the transition to land.

At the base of the female cones are two cells that will undergo meiosis, each producing four cells. Three of these will degenerate, leaving a single spore that will develop into the female gametophyte. Since there are two cells undergoing meiosis, two female gametophytes will be produced at the base of each scale. Then the nucleus of each repeatedly

FIGURE 14.6 Alternation of generations in a conifer. With the exception of cells hidden in their reproductive structures and an occasional yellow cloud of windborne pollen, all of the conifer we see is the sporophyte. The conifers have no truly separate gametophyte generation. Cells in the male cones, known as microspore mother cells, enter meiosis and form haploid microspores. These microspores form a winged pollen grain and, later, the male gametes. Cells in the female cone, the megaspore mother cells, enter meiosis and produce haploid megaspores. These produce the female gametophyte with two archegonia. When pollen (the male gametophyte generation) lands on the female cone, a pollen tube grows into the tissue surrounding the egg cell within an archegonium. Fertilization occurs when a sperm from a pollen tube reaches an egg cell.

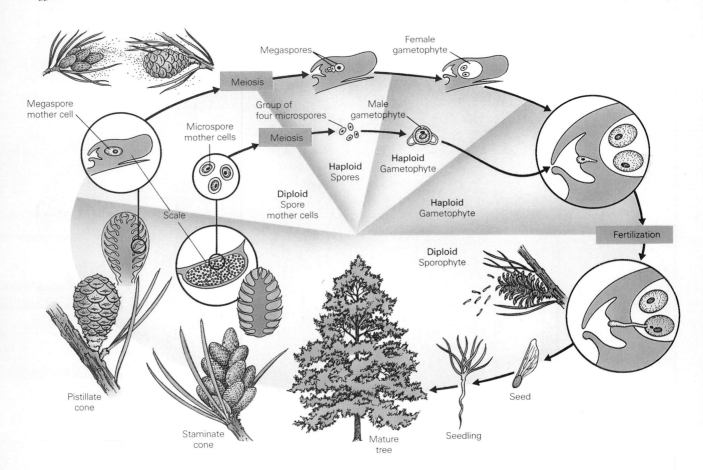

(A) (B) (C)

FIGURE 14.7 The incredible diversity of flowers is illustrated by the differences in sizes of these flowers: (A) the tiny ladies tresses, (B) black-eyed Susan, and (C) one of the world's largest flowers, Rafflesia.

divides, producing gametophytes with multinucleated cytoplasm. Only one of these nuclei will become the **egg nucleus** (it is not known how one comes to be "chosen"). The egg nucleus will be fertilized by the nucleus of a sperm in a pollen grain.

In tracing the route of the pollen grain, we find that when a pollen grain reaches the base of a scale, it usually encounters a moist pool through which it can move until it reaches one end of the female gametophyte. There it germinates and produces a **pollen tube** that slowly elongates, reaching, finally, the archegonium of the female gametophyte. As the tube grows, the pollen grain begins nuclear division, producing the **sperm nucleus,** which will ultimately join with the egg nucleus. After the pollen tube has grown deep into the female tissue, the sperm nucleus is released and makes its way to the egg nucleus, where the two fuse in fertilization. The zygote, or fertilized egg, undergoes repeated mitotic divisions until it forms an embryonic plant that is surrounded by both protective and nutritive tissue. This entire structure is borne exposed (naked) to the elements and is called a **seed.**

Flowering Plants

It sometimes seems that flowers are nature's gift to humans. We make so much of their beauty and fragrance that they have become symbols of grace and elegance in our lives. However, those flowers are not "for" us. We have interjected ourselves into a functioning natural unit. Flowers appeal to us quite coincidentally. In the wild, it might be said, flowers are simply nature's way of making more flowers.

No matter what the role of flowers in nature, a woodland stroll is certainly enhanced by the colorful accents they provide. A spot of blue here, yellow there, then red attests to the great diversity in flowering plants. However, all the flowers, from tiny, quivering bells in a meadow to forest giants (Figure 14.7), are variations on a theme. Therefore, we

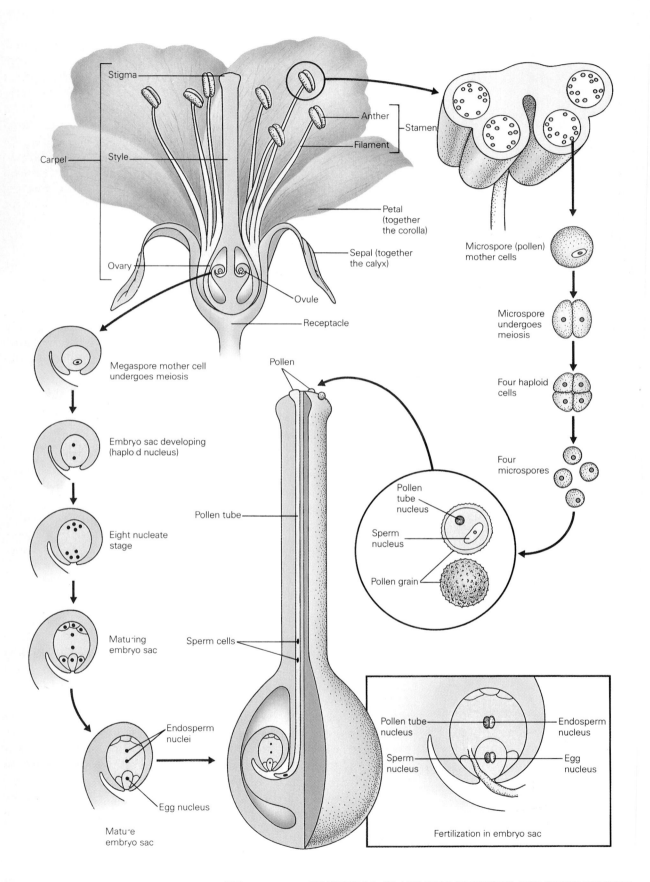

Stigma

Anther

Stamen

Filament

Carpel

Style

Petal
(together
the corolla)

Ovary

Sepal (together
the calyx)

Ovule

Receptacle

Megaspore mother cell
undergoes meiosis

Pollen

Microspore (pollen)
mother cells

Microspore
undergoes
meiosis

Four haploid
cells

Four
microspores

Embryo sac developing
(haploid nucleus)

Pollen tube

Eight nucleate
stage

Pollen
tube
nucleus

Sperm
nucleus

Pollen grain

Maturing
embryo sac

Sperm cells

Endosperm
nuclei

Pollen tube
nucleus

Endosperm
nucleus

Sperm
nucleus

Egg
nucleus

Egg nucleus

Mature
embryo sac

Fertilization in embryo sac

FIGURE 14.8 Overall structure of a flower and fertilization in a flowering plant. The development of the male gametophyte begins when a microspore mother cell divides through meiosis, producing four microspores. Each will become a mature pollen grain when its coat develops and a mitotic event produces a pollen tube nucleus and a tube nucleus.

The female gametophyte, the embryo sac, begins its development when meiosis occurs in the megaspore mother cell, producing four haploid megaspore nuclei (not shown). Three of these degenerate and the surviving haploid cell divides mitotically three times to produce the eight-celled embryo sac within the ovule. The endosperm cell contains two nuclei that will later fuse with the pollen tube nucleus to form the endosperm.

Before entering the ovule, the sperm nucleus divides to form two sperm cells. Since one fertilizes the egg nucleus and the other fertilizes the endosperm nucleus, the result, then, is a diploid zygote and a triploid endosperm.

can make certain generalizations, keeping in mind that we are dealing with a convenient fiction.

Our generalized flower (Figure 14.8A) is composed of four regions, each a **whorl** (or circle) of highly modified leaves. The whorls arise from a widened area, the **receptacle,** at the base of the flower. The whorl closest to the stem is the calyx, formed from leaflike **sepals** that were once the protective covering over the developing bud. The second whorl is the **corolla,** composed of the **petals.** In species that must attract animal pollinators (such as insects, birds, or even bats, see Essay 14.1), the petals may be very bright and attractive, some with just the qualities to particularly attract certain species of pollinators. The third and fourth whorls contain the reproductive organs. The male part is the **stamen,** the female part, the **carpel.** Each stamen consists of a slender **filament** capped by an **anther** where, following meiosis, the male gametophyte is produced and released as **pollen.** Each carpel has three parts: the **ovary, style,** and **stigma.** Carpels are often fused.

The ovary is the widened base of the carpel. It will later form the fruit. In its early developmental stages, the ovary contains the cells that will undergo meiosis to produce the female gametophyte. After fertilization, it will house the growing seed and then the developing embryo. The style is the slender stalk that arises from the ovary. At its tip is the stigma, a sticky or hairy structure that receives the pollen.

How Flowers Work

Within the flower's young ovary are the **ovules.** Ovule technically means "little egg," but these are not eggs. They contain the sporophyte cells that will produce the female gametophyte. Each ovule consists of a megaspore mother cell surrounded by nutritive and protective tissues. The megaspore mother cell is diploid, but it will produce the haploid female gametophyte by meiosis. The meiosis results in four haploid cells, but three disintegrate, leaving a functional megaspore cell that goes through three rounds of mitosis to produce eight haploid nuclei. This eight-nucleate structure is the mature female gametophyte, now called the **embryo sac.** Cell walls then form in such a way that all but two of the nuclei (the **endosperm nuclei**) are isolated. There are now three isolated nuclei at either end of the embryo sac, and one of these begins changes that will produce the egg cell.

While this is going on, the male gametophyte is also developing in preparation for that glorious union of sperm and egg. Within the anthers are chambers called **pollen sacs** that contain numerous diploid cells. There are the flower's microspore mother cells. Through meiosis, each produces four microspores. Each of these then doubles by mitosis to produce two-celled male gametophytes. One of the cells is the **sperm cell,** the other, the **pollen tube cell.** The pollen tube cell then expands to enclose the sperm cell. A tough, resistant coat forms over the two. This is the **pollen grain.** So we see that the gametophyte stage, which was so prominent in more primitive plants, is now represented only by the pollen grain and embryo sac.

Pollination occurs when the pollen grain lands on the stigma, but actual fertilization does not occur until later (Figure 14.8B). On the moist stigma, the pollen grain opens and sends a long tube down the

ESSAY 14.1 THE COEVOLUTION OF POLLEN AND INSECTS

We have seen the evolution of tiny, resistant pollen that requires no water and can even be carried about by light winds. This development, then, set the stage for the development of another means of pollen transport. As flowering plants were evolving on earth, so were other species. Many of these developing forms began to react to each other and to utilize one another in increasingly intricate and complex ways. As the insects were spreading out, invading new niches, taking advantage of what the environment offered, some began to exploit the food in flowers. These insects, taking food from one flower after the next, became vehicles for pollen transport. As time passed, some flowers would have developed an increasing dependency on the visits of pollen-bearing insects.

A bee pollinating an orchid. Each is well-adapted to the other.

In time, plants began to compete for the attentions of insects. They would have developed increasingly attractive colors that insects can see (such as those with shorter wavelengths, like violet), sweet nectar, and shapes that are somehow appealing to insects (such as broken margins—as are produced by petals). Those that attracted more insects left more offspring, and some lines of flowers became true specialists in attracting certain insects. Insects, on the other hand, became increasingly specialized so that they could tap the offerings of the specific sorts of plants that yielded the nutrients most closely fitting their specific needs. The insects and the flowers changed in ways that enhanced their interdependency and specializations in a steady progression toward finer attunement to each other. Such reciprocal influences on the development of interdependent species is referred to as **coevolution.**

style, with the pollen tube nucleus near the tip and the sperm cell nucleus lingering farther back. The sperm cell then undergoes mitosis once to produce two identical sperm.

The pollen tube finally penetrates the ovule through a tiny opening called the **micropyle,** and the two sperm enter the embryo sac. Only one of these will fertilize the egg cell. The resulting zygote then begins the next sporophyte generation. Meanwhile, the other sperm penetrates the large binucleate cell in the center of the embryo sac. The three nuclei fuse to form a special nutritive tissue called the **endosperm.** (The endosperm tissue also provides us with the flour and meal from wheat, corn, rice, rye, millet, and oats.

As the embryo develops within the seed, it quickly forms either one or two wings of tissue. These are the cotyledons (see page 274). In many monocots, the single cotyledon absorbs food from the surrounding starchy endosperm, while in dicots, the cotyledons themselves contain food (Figure 14.9).

FIGURE 14.9 **FIGURE 14.9** Two reproductive developments helped the gymnosperms adapt to the new conditions. The pollen grain carried the male gamete safely in a tough resistant case, and the embryo came to be protected in an equally tough container of its own, the seed. The pollen shown here in the SEM photograph is from the northern white cedar. The seed contains the embryo of the plant, some stored food, and a surrounding seed coat, which is hard and water-resistant. The seed of the pine is illustrated here. The protective seed coat makes it possible for the embryo to remain intact for considerable periods of time, until conditions are right for growth.

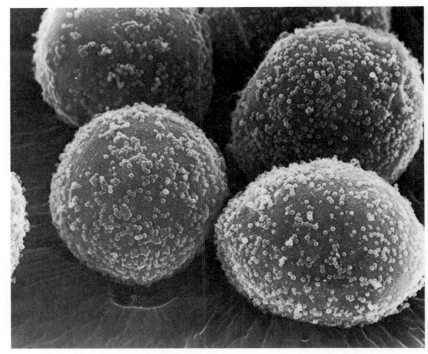

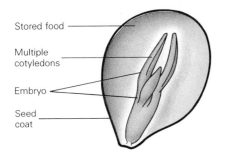

Stored food

Multiple cotyledons

Embryo

Seed coat

The seed is well-adapted to the terrestrial life, as evidenced by its tough, protective **seed coat.** Once the seed is released into the environment it will remain dormant, germinating (beginning growth) only when triggered by the proper stimuli. This waiting period varies enormously among different kinds of seeds. For example, the seeds of the mangrove may start to germinate while the fruit is still on the tree, whereas Indian lotus seeds have been known to lie dormant for as long as 400 years, undoubtedly frustrating generations of gardeners.

Unlike the "naked" seeds of gymnosperms, the angiosperm seed develops within an ovary. As the seed develops, the ovary undergoes rounds of mitosis and grows to form a structure called the **fruit.** You may have an idea of what a fruit is, but technically, a fruit is any structure that develops from the ovary and encloses ovules or developing seeds. Fruits come in all sorts of sizes and shapes, ranging from grains and nuts (dried fruits) to pineapples and raspberries (clusters of small fruits) to fleshy tomatoes and apples. (You thought a tomato was a vegetable, didn't you?)

These, then, are the events of reproduction in plants. Reproduction is followed by a remarkable chain of developmental events that eventually produces the familiar kinds of plants that grace our world. As one might expect, different kinds of plants develop in different ways, but because of a common heritage, they all progress along a basic developmental theme.

PLANT DEVELOPMENT

Plants develop differently from animals in two basic ways. First, because of the rigid walls of plant cells, new cells are prevented from developing in the interior tissue, but this is not the case in animal tissue. Plants

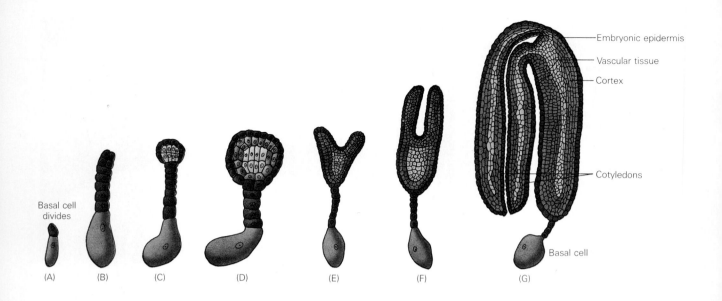

Embryonic epidermis
Vascular tissue
Cortex
Cotyledons
Basal cell divides
Basal cell
(A) (B) (C) (D) (E) (F) (G)

FIGURE 14.10 Development of *Capsella,* or shepherd's purse. Note the uneven first division. The larger daughter cell will grow and ultimately become the basal cell. The smaller cell will continue dividing linearly for a few divisions until the terminal cell begins forming smaller cells that will produce a globular embryo. The embryonic tissue will grow increasingly specialized until, as we see in part G, several distinct tissues have formed. A micrograph of a developing *Capsella* shows a stained specimen and the nuclei appear as dark spots in each cell. The micrograph indicates some of the problems with relying on the light microscope to study cell detail. The cell walls are indistinct and many of the organelles are difficult to distinguish. However, compare the photograph with the artist's renderings in part G.

therefore grow by adding new cells to the **periphery** of the established cells. Second, whereas most animals stop growing at some point (maturity), plants continue to produce new cells and to grow throughout their lives.

Let's begin by considering the early development of the angiosperm, starting with the seed. Before the seed germinates (sprouts), one of the enclosed cells, the endosperm nucleus that will nourish the seedling, begins a period of rapid division followed by the mitotic divisions of the zygote itself. As we see in Figure 14.10, the first division of the zygote is uneven, so one daughter cell is larger than the other. The smaller cell continues to divide, forming a chain, or stalk, of cells. Cells near the end of the stalk then begin to divide in every direction, producing a globular **embryo.** As the divisions continue, the cells in the embryo begin to change and become different from one another. Such a developmental process is known as **differentiation,** or **specialization.** Some will eventually produce the **epidermis,** or protective "skin;" others will form the **vascular** tissue, which transports nutrients; and still others will form the **cortex,** or structural tissue.

As differentiation continues, the globule flattens and elongates, forming the embryonic leaves, or **cotyledons.** Seeds of cone-bearers, such as pines, have many cotyledons, but those of flowering plants have only one or two. Those with one are called **monocotyledons;** those with two, **dicotyledons.** In general, broad-leaved plants are dicotyledonous (or **dicots**) and narrow-leaved plants such as grasses are monocotyledonous (or **monocots**). Some differences in monocots and dicots are shown in Figure 14.11.

There are areas of the plant in which the cells do not mature or differentiate; instead, they remain permanently embryonic. The division of cells in two of these areas is responsible for growth in the length of the roots and stems. They are called **apical meristems** and are located at the root tips and the stem tips. As the roots and stems lengthen, new embry-

FIGURE 14.11   Summary of differences in monocots and dicots.

Monocot

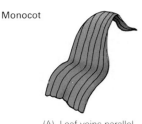

(A) Leaf veins parallel.

(B) One cotyledon (as in corn).

(C) Flower parts in threes or multiples of three.

Vascular bundles

(D) Vascular bundles scattered. No vascular cambium.

Dicot

(A) Leaf veins branching (netlike).

(B) Two cotyledons (as in beans).

(C) Flower parts in fours or fives or mu tiples of four or fives.

(D) Vascular bundles in cylindrical arrangement. Vascular cambium present.

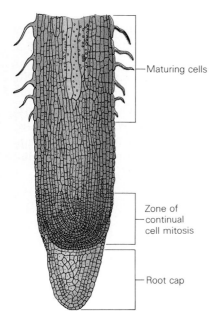

- Maturing cells

Zone of continual cell mitosis

- Root cap

FIGURE 14.12 The development of the root. The tip is shielded by relatively large cells comprising the root cap. Such protection is necessary as the delicate root pushes its way betweeen soil particles. The area it protects is the site of intense mitotic activity, hence these cells, not yet mature, are smaller. As they mature, they increase their cytoplasm and inclusion bodies, eventually elongating. The most exterior of these may send out delicate root "hairs" through which water is readily absorbed.

onic tissue is formed near the tips as the older meristematic cells are left behind to mature and specialize. (The root meristem is seen in Figure 14.12.)

The apical meristem of the stem differs slightly from that of the root in that the stem tip lacks a protective cap, and it may have leaves. Also, the stem apex is divided by **nodes** that are the future sites of new leaf formation. The stem tip is not protected by a cap, but is safely shielded deep in a **terminal bud.** Thus, the small, mitotically active cells behind the tip can carry out their complex functions with some degree of protection from disruptive influences (Figure 14.13).

In the trees and shrubs of the temperate zones, the terminal bud opens in the spring, giving rise to a new area of tissue. Then, in late summer or fall as growth ceases, a new terminal bud forms. This means that one can determine how much a stem has grown in any year by measuring the distance between the scars. If the weather was unfavorable in a certain year, the distance will be small.

The lateral buds on stems end up doing nothing if their growth is suppressed by hormones from the terminal bud at the tip. If the buds are not suppressed, they grow into branches. Such growth occurs, for example, if the terminal bud has been removed. Gardeners sometimes snip off the terminal bud in order to encourage a plant to be more bushy.

Just as apical meristem allows growth in length, **lateral meristems,** also called **cambia** (singular, **cambium**), are the tissues that enable growth in girth or diameter. This kind of growth, however, presents plants with a special problem. As a woody plant grows in diameter, the vascular system must be enlarged by adding cells to the **xylem,** which carries water up from the roots, and to the **phloem,** which carries food down from the leaves. Since these areas lie deep within the plant, it

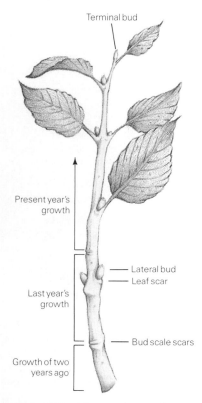

FIGURE 14.13   The growth area of a stem.
One can read a bit of the plant's history by
noting the distance between bud scale scars,
the distance being greater for good years
when the stem has been able to grow rapidly.

FIGURE 14.14   Section from the stem of a
dicotyledonous plant. The woody xylem
shows two years' growth, the winters being
recorded in the areas of the smaller cells (and
hence less growth). The vascular cambium
produces xylem on its inner surface, phloem
on its outer.

might be asked, how can new cells be added without rupturing the
plant's surface? The answer is that there are two kinds of cambium. The
vascular cambium produces new xylem and phloem and the **cork cambium** produces new cork, a protective tissue that functions as a kind of
epidermis, or skin, which protects the tissue below that would otherwise
be exposed as growth caused the surface layers to rupture. The vascular
cambium lies between the xylem and phloem, and as it produces new
cells, those toward the inside of the stem form xylem and those toward
the outside form phloem (Figure 14.14).

Plants are able to carry out a number of very specialized functions that
belie their simple and seemingly inactive appearance. For example, the
thick-walled tissues called **sclerenchyma** and **collenchyma** are structural supports (Figure 14.15). **Parenchyma,** on the other hand, is composed
of thin-walled cells with large vacuoles. Some parenchyma cells contain
chloroplasts.

Plant development, we see, is very complex and coordinated. Obviously, there must be some means of regulating such activity. That regulation, we will see now, is largely due to the activities of special kinds of
molecules to which the plant is particularly sensitive.

PLANT HORMONES

The seeming simplicity of plants sometimes leaves us unprepared for
their complexities. As a case in point, researchers have for years been
trying to unravel the mysteries of plant hormonal control. **Hormones** are
substances formed in one part of an organism that are transported to
other places where they cause changes. Many aspects of plant growth
and development are under the control of these itinerant chemicals, and
so the search is not only fascinating but increasingly important as we
grow more reliant on these green, intermediate links with our life-giving
sun. Let's briefly consider three of the better known groups of plant
hormones involved in regulating growth and development: **auxins, gibberellins,** and **cytokinins.**

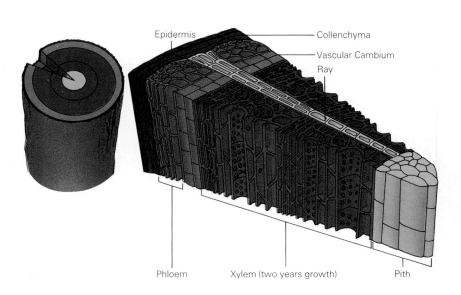

FIGURE 14.15 Sclerenchyma (left) and collenchyma (right). Both lend support to plants. The sclerenchyma dies at maturity, its thickened walls remaining. The collenchyma is alive when functioning, and its strong, supple walls are prevalent in growing plants.

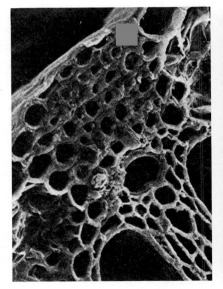
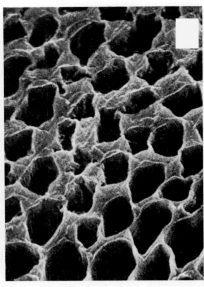

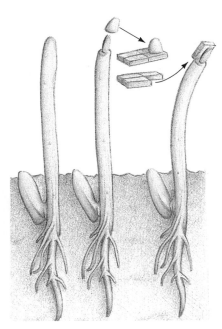

FIGURE 14.16 In this classical experiment, the tip was cut from an oak seedling and placed on an agar cube which absorbed any fluids that might seep from the tip. Then the block was placed off center on a decapitated seedling. As the seedling grew it bent away from the side with the agar block (meaning that cells on this side were growing faster). The experiment demonstrated that the fluids from the plant tip contained substances that enhanced growth. Those substances, it was discovered, were plant hormones.

Auxins

Some of the first experiments on plant hormones were conducted by one Charles Darwin and his son, Francis, and reported in a 1881 publication, *The Power of Movement in Plants.* After noticing that plants tend to bend toward light, the Darwins wondered if the plants were responding to some sort of signal from the stem tip. So they covered the shoot tips of growing plants and exposed them to light from the side. The plants now did not bend toward the light. However, they found that if the tips were capped with transparent glass, normal bending occurred. They wrote, "We must therefore conclude that when seedlings are freely exposed to a lateral light some influence is transmitted from the upper to the lower part, causing the latter to bend."

In 1926, the Dutch physiologist, Fritz Went, discovered something else about that "influence." He cut off the tips of emerging oat seedlings and set them on a block of gelatinous agar for about an hour. The decapitated plants were kept in a dark place. Then they were removed and pieces of the agar on which the tips had rested were placed along one side of the stump. Within an hour, the plant began bending in the opposite direction (Figure 14.16). Thus, the "influence" was believed to be chemical and was called **auxin** (from the Greek *auxein,* to increase). It was later found that auxin is generally produced in the stem tips and seeps through the cell tissue to the rest of the plant, instead of travelling in pipelines of xylem and phloem.

How does auxin control elongation in plants? No one knows. It is known that auxin softens the rigid cell walls of plants and allows water to swell the cells. Thus, growth may be accomplished by cell enlargement rather than cell division.

A synthetic form of auxin, called 2,4-D, is a powerful weedkiller (as is auxin at high concentrations) and was used by the military in Southeast Asia to expose enemy movement through the jungle.

FIGURE 14.17 Chemicals such as auxin are used as defoliants for a variety of crops to increase the ease of machine-harvesting.

Small amounts of auxin may stimulate the growth of roots, and although it doesn't affect the growth of leaves, it plays a role in the leaf drop and is often used as a defoliant (Figure 14.17). With the approach of fall, the plant draws certain ions, amino acids, and sugars from its leaves, and auxin helps to break down the cells that hold the leaf to the stem. The application of auxin also stimulates fruit to grow without the flower having ever been pollinated, thus growers can produce seedless varieties.

Gibberellins

While Went was performing his agar experiments, E. Kurosawa in Japan was looking for the cause of the "foolish seedling disease" of rice. The diseases caused the plants to become spindly and pale before finally collapsing. It turned out that the culprit was a parasitic fungus that contained substances called **gibberellins.**

Gibberellins are found in most, if not all, plants, and they have some rather surprising properties. For example, they cause dramatic increases in stem length by stimulating cell division and cell elongation in both leaves and stems. I recall that in my first garden, the lettuce, to my astonishment, grew to about five feet tall. It was very embarrassing. However, it turns out that in some plants, certain weather patterns can produce effects that are similar to the effects of gibberellin (Figure 14.18), and the weather in Santa Barbara was right for tall lettuce.

Finally, gibberellins stimulate pollen germination and the growth of pollen tubes in a number of plant genera and can break the dormancy of many kinds of seeds that normally can be aroused only by cold or light.

Cytokinins

In 1941, a Dutch physiologist, J. van Overbeek, found that coconut milk (which is really a liquid endosperm, or "seed food") contained a peculiar growth factor unlike anything known. The factor, whatever it was, not only accelerated the development of plant embryos, but it increased the growth rate of isolated cells in a test tube. Also, a drop of coconut milk could stimulate mature, nonmitotic cells to begin dividing again.

What, exactly, in the coconut milk was causing such changes? After all, coconut milk is rich in a variety of substances. Years of research efforts to isolate the growth factor proved futile. Then researchers began looking for the factor in something besides coconut milk, something easier to work with. Somehow a graduate student discovered that old herring-sperm DNA could make tobacco cells divide. (One wonders what possessed him to investigate the effect of aging fish sperm on tobacco.) Others picked up on the line of investigation, and it was soon found that any stale DNA would provide the factor. Apparently, it was a breakdown product of nucleic acid. The product was finally isolated. It was called **kinetin** and was found to be in the chemical group called **cytokinins** (from "cytokinesis," cell division).

Cytokinins can react with auxin to produce a variety of growth effects, such as rapid cell division. It can also work alone to enhance germination once it has started. In addition, cytokinin can somehow

FIGURE 14.18 Treatment with gibberellins may cause bizarre growth in plants. The normal cabbages on the left are shown with gibberellin-treated cabbage plants that have "bolted."

keep leaves from turning yellow after they are removed from the tree. Apparently, it keeps the DNA sequences that function in young, healthy leaves from ceasing activity. In general, however, the workings of cytokinins remain a mystery.

Reproduction and development in plants is highly regular, specialized, and coordinated. Cells divide, grow, and move in a silent ballet that moves strangely, and often in complex ways, to their own cellular choreography. Their dance may mystify us and leave us a bit wide-eyed at their dexterity. The ritual is so compelling that we may, if even for a moment, think it is all done just for us. But it is not so. Even when it seems plants are at rest, they are not. On a peaceful forest walk, those plants that shade us and soften our steps are not resting with us. They, at least, are working. They are alive, after all. So even as we sit among them, they are reproducing, growing, changing, reacting, and struggling to maintain themselves. They are carrying out life's tasks, often in enormously complex ways—ways tested by time and the relentless processes of natural selection.

SUMMARY

The disadvantages of sexual reproduction include dilution of genes in one's offspring (by mixing with the genes of the other parent) and the element of chance in getting gametes together. The advantages include increased variation in the offspring and the chance of fortuitous new combinations of genes. Reproduction varies widely in different sorts of plants. In green algae, the gametophyte is dominant. In some species, the gametes are quite distinct. Mosses are better adapted to the terrestrial life but their flagellated sperm must still swim to the egg. Ferns have a pronounced sporophyte stage with a relatively vulnerable gametophyte that (unlike the sporophyte) lacks conducting tissue and must remain in contact with water (through which the sperm must swim). Conifers bear male and female cones. Pollen released from the male cones contains a sperm nucleus that will fertilize the egg nucleus produced from the female gametophyte. The resulting seed is naked. Flowering plants are reproductively varied, but the sporophyte stage is highly dominant. The seed is enclosed in a protective seed coat and develops within an ovary.

Monocots and dicots develop differently, but along somewhat similar lines. Growth takes place in the meristematic areas. As they grow, they add vascular tissue, the xylem and phloem. Structure is largely maintained by strong collenchyma and sclerenchyma cells. Parenchyma is thinner-walled and often contains numbers of chloroplasts. Plant development is largely under the control of hormones, such as auxins, gibberelins, and cytokinins.

Chapter 15

ANIMAL REPRODUCTION

It struck me as strange a few years ago when I was writing a book called *How They Do It*, just how little we know about animal sex. It seemed strange because we are obviously fascinated by practically anything that has to do with sex, and the weirder the better. I learned there is indeed weird sex going on out there (Essay 15.1), but most of us really don't know much about it. One reason is because the information is not easy to find. I was forced to scour reams of dusty journals to dig out bits and pieces of information that had somehow failed to make its way to the public view. After the book was published, I learned that many people (even urbane, educated people) were somehow embarrassed to take the book to the cashier. (Many of the books, I was told, were *stolen!*)

Why were people embarrassed? Why is the subject such a delicate one? You undoubtedly have your own ideas about such matters, but it cannot be denied that sex is a touchy subject, even that of other species. Nonetheless, this chapter is about animal reproduction, so it seems that we will be travelling over some hallowed ground. Of course, all due effort will be made to maintain our usual humble reverence.

ASEXUAL REPRODUCTION

Some animals avail themselves of the great advantages of asexual reproduction that we discussed in the last chapter. As you might expect, these are generally rather primitive species, but the processes can involve some rather elaborate and complex patterns. We will consider two rather distinct cases here.

Binary Fission

A number of organisms reproduce by simply splitting in two, a process called **binary fission** (Figure 15.1). The planarian flatworm can also reproduce itself in this way. (It can also reproduce sexually, but we'll ignore that small inconvenience for now.) In the process of fission the flatworm simply begins to constrict about midway along its length until it pinches in two. Cells from each part then move into the severed area and regenerate the missing parts. The result is two new flatworms.

It is apparent that the cells that rebuild the missing structures in such cases must undergo some sort of fundamental reorganization. After all, these cells had already differentiated and specialized and were functioning in some specific role. Now they must somehow change and take on a new role. In some cases, cells can reverse their development; that is, after proceeding along one very specific developmental pathway, they reverse the process, revert to an earlier stage, and proceed along a different developmental pathway. (If a flatworm is cut in half, each half can regenerate the missing parts, presumably by the same reorganizational pathway.)

Reproduction through regeneration is possible in some species that do not normally reproduce by such means. Consider, for example, various starfish, a group that normally reproduces sexually. Only a few years ago, oystermen hauling up a starfish in a dredge would chop up the unfortunate creature and throw the pieces back in the water, quite sure that the starfish would eat no more oysters. However, they were wrong. The oystermen didn't know that an entire starfish can regenerate from a single small piece. As a result, the starfish population flourished, and the oyster catches dwindled, causing the oystermen to chop away with even greater diligence until biologists were able to convince them to axe their behavior instead of the starfish.

FIGURE 15.1 Paramecium reproduce through binary fission. Basically, the body pinches in two roughly halfway along its length. Each end then regenerates the missing parts. (Photos from Carolina Biological Supply Company.)

Budding

Another method of asexual reproduction is **budding,** a process in which a new organism is produced as an outgrowth of the parent organism. The new appendage develops while attached to the parent, then pinches

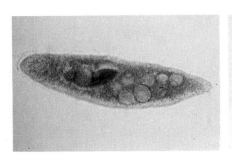
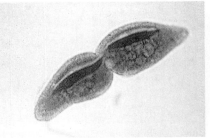
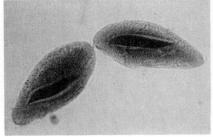

off and moves away on its own. Many of the simple invertebrates reproduce this way. An example is found in the tiny Hydra, a freshwater relative of the sea anemone (Figure 15.2).

SEXUAL REPRODUCTION

Sexual reproduction may not involve the intimacy that many of us associate with the process. This is because fertilization may be accomplished **externally,** outside the body, as frogs do (Figure 15.3), as well as **internally,** as we see in mammals. The two methods involve quite different evolutionary strategies. In external fertilization, the eggs and sperm are scattered into the environment, leaving the rest to chance. External fertilization apparently evolved first, and whereas there is great wastage of gametes, there is little cost involved in the production of each.

In some cases, external fertilization can involve rather complex and even moving rituals. For example, the Caribbean fireworm is about an inch long and spends its time burrowed in the bottom of bays in the West Indies. On the fifth night after an August full moon, the worms crawl out of their burrows and swim to the surface of the warm tropical seas. In a remarkable timing feat, they reach the surface after sunset, about an hour before the moon rises and in the blackness of night, so that nothing can interfere with what they are about to do. Then, for the only time in their lives, they become phosphorescent. The water sparkles and dances with greenish blinking lights. This is their time of reproduction. The water seems to be alive with their shimmering bodies. Then, suddenly, in a tiny phosphorescent explosion, clouds of eggs and sperm fill the water, and the ruptured bodies of the fireworms sink slowly to the bottom, leaving the gametes to play out a game of chance in the dark waters above.

The method obviously works for fireworms and for the salmon that shed their gametes in northern streams each year. But, one might ask,

FIGURE 15.2 The *Hydra* can reproduce by a variety of mechanisms, one of which is asexual budding. In this process, buds appear along the trunk of the body. They grow into tubelike structures complete with developing arms. Finally, the structure, resembling a tiny adult, breaks off and swims off alone.

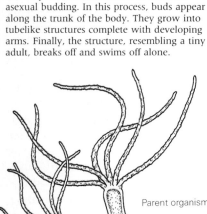
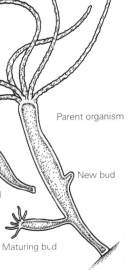

Parent organism

New bud

Genetically identical new organism

Maturing bud

FIGURE 15.3   Insemination in frogs. The male clasps the female from the back, gripping her behind her forelegs. As she releases her eggs he pours sperm over them, and the eggs are then on their own, protected only by their gelatinous coating. This method of insemination is somewhat more intimate than simply releasing eggs and sperm randomly into water, but less "personal" and less efficient than the male injecting sperm directly into the reproductive tract of the female. Other frogs have rather interesting ways of getting their gametes together, such as pouring them down a leaf funnel over a body of water.

isn't there a better way? (I'm sure you have an answer, but let's think in evolutionary terms for a moment.) Why risk having the sperm or eggs exposed to such a hostile environment? Why not just put the sperm inside the female where the eggs are? (I can hear the cheering.) With the sperm safe in her body, there would be less risk of unsuccessful matings and fewer metabolically expensive eggs and sperm would have to be produced. Thus, internal fertilization arose. Probably at first, the reproductive orifices were simply pressed together as the male ejaculated, as most birds do today. In time, however, many species would have developed specific reproductive structures and behavioral patterns that would have increased the probability of successful matings. The array of specialized male and female sex organs of today's creatures attests to the pervasiveness of this evolutionary direction. So today, sexual reproduction in many animals is internal, accomplished by **copulation,** the insertion of the penis into the vagina where sperm is deposited. In many species, the timing of the copulation is critical to maximum reproductive output. For example, mammals copulate only at specific periods—usually when the female is physiologically and behaviorally receptive (in **estrus,** or heat). Such timing is likely to increase the likelihood of the young being born at the most opportune time of year, such as when the weather is warm or food is most abundant.

In many mammalian species, including dogs and horses, the timing is controlled by the receptivity of the female. Her condition, in turn, may be controlled by environmental conditions, such as the changing of the seasons. In many species, the female advertises receptivity through odors associated with vaginal discharge and changes in her behavior. In some primates, receptivity is also communicated by a reddened and enlarged area around the rump. (Receptive female chimpanzees are sometimes referred to as "pink ladies.") Obviously, sexual signaling can be accomplished through a variety of means.

Many people seem to be under the impression that sex is sex, that basically all the species must do it the same way. Nothing could be further from the truth. Some animals do behave in ways that humans would recognize, but most of them don't. For example, bedbugs are often homosexual, however neither male nor female bedbugs can survive many matings because the male pierces his mate's back with his sharp penis and ejaculates directly into the body cavity. Special cells then capture and ingest many of the sperm as they roam the recipient's tissues. Thus, the recipient is nutritionally rewarded. Other species may also surprise us. In some mites, brothers and sisters copulate before they are born, so the little females are born pregnant. Certain snails have an enormous penis just over the right eye. They are hermaphroditic (each possessing the organs of both sexes), but they can't exchange sperm until they have pierced each other with a chalky dart that usually acts as a

sexual stimulant but can also kill. Other snails begin life as wandering males, but evenutally settle down and become sedentary females. If a wandering male mounts a female, he must copulate before his masculinity fades. The female praying mantis sometimes devours the male even while he's copulating with her. In fact, after she eats his head and brain, his sexual behavior becomes more vigorous (but we'll try not to extrapolate from that).

Some snails, fish, and lizards change gender, so they may be a father at one time and a mother at another. In yet other fish, there are no males at all. The egg is stimulated to develop when the female mates with a male of a different species. His sperm only activates the egg; it doesn't join with it.

On the kinkier side, by our standards, we find geese that form ménages à trois (a threesome, usually with two males). Rape, by the way, is common among lobsters, skunks, and orangutans. Furthermore, the female of many species

Animals mate in a variety of ways. Bighorn sheep rut in a fairly traditional way. The male dragonfly deposits sperm in a receptacle near the female's neck. With lions, conflicts may arise if the female is not receptive. And wood ducks mate by pressing cloacas together.

must be subdued before they will allow mating. These include camels and rhinos. Rhino females are apparently stimulated by a good fight.

The reproductive structures of many animals can be somewhat surprising. Males of many insects have penises that look like instruments of torture with points, hooks, barbs, and impossible angles. That of each species is so distinct that interbreeding is usually not possible. Male opossums have split penises, with grooves instead of tubes, that match the divided vagina of the females. Pigs have a corkscrew-shaped penis that locks tightly into the female's vagina. Snakes and alligators can copulate with either of two penises that evert from the cloaca like turning the finger of a glove inside out. Each is covered with backwardly-directed barbs, as are the male organs of cats and skunks. And, finally, as the old joke goes, how do porcupines do it? They, indeed, do it very, very carefully.

HUMAN REPRODUCTION

In humans, theoretically, there is no period when the female is not sexually receptive or when she is not sexually attractive to males, your personal experience notwithstanding. Women are physiologically able to copulate at almost any time and use no particular signal to attract males. (Although several sure-fire ones may come to mind.) It has been suggested that the continual receptivity of women may have evolved as a means of enticing larger and stronger males to remain with a specific female, one that may be bearing his children. In such an association, the female might have greater reproductive success if she can elicit the male's assistance, and the male might also leave more offspring by assisting his reproductive partner, who is usually smaller and also burdened with the responsibility of nursing the young. Traditionally, the male has been useful both in protection and in bringing home protein-laden meat, while the female gathers plant material (she actually supplies more calories than the male among today's hunters and gatherers). Of course, the whole thing has not been a cold business arrangement (at least historically it hasn't been). It is simply the system that has produced the most offspring, and therefore it has become woven into our social fabric.

Considering such matters in scientific terms admittedly puts some people's teeth on edge. We should hasten to say that such theorizing does not neglect those more tender emotions of which we are so proud. A certain bonding, feeling, or attraction seems to develop between reproductive partners, emotions that serve to strengthen a union—even if it was originally established on a purely physical basis. Perhaps this kind of bond provided the basis for what is called "love," or perhaps such bonding *is* love. It's hard to say since no universally acceptable definition exists. In any case, the human bond often develops and, in fact, people seem to have a need for strong attachments with others. Attraction between people, as we all know, may be long-term or short-term. The resulting bonds may also be of intermediate length, or intermittent, and even retractable. Some long-term relationships among humans are sometimes formalized and referred to as "marriage," a system that, in our culture, discourages the breaking of the relationship. Interestingly, other species may also build long-term bonds. For example, geese often mate for life, and a member of a pair may never mate again if its mate is lost.

Copulatory behavior in many species is rather easy to predict. Naturalists are often generally aware of when the wild animals of their area will begin breeding, and animal breeders usually have a keen sense of when copulation is likely to occur between their charges. Humans, however, are highly variable in this regard. There is no specific point at which a man and woman are likely to initiate copulation. It can happen any time the coast is clear. Furthermore, the events preceding copulation in humans may vary widely. In certain other animals, specific signals always precede the copulatory act. such as a female bird "soliciting" by lowering her fluttering wings, or a female chimpanzee's "presentation" of her livid rear end to a male (although the blasé males may not mount the soliciting temptress). In humans, under normal conditions the physical prerequisites of copulation or **intercourse** are simply an erect **penis** (Figure 15.4) and a lubricated **vagina** (Figure 15.5). These conditions are usually achieved during the course of precopulatory sexual behavior, or

FIGURE 15.4 The male reproductive system includes the testes, scrotum, seminal vesicles, accessory glands, various ducts, and the penis. The testes are the glandular organs in which sperm are produced. Before birth, the testes descend from the abdominal cavity into a pouch behind the penis called the scrotum. One testicle (usually the left) hangs lower than the other so the two cannot be crushed together. There are two testes (or testicles) that contain hundreds of coiled seminiferous tubules in which the sperm develop. These tubules lead into the epididymis, which appears to be a large, short tube but is actually a highly coiled, very thin tube about 20 feet long. The epididymis acts as a filter to remove any cell debris or pigment accompanying the sperm and to propel the sperm into the sperm duct. If more sperm are produced than can be utilized, the epididymis may "digest" some of them and return their nutrients to the body. The epididymis also acts to increase the fertility of the sperm as they pass through it. The sperm duct, or vas deferens, transports the sperm over the pelvic bone to the back of the urinary bladder where it joins the sperm duct from the other testis. The seminal vesicle is a tightly coiled tube that lies in a pouch at about the point of juncture of the seminal ducts. Fluid from the seminal vesicle is discharged at ejaculation. The prostate gland lies just below the bladder and secretes prostatic fluid that, together with seminal fluid from the seminal vesicle, activates and maintains the sperm. Below the prostate gland the Cowper's gland, or bulbourethra, also contributes a fluid to the semen. The penis is composed of three long areas of spongy tissue bound together by fibrous tissue. Two of these long areas of spongy cavernous areas are attached to the base of the Cowper's gland and at the penis tip to form the glans penis. This structure may be covered by the foreskin, the fold that is removed at circumcision.

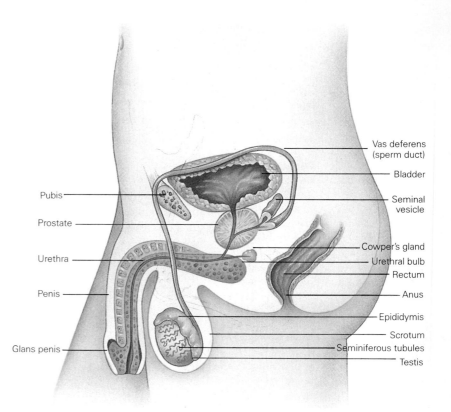

Labels: Vas deferens (sperm duct) — Bladder — Seminal vesicle — Cowper's gland — Urethral bulb — Rectum — Anus — Epididymis — Scrotum — Seminiferous tubules — Testis — Pubis — Prostate — Urethra — Penis — Glans penis

foreplay. Foreplay, as well as the copulatory act itself, varies widely in its expression between cultures, from one individual to the next, and even between established sexual partners from one time to the next. Let's examine some of the changes that accompany sexual intercourse in our species.

The human penis is supplied with a great number of blood vessels, some of which open into large blood chambers. Parallel veins and arteries service the penis, and during sexual arousal, the arteries relax and enlarge, allowing an increased blood flow into the penis. The veins do not expand, however, and in fact are pressed by the expanded arteries. Thus, the penis becomes filled with blood, causing it to grow stiff and lengthen, sometimes to a surprising degree—perhaps more surprising to some than others. The penis stands erect at such times and is said (by the more delicate among us) to be in the **tumescent** condition.

In women, foreplay usually causes the cells in the vaginal wall to secrete a lubricating substance that aids in the insertion of the penis. The lubricating may begin, however, well before she is emotionally ready for intercourse. With more extensive foreplay, the **labia minora** (small lips) and **clitoris** may enlarge and redden as blood rushes to those areas. About this time, the nipples may also harden and enlarge. As foreplay continues, the shoulders and chest may mottle and the breasts may enlarge and grow more sensitive.

Copulation involves the insertion of the erect penis into the vagina. After a few or many pelvic thrusts (depending on the species, and, in the case of humans, depending also on the individual, the degree of arousal, and one's schedule), the male ejaculates.

FIGURE 15.5 The reproductive system of the human female includes the ovaries, oviducts, uterus, vagina, and external genitals. The ovaries are located at the sides of the abdominal cavity and are supported by ligaments. Eggs released from the ovaries move into the tubelike oviducts that extend to join to either side of the uterus where the embryo develops. The oviduct is lined with cilia that sweep the egg toward the uterus. The uterus is pear-shaped, muscular, and thick-walled, with the lower end opening into the vagina. The inner lining or endometrium of the uterus is glandular and is the site where the fertilized egg will develop. The lower part of the uterus, just above the vagina, is called the cervix. The vagina is a muscular tube, 3 or 4 inches long, that leads from the exterior to the cervix. The vaginal muscles are thin-walled, and the epithelial lining may have a corrugated appearance. The external genitals, or vulva, are comprised of the mons veneris, a fatty mound that becomes covered with hair at puberty. The labia majora are thick folds that lie on either side of the vagina and are also covered with hair. The labia minora are smaller folds of skin that lie between the labia majora and the vaginal orifice. Where the labia minora meet above the vaginal orifice is the clitoris, a small erectile organ that is highly sensitive and that corresponds developmentally to the penis of the male. The urethra (urinary tract) opens between the clitoris and vagina. There may be a mucous membrane called the hymen stretched across the vaginal opening.

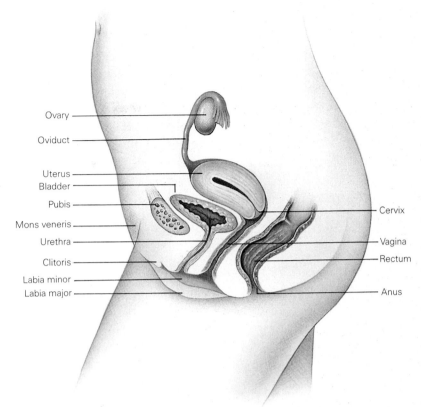

Ovary
Oviduct
Uterus
Bladder
Pubis
Mons veneris
Urethra
Clitoris
Labia minor
Labia major
Cervix
Vagina
Rectum
Anus

Ejaculation takes place in two stages. First, the entire genital tract contracts, including the epididymis, sperm ducts, and seminal vesicles. Fluid from the seminal vesicles flows into the upper urethra (urinary tract), where another fluid from the prostate gland is added. The bulbourethral gland (also called Cowper's gland) secretes a slippery fluid that may aid in penetrating the vagina. As the urethra becomes filled with semen, the urethral bulb at the base of the penis expands to accommodate the influx. The second stage begins as a sphincter muscle (a ring of muscle under involuntary control) closes off the urethra at the bladder. A sphincter at the base of the prostate then relaxes, allowing semen to move into the urethral bulb and then into the urethra as it extends into the penis. This is followed by contractions of the muscles in the area behind the testicles, which pump the accumulated fluid out through the penis. Several contractions may follow, but most of the semen is expelled in the first. The semen usually ejects from the penis with considerable force in the first few contractions. Also, it is in this first fluid that the greater part of the sperm is concentrated.

Ejaculation is ordinarily accompanied by **climax** or **orgasm,** which is the culmination of the copulatory act. Accompanying physiological changes include an increase in blood pressure and a quickened pulse. The postclimatic period is marked by a feeling of relaxation, sometimes to the point of drowsiness. The blood vessels return to their original state, and the penis again becomes flaccid or limp.

The physiology of orgasm in women is not as well documented. In fact, the very existence of the female orgasm was argued until 1966, when a pioneer study of human sexual response was published by the physician-psychologist team of Virginia Masters and William Johnson. On the basis of many carefully monitored observations of various sexual activities, they demonstrated that the female orgasm differs little from that of the male. However, orgasm among women is not so universal a response as among men, and there is a much wider range of individual differences. Some women rarely or never reach climax, while others may climax frequently and easily. (The whole topic was once quite fashionable in suburbia.) The inability to reach orgasm, by the way, has absolutely no effect on the ability to reproduce.

Sex and Society

The act of copulation in many societies has been regulated by a number of religious, legal, and social restrictions. In fact, in our culture, "morality" has come to refer primarily to sexual matters. To many people, an "immoral" person is one who is sexually promiscuous, period.

It has been suggested that such values may stem from the idea that copulation must be reserved for child-bearing activities. One argument is that in nature, copulation is reserved solely for the purpose of procreation and that copulation for any other reason is "unnatural," and therefore wrong. Logic aside, the premise is erroneous; copulation has many functions in nature. For example, among rhesus monkeys, subordinate males often present their rears to dominant males as an act of submission—an act that has the effect of reducing the dominant individual's aggression. The dominant male may follow up with a demonstration of his authority by mounting the subordinate and making several pelvic thrusts, although there is no actual penetration. Furthermore, among chimpanzees, a pink lady may copulate with every adult male in the group, as well as some adolescents. Since only one male is needed to impregnate the female, the act of copulation is believed to be a means of developing social bonds within the group. Among baboons, the females in the earlier phase of their estrus period copulate with males of any rank, but at the time they are most likely to conceive, they may accept only high-ranking males. Baboons are also highly social, and in this case, too, the earlier acts of copulation may serve to reinforce group bonds. Female monkeys have been known to present themselves to a male, distracting him long enough for her to swipe his bananas. This is a "natural" act, but not many people emulate it, at least in the better social circles.

Copulation may be an important bond builder. If this is so, it would help to explain our caution regarding what seems, on the surface, to be a rather mechanical process. In other words, it may not be advantageous to build bonds indiscriminately.

An indication that copulation between humans may have functions other than procreation is provided by evidence that whereas women may be prepared to copulate at any time, pregnancy can occur only at a specific time in the **menstrual cycle.**

The Menstrual Cycle

The menstrual cycle begins in females at puberty, ordinarily around the age of twelve to fourteen. (The onset has become progressively earlier in

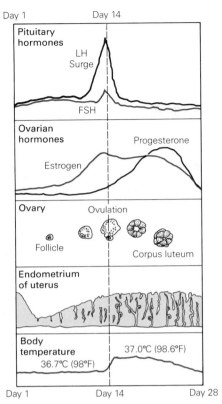

Day 1   Day 14

Pituitary hormones

LH Surge

FSH

Ovarian hormones

Progesterone

Estrogen

Ovary   Ovulation

Follicle

Corpus luteum

Endometrium of uterus

Body temperature   37.0°C (98.6°F)

36.7°C (98°F)

Day 1   Day 14   Day 28

FIGURE 15.6 The relationship of female reproductive hormones and the events in the ovary and uterus during the menstrual cycle. Note that FSH and LH are released simultaneously from the anterior pituitary, but that LH is much more abundant. In the meantime, the maturing follicles are producing estrogen, but progesterone will not begin to rise until after ovulation, when the corpus luteum is formed. The sharp rise in FSH, LH, and estrogen marks the rupture of the egg, but by this time the uterine endometrium is halfway through its period of development.

Note that estrogen peaks just before day 14, about the time that progesterone levels begin to rise, peaking from day 21 to 24. The follicle reaches maturity at day 14, when ovulation occurs, leaving the corpus luteum, which continues secreting some estrogen, but particularly progesterone, for about 10 more days. The endometrium (lining) of the uterus undergoes growth and repair during the first 14 days, reaching full development about a week later. At day 14, the body temperature elevates slightly, a signal that ovulation has occurred.

American girls, probably as a result of increasingly good health practices.) The cycles will continue for about the next thirty to forty years of the woman's life, usually interrupted only by pregnancies and terminated finally by menopause.

What exactly is this cycle? How does it work? **Menstruation** is the result of the shedding of the blood-rich **endometrium,** the lining of the uterus, when the egg released from the ovary has failed to be fertilized. In preparation for receiving a fertilized egg, the endometrium becomes engorged with blood-carrying, life-sustaining food and oxygen. If the egg reaches the uterus unfertilized, the endometrial preparations have been for nil, and the uterus reverts to its previous condition. The old endometrial lining breaks loose and slides out the vaginal opening, accompanied by blood from ruptured vessels. This marks the menstrual period.

An egg is released from an ovary about every 27 to 30 days. If it is not fertilized, menstruation will begin about 14 to 16 days after that. The woman's "period" will last from 3 to 6 days. However, there is a wide variation in the length of both the cycle and the menstrual period, and almost any pattern is considered normal as long as it recurs on a rather regular basis. Figure 15.6 describes the relationship of the ovarian, uterine, and hormonal cycles from the start of one cycle to its termination. **Hormones** (Chapter 21) are "chemical messengers," produced in certain parts of the body, that regulate events in other parts of the body.

At the end of a menstrual period, the cycle begins anew, deep within an ovary, where other follicles are developing. Follicles are rounded bodies, each containing an egg that protrudes inward from the follicle wall into a large fluid-filled space. As the follicles mature, they grow larger and migrate to the surface of the ovary. Their development is stimulated by rising blood levels of a hormone simply called the **follicle stimulating hormone** (FSH), which interacts with another hormone called the **luteinizing hormone** (LH). Both are produced in a small but remarkable structure called the **pituitary gland,** located at the base of the brain. (We'll look more closely at this important gland later.)

As the follicles mature, they begin to secrete a hormone of their own called **estrogen,** which induces a thickening of the endometrium. For some reason, one bubblelike follicle begins to outgrow the others, and as soon as its ascendancy is established, the others stop developing. This larger follicle then moves to a position just below the surface of the ovary. In response to the pituitary's luteinizing hormone, the follicle ruptures on about the thirteenth or fourteenth day after the follicle-stimulating hormone initiated its growth. The egg is released, leaving behind the yellowed body of the follicle, now called the **corpus luteum** ("yellow body"). However, the corpus luteum is far from being a corpse, for now it begins to secrete another hormone, **progesterone,** which stimulates the uterus to prepare for a fertilized egg. The endometrium thickens and becomes engorged with tiny blood vessels that will carry food and oxygen to the embryo in the event that fertilization occurs. The interaction of the progesterone and estrogen also inhibits the production of the pituitary's follicle-stimulating hormone at this time and, thus, prevents the start of a new cycle. These inhibiting properties of estrogen and progesterone are the basis for their use in birth control pills, as we will see later.

When fertilization does not occur, the corpus luteum stops its secretion of progesterone and estrogen, and the swollen uterine lining begins to shrink. Deeper within the uterus, blood vessels contract, reducing the supply of nutrients and oxygen to the endometrium. The tissue then breaks down, rupturing the vessels there, and bleeding begins. The blood flow carries bits of cell tissue out of the uterus and through the vaginal opening, marking menstruation.

The pressure of blood-engorged tissue often causes discomfort and cramping. (In some cases, the changes in hormone levels associated with menstruation also cause depression and edginess.) As soon as the uterus has returned to its original state, the pituitary stimulates the development of more follicles, and the cycle begins again.

If copulation has been successful (depending on your point of view), and the egg that reaches the uterus is fertilized, the zygote will implant in the uterine wall and begin its growth, taking its necessities from its mother's blood in a parasitic fashion. (Some say it's just the beginning of a long parasitic existence.)

Conception

Although only one egg is produced during each menstrual cycle, the semen released at each ejaculation may contain millions of sperm, each with its own genetic makeup. Of course, there is no way to predict which sperm will fertilize the egg. During copulation, the sperm are usually ejaculated into the upper reaches of the vagina, near the cervix. Each sperm cell has a whiplike "tail" that can propel it along at a rate of about $\frac{1}{2}$ centimeter per minute (Figure 15.7). Surprisingly, however, sperm may reach the cervix only $1\frac{1}{2}$ to 3 minutes after ejaculation. This is probably because they are assisted by beating cilia that line the **oviducts,** the tubes leading from each ovary to the uterus, and currents created by vaginal contractions.

The sheer number of sperm in each ejaculation is an adaptation to the extraordinary hazards they face in making their way to the egg. Some may die of "natural causes," as a result of their own physiological changes. Others may be rendered immobile by the chemical environment of the vaginal tract. Some are devoured by the woman's roaming white blood cells, which treat them as the foreign bodies they are. Furthermore, semen, the fluid in which the sperm are carried, is alkaline (basic), while the vaginal environment is slightly acid, and this abrupt change in conditions can render sperm inactive.

In spite of the dangers, some sperm are able to make it into the relatively hospitable alkaline environment of the uterus. Through ways that largely remain a mystery, a few manage to find their way through the uterus and into the tiny pore leading into the oviduct (also called fallopian tube) and up the tube nearly to the ovary, where they may encounter an egg on its way down. Conception actually takes place, then, not in the uterus, but far up in the oviduct.

Several sperm may reach the egg at once, but for some reason, only one will penetrate the egg's membrane and effect fertilization. The mechanism that prevents fertilization by all but one of the sperm is not completely understood, but the initial penetration causes dramatic changes in the membrane surrounding the egg.

Once fertilization has taken place, the zygote (fertilized egg) will continue down through the oviduct to be received by the ready uterus. It will quickly begin to draw sustenance from the uterine fluids. Then, it systematically dissolves the uterine wall and sinks deeper into the maternal tissues until it is completely embedded. It soon begins to receive sustenance as oxygen and nutrients filter through from the blood of the mother. In another 266 days, a new individual will join the population.

There has never been another individual quite like this one, nor will there be. The genetic combination cannot be duplicated, and neither can the experiences that will be superimposed on that constitution. In some cases, its existence will be brief, the time counted in days or even minutes. Most, though, will join the planet's teeming human population where it may become anything from a Sikh to a Republican. Whatever it will call itself, though, it will be unique.

CONTRACEPTION

Now let us consider a problem at the other end of the reproductive spectrum: the avoidance of parenthood.

Methods of Contraception

Contraception, or the avoidance of pregnancy, can be achieved in many ways. Some methods are employed by the female, others by the male, and some by both partners. Unfortunately, none, except abstinence and sterilization, is 100 percent effective (Table 15.1).

Historical Methods

Historically, contraception has been attempted in a variety of ways, some effective, some worthless. (One might wonder just how our ancestors must have reacted when they began to discover what causes pregnancies. It's still a bit hard to believe.) The ancient Egyptians, for example, blocked the cervix with leaves, cotton, or cloth. The encouragement of homosexuality among the ancient Greeks probably reduced the rate of conception in that culture, although the practice was undoubtedly not instigated for that reason. In the Middle Ages, condoms that fit, sheath-like, over the penis were fashioned from such materials as linen and fish skin. Douching has long been practiced as a means of flushing sperm out of the vagina. The finger has also been used to remove semen after it has developed a stringy quality through being exposed to the vaginal environment. The Old Testament refers to removing the penis just prior to ejaculation. This is unreliable for reasons we will see shortly. The most effective method, of course, has always been total abstention, a method that has met with little applause in most circles.

There are also a number of "folk" devices that are employed even today. These range from regulating intercourse according to the phases of the moon, to stepping over graves, to using cellophane sandwich-wrappings as condoms, to douching with soft drinks (the carbonic acid and sugar are believed to be spermicidal, and a shaken drink provides the propulsion for the agents to reach far into the vagina). Since the effectiveness of most folk methods is unknown, we might consider a few other contraceptives that we know more about.

The Rhythm Method

As an old joke goes, people who practice the rhythm method are called "parents," although the joke is funnier to some than others. Basically, the **rhythm method** involves periodic abstention. It is based on the notion that a woman's fertile period can be predicted or revealed by tests (such as testing secretions for pH or glucose levels), and that by refraining from intercourse during this time, pregnancy cannot occur. The couple must therefore avoid intercourse at least two days before and one day after ovulation, although, for safety, this interval is usually increased. Interestingly enough, although the method requires that couples refrain from intercourse at the very time when the woman may be psychologically and emotionally most receptive, and although it requires the use of thermometers, calendars, paper and pencils, the claim is made that this is the only "natural" means of birth control. In any case, the method is not particularly reliable (about 80 percent under ideal conditions), partly because the menstrual periods of many women are highly irregular. One way of interpreting these numbers is that for every 100 women using this method for a year, 20 are likely to become pregnant.

Coitus Interruptus

In spite of the obvious inconvenience of withdrawing the penis from the vagina immediately before ejaculation, **coitus interruptus** is one of the most widely used contraceptive practices in Europe today. There are little data on the effectiveness of this method, but it is probably not very reliable (probably about 70 percent), because the man must do precisely the opposite of what he wants to do. The practice is also risky because sperm may unobtrusively leave the penis before it is withdrawn. In addition, because of residual semen in the penis after the first ejaculation, there is higher risk of conception for those who are able, and have the time, to immediately copulate again.

The Condom

The **condom,** or rubber, is one of the most widely used contraceptive devices in the United States. Basically, it is a balloonlike sheath made of rubber, or other material, that fits tightly over the penis and traps the ejaculated sperm. There are many grades of condoms, ranging from the two-for-a-quarter specials sold in restroom vending machines to the expensive types that are made from animal membranes.

Condoms are only about 70 to 93 percent effective (partly because of differences in quality and the fact that they are often incorrectly used). The disadvantages of using condoms include the frantic effort to find one and put it on, and the reduced level of the man's sensation. One reason it remains popular is the protection it affords against venereal disease.

Diaphragm

The **diaphragm** is a dome-shaped rubber device with a flexible steel spring enclosed in the rim. It is inserted into the vagina and fitted over the cervix to prevent sperm from entering the uterus. Unlike the condom, the diaphragm must be individually fitted. It is usually coated with a spermicidal cream or foam and left in place for several hours after intercourse. The device is about 90 to 98 percent effective if well-fitted and used correctly. With a certain element of anticipation on the part of the woman, it need not interrupt sexual activities (Figure 15.7).

The Cervical Cap

The **cervical cap** functions similarly to the diaphragm, but has become less popular in the last two decades. However, it is presently being redesigned and retested and, in its new form, is strongly favored by some women. Essentially, it is a small plastic or metal cap that fits tightly over the tip of the cervix. It has traditionally been used with chemical spermicides. One of its advantages is that it can be left in place for days. In fact, some women remove it only for menstruation. Whereas its effective period when used with the spermicide is presently being reinvestigated, it may be more effective than the diaphragm since it ordinarily cannot develop leaks. One problem is that it can move out of position without being noticed.

Spermicides

A number of sperm-killing foams, jellies, aerosols, suppositories, and creams are available as contraceptives or to increase the effectiveness of other contraceptives such as the diaphragm or cervical cap. The spermicides are usually quickly and easily applied inside the vagina and require neither fitting nor prescription by a doctor. Used alone, they are 75 to 90 percent effective, depending on whether they are used properly.

The Fallopian Plug

The **fallopian plug** is also called the **tubal occlusion.** The process essentially consists of simply inserting silicone rubber plugs into the oviducts (fallopian tubes). The egg, unable to reach the uterus, simply withers and is reabsorbed.

FIGURE 15.7   Left, diaphragm in place. Right, cervical cap in place. They both operate on the same principle—that is, blocking the cervix. However, the diaphragm is normally inserted each time before intercourse, while the cervical cap may be worn for weeks at a time. The diaphragm is far less effective unless it is used with a spermicidal jelly.

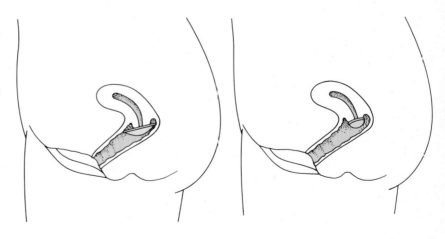

A fiber-optic device helps guide a tube to a place near the opening of the oviduct and liquid silicone, with a hardener, is then slowly pumped into the oviduct. The liquid hardens in about six minutes and x-rays determine if the plug is in place in both oviducts.

The woman is free to leave right away, but her activity may be governed by cramps at first. Whereas the process is a bit difficult for the physician, it is easier on the woman than any surgical techniques. The effectiveness of the technique approaches 100 percent. At present, the process is considered irreversible, but new methods are being devised that permit retrieval of the plug.

The Intrauterine Device

The **intrauterine device,** or IUD, was recognized as an effective contraceptive for many years. The principle has apparently been utilized for a long time. In fact, it is said that Arab camel drivers on long, hard caravan marches prevented pregnancies in their charges by placing apricot seeds in the camels' uteruses. Today, though, IUDs are made from several materials and come in a variety of sizes and shapes. No one knows exactly how the IUD works, beyond the fact that it seems that a foreign body in the uterus somehow prevents pregnancy. It is theorized that the IUD prevents the implantation of the zygote, or that it may cause the egg to move too rapidly through the oviduct, and even that it may alter the condition of the uterine endometrium.

The IUD has several major advantages. Once it has been installed in the uterus, it requires little attention. Also, it is inexpensive and about 97 percent effective.

The IUD has recently fallen into disrepute, however, because of the many problems it causes. One problem is that the device may be expelled by the uterus in some women, perhaps without being noticed. For some reason, IUDs are more likely to be rejected by the uterus in women who have never been pregnant. There may also be side effects just after insertion, including bleeding and pain. If these persist, the IUD must be removed. The IUD has also been associated with a number of ectopic pregnancies (those that occur in the body cavity, outside the reproductive tract). More important, however, the IUD has been implicated in a number of serious disorders of the pelvic region, perhaps brought on as the string hanging from the uterus into the vagina serves as a "wick" for vaginal bacteria, permitting them to ascend into the uterine area and on into the oviduct. The IUD is, therefore, steadily being withdrawn from use at this time.

The Birth Control Pill

The **birth control pill** contains the hormone estrogen, either in natural or synthetic form, usually administered with an "artificial hormone" called progestin, which is similar to progesterone. The combination of the two apparently inhibits the formation of egg follicles in the ovary. The pill is taken each day for 20 or 21 days, beginning on the fifth day after the onset of menstruation. The next menstrual period will begin when this series of pills is finished, so the menstrual period occurs precisely every 28 days. Apart from the convenience of knowing the day of the week on which the periods will begin, in many cases, the pill may alleviate the pain accompanying menstruation. However, the pill's greatest advantage is that it is probably over 99 percent effective when used properly. There are reports that regular use may also delay menopause and increase the libido of the middle-aged user, although the bases for these effects are not known.

You should be aware that birth control pills may cause undesirable side effects similar to those of pregnancy, especially during the first few months of use. They include weight gain, fluid retention, nausea, headaches, depression, irritability, and increased facial pigment in certain areas. Another side effect may be an enlargement of the breasts. Some or all of these symptoms may diminish after the first few months of use.

There may also be more serious problems associated with birth control pills. You may have heard that the pill may be associated with cancer. Early results of a study sponsored by Planned Parenthood of New York were inconclusive; while such studies did indicate a higher incidence of "precancerous" areas in the cervix among women who use the pill, such areas do not invariably develop into cancer, and they are easily and totally curable.

Although the pill has produced no serious medical problems for the overwhelming majority of its users, some argue that it is too early to assess its long-term use. Some recent results are rather interesting, however. A study of about 5,000 women revealed that those who have used "combination" pills (pills combining synthetic estrogen and progesterone) for at least a year run half the risk of cancer of the cervical endometrium as those women using the "sequential" pill (containing only estrogen). The results indicate that the longer the combination pill is used, the more protection it confers, and the protection may last up to five years after the woman stops using the pill.

The relative amounts of estrogen and progesterone in the pill are critical because of the powerful potential side effects of estrogen. Currently, a triphasic pill is in use that varies the relative amounts of these hormones at three levels as the cycle progresses.

A great deal of research attention has been directed toward developing a contraceptive pill for men, but so far the search is unsuccessful.

The Sponge

A recently-developed contraceptive that is gaining popularity is the **sponge** ("Held tightly between the knees," someone has said). The sponge is a small absorbent polyurethane sponge that is saturated with a spermicide. It is inserted into the upper area of the vagina up to 24 hours before intercourse and is left there for about 8–10 hours afterward. Its success rate approaches 98 percent.

Sterilization

Sterilization, as you might guess, works, and it's 100 percent effective. Furthermore, either the male or female can be sterilized, although it is a much simpler matter in the male.

A **vasectomy** for a male normally takes about 15 minutes and can be performed in a doctor's office. A small incision is made in the side of the

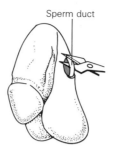
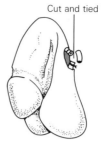
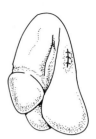

Sperm duct   Cut and tied

FIGURE 15.8 The simple procedure of a vasectomy. The operation has been described as similar to a visit to the dentist—with a few differences. Normally, under local anesthesia, the vas deferens are exposed by a small cut in the scrotum, and a section is removed from the tube. Both ends are tied in case they show a tendency to rejoin. Although surgeons have had some success in reversing their results (particularly in California where the operation is rather common), vasectomy is recommended only for those who have made up their minds.

scrotum, a short section is cut from the seminal duct (the vas deferens), and then the incision is closed (Figure 15.8). The operation can be reversed in 50 to 80 percent of the cases in those places where doctors have had the greatest experience with the operation. Another method, in which a removable plastic plug is inserted in the seminal duct, shows promise of being reversible in almost all cases.

Sperm continue to be produced after a vasectomy, but since they cannot be ejaculated, they are reabsorbed by the body. Seminal fluid is ejaculated as before, but it contains no sperm. Since seminal fluid makes up about 95 percent of the normal ejaculate, and hormonal levels are not affected, there is no noticeable difference in the male's sexual performance.

Vasectomy is, in itself, a simple procedure, but other factors must be taken into account. For example, in many cultures, including that of the United States, men tend to identify rather strongly with the "male role." The realization that they are no longer able to impregnate women may have a rather marked psychological effect in some individuals. For this reason, the procedure is ordinarily advisable only for those men who are fairly secure in their own sexual identity.

Sterilization in women is usually accomplished by **tubal ligation** (Figure 15.9). The oviduct is cut and both ends are tied back. The oviduct can also be cauterized, or seared, causing the opening to seal shut as it heals, or it can be pinched together with a tiny clip or rubber band. This operation is much more complicated than a vasectomy since the abdominal wall must be opened. In some cases, the abdomen can be entered through a small incision at the navel. It is also possible to enter through the vaginal opening, making the ligation a far simpler procedure. As with vasectomy, there is no evidence that, in humans, tubal ligation causes any change in hormone production, sex drive, or sexual performance.

Abortion

Abortion, the artificial termination of a pregnancy, has been one of the more common forms of birth control throughout recorded history. In the United States, restrictions on abortions have been relaxed, but this has not necessarily resulted in more abortions. Instead, it very likely means that *illegal* abortions may now be less common. When abortion was illegal, there was estimated to be one illegal abortion for every two live births in the United States.

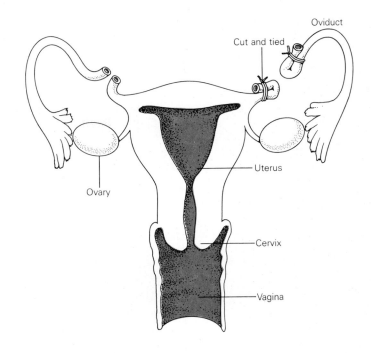

FIGURE 15.9 Tubal ligation, or "tying the tubes," is a means of sterilization in women. It is a more serious operation than the male's vasectomy. It has traditionally been accomplished by opening the body wall, cutting the oviduct and tying back the ends. Newer methods, however, include tying with a device that is pushed up the oviduct from the uterus. Ligation does not affect the ovaries, thus they continue to produce their hormones, so there is no change in the physiology of the woman other than totally negating the possibility of pregnancy. Some women have reported increased sexual satisfaction with the knowledge that they cannot become pregnant, but such reactions are strictly a personal matter and cannot be predicted.

Illegal surgical abortions and do-it-yourself methods have resulted in innumerable deaths and permanent physical and psychological damage as well. Some of the methods employed under illegal or "back alley" conditions have been as barbaric as they are hazardous. Furthermore, abortions by untrained people may not even be successful. Perhaps the most important point here, however, is that it is probably safer to deliver than to attempt to abort under these conditions.

When done properly, under medical supervision, an abortion during the first three months of pregnancy is a simple procedure. The most common method of aborting a fetus is called **dilation and curettage,** or **D and C.** The vagina and cervix are dilated, and the lining of the uterus is scraped with a curette, a steel loop at the end of a handle. Other methods are used quite successfully also, such as **vacuum curettage,** an increasingly popular technique in which a suction device pulls the embryo from the uterine wall. Another common procedure is called **salting out.** It involves the injection of salt solutions into the uterine cavity.

Abortion first became illegal in the United States in the nineteenth century. At that time, it was ruled dangerous to the prospective mother. Recently, courts have upheld the right of a woman to undergo abortion on request, but the decision has caused heated debates, animosity, and even violence. The groups opposing abortion have coined the term "right to life," but others argue that pregnancy is a very personal matter and that the decision to accept this condition and the ensuing years of responsibility are primarily the woman's since she must bear the brunt of any pain, risk, and responsibility. One argument has revolved around the question of when a developing embryo becomes a "person." Some believe that the fertilized egg is a human being and has a "right" to develop and be born. Others argue that abortion does not actually destroy "life" until later, such as when the heart (still a tubular muscle) begins to beat, or when the embryo begins to stir or "quicken," or the time when it becomes capable of surviving after birth.

The question basically boils down to opinions, and emotional ones at that. Logic has rarely found a place in these discourses, nor have decisions based upon the good of all. For example, many antiabortionists have stated their support for capital punishment or military enterprises that involve loss of life. It almost seems that life's sacredness is related to one's birthplace, ancestors, or voting record. Obviously, the question of abortion is not likely to be resolved on any rational or scientific basis.

It should be mentioned that abortion has various effects on different women. Some take abortion very lightly and measure it in physical or economic terms. Others, however, suffer after-effects such as extreme guilt, a feeling of loss, or some other severe emotional trauma. People who do not want to be parents may be forced to make some very difficult decisions. Some are prepared for the effects, but others are surprised at their own anguish. Furthermore, the repercussions may be severe to both the man and the woman, a point often overlooked.

Abortion is also opposed on the assumption that it encourages sexual freedom, a possibility that is viewed with more alarm in some quarters than others. More than one legislator has publicly stated that if a woman has sex, it is only fair that she "pay the piper" and accept her punishment in the form of "compulsory pregnancy." Apart from the tacit assumption that moral responsibility should be expected only of women, children being considered a form of punishment is interesting, but probably rarely acknowledged by its *de facto* proponents. These are only a few of the arguments regarding abortion, but they may provide food for thought or perhaps launch a discussion or two. If you would like to hear some other arguments, just bring up the subject at your next family gathering.

As a final point of consideration, Swedish scientists attempted to find out what usually happened to unwelcome children whose mothers were denied abortion. Such a group was compared to a matched set of children who were wanted. They found that twice as many of the unwanted ones grew up labelled as "illegitimate," or in broken homes or institutions. In addition, twice as many unwanted ones had records of delinquency, twice as many were declared unfit for military service, and twice as many had required psychiatric care. Finally, five times as many had been on social-welfare programs even in their teens.

SUMMARY

Animal reproduction can be asexual or sexual. Two forms of asexual reproduction are binary fission and budding. Sexual reproduction may be external or internal. Internal reproduction involving copulation gives a relatively good chance of the gametes encountering each other. Human sexuality may be adaptive, not only in reproduction, but in bonding. Theoretically, humans are physiologically capable of copulating at any time, but the woman's reproductive state may depend on her stage in the menstrual cycle. Menstruation is the shedding of the endometrium after an egg has failed to be fertilized. Menstruation begins 14 to 16 days after an egg is released from the ovary. Follicles, containing eggs, are stimulated to develop by FSH. The follicles then begin to secrete estrogen, which causes the endometrium to develop. The corpus luteum, left when the egg is released, secretes progesterone which also helps prepare the endometrium to receive a fertilized egg. If fertilization does not occur, estrogen and progesterone cease being formed and menstruation begins soon after. Fertilization takes place in the upper oviduct and the zygote implants in the uterine wall. A number of means of contraception are practiced with varying rates of success.

Chapter 16

ANIMAL DEVELOPMENT

The story of development encompasses all those things that happen to us from the moment of conception (which most of us hardly remember) to that final indignity that mocks us all. In its broadest sense, then, it includes not only the events of the womb, but our birth, growth, aging, and perhaps even our death. (Some "primitive" tribes consider death just another phase of life, but that philosophy may take a bit of adjustment on our part.) Any such broad treatment of development would involve as much philosophy as science, though, so we will take a more modest approach.

We will focus primarily on the events from fertilization to birth, and we will concentrate on humans. However, even this narrowing leaves us with some rather complex issues, so we will begin by considering two other kinds of animals and see how they develop. Different kinds of species, we will see, may develop from quite different kinds of eggs.

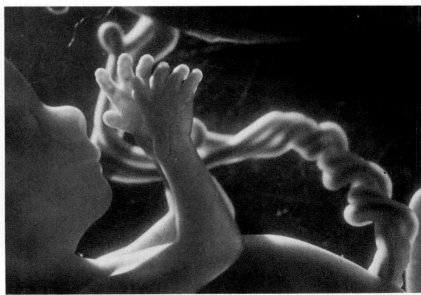

TYPES OF EGGS

The three basic types of animal eggs are roughly categorized according to the amount of **yolk** they have. The amount of yolk is critical since it is the embryo's food supply, at least for a time. In some species, the embryo needs only a small supply of yolk since it soon switches to nutrients derived from the mother's blood, as is the case with humans. We only need enough to last until the embryo has implanted in the wall of the uterus. In contrast, birds leave their mother's body at a very early developmental stage, and they must carry their entire embryonic food supply with them. So, whereas a human egg is smaller than the period at the end of this sentence, it would be hard to hide an ostrich egg with this whole book. Interestingly, both the young and the adults of these two species are about the same size.

Other kinds of animals have a moderate yolk supply. In these, the young must begin to find its own food long before it has reached its final body organization. The frog is an example. The frog egg has just enough yolk to get the developing embryo to the tadpole stage; after that, the tadpole can survive on food stored in its tail for a time, but it must soon begin to eat on its own.

EARLY DEVELOPMENT IN THE FROG

You may recall that at fertilization, the zygote receives all the genetic information it is ever going to get. No matter what manner of creature will ultimately be formed, all the genetic instructions are in that first cell. Here, then, let's see how those instructions tell a cell to build a frog.

The First Cell Divisions

In frogs, as in other animals, the egg begins drastic changes as soon as it is fertilized. First, the egg becomes unresponsive to other sperm, even as the two sets of chromosomes draw together for the first of countless mitotic divisions. The genetic material doubles, and virtually identical chromosomes move to opposite poles. Then the cell divides, forming two daughter cells. The embryo is on its way.

The first two cell divisions are usually at right angles to each other, along the same vertical axis; the third line of division is perpendicular to these. (Figure 16.1 compares the developmental sequence of frogs, chickens, and humans.) In the case of the frog, this third cleavage (division into smaller cells) is somewhat closer to one pole. This displacement marks the **animal pole.** The displacement is due to the relatively large supply of inert, nondividing yolk at the other, **vegetal pole.** The poles are determined by the point at which the sperm penetrated the egg at fertilization.

As the cells of the embryo continue to divide, they form a cluster, or ball, of cells called a **morula.** A cavity, the **blastocoele,** appears within the ball, which itself is now called a **blastula.** Because of the unequal distribution of yolk, the cells are smaller and of more similar size at the animal pole.

FIGURE 16.1 A comparison of the development of three vertebrates. In the early cell divisions (1, 2, 3), differences are obvious. The chick embryo, perched atop a massive and inert yolk, first undergoes incomplete cleavages. Frog and human begin similarly, but soon cell divisions lag in the larger, yolk-laden vegetal region of the frog, resulting in larger cells there. All three embryos produce a hollow blastula stage (4). Frog and human blastulas are spherical with prominent cavities. In the chick embryo, a streak of cells rises up slightly, forming only a minute cavity above the yolk. Gastrulation (5) produces new cavities and the three germ cell layers. From these layers, future tissues and organs will be molded. About this time, the human embryo implants in the uterine wall, its meager yolk reserves depleted. Its chorion grows into the uterus, seeking nourishment. The bird, too, will produce membranes that assist in bringing food from the yolk. The frog soon incorporates its food supply into its body.

In (6), all three produce the rudiments of a nervous system, ectoderm rising up in folds to outline the system. Subsequently, discrete pockets of cells contribute to organs as systems are built. Heart and blood vessels in the chick and human form early. In each embryo, mesodermal blocks called somites contribute to vertebrae, muscle, dermis, and other structures. Interestingly, each species forms pharyngeal gill pouches, a primitive vertebrate feature. In land creatures, these contribute to structures other than gills.

With continued refinement of form (10, 11), the species become easier to recognize. The tadpole will be eating long before the chick has used up its food supply, and the chick will be an adult by the time the human is born.

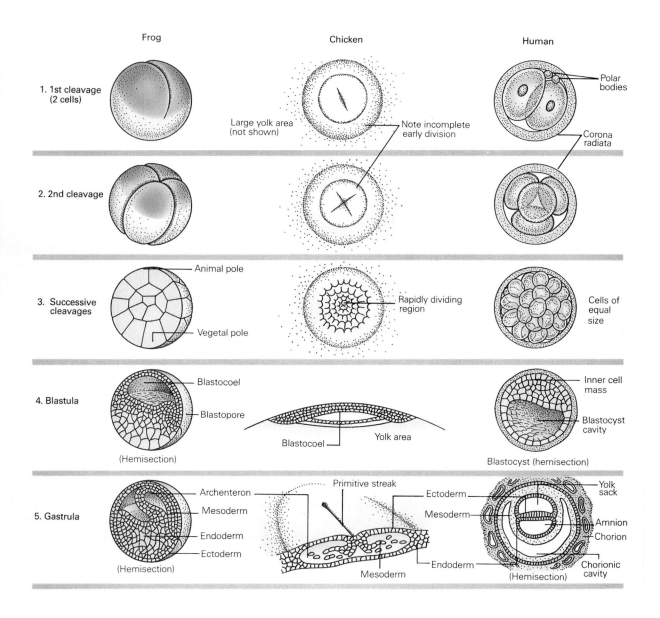

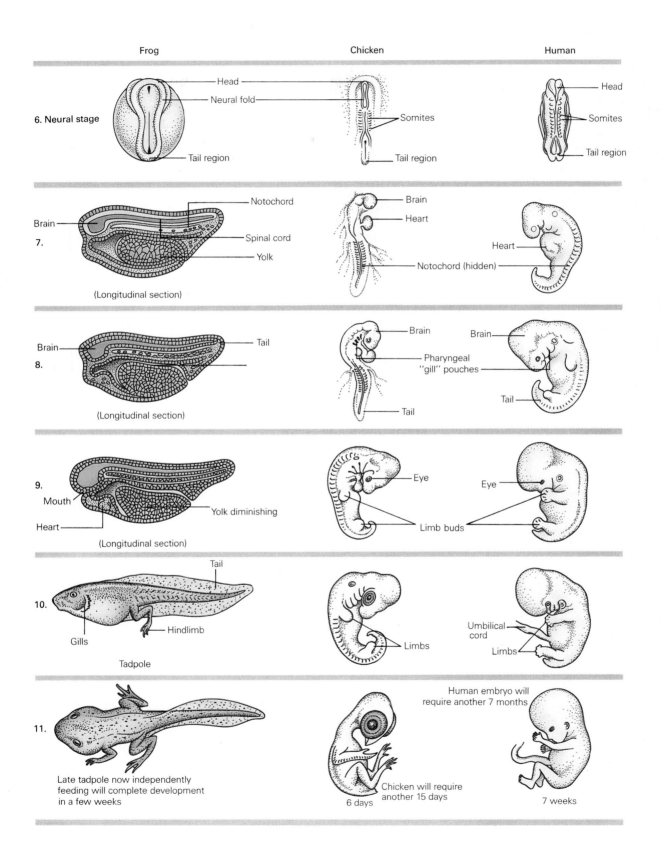

	Frog	Chicken	Human

6. Neural stage

Head
Neural fold
Somites
Tail region

Head
Somites
Tail region

7.

Brain
Notochord
Spinal cord
Yolk

(Longitudinal section)

Brain
Heart
Notochord (hidden)

Heart

8.

Brain
Tail

(Longitudinal section)

Brain
Pharyngeal "gill" pouches
Tail

Brain
Tail

9.

Mouth
Heart
Yolk diminishing

(Longitudinal section)

Eye
Limb buds

Eye
Limb buds

10.

Tail
Hindlimb
Gills

Tadpole

Limbs

Umbilical cord
Limbs

11.

Late tadpole now independently feeding will complete development in a few weeks

Human embryo will require another 7 months

Chicken will require another 15 days

6 days

7 weeks

Gastrulation

Early on the second day after fertilization, another important change occurs. At this time, certain cells migrate toward an area in the vegetal hemisphere. Here, they converge and then begin to roll under the surface of the embryo, marking the stage called **gastrulation.** The indentation formed by their rolling under (or **involution**) is called the **blastopore.** The blastopore forms a curved line over the yolky area at what will be the back, or dorsal, side of the embryo. The involution continues, extending its margins, until the crescent finally becomes a circle. The vegetal area inside this circle is called the **yolk plug** and will eventually be enclosed by the expanding layer of surface cells around it.

It is at this stage that we can clearly distinguish three distinct types of cells, called **germ layers.** They are the **ectoderm, endoderm,** and **mesoderm.** Each will contribute to specific structures as the body slowly forms. The ectoderm lies on the outside, an inner endoderm is formed from cells of the yolky vegetal area, and the mesoderm is derived from the cells that rolled under from the outer layer. The mesoderm now lies between the ectoderm and endoderm.

Few structures of the vertebrate body are formed entirely from any one germ layer, but it is traditional, and simpler, to refer to them as if they are. For example, we will say that the ectoderm forms the outermost layer of the skin, the sense organs, parts of the head and neck, and the nervous system. The ectoderm is held responsible for hair and feathers. The mesoderm forms tissues associated with support, movement, transport, reproduction, and excretion (for example, muscle, bone, cartilage, blood, heart, blood vessels, gonads, and kidneys). The endoderm is also largely responsible for the structures associated with breathing and digestion (including the lungs, liver, pancreas, and other digestive glands).

Embryonic Regulation

It must be admitted that at this stage of development, it is hard to see just how this peculiar ball is going to become a frog, but stranger things have happened. Interestingly, even at this early stage (the second day), the fate of some cells has been sealed for hours. In other words, they are firmly set on particular paths that lead to the formation of specific cell types.

However, other cells in the embryo have not yet specialized, and these can presumably take any of a number of developmental routes. They are so "flexible" that if they are transplanted to other parts of the embryo, they will form whatever type of cells their new neighbors happen to be and become something entirely different from what they would have formed had they not been moved. As each hour goes by, though, more and more cells become committed as their changes take them past the point of no return.

INFLUENCES ON ORGANIZATION

At this point, we will leave our friend the frog and consider a few general principles about what causes an embryo to take its final form. Then we will see how these principles apply to embryonic birds and mammals.

A few years ago, a young boy was admitted to the Children's Hospital in Sheffield, England. He had accidentally cut off the end of his finger. Ordinarily the part would have been reattached by a plastic surgeon in the hope that it would grow back and that the feeling would be restored. However, due to a clerical error, the stub was simply bandaged and the boy was ignored for several days. When the error was discovered, anxious physicians unwrapped the finger to see how extensive the damage had become. They were stunned to see that the finger was regenerating the missing part. They carefully monitored the progress of the finger over the next few weeks until the damage had, for all practical purposes, repaired itself and the finger was restored. The technique is now routinely used for young children, and there are several cases of regrown fingers, complete with nails and fingerprints. But how? This was contrary to all medical expectations. Should it have been so unexpected? After all, identical twins are formed when a developing human breaks apart at the two-cell stage, each part regenerating whatever was lost. So there are some regenerative powers in humans. The phenomenon is apparently associated with youth. It has not been found in adults, other than to a limited degree in wound healing. Subsequent research has shown that age does, in fact, have something to do with the regeneration. It seems that children under the age of eleven have marked powers of regeneration, and these rapidly dissipate after this time. As far back as the early 1970s, it was demonstrated that electrical charges surrounding the tissues in young people become reversed about the time that regenerative powers dissipate. Research attention is now focusing on maintaining particular (young) electrical fields around those areas in which regeneration is being attempted. The time when humans can regenerate lost limbs is probably not near, but scientists have new reasons for optimism.

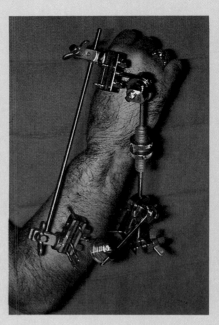

This arm was severed by an alligator. The arm was reattached and the above apparatus was used to stabilize it after a certain amount of healing had occurred.

The Flexible Fate

We are aware that the fate of some cells is never sealed. Otherwise, how could a flatworm that is cut in half regenerate its lost parts? Or how could a new starfish form from only a piece? (See Essay 16.1.) Furthermore, if sponge cells are separated by being passed through a sieve, they will crawl back together, ameboid fashion, and form a new sponge. (I suspect that if *my* cells were separated, they would take advantage of the situation and clear out.) Such cell flexibility is also present to a degree in many vertebrates. For example, in some kinds of salamanders, when a leg is amputated, the cells in the area of the cut despecialize and revert back to a more primitive state that is usually associated with embryos. Then they begin to specialize again, taking new developmental routes, moving about and changing until a new leg is formed.

What about mammals? If the leg of a mouse is amputated, it doesn't grow back. Cell differentiation in mammals is more permanent; the cells are less able to revert to an earlier condition. There are some interesting exceptions to this rule, however. For example, certain cells in the repro-

Centuries ago, Egyptians covered open wounds with raw meat, and today in Brooklyn, people are known to place raw steak on black eyes. Both habits may be regarded as a bit peculiar in some circles, but perhaps both have merit. In fact, it has recently been discovered that something in muscle tissue indeed promotes healing. Healing, in general, is a remarkably orchestrated sequence, especially when open flesh wounds are involved.

Flesh wounds are common and dangerous. After all, the protective covering around the body has been broken, leaving an opportunity for invasion by dangerous organisms. Thus, the body acts quickly by first sealing blood vessels in the process called **clotting.** Platelets—tiny bodies in the bloodstream—rush to the site, disintegrate, and, as fibrous proteins begin to form, the blood

that has already escaped hardens and forms a protective scab. Once the blood flow has stopped, the body begins to respond to other chemicals **(pyrogens)** that have been released in response to the trauma. The area becomes warm (perhaps too warm for any invading bacteria) and blood vessels open up (bringing in not only nutrients and oxygen for stepped-up metabolism as new tissue forms, but also perfusing the area with white blood cells). In the meantime, epidermal (skin) cells begin to multiply in the area of the wound, extending underneath the scab and growing together from all sides. Capillaries in the area begin to grow side branches and to penetrate the regenerating area, bringing in yet more fresh blood laden with oxygen and nutrients.

Here, a netlike structure called fibrin forms, holding red blood cells as a clot develops.

ductive tract of rats are able to dedifferentiate and take a new route, becoming different sorts of cells, replacing those experimentally removed. We are aware that, in mammals, epidermal cells grow over a wound as the cells underneath are engaged in the healing processes (Essay 16.2). In general, though, among mammals, cellular deorganization and regeneration is very limited. In most cases, once a cell has developed along a certain route, its fate is sealed. So what determines that route?

Sequencing

Embryonic regulation demands stringent controls on the sequence of development. For example, before those first trembling palpitations in the developing heart begin, signalling the onset of a rhythmic beating, there must be some place for the blood to go. So, great, irregular channels appear even earlier in development, formed along extensions of loose-knit, undifferentiated first-blood pools. What sorts of mechanisms determine such sequencing?

The details of most such controls remain a mystery, but ultimately the control must reside in the chromosomes. In the tightly choreographed sequence of developmental processes, some cells slow down, or even cease, mitotic activities just as others begin a burst of reproduction, accompanied by a surge in the growth of the tissue they comprise. Some cells must even die as part of the developmental process. It is essential

In the meantime, cells called **fibroblasts** (comprising a kind of connective tissue) are rapidly multiplying, forming "scar tissue," and filling in the depression caused by the wound. The scar tissue itself is inordinately strong because the fibroblasts produce cablelike fibers with the ability to stretch and contract. As they grow over the area and contract, they help pull the edges of the wound together. By now, nerve cells have begun to sprout side branches that invade the injured area. After a time, the scab loosens and falls away, exposing the layer of epidermis underneath. Below this, the collagen fibers grow along the lines of stress that produced the injury in the first place, further strengthening the healed area.

that some cells go through the process of formation and then vanish, such as those that lie between what will be the fingers of a developing hand.

The question arises, then, how chromosomes control the timing of developmental events. Such timing can be accomplished because a particular segment along the length of a chromosome may have a shorter active period than another segment. Thus, after an initial period of activity, one set of genes might cease its activity while another continues to turn out its enzymes. Obviously, the real questions of control involve the signals that turn genes on or off, and the challenge lies with identifying those signals and perhaps learning to control them.

Stress

Interestingly, stress can act as a regulatory factor in embryos. Just as karate practitioners build up the tissues around their knuckles by pounding makiwara boards, so an embryo builds up tissue along its lines of stress.

Embryos are stressed in a number of ways during normal development—for example, by bending and twisting that places tension on the back area and encourages the development of the sturdy muscles in that area. This happens when a type of mobile, undifferentiated tissue called **mesenchyme** moves into the areas of tension or pressure, settles there, and forms new supportive tissue.

Induction

Perhaps the most fascinating form of regulation involves one type of tissue influencing the development of another, a process called **induction.** As an example, after the mesodermal layer is formed, the tissue just under the area that will ultimately become the spinal cord begins to thicken, forming that rodlike structure, the **notochord.** The notochord itself doesn't form the spinal cord; instead, it causes the overlying ectoderm to do so.

Under the influence of the notochord, the overlying ectoderm begins to thicken, forming a **neural plate.** Soon tissue starts to build up along each side of the long, thickened plaque (Figure 16.2). After a period of enlargement, these two ridges, called the **neural folds,** lying along either side of the **neural groove,** begin to grow toward each other. At this stage, the folds form the prominent **neural crest.** The margins of the crest grow

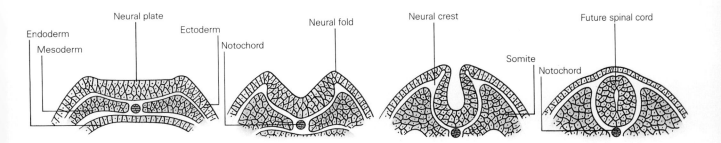

together until their tips touch and finally join and fuse. The result is a hollow tube lying along the dorsal surface of the embryo. This is the early **spinal cord,** and it is soon overgrown by ectodermal cells from the surrounding area so that it comes to lie beneath the skin.

In most vertebrates, the formation of the spinal cord marks the partial or complete disintegration of the notochord that triggered it. Any supportive function it might have had is taken over by bones that first form largely from **somites,** segmented blocks of mesoderm, lying alongside the spinal cord. These will form the **vertebrae** (backbones) and eventually enclose the spinal column in protective bone. The role of the notochord as the **inducer** of the spinal cord has been demonstrated by transplanting segments of the notochord to other areas of the embryo. If this is done before the ectoderm has become too differentiated, it will form incipient spinal cords almost anywhere on the surface of the embryo. For example, a spinal cord could be induced to grow along the belly. Such freaks, as you might suspect, usually do not survive.

Sometimes different inducers operate together in a coordinated fashion. For example, in some cases, where two tissues have developed from the same germ layer, one may have induced the other. Perhaps the best example is found in the induction of the lens of the eye from ectodermal cells by an underlying area of the brain, a tissue also formed from ectoderm (Figure 16.3).

One of the key developmental processes in vertebrates involves the formation of various membranes that sustain the embryo in various ways. Their formation and roles are best illustrated in the development of bird embryos.

FIGURE 16.3 Lens induction in the vertebrate eye. By the fifth to sixth week (below), outpocketings of the highly lobed brain (called optic vesicles) come to lie under the head ectoderm, inducing that ectoderm to thicken and roll, forming the lens of the eye by the eighth week (bottom). The underlying brain tissue will form elongated cells that become the light-sensitive retina of the eye. Thus, part of the eye is directly confluent with the brain and both the lens and the retina are ectodermal in origin. By the eleventh week (right), the eye is covered by a cornea.

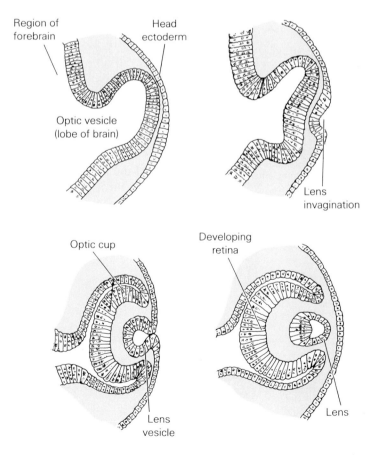

DEVELOPING BIRDS AND THEIR MEMBRANES

Although a bird's egg is very large, the early embryo is very small; in fact, the zygote is at first only a disklike **germinal spot.** Because a bird embryo rests on the bulky and inactive yolk, cell division is limited to the tissue comprising the germinal spot. The blastopore in the chicken embryo does not develop as a crescent, but as a slit along what will be the body axis (Figure 16.4). As a result, the mesoderm in birds is formed somewhat differently than in species with less yolk. In birds, the endoderm first splits away from the underside of the ectoderm, then the mesoderm is formed as cells from the upper ectodermal layer roll under along the midline of the embryo.

After the spinal cord has formed, the developing head and tail areas grow so rapidly that they overlap the tissue below. Each area is composed of an outer layer of ectoderm and an inner layer of endoderm. The two layers then form pockets, one at the head end and one at the tail end. These thinned areas then rupture and form the mouth and excretory openings. The characteristic tube-within-a-tube structure of vertebrates (Figure 16.5) is now complete.

In the meantime, a membrane composed of endoderm and mesoderm grows outward from the embryo over the yolk. Blood vessels develop in the mesoderm and will soon be able to carry food from the yolk to the embryo. Another membrane, called the **allantois,** begins to grow out of

FIGURE 16.4 Top, a cutaway view of a hen's egg with a developing embryo (the germinal disc). The yolk will provide food for the embryo, food that will be transported through embryonic vessels as membranes grow and enclose the nutrient supply. The yolk and embryo are held in place by the chalazae and cushioned by the albumen (or egg white). Later, membranes will come to lie pressed against the inside of the shell, picking up oxygen from the environment and losing CO_2 to it.

Center, the blastoceol stage of a developing chick. Compare the developmental stage with the sequences in Figure 16.1.

Lower, the embryo proper is still only a rather flattened area sitting atop the yolk. The disk is now elongated with the outline of developing membranes growing outward from the embryo. The primitive streak is an indentation above the notochord that indicates the axis along which the spinal cord will develop. Later, the head and tail regions will grow so fast as to overlap the yolk as is shown in the cross-section of the head in Figure 16.1.

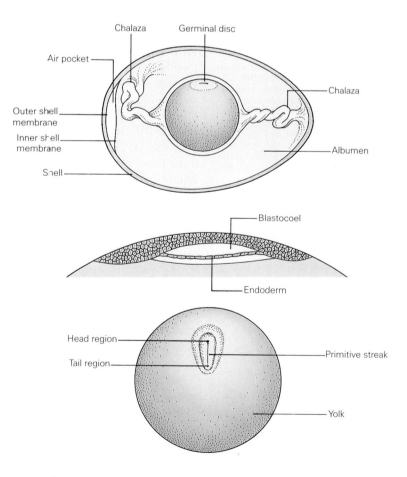

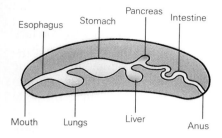

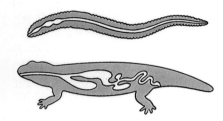

FIGURE 16.5 The basic tube-within-a-tube structure of vertebrates and most invertebrates. This general body plan is permitted by the events at gastrulation. Note the number and variety of organs that essentially develop as pouches from the embryonic gut.

FIGURE 16.6 Later development of membranes in the chick embryo. The head of the darkened embryo is at the right of both figures. Note that as the yolk mass decreases, the membrane complex becomes more extensive. As extraembryonic membranes grow upward and over the embryo, finally to fuse, they form a four-walled membrane. The inner two form the protective amnion. The outer two form the chorion. The allantois first appears as a saclike extension near the rear of the embryo and is a storage site for embryonic waste. Later it will become compressed with the chorion against the eggshell and exchange between the egg and the environment will be accomplished over this chorioallantoic membrane.

the hind area of the developing gut, forming a receptacle for nitrogenous wastes. The waste-laden allantois will be left behind when the chick hatches (See Figure 16.6).

As the extra-embryonic membranes develop, they arc upward over the embryo and grow toward each other, finally fusing and enclosing the embryo in a new, four-walled membrane. The inner two layers will become the **amnion** and the outer two the **chorion,** each consisting of both a layer of mesoderm and a layer of ectoderm.

The amnion fills with a slippery fluid that acts as a protective shock absorber and as a lubricant preventing the appendages of the embryo from fusing to the body. The mesoderm on the outside of the allantoic membrane later fuses with the mesoderm on the inside of the chorion, forming a three-layered **chorioallantoic membrane.** This membrane will come to surround the **albumen,** the egg-white that lies just under the porous shell. This membrane is then in a position to pick up oxygen diffusing in through the porous shell and transport it to the embryo. Carbon dioxide from the respiring embryo can leave the egg through the same membrane.

As the embryo continues to develop, it eventually is connected to the yolk only by a thin stalk through which food passes. In time, the anterior, or head, part of the spinal cord begins to bulge, forming the large lobes of the brain. The mesoderm on either side of the embryo forms blocks of somites, which will soon form the vertebrae and large trunk muscles. The mesodermal membranes, in fact, were formed from broad, lateral extensions of these somites.

About this time, a large blood vessel underneath the embryo begins to twitch irregularly and then to pulsate more and more rhythmically. Later, it will loop and fuse to itself, forming the heart. Tiny flaps are now forming that will become the limbs. The brain has continued its rapid growth and now outpockets on either side to form the great orbs that will be the eyes. Meanwhile, the endoderm begins to form pockets here and there that will become the highly complex glands and organs associated with digestion.

A crucial change occurs on the eleventh day. This is a time of rapid transition, marked by a great eruption of enzymes. Many embryos whose systems are not quite functioning properly die at this time. After this, if all goes well, the various systems of the organism will coordinate their activities and begin to function in even greater harmony.

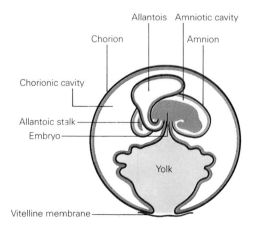

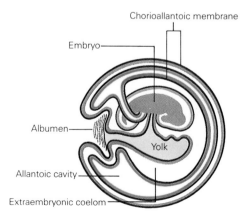

After about three weeks, the chick will begin the first feeble pecks that will eventually enable it to break free of its protective shell. When it finally emerges, it will begin to react to strange and fascinating new experiences. Some responses will be innate, others will develop as the result of experience. As it grows, it will constantly adjust to its world, adapting, changing, and learning. Then someone will eat it.

HUMAN DEVELOPMENT

Many of the principles we have considered apply to a wide range of animals, including us. That brings us to us. As you are doubtless aware, it is quite easy to plunge into any number of controversial areas by this simple shift in species. There is an understandable tendency for us to assume a certain "specialness" and to minimize the idea of ties with other, less "noble," species. Nonetheless, many commonalities remain, beginning with fertilization.

Descriptions in bus-station novels notwithstanding, the prospective mother has no way of knowing if fertilization has occurred—at least not at first. If there are sperm cells thrashing around in her genital tract from about forty-eight hours before ovulation to about twelve hours after, the odds are good that pregnancy will result. (The whole story of copulation, pregnancy, and birth may be a bit hard to believe, but let's assume it's true and go on.) As soon as the egg is touched by the head of a sperm (see Essay 16.3), it begins to pulsate violently, uniting the twenty-three chromosomes of the sperm with the same number of its own. From this single fertilized cell, now only about 1/175th of an inch in diameter, a baby weighing several pounds and composed of trillions of cells will appear about 266 days later. It may one day vote.

For convenience, we will divide the 266 days, or nine months, of pregnancy into three periods of three months each and consider these trimesters separately.

The First Trimester

In those delicate days of the first trimester, the misshapen embryo begins the coordinated changes that will lead to its final form. During this period, the embryo is particularly susceptible to any number of factors that might influence its development. In fact, human embryos often fail to survive this first critical stage.

The first few divisions of a human embryo produce very similar cells with roughly the same potentials. In other words, at this stage the cells are, theoretically anyway, interchangeable. There are very few cells produced in those first days. In fact, only sixteen such cells will exist seventy-two hours after fertilization. (How many divisions will have taken place?) The cells in early embryos grow progressively smaller because, since there is no input of food to fuel growth, each cell divides before it reaches the size of the cell that has produced it. The rate of mitosis begins to increase, so that by the end of the first month, whereas the embryo will be only 1/8 inch long, it will consist of millions of cells.

The First Month

The first cleavages take place in the oviduct before the zygote reaches the uterus. In fact, after the zygote reaches the uterus, a period of a few days, it does not implant for about another three days, even though the uterine wall is already swollen and ready to receive it. The result is that by the time it does implant, perhaps a week after fertilization, it has already reached the hollow-ball, or **blastula**, stage. This blastula is lopsided, somewhat like a class ring. One side of the blastula is thin-walled and forms the **trophoblast**. Later, this will produce the complex membranes that surround the embryo. The thicker-walled side (the side of the ring with the setting), forms the **inner cell mass**. It will become the embryo proper.

At the time of implantation, the trophoblast secretes a digestive enzyme that breaks down some of the blood vessels in the uterine wall around it. The embryo sinks into this cavity and comes to lie in a pool of the mother's blood. The enzymatic breakdown of the uterus produces some bleeding that is sometimes mistaken for menstruation. (You might like to remember that.) The embryo continues secreting these enzymes for about a week as it sinks deeper into the blood-engorged uterus.

The embryo's early days, then, are spent in a pool of bloody, glycogen-laden fluid that provides it with sustenance. By about twelve days, the injured uterine wall will have repaired itself, and the embryo, buried deep with the flesh of the uterine wall, silently continues its mysterious changes.

By this time, the trophoblast has begun extending fingerlike projections called **chorionic villi** into the uterine wall, where they probe deeply into the mother's tissues, which are also undergoing drastic changes. Where the embryonic and maternal tissues press together, the **placenta** is formed. Through this placenta will pass food and oxygen from the mother and waste and carbon dioxide from the fetus.

Surprisingly, after a month, the embryo is still smaller than a pea (Figure 16.7), but it has nevertheless undergone momentous changes. The anterior end of the spinal cord has begun to develop the bulges and lobes that will be the brain. The mesoderm has now divided into thirty-two pairs of dense blocks of somites that lie on either side of the devel-

FIGURE 16.7 The development of the human embryo showing relative sizes at different ages. Note that at fourteen days, very few features are clearly distinguishable. However, its body axis is established and the major organs have begun to form, unlikely as it seems. The embryo at the third week is strangely vulnerable to a variety of dangers from drugs to radiation. You probably can't make out the developing eyes or the tubelike heart at the fourth week. At seven-and-a-half weeks and almost an inch in length, the embryo still weighs less than an aspirin tablet and has a short tail. At the end of eight weeks it will be called a fetus, and by nine weeks it will take on the general appearance of a human. The head will be oversized through the embryonic period and on into childhood. By eleven weeks fingers and toes have formed as well as such refinements as the ridges of the ear. By fifteen weeks the fetus is moving frequently and has taken on the facial features of its species.

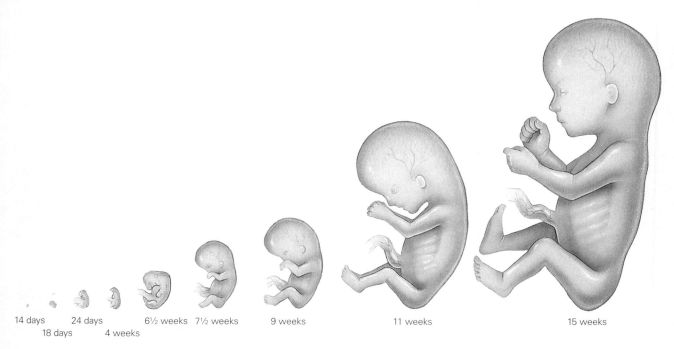

14 days   24 days   6½ weeks 7½ weeks   9 weeks   11 weeks   15 weeks
        18 days   4 weeks

In the developing female embryo, there are about seven million eggs in the ovaries by the twentieth week. By puberty, many of these have been lost (though the means of selection is unknown) so that there are by then only about 400,000 eggs. At about the age of sexual maturity (usually around 12 to 16), there may only be 300,000 eggs. Of these, only one will escape a ruptured follicle every 28 days. This goes on for about 40 years, so a total of about 400 eggs are released. Each woman could actually produce a maximum of about 40 children in a lifetime (the record is 69). So, of millions of potential ova, only 25 to 40 can contribute their genes to the next generation.

If no type of developing germ cell has an advantage, the probability of any one developing to maturity is remote, and the probability of ova with identical sets of genes approaches nil.

Males, after puberty, normally produce millions of sperm each day and the production may continue throughout life. The chromosomes of each sperm and egg have undergone meiosis and crossing over, and there is almost no likelihood of two identical sperm being produced. Of millions of sperm in each ejaculation, only one will fertilize an egg. One reason so many sperm are necessary is that the female reproductive tract is essentially a hostile environment, and most sperm will quickly perish there. Also, a mucous plug may block the cervix, and enzymes carried by sperm are necessary to break it down to open the way for other sperm. These same enzymes act to break down the follicle cells surrounding the egg so that a sperm can get through to fertilize it. If no particular type of sperm (in terms of genetic constituents) has an advantage, you can see that the likelihood of identical individuals being produced from the same parents is very slim. In fact, the odds of your having children with identical genetic constituents have been calculated at about 1 in 14 trillion.

oping spinal column. Later, the somites will give rise to much of the skin, as well as bone and voluntary muscles. Three pairs of arterial arches have appeared in the neck region. In humans, these disappear, but in fish they become part of the gills. The opening that will be the mouth has broken through, although no nourishment will enter for some time yet. Even large sightless eyes have begun to form. By the end of the third week, the weak, developing heart begins its first twitching pulsations.

Also during this first month, undifferentiated cells roaming the yolk sac and chorion begin to form blood cells. These, by day twenty-four, will be pumped by a tubular and incomplete heart through unfinished blood vessels. Nonetheless, they are able to pick up oxygen that has diffused across the placenta from the mother's blood. Near the end of the first month, primitive kidneys have formed, but they will function for only a few days before they are replaced by a more specialized permanent kidney. By now, the first cells of the endocrine glands are apparent. Tiny buds are visible that will eventually form the arms and legs.

The Second Month

In the second month, the features of the embryo become more recognizable. Bone begins to form throughout the body, first in the jaw and shoulder areas. The head and brain are now developing at a much faster rate than the rest of the body. Ears appear. Open, lidless eyes stare blankly into the amniotic fluid. The tubular heart has looped back on itself, setting the stage for its final four-chambered form. The circulatory system continues to develop, and blood is continuously pumped through the umbilical cord and out to the chorion where it receives life-sustaining nutrients and deposits the poisons it has removed from the developing embryo. The nitrogenous wastes and carbon dioxide filter into the mother's bloodstream, where they will be circulated to her own kidneys and lungs for removal. About day forty-six, the reproductive organs begin to form, either as testes or ovaries, and now, for the first time, the sex of the embryo is apparent. Near the end of the second month, fingers and toes begin to appear on the flattened paddles that have formed from the limb buds. By this time, the embryo is about two inches long and has the general appearance of a human. From now on it is called a **fetus**.

The Third Month

The fetus continues to grow and change during the third month, and now it may begin to move. It may breathe the amniotic fluid in and out of bulblike lungs, and from time to time it swallows. Even at this stage, fetuses begin to show individual differences, especially in their facial expressions. Some frown a lot; others tend to smile or grimace. It would be interesting to correlate this early behavior with the personality traits that develop after birth.

At the end of the first trimester, the reproductive and excretory organs show marked development as the embryo's urine appears in the amniotic fluid, from where it is filtered into the mother's blood to be removed by her kidneys. Bone continues to form throughout the body, much of it replacing scaffoldings of cartilage. The bones arise independently of each other and grow outward to meet other bones at what will be the joints. All the organ systems have been formed by this time, but if the fetus is removed from its mother it will die.

The Second Trimester

In the second trimester the fetus grows rapidly, and by the end of the sixth month it may be about a foot long, but it will weigh only about a pound and a half. Whereas the predominant growth of the fetus during the first trimester was in the head and brain areas, during the second trimester, the rapid growth of the body begins to catch up with the head.

The Fourth and Fifth Months

By the fourth month, the fetus is moving vigorously, kicking and thrashing in its amniotic fluid, movements clearly felt by the mother. Interestingly, it must sleep now, and the thankful mother can also get some rest. As time passes, it becomes increasingly sensitive to more types of stimuli. For example, by the fifth month the eyes are sensitive to light, although there is still no hearing. Other organs, such as the lungs, seem to be complete, but are still nonfunctional. The digestive organs are present but cannot digest food. The skin is well-formed, but it cannot adjust to any temperature changes. By the end of the fifth month, the skin is covered by a protective, cheesy paste consisting of wax and sweatlike secretions mixed with loosened skin cells. The fetus is still incapable in nearly all instances of surviving alone.

The skeleton has been developing rapidly during the second trimester, with some bones arising anew from undifferentiated embryonic cells, and others forming through the gradual replacement of cartilage cells by bone cells. Now the mother must supply large amounts of calcium and other bone constituents as building materials for the fetal skeleton.

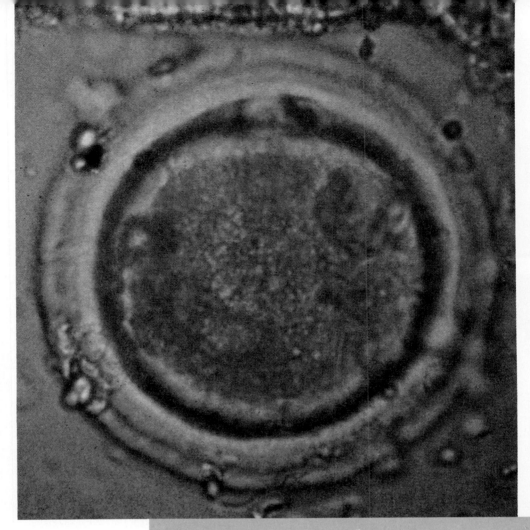

Human Development

Humans are indeed a remarkable form of life. However, we are only *one* form of life. We almost intuitively try to set ourselves apart from the other forms, but we seem to base the differences primarily on the stages that occur after our first few years of life and what we presume lies in store after life ends. As these photographs show, even if we do differ from other forms of life in both accomplishment and destiny, we are clearly tied to the others, from pigs to pike, in our earliest days. The drama of the beginnings of human life is no more complex, fascinating, and poignant than are the first stirrings of life in other creatures. Every guru was once a gastrula, and perhaps a clearer understanding of our individual beginnings will help us to more comfortably take our place in the grand parade of life.

This remarkable photograph captures the moment of fertilization of a human egg. Shown faintly in the darkened center are the chromosome-laden male and female nuclei approaching each other. At the top are two polar bodies formed during meiosis. One might wonder how much of our behavior and even our values is determined at this moment. It seems strange, but we really don't know just what those chromosomes carry. We know the limits of height are more or less determined genetically. How about the limits of empathy? In any case, about thirty-six hours after this event, the zygote will undergo its first cleavage, all the while wending its way slowly toward the uterus that will be its next home.

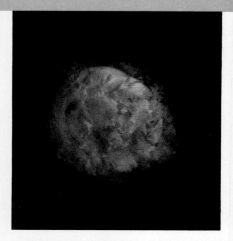

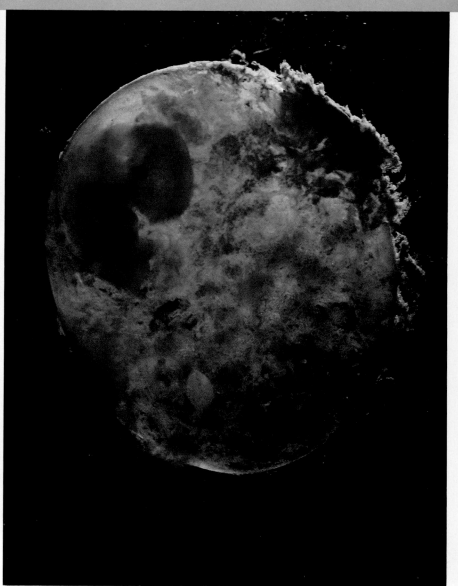

At the beginning of the second week, the embryo has developed extensive membranes that lie in close contact with the mother's tissues (top left). The delicate projections of the chorion have penetrated the mother's tissues in all directions as they take nutrients and oxygen from her blood and deposit their own metabolic wastes for her system to carry away.

Well into the third week (bottom left) the chorionic membrane has continued to penetrate the mother's endometrium. The chorion, shown radiating outward, carries blood vessels that lie closely intertwined with those of the mother, but the two do not join. The balloon-like structure is the yolk sac. The human embryo need carry only enough food to last through its first few weeks since food will later be derived from the mother's blood.

The embryo at the fourth week (right). It lies protected in its amniotic sac. The dark eye is prominent and the enormous brain lies tucked against the embryonic heart. The embryo by this time has already developed primordial cells that will form its own gametes. By now the tubular heart has begun its first timorous beats.

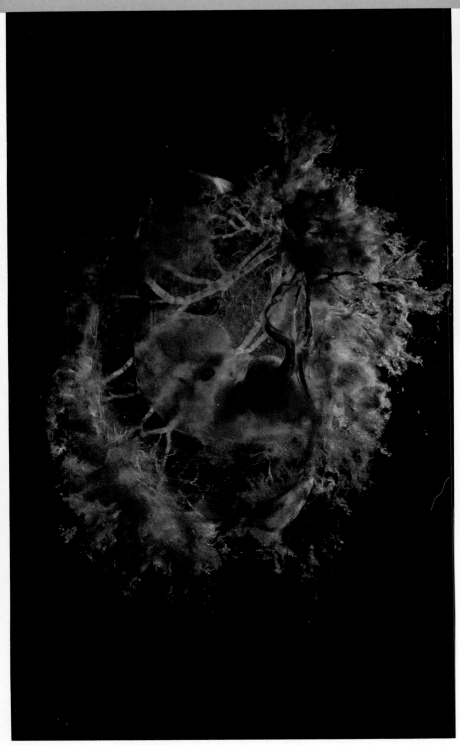

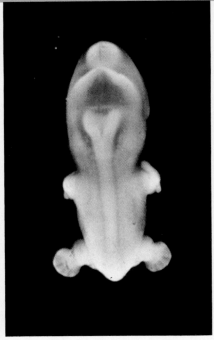

The human embryo is here shown at forty-two days with the surrounding membranes removed (above). It is about half an inch (or sixteen millimeters) long. This is a dorsal view showing the enormous head (which helps direct the growth of the rest of the body) and the spinal cord extending to the rump. Notice the paddlelike appendages. Fingers and toes are already apparent as the tissue between them dies as a result of a mysterious chromosomal timing mechanism.

At about six weeks (left) the extensive vascular system is clearly visible leading from the projections of the embryonic chorion to the embryo itself. The organ below the eye is the now-looped heart. The embryo is still so water-laden that its tissues are virtually transparent.

At six weeks (right), with the amnion removed, the fingers are apparent and the bulbous brain still dominates the embryo. Notice the "tail" tucked under the abdomen. (It is destined to disappear.) The tiny pit above the arm will become the ear.

At about this time the embryo is extremely vulnerable to all sorts of chemical agents. An increasing number of drugs have been found to produce congenital abnormalities, for example, in the growth of limbs. Even x-rays, at this time, may endanger the development of the embryo. Certain diseases are also particularly dangerous. For example, if a mother contracts German measles during the fourth to twelfth week of her pregnancy, the result may be deformities in the offspring's eyes, heart, and brain.

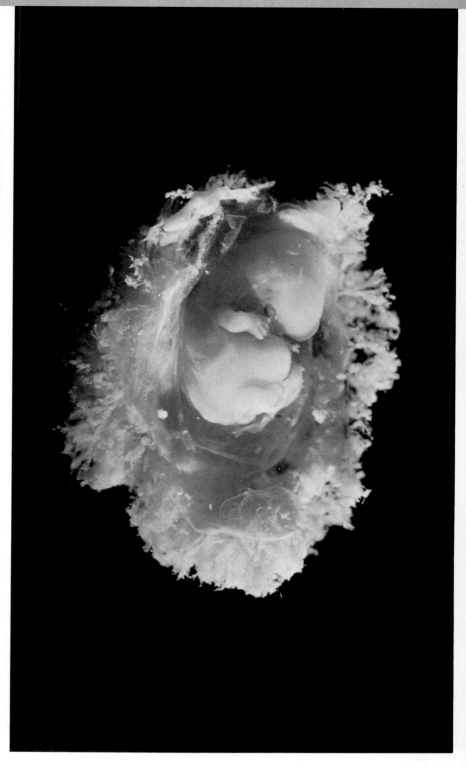

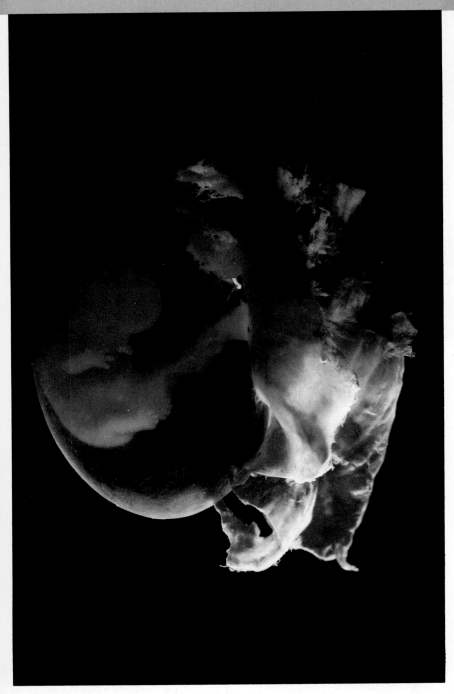

At about seven weeks (left) the embryo, afloat in its amniotic fluid, is clearly anchored to its placenta by the twisted umbilicus through which great blood vessels pass. The abdomen is swollen due to the rapid growth of the liver, the main blood-forming organ at this time.

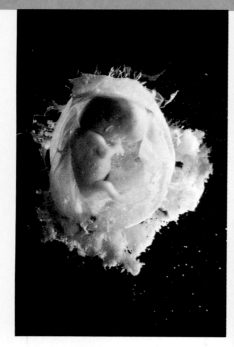

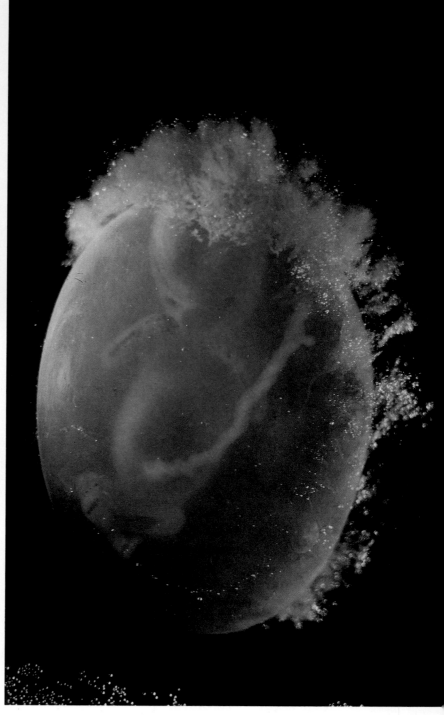

The eight-week embryo is shown here (right) in front view. The organs are now more or less complete after a rapid period of growth and development. From here on, development will primarily consist of refinements of existing structures. The skeletal system is among the last to form, but bones are now evident in the arms and legs.

At nine weeks (above) lids have begun to grow down over the eyes, and the outer ear begins to form. Because the plates of the skull have not fused, the head is rather flexible. During this third month, the fetus may begin to move, wave its arms and legs, and may even suck its thumb. It is now beginning to fill its amniotic space and will soon assume the typical, upside-down fetal posture.

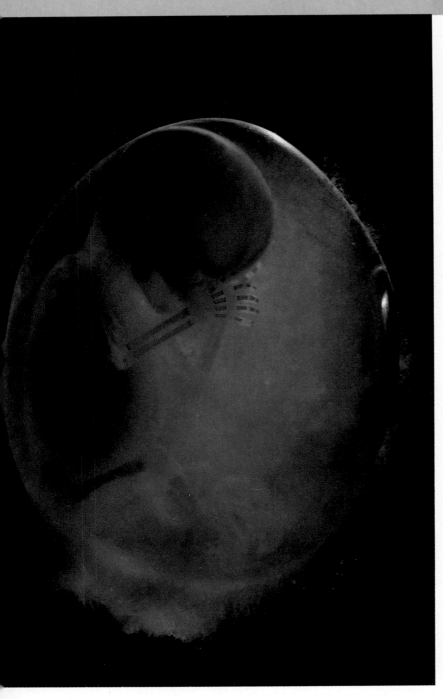

At ten weeks (left) the skeleton is well along in its development. The long bones begin developing independently, growing from areas near their ends. They will join, forming joints, later. In fact, the joints may not be firmly abutted by the time the baby is born. The head is still disproportionately large and will remain so, to a decreasing degree, through childhood. Notice the coiled umbilicus lying near the cuboid ankle bones showing as dark spots. The wrist bones have not yet begun to form and the jaw structure is weak indeed.

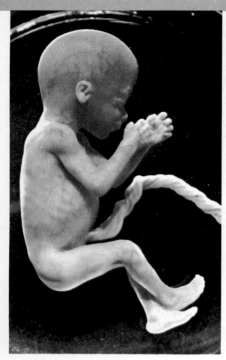

At fourteen weeks (right) the fetus is fist-sized. Ribs and blood vessels are visible through the translucent skin. The vigorous movements of the fetus can now be felt by the mother. The delicate skin is actually covered with a cheesy protective coating. Refinements such as fingerprints and fingernails have not yet developed.

By the end of five months (above) the fetus is covered with fine, downy hair and its head may have already started to grow its own crop. It has already started the lifetime process of discarding old cells and replacing them with new ones. The heart is beating now at a rate of 120 to 160 times per minute.

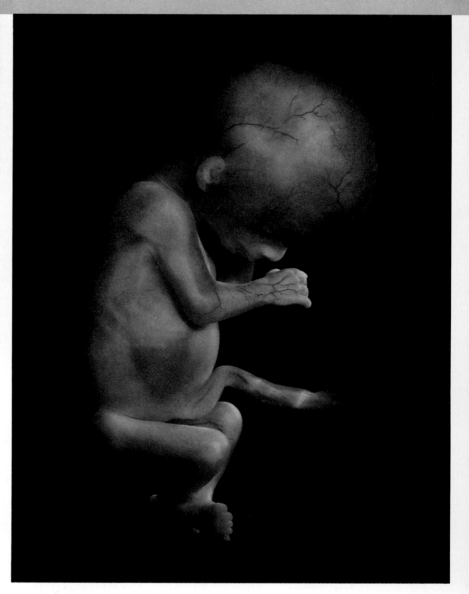

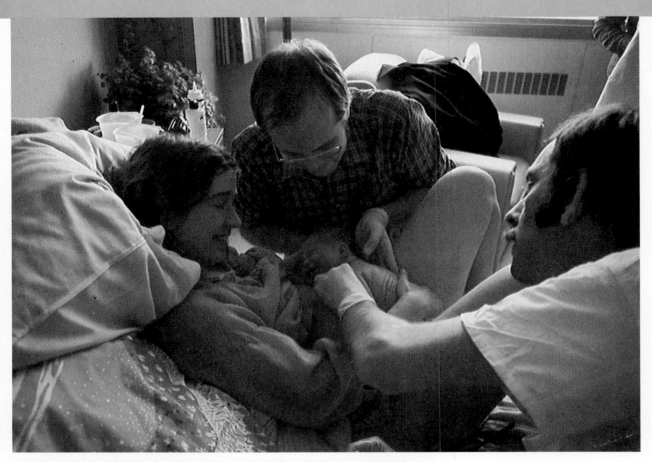

Delivery is a much too clinical term for what
the mother is experiencing both physically
and emotionally. It is a culmination of long
months of changes. The infant is still not completely developed, but it is able to survive
apart from the mother.

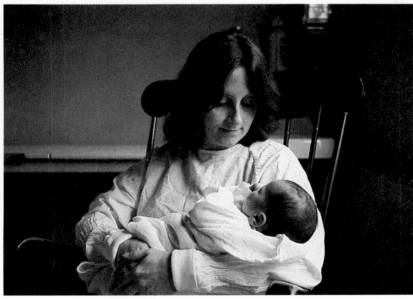

The Sixth Month

By the sixth month, the fetus is kicking and turning so constantly that the mother often must time her own sleep periods to coincide with its schedule. The distracting effect has been described as feeling somewhat similar to being continually tapped on the shoulder, but not exactly. The fetus now moves so vigorously that its movements can be seen clearly from the outside. To add to the mother's distraction, the fetus may even have periods of hiccups. It is now so large and demanding that it places a tremendous drain on the mother's reserves.

At the end of the second trimester, the fetus is clearly human, but it resembles a very old person because its skin is loose and wrinkled at this stage. In the event of a premature birth around the end of this trimester, the fetus may be able to survive.

The Third Trimester

During the third trimester, the fetus grows until it fills its available space and is no longer floating free in its amniotic pool, and, even in the greatly enlarged uterus, its movement is restricted. In these last three months, the mother's abdomen becomes greatly distended and heavy, and her posture and gait may be noticeably altered in response to the shift in her center of gravity. (However, some markedly overweight women may go through much of this stage without realizing they are pregnant at all.) The mass of tissue and amniotic fluid that accompanies the fetus ordinarily weighs about twice as much as the fetus itself. Toward the end of this period, milk begins to form in the woman's mammary glands, which in the previous trimester may have undergone a sudden surge of growth.

At this time, the mother is at a great physical disadvantage in several ways. About 85 percent of the calcium she eats goes to the fetal skeleton, and about the same percentage of her iron intake goes to the fetal blood cells. Much of the protein she eats goes to the brain and other nerve tissues of the fetus.

Some interesting questions arise here. If a woman is unable to afford expensive protein-rich foods during the third trimester, can it affect the brain development and intelligence of her offspring? On the average, poorer people in this country show lower I.Q. scores. Are they poor because their intelligence is low, or is intelligence low because they are poor? Is there a self-perpetuating nature about either of these alternatives?

During the third trimester, the fetus grows quite large. It requires more food each day, and it produces more poisonous wastes for the mother's body to carry away. Her heart must work harder to provide food and oxygen for two bodies. She must breathe, now, for two individuals. Her blood pressure and heart rate rise. The fetus and the tissues maintaining it form a large mass that crowds her internal organs. In fact, the fetus pressing against her diaphragm may make breathing difficult for her in these months. Several weeks before delivery, however, the fetus will change its position, dropping lower in the pelvis (a process called *lightening*), and it thus relieves the pressure against the mother's lungs.

The "finishing" of the fetus proceeds rapidly in the last three months. Such changes are reflected in the survival rate of babies delivered by cesarean section (an incision through the mother's abdomen). In the seventh month, only 10 percent survive; in the eighth month, 70 percent; and in the ninth, 95 percent survive.

Interestingly, there is a change in the relationship of the fetus and mother the last trimester. In the first trimester, measles and certain other infectious diseases would have affected the embryo. However, during the third trimester, the mother's antibodies confer an immunity to the fetus, a protection that may last through the first weeks of infancy.

About 255 to 265 days after conception, the life-sustaining placenta begins to break down. Parts of it begin to shrink and change, and the capillaries begin to disintegrate. The fetal environment becomes rather inhospitable, and premature births at this time are not unusual. At about this time, the fetus slows its growth and changes its position so that its head is directed toward the bottom of the uterus. Its internal organs undergo some final changes that will soon enable it to survive in an entirely different kind of world. So far, its home has been warm, sustaining, protected, and confining. It is not likely to encounter anything quite so secure again.

Birth

The signal that there will soon be a new customer on the planet is the onset of **labor,** a series of uterine contractions that at first appear at about half-hour intervals and gradually increase in frequency. Meanwhile, the sphincter muscle around the cervix dilates, and, as the periodic contractions become stronger, the baby's head pushes through the extended cervical canal to the opening of the vagina.

Once the baby's head emerges, the pattern of uterine contractions changes, becoming milder but more frequent. The small shoulders come into view, and then the body appears. With a rush, the baby slips into a new world. A few minutes later, the umbilical cord that had connected the fetus to its life-sustaining placenta is tied off and cut. The contractions continue until the placenta is expelled as the **afterbirth.** The mother recovers surprisingly rapidly, in humans and wild mammals as well. In many other mammals, the mother immediately chews through the umbilicus and eats the afterbirth. She thus regains nutrients and prevents the tissue from advertising to predators the presence of a helpless newborn.

Cutting the umbilicus halts the only source of oxygen the infant has known. With no oxygen, carbon dioxide quickly builds up in the blood. It triggers a breathing center in the brain that causes a nerve impulse to be fired to the diaphragm. The contraction of the diaphragm causes the baby to gasp its first breath. An exhaling cry means that the baby is breathing on its own.

In American hospitals, a newborn baby is given the first of the many tests it will encounter in its lifetime. This one is called the **Apgar test series,** in which muscle tone, breathing, reflexes, and heart rate are evaluated. The physician then checks for skin lesions and evidence of hernias. If the infant is a boy, it is checked to see whether the testes have properly descended into the scrotum. A footprint is then recorded as a means of identification, since the new individual, despite the boasts of proud parents, does not yet have many other distinctive features. (There have been more than a few cases of accidental baby-switching. I'm *convinced* it happened to me!)

Miscarriage

Unfortunately, not all pregnancies result in the delivery of a healthy offspring. The route between conception and birth is fraught with risk, perhaps largely because of the very complexity of living things. A phenomenal array of physical and chemical interactions occur within a developing body, and, if somewhere along the line a simple component fails in its function or its timing, the embryo may be lost. Should the problem arise early in the pregnancy, the embryo may be broken down—digested—and reabsorbed by the mother's body, perhaps without her ever knowing she was pregnant. It is interesting that if failure should occur later in the pregnancy, the mother's body often mysteriously seems to recognize that something has gone wrong, and the fetus is aborted as a "miscarriage." About half of all miscarriages involve abnormal fetuses.

A miscarriage can arise from a number of factors. For example, it may occur if an embryo does not implant correctly, or if it invades the uterine wall too soon in its development. The mother's activities also have effects. Alcoholism or heavy smoking may cause premature birth of otherwise healthy babies. Drugs, legal or otherwise, may impair the normal development of the embryo. A weak cervix may not be able to support the weight of a fetus and produce a spontaneous abortion. Other factors may also interfere with a successful pregnancy, such as disease, stress, or malnutrition. In fact, so many things can go wrong that one might well marvel at the successful delivery of any healthy babies—or so it seems when reviewing the list of potential hazards to embryos.

SUMMARY

Animal development, the events between fertilization and birth (or hatching), is largely dependent on the type of egg as determined by yolk abundance. As examples, the human egg has little yolk, the frog moderate yolk, and birds large amounts of yolk. The early cell divisions are easily seen in the frog as the metabolically inert yolk displaces the cleavages toward the animal pole. The three germ layers (ectoderm, endoderm, and mesoderm) are established at gastrulation. In the early stages, the fate of the cells is not completely fixed. Continuing development is not only a function of genetic control (especially through timing), but external influences such as stress and induction play a role as well. Developing membranes are best seen in birds' eggs, especially the amnion, chorion, and chorioallantoic membranes. Human development takes about 266 days, usually divided into three trimesters. The embryo is several days old by the time it implants as a blastula. Nutrients and oxygen reach the embryo across the placenta. Growth is rapid in the second trimester as various organs begin to take form. The third trimester marks a great drain on the mother. Birth occurs as the placenta breaks down. Sudden changes occur in the fetus that allow it to exist independently of its uterine home.

Part Five

SYSTEMS
AND
THEIR
CONTROL

Chapter 17

HOMEOSTASIS AND THE INTERNAL ENVIRONMENT

About 200 years ago, Dr. Charles Blagden, then secretary of the Royal Society of London, proved that he was one of the most persuasive people on earth. He talked some friends into joining him, a small dog, and a steak in a room in which the temperature had been raised to 126°C (260°F). In fact, he managed to persuade his friends to stay in there for 45 minutes. (The dog and the steak had no choice.) At the end of this time, the men and the dog emerged unharmed, but the steak was cooked!

In addition to demonstrating his polemic powers, Blagden also showed that the bodies of animals are able to compensate for extreme physiological conditions and, in particular, that some living things can control their internal temperatures in the face of extreme external conditions. Since this early experiment, it has been found that many kinds of animals can regulate a host of other internal physical and chemical states. Such an ability, to some degree, should not be unexpected since the delicate processes of life would not be possible under wildly fluctuating conditions.

HOMEOSTASIS AND THE DELICATE BALANCE OF LIFE

Homeostasis refers to the tendency of living things to maintain a constant internal environment (from the Greek *homios*, same, and *stasis*, standing). The term is a bit misleading, however, because no living thing strictly maintains a constant internal environment. One reason is that maintaining such rigid constancy would place a great demand on the organisms's metabolic machinery. Another reason is that some change must occur in order for things to stay the same. That is, as the environment changes, the body's internal processes must also shift in order to counteract the outside changes and keep the internal conditions stable. So cells do change as they constantly monitor, metabolize, and adapt. The result is a certain internal constancy, and that is what homeostasis is all about.

FEEDBACK SYSTEMS

Often, the steady state is maintained by **feedback** mechanisms. Feedback occurs when the product of an action influences that action. Biologically, the most important kind of feedback action is called **negative feedback.** Negative feedback occurs when an increase in a system's product causes a slowdown of that system, or when a reduction of the product stimulates the system to increase its activity. One kind of "cruise control" on an automobile engine (a device that keeps the car moving at a certain speed) works because as the motor runs faster, valves are closed, causing the car to slow down. As it slows, the valves are reopened, and the car accelerates, keeping its speed within certain limits. Because of negative feedback mechanisms, you don't have to constantly nibble and fast to keep your blood sugars at the proper level. When blood sugars are low, the liver simply breaks down some of its stores of glycogen and releases glucose into the blood. As blood sugars rise, the liver releases less glucose.

Positive feedback works on the opposite principle: the product of a system increases the activity of that system. Using the example of an automobile, with positive feedback, accelerating the motor would tend to open the carburetor and cause the engine to run even faster. Such an engine might be revved to such limits that it would explode. Obviously, homeostasis would operate primarily through the more delicate mechanisms of negative feedback, but this is not to say that living things are not sometimes subjected to positive feedback.

Positive feedback mechanisms in humans are sometimes associated with severe health problems. For example, if a person's temperature begins to rise above the normal 37°C, the body will activate corrective devices such as sweating and the opening of peripheral blood vessels (producing a heat-dissipating "flush"). However, at some point (usually at about 42°C), the negative feedback system breaks down, and a positive feedback begins. The high temperature begins to cause an increase in metabolic activity, which raises the heat, which increases metabolic activity, which can kill the unfortunate soul. (Positive feedback is the basis of the famed "vicious circle.") To better understand the principles of the two forms of feedback control, consider the operation of two simple storage tanks (Figure 17.1).

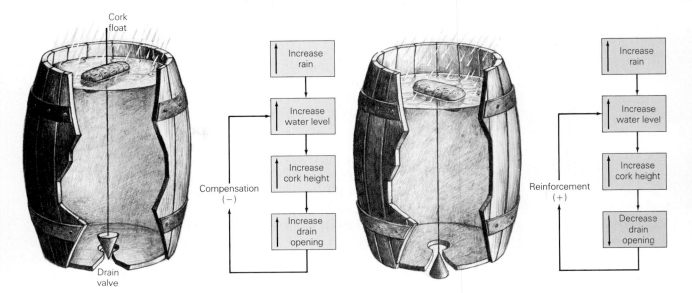

FIGURE 17.1 Two types of feedback systems. In the figure at left, a rather constant water level is maintained as long as there is enough rain. As the amount of water increases, more is allowed to run out; as it decreases, the drain is plugged. This, then, is a typical negative feedback system. In the figure at right, as the water level rises, less is allowed to escape, so the water rises out of control. As it falls, even more is released until the barrel is empty. This is a positive feedback system, a type usually not associated with the normal functioning of living things.

FIGURE 17.2 The pupfish is a denizen of hot, briny desert springs. In natural habitats, they are able to live under a temperature range of 38°F (3.3°C) to 108°F (42.2°C). The physiological system that permits such abilities is not completely understood.

TEMPERATURE REGULATION

Many of the processes of life can occur only within a narrow range of temperatures, and the mechanisms by which this stabilization occurs may serve as a good example of homeostasis. First, we should note that an increase in temperature of only a few degrees can cause great leaps in the rate of chemical reactions. In nonliving material, the rule of thumb is: the rate of chemical reactions doubles for every 10°C increase in temperature. However, in living cytoplasm, such rules may not apply, and a change in temperature may have less effect.

Why is temperature so important to living things? Because of the severe effects of temperature extremes. At about -1° to -2°C, the water in cells freezes, causing ice crystals to form that may rupture delicate membranes. Also, with water tied up as ice, the remaining cell constituents may become so concentrated that they are unable to function properly. The result of such disruptions may be death. At the other extreme, the upper temperature limit that life can withstand is apparently largely set by the temperature at which the hydrogen bonds holding proteins in their tertiary structures begin to break, thus unwinding (or denaturing) the protein. Because of the effects of temperature, most animals live in places that are not much colder than freezing or warmer than about 40°C (Figure 17.2). (Exceptions to such rules include certain algae, fungi, and bacteria that are able to live in hot springs at temperatures up to 80°C.)

Ectotherms

The **ectotherms,** or the so-called cold-blooded animals include virtually all the vertebrates except birds and mammals (*ecto,* outside; *therm,* temperature). These are the animals that do not physiologically regulate their body temperatures to any great extent. Although they do produce metabolic heat, they have no efficient means of conserving it or of increasing or decreasing its production. If it is necessary to change their body temperatures, they may respond behaviorally simply by moving to a cooler or warmer place. Because they lack metabolic temperature con-

FIGURE 17.3 A desert-dwelling lizard. Note the long appendages that not only dissipate heat quickly, but enable the lizard to move quickly to find cooler areas. Even desert lizards cannot stand the direct desert sun for long, and so they hunt in the shade, being most active at dusk and dawn.

trol, fish, amphibians, and reptiles have specific problems in adapting to their environments.

In general, the saltwater fish have no great problems because the oceans rarely change in temperature more than a few degrees. However, freshwater fish living in the shallows are much more at the mercy of the elements. Furthermore, their efforts to thermoregulate behaviorally can put them at risk. As they move from place to place, they may find themselves in danger of becoming landlocked. This means they risk overheating or drying out if they are unable to return to deeper water. Of course, freshwater fish that remain in deeper water suffer no such threat, but because of the low temperature at these depths, their metabolic rate is slow and so must be the pace of their lives.

Among land dwellers it's a different story. They face unbuffered temperatures and dry air with its rapidly changing temperatures. Somewhat surprisingly, amphibians have been able to adapt to such arid environs. One way they adjust is by simply moving overland to places with more agreeable temperatures. Some species may also bury themselves and thereby escape the drying air. Other species that are exposed to extreme heat and drought have the ability to **estivate**—that is, to enter a form of summer hibernation until reactivated by cool temperature and moisture. Toads in certain arid parts of Australia, in fact, have been known to remain buried for as long as two years while waiting for a good rain. The estivating toads have low metabolic rates, just as do hibernating animals, and therefore their food and oxygen requirements are low.

Land reptiles strongly rely on behavioral regulation. As do amphibians, they simply move to areas that are more appropriate to their needs. As with many other ectotherms, they also allow their body temperatures to drop with the cool of night. This results in a certain sluggishness in the morning because their metabolic rates have slowed accordingly. If they are going to catch any food, however, they've got to warm up, and they quickly do so by utilizing the warmth of the sun. Basking ectotherms are able to absorb the heat of the sun even when the air temperature is near freezing. As the day wears on, the animals may seek shade or turn to face the sun, which reduces the surface area exposed to its rays. During the hottest part of the day, some species may retreat underground, reappearing only in the cooler afternoon (Figure 17.3).

Endotherms

Birds and mammals comprise the **endotherms** (*endo*, inside), the so-called warm-blooded creatures. They have evolved physiological methods of keeping their body temperatures within very narrow limits. Their bodies can carry out biochemical activities more efficiently by having specialized to within a range of temperatures that is conducive to biochemical reactions.

Interestingly, both birds and mammals are descended from the reptiles. Mammals appeared at about the time the dinosaurs made their entrance, about 150 million years ago (the birds evolved a little later). While the great dinosaurs were thundering around the earth terrorizing everything in sight, the mammals were existing as tiny mouselike forms that probably terrorized only insects (Figure 17.4).

After the disappearance of the dinosaurs, the numbers of birds and mammals burgeoned. As they radiated out over the earth, invading

FIGURE 17.4 The tiny primitive mammals scurrying about under the leaves at the lower right and the low-flying Archaeopteryx were unlikely candidates to inherit the earth, but they survived the ruling reptiles. One reason may have been due to a greater ability to physiologically regulate their temperatures. Such surmising is based on the fact that most modern reptiles are not able to regulate their internal temperatures very effectively, unlike birds and mammals.

newly-vacated niches, the endotherms began to change and specialize in innumerable ways. Nonetheless, both birds and mammals, having sprung from the same distant stock, had certain critical traits in common—traits related to temperature control. First, they were well-insulated creatures, with their fur and feathers. Second, they had a more efficient four-chambered heart.* Theirs was a powerful organ (described in Chapter 19) able to pump enough fuel and oxygen throughout the body to stoke the cells' metabolic furnaces. Remember, the cost of struggling against the environment to maintain a constant internal temperature would have been metabolically expensive and would have required increasingly efficient physiological mechanisms.

Since in endotherms, metabolic heat (a by-product of certain chemical reactions) travels from the inside out, bodies tend to be warmer toward the inside. Thus, if you should touch something that is 37° C (your body's internal temperature), it will feel warm because your skin is cooler than that. So, there is some variation from place to place in your body, and there is even some variation from time to time. (Your temperature usually falls a bit in the wee hours of the morning and rises in the early afternoon.) Because of the greater constancy in the deeper tissues, if you measure someone's temperature with a rectal thermometer, you will get not only his or her undivided attention, but a more precise reading as well.

Here I should add that there is some recent evidence suggesting that for those who are ill, aspirin or other fever-reducers may do more harm than good. It has been suggested that a slight rise in your body's temperature can render it inhospitable to temperature-sensitive viruses and bacteria. Physical exercise can also raise the body's temperature and, some say, thereby stave off certain illnesses by rendering the body unsuitable for habitation by pathogens.

How does your body "know" its own temperature? The temperature is monitored by a delicate thermostat in the **hypothalamus** (an ancient part of the brain—see Chapter 22). Certain receptor cells monitor the temperature of the blood reaching the brain. If the blood is too warm, the hypothalamus initiates a chain of events that brings down the temperature. This is done in different ways in different animals, and both physiological and behavioral responses may be involved. As body temperatures rise, humans and most other large animals sweat, and the evaporation cools them. Dogs pant and rapidly move cooling air over their large tongues and through their lungs. (They may also lie down under a tree—a behavioral response.) Cats are stimulated to lick themselves and are cooled by the evaporation of the saliva. In some animals, including humans, peripheral blood vessels dilate, bringing warm blood to the surface of the skin where heat can be exchanged.

As the blood cools, the surface vessels contract, reducing heat loss over the skin's surface. The hypothalamus then signals the pituitary to direct the thyroid glands, located in the throat area (Chapter 21), to release a chemical (called thyroxine) that increases metabolic rate. In addition, if the temperature of the blood in the brain should drop to

---

*Since some modern reptiles, notably alligators and crocodiles, have four-chambered hearts, theorists have suggested that the dinosaurs may have had such an advantage as well.

dangerous temperatures, the hypothalamus will direct the adrenal glands to secrete epinephrine (adrenalin), thereby increasing the metabolic rate and raising the body's temperature. If the temperature continues to drop, shivering begins, causing the surface muscles to work and to increase their metabolism, thereby producing more heat and using more energy. Have you noticed that you eat more in the winter? Why do you suppose that is?

Occasionally the body's thermostat is "turned up," and fever results. No one really knows how fever occurs, but infection seems to trigger the release of **pyrogens** ("heat causers") that elevate the body's temperature. Before fever, when the temperature is still normal, the body begins to behave as if it were cold. (The hypothalamus responds as if it were detecting cold.) Blood vessels in the skin contract ("You look pale; do you feel well?"), and shivering may begin as the rate of metabolism increases. All these things can act together to bring on higher temperature. Finally, the temperature temporarily stabilizes at its new, higher setting. When the fever breaks, the opposite responses occur. The skin becomes flushed and sweating occurs until enough heat is lost to restore the body to normal temperature.

Just as your body's temperature is kept within narrow limits, so are other physiological factors. In fact, every activity involving growth and maintenance is controlled by complex regulatory mechanisms. One of the clearest examples of such control involves water balance. You have probably heard that the body to which you are attached is composed mostly of water, and physiologists have glibly quoted rather precise numbers to tell you just what the percentage is. How can they be so sure? Do they include the bodies that have just completed marathons, or those that have been drunk for three days? Not normally, but still they are confident because they are aware of the delicate and intolerant mechanisms of the kidneys.

THE EXCRETORY SYSTEM

All animals have the problem of getting rid of metabolic wastes, or "cell garbage." Much of this waste is in the form of excess nitrogens that are left over when proteinaceous foods are metabolized. In this process, the proteins are stripped of their nitrogens and the remainder of the molecules are converted into whatever is needed, such as glucose. This leaves the nitrogens to combine with other elements, forming poisonous nitrogenous by-products.

Different species may dispose of nitrogenous metabolic wastes in different ways. In protistans, such as the amoeba, nitrogenous wastes simply diffuse across the cell membrane and out into the surrounding water. However, since protoplasm is hypertonic to pond water, water constantly enters the body by osmosis (Figure 17.5), threatening to cause the organism to swell and rupture. (You may recall our earlier discussion of osmosis.) Some protozoa must constantly expel water by the continual pumping of a contractile vacuole. In the case of freshwater protista, then, the problems of metabolic waste and internal water regulation are solved by different mechanisms. In other animals, the two problems may be solved by the same mechanisms.

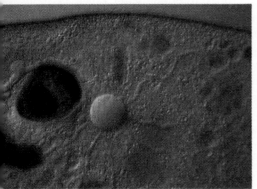
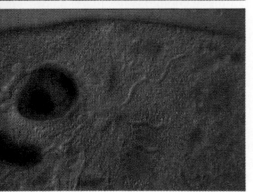

FIGURE 17.5 Freshwater protists have a marked problem with maintaining internal water balance. They excrete the water that moves in osmotically by collecting it in a contractile vacuole and pumping it back out again. Here, a vacuole fills and then contracts, expelling water through the membrane.

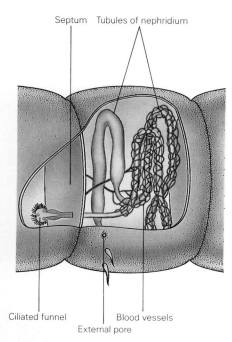

Septum   Tubules of nephridium

Ciliated funnel   Blood vessels
External pore

FIGURE 17.6 Nephridium of an earthworm. Most segments boast a pair of nephridia. The tubelike nephridium passes through the body septum. Beating cilia set the waste-laden fluid of the body in motion so that it flows into the funnel and past blood vessels lying tight against the nephridium that shares that segment. Water and other desirable substances move from the fluid into the blood vessels, concentrating the waste in the nephridium as it moves along. Metabolic wastes from the blood are finally excreted.

As an example of a single mechanism adjusting the body's level of both metabolic wastes and water content, consider the excretory system of the earthworm. Each segment of an earthworm has a pair of coiled tubes called **nephridia,** connecting the body cavity to the outside (Figure 17.6). The tubes opening inside the body cavity are rimmed with beating cilia that circulate the fluids there. Capillaries lie tightly coiled around the outside of the nephridia, and as the fluid moves into and through the tube toward the outside, water, salts, and minerals are drawn back into the capillaries. The concentrated waste, then, is excreted through the external pore. The amount of water the animal excretes with the wastes (primarily urea and ammonia) depends on how much water has entered the body cavity. Worm urine, in case you've ever wondered, is very dilute, and the amount excreted each day equals about 60 percent of the worm's body weight.

SOLUTIONS TO THE WATER PROBLEM

One of the great problems in the development of excretory systems was how to flush metabolic wastes from the body without losing too much water. The evolutionary solutions, we shall see, have been quite diverse.

Vertebrates have evolved a number of ways of ridding the body of metabolic wastes and maintaining proper internal water levels. For example, freshwater fish, like freshwater protozoa, live in a hypotonic medium; that is, the concentration of particles in their body fluids is higher than that in the surrounding water. Hence, water tends to move into their bodies by osmosis. One means by which they rid themselves of excess water is through a great number of tiny blood vessels in their kidneys. These blood vessels are coiled into tiny clumps that keep the blood in one area for an extended time and are called **glomeruli** (singular, **glomerulus**). We'll consider the structure of the glomeruli shortly, but suffice it to say here that, in general, they provide a large surface area through which fluids leave the blood to be collected in the kidney. The metabolic waste in freshwater fish is also highly dilute. This is to be expected since they have no problem finding enough water to flush out their wastes.

Saltwater animals, interestingly enough, have the same problem as desert animals: conserving water. Seawater is a hypertonic medium; that is, its concentration of solutes is higher than the fluids in the marine animal's body. Hence, these animals tend to lose water by osmosis. As a result, many have developed smaller glomeruli, so that there is less opportunity for water to filter out as it passes through the kidney. It is also in the kidney that excess salt taken in with the seawater is removed. In other animals, specialized gills or other structures may also eliminate nitrogenous waste as blood passes through them.

Most sharks and rays in the ocean have a different strategy: they retain some urea, or waste, in their blood. This raises the osmotic pressure of their body fluid so that it is closer to that of seawater. Thus they sidestep many problems of the excretion of both water and metabolic wastes.

All terrestrial mammals, including humans, face the problem of how to flush nitrogenous wastes out of their bodies without losing too much

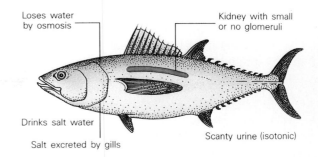

Fresh-water fish

Gains water by osmosis

Kidney with large glomeruli

Drinks no water

Copious urine (hypotonic)

Salt-water bony fish

Loses water by osmosis

Kidney with small or no glomeruli

Drinks salt water

Salt excreted by gills

Scanty urine (isotonic)

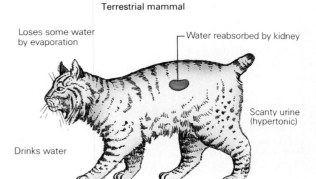

Shark

Gains water by osmosis

Kidney with large glomeruli

Some salt taken in with food

Salt-excreting gland

Urea retained by gills

Copious urine (hypotonic)

Terrestrial mammal

Loses some water by evaporation

Water reabsorbed by kidney

Drinks water

Scanty urine (hypertonic)

FIGURE 17.7 Regulation of water and salt balances in a representative group of animals. Water enters the freshwater fish by osmosis. Great amounts of fluid cross its glomeruli daily, and it excretes large amounts of water in its urine. The saltwater fish loses water by osmosis, so it saves the remainder by filtering little fluid through its kidneys and urinating little. The shark gains water by osmosis and passes large amounts through its kidneys, urinating copiously. Land mammals drink water and lose some by evaporation. They save water by reabsorbing much of what passes through the kidneys and urinate only a little hypertonic urine. Look at the cases individually and compare their problems and solutions. Why do the fish solve their problems three different ways? What can you say about their cytoplasm?

water. A large part of human food is protein (for those who can afford it), and the excess nitrogens from protein are, as we have seen, a primary source of our excretory problems. (Various solutions to different water and nitrogen problems are summarized in Figure 17.7. See also Essay 17.1.)

The first step in deriving energy from an amino acid is to get rid of the nitrogens. Remember, nitrogen is not utilized in either the anaerobic or aerobic energy cycles. So the nitrogens are stripped away, and once free, they usually pick up three hydrogens—and NH_3 is **ammonia**, a deadly poison. The problem, then, is how to dispose of the ammonia. Since it can't pass out through the lungs like carbon dioxide, it must be excreted by the kidneys. However, ammonia is so poisonous that it must be highly diluted in order to be handled by the body, and in water-conserving species, any such dilution would result in the loss of too much water. The problem is solved by the addition of carbon dioxide to the ammonia, forming **urea**, a less toxic molecule that the body can handle in higher concentrations. A moderately dilute urea is then passed out of the body as **urine**.

Many terrestrial species, notably the egg-laying birds and insects, have a particular need to conserve water, so they have developed the means to convert ammonia to an **insoluble** nitrogenous product called **uric acid**. They withdraw so much water from this product that it is

A quick look around the animal world reminds us that living things solve the problems of life in all sorts of ways. Some of the problems are monumental, indeed, especially those relating to conserving water. For example, some insects are able to survive for years enclosed in jars of dry pepper. They do this by utilizing metabolic water and excreting almost dry uric acid.

Few other species face such severe problems, but water conservation nonetheless remains a problem for many forms of life. Perhaps the most famous water conserver is the camel. A camel can tolerate dry conditions by adopting a number of tactics. At night it drops its body temperature several degrees so that bodily processes slow down. In the heat of day, it doesn't begin to sweat until its body temperature reaches about 105°F. In addition, the camel can lose twice as much of its body water (40 percent) without ill effect than can most other mammals. Interestingly, the thick coat of a camel acts as insulation to keep the heat out. Also, when the camel does drink, it can hold prodigious amounts of water (but not in its fatty hump).

The camel's desert colleague, the tiny kangaroo rat, doesn't drink at all, and it lives on dry plant material. It survives because it doesn't sweat and is active only in the cool of night. Its feces are almost dry, and its urine is highly concentrated. Most of its water loss is through the lungs. Fats produce more metabolic water than other foods, so it prefers fatty foods. On the other hand, proteinaceous foods, such as soybeans, produce a lot of nitrogenous waste for which a lot of water is needed—so, if a kangaroo rat is fed only soybeans, it will die of thirst.

Some seabirds and turtles have salt removal glands. The huge tears in the eyes of sea turtles have nothing to do with the realization that we are poisoning the oceans. The tears are flowing from special salt-removing glands that remove sodium chloride from the blood.

Marine iguanas expel excess salt from their bodies by blowing salty excretion out of their noses. Part of the white on the nose is evaporated salt.

excreted almost as a dry paste in some species. The major waste products of metabolism in fish (ammonia), birds (uric acid), and mammals (urea) are:

$$NH_3 \qquad \begin{array}{c} NH_2 \\ | \\ C=O \\ | \\ NH_2 \end{array}$$

Ammonia Urea Uric Acid

The Human Kidney

Humans, as do the other mammals, have two kidneys (although we can live with one). These lie in the dorsal area (at the back) and extend slightly below the protective rib cage (which is one reason blows to the kidney area are so dangerous). Actually, they lie outside the body cavity behind the membrane that lines the cavity. Each kidney contains about a million **nephric units**, each unit consisting of a **glomerulus**, a tightly convoluted blood vessel, a cuplike **Bowman's capsule** surrounding the glomerulus, and the connecting **tubules** leading from the capsule to the collecting duct (where urine collects on the way to the urinary bladder). The structure of these units is shown, along with the position of the kidney itself, in Figure 17.8.

The exit from the glomerulus is smaller than the entrance, so that blood is placed under high pressure as it enters from the **renal artery**. As the blood wends its way through the tortuous route of the glomerulus, much of its fluids are forced out through the thin-walled vessels and into the cuplike Bowman's capsule. Most larger particles, such as blood cells and certain proteins, cannot pass through the glomerular walls.

The filtrate received by the Bowman's capsule contains a high concentration of waste products, but also water, salts, and nutrients. So, as the filtrate begins to move through the tubules on its journey to the urinary bladder, it passes through the twisted proximal convoluted tubule, down through the hairpin turn of the **loop of Henle**, up to the distal convoluted tubule, and on to the collecting ducts. Blood vessels from the glomerulus head off the filtrate at the convoluted tubules and the loop of Henle. The blood vessels wind around the tubules and the loop and retrieve many of the usable products, both through diffusion and by active transport, allowing the urea and some water to continue on their way. (See Essay 17.2 for a description of the forces that move these materials.)

The remaining fluid in the tubules proceeds to a larger collecting tubule, and then to the **ureter**, which leads from each kidney to the **urinary bladder**. From the urinary bladder, the urine passes through a

FIGURE 17.8 The human kidney. Note the position of the kidneys in the body. The lower part actually extends below the protective rib cage. Also note the vast maze of blood vessels penetrating the kidney. Blood is brought into the kidney under considerable force, and many of its constituents are filtered through the glomeruli. From there they enter a tubular system from which many of the constituents (including some water) pass back into the bloodstream. Those that are not recollected are passed into increasingly larger tubes until they enter the urinary bladder as urine. Here the urine is held until, in long lines halfway to the box office, it demands to be released back to the environment. Because of its intensive filtering, urine is largely germ-free and has been used to wash wounds under field conditions when the available water was known to be impure.

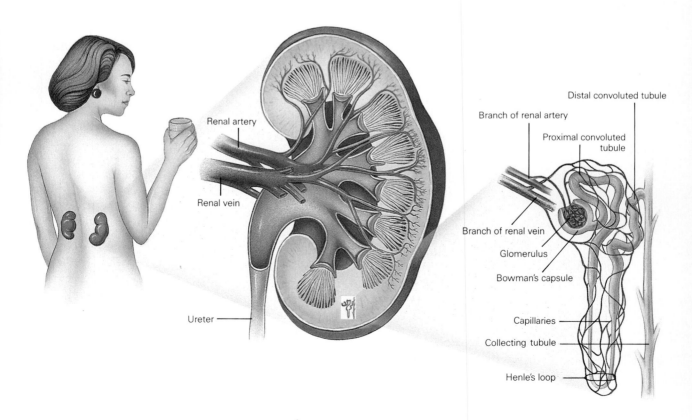

There are a number of mechanisms producing hypertonic urine in the human nephron. First, body fluids filter through the capillary wall into Bowman's capsule and from there through the proximal convoluted tubule (here straightened), the long loop of Henle, the distal convoluted tubule, and then into the collecting duct. As the urine passes through the ascending loop of Henle (on the right), cells lining the loop pump chloride out of the urine into the tissue surrounding the loop. These chloride ions then diffuse over to the descending loop (on the left) where they enter the loop. From there they flow to the ascending loop where they are pumped out again. This constant circulation means that the loop of Henle and the adjacent collecting duct are always bathed in a salt solution, and it is this solution that sets up the osmotic gradient that withdraws water and solutes from the urine. The ascending loop is apparently impermeable to water since no water passes out with the sodium. The collecting duct, however, is permeable to water, and water in the wastes reaching it freely flows out in response to the high salt concentrations in the surrounding tissue. This water is picked up by the blood vessels that come from the glomerular region, and the hypertonic urine passes to the bladder.

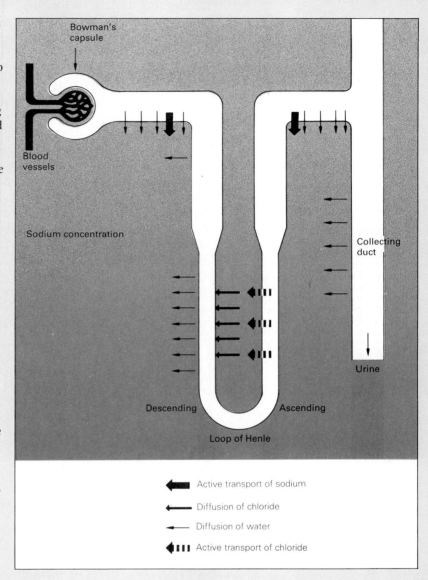

single opening, controlled by sphincter muscles, that leads into the **urethra** and on to the outside.

Have you ever noticed how often your rowdy, beer-drinking friends must urinate once the evening gets underway? You may have heard them comment that more seems to be going out than was coming in. The next day they may complain of being thirsty. In your stern lecture to them you should say something about the diuretic effects of ethyl alcohol. You will find them particularly interested when you say that the reason for their thirst is that reabsorption of water at the tubules is controlled by an **antidiuretic hormone** (ADH) that is secreted by the posterior lobe of the pituitary. Go on to say that alcohol suppresses the secretion of this hormone, resulting in less water being reabsorbed in the loop of Henle. The urine thus becomes hypotonic, or more dilute than the body fluids, and the body actually becomes somewhat dehydrated. Point out that, on the other hand, ADH secretion also decreases when you drink a great deal of water, since the osmotic concentration of the blood decreases (so drinking water can temporarily "thin" the blood). Drinking water, you can say, also results in the formation of a urine that is hypotonic to the blood. They will appreciate this information, especially early the next morning.

If your friends are so appreciative that they buy you a sailboat, you may be interested in another bit of practical physiology. If you discover, far at sea, that for some strange reason the boat sinks, you should know that a person on a life raft cannot get the water he needs by eating fish, contrary to popular rumor. The reason is that human urine cannot exceed a salt concentration higher than about 2.2 percent. Fish fluids are not actually this concentrated, but fish are high in protein. This means that in order to get rid of the excess nitrogen, you would have to excrete a lot more water than you could get from eating the fish. Of course, you would do better eating fish than drinking seawater, since the salt concentration in seawater is about 3.5 percent. Some shipwreck survivors claim that the spinal fluid of fish is a good supply of water, but all things considered, you would be better off to have remembered to take water along.

SUMMARY

A sensitive and responsive system of balancing homeostatic balancing mechanisms should not be unexpected in living things. After all, life is possible under only a very narrow range of conditions on the planet, and its delicate processes must be kept within rather precise limits. Nevertheless, the unfolding story of just how the constancy of life is maintained is fascinating. As new stories come along, describing yet other kinds of regulation, we will undoubtedly be reminded of the complex, interactive nature of even the simplest living thing.

Homeostasis involves the mechanisms that keep the internal environment of living things relatively constant. It operates primarily through negative feedback systems. Homeostasis is particularly important to endotherms (birds and mammals), those animals that metabolically (as well as behaviorally) thermoregulate. Ectotherms (the other animals) thermoregulate mainly behaviorally, and their internal temperatures tend to fluctuate with that of the environment. The excretory system keeps the body's water levels within certain limits, even as it rids the body of excess nitrogens. The nitrogens are handled differently in different species according to the amount of water the animal can afford to lose. The human nephric unit is composed of the glomerulus, Bowman's capsule, proximal and distal convoluted tubules, the loop of Henle, and the collecting ducts. Filtrate passes from the glomerulus into Bowman's capsule. As the filtrate continues to the collecting duct, water and minerals pass back into the body through diffusion and active transport. The result is a dilute concentration of urea, called urine.

Chapter 18

SUPPORT AND MUSCULAR SYSTEMS

There are few things that one can be sure of these days, but here's one: There is no such thing as an amoeba large enough to eat your car. The largest amoeba in the world can't even eat your cat. Amoebas are small, and they are destined to stay that way essentially because of one reason: They are composed of a single cell. There are a great many repercussions to this seemingly insignificant trait, as we saw earlier. For example, you may recall that as an object becomes larger, its volume increases at a proportionately greater rate than its surface area. Thus, a huge amoeba would have an unsatisfactory surface-to-volume ratio. In other words, its cell membrane would be too small to accommodate the cell's mass. Put another way, the cytoplasm would be served by an inadequate structure with which to communicate with the environment.

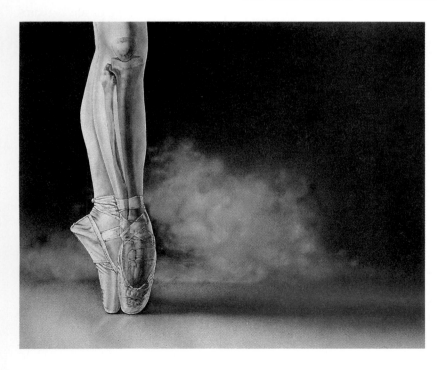

In addition, the unsupported weight of a huge amoeba would create such great internal pressures that the creature would likely rupture, and you can imagine the mess. Finally, oxygen and carbon dioxide can move only across wet membranes, and a giant amoeba would have no way of keeping its membrane wet. Even if oxygen were able to move into the amoeba's body, once inside, the only way it could reach the cytoplasm at the center of the beast would be by diffusion, perhaps aided by the cytoplasm slowly shifting due to the amoeba's sluggish movements. The movement of carbon dioxide and oxygen through the amoeba indeed would be slow and inefficient. Because of such problems, your car is safe and so is your cat.

While there are few creatures on earth that could or would eat your car, there are a number that could legitimately threaten your cat. Many of these are quite large, and we can assume they are composed of many cells. Since evolution has produced so many kinds of multicellular creatures, we can assume the condition has certain marked advantages. So, let's look at multicellular animals now and see how those cells are arranged so as to help the creature succeed in its world. We can first remind ourselves that groups of cells existing together, and with the same functions, are called tissues. Groups of tissues with the same function are called organs, and groups of organs acting together are called organ systems. In this chapter, we will concentrate on the skeletal (supportive) and muscular (contractile) tissues and systems. Later we will consider other kinds of systems, keeping in mind that they do not act independently, but generally in a highly coordinated manner as the organism responds to its environment.

It would seem that bones and muscles would be of such interest to people that we would, as a matter of course, know a great deal about them. But most people probably don't know the answers to some of the most fundamental questions about these great tissues. For example, how many bones are there in the human body? (You don't know?) Why can't the knee flex forward? How do muscles enlarge? What is a bone bruise? What are "fallen arches"? Why is our back so strong and our belly so weak? If you are allowed to grow old, why should you worry about a broken hip? What we cover here should allow you to deduce the answers to such questions and to gain a better perspective on how we are supported and protected by our skeletons and how that body framework moves.

SUPPORT SYSTEMS

Trees are able to stand erect in spite of their great weights because of the combined strength of their hardened cell walls. Animals such as ourselves, though, lack cell walls. Still, we do not flounder about, collapsing here and there in quivering heaps, except perhaps on very special occasions. Generally, animals maintain their form and structure because of various sorts of support systems. The most common of these are **skeletal systems.** These are formed as special cells secrete a fluid that then hardens to become the skeletal material. In many species, including humans, the skeletons are on the inside **(endoskeletons),** but certain other kinds of animals, such as insects and clams, wear their skeletons on the outside **(exoskeletons).** Some species utilize other forms of support systems (several are mentioned at the end of this chapter). For example, in certain soft-bodied animals that lack skeletons, such as hydra, flatworms, and earthworms, rigidity is maintained as muscular contractions compress cell fluids. The resulting pressure forms a type of hydrostatic skeleton.

Types of Skeletal Systems

One of the most primitive of skeletal systems is found in the sponges (Figure 18.1). They require little support, of course, because the body is largely supported by the buoyancy of the seawater. Their skeletons consist largely of tiny **spicules** scattered throughout the tissue. The 10,000 species of sponges can be divided into groups according to the type of spicules they produce.

Species with exoskeletons include not only insects and clams, but lobsters and snails. Exoskeletons efficiently protect internal parts (just as do our skull and rib cage). Exoskeletons, however, present special problems of growth that are solved in different ways by different species. Snails and clams simply secrete extensions to their shells as they grow, so their shells simply grow with them. Lobsters, crabs, and crayfish, however, discard their old shells and grow new, larger ones. They do this by first withdrawing the valuable minerals from their shells, causing the shell to soften. The weakened shell, stressed by the force of the growing animal, then splits down the back. The animal crawls out, and, soft and vulnerable, it usually retreats to some darkened corner where it swells by absorbing quantities of water. It grows rapidly for a time and then begins to secrete a new shell into which it redeposits its hoarded minerals (Figure 18.2).

FIGURE 18.1   A group of sponges, part of a colony (left). At right is a longitudinal section through a simple sponge. Note the various types of cells that make up the sponge tissue. The delicate choanocytes, or collar cells, on the interior have flagella that move nutrient-laden water through the porocytes and on through the body cavity. The mysterious wandering mesenchyme cells, although rather undifferentiated themselves, can work together to form very specific structures such as the three-pronged spicule at lower left. In sponges, spicules of rather specific shapes give some rigidity to the sponge and, hence, act as a kind of skeleton.

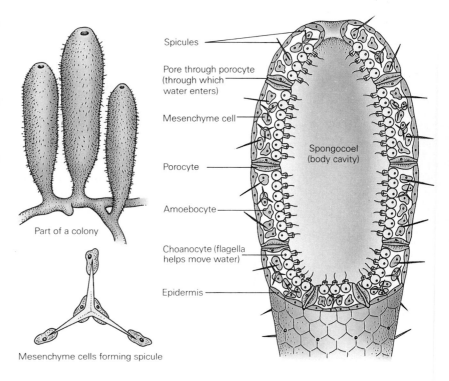

Spicules

Pore through porocyte (through which water enters)

Mesenchyme cell

Spongocoel (body cavity)

Porocyte

Amoebocyte

Choanocyte (flagella helps move water)

Epidermis

Part of a colony

Mesenchyme cells forming spicule

FIGURE 18.2   A crayfish completing molting. Crayfish molt at rather regular intervals throughout the year. Their growth is restricted by a hardened and protective exoskeleton, so growth can only be accomplished in the brief interludes between the development of new skeletons. The crayfish that has recently lost its exoskeleton may compensate behaviorally by becoming more furtive and hiding under rocks until it can venture forth in its new coat of armor.

All vertebrates possess an endoskeleton, a skeleton inside their bodies. In the most ancient and ancestral lines of vertebrates, such as sharks and rays, the skeleton consists entirely of cartilage, similar to that composing the bulb of your nose. Even in most other vertebrates, parts of the skeleton are formed of cartilage during embryonic development, but these are largely replaced by bone as the body develops. Some parts of the skeleton, however, form directly as bone.

Types of Connective Tissue

Bones are only one form of **connective tissue,** which is a supportive tissue that also includes *fibrous connective tissue, tendons, ligaments,* and blood, which we will consider later. Connective tissue occurs throughout the body and generally functions as a binding material—for example, holding glands in position or binding bone to muscle or bone to bone. Generally, cells comprise only a small part of connective tissue; most of the mass consists of substances secreted by the cells. These secreted, nonliving substances are called **matrix.** The specific qualities of each type of connective tissue are largely dependent on the nature and composition of the matrix (Figure 18.3).

Bone is heavy material with a matrix of calcium salts, which make the bone hard, and a protein called **collagen,** the most abundant material in most connective tissue, which gives the bone some degree of flexibility. (The decrease in collagen-forming cells that occurs with age causes a certain brittleness in the bones of older people.) The larger bones are made somewhat lighter by being hollow. The tissue inside these hollow cavities is called **marrow. Yellow marrow** consists mostly of inactive fat cells, while **red marrow** is active and forms red blood cells as well as some white blood cells. Running throughout the hard, bony matrix are **Haversian canals,** small channels containing the blood vessels and nerves that serve the bone cells. The bone cells and the Haversian canals are connected by even smaller canals, called **canaliculi,** through which the bone cells receive food and oxygen and expel waste, including carbon dioxide.

The matrix of bone also contains cells that have the peculiar ability to withdraw calcium salts from the matrix and to redeposit them in other places that may be more severely stressed. Because of these cells, bone

FIGURE 18.3   At left, a Haversian system of a bone. Through the large opening in the center, the Haversian canal, run the nerves and blood vessels that serve the bone. The bone-forming cells, (or osteocytes) lie in concentric rings around the canal and communicate with each other by the tiny canals, or canaliculi, radiating from them. At right, a cartilage cell. For some reason, these usually occur in pairs, and they secrete the fibrous mat that is the matrix. Some bone contains cartilage cells, giving bones elasticity and resilience, but, with age, these may be replaced by bone cells. Thus, old people have a high risk of breaking brittle bones by a simple fall. (Photo at left from *Tissues and Organs—A Text-Atlas of Scanning Electron Micrography,* by Richard G. Kessel and Randy H. Kardon. W.H. Freeman and Company, copyright © 1979. Photo at right from Carolina Biological Supply Co.)

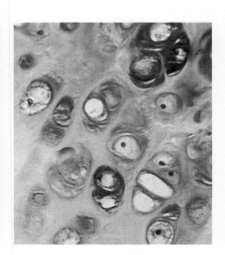

has the ability to change in response to the stresses placed on it. As an example, in certain African tribes, the heads of women are bound in a tight metal ring that drastically changes the head shape. It is also interesting that experts can determine the muscular development of a person by skeletal examination only. Larger muscles place greater stress on the bones where they are attached, and the bone thus enlarges to accommodate the tension. As a general rule, the larger the area of attachment, the larger the muscle. Incidentally, training with heavy weights not only changes muscle development but bone structure as well. (We will see later how calcium is withdrawn from the bones of pregnant women to be deposited in the fetuses they carry.)

In **fibrous connective tissue,** the matrix consists of a heavy, interlaced mat of microscopic fibers interspersed with the cells that secreted them. Leather is formed by tanning the fibrous connective tissue that lies in the deeper layers of the skin of certain animals.

Tendons and **ligaments** are specialized types of fibrous connective tissue, each with a specific function. Ligaments are fairly elastic and serve to bind one bone to another. Tendons are not very elastic, but they are flexible and cordlike, connecting muscle to muscle or muscle to bone (Figure 18.4).

The Human Skeleton

The adult human skeleton consists of about 200 bones (the number decreases with age as some bones fuse). In humans, as well as in other vertebrates, the skeleton is generally divided into two major groups called the axial and appendicular skeletons (Figures 18.5 and 18.6).

The **axial skeleton** is the part of the skeleton that forms the axis of the body—lengthwise—running along the direction of the backbone. It

FIGURE 18.4   The tendons and ligaments of the knee form a very complex array of supporting and binding structures. The entire system is very vulnerable to twisting injuries, as can be attested by former great skiers and football players who are now virtual cripples because of injury to this delicate area.

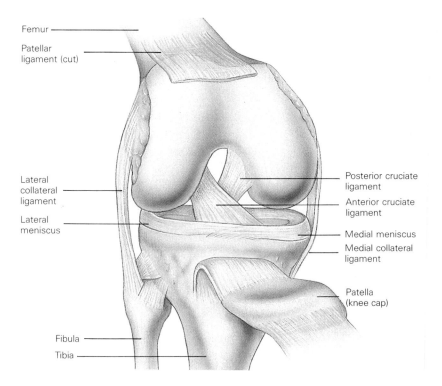

Femur

Patellar ligament (cut)

Lateral collateral ligament

Lateral meniscus

Posterior cruciate ligament

Anterior cruciate ligament

Medial meniscus

Medial collateral ligament

Patella (knee cap)

Fibula

Tibia

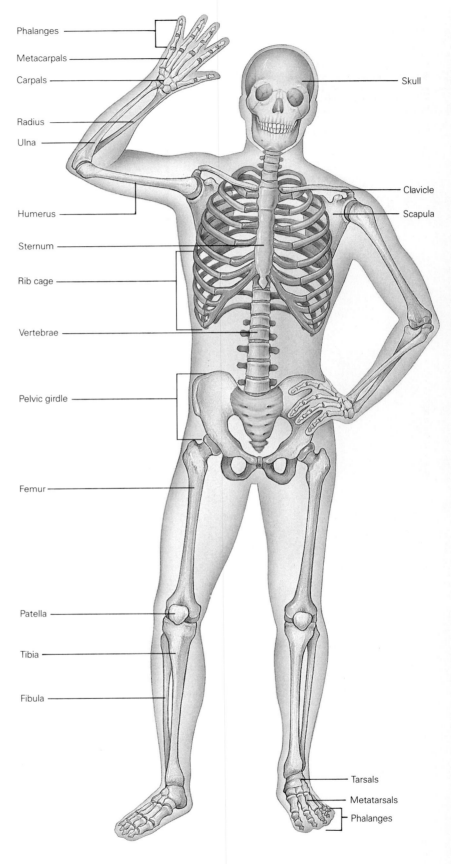

FIGURE 18.5 The axial skeleton (shaded orange) and the appendicular skeleton (white). The axial skeleton is essentially comprised of the skull, vertebral column, and rib cage. In its protective function it houses the most critical parts of the human body (the central nervous system and parts of the viscera such as the heart, lungs, and kidneys). The appendicular skeleton is primarily composed of the long bones (those of the pectoral and pelvic girdles). Other animals, of course, may differ substantially in their skeletal makeup. An astute observer can tell a great deal about an unfamiliar animal's habits simply by looking at the skeleton. Would shortening the upper leg and lengthening the lower leg make humans faster or slower runners?

Phalanges

Metacarpals

Carpals

Radius

Ulna

Humerus

Sternum

Rib cage

Vertebrae

Pelvic girdle

Femur

Patella

Tibia

Fibula

Skull

Clavicle

Scapula

Tarsals

Metatarsals

Phalanges

FIGURE 18.6 The axial (orange) and appendicular (white) skeleton of the frog. The area between the large eyes houses the tiny brain. Some might deduce that what it sees is obviously more important than what it learns. Note the elongated transverse processes of the vertebrae and the absence of ribs. With careful attention, you might discern other differences between the skeleton of a frog and a human. Note also the basic similarities in the body plan. What special features of the frog skeleton adapt the animal for its particular way of life?

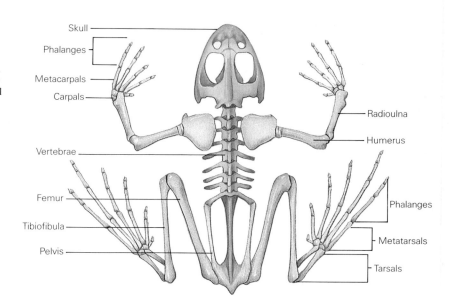

Skull
Phalanges
Metacarpals
Carpals
Radioulna
Humerus
Vertebrae
Femur
Tibiofibula
Pelvis
Phalanges
Metatarsals
Tarsals

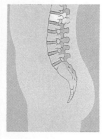

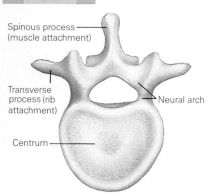

Spinous process
(muscle attachment)

Transverse
process (rib
attachment)

Neural arch

Centrum

FIGURE 18.7 A lumbar (lower back) vertebra (shown from above) showing the centrum that articulates with vertebrae above and below the neural arch through which the spinal cord passes and the various projections and processes that provide surfaces for the attachment of muscles and ribs. Inset, side view of the same vertebra.

FIGURE 18.8 Anatomical landmarks in the human.

includes the skull, the spinal column, the **sternum** (breastbone), and the ribs. The skull consists of the fused bones of the braincase, or **cranium,** and the bones of the face. The backbone is made up of thirty-three separate **vertebrae,** which differ according to their position along the spine (Figure 18.7).

The ribs are attached dorsally (toward the back—see Figure 18.8) to the vertebral column and ventrally (toward the belly surface) to the sternum or to cartilage that is, in turn, attached to the sternum, except for the last pair, called floating ribs, which is not attached ventrally. The rib cage is slightly moveable, which enables it to raise and lower during breathing, and it is flexible enough to absorb powerful blows, thereby protecting the heart, lungs, and other vital organs.

The **appendicular skeleton** includes the arms, the legs, and two girdle complexes (Figure 18.9). The **pectoral girdle** consists of two collar bones

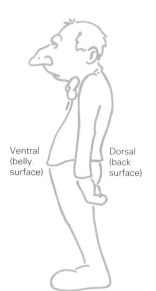

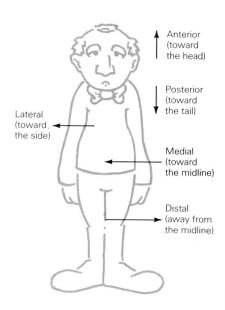

Ventral
(belly
surface)

Dorsal
(back
surface)

Anterior
(toward
the head)

Posterior
(toward
the tail)

Lateral
(toward
the side)

Medial
(toward
the midline)

Distal
(away from
the midline)

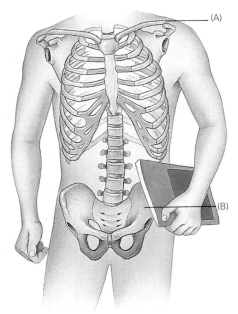

FIGURE 18.9 The pectoral (A) and pelvic (B) girdles. In humans, the bones of the pectoral girdle are loosely joined and can be flexed. In leaping animals, such as cats, the collarbone is greatly reduced so as not to interfere with the forward movement of the forelegs.

(clavicles) and two shoulder blades **(scapulas)** meeting the sternum. The **pelvic girdle** is composed of three bones fused to form the hips. Many fish, by the way, have rather primitive pectoral and pelvic girdles. Fossil evidence indicates that the arms and legs of land animals evolved from the fins of early fish, and that these girdles enlarged to support the ancient creatures as they invaded the terrestrial realm.

The arms and legs are comprised of upper and lower parts that join at the elbow and knee respectively. The upper bones of both the arms and legs are single bones that join two bones of the lower limbs (Figure 18.10). Both the arms and legs terminate in a complex arrangement of several bones. The arms end with the wrist bones **(carpals)** and the legs with the ankle bones **(tarsals).** These bones are able to slide past each other, permitting a rather wide range of movement for both the hands and feet. The hands and feet themselves are made up of the **metacarpals** and **metatarsals,** respectively. Both the hands and feet terminate in five **digits** (or phalanges), the fingers and toes. Although the hands and feet of humans are structurally similar, they differ greatly in their dexterity and maneuverability. In fact, humans are remarkable for a highly-specialized hand existing on the same body with a much less talented foot. Both the human arms and legs are rather primitive structures as vertebrate limbs go. A much more specialized limb, for example, is that of the horse, adapted for speed. However, the legs of horses are not particularly well suited for climbing trees, to the relief of picnickers in the shade.

Joints

Some bones move freely, while others are only slightly moveable, and yet others are solidly fused together. Bones are able to move with respect to each other at points of articulation, or **joints** (Figure 18.11). The greatest freedom of movement is provided by **ball-and-socket joints,** such as those that connect the humerus to the scapula at the shoulder and the femur to the pelvis at the hip. **Hinge joints** at the knee and the elbow also permit great movement, but primarily in an arc, back and forth. The ankle is formed from both a hinge and a rotating joint and is quite moveable, while the wrist is a hinge and does not rotate (although it seems to). An example of a joint with little flexibility is found where the pelvic bones meet at the pubic symphysis. These can separate slightly during childbirth. The bones of the skull develop separately and later fuse along **suture lines** that are completely immobile joints. Therefore, if you try to teach a person to shrug his or her head, you will only waste your time and annoy the person.

MUSCLES

Muscles are tissues with the remarkable ability to contract. There are different kinds of muscle tissue, each with specific abilities and distinctive appearances.

Types of Muscles

Human muscles can be divided into three groups according to their structure and the nerves that activate them. **Smooth muscle** is found in a number of internal structures, such as the walls of the digestive tract,

FIGURE 18.10 The human arm (right) and leg (left). Note that they are very similar in their basic design. The lower arm, however, is free to rotate, but not the lower leg. This is one reason knee injuries can result from severe twisting. The shoulder and hip joints are similar, as are the ankles and the wrists. The knee joint is protected by a kneecap, but the elbow, less prone to impact, has no such protection.

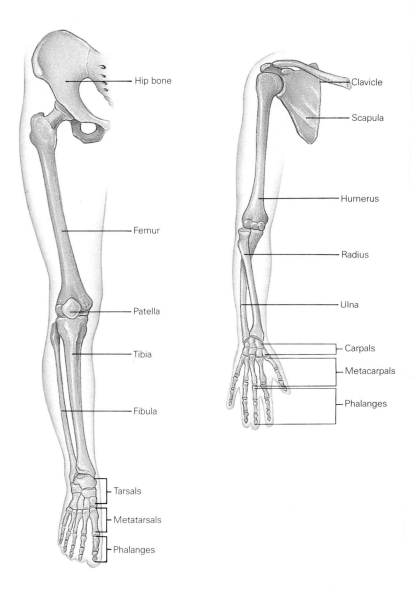

Hip bone
Clavicle
Scapula
Humerus
Femur
Radius
Ulna
Patella
Carpals
Metacarpals
Tibia
Phalanges
Fibula
Tarsals
Metatarsals
Phalanges

FIGURE 18.11 Representative joints in the human body. Some joints are sutured and do not flex at all. Others almost never move, while yet others move in very limited directions. Ball-and-socket joints give the greatest movement, although complexes of joints such as those found in the wrists also provide great freedom of movement.

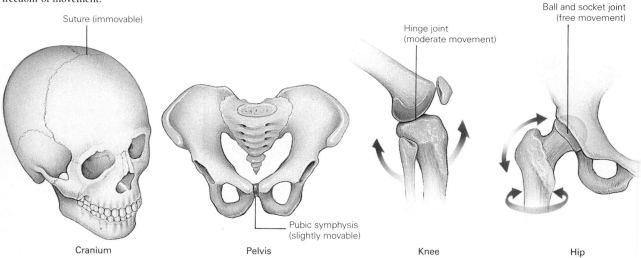

Suture (immovable)

Hinge joint
(moderate movement)

Ball and socket joint
(free movement)

Pubic symphysis
(slightly movable)

Cranium

Pelvis

Knee

Hip

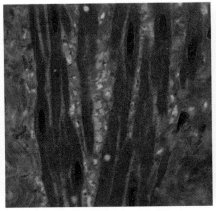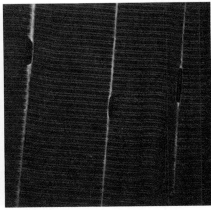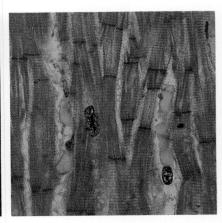

FIGURE 18.12 Three types of muscle tissue. At left is smooth muscle, unstriated and spindle-shaped (although the shape is not apparent here). The nuclei are elongated and prominent. At center is striated, or skeletal, muscle. The striations are due to the banding as shown in Figure 18.15. At right is cardiac muscle. Note the striations.

around some blood vessels in the diaphragm, and in certain internal organs. **Cardiac muscle** is found in the walls of the heart. Both of these are labelled "involuntary," since they function without conscious control. Thus, you don't have to lie awake nights keeping your heart beating or remembering to breathe. There is some fascinating evidence that "involuntary" responses can be voluntarily controlled to a degree, as we will see later. **Skeletal muscles** are the voluntary muscles. We will consider these in more detail next.

Both the skeletal muscles and the cardiac muscles have several nuclei in each cell or **muscle fiber,** whereas each smooth-muscle fiber has only one nucleus (Figure 18.12). In cardiac muscle, the nuclei are scattered throughout the cell, but in skeletal muscles, the nuclei lie just under the membrane of the fiber. The cells of skeletal muscles are also distinctive in that they may be very long, some even extending the entire length of the muscle. Microscopic examination shows that both skeletal and cardiac muscles have alternating light and dark bands, called **striations.** The three muscle types are compared in Table 18.1.

Skeletal Muscles

Typically, when we think of a "muscle" we conjure up images of skeletal muscle. A single such muscle is actually a bundle of millions of individual muscle fibers bound together by connective tissue. (This connective

Table 18.1 Comparison of Muscle Types

	Skeletal	Smooth	Cardiac
Location	Attached to skeleton	Walls of viscera, around arteries	Wall of heart
Shape of fiber	Elongated, cylindrical	Elongated, spindle-shaped	Elongated, cylindrical, branched
Number of nuclei	Many	One	Many
Position of nuclei	Peripheral	Central	Central
Cross striations	Present	Absent	Present
Speed of contraction	Rapid	Slow	Intermediate
Ability to remain contracted	Least	Most	Intermediate
Common type of control	Voluntary	Involuntary	Involuntary

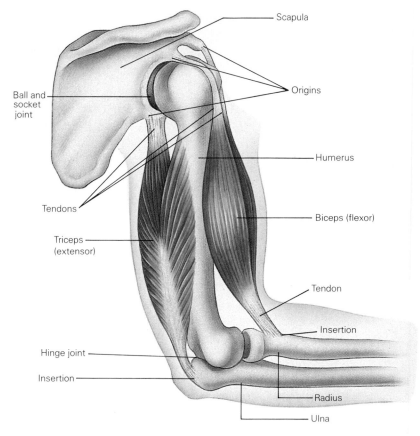

FIGURE 18.13 The attachment of the upper arm muscles. The biceps brachii lies on the forward part of the humerus. On the back is its antagonistic muscle, the triceps brachii. Shortening of the biceps has what effect on the lower arm? Note that the origins are attached to bones that will move less than the bones with the insertions when the muscle is shortened.

tissue is what makes some meat so tough.) The muscle is surrounded by a tough sheet of whitish connective tissue called the **fascia.** The slippery fascia minimizes friction as the muscles move past each other.

Each end of most skeletal muscles is attached by cordlike tendons to different bones. When a muscle of this type contracts, it shortens and thickens, drawing the areas of attachment closer together, and thereby moving the bones. (Note, then, that muscles "pull" and do not "push.") Ordinarily one of the bones moves less than the other and, in some cases, is completely stationary. The bone that moves less provides the area of attachment for the end of the muscle called the **origin;** the more moveable bone provides the surface for the muscle's **insertion.**

Consider an example. If you ask a fellow to show you his muscle, if he is a gentleman he will display his **biceps brachii** (Figure 18.13). Note that the origin of this muscle is at the joint of the humerus and scapula (**biceps** means "two-headed," so note also the double origin). The muscle inserts on the radius, the arm bone that terminates behind the thumb. (As the forearm is rotated, the radius describes an arc, hence its name.) As the biceps is flexed, the lower arm forms an increasingly sharp angle with the upper arm. When the biceps is fully shortened, the muscle can then be admired by those interested in such things.

Antagonistic Muscles

On the other side of the humerus, opposite to the biceps brachii, is another large muscle. This one originates on the scapula and at two places on the humerus and is called the **triceps brachii.** It inserts on the elbow, which is actually the part of the **ulna** (the bone lying behind the little finger) that extends past the joint. As the biceps is contracted, the triceps must relax and lengthen. When the triceps is contracted, extending the arm, the biceps must then relax and lengthen. Thus the biceps increases the angle between the upper and lower arms. Muscles that increase the angles between bone are called **flexors.** The triceps decreases the angle and so is called an **extensor.** Most skeletal muscles have such opposing muscles that reverse their action. Such pairs are called **antagonist** muscles.

In some cases, antagonistic muscles must partially contract simultaneously in a highly coordinated manner and, thereby, keep a certain tension between them. For example, standing requires simultaneous contractions of opposing leg muscles. The flexor (in the back of the leg) and the extensor muscle (in the front) pull against each other and keep the leg straight and the knees locked. The tension, of course, must be highly regulated to keep us from leaping about and sprawling in such a manner as to arouse the curiosity of our fiancée's parents. Walking, by the same token, is not a simple matter; it requires an amazing degree of intricate coordination between a number of opposing muscles. Balance is maintained by constantly changing tensions as our weight shifts from one leg to the other.

Types of Movement

Just as the biceps brachii is a flexor and the triceps brachii is an extensor, other muscles also produce specific forms of movement. **Adductors** and **abductors** move parts of the body toward or away from the central axis of the body respectively, as when the arm is pressed against the body or swung outward. **Levators** and **depressors** raise or lower parts of the body, as in shrugging and lowering the shoulders. **Pronators** rotate body parts downward, as when you turn your palm downward to look at your watch, and **supinators** rotate those parts upward, as when accepting money for the watch. **Sphincters** and **dilators** are rings of muscle that open and close body openings, such as the muscles of the lips. Most muscles in this last group are involuntary, such as those around arteries and at the juncture of the stomach and small intestine. However, some, such as the urinary bladder and anus, are under partial voluntary control, the level of which may influence social acceptability.

The larger superficial muscles of the human body are shown in Figure 18.14. Look this over, and try to find each of these muscles on yourself. Contract the muscle, identify its action, but not while waiting for a job interview. Does anything about the arrangement surprise you? Why is the gluteus maximus of humans so much larger than that of other species? What muscles are involved in your shrug?

Muscle Contraction

The striated muscle fiber, or cell, has several nuclei, as was mentioned. It is also surrounded by a rather tough cell membrane. Closer examination (Figure 18.15A and B) reveals that each fiber if made up of yet smaller fibers. These are composed primarily of protein and are called **myofibrils.** The cross banding, or striations, of the myofibrils gives the fiber its striated appearance. Since muscle tissue is very active, we should not be surprised that it contains a rather larger number of mitochondria, those

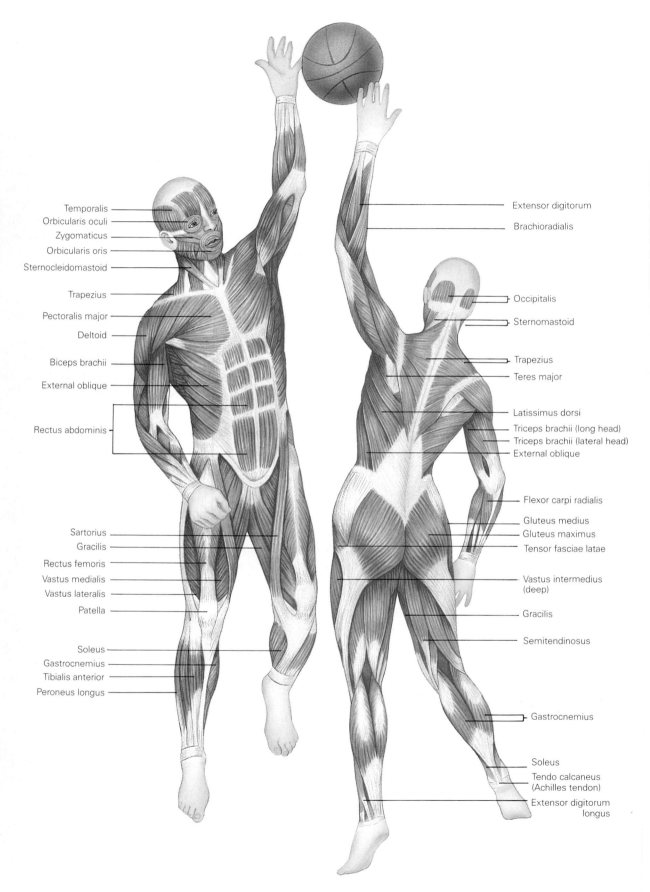

Temporalis

Orbicularis oculi

Zygomaticus

Orbicularis oris

Sternocleidomastoid

Trapezius

Pectoralis major

Deltoid

Biceps brachii

External oblique

Rectus abdominis

Sartorius

Gracilis

Rectus femoris

Vastus medialis

Vastus lateralis

Patella

Soleus

Gastrocnemius

Tibialis anterior

Peroneus longus

Extensor digitorum

Brachioradialis

Occipitalis

Sternomastoid

Trapezius

Teres major

Latissimus dorsi

Triceps brachii (long head)

Triceps brachii (lateral head)

External oblique

Flexor carpi radialis

Gluteus medius

Gluteus maximus

Tensor fasciae latae

Vastus intermedius (deep)

Gracilis

Semitendinosus

Gastrocnemius

Soleus

Tendo calcaneus (Achilles tendon)

Extensor digitorum longus

FIGURE 18.14 **FIGURE 18.14** Ventral and dorsal views of the superficial muscles in the human. The darker areas are muscle bellies. The lighter areas are connective tissue, mostly tendons. Note the difference in the size of various muscles in the two sexes (male, left; female, right). You should be able to infer the action of any muscle by noting its position and attachments. For example, the deltoid is largely responsible for raising the arm toward the ball. What do you suppose the sartorius does? Why do you suppose a sprain of the lower back sometimes takes so long to heal? (Keep in mind the white connective tissue has fewer blood vessels.) In which areas of the body are the bones covered with little or no muscles? (Note the back of the pelvis and the sternum.) Why do you suppose the gluteus maximus is so much larger in humans than most other mammals? (Keep in mind our erect posture and type of leg movement.)

tiny organelles so intimately involved in the production of energy. The mitochondria lie snugly against the myofibrils, the actual contractile units.

As we see Figures 18.15C and D, the tiny myofibrils are made up of still smaller units, called **myofilaments**. There are two kinds of myofilaments; the thicker ones are about 10 nanometers in diameter, and the thinner ones are about half that size. We can now see that the banded appearance of the myofibril is due to how two kinds of myofilaments are arranged with respect to each other.

The myofilaments themselves consist primarily of two proteins, **actin** and **myosin**. The thinner filaments are composed of actin and the thicker ones of myosin. Note that the various areas along the fiber are designated by letter. The thick myosin filaments, we see, are found only in the darker A band, while the I band contains only the thin actin filaments, which may extend into the A band.

During muscle contraction (are you ready for this?), the A band remains constant in length, while the I band shortens, and the length of the H zone within the A band also decreases. Such changes occur because the filaments themselves do not actually contract; instead, they slide past each other. In other words, during contraction, the actin filaments on the opposite ends of the H zone (the area free of actin filaments) move toward each other, reaching into the A band. As this happens, the H zone and the I band become narrower as the actin filaments move over the larger and stationary myosin filaments and approach what is termed the Z line.

The precise mechanism by which muscles contract is not entirely understood, but the prevailing explanation is called the **sliding filament theory**. Its acceptance is partly due to the fact that it accounts for the charges in banding patterns that result in the I band becoming shorter and the H zone disappearing. The basic idea is diagrammed in Figures 18.15E and F. Other parts of the puzzle are better understood. For example, it seems clear that the sliding movement is due to the activity of myosin. Myosin filaments have projections called **cross bridges** (Figure 18.15E and F). These attach to the actin fibers and pull them toward the center of the contractile unit while the myosin itself remains in place. The cross bridges attach, pull, release, reach and attach again to the actin, thereby pulling it along in a rachetlike fashion.

The myosin cross bridges can only attach to actin in the presence of calcium ions (Ca^{++}) and ATP. The calcium ions react with protein molecules in the actin filament, causing "raw" sites along the filament to which the cross bridges of myosin can attach.

After the bridges have attached and pulled the actin along, the myosin takes on the role of an enzyme and breaks down ATP to ADP, releasing energy that can then be used to break the cross bridge, freeing it to reach further along the actin fibril, attach, and pull again. This all sounds rather tedious and plodding, but in actuality it happens with astonishing swiftness. The sequence may occur hundreds of times each second.

The efficiency and swiftness of muscle contractions illustrates the conservative efficiency of living systems that have developed over time to meet certain environmental needs. The environment has traditionally been a demanding, unforgiving, and unprejudiced place, but life has had the time to meet its challenges. We are well adapted for living on this planet, and one prevailing theme has been conservation, not only of

FIGURE 18.15   How skeletal muscles are believed to work. The sequence shows increasingly enlarged areas of a muscle. (A) Illustrated here is the sliding filament hypothesis of how such voluntary muscles work at the subcellular level. A and B give you some perspective on precisely where the Z lines are in a muscle. D–F indicate what happens when a muscle contracts. Notice that the length of the A band remains constant. The fibers of the H zone give rise to tiny "bridges," which the I fibers can move along in a progressive fashion. The converging I bands, then, decrease the length of the H zone as the muscle shortens. The I bands are believed to be composed of actin, the H zone of myosin.

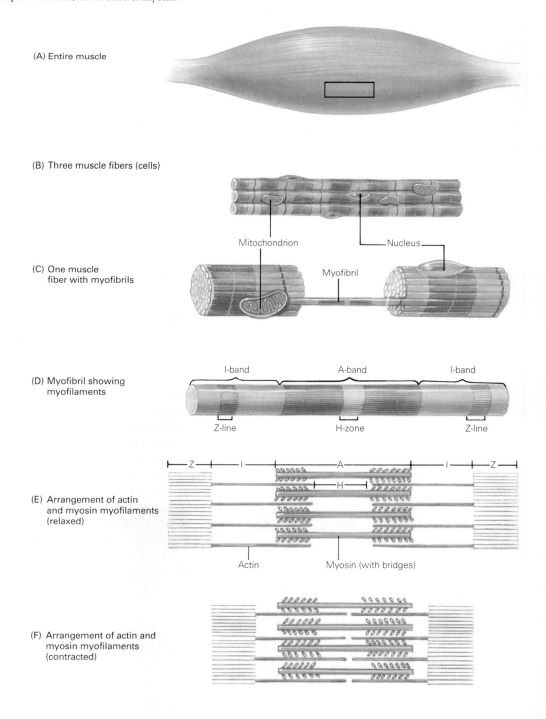

(A) Entire muscle

(B) Three muscle fibers (cells)

Mitochondrion   Nucleus

(C) One muscle
fiber with myofibrils

Myofibril

I-band   A-band   I-band

(D) Myofibril showing
myofilaments

Z-line   H-zone   Z-line

(E) Arrangement of actin
and myosin myofilaments
(relaxed)

Actin   Myosin (with bridges)

(F) Arrangement of actin and
myosin myofilaments
(contracted)

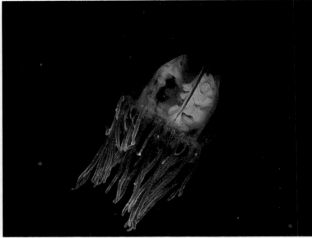

FIGURE 18.16 Coelenterates lack mesoderm, hence they do not form true muscle.
They can move slowly by contracting their
outer ring of contractile tissue formed from ectoderm and an inner radially arranged set
formed from endoderm.

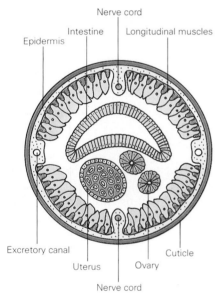

FIGURE 18.17 Cross-section of a roundworm, *Ascaris*, a parasite of humans. Note that
the nerve cords lie both dorsally and ventrally.
(What is the case in insects and in vertebrates?) Note that in the ectoderm, the nuclei
do not lie in separate cells. What kind of
movement is possible with only longitudinal
muscle fibers? Why does the worm have such
a heavy protective cuticle? Keep in mind that
it lies bathed in digestive juices within the intestines. The body cavity that lies inside the
body wall and outside the gut is incompletely
lined with mesoderm and so is called a pseudocoelom. In animals in which the cavity is
completely lined with mesoderm, it is called a
coelom.

energy, but of time. Imagine a woman lying on the grass. She is quietly
at rest, enjoying the peaceful inactivity. However, she is actually not at
rest in the most profound sense. Her body can't spare the time. Even as
she calmly describes her inner peace, her cytoplasm is abuzz with activity—shuffling molecules, shifting energy about, saving both time and
resources at every turn, doing the things necessary to keep her alive—
and at a dazzling rate.

VARIOUS INTERACTIONS BETWEEN SUPPORT AND
CONTRACTILE SYSTEMS

Whereas humans can provide a reasonably good example of the support
and contractile systems of vertebrates, the invertebrates may vary greatly
from this scheme, as you might imagine. Let's briefly review a few other
types of support and movement in some of these groups.

Coelenterates, you may recall, are organized rather like a hollow sac.
The group includes the jellyfish, hydra, and sea anemones (Figure
18.16). Since coelenterates lack mesoderm, they cannot develop true
muscles. However, their double-walled bodies contain two sets of rather
efficient contractile fibers. The fibers in the outer wall (which was
formed from ectoderm) extend along the length of the animal, and those
in the inner wall (formed from endoderm) run circularly. Because
anemones are not restricted by some sort of rigid skeleton, they are able
to perform some rather remarkable contortions.

Beginning with the roundworms, (Figure 18.17), we see true muscles
and, hence, we are among animals with three germ layers, including
mesoderm. Flatworms and segmented worms go a step further. Whereas
the roundworm has only longitudinal muscles, the flatworms and segmented worms also have longitudinal muscles and some species have
yet other muscle arrangements, such as diagonal muscles (running at an
angle between the dorsal and ventral body walls) or dorsoventral muscles (running directly across the body cavity between the dorsal and
ventral walls).

Arthropods, the joint-legged group that includes the insects, are supported by an exoskeleton. This system entails a distinct arrangement of
contractile fibers, quite different from any we've considered so far. This is

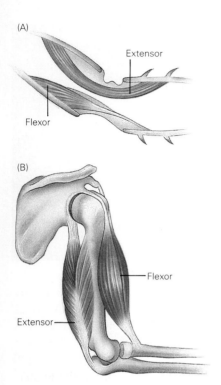

(A)

Extensor

Flexor

(B)

Flexor

Extensor

FIGURE 18.18   A comparison of the musculature of an insect (A) and a human (B). Note, particularly, the relationship of muscles and joints. Insects are much stronger than humans for their weight, so why haven't we evolved internal musculature? What are its disadvantages? Does shortening the insect's upper muscle here have the same effect as flexing the human bicep?

because the muscles lie *within* the skeleton. The joints are also unusual in that they are simply thin, flexible places in the exoskeleton. A muscle may extend across a joint, or it may lie entirely within a single segment of an appendage and insert on the next appendage by an **apodeme**—a long, thin filament.

The musculature of animals with exoskeletons lends them remarkable strength for their size. You have undoubtedly noticed tiny ants carrying loads of food many times their weight or dragging along the corpses of huge insects. Their strength is largely due to the mechanical advantage inherent in the arrangement of their muscles (Figure 18.18). However, the same sort of physical principles that makes them so strong for their size also limits their size. Because of some complex biomechanical principles, an insect the size of a human would have greatly-diminished relative strength. Thus, if you should encounter a 150-pound insect on your way home some night, don't be afraid. You're stronger than it is.

The insect muscle-skeleton arrangement is also responsible for that lovely whine of the mosquito. Physiologists once puzzled over how mosquitoes and other insects could move their wings so fast. It would seem that the nerves that activate the flight muscles simply couldn't fire so rapidly. However, it has now been found that a single muscular contraction can set up a reverberation of the insect's exoskeleton. As the thorax of the exoskeleton vibrates, it moves the wings. Thus, a single muscle contraction flexes the exoskeleton and produces multiple wingbeats.

The problems of support and movement, we see, beset most of the animal life on the planet. Because natural selection would have necessarily responded to the problem early in the evolutionary development of animal life, we see a few general themes within related groups. Since the earth has set other fundamental problems for life, we will continue our survey of systems.

SUMMARY

Systems of support are important in multicellular animals, particularly terrestrial species that are not buoyed by water. Support is usually accomplished by exoskeletons (outside the body) or endoskeletons (inside the body). Bones are formed of connective tissues that are largely a matrix of calcium salts and collagen. Haversian systems of bone center around an opening through which nerves and blood vessels run. The bone cells lie in a concentric arrangement around these channels. The human skeleton is divided into the more central axial groups (the skull, vertebral column, sternum, and ribs) and the more peripheral appendicular group (the pectoral and pelvic girdles and the limbs). The various body parts move by a complex contractile system basically formed of three kinds of muscles: cardiac (heart), smooth (such as those around the arteries), and skeletal. Skeletal muscles often work in antagonistic pairs. Their actions are defined according to the movement they produce. The skeletal muscles are composed of myofibrils that are composed of still smaller units called myofilaments. The thicker myofilaments are composed of myosin, the thinner ones of actin. During contraction, the myosin essentially remains stationary while the actin bands move toward each other, sliding rachetlike along the myosin. They move by rapidly forming and breaking cross bridges with the myosin.

Chapter 19

THE RESPIRATORY, CIRCULATORY, AND DIGESTIVE SYSTEMS

A whole host of problems rained down on the first struggling life on a swiftly changing planet. There was the devastating weather, the withering sunlight, competition at every turn, and there were predators. One solution to survival in such a place was the specialization and efficiency that comes with multicellularity. With multicellularity came larger size, and with larger size came a whole host of new challenges. One of those challenges involved acquiring certain critical molecules and then moving them to where they were needed within the organism. Here, then, we will take a look at some of the systems that helped solve these sorts of problems.

RESPIRATORY SYSTEMS

Life on the planet evolved in a sea of gases, one of which was quite deadly. When the first forms of life appeared, the deadly gas was quite rare, and so it wasn't a threat. But, in time, it accumulated until it posed a very real threat to the continuance of life.

The gas corroded all sorts of hopeful experiments; the air seemed to rot, digest, and break down everything it touched. But life ebbed and flowed, this way and that, in mindless shifts and evolutionary eddies, and in time, there appeared forms that could actually utilize this dangerous gas that we now call oxygen. Many life forms succumbed to its corrosive effects, but others not only survived, they developed ways to put this ethereal spirit of decay to good use—and so we breathe.

Breathing, or **ventilation,** is often equated with respiration, but actually, breathing is only one method of **external respiration** (getting oxygen to the cells). There are also other ways of oxygenating cells, as we shall see. We might also keep in mind that **internal** (or **cellular**) **respiration** (Chapter 5) involves the use of oxygen in cellular metabolism. Here, then, we will concentrate on external respiration.

The Various Ways Species Get Oxygen

Because life is so splendidly varied across the far reaches of the planet, and because so many life forms depend on oxygen, it is not surprising that living things have developed a variety of means to get oxygen. We will consider only a few representative cases, but we will again see that natural selection can take many paths to the same adaptive end.

Aquatic Animals

Small aquatic animals have relatively little problem in getting oxygen. They can simply absorb it through their body walls. Planarian flatworms, for example, have no organs for taking in oxygen, but their thin, flattened shape provides them with a large surface area in relation to their mass. Diffusion of oxygen from the water into their bodies gives them all the oxygen they need for their slow-paced lives.

Larger aquatic animals have proportionately less surface area relative to their mass; hence, they need some specialized system of obtaining and delivering oxygen. Consider the chiton, a mollusk of both the Atlantic and Pacific coasts (Figure 19.1). You may have seen them clinging tenaciously to rocks along the shoreline and commented on their lack of personality. The Atlantic species may be less than two centimeters long; the Pacific species run as long as twenty-five centimeters. They are large enough, though, to require a respiratory system. The problem is solved by gills projecting from a concavity around the body, each gill covered with tiny beating cilia that circulate oxygen-laden water over it. The gills are very thin-walled with a large surface area and are permeated by tiny blood vessels; thus, oxygen can enter the bloodstream easily and be carried quickly to the rest of the body.

Although fish are much more complex creatures than chitons, they also acquire oxygen through gills. The gills of fish are also essentially comprised of protrusions or shelves (lamellae) that provide a very large surface area (Figure 19.2). The cell membranes are very thin and readily permit the passage of gases across them. Lying in close proximity to these

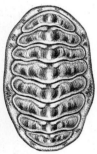

Dorsal view

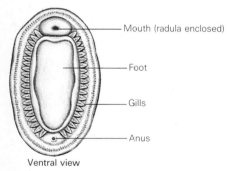

Mouth (radula enclosed)

Foot

Gills

Anus

Ventral view

FIGURE 19.1 The external structure of the chiton, dorsal and ventral views. The animal attaches to rocks by suction produced by muscles of the foot. It feeds by rasping algae off rocks with a horny-toothed structure called a radula. Between the foot and the upper body is a concavity in which lie eleven to twenty-six leaf-shaped gills. From the gills extend tiny cilia that, by their beating, keep oxygen-laden water flowing past them.

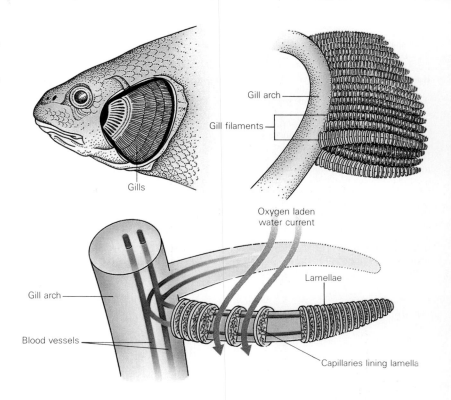

FIGURE 19.2 The gill structure of a bony fish. The lamellar projections increase the surface area of the gill. Note the blood vessels within each lamella. Blood passes from the gill arch into the lamellae of the gill filaments. Here, gases are exchanged. The oxygenated blood then returns to the gill filaments, then through the gill arch, back to the body.

Labels in figure: Gills; Gill arch; Gill filaments; Oxygen laden water current; Lamellae; Gill arch; Blood vessels; Capillaries lining lamella

cell membranes are tiny, thin-walled capillaries that exchange carbon dioxide produced in the body with oxygen from the water. The oxygen is then transported to the rest of the body. Remember that there are normally higher carbon dioxide levels within the body and higher oxygen levels outside it, and it is these differences that determine the direction of flow of the dissolved gases according to the principles of diffusion. Thus, the higher level of carbon dioxide within the body results in carbon dioxide moving outward, and the lower level of oxygen in the body means oxygen moves inward from the surrounding water.

Whereas the chiton moves oxygenated water over its gills by movement of its cilia, the fish pumps water over its gills as it enters the mouth and leaves from under the gill covering. Some fish, such as sharks, lack the muscles for pumping water along this pathway and must swim to keep water flowing over their gills. It was long thought that all sharks were destined to swim virtually every minute of their lives—that if they stopped, they would suffocate. Whereas this may be true for some sharks (Figure 19.3), divers have recently found sharks sleeping motionless at night. No one yet quite understands how this can be, but it has been suggested that they can supersaturate their blood with oxygen before sleeping and that they lower their metabolic rate at such times so that less oxygen is required.

Insects

Insects generally solve the oxygen problem by breathing through a **tracheal system** of tiny tubes that opens to the air and permeates their body tissue. The openings to the system are called **spiracles.** As shown in Figure 19.4, oxygen enters the tracheal system directly from the air and, aided by the insect's bodily movement, moves through the tracheae (plural) deep into the body tissue. Each trachea ends amidst groups of

FIGURE 19.3 **FIGURE 19.3** It was once assumed that since sharks lack a water-pumping system, they must constantly stay in motion to keep oxygen-laden water flowing over their gills. However, divers have recently reported that some species of shark sleep, lying dormant on the bottom. If this is true, their physiology must be much more complex than we had thought. It may be that their constant motion is more directly related to their incredible appetites. A list of things found in the stomachs of sharks over the years would probably astound, amuse, and perhaps dismay even the most jaded biologist. On an expedition in which I participated, we found a Polaroid negative in the stomach of a twelve-foot tiger shark. The picture was taken of three of us and the negative thrown overboard 200 miles and two days previously. Were we being shadowed? It is most disconcerting to find a photograph of one's self in the stomach of a shark. Here we see a sleeping nurse shark.

cells or in expanded air sacs. The tracheal network is so extensive that no cell lies far from an oxygen source.

The insect's system of getting oxygen has probably placed a limit on how large insects can become. This is because a tracheal system cannot deliver oxygen over a great mass. Actually, a tracheal system can work only for organisms no larger than, say, about five centimeters. Thus, an evolutionary direction taken eons ago may be responsible for the fact that we don't have to contend with 800-pound mosquitoes.

Mammals

Among mammals, external respiration takes place by means of special pouches called **lungs.** External respiration is accomplished as oxygen-laden air is brought into the lungs by **ventilation,** or breathing, the rate of which is controlled by the brain's medulla (Chapter 20) as it receives input on blood CO_2 levels from sensors throughout the body. Air passes from the large **trachea** into the branched **bronchi** (singular, **bronchus**) and on into increasingly smaller **bronchioles** that terminate in the tiny **alveoli,** which are so numerous that they give a spongelike quality to the lung (Figure 19.5).

FIGURE 19.4 The respiratory system of an insect. Note the spiracles in the animal's side through which air enters, eventually moving into a highly branched tracheal system and, in some species, into large air sacs. The insect's movement hastens the flow of air through these channels. It has been suggested that because of the limitations of such a system, insects have never been able to become very large.

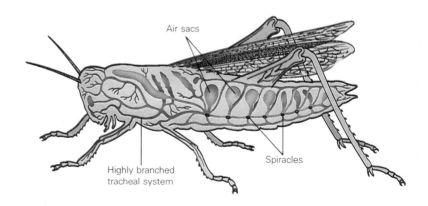

Air sacs

Highly branched tracheal system

Spiracles

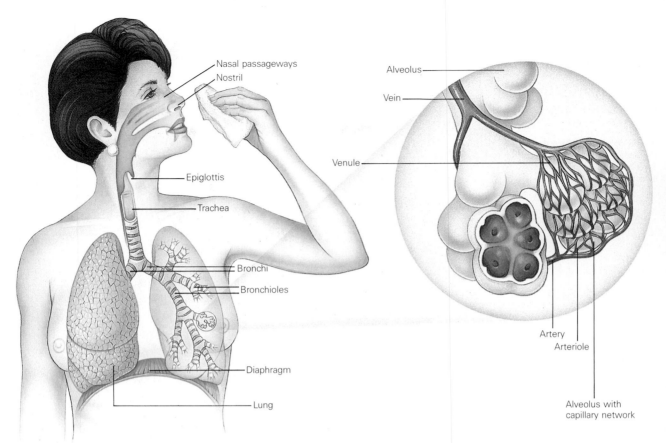

Nasal passageways
Nostril
Epiglottis
Trachea
Bronchi
Bronchioles
Diaphragm
Lung

Alveolus
Vein
Venule
Artery
Arteriole
Alveolus with capillary network

FIGURE 19.5 The human respiratory system. The contraction of the diaphragm creates a partial vacuum in the chest and causes oxygen-filled air to rush in to fill the void. The air is warmed and moistened as it travels through the upper air passages. It passes through increasingly smaller passages until it reaches the very thin-walled, saclike alveoli, which allow oxygen to pass into the bloodstream and carbon dioxide to pass into the lungs, from where it is expelled.

The surface of the lungs is moist, as is the respiratory surface of any animal. The moistness is necessary because oxygen must dissolve before it can cross these delicate membranes. Once in solution, it can diffuse across the thin cell membranes of the saclike alveoli where the respiratory passages terminate and enter the bloodstream. The blood transports the oxygen to the body's tissues where it diffuses into the cells. In the cells, the oxygen has the humble but critical role of picking up spent hydrogen atoms from the electron-transport chain, thereby producing water.

Meanwhile, the carbon dioxide that has been produced in metabolic reactions throughout the body diffuses from the cells into the capillaries, and from there it is carried to the lungs. In the lungs, CO_2 diffuses through the tiny vessels out of the blood and into the interior of the lungs. From there it is exhaled into the environment. (Breathing rate is controlled by the brain's medulla, which responds to several sensors throughout the body, as is shown in Figure 19.6.)

CIRCULATORY SYSTEMS

Not only must oxygen and carbon dioxide be transported throughout the body, but a host of other substances must move through the body as well. These include nutrients, various wastes, and hormones. Thus, the development of an efficient circulatory system has been critical. Of course, some species have less need of such a system, and those that have them have developed a variety of mechanisms for moving substances about.

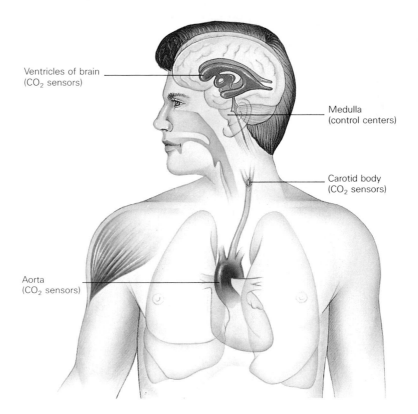

FIGURE 19.6 The primary respiratory control center is located in the medulla of the brain. It receives neural input from carbon dioxide-measuring centers elsewhere and responds by altering the rate and depth of breathing. The carbon dioxide-sensing centers are located in the ventricles (openings) of the brain, the carotid arteries leading to the head, the aorta (the large artery leaving the heart), and, to some extent, in the voluntary muscles. Because of the action of the respiratory center, the body's oxygen demands can be met under a variety of conditions, from rest to strenuous activity.

Ventricles of brain (CO_2 sensors)

Medulla (control centers)

Carotid body (CO_2 sensors)

Aorta (CO_2 sensors)

Circulation in Small Animals

A number of factors may influence the type of circulatory systems in living things. For example, size may be important. In one-celled organisms and small multicellular species, there may be no need for a circulatory system at all since materials can easily diffuse from one part of the body to another and a large part of the total membrane area of the body is exposed to the outside environment.

The problem becomes more critical in increasingly larger animals, and it has been met in a variety of ways. Coelenterates, for example, have an unusual kind of circulatory system among animals. In this group, some of their cells are **amoeboid.** That is, they can crawl and squirm their way through the body's tissues. When they reach the inner layer of the pouchlike animal, they surround and engulf food particles just as an amoeba does. Then they withdraw into the body with the captured nutrients and distribute the food.

Flatworms also have saclike guts with a single opening used for food intake and the elimination of waste. Since the body is flattened, no cell is very far from the gut where digestion takes place; thus, nutrients can easily diffuse to the areas where they will be utilized. Small animals, we see, have little need for the complex systems of their larger colleagues. But this section is supposed to be about circulation, so let's move to the larger creatures.

Circulation in Larger Animals

We can begin by noting that circulatory systems may be classified according to whether they are **open** or **closed.** In open systems, the

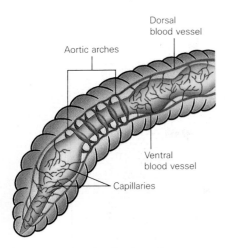

FIGURE 19.7 The closed circulatory system of the earthworm. The pulsating dorsal vessel sends blood downward through the tubular rings of the aortic arches backward through the ventral vessel. Note the branches of the major vessels that lead into other parts of the body. The blood remains enclosed in vessels, thus the system is "closed."

Aortic arches

Dorsal blood vessel

Ventral blood vessel

Capillaries

blood does not remain enclosed in vessels. At certain places, it leaves the vessels to seep through the tissues. For example, in mollusks, such as clams and snails, and in arthropods, such as insects and crayfish, a tubular heart pumps blood through vessels for a way, but then the blood empties into open cavities, or **sinuses.** From there it seeps or percolates through tissues and then collects again in a large sinus around the heart. When the heart relaxes, blood flows in through tiny, one-way pores in the heart wall. When the heart contracts, the pores close, and blood is pumped out through the vessels. In open circulatory systems, the blood slowly seeps through the tissues, so the process is generally too inefficient for larger animals in which the more rapid movement of channeled blood would be favored.

In closed circulatory systems, blood remains enclosed in vessels (with a few inevitable exceptions as we'll see) and can therefore move more swiftly with its life-giving burden. Among the most primitive of the closed systems are those of the annelids, the segmented worms. In the earthworm, for example, the heart consists of a pulsating dorsal aorta that sends the blood forward and then down through a series of five tubular rings, the aortic arches, from where it flows backward through a large ventral vessel (Figure 19.7).

Vertebrates also have a closed circulatory system, but it is in this group that we most clearly see an exception to the rule that blood in closed systems always flows in vessels. In vertebrates, blood, indeed, leaves vessels and flows through small open cavities of the liver and certain bones.

Like most vertebrates, humans have a large muscular heart that pumps blood into large muscular **arteries,** which in turn branch into increasingly smaller **arterioles** and finally into **capillaries.** The walls of the capillaries are so thin that oxygen, carbon dioxide, nutrients, and metabolic waste are able to pass through them. From the capillary beds the waste-laden, oxygen-depleted blood flows into larger vessels called **venules,** and then into increasingly larger **veins** that return the blood to the heart. (Arteries, then, carry blood from the heart, and veins return it to the heart.) In humans, the length of the entire system is estimated to be between 50,000 and 60,000 miles, 70 percent of which is capillaries.

The body faces the problem, of course, of how to get rid of various wastes, primarily carbon dioxide and various nitrogen compounds. This process begins with a series of chemical reactions that result in waste products being produced in a form that can be handled by the body's excretory systems, as we saw in the case of nitrogenous waste in Chapter 17.

Carbon dioxide is handled in quite a different way. Dissolved carbon dioxide gases in the blood are in equilibrium with carbonic acid and ionic bicarbonate. That is,

$$CO_2 \;+\; H_2O \;\longleftrightarrow\; H_2CO_3 \;\longleftrightarrow\; H^+ \;+\; HCO_3^-$$

| Dissolved carbon dioxide | Water | Carbonic acid | Hydrogen ions | Ionic bicarbonate |

For these wastes to be excreted, they must reach the lungs as CO_2, the form that most easily passes through the thin membranes of the lung. This is accomplished by the activity of an enzyme in the blood, **carbonic**

anhydrase, which moves the reaction to the left. The high levels of CO_2 inside the bloodstream, relative to the low levels inside the lung, cause this waste gas to move out of the body and into the lung, from where it can be exhaled.

The Heart

The muscular pump called the **heart** varies widely in invertebrates and is related to the life style of the organism. There is less variation in vertebrates, but there is an important progression regarding the number of chambers.

The fish heart has only two chambers (Figure 19.8). One is a thin-walled **atrium** that accepts blood after it has been circulated through the body. Contraction of the atrium forces blood (see Essay 19.1) into the other chamber, the **ventricle.** The atrium doesn't have to be very strong because it doesn't take much work to pump blood only to the next chamber (see Figure 19.9). The ventricle, however, must generate tremendous pressure, since it must force the blood through the large artery called the **aorta** into the tiny mesh of the gill capillaries where gases are exchanged, and then through the rest of the body. From the major tissues of the body it reenters the veins. The fish ventricle, then, must be able to pump blood through two capillary beds—one in the gill and one in the body. Friction is high in these beds due to the enormous surface area of the interior of the tiny vessels of such an expansive network. Not surprisingly, the ventricle is extremely thick and muscular.

The amphibian heart is a bit more complex in that it has two atria. One receives oxygenated blood (blood high in oxygen) from the lungs, and the other receives deoxygenated blood (high in carbon dioxide) from the rest of the body. The two are mixed in the single ventricle, so the system is not particularly efficient; but it is efficient enough for the rather lethargic frog.

Reptiles go one step better and have two atria and two ventricles, although in most species the ventricles are not completely divided (exceptions include crocodiles and alligators). Because of the incomplete ventricular septum (wall), there is some mixing of blood from the two sides of the heart. Still, this system is more efficient than the frog's, since one side of the heart pumps blood through the lungs and the other side pumps it through the rest of the body.

In birds and mammals, the right and left sides of the heart are completely separated. These groups require an efficient circulatory system because they are endothermic ("warm-blooded"). The need to maintain a constant temperature requires a means of moving blood to the heat-generating cells.

Heart Control and Variation

The heart must function in a precise and coordinated manner (but see Essay 19.2), so we are not surprised to learn that it generally operates under a complex system of controls. The control centers, shown in Figure 19.6, respond to a variety of stimuli. For example, heart rate can increase or decrease according to whether arteries are being stretched by

FIGURE 19.8 Various kinds of vertebrate hearts, ranging from the two-chambered to the four-chambered. The simple to complex arrangement is assumed to trace the evolutionary development of the "warm-blooded" heart (that of birds and mammals). Note the increased efficiency in separating oxygenated and deoxygenated blood. These schemes are highly generalized and exceptions do exist. For example, among reptiles, the crocodile heart is more strongly divided than that of snakes. In birds and mammals, the developing tubular heart forms a loop, resulting in the atria coming to lie headward. In our illustration, however, the heart appears as if the looping has not occurred.

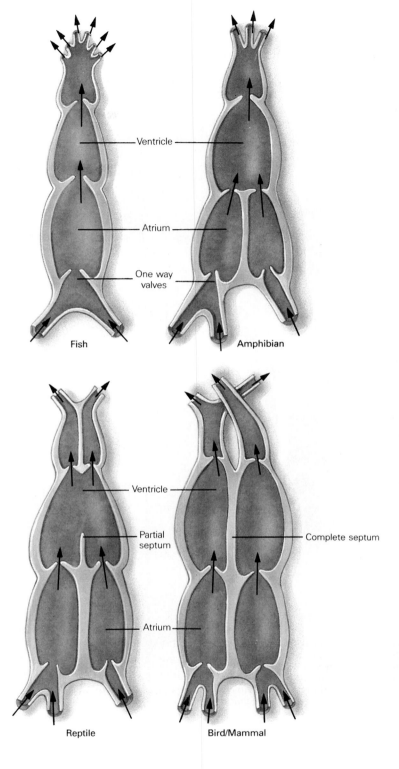

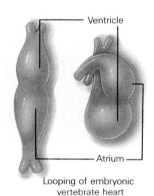

Looping of embryonic vertebrate heart

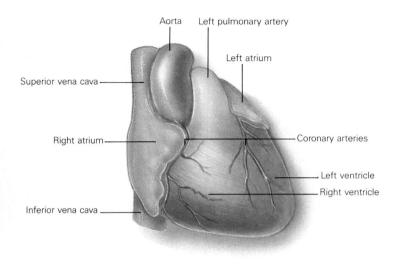

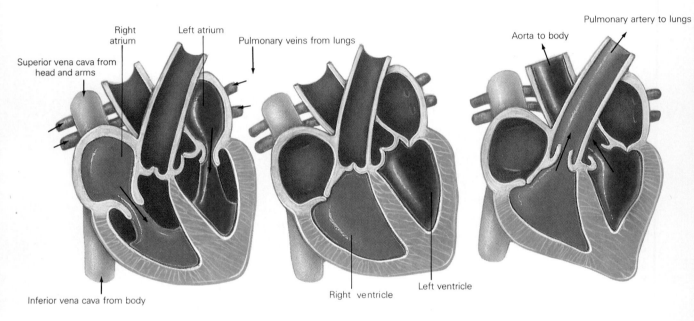

Atrial contraction   Ventricles filled   Ventricle contraction

FIGURE 19.9 The flow of blood through the human heart. Notice the coronary arteries that service the heart itself. Blood from the right atrium (deoxygenated and deep red) is pumped to the right ventricle and from there through pulmonary arteries to the lungs. It returns oxygenated and bright red to the left atrium and from there to the left ventricle to be pumped to the body via the aorta. Notice that the two atria contract simultaneously, as do the ventricles.

the volume of blood being pumped into them (which decreases the heart rate and lowers that volume) or veins are being stretched (which increases the heart rate, thereby reducing the pressure within the veins). Heart rate can also change according to the relative levels of carbon dioxide and oxygen in the blood (monitored by sensors that detect carbon dioxide) and by emotional stress, as we will discuss later. The system is not a simple one. For example, the nervous system can expand certain arteries while constricting others. Thus, after a meal in a fine restaurant, blood is shunted away from muscles toward the stomach and intestines where it aids in digestion. By the same token, the sight of the bill can cause blood to drain from the head, but the image of a shadowy figure in an alleyway outside the restaurant can cause blood to be shunted back to the muscles.

What does the separation of the two halves of the heart have to do with efficiency? The two halves are essentially two hearts, one serving the lungs, the other serving the body. When CO_2-laden blood enters the right atrium from the large veins (the **superior vena cava** from above and the **inferior vena cava** from below), it is pumped from the atrium to the right ventricle. Contraction of the right ventricle sends the blood through the **pulmonary arteries** to the lungs. Here the blood picks up oxygen and releases carbon dioxide before returning to the left atrium via the **pulmonary veins.** The left atrium pumps the blood into the left ventricle, which then contracts to send the blood into the large aorta, which immediately branches before looping downward, carrying the blood on the first leg of its long journey through the body, bearing its gift of oxygen. The left ventricle, then, is very thick-walled and muscular; hence, the left side of the heart is larger. For this reason, you may have the notion that your heart is on the left side. It isn't. It's right in the middle of your thoracic cavity and is about the size of your fist. Thus, when you pledge allegiance to the flag, your hat is actually over your left lung.

Among invertebrates, the circulatory fluid, or blood, is quite diverse. It may be a colorless fluid, it may be green (copper-containing), or even red (iron-containing). In vertebrates, the **erythrocytes,** or red blood cells, float in a colorless fluid called **plasma.** The plasma also contains proteins, hormones, and various ions, as well as food and waste material. The plasma also carries cells in addition to erythrocytes, such as **leukocytes,** or white blood cells, which function mainly in combating invasion or infections, and cellular fragments

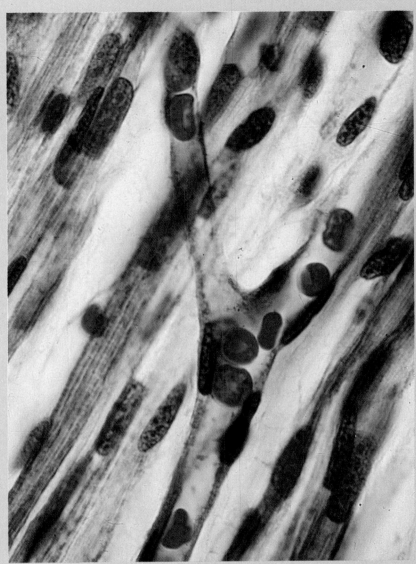

called **platelets,** which function in clotting.

The red blood cells of vertebrates contain a protein-iron compound called **hemoglobin,** which is what makes blood red. Hemoglobin is able to combine reversibly with oxygen. The result is that human blood is able to carry far more oxygen than could be dissolved in plasma alone. Hemoglobin readily combines with oxygen in body parts where the oxygen concentration is high (such as in the lungs). Where

oxygen is low (such as in the tissues), hemoglobin tends to give up its oxygen.

Hemoglobin is comprised of four subunits, each of which has a protein chain (globin) and **heme** (a complex carbon ring structure that contains iron). Each of these subunits can hold one molecule of oxygen.

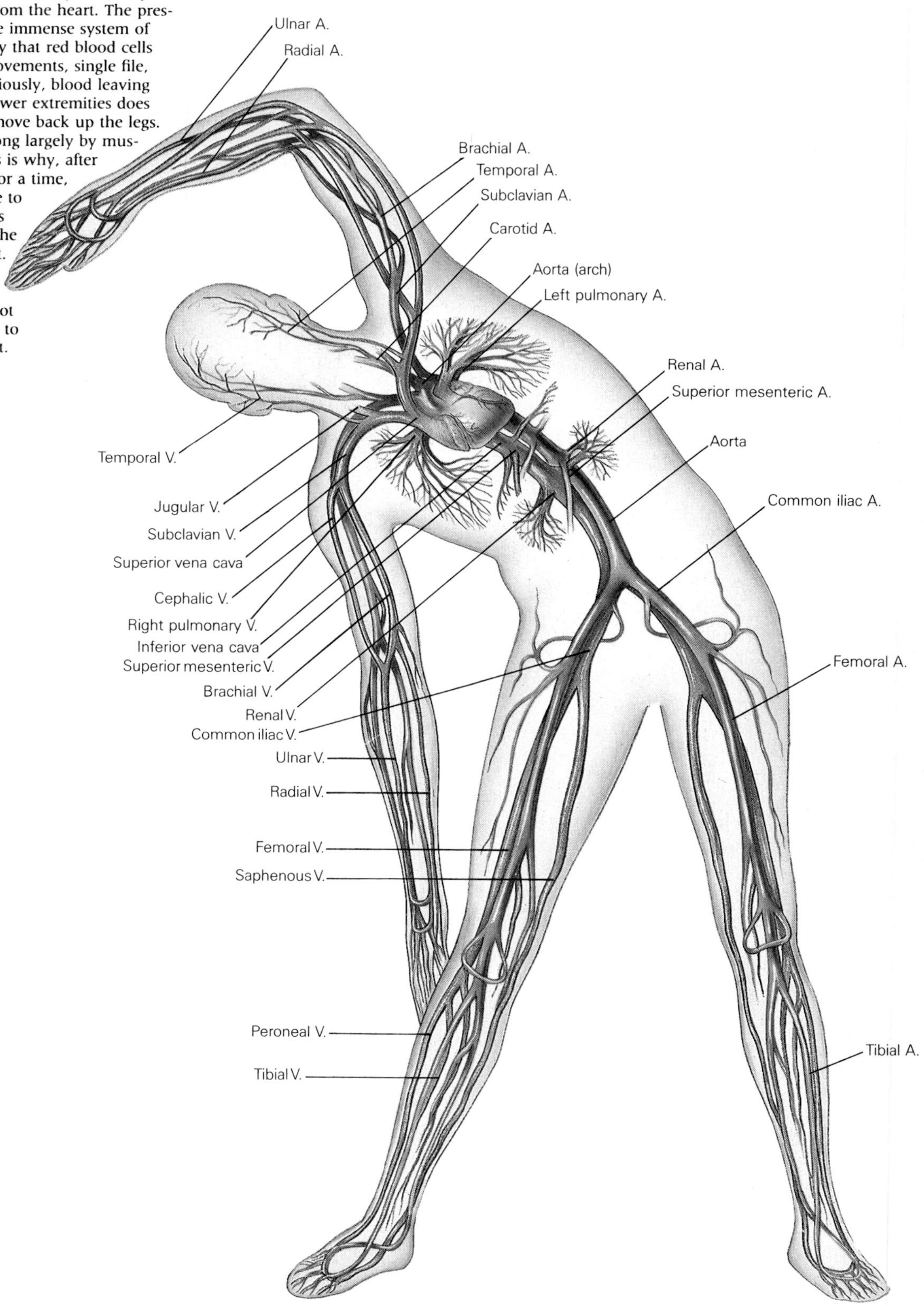

FIGURE 19.10 The major blood vessels of the human body. The arterial (red) and venous systems (blue) are connected by capillaries. Generally, the arteries are thicker-walled because they are under intense pressure by blood being pumped from the heart. The pressure is dissipated in the immense system of capillaries, some so tiny that red blood cells must move in jerky movements, single file, between the cells. Obviously, blood leaving the capillaries of the lower extremities does not have the force to move back up the legs. This blood is forced along largely by muscular contractions. This is why, after being in one position for a time, you may feel the desire to stretch, which contracts muscles and squeezes the blood back to the heart. Without such movement, people standing (as at attention) may not receive enough oxygen to the brain and may faint.

Ulnar A.
Radial A.
Brachial A.
Temporal A.
Subclavian A.
Carotid A.
Aorta (arch)
Left pulmonary A.
Renal A.
Superior mesenteric A.
Aorta
Common iliac A.
Femoral A.
Tibial A.

Temporal V.
Jugular V.
Subclavian V.
Superior vena cava
Cephalic V.
Right pulmonary V.
Inferior vena cava
Superior mesenteric V.
Brachial V.
Renal V.
Common iliac V.
Ulnar V.
Radial V.
Femoral V.
Saphenous V.
Peroneal V.
Tibial V.

Figure 19.10 shows the major blood vessels of the human body. Where are the pulmonary vessels? What is the large network below the heart on the right side of the body? You might now wonder why the good guys survive so many shoulder wounds in so many bad Westerns. According to the figure, where is a better place to be shot? Some of these vessels, especially the larger ones, are likely to be found in about the same place in any human body. This element of predictability is undoubtedly appreciated by surgeons. But other vessels are extremely variable, as you can demonstrate with an indelible pen by tracing the veins in the hands of the people sitting around you. If they object, point out that you only want to show that theirs is different from yours. They will understand. You can always refer them to the discussion of how the circulatory system forms in Chapter 16.

The Lymphatic System

When we think of a circulatory system, we usually think of blood. However, there is a circulatory system of another kind, the one that circulates **lymph.** This one, logically enough, is called the **lymphatic system** (Figure 19.11). Lymph is part of the fluid constituent of whole blood. When it moves as part of the blood, it is called **plasma.** This blood plasma,

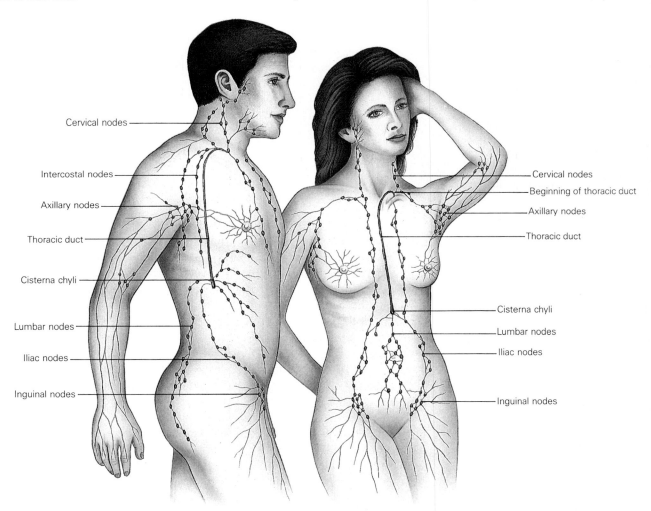

FIGURE 19.11 The human lymphatic system, sometimes referred to as the "other" circulatory system. Lymph nodes are scattered throughout the body, but concentrated in specific areas. These nodes harbor large numbers of one kind of white blood cell, the lymphocytes, which attack any bacteria trapped there. Lymph nodes tend to swell and become sore if they are involved in fighting an infection near them, thus they may signal infections that might otherwise go unnoticed. The axillary lymph nodes are often affected by cancer of the breast and are often removed with the breast and its underlying muscles. Mastectomies of this sort are currently being reexamined to see whether they are merited or effective in most cases.

Cervical nodes

Intercostal nodes

Axillary nodes

Thoracic duct

Cisterna chyli

Lumbar nodes

Iliac nodes

Inguinal nodes

Cervical nodes

Beginning of thoracic duct

Axillary nodes

Thoracic duct

Cisterna chyli

Lumbar nodes

Iliac nodes

Inguinal nodes

We have all heard the old story about the faint-hearted guard dog who, upon being told "Attack," had one. *Attack* is now a common word in our vocabulary, as well it should be since heart attacks kill over half a million Americans each year. The incidence of heart disease among Americans is one of the highest in the world; about 30 million people are affected. (Some countries have a higher rate, as do the French, or a lower one, as do the Japanese.) But what *is* a heart attack? Technically, it is a the result of a **myocardial infarction**—that is, a blockage of the arteries that feed the heart. (The lighter areas in the photo indicate blockage.)

When such an artery is blocked, the oxygen-starved muscles of the heart begin to die. Depending on how much of the heart is damaged and how badly, the results can vary from almost complete recovery to death. Blockage of the coronary vessels that feed the heart is usually caused by one of three things: a clot lodged in the vessel, a prolonged contraction of the walls of the vessel, or atherosclerosis.

Atherosclerosis is the result of the buildup of a number of sub-stances, such as fat, fibrin (formed in clots), parts of dead cells, and calcium. These substances reduce the elasticity of the vessel, and by decreasing the diameter of the vessel, they raise blood pressure, just as you would raise the pressure in a garden hose by holding your thumb over the end. No one knows what causes atherosclerosis, but a number of things can speed its development, such as smoking cigarettes and eating animal fat and cholesterol. Other factors include age, hypertension, diabetes, stress, heredity, and gender (males have more heart attacks).

The warnings of a heart attack are often (but not always) (1) a pain that spreads along the shoulder, arm, neck, or jaw; (2) sudden sweating; (3) a heavy pressure and pain in the center of the chest; and (4) nausea, vomiting, and shortness of breath. The symptoms may come and go.

People who have not developed a strong and efficient cardiovascular system through exercise are particularly susceptible to **angina pectoris** ("chest pain"), which occurs when the heart fails to receive enough blood, particularly during

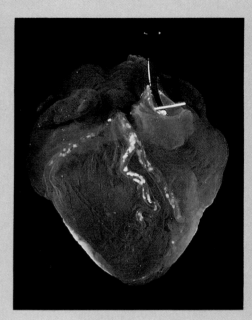

times of stress or exercise. It should not be confused with a true heart attack. The pain may be relieved by stopping the unusual exercise or by reducing the levels of stress. Blood flow to the heart can be increased by a program of exercise or by surgically inserting vessels from other parts of the body (a coronary bypass). Certain chemicals, such as nitroglycerin, also dilate the heart's

however (minus its dissolved proteins), can also move on its own, as when it filters out into the tissues through the walls of tiny capillaries. Here it bathes the cells, moves between them and finally winds up in small channels called **lymph capillaries.** The lymph may then move into larger channels which may eventually bring the fluid to one of the many **lymph nodes** scattered throughout the body. In the nodes it is filtered, thus removing any cellular debris or bacteria. Frogs and some other vertebrates have "lymph hearts" that move the lymph along, but mammals rely chiefly on muscular movements to squeeze the vessels and circulate the fluid. (Again, we see that people are adapted to being on the move.)

White blood cells called **lymphocytes** are found in great numbers in the lymph nodes. If bacteria have entered the system, they are likely to eventually find themselves at one of these nodes, where they will be immediately attacked and destroyed by the lymphocytes. When such a battle is raging, the lymph nodes enlarge (a sign of infection). As the vessels and increase the circulation of blood there.

Another form of heart attack results in a phenomenon called **sudden death.** The death may be due to chaotic and uncoordinated contractions of the ventricles. The contractions do not move blood along and, after a few spasms, the heart may stop entirely. Many people afflicted in such a way mysteriously fall dead in their tracks. Some, however, could have been saved if they had been helped in time.

In fact, victims of any form of heart attack stand a much greater chance of surviving if they are treated immediately. In many metropolitan areas, citizens are being trained in cardiopulmonary resuscitation (CPR) to help restore circulation in such emergencies. Their efforts continue the flow of blood to the brain, where sensitive tissues die quickly without oxygen. In Seattle, Washington, with an extensive citizen-training program, passers-by have performed about one-third of the city's resuscitations. Their success rate is higher than that of professionals because they usually reach victims sooner.

lymph moves along, it flows into increasingly larger channels, or **lymph ducts,** until, finally, it enters large veins near the heart, rejoining the blood, whereupon it is again referred to as plasma.

DIGESTIVE SYSTEMS

There is a tired old cliché about our bodies being internal-combustion engines, with food being the fuel that runs them. However, the idea can still be squeezed for elements of truth. The point, more precisely stated, is that energy is required to do the work that must be done to minimize entropy, or disorganization, within our bodies. The energy for this work is found in food and is released after the food is broken down into its constituents (digestion) and the product transported to the cells (circulation), where it enables the cells—which are actually the engines, if the analogy is to be valid at all—to burn the food for energy (metabolism). Of course, not all ingested material is "burned;" some of it is used as building blocks for the repair and maintenance of the machinery and the growth of the organism.

Digestive Arrangements

In a nutshell, **digestion** reduces the molecular size of nutrients so that they can pass through cell membranes and enter the bloodstream. Since the process is so fundamental, it may seem that the basic scheme for digestion is the same in all animals, except for a few details. This may be true, but these "details" vary widely, as we see in Figure 19.12. An amoeba, for example, simply moves its flexible body around a food particle and engulfs it (phagocytosis), so that the particle becomes contained in a **food vacuole.** Acidic digestive enzymes are then secreted into the vacuole until the food is digested. Any indigestible particles are brought, still enclosed within the vacuole, to the surface of the protist and squeezed out through an opening that appears in the membrane.

The hydra and the flatworm both have pouchlike digestive systems— that is, with a single opening to the outside. Whereas the hydra stings its prey with special cells on its tentacles and then hauls the prey into its digestive cavity, the flatworm is able to extrude its **pharynx,** or "throat," to capture food. In both these animals, digestion begins in the **gastric cavity,** or "stomach," but before the process is completed, the particles of partly digested food are engulfed by cells of the stomach wall and digestion is completed within these cells. Undigested particles are excreted from the cells and expelled through the single opening. These animals, then, digest food by both extracellular and intracellular processes. (Which of these two processes typifies human digestion?)

In more complex animals, the digestive system is a tube rather than a pouch, with one opening through which food enters and another through which it exits—a much more civilized arrangement, to be sure.

Although both humans and earthworms have tubelike digestive systems, you will notice in the earthworm several structures that are absent in humans, such as the crop and gizzard (also present in birds). The **crop** is where food is stored, and the muscular **gizzard** is where food is ground

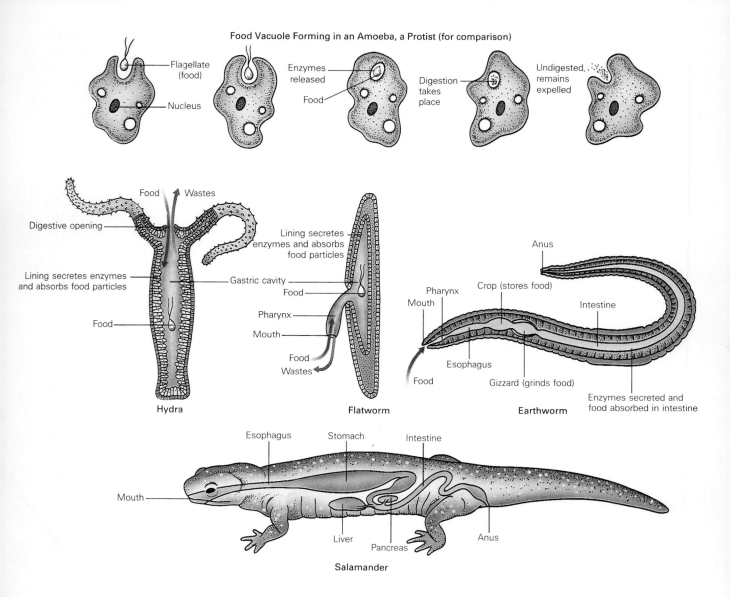

Food Vacuole Forming in an Amoeba, a Protist (for comparison)

Flagellate (food)

Nucleus

Enzymes released

Food

Digestion takes place

Undigested, remains expelled

Food   Wastes

Digestive opening

Lining secretes enzymes and absorbs food particles

Food

Hydra

Lining secretes enzymes and absorbs food particles

Gastric cavity

Food

Pharynx

Mouth

Food

Wastes

Flatworm

Anus

Pharynx

Mouth

Crop (stores food)

Intestine

Esophagus

Food

Gizzard (grinds food)

Enzymes secreted and food absorbed in intestine

Earthworm

Esophagus   Stomach   Intestine

Mouth

Liver   Pancreas   Anus

Salamander

FIGURE 19.12 Digestive systems in a variety of animals. Here, a protist, an amoeba, has an unspecialized arrangement, ingesting food through any part of its surface, simply ejecting waste through its membrane. The hydra and flatworm have sac-like digestive tracts, food exiting over the same route it entered. The earthworm has an essentially tubular system with a food-grinding gizzard to prepare the food for digestion. The tubular system of the salamander is embellished by a number of glands and organs that alter the food as it passes. Here, notice that, technically, food always remains outside the body until its digestive products pass through the intestinal membranes. Thus, "I would like to get on the outside of a hamburger," is a proper statement.

against small stones that have been ingested. There is now no reason for you to lie awake night after night wondering how an earthworm chews.

As another fascinating aside, earthworms have amoeboid cells in their bodies that ingest particles moving through the digestive tract as the worm literally eats its way through the soil. These cells then migrate through the body and come to rest just under the skin, giving the earthworm a color similar to the soil in which it lives and causing the early bird to come up short.

In vertebrates, such as the salamander, food enters a well-defined mouth where some basic processes of digestion may begin. The food is swallowed by voluntary action and then moves down the smooth-walled **esophagus** by a wave of involuntary muscular contractions of the tube. The contractions resemble a ring being slid along the tube. The process is called **peristalsis** and may appear along the entire gastrointestinal tract. The food then enters the stomach, where digestion continues.

From there it passes to the intestine, where it is further broken down by the action of intestinal secretions along with those from the pancreas and liver. When the food is finally digested, the products move through the intestinal wall into the blood vessels that line it. These nutrients are then carried by the blood to the tissues where they will be used. The indigestible particles move on through the intestine, to be eliminated through the **anus** or the **cloaca.** (The cloaca is a common opening for the intestine, kidneys, and reproductive organs. It is found in amphibians, reptiles, and birds—a primitive plumbing arrangement that has been "improved upon" in mammals.)

One of the most important chemical processes in digestion is hydrolysis. Hydrolysis, as mentioned in Chapter 4, involves breaking down a substrate by the chemical addition of water. In digestive processes, large molecules of proteins, fats, and carbohydrates are thus broken down into their simpler components, as we shall see shortly. These smaller molecules are then absorbed through the intestinal wall and enter the bloodstream.

The Human Digestive System

In the human digestive system (Figure 19.13), the breakdown of food particles begins in the mouth, where chewing breaks food apart and increases the surface area on which enzymes can act. Chewing also has another function in those animals that eat plants, since it ruptures the cellulose-laden cell walls that cannot be broken down by digestive enzymes. Since animal cells lack cell walls, most animal tissue *can* be entirely broken down by digestive enzymes. This is why cats don't have flattened molars with which to grind food, and why you never see them lying around chewing their cud. Humans, being omnivores ("all eaters") have teeth that are adapted to handling both types of food.

In the human mouth, the food is broken up and lubricated by saliva (this is not my favorite part of biology, either). Also, some chemical digestion begins there as starches are changed to disaccharides, or double sugars. The food is then swallowed and moves to the stomach by involuntary contractions of the smooth-walled esophagus, which also moistens the food by secretory glands in its lining.

The **stomach** is a muscular sac that churns the food as it secretes mucus, hydrochloric acid, and enzymes. The food is meanwhile sealed in the stomach by two sphincters, or rings of muscles, one at either end of the stomach. After the mixing is completed, the lower sphincter opens and the stomach begins to contract repeatedly, squeezing the food into the small intestine. A fatty meal, by the way, slows this process and makes us feel "full" longer. This is also why we're hungry again so soon after a low-fat Chinese dinner.

The **small intestine** is a long convoluted tube in which digestion is completed and through which most nutrient products enter the bloodstream. Its inner surface is covered with tiny, fingerlike projections called **villi,** which greatly increase the surface area of the intestinal lining. Furthermore, the surface area of each villus is increased by about 3000 tiny projections called **microvilli.** Within the villi are numerous minute lymph vessels, as well as tiny branches of the **mesenteric arteries** and **veins,** the blood vessels that surround much of the viscera. Many of the

FIGURE 19.13   The human digestive system. Once food enters the esophagus, it is moved along the digestive tract largely by involuntary muscular movements. Various enzymes and other juices are added along the way to break down the food molecules so that by the time they reach the small intestine, they can be absorbed into the bloodstream. Undigested or in-digestible matter then moves to the large intestine where it waits for the 7:05 out. The diges-tive process is curiously involved with emotion somehow, so when we are stressed, the stomach may overproduce acid that, when it moves back up the esophagus a short way, causes heartburn. In addition, anxiety may reduce the efficiency of digestion or may cause sphincter muscles to remain closed, stopping the movement of food through the tract.

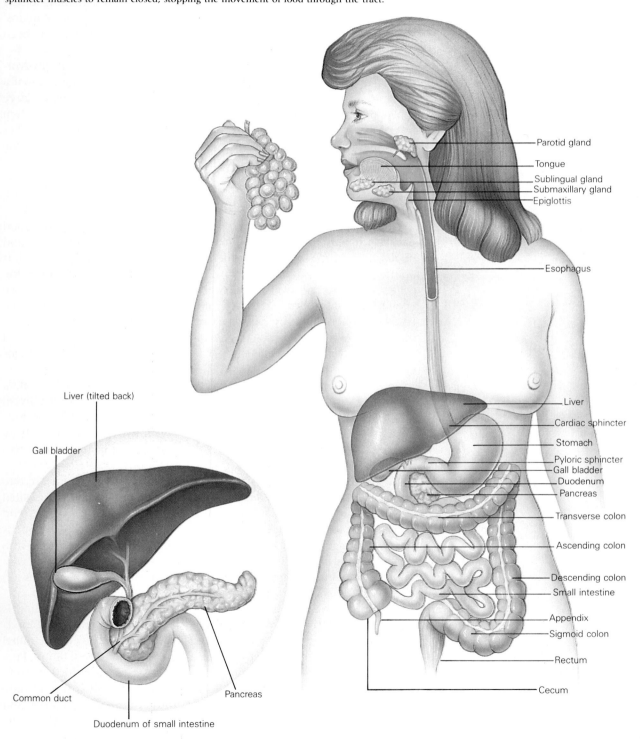

Parotid gland
Tongue
Sublingual gland
Submaxillary gland
Epiglottis

Esophagus

Liver (tilted back)

Gall bladder

Liver
Cardiac sphincter
Stomach
Pyloric sphincter
Gall bladder
Duodenum
Pancreas
Transverse colon
Ascending colon
Descending colon
Small intestine
Appendix
Sigmoid colon
Rectum
Cecum

Common duct

Pancreas

Duodenum of small intestine

products of digestion move directly into the lymph vessels and the capillaries that spring from the mesenteric arteries.

Fat absorption is a rather complex process. However, once the products of fat digestion enter the bloodstream, they are quickly restored as fats. In fact, this process happens so quickly that after a meal high in fat, the blood may take on a weird, milky appearance.

The first ten inches of the small intestine comprise the **duodenum.** The enzymes produced by the pancreas enter the gut in this area. These are powerful enzymes that break down proteins. One might wonder, then, why doesn't the pancreas digest itself? Largely because the enzymes are stored in an inactive form in the pancreas and are activated by the duodenal environment. Well then, the sharp-witted reader will ask, why isn't the duodenum digested? As a matter of fact, it would be if it were not "alive" with a membrane system that actively excludes harmful substances, such as digestive enzymes, from entering the cells. This is also why, in those unfortunate enough to be "wormy" (and many people are), when a worm dies, it is immediately digested—the body's ultimate vengeance. In addition, the intestine protects itself by secreting a protective shield of mucus, and the high level of sodium bicarbonate from the pancreas helps to neutralize the disruptive acids entering the intestine from the stomach. When the system fails or extreme amounts of acid are produced, perhaps due to emotional stress, ulcers (localized lesions of the digestive tract) can result.

Another substance secreted into the duodenum is **bile,** which is stored in the gall bladder after it is produced by the liver. Bile contains certain sodium salts that act as detergents, breaking up the large fat droplets and making them more water-soluble so that the pancreatic enzymes can work on them.

Once the products of digestion move into the tiny capillaries within the villi, the nutrient-laden blood is carried to the **hepatic portal vein,** which leads to the liver (a **portal** vessel is one that lies between two capillary beds). The blood then filters through tiny vessels and sinuses in the liver. If the blood contains excess carbohydrates, some are removed by the liver and stored as glycogen. If the blood is low in carbohydrates, stored glycogen is broken down into its glucose subunits, which are then released into the bloodstream. Thus, proper glucose levels are maintained in the blood despite variations in food intake. Such shifts in blood glucose levels can result in various changes, including behavioral ones. For example, as the body's reserves of glucose drop below the amount needed to maintain a constant blood level, one has the sensation of hunger and is motivated to go out and find some glucose (in a variety of forms from rabbits to apples).

As food passes from the duodenum through the next two sections of the small intestine, the **jejunum** and the **ileum,** further digestion and absorption occurs (as we see in Table 19.1). These sections compose most of the small intestine. The remaining matter, mostly composed of waste products, then moves on into the **large intestine.** The large intestine consists of two portions, the **colon** and the **rectum.** The colon has three distinctive areas, the **ascending, transverse,** and **descending** parts that lead to the rectum. The inner surface of the large intestine is rather smooth, and no digestion, or absorption of digested food, occurs here. However, it is a veritable hotbed of *E. coli* bacteria. The *E. coli* produce a

Table 19.1. Digestive enzymes and their functions

Sources and Enzyme(s)	Substrate	Product
Salivary glands		
Salivary amylase	Starch	Glucose
Stomach lining		
Pepsin	Protein	Polypeptides
Renin	Casein	Insoluble curd
Gastric lipase	Triglyceride	Fatty acids + glycerol
Pancreas		
Trypsin	Peptide linkage	Shorter polypeptides
Chymotrypsin	Peptide linkage	Shorter polypeptides
Ribonuclease	RNA	Nucleotides
Deoxyribonuclease	DNA	Deoxynucleotides
Pancreatic amylase	Starch	Glucose
Pancreatic lipase	Triglyceride	Fatty acids + glycerol
Carboxypeptidase	C-terminal bond	Shorter peptide and one free amino acid
Intestinal lining		
Aminopeptidase	N-terminal bond	Shorter peptide and one free amino acid
Tripeptidase	Tripeptide	Dipeptide + amino acid
Dipeptidase	Dipeptide	Two amino acids
Nuclease	Nucleotide	Pentose + nitrogen base
Maltase	Maltose	Two glucose units
Sucrase	Sucrose	Glucose + fructose
Lactase	Lactose	Glucose + galactose

waste that is high in certain vitamins. (The presence of these bacteria in the environment is an indication of fecal contamination, an important factor in the spread of some diseases, such as infectious hepatitis.) The large intestine also extracts water from the solid waste product, or **feces.** This is why constipation results if one puts off nature's call and why diarrhea robs the body of both water and nutrients. Finally, let us note that the feces are stored in the rectum before elimination through the anus.

We end our discussion on that poetic note. We will next consider the immune system, but we must keep in mind that although we are considering each system separately, each is but a fixture in a vast constellation of processes. Ultimately, the various systems must be considered in terms of their precise and balanced interactions if we are to begin to understand this complex phenomenon of life.

SUMMARY

With increasing size came a variety of interacting systems that met the demands of size. The respiratory system delivered oxygen to the cells and carried away carbon dioxide. Two primary exchange organs are gills and lungs, both with very large surface areas. Some species rely on neither, including insects, with their system of tracheae. In ventilation, or breathing in vertebrates, air passes through increasingly smaller tubes that end in blind pouches, or alveoli.

Circulatory systems move fluids through bodies, particularly blood. Smaller animals have less need for such systems since materials can move through them by diffusion. In larger animals, fluids are moved by pumps called hearts through systems that may be open (the blood not always enclosed in vessels) or closed (the blood virtually always enclosed in vessels). The blood carries oxygen, carbon dioxide, nutrients, and wastes. The heart's efficiency was increased by separating oxygenated and deoxygenated blood. Such systems culminate in the four-chambered heart. Heart rate is controlled by the medulla responding to several sensors throughout the body. In vertebrates, the lymphatic system moves blood plasma through ducts by muscular contraction. The lymphatic system is important to the body's defense against infection.

The digestive system breaks down food, primarily through hydrolysis, and allows the nutrient products to enter the bloodstream. In simpler animals, the gut may be saclike; in other animals it is tubular. In the human system, as in many other animals, the digestive tract has highly specialized regions (such as the mouth, esophagus, stomach, and intestine), and accessory glands and organs (such as the liver, gall bladder, and pancreas), all specialized to play particular roles in making food available to cells.

Chapter 20
THE IMMUNE SYSTEM

The earth is really not a very friendly place. You can get yourself killed here. And the dangers are of many stripes, some much more apparent than others. We occasionally hear a story about someone encountering travellers in an alien spaceship that has landed on our planet. If such beings did land here, their greatest concern need not be a trigger-happy farmer. Their greatest risks may be of a far subtler sort, such as an agonizing corrosion from our oxygen. But in addition to our deadly atmosphere, they might find themselves exposed to innumerable chemical and microbial agents, many of which are able to penetrate the bodies of living things and disrupt their delicate internal balances, bringing life to an end. Of course, we live in this deadly sea, and most of us, most of the time, are able to withstand the dangers. After all, we evolved on this planet, and our presence here attests to the fact that natural selection has endowed us with certain defenses.

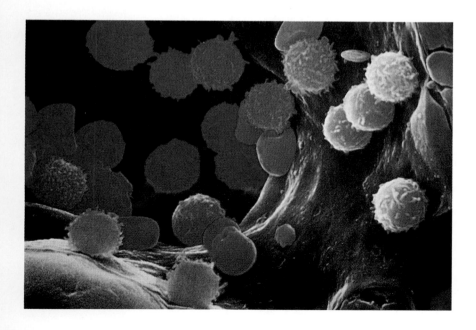

IMMUNITY AND THE CIRCULATORY SYSTEM

The sort of internal defense that we will consider here is loosely referred to as the **immune system.** The system is a battery of protective devices against such intruders as bacteria, viruses, and even the increasing list of agents that cause cancer. Some of the most important factors of our immune systems are associated with the circulatory system. Among these are the white blood cells, or leukocytes, two kinds of lymphocytes called T and B, and those called phagocytes, as we will see.

Leukocytes are produced in the same bone marrow tissues that produce erythrocytes, or red blood cells. However, unlike the red blood cells, the leukocytes retain their cell nuclei and mitochondria. Later, some of them will even divide and produce daughter cells like themselves, a trait not shared with the red cells. In general, leukocytes are larger than the red blood cells. Several types are shown in Figure 20.1.

There are far fewer leukocytes than there are red blood cells. Generally, there are about 5,000–9,000 leukocytes per mm^3 (cubic millimeter) of blood. Counts of over 10,000 per mm^3 indicate infection, and figures approaching 100,000 mm^3 indicate a severe condition, even a cancerous condition, such as leukemia.

Leukocytes are amoeboid, that is, they can crawl around like an amoeba. They have directional abilities that permit them to move toward an injury in response to certain hormones released at such places or to toxins released by the invaders. They can squeeze between the cells of capillary walls and out into the tissues as they roam throughout the body, tiny vigilantes, prowling in search of a violation (Figure 20.2).

FIGURE 20.1 The white blood cells (leukocytes) are far less numerous than red cells (marked *rbc* in the photo). Whereas there are about 5 million red blood cells per cubic millimeter in the adult, there are only 5,000–9,000 white cells. These are of five known types. Neutrophils and monocytes (marked *n* and *m*) are phagocytic, engulfing foreign matter, bacteria, and cellular debris from cells destroyed by infection. Lymphocytes (marked *l*) are undifferentiated; they can differentiate into other white cells and even into red cells, and under certain conditions they can produce antibodies. Eosinophils increase in number in the presence of foreign proteins, but no one knows why. Basophils are known to secrete serotonin (which constricts blood vessels in a damaged area), histamine (which causes damaged areas to swell with fluid), and heparin (an anticoagulant). Neutrophils and lymphocytes make up about 95 percent of the white cells.

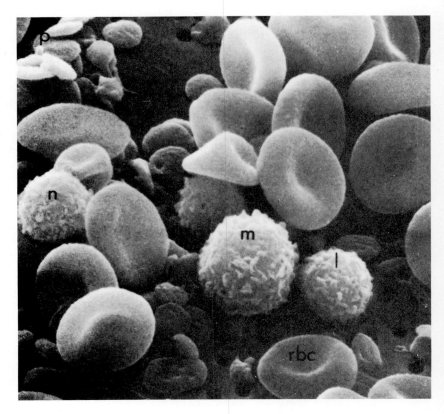

FIGURE 20.2 Some leukocytes roam the bloodstream and tissues of the body, attacking invading bacteria. Here streptococci (spherical bacteria, often occurring in rows) are being engulfed by a monocyte, a large amoeboid leukocyte.

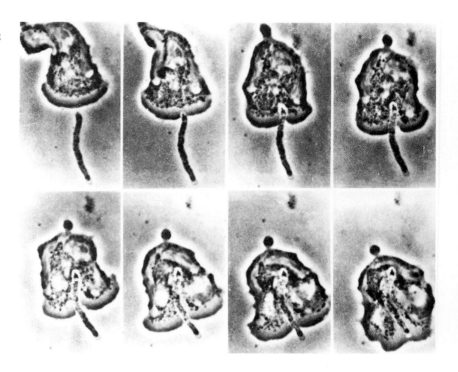

Each kind of leukocyte functions in its own way. Essentially, they fall into three groups, *eaters, helpers,* and *killers.*

The Eaters

The eaters are the amoeboid cells called **phagocytes** (eating cells), which engulf foreign particles and destroy them. The larger phagocytes are called **macrophages** (a monocyte). Whereas phagocytes can wander anywhere throughout the body, they tend to accumulate at the site of an infection or wound. If bacteria are numerous at the site, the phagocytes can literally eat themselves to death by ingesting so many bacteria that the accumulation of bacterial poisons kills them. The accumulation of dead phagocytes forms **pus.** (Lunch, anyone?)

Phagocytes have several roles in the healthy body. They not only engulf and digest bacteria and viruses, but they also dispose of certain kinds of wastes. They will eat dead cells, worn-out connective tissue, old red blood cells, and cellular debris. They also signal the immune system to swing into action.

The Helpers

The helper T-cells are a remarkable group. There are literally millions of kinds, each carrying, embedded in its membrane, a mirror image of some sort of antigen. (Some fit antigens that are likely never to be encountered, some having been manufactured only recently in the laboratory.) No one knows how the body comes to produce so many kinds of helper T-cells.

The Killers

The killer cells are a fascinating lot. There are two main types, the **natural killer cells** and the **cytotoxic killer cells.** (Together, they are sometimes referred to as killer T-cells since they are localized in the thymus gland.) The natural killer cells roam the body, touching other cells along the way, checking the identity of each and then usually moving on. They behave much as security guards frisking everyone at a rock concert. Normal cells are allowed to continue on, but cancerous cells, or those infected with a virus, are immediately attacked. The battle, we know, is not always won, and we may fall ill.

The cytotoxic killer cells (*cyto,* cell; *toxic,* poison), like the natural killer cells, do not tolerate cancerous or infected cells. They immediately attack them. However, these cells must "learn" which cells to kill. They acquire the information through a remarkable process as new antibodies are formed. Let's have a look at antibodies in general, and then see how the cytotoxic killer cells get their instructions.

ANTIBODIES: THE SILENT WEAPONS

Antibodies are globular proteins produced by lymphocytes. They are found in the bloodstream and are among the body's most important defenses. They are produced in response to **antigens** (foreign substances), such as bacteria or viruses. An antigen, then, is something that stimulates the production of an antibody, and an antibody is something produced in response to an antigen. (Some of our more profound thinkers have suggested a certain circularity here. Nonetheless, these definitions will suffice for now.) Antibodies tend to remain in the bloodstream for long periods once they are produced, either inactivating the antigen or actually destroying it, as we shall see. Various cells have their own specific proteins embedded in their membranes and are recognized by other cells on the basis of these proteins. Thus, an introduced cell or tissue can be recognized as "foreign" and initiate the production of antibodies. (This is a basic problem in organ transplants.)

The most familiar antibodies are called **immunoglobins** (Figure 20.3). Each immunoglobin molecule is comprised of four polypeptide chains—two identical light chains and two identical heavy chains. The four chains are bound together by disulfide (S-S) linkages, and together they form a general molecular "stalk."

The stalks of many immunoglobins are much the same, and, thus, they have little "specificity" when responding to antigens. In certain types, however, the stalk divides at one end, forming a Y. It is in these separated areas that the immunoglobins take on very specific characteristics and differ from each other in critical ways. These differences arise as the Y-regions react to various antigens, taking on forms that "fit" the molecular structures of the antigens. These variable parts are called the **antigen recognition regions** (Figure 20.4).

Once formed, some immunoglobins are released into the blood plasma while others remain embedded in the surface of the lymphocytes that produced them. No matter where they exist, however, when an antibody encounters its antigen, it binds to it, the two fitting each other like a lock and key. Furthermore, the resulting complex becomes essentially

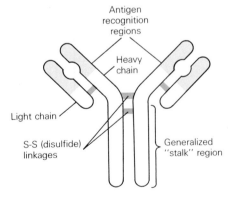

Antigen recognition regions

Heavy chain

Light chain

S-S (disulfide) linkages

Generalized "stalk" region

FIGURE 20.3 Immunoglobins are proteinaceous antibodies that consist of two light and two heavy polypeptide chains, connected by disulfide linkages. Each chain has two general regions common to a number of antibodies and two highly specific antigen recognition regions that bind only to specific antigens.

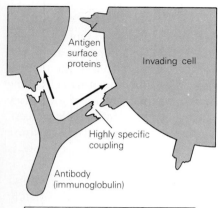

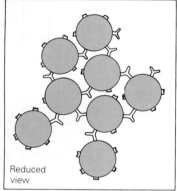

FIGURE 20.4 When a specific antigen is encountered, the recognition regions of the antibody molecules attach to specific binding sites on the antigen, eventually forming an immobile mass that can be engulfed by phagocytes. In other cases, the antibody may simply destroy the antigen.

"sticky" and may join others. As they join, they form unwieldy clumps. The resulting large, immobile mass is then easily devoured by phagocytes.

Identifying the Enemy

Now let's see what happens after a macrophage has engulfed and ingested some foreign protein, cell, or virus. Essentially, what we find is that as soon as digestion is nearly completed, the remaining bits and pieces become embedded in the membrane of the macrophage. The stage is now set for the next role of this large white blood cell.

The macrophage, with its embedded particles, then begins to move about through the body. It wanders tirelessly, touching surfaces with any **lymphocyte** (undifferentiated small, white blood cell that can take any of several developmental routes) it encounters. Thus, the foreign substances embedded in its membrane are presented to countless numbers of lymphocytes. Usually there is no response. But eventually—after perhaps thousands of uneventful encounters—the macrophage may find a particular kind of lymphocyte called a **helper T-cell** whose surface antibodies match the structure of the captured foreign molecule.

There are millions of lymphocytes in the body and, perhaps, millions of kinds of molecules on their membranes. Thus, an encounter between a macrophage and a helper T-cell that bears a matching membrane is entirely a chance event. However, when such a match does occur, the lymphocyte begins to divide. Each daughter cell divides in turn, and within 10 days or so, the lymphocyte with the matching molecule will have produced literally thousands of descendants, all making that same antibody (Figure 20.5). Some helper T-cells will rush to the spleen and lymph nodes and cause the production of killer cells and B-cells. Some of the killer cells, the cytotoxic killer cells, will wear the antibody on their surfaces, destroying any invader bearing a matching antigen. (This is sometimes called cell-mediated immunity.) In this manner, then, the cytotoxic killer cells get their instructions. The B-cells will furiously produce antibodies and release them into the bloodstream (antibody-mediated immunity). These antibodies can attach to the antigenic proteins that coat the invading cells, making them "sticky," as was just mentioned, so that they will clump together, making it easier for the phagocytes to devour them. (As the battle is won, the immune process is halted partly by the production of a third kind of T-cell, called the **suppressor T-cell**.)

The developmental routes of T-cells and B-cells are quite different, although their origins are the same. Both arise in special cells of the bone marrow, from which they are released into the bloodstream. The T-cells are not yet functional. They must first migrate to the thymus gland (hence their name) where they mature. Here, their cell surfaces change in a way that prepares them for their special role. When a T-cell is finally mature, it leaves the thymus and begins roaming the bloodstream, where by chance it could meet a macrophage cell and enter the immune process.

The development of B-cells is a bit more mysterious. They, too, are formed in the bone marrow, but no one knows where they mature or how the process takes place. Furthermore, they, too, are preprogrammed to react only to certain antigens, but how they learn to recognize

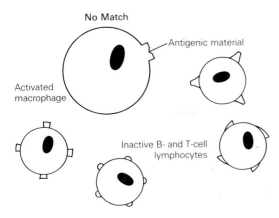

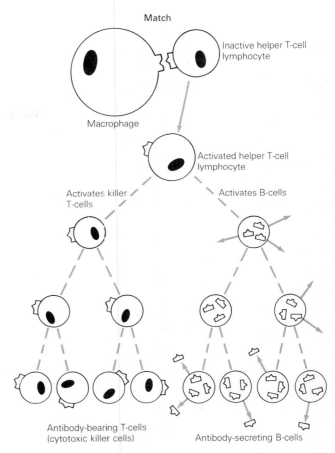

FIGURE 20.5 In the ongoing combat against invaders, macrophages bearing captured cell-surface antigens seek out lymphocytes (helper T-cells) with matching membrane surfaces. Once a match is made, the macrophage induces the lymphocyte to begin to reproduce its specific line. These cells then trigger the production of cytotoxic killer cells or B-cells, which secrete antibodies.

these antigens remains unknown. However, it is known that in chickens, the B-cells mature in the Bursa of Fabricius, a lymphoid organ for which the B-cell was named.

We don't know how many specific types of lymphocytes exist, or even *why* they exist. After all, the human body can even make antibodies against, say, crocodile blood. Why would we have evolved antibodies to fight crocodile blood? However it is done, the immune system somehow covers all bets in its preparation for trouble.

Attack Against Nonself

If the body must mobilize against an attack, certain changes occur that cause that unmistakable feeling of illness. This all starts when an antibody binds to an antigen, the resulting complex interacting with a group of proteins in the plasma known as the **complement.** When the antigen/antibody complex joins with the proteins of a complement, those proteins begin a series of reactions, some of which cause the feeling we associate with illness. When you catch a cold, the complement stimulates the secretion of **histamine,** dilating blood vessels, increasing local blood flow, and flooding the affected area with white blood cells and oxygen. The histamine also causes redness, swelling of mucous membranes, and an increase in the flow of mucus. The complement may also cause the release of certain hormones that set even more white blood cells into action.

The complement also induces molecules called **pyrogens** to increase body heat, bringing on fever. By the way, it is sometimes a mistake to reduce a low-grade fever by taking aspirin, because most harmful bacteria and viruses are inactivated or killed by even a slight increase in body temperature.

Then the immune response itself begins. The immune response involves two distinct steps. The **primary immune response** occurs when a foreign "nonself" substance (or invader) is encountered for the very first time. The reaction normally is a bit slow, but usually the invader is finally overcome. Should this sort of invader appear again, the response will be astonishingly swift because of the efficient **secondary immune response.**

The primary immune response is a little slow because it depends on the macrophage first finding a few specific lymphocytes that must then build an army. Of course, while the army is mobilizing, the invaders may be on a rampage. In time, however, the primary response is usually successful, and the invader is defeated. Once the battle is over, the lymphocytes decrease in number as they begin to die off since they are no longer needed. However, when the helper T-cells and killer cells were multiplying in response to the invasion, some of the cells produced were programmed for long lives. These cells and their descendants, then, continue to roam the body in greater numbers than before, no matter how inconspicuously. Furthermore, because of their numbers and their "sensitization," they are primed and ready for the next invasion. They will quickly be activated the next time that same type (or a very similar type) of invader appears, and they are likely to knock out the invader almost immediately.

This rapid reaction, then, is the secondary immune response. It is the reason you don't get measles more than once. And if you've been vaccinated for measles, you may not get the disease at all. This is because being vaccinated with a harmless form of the measles virus triggers a primary response, so if a real invader follows later, it will meet the secondary response. (For some vaccines, booster shots are given just to jog the memory of the lymphocytes and keep the numbers of the antibodies relatively high.)

Attack Against Self

Leukocytes have been called the body's "police force." In some ways, the metaphor is distressingly apt. We do, indeed, need police officers for protection, but we also know that on rare occasions, they may overreact and harm the very group they normally protect. Similarly, our protective immune system sometimes may go awry—overreacting not only to dangerous (nonself) foreign antigens, but to harmless nonself antigens such as pollen (resulting in hay fever), as well. More seriously, leukocytes may misidentify the body's own tissues. In such cases, normal body tissues are erroneously identified as foreign and are attacked by the misguided leukocytes.

These attacks can produce a number of autoimmune (self) diseases with rather esoteric-sounding names, such as rheumatoid arthritis, systemic lupus erythematosus, pernicious anemia, and thyroiditis. Researchers are also taking a second look at familiar illnesses such as certain kinds of diabetes, Addison's disease, and myaesthenia gravis, each of which may be linked to autoimmune problems.

INTERFERON, A NEW PROMISE

The term **interferon** comes from **interference phenomenon,** which refers to a group of antiviral substances manufactured by the cells of most vertebrates in response to viral attack. Interferon causes cells to become resistant to attacks by other viruses. As we saw in Chapter 11, an attacking virus tends to alter a cell's own replicating mechanisms, using them to make, instead, more viruses that can infect other cells. Interferon helps to block this deadly geometric increase. Interferon does not act against specific viruses, but will inhibit any viral attack. Interferon from one species, however, cannot increase resistance to viruses in another species.

When it recently became known that interferon could be manufactured by recombinant techniques, hopes in the medical community soared. Interferon seemed to be the answer to everything from cancer to the common cold. But the promises were apparently premature; interferon was simply not the magical cure-all we hoped. However, research is progressing, and there are promising signs. In one experiment, not one of eleven volunteers given interferon in a nasal spray caught cold after being exposed to cold viruses, while in control groups, eight of eleven people exposed and given plain water spray did catch cold. Interferon has also reduced tumor size in a number of patients who did not respond to other treatment. In one case, two of three separate cancers discovered in one man completely disappeared after treatment with interferon.

Interferon does have side effects. In some people, it triggers cardiac arrhythmias (irregular heartbeats). It may also complicate liver or kidney problems. In high doses, interferon can cause confusion, change brain waves, and bring on seizures. Nonetheless, it is still considered a potentially useful substance, based on what we know now, and may become a superb form of treatment for some ills, when we learn more about it.

AIDS: A DEVASTATING NEW PROBLEM

Not long ago, an accused murderer was led into a courtroom by a sheriff's deputy who was wearing rubber gloves. The jurors facing the accused did not include the fourteen people who were asked to be excused because of the medical condition of the accused. The unusual circumstances arose because the defendant was guilty of having AIDS.

The fear of AIDS now probably surpasses the fear of flying. People with the disease, or those in high-risk groups, are often shunned, even by those in the health services community. The argument regarding the contagion of the disease seems to be more vigorous outside the medical community, however, because most researchers in the area seem to agree that the disease is not particularly contagious *if* certain simple safeguards are taken. But what is AIDS? What is the problem? And what are the safeguards?

AIDS is an acronym for **acquired immune deficiency syndrome.** Essentially, it acts by suppressing the victim's immune system. People with AIDS are susceptible to virtually any disease. In fact, the appearance of rare diseases such as Kaposi's sarcoma (a skin cancer) and pneumocystic pneumonia frequently occur with the disease. Early signs of

The reason the AIDS virus is so deadly is that it can hide within, and then kill, the one lymphocyte critical to initiating an immune response, the helper T-cell. Like soldiers within a Trojan horse, the AIDS virus usually enters the body concealed within a helper T-cell in the blood or semen of an infected host. The defending T-cells immediately detect the infected T-cells, but when the defender moves toward the infected cell, the virus slips into the defender and immobilizes it before it can summon the killer cells and the B-cells.

The AIDS virus is a retrovirus, which means that instead of storing its genetic program in DNA, as most viruses do, it uses RNA. And though the AIDS virus is frail outside the body, it's remarkably immune to drugs once it's inside. The

virus may remain within its inactivated helper T-cell for perhaps years until some unrelated infection triggers the T-cell to divide. As it begins to divide, the viral DNA instructs the T-cell's machinery to make copies of the AIDS virus, which then bud from the T-cell's surface, killing the cell and spreading copies of the virus through the body and invading new cells until the body loses its army of tiny sentinels. Phagocytes, killer cells, and B-cells remain ready and able to defend, but they are not alerted, and the invaders run free in a sickening body.

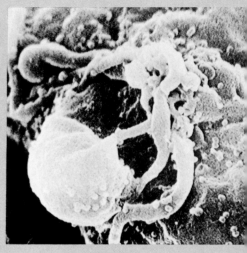

The AIDS virus.

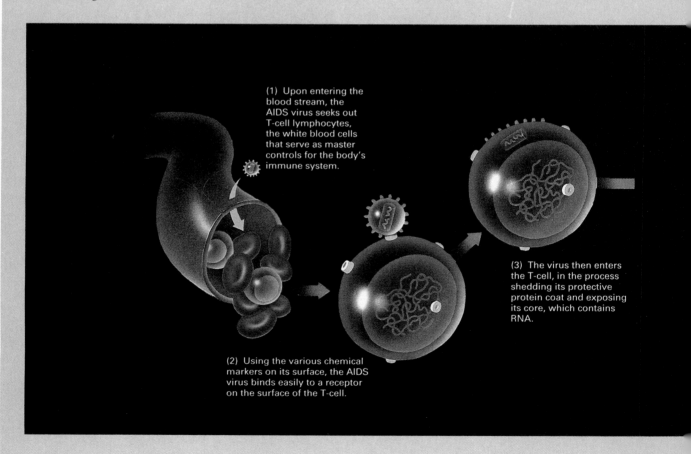

(1) Upon entering the blood stream, the AIDS virus seeks out T-cell lymphocytes, the white blood cells that serve as master controls for the body's immune system.

(2) Using the various chemical markers on its surface, the AIDS virus binds easily to a receptor on the surface of the T-cell.

(3) The virus then enters the T-cell, in the process shedding its protective protein coat and exposing its core, which contains RNA.

The Micro-anatomy of an AIDS Virus

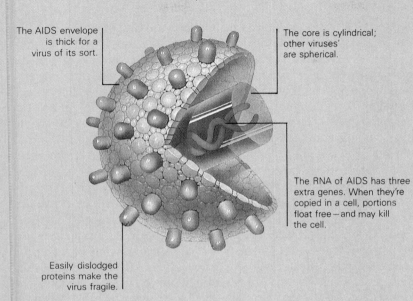

The AIDS envelope is thick for a virus of its sort.

The core is cylindrical; other viruses' are spherical.

The RNA of AIDS has three extra genes. When they're copied in a cell, portions float free—and may kill the cell.

Easily dislodged proteins make the virus fragile.

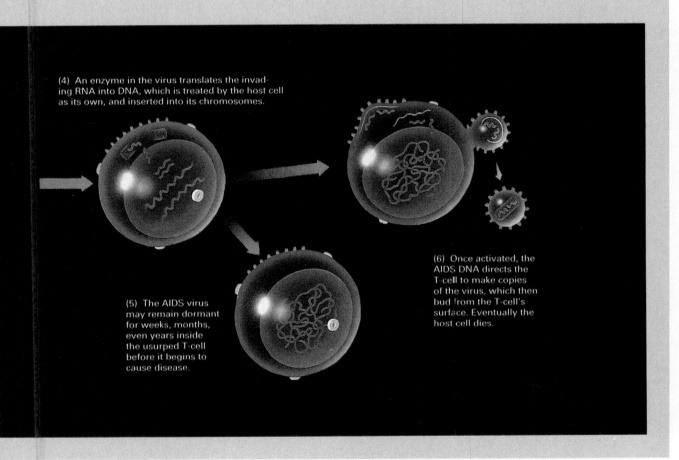

(4) An enzyme in the virus translates the invading RNA into DNA, which is treated by the host cell as its own, and inserted into its chromosomes.

(5) The AIDS virus may remain dormant for weeks, months, even years inside the usurped T-cell before it begins to cause disease.

(6) Once activated, the AIDS DNA directs the T-cell to make copies of the virus, which then bud from the T-cell's surface. Eventually the host cell dies.

AIDS include a series of lingering, simple colds, "night sweats," persistent fever, swollen glands, and coughing. (Immediately upon learning this, of course, everyone detects just those symptoms in themselves.) More serious conditions follow, including at least three forms of cancer and destruction of the lungs and brain.

By some accounts, the first case of AIDS in the United States appeared in 1979, followed by a half-dozen cases reported in Los Angeles in 1981. By early 1986, the number of AIDS victims had passed 15,000 in the U.S., with the number of known victims doubling every nine months. Many people carry antibodies to AIDS, showing that they have been exposed to it, and some may carry the disease in its early stages without developing the symptoms. It is feared that such people may be able to transmit the disease, nonetheless.

The disease, once full-blown, is believed to be incurable and to virtually always cause death within a few years (fewer than fourteen percent of victims survive past three years). Because so much is unknown about the disease, much of what is known is misconstrued, often by sensationalist media. Who, then, is at risk, and how does the disease progress?

The disease is largely confined to homosexual men and intravenous drug abusers. The agent, a virus (see Essay 20.1), is transmitted in the blood and semen. The primary means of contagion is believed to be anal intercourse, when the delicate tissues of the bowel are likely to be injured, allowing the virus to enter the blood through broken vessels. The drug abusers may pass the virus along by sharing needles. (There have been cases of people contracting the disease through medical blood transfusions, but careful screening and processing of blood is reducing this risk.) The risk of contracting the virus through heterosexual vaginal intercourse is extremely low but on the rise.

The geographical source of AIDS is not entirely established, but some researchers believe the virus is a mutant of a strain that infects the African green monkey, a species that lives in close contact with humans in West Africa. The disease is widespread in certain areas there. Certain French-speaking and AIDS-ridden nations of West Africa have developed exchange programs with Haiti, a favorite vacation area of American homosexuals, and the disease may have spread to America in this way. One problem with tracing the movement, sources, and modes of transmission of the disease is that people tend to be less than honest about their sexual behavior. For example, almost all the hundred or so AIDS victims in the American military claim to have contracted the disease from prostitutes. Of course to say otherwise would be grounds for prosecution and discharge. At present, the American Centers for Disease Control maintain that the disease is not likely to sweep through the general population, but is likely to continue to be transmitted through its present means. In West Africa, the disease affects men and women in roughly equal numbers, but it has been pointed out that heterosexual anal intercourse is common there (often after ritual clitorectomy).

The problem of AIDS all too clearly illustrates the challenges presented to our immune systems. Not only must our own bodies fend off the usual agents that have attacked our delicate systems over evolutionary time, but they must stand ready to meet the new challenges that can be expected in the changing world of living things.

Most of us wish to live long and well. But that implies a continuing existence on an essentially hostile planet. Furthermore, that hostility may be increasing, largely because of our own behavior. Not only are our increasing numbers on the planet threatening our individual access to resources that contribute to good health, but each additional person might be considered a potential reservoir for some mutant threat. Crowding, of course, goes hand-in-hand with contagion, and we grow more crowded daily. In addition, we must rely on new technologies to help us solve our immediate problems, and the earth is becoming permeated with technology's by-products. We are forced to stand against a tide of chemical agents that are totally new to the environment, and that tide rises daily. Indeed, our immune systems, our ability to withstand, may soon be tested in ways we can now only imagine.

SUMMARY

Living things have developed complex systems that confer immunity against environmental hazards, including pathogenic bacteria and viruses. Among the most important such defenses are leukocytes, or white blood cells, that attack invading life forms. Leukocytes are of two basic types: phagocytes (including large macrophages) and lymphocytes (specifically, T-cells and B-cells). There are two kinds of T-cells—helper T-cells and killer T-cells. When a macrophage devours an invader, parts of the invader become incorporated into the macrophage's membrane. The macrophage then presents its trophies to countless lymphocytes until it encounters a helper T-cell with a matching configuration in its own membrane. The helper T-cell then divides rapidly and triggers the production of killer T-cells and B-cells in the lymph nodes and spleen. The killer T-cells attack the invaders bearing that particular membranal insignia and the B-cells produce antibodies against that invader. The most familiar antibodies are immunoglobins, composed of four polypeptide chains, some of which diverge at one end to form variable areas called antigen recognition regions that give the immunoglobins their specificity. Some immunoglobins are released into the bloodstream while others remain embedded in the surface of the lymphocytes that produced them. Antibodies bind to antigens and cause them to clump in such a way that the phagocytes can devour them.

The immune response has two steps: the slow, primary immune response and the rapid, secondary immune response. The primary response involves the antigen-bearing macrophage finding its matching antibody-bearing lymphocyte. The secondary response occurs after the lymphocyte has been triggered to reproduce many descendants, some of which are very long-lived, so that the next time a macrophage has detected and attacked that same kind of invader again, it will quickly encounter a matching lymphocyte. Leukocytes can sometimes mistake the body's own materials for harmful substances, and they can overreact to harmless foreign substances.

The cells of most vertebrates can produce an antiviral substance called interferon that blocks the virus's replicating mechanism. One of the most deadly new diseases is AIDS, which is largely confined to male homosexuals and intravenous drug users, but which can spread into the general population.

Chapter 21

HORMONES
AND NERVES

If life can be thought of as a symphony, then the body itself must be the orchestra. The orchestra is a physical thing, composed of many parts, each contributing to the music of the others, and all indispensable. Each part, however, must act in concert with the rest; each must contribute at just the right time and for the proper duration to produce symphonic harmony.

This coordination is possible because of two major means of communication: chemical messengers called hormones and electrochemical devices called nerves.

HORMONAL REGULATION

A **hormone** is a chemical that is produced in one part of the body and carried by the blood to another part of the body where it ultimately influences some process or activity there. Hormones are produced by **endocrine** (ductless) **glands** that are found in various parts of the body. (Since hormones are released directly into the bloodstream, there is no need for ducts.)

Hormonal regulation is common throughout the animal world, in invertebrates as well as vertebrates. In fact, a great deal of hormone research has been carried out on arthropods, such as insects and crayfish. For example, hormones have been found to regulate the change in the body organization of insects as they pass through their larval and pupal stages. The importance of hormones in insect development is illustrated in Figure 21.1.

An enormous amount of research attention has been focused on hormones, but the work can be quite difficult. One of the greatest problems in doing hormone research is acquiring the hormones in the first place. Animal bodies contain very little hormone substances because only tiny amounts are needed to cause great changes. For example, a woman produces only about a teaspoonful of the female hormones called estrogens in an entire lifetime. A source of the estrogen called estradiol has been pig ovaries. However, more than *two tons* of pig ovaries are required to extract only a few *milligrams* of the hormone. (It may have occurred to you that perhaps recombinant DNA techniques can play a role here.)

Hormones, or *chemical messengers,* as they are sometimes called, basically function in regulating various bodily processes. Some keep those processes within their proper daily limits, as does the antidiuretic hormone associated with urine formation (Chapter 17), while others may cause permanent changes, such as growth or sexual maturation. They may also function in emergency situations by causing the body to prepare for stress. And they may help adapt the organism to its environment in other ways, as when amphibians or reptiles change colors to match their surroundings. Hormones also play a part in nature's chemical war-

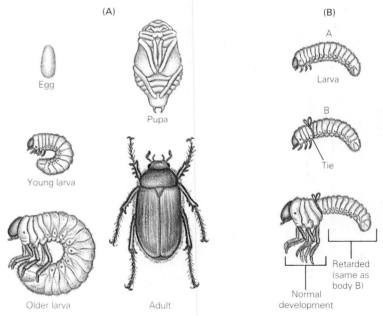

FIGURE 21.1 (A) Development in a beetle. Not all insect species go through the sequence of egg, larva, pupa, adult, but whatever their pattern, each stage is brought on at a specific time by a different hormone.
(B) An experiment showing the importance of hormones in insect development. When growth hormones produced in the head are kept from reaching other parts of the body by tying off those parts, the hormone-deprived areas do not mature.

(A)

Egg

Young larva

Older larva

Pupa

Adult

(B)

A

Larva

B

Tie

Normal development

Retarded (same as body B)

FIGURE 21.2 The major endocrine glands in the human. Because of the immense effect of small amounts of secretions from these glands and because they often interact in complex ways, their release must be very closely regulated. The pituitary is about the size and shape of a bean and is located in the geometric center of the skull. It is the source of several hormones, some of which affect other glands, but it, in turn, may be affected by them. You should be aware of the role of each of these glands and the results of their malfunction. Hormones are formed in one part of the body and operate in another, usually after being carried there by the bloodstream. Some diseases can be diagnosed by variations in the amounts of certain hormones in the blood. In addition, some hormone levels may be changed by experience. For example, there is preliminary evidence that dominance in males causes a rise in testosterone levels (but it might be argued that the cause-effect relationship here is reversed).

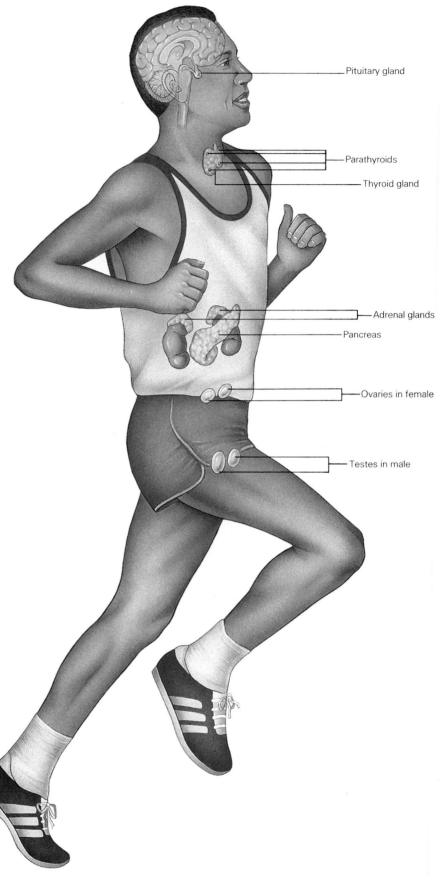

Pituitary gland

Parathyroids

Thyroid gland

Adrenal glands

Pancreas

Ovaries in female

Testes in male

fare. Some plants, for example, manufacture substances that mimic the growth hormones of insects, and thus they disrupt the normal growth of insect larvae that feed on them. Perhaps the greatest research attention, though, has been focused on our own species, and it is here that we will begin our closer look at these remarkable chemicals.

THE HUMAN ENDOCRINE SYSTEM

That delicate balancing act called homeostasis is strongly reliant on the constellation of endocrine glands as their powerful secretions ebb and flow through our sensitive bodies. We will consider each gland separately, but we must keep in mind that any is meaningless alone. Their interaction is critical and life's picture is not complete if any of these stars should fade.

In the human endocrine system (Figure 21.2) the *pituitary* has been called the "master gland," but this flattery is misleading, since it is regulated by some of the glands it influences and since many of its activities are also regulated by a part of the brain called the **hypothalamus.** The pituitary does have a number of important functions, however, and it indeed does affect some other glands. Structurally, it is composed of two major parts, the **anterior** and **posterior lobes,** with a central **midlobe** area. Embryologically, the posterior lobe develops from nerve tissue and the anterior lobe from an outpocketing of the embryonic mouth. (What germ layers are involved?)

The pituitary is located at the geometric center of the skull, where it lies shielded by heavy layers of surrounding bones. If you point one finger directly between your eyes and stick the other in your ear, you'll point right to it, as well as gain the attention of other people on the bus. The pituitary is only about the size of a bean, but its size belies its physiological complexity. In all, the pituitary secretes nine known hormones. The anterior pituitary alone secretes seven hormones, six of which influence other endocrine glands. The seventh stimulates body growth.

The growth hormone is called **somatotropin,** and it acts by stimulating bone and muscle growth. If too little somatotropin is present during childhood, the result is a pituitary dwarf. Too much somatotropin results in a giant. Excessive somatotropin in an adult increases the size of only those bones that can still respond to the hormone—the jaw and the bones in the hands and feet. This condition is called **acromegaly** (Figure 21.3).

When the **thyroid** gland is stimulated by the pituitary, it produces **thyroxine,** the hormone that controls the body's metabolic rate. Deficiency of thyroxine in childhood results in a characteristic physical appearance and mental retardation called **cretinism** (because the condition was originally associated with the population of the Isle of Crete). It is now possible to recognize thyroid deficiency early in life, so that thyroxine can be administered artificially to promote normal development. Excess thyroxine causes hyperactivity and low body weight. The thyroid also produces **calcitonin,** which inhibits the release of calcium from bone into the bloodstream. It operates with the **parathyroid hormone** to regulate levels of calcium ions in the bloodstream. The **parathyroid glands** are the smallest endocrine glands, located behind or within the thyroid. They produce the parathyroid hormone that increases the

FIGURE 21.3 In 1914 these were the tallest and shortest women in Europe (top). The giant is seven feet four inches tall and weighs 294 pounds. She is sixteen years old. The midget is eighteen, is twenty-five inches tall, and weighs thirteen pounds. Both conditions arise from abnormal pituitary secretions. Sudden increases in the production of growth hormone after maturity produces growth in only certain body parts, such as those of the face and hands (bottom). The condition is called acromegaly.

absorption of calcium ions from the intestine and reduces its excretion from the kidneys. It can also help withdraw calcium from the bones. It thus raises the calcium concentration in the blood. Calcium ions are important in muscle contraction and nerve conduction.

The testes and the ovaries predominantly produce **testosterone** and **estrogen,** respectively. Both hormones are responsible for those secondary sex characteristics we all love and admire. In males, testosterone causes deepening of the voice, broadened shoulders, hairy chests, facial hair, muscular development, sperm production, and increased production of red blood cells (thus, "red-blooded he-man"?). In females, estrogen causes widening of the hips, changes in the distribution of body fat, production of body hair, and development of the breasts. **Progesterone,** produced by the corpus luteum (the secretory "yellow body" of the ovary), stimulates the development of the uterine lining and causes further breast development during pregnancy.

Ad- means "upon" and *renal* refers to the kidney, so guess where the **adrenal glands** are located. The adrenals have two distinct parts, the outer **cortex** and the inner **medulla.** About fifty steroids have been isolated from the adrenal cortex of various mammals, and it is believed that some of these may be used in the production of hormones, but it is not yet known just what role, if any, these steroids play. One hormone produced in the cortex is **cortisone** (manufactured when the cortex is stimulated by adrenal corticotropic hormone (ACTH), from the anterior pituitary). Cortisone is well known as an agent that reduces inflammation. **Aldosterone** is also produced in the adrenal cortex and functions in the regulation of salt and water in the body. It works in the cells of the kidneys' distal convoluted tubules where it increases their rate of sodium reabsorption. (When sodium is reabsorbed, you may recall, water must follow because of the osmotic gradient that is set up, so aldosterone helps the body recover water that is moving through the kidney.) The adrenal cortex also produces some male sex hormones. This function is greatly diminished in women, but if it should be prompted, for example, by a tumor, it will cause such characteristic masculine qualities as beard growth.

The adrenal medulla secretes the hormones **norepinephrine** and **epinephrine** (sometimes called **adrenaline**), as we will see shortly. Other endocrine glands such as the parathyroids, the hormones they produce, and their principal effects are listed in Table 21.1.

How Hormones Work

Actually, we don't know how hormones work. Although we have known the composition of some hormones for a long time, we still aren't sure how any of them function. One hypothesis, however, is that when hormones come in contact with the membranes of "target cells," they alter certain sites on the membranes of these cells and trigger the production of a group of substances called **prostaglandins** within the membrane. Prostaglandins are involved in the production of the enzyme **adenyl cyclase,** which helps to change ATP to a form of adenosine monophosphate, or **cyclic AMP,** in the cytoplasm. The cyclic AMP may act by altering enzyme activity (think of the implications) or by stimulating the production of certain types of *m*RNA that would cause the production of

TABLE 21.1 Major Vertebrate Endocrine Glands and Their Hormones

Gland	Hormone	Principal Action
Anterior lobe of pituitary	Thyrotropic hormone (TH)	Stimulates thyroid
	Follicle-stimulating hormone (FSH)	Stimulates ovarian follicle
	Luteinizing hormone (LH)	Stimulates testes in male, corpus luteum in female
	Growth hormone (somatotropin)	Stimulates growth of bones and muscles
	Adrenocorticotropic hormone (ACTH)	Stimulates adrenal cortex
	Prolactin	Stimulates secretion of milk, parental behavior (such as nest building in birds and fish)
	Interstitial cell-stimulating hormone	Stimulates testosterone production
Midlobe of pituitary	Melanocyte-stimulating hormone	Regulation color of skin in reptiles and amphibians
Posterior lobe of pituitary	Oxytocin	Stimulates contractions of uterine muscle, milk release
	ADH or vasopressin	Increases water absorption
Thyroid	Thyroxine	Increases metabolism
	Calcitonin	Inhibits the release of calcium from bone
Parathyroid	Parathyroid hormone	Controls calcium metabolism
Testes	Testosterone	Stimulates production of sperm and secondary sex characteristics
Ovary, follicle	Estrogens	Stimulates female secondary sex characteristics
Ovary, corpus luteum	Progesterone	Stimulates growth of uterus
Adrenal medulla	Epinephrine (adrenalin), norepinephrine	Activates sympathetic nervous system
Adrenal cortex	Cortisone, cortisone-like hormones, aldosterone	Controls carbohydrate metabolism, salt, water, and sugar
Pancreas	Glucagon	Stimulates breakdown of glycogen into glucose
	Insulin	Lowers blood sugar levels, increase formation and storage of glycogen

certain enzymes. Enzymes, we know, direct the behavior of cells. Changes such as these could theoretically produce all the effects that are attributed to hormone action.

Feedback Systems

Hormone production is regulated, in part, through a negative feedback system in which the product of a process slows the process. As an example, consider the interaction of the tropic hormones and their target endocrine glands. The anterior pituitary gland secretes a thyroid-stimulating hormone (TSH) that causes the thyroid gland to secrete the hormone thyroxine. As the thyroid responds and the thyroxine level rises in the blood, the thyroxine itself suppresses the secretion of TSH, which lowers the level of thyroxine. As a result, more TSH is released. The levels of both hormones are thus kept within tight limits because of the influence they exert on each other. Such feedback mechanisms are common throughout the endocrine system and furnish some of the clearest and most fascinating examples of homeostasis in living things.

THE INTERACTION OF HORMONES AND NERVES

Nerves may regulate bodily processes in much the same way hormones do, but their effect is more immediate, as we shall see. In some cases, neural regulation may duplicate the work of hormones. In other cases, the nerves act in coordination with hormones or independently of them.

To underscore the indistinctness of the line separating regulation by nerves and hormones, consider the operation of the pituitary gland. It is indeed an endocrine gland, but it is partly formed from embryonic neural tissue. Also, it lies directly under the hypothalamus (a part of the brain), which releases hormones that partly control the anterior pituitary, and which sends regulatory neurons to the posterior lobe of the pituitary. In addition, two hormones associated with the posterior pituitary are known to be manufactured in the hypothalamus and only stored in the posterior pituitary.

Epinephrine (or adrenaline) produced by the adrenal glands, provides another example of the close association of nerves and hormones. It is a hormone largely responsible for those astonishing physical feats we hear of people performing in emergencies (the story about a little old lady lifting a car off her trapped husband crops up periodically). The hormone is also a **neurotransmitter.** That is, it is produced at certain nerve endings, enabling an impulse to pass, say, from one neuron to another, or from a neuron to a muscle.

THE NEURON

The **neuron** is simply a nerve cell—although you may reach the conclusion that "simply" is hardly the word for it. Neurons come in a variety of sizes and shapes (Figure 21.4), and there are billions of them in the human body.

A neuron is composed of a cell body, dendrites, and the axon. The **cell body** contains the nucleus of the cell and most of its cytoplasm, and the usual cellular structures, such as the ribosomes and endoplasmic reticulum. The neuron's highly branched **dendrites** receive stimuli from other neurons and conduct impulses toward the cell body. The elongated **axon** conducts impulses away from the cell body. The profusely branching dendrites provide numerous points of contact with other neurons. Most neurons have only one long axon, but in some cases, it may branch.

We can differentiate between dendrites and axons on the basis of their interaction with other cells. In general, dendrites can be stimulated by other cells, while axons cannot be. Conversely, axons, but not dendrites, can stimulate other cells. (As we might expect, there are certain exceptions to these rules.)

Some vertebrate axons are enveloped in a fatty **myelin sheath,** which essentially consists of the tight membranes of flattened **Schwann cells** wrapped around the fiber (Figure 21.5). The myelin sheath serves to speed up the transmission of impulses along the axon (Figure 21.6). It is the myelin that gives the tissues of the **central nervous system** (that is, the brain and spinal cord) a whitish appearance. Nonmyelinated fibers are usually gray. The white myelinated fibers lie on the outside of the spinal cord, but in the brain they retreat to the inside, leaving gray matter visible on the outside. (Thus, "gray matter" has come to refer to the brain.) In most cases, only myelinated fibers are capable of regeneration

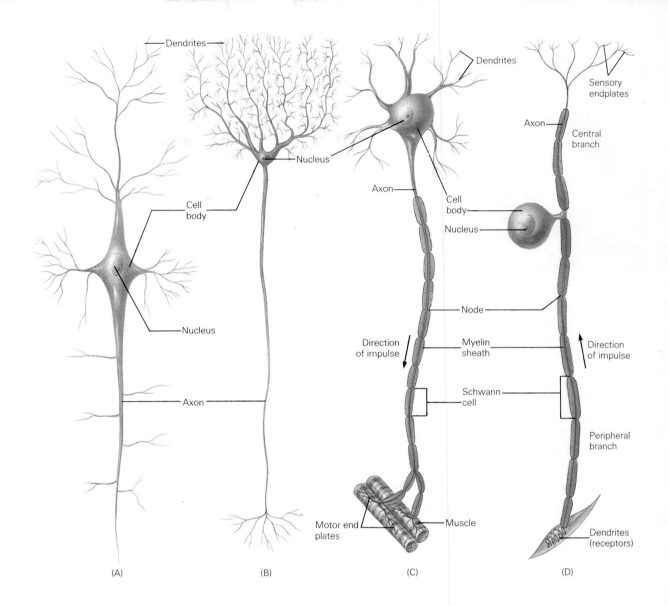

Dendrites

Nucleus

Cell body

Nucleus

Axon

(A)

Nucleus

(B)

Axon

Cell body

Nucleus

Node

Direction of impulse

Myelin sheath

Schwann cell

Motor end plates

Muscle

(C)

Dendrites

Sensory endplates

Axon

Central branch

Nucleus

Direction of impulse

Peripheral branch

Dendrites (receptors)

(D)

FIGURE 21.4 Four neurons found in human beings show the diversity of these cells. (A) and (B) show clear differences in cell bodies, axons, and dendrites. (C) is a motor neuron with axons that run from the nervous system to the effector (in this case, a muscle). (D) is a sensory neuron that runs from the receptor to the spine. Note that the sensory neuron has no true dendrites. The nodes in the myelin sheath are where one Schwann cell ends and another begins. A single nerve cell may be nine feet long, such as those that run from the base of a giraffe's spine down its hind leg.

if they are severed (as long as the cell body is not destroyed); the gray areas of the spinal cord normally do not regenerate.

Nerves are tracts of neurons. They are white, glistening strands, some quite large, ranging up to about the diameter of your finger. Although a nerve may contain many parallel neurons, each neuron can transmit an impulse independently of the others.

Another kind of cell in the vertebrate nervous system is the **glial cell.** No one is sure just what glial cells do, although they outnumber nerve cells by ten to one in the nervous system. Their branches often permeate the nerves, weaving between the neurons, and it is thought that perhaps they somehow support or nourish neurons. It is also suggested that they may store certain kinds of molecules necessary for impulse transmission.

Whereas neurons conduct impulses from one part of the body to another, the term "conduct" is perhaps misleading. This is because neurons are far more than simple conductors; each neuron not only conducts its impulses, but *generates* them as well.

(A)

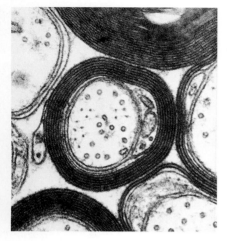

(B)

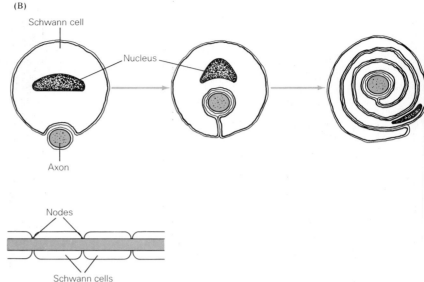

FIGURE 21.5 (A) In this electron micrograph, a cross section through a myelinated axon reveals several layers of wrappings surrounding the axon itself. The wrappings are encircling membranes of Schwann cells, whose cytoplasm and nucleus are seen as an enlarged region. The membranal wrappings produce an effective insulation, which plays an important role in impulse propagation. During development (B), a sheath-forming Schwann cell begins its enveloping action by simply surrounding part of an axon. Then one free end of the plasma membrane grows under the other, advancing along in a "burrowing" action until it has wrapped itself about the axon several times. This growth must be completed by numerous cells for a complete sheath to form.

Impulse conduction in neurons has been compared to electrical impulses in, say, a copper wire, but the analogy is not a good one. An electrical current in a copper wire diminishes with time and distance, but neural impulses, once started, do not diminish along the length of the neuron. They are as strong at the farthest branch of the axonal tree as they were at their origin in the dendrite. Therefore, instead of being like a current of electricity in a copper wire, the nerve impulse is more like a line of falling dominoes. Each domino triggers the fall of the next, but once triggered, each domino falls with the same energy. And just as dominoes must be set back up before they can repeat their performance, so must the potential energy of a neuron be restored before it can be fired again.

Impulse Pathways

If you step on a tack and are a normally sensitive and intelligent person, you may wish to take your foot off it. For you to be aware of your predicament, however, the condition of your foot has to register in your brain, which is some distance away. This message travels over neurons as an **impulse.** Let's start at the bottom.

To begin with, the tack stimulates certain **receptors** in the sole of your foot. Receptors come in a variety of types, and each type specializes in a particular sensation, such as cold, pressure, or pain (Chapter 23). In some cases, bare nerve endings function as receptors. In any case, the stimulation of a receptor results in the transfer of that impulse to an **afferent** (sensory) **neuron,** which is one that carries impulses *toward* the central nervous system. The impulse may be relayed from one neuron to the next until it enters the spinal cord and is transferred to ascending neurons that reach the brain with the message, "Sharp pain in foot!" If the foot has not already been lifted by a reflex action (which we'll discuss later), you may say to yourself, "I really must remove my foot from the tack," whereupon an impulse is sent from the brain, down the spinal

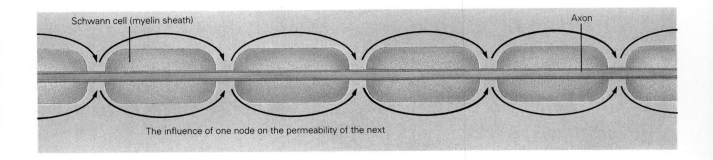

Schwann cell (myelin sheath)

Axon

The influence of one node on the permeability of the next

FIGURE 21.6 In myelinated neurons, action potentials occur only at the nodes. The neural impulse leaps from node to node down the axon. This type of transmission, known as saltatory propagation, is considerably faster and requires less energy in terms of ATP than transmission in nonmyelinated neurons. Because of saltatory propagation, myelinated neurons transmit impulses up to 20 times faster than the fastest nonmyelinated neurons.

cord, and out along the proper **efferent** (motor) **neuron;** that is, one that carries the impulse away from the spinal cord toward the **effector.** The effector here would be the muscle that raises the leg.

The junction of two neurons—the point at which the axon of one neuron transmits the impulse to the dendrite of another neuron—is called a **synapse.** Whereas, theoretically, an impulse can travel in either direction along a neuron, it can cross the synapse in only one direction.

The Mechanism of the Impulse

When a nerve fiber is at rest, certain chemicals are found in greater abundance within the cell, while others are more concentrated outside the cell. Furthermore, some of these chemicals are ionic, or electrically charged. For example, the resting neuron is literally bathed in sodium ions (Na^+), but these ions are relatively scarce inside the cell. This difference results in a net positive charge outside the cell. The cytoplasm of the neuron, however, is negatively charged. This is because, in spite of the positively charged ions such as potassium (K^+) in the cell fluid, they are overmatched by the negatively charged ions found there, especially chlorine (Cl^-) and negatively charged proteins.

The electrical potential—that is, the difference between the net electrical charges of the cell's interior and its surroundings—that is set up by this peculiar distribution of charged particles is called the cell's **resting potential.** Typically, this potential difference is about 70 millivolts, or about 5 percent as much electrical energy as is found in a regular flashlight battery.

The difference in the ion concentrations inside and outside the cell is maintained by considerable work on the part of the cell. The membrane is not very permeable to the positively charged Na^+ ions, but as fast as they diffuse into the cell, they are pumped right back out again by a mechanism called the **sodium/potassium ion exchange** pump (described shortly). The pump is actually an assemblage of proteins that moves sodium and potassium ions in opposite directions simultaneously. So, as sodium is pumped out, potassium is pumped in. The energy for this pumping activity is provided by ATP.

Now, when an impulse travels along a neuron, the first thing that happens is that the cell membrane suddenly becomes leaky to sodium ions, so that they rush in faster than they can be pumped out. The result of this shift in ions is that there is an interruption in the resting potential of the cell. However, the change doesn't occur simultaneously along the entire length of the cell; it is sequential. Beginning at the point of stimulation, a wave of inrushing sodium ions sweeps along the length of the cell, and as it passes, the cell fluid behind it immediately restores itself to its normal resting potential. Since areas along the membrane are no longer separating positively and negatively charged particles, the neuron is said to be **depolarized** at these areas. This depolarization sweeps along the neuron at a regular rate of speed, depending on the size and type of neuron.

The inrush of sodium ions at any point along the membrane results in a momentary net positive charge inside the cell at that point, so that there, the polarity is reversed. In other words, for an instant the outside becomes slightly negatively charged and the inside slightly positively charged. The change in the distribution of charges that sweeps along a neuron as it carries an impulse is called an **action potential** (Figure 21.7).

FIGURE 21.7 The action potential occurs as the inside of the cell gains a positive charge. The charge goes from about −70 to +20 millivolts as ion concentrations inside and outside the cell change. Changes in net charges inside and outside the membrane occur as an impulse passes. Note that the impulse travels from left to right. The absolute refractory period designates the period at which the nerve cannot respond to another stimulus. The relative refractory period designates the period when the cell can respond only to an unusually strong stimulus.

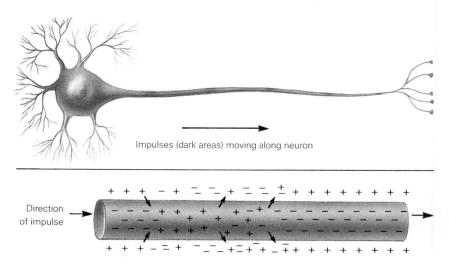

Impulses (dark areas) moving along neuron

Direction of impulse

Movement of charges (ions) across segment of neuron

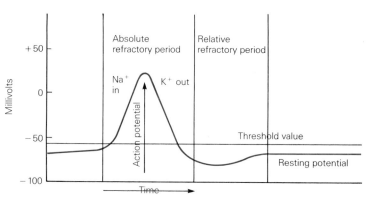

Action potential (oscilloscope tracing)

FIGURE 21.8 When not conducting neural impulses, a neuron maintains a high sodium ion (Na^+) gradient outside the axon and a lesser potassium ion (K^+) gradient inside. This gradient represents a considerable amount of potential energy that will be released when a neural impulse is generated. The sodium/potassium gradients are produced by special membranal carriers known as sodium/potassium ion exchange pumps. In each action of the pumps, three sodium ions are pumped outside and two potassium ions are pumped in. The channels are opened by the release of energy from ATP, which engages in special receptors in the complex.

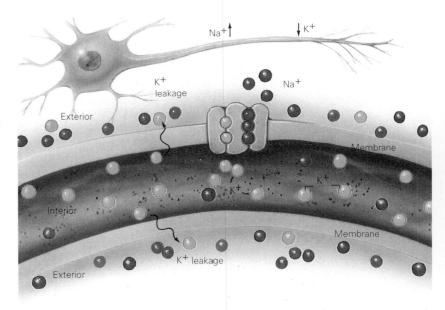

The **repolarization** of the cell begins as positively charged potassium ions rush out of the cell faster than the sodium can move in. This shift sets up a net positive charge outside again. The membrane becomes less permeable to sodium as the sodium/potassium ion exchange pump begins to move these ions out again and moves potassium back in. For each molecule of ATP used, the pump moves three sodium ions out of the cell and two potassium ions into the cell (Figure 21.8). Thus, the membrane potential of the cell is restored. The description may sound a bit tedious, but the whole process takes place in an instant. Most neurons can be depolarized and repolarized several times each second.

Two important qualities mark neural impulses. One, as was mentioned, is that an impulse does not increase or decrease in strength as it moves along a neuron. The other is that impulses operate on an "all-or-none" basis. This means that if a stimulus reaches **threshold value** for a neuron, the lowest intensity at which a stimulus can be registered for that neuron, the cell fires, and it fires at only one intensity no matter what the strength of the stimulus. This "go" or "no go" principle has been likened to firing a gun. The velocity of the bullet is not changed by how hard you pull the trigger.

If neurons operate in this all-or-none manner, how is it that we can detect various intensities of stimuli? If you put your hand on a stove, how do you know whether it is simply nice and warm or dangerously hot? Actually, we can make such discriminations in a number of ways. First, the more intense the stimulus (the hotter the stove), the higher the rate of impulses that pass along the neuron (although there are maximum limits). Also, different neurons in the same nerve may have different threshold levels. If some are easily stimulated and others harder to fire, then the stronger the stimulus, the more neurons will be stimulated. Hence, the brain can interpret either the *frequency* of the impulses or the *number* of neurons stimulated to tell you whether you should remove your hand from the stove before you begin to see smoke. Again, fortunately, the process doesn't take as long as the explanation.

The Synapse

Now let's consider how an impulse travels from one neuron to the next. Actually, we are concerned with how one neuron activates another so that the impulse continues along a chain of nerve cells. The story is an interesting one that centers on the secretion of certain remarkable chemicals.

When an impulse traveling along a neuron reaches the axon tip, it causes the secretion of a chemical into the space between itself and the dendrites of the next neuron. The axon is enmeshed in the highly branched dendrites of the next neuron so that there is a large total surface area over which the two neurons communicate. Any chemical secreted by the axon endings of one neuron, then, can be expected to immediately affect the next neuron. And this is just what happens.

In many of the neurons in our body the transmitting chemical is **acetylcholine.** The transmitting chemical is stored in small packets (or "buttons") in the tips of the axon endings (Figure 21.9). When an impulse reaches the end of the axon, some of these packets move toward the membrane and rupture, thus releasing their transmitter substance into the synapse. The chemical quickly diffuses across the gap and contacts the dendrites of the next cell, where it sets up another action potential in that neuron.

After a few impulses have traveled down a neuron, causing release of the transmitter chemical into the synaptic space, it might seem that the chemical would build up in the synaptic area and cause the next neuron to continue firing in the absence of a real stimulus. But do not despair. Almost all types of dendrites secrete a substance into the synaptic space

FIGURE 21.9 Synaptic transmission. The axon endings of one neuron lie very near the dendritic endings of the next neuron. As the impulse reaches the end of its axon, it stimulates the release of a neurotransmitter, such as acetylcholine. The process requires the expenditure of energy, thus the axonal endings are peppered with mitochondria. The neurotransmitter is released through the terminal portion of the axon into the space between the axon and the dendrite of the next neuron. The neurotransmitter thus stimulates that dendrite to initiate an impulse. The neurotransmitter is quickly destroyed by an enzyme, such as acetylcholinesterase, so that the second neuron does not continue firing in the absence of a real impulse.

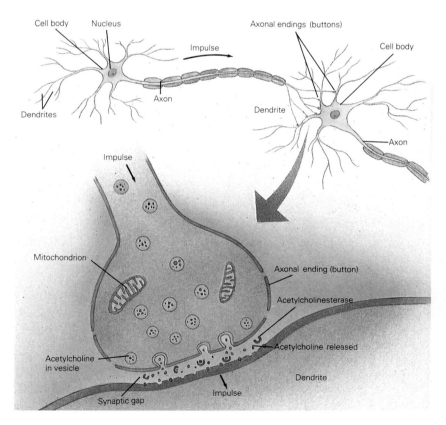

that quickly neutralizes any residual transmitter substance. For example, the neutralizer for the transmitter substance acetylcholine is the enzyme **acetylcholinesterase.**

Humans sometimes accidentally come into contact with substances that inactivate these important neural enzymes, with horrible consequences. For example, certain insecticides, such as Malathion or Parathion, which are sprayed on crops, have the advantage of being short-lived, unlike DDT, which remains intact in the environment long after it is applied. Within a few days after spraying, the short-lived chemicals are believed to be relatively harmless. However, while they are active, they can block the effects of acetylcholinesterase. The enzyme functions not only to keep chains of neurons from firing blindly in response to accumulating transmitter substance, but also to keep acetylcholine from diffusing over to nearby neurons and causing them to fire as well. With this in mind, what do you think might be the behavioral effect on a farmworker exposed to the insecticides? Already there have been several deaths among farmworkers attributed to these chemicals.

In addition to excitatory transmitter substances such as acetylcholine, there are also numerous **inhibitory transmitter substances** manufactured by **inhibitory neurons** and released into **inhibitory synapses.** Inhibitory neurons converge with the axons of excitatory neurons at the dendrites of the next cell where they counteract the effect of the excitatory neurons. This opposing system has been found at every central nervous system synapse that has thus far been extensively studied. Inhibitory synapses may function as a means of keeping neural excitation under control, so that a chain reaction of neural activity is not set up throughout the nervous system (Figure 21.10). It is believed that uninhibited neural activity of this kind may be responsible for the convulsions that occur in epilepsy.

SUMMARY

Coordination between the various body parts is generally maintained by hormonal and nervous systems. A hormone is a chemical produced by a ductless endocrine gland in one part of the body; it is carried by the blood to another part where it causes changes. Hormones may cause short-term changes or permanent ones. The major glands of the human endocrine system include the pituitary (which secretes nine hormones), the thyroid, the testes and ovaries, the adrenals, and the pancreas. The hormones they secrete have very specific chemical and physiological activities. The method of hormone function is not understood. One hypothesis suggests that hormones alter target cells on membranes and trigger prostaglandins that act in the production of adenyl cyclase. This enzyme then changes ATP to cyclic AMP, which, in the target cell's cytoplasm, alters enzyme activity or stimulates specific *m*RNA production, thereby producing certain enzymes. Many kinds of hormone production are regulated through feedback mechanisms, and hormone function may be tightly interwoven with neural activity.

A neuron is composed of a cell body, a dendrite, and an axon. The impulse travels along the axon, which may be wrapped in Schwann cells forming a myelin sheath. Myelinated fibers can regenerate; the gray, unmyelinated fibers normally do not. Nerves are tracts of neurons (nerve

FIGURE 21.10 Excitement and inhibition among neurons. Here, two excitatory neurons and one inhibitory neuron synapse with the dendrites of another neuron. The darker axons are firing. In (A), the neuron is at rest with no axonal action; in (B), one excitatory neuron fires, but it does not stimulate to threshold levels; in (C), both excitatory neurons fire and the second neuron is stimulated to fire; in (D), the inhibitory neuron is fired at the same time so that the threshold levels of the second neuron are raised above the effects of even both excitatory neurons; and in (E), the inhibitory neuron firing alone will, of course, fail to excite the second neuron.

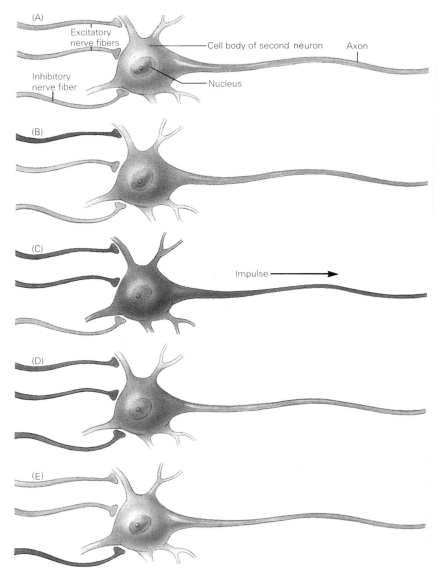

cells). The nerve cells of the vertebrate nervous system are outnumbered by glial cells whose function is unknown.

The resting neuron has a positive charge outside the cell and a net negative charge inside. In the resting cell, positively charged sodium ions (Na^+) tend to be kept outside the cell and positively charged potassium (K^+) ions kept inside the cell by a sodium/potassium ion exchange pump. The negative charge inside is maintained by the accumulation of chlorine (Cl^-) ions and negatively charged proteins. As an impulse passes, the sodium rushes in, producing a net positive charge there. Then the potassium begins to leak out, until the pump can reestablish the resting potential. Neurons operate on an all-or-none principle. The intensity of a stimulus is measured as the brain decodes either the frequency of stimulation or number of nerves stimulated.

Impulses pass from one neuron to the next by synapses—gaps into which neurotransmitters are secreted. These chemicals are quickly broken down by specific enzymes.

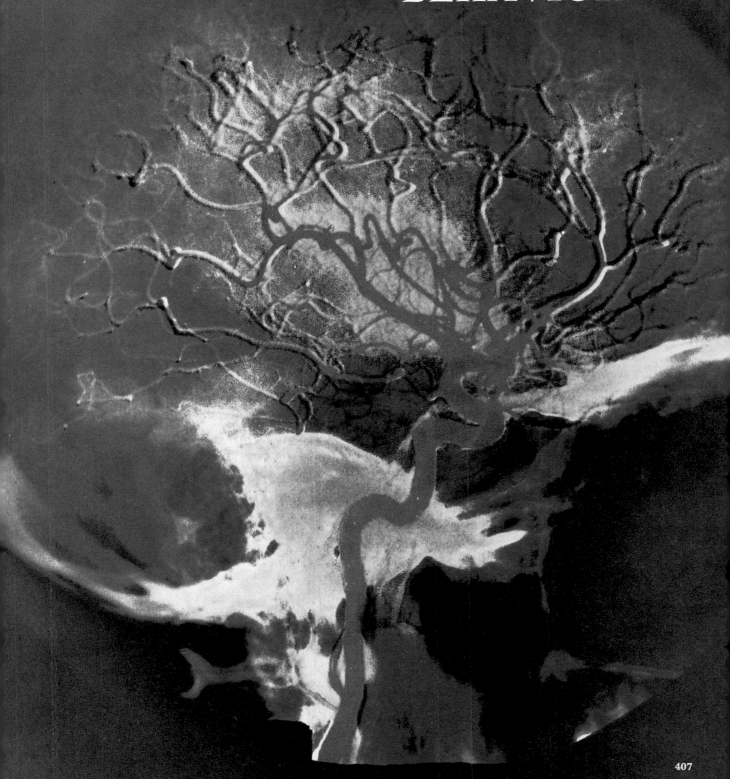

Part Six

BRAIN
AND
BEHAVIOR

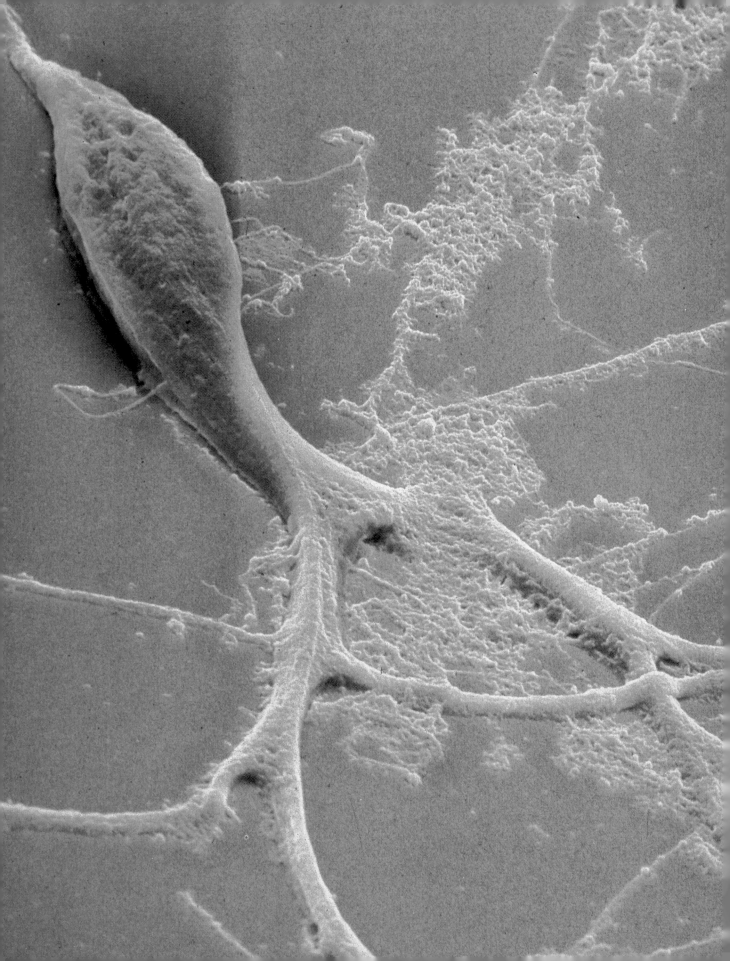

Chapter 22

THE CENTRAL AND AUTONOMIC NERVOUS SYSTEMS

Humans almost seem to worship intelligence and that great, gray orb from which it stems—the brain. However, a certain irony arises here because we don't even know what intelligence is. People have been wrestling for years to define it and measure it. The result has been sharp disagreement and a complete lack of consensus. And when we try to broach the idea of intelligence in other species, the conversation becomes a shambles. Yet we believe that intelligence, whatever it is, is good—at least for us. And we believe that it somehow resides in the brain.

In our consideration of the brain and other neural structures, we will first review various types of nervous systems, from simple to complex. We could say from "lower" to "higher," except that the terms lead to unfortunate misunderstandings. "Higher," in the minds of many people, implies any characteristic similar to those of humans, such as a large "thinking" center. The implication of this usage is that all species tend to evolve toward humanlike characteristics, including higher intelligence. Nothing could be further from the truth.

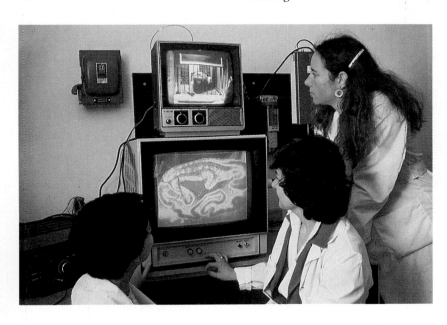

THE EVOLUTION OF NERVOUS SYSTEMS

Clues to the evolutionary development of nervous systems can be gleaned from a cross-species survey from simple to complex (Figure 22.1). One of the simplest nervous systems is that of the freshwater *Hydra*. It consists simply of a two-dimensional net of interconnecting neurons, or nerve cells, spread throughout the outer body layer. The entire surface of the animal is about equally covered. There is no part that controls the rest; no nerve center that functions in the regulation or coordination of the nerve net. Thus, if one part of the animal is stimulated, the entire body responds and shows awareness of the stimulus.

The flatworm has a somewhat more specialized nervous system (Figure 22.1B). The neurons are arranged in two longitudinal nerves connected by **transverse nerves,** producing a "ladder" as opposed to the *Hydra*'s "net." Your keen eye will undoubtedly also have noted the aggregation of nerves in the head region. These are the **cephalic** (head) **ganglia,** which are composed of clumps of neural cell bodies. As these become relatively larger and more complex in other species, they eventually are referred to as the brain.

As a brief aside, you may have wondered why the brain is located in the head. There have been exceptions, such as the huge, herbivorous dinosaur, the Brontosaurus, which had a second "brain" at the base of its tail to help direct its immense body as it browsed in prehistoric lakes. But the evolutionary reason for nerve centers in the anterior, or head, region may have been to permit quick analysis of the environment into which the animal would be moving. These centers would in all probability have come to be associated with the organs of perception, the

FIGURE 22.1 Examples of animal nervous systems. It is likely that the evolutionary route of vertebrate nervous systems was much the same. (A) The *Hydra,* with its nerve net and no coordinating center; (B) the planarian, or flatworm, with its longitudinal nerves connected by transverse nerves and a nerve concentration in the head; (C) the earthworm, with its single ventral nerve cord and well-defined cerebral ganglia; and (D) the frog, with its dorsal hollow nerve cord and the well-developed brain protected by bone. The trend is toward condensing the nervous system into a longitudinal arrangement (arising with the development of bilateral symmetry). A segmented body innervated by a segmented nervous system sets the stage for specialization along the nerve length. The nerve cord moves to a dorsal position in vertebrates as the brain becomes more complex and the segmentation becomes less regular.

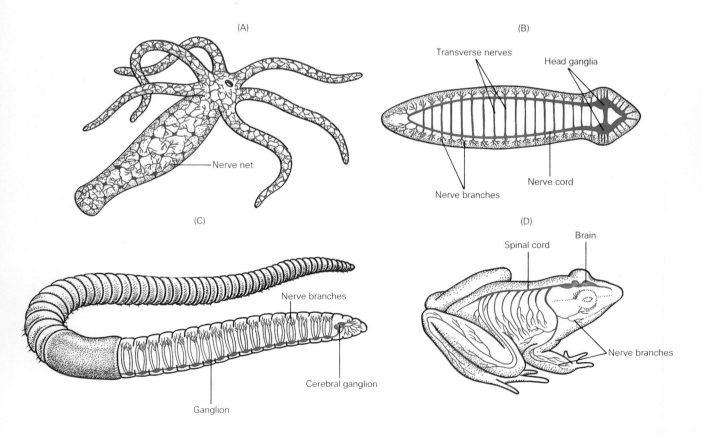

specialized receptors we refer to as the **senses.** And as we know, the senses of sight, sound, smell, and hearing are commonly located in the head region. Thus, the environment into which the animal is moving may be quickly assessed. If the brain and these special receptors were located at the posterior end, the animal might find itself in an inhospitable environment by the time it realized its predicament.

A somewhat more complex nervous system is found in the earthworm (Figure 22.1C). The earthworm has a single longitudinal nerve, but it shows vestiges of a paired arrangement in that it is two-lobed, much like two cords pressed together. The nerve is ventral, with the heart and digestive tract lying dorsal to it. Note the distinctness of the cerebral ganglia and the obvious nodes along the nerve cord, each with paired nerves reaching into a segment of the body.

The frog nervous system (Figure 22.1D) is relatively primitive for that of a vertebrate, but it can be used to illustrate the basic neural plan in vertebrates. In this group, we find a distinct brain and spinal cord that together form the **central nervous system.** In frogs, as in all vertebrates, the longitudinal nerve (the spinal cord) is dorsal, hollow, filled with fluid, and protected by bone. The anterior end is marked by a brain, an elaboration of the primitive ganglionic mass of ancient forebears. The vertebrate brain shows marked specialization; that is, different parts of it are associated with very specific functions. The central nervous system of vertebrates shows traces of the paired and segmented neural arrangements of their distant ancestors. For example, the brain is two-lobed and paired nerves extend from it and from the spinal cord. However, the paired branching is no longer so regular and apparent because of specialization along the spinal cord as the vertebrate body plan became more complex.

The Vertebrate Brain

Actually, there is no such thing as the vertebrate brain, because the vertebrates include widely diverse groups, each of which is highly specialized and distinctive. In the midst of such diversity, however, it is possible to detect trends that give some clues to the general pattern of brain development in animals with backbones.

Figure 22.2 illustrates relative differences in parts of the brain from fish to reptile to mammal. The **medulla** is simply the specialized anterior end of the spinal cord. The **cerebellum** is associated with various lower, or unconscious, forms of behavior, and the **cerebrum** is the "gray matter," the thinking part of the brain. (Regarding the brain, the higher centers are those most recently evolved—"advanced"—and usually refer to the cerebrum. The lower centers are more ancient and generally refer to areas nearer the brain stem.) We will discuss all these structures shortly, but there are a few preliminary points you might find interesting. First, not only is there an increase in general brain size (in relation to body size and, particularly, to spinal cord size) as we go from fish to mammal, but there is also an increase in the size of the cerebrum in relation to other parts of the brain. However, there are notable exceptions to this trend. For example, the olfactory lobe doesn't follow this pattern. Olfaction has to do with the sense of smell, so which of these

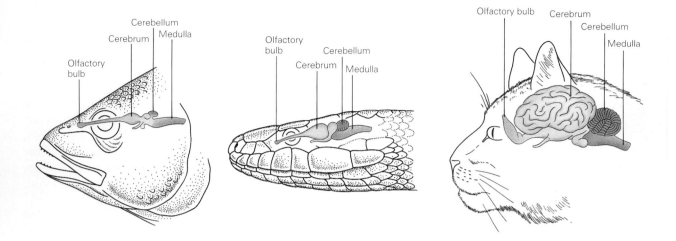

FIGURE 22.2 A comparison of the brains of a fish, a reptile, and a mammal. Notice the diminutive size of the cerebrum relative to the olfactory bulb in the fish. Also note the relative mass of the lower brain (here, the cerebellum and medulla) compared to that the cerebrum. Reptiles have a somewhat larger, but still smooth, cerebrum and a reduced olfactory area. The brain of the cat is dominated by the convoluted cerebrum. The cerebellum, involved in coordination, is well-developed in the cat, as is the olfactory bulb. It is important to realize that all areas of intelligence do not necessarily increase as one moves toward cerebration. Fish, for example, can learn some things easier than reptiles can, even though reptiles are generally more behaviorally adaptable and generally believed to be more intelligent.

animals do you suppose would rely more on a sense of smell? (Remember to consider the olfactory lobe in relation to total brain size.) Other sensory lobes could also be singled out, such as the optic lobe, which has to do with vision. Which animal do you suppose would have a larger optic lobe with respect to its brain mass, an eagle or an elephant?

I do not mean to imply that as one moves up the evolutionary family tree, each species is "smarter" than the ones below it. There may be such a trend toward higher intelligence within certain groups, but there are also many exceptions. For example, the octopus, which is a mollusk like the snail, is more intelligent than many vertebrates. The octopus brain, in fact, is similar to the mammalian brain in terms of its complexity and organization.

The evolutionary development of the mammalian cerebrum, which culminates in the human brain, undoubtedly has been one of the most crucial events in the history of life on earth. Such a statement is admittedly somewhat grandiose, but its validity becomes apparent when we consider the impact of the human species on the fragile life-support system that so thinly covers the planet. It would be interesting, therefore, to know how such a brain came to be. What spurred its development? And, as usual, no one is certain. One prevailing idea, however, is that an increasingly developed cerebrum aided in the survival of one small, seemingly insignificant group of reptiles that lived in an exceedingly dangerous world dominated by the great dinosaurs. This group, which was to give rise to mammals, branched off from other reptiles 180–200 million years ago.

These small premammals were certainly no match for the speedy, incredibly powerful, and voracious dinosaurs. Since they couldn't outrun or outfight the "ruling reptiles," it is believed that a premium was placed on their mental agility. In order to survive, the premammals were forced to *outthink* the dinosaurs. The dimmer members among them would rapidly have fallen to predators and, thus, would have failed to leave their "dim genes" in the next generation. So, with their very existence at stake, the premammals must have rapidly increased their mental agility, starting the cerebrum on its way toward dominating the brain of at least one line of animals.

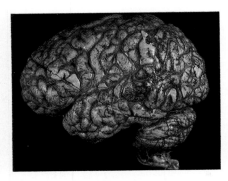

FIGURE 22.3 Under the wrinkled exterior of the human brain resides an incredibly complex array of neurons. We know very little about the precise mechanisms of brain functioning. The problem is compounded by the uncountable neurons, the innumerable neural tracts and neural bodies, and the maze of chemical and physical interactions among neurons.

The rapid development of the cerebrum and increasing reliance on intelligence would have been accompanied by other changes among the early mammals. For example, since learning requires experience, the brainier animals would have had to develop a life pattern that gave them a chance to learn *before* they were exposed to the dangers of the world. On this basis, some believe parental care developed. Even today the offspring of "learning animals" stay with their parents for extended periods. (How old were you when you left home?) During this time, they are cared for and protected by their parents until they gain enough experience to cope with the world. Also, by associating with parents or others of their kind, they can learn from them. (Most of the more brainy species are social animals, although there are notable exceptions.)

Even in a high-risk world, however, increased braininess is not the only course open to natural selection. As an example of another evolutionary route, consider the "strategy" of birds. The noted psychologist Oscar Heinroth is said to have commented, "Birds are so stupid because they can fly." The idea is that a bird doesn't normally "outfox" its enemies through intellectual maneuvering. It simply flies away. Even in escaping airborne predators, such as hawks, most birds do not rely on their cunning. Instead, they employ a few stereotyped behavior patterns, such as attacking in mass, aggregating to confuse predators, taking rather specific evasive maneuvers, or giving warning cries. Birds just never have had to develop great mental capacities, so don't be misled by the discerning frown on an eagle's face.

THE HUMAN BRAIN

Now we come to the rather interesting notion of the human brain talking about itself. Perhaps it is because of this strange twist that we encounter so many endless accolades about this great organ (Figure 22.3). Let's begin with some basic descriptions.

The human brain is divided into three parts: the hindbrain, the midbrain, and the forebrain. The **hindbrain** consists of the medulla, the cerebellum, and the pons. The hindbrain is sometimes called the "old brain" because it evolved first. These structures still dominate the brain of some animals, as we have seen. The **midbrain,** logically enough, is the area between the forebrain and hindbrain and connects the two. The **forebrain,** or "new brain," consists of the two cerebral hemispheres and certain internal structures (Figure 22.4).

The Hindbrain

The Medulla

As a rough generality, the more subconscious, or mechanical, processes are directed by the more posterior parts of the brain. For example, the hindmost part, the medulla, is specialized as a control center for such basic functions as breathing, digestion, and heartbeat. In addition, it is an important center of control for certain social activities such as swallowing, vomiting, and sneezing. As we have already seen, it connects the spinal cord and the more anterior parts of the brain.

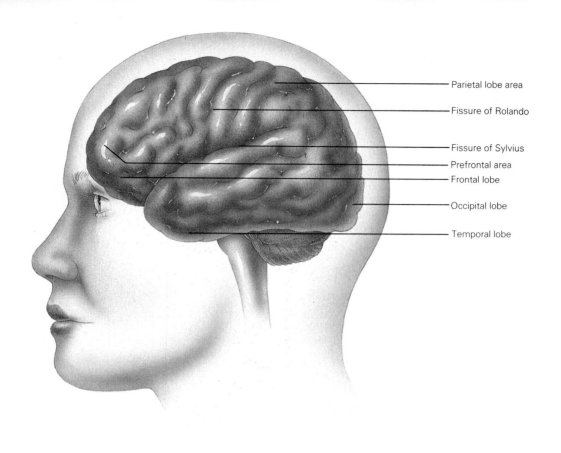

Parietal lobe area

Fissure of Rolando

Fissure of Sylvius

Prefrontal area

Frontal lobe

Occipital lobe

Temporal lobe

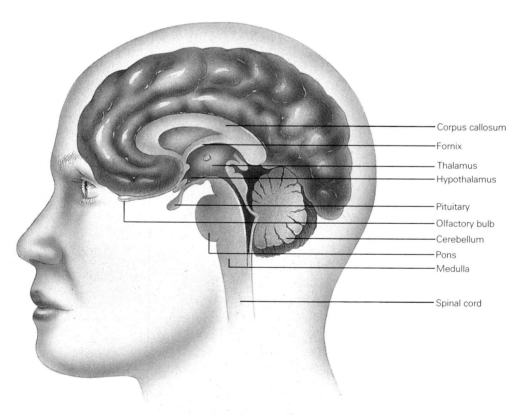

Corpus callosum

Fornix

Thalamus

Hypothalamus

Pituitary

Olfactory bulb

Cerebellum

Pons

Medulla

Spinal cord

The Cerebellum and Pons

Above the medulla and more toward the back of the head is the **cerebellum,** which is concerned with balance, equilibrium, and coordination. Do you suppose there might be differences between athletes and nonathletes in this part of the brain? (Apparently, there are, but the differences are slight.) Do you think this "lower" center of the brain is subject to modification through learning? (Can you improve your coordination through practice?) The **pons,** which is the portion of the brainstem just above the medulla, connects the cerebellum and the cerebral cortex, accenting the relationship between the cerebellar part of the hindbrain and the more "conscious" centers of the forebrain.

The Midbrain

The **midbrain** connects the hindbrain and forebrain by numerous tracts. In addition, certain parts of the midbrain receive sensory input from the eyes and ear. In vertebrates, sound is processed here before being sent to the forebrain. The midbrain has a more complex role in fishes and amphibians than in reptiles, birds, and mammals because in the latter group, many of its functions are taken over by the forebrain.

The Forebrain

The Thalamus and Reticular System

The thalamus and hypothalamus are located at the base of the forebrain. The **thalamus** is rather unpoetically called the "great relay station" of the brain. It consists of densely packed clusters of nerve cells that presumably connect the various parts of the brain—between the forebrain and the hindbrain, between different parts of the forebrain, and between parts of the sensory system and the cerebral cortex.

The thalamus contains a peculiar neural structure called the **reticular system,** an area of interconnected neurons that are almost feltlike in appearance. These neurons run throughout the thalamus and into the midbrain. The role of the reticular system is still a bit mysterious, but several interesting facts are known about it. For example, it bugs your brain. Every afferent and efferent pathway to and from the brain sends side branches to the reticular system as it passes through the thalamus. So all the brain's incoming and outgoing communications are "tapped." Also, these reticular neurons are rather unspecific. That is, the same neuron may respond to stimuli from, say, the hand, foot, ear, or eye. It has been suggested that the reticular system serves to activate the appropriate parts of the brain upon receiving a stimulus. In other words, it acts as an arousal system for certain brain areas. Furthermore, the more messages it intercepts, the more the brain is aroused. Thus, the reticular system is important to sleep. You may have noticed it is much easier to fall asleep lying on a soft bed in a quiet, darkened room than on a pool table in a disco. With the quietness, there are fewer incoming stimuli; as a result, the reticular system receives fewer messages and the brain is lulled rather than aroused. On those nights when you have the "big eye" and just can't sleep, the cause may be continued (possibly spontaneous) firing of reticular neurons (Figure 22.5).

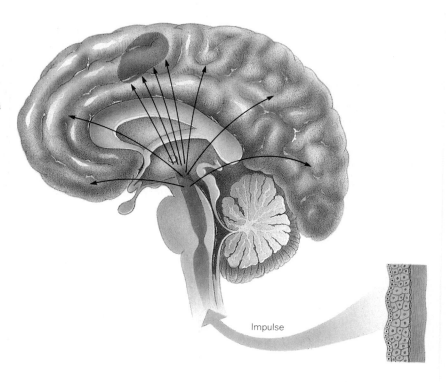

FIGURE 22.5 The reticular system of the human brain. The impulse originating at the lower right passes through the reticular system, with its untold millions of neurons. The smaller arrows indicate that, in this case, the entire cerebrum has been alerted, but a specific part (the shaded area) is the target of most of the impulses. It is likely, then, that this area of the cerebrum will be required to deal with whatever initiated the stimulus in the first place.

Impulse

The reticular system may also regulate which impulses are allowed to register in your brain. When you are engrossed in a television program, you may not notice that someone has entered the room. But when you are engaged in even more absorbing activities, it might take a *general* stimulus on the order of an earthquake to distract you, whereas the *specific* stimulus of a turning doorknob would immediately attract your attention. Such filtering and selective depression of stimuli apparently takes place in the reticular system.

The Hypothalamus

The **hypothalamus** is a small body, densely packed with cells. It helps regulate the internal environment as well as some aspects of behavior. For example, the hypothalamus helps to control heart rate, blood pressure, and body temperature. It also plays a part in the regulation of the pituitary gland, as we learned earlier. And it controls such basic drives as hunger, thirst, and sex. So now you know what to blame for all your problems. Experimental electrical stimulation of various centers in the hypothalamus can cause a cat to act hungry, angry, cold, hot, benign, or horny.

The Cerebrum

For many people the word *brain* conjures up an image of two large, deeply convoluted gray lobes. What they have in mind, of course, is the outside layer of the two cerebral hemispheres, the dominant physical aspect of the human brain. The cerebrum is present in all vertebrates, but it assumes particular importance in humans. In some animals, it is essentially an elaborate refinement that implements behavior that could be performed to some degree without it. It has a far greater importance in other animals.

For example, if the cerebral cortex of a frog is removed, the frog will show relatively little change in behavior (*cortex*, rind; the cerebral cortex is the outer layer of the forebrain in which a great deal of the active neural tissue is found). If the frog is turned upside down, it will right itself; if it is touched with an irritant, it will scratch; it will even catch a fly. Also, sexual behavior in frogs can occur without the use of the brain—but we'll try not to extrapolate from that. Rats, on the other hand, are more dependent on their cerebral cortex. A rat that is surgically deprived of its cerebrum can visually distinguish only light and dark, although its body movement seems unimpaired. A decorticated cat (that is, with its cerebral cortex removed) can meow, purr, swallow, and move to avoid pain, but its movements are sluggish and robotlike. A monkey whose cerebral cortex has been removed is severely paralyzed and can barely distinguish light and dark. In humans, the destruction of the cortex causes total blindness and almost complete paralysis. Although such persons can breathe and swallow, they soon die.

It seems apparent that the cerebrum is more than just the center of "intelligence," and that from an evolutionary standpoint, as the cerebrum enlarges, more and more of the functions of the lower brain are transferred to it.

Hemispheres and Lobes

The human cerebrum consists of two hemispheres, the left and the right, each of these being divided into four lobes. At the back is the **occipital lobe,** which receives and analyzes visual information. If this lobe is injured, black "holes" appear in the part of the visual field that is registered in that area.

The **temporal lobe** is at the side of the brain. It roughly resembles the thumb on a boxing glove, and it is bounded anteriorly by the **fissure of Sylvius.** The temporal lobe shares in the processing of visual information, but its main function is auditory reception.

The **frontal lobe** is right where you would expect to find it—at the front of the cerebrum, just behind the forehead. This is the part that people hit with the heel of the palm when they suddenly remember what they forgot. One part of the frontal lobe is the center for the regulation of precise voluntary movement. Another part functions importantly in the use of language, and damage here results in speech impairment.

The area at the very front of the frontal lobe is called the **prefrontal area,** if you follow that. Whereas it was once believed that this area was the seat of intellect, it is now apparent that its principal function is sorting out information and ordering stimuli. In other words, it places information and stimuli into their proper context. The gentle touch of a mate or the sight of a hand protruding from the bathtub drain might both serve as stimuli, but they would be sorted differently by the prefrontal area. Up until a few years ago, parts of the frontal lobe were surgically removed in efforts to bring the behavior of certain aberrant individuals more into line with what psychologists had decided was the norm. The operation was called a frontal lobotomy, and it resulted in passive and unimaginative individuals. Fortunately, the practice has been largely discontinued, largely because chemical treatments now meet the same objectives.

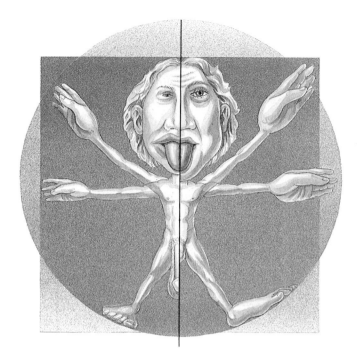

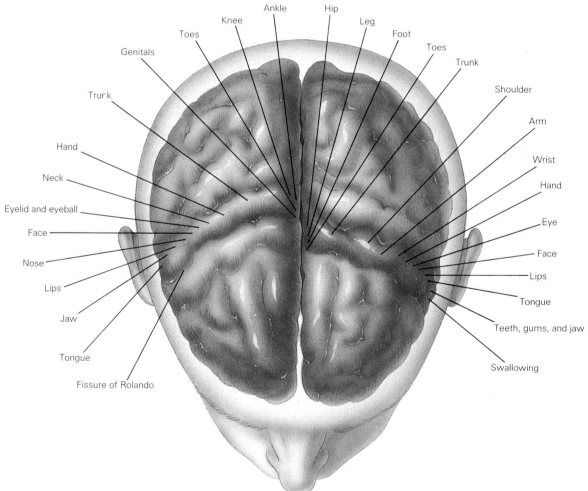

Genitals
Trurk
Hand
Neck
Eyelid and eyeball
Face
Nose
Lips
Jaw
Tongue
Fissure of Rolando

Toes
Knee
Ankle
Hip
Leg
Foot
Toes
Trunk
Shoulder
Arm
Wrist
Hand
Eye
Face
Lips
Tongue
Teeth, gums, and jaw
Swallowing

FIGURE 22.6 Function maps of the cerebral cortex showing sensory areas (top) and motor areas (bottom). The sensory section is taken posterior to the fissure of Rolando, the motor section anterior to it. Note that a large part of the human brain is devoted to face, hands, and genitalia. Also, the eyes and hands are given more motor than sensory space in the brain in spite of our great dependence on them for sensory information. Our lips and genitalia have a rather surprising amount of brain tissue allocated for their sensory input. Perhaps we have so much of the brain devoted to facial areas because the face is so important in the subtle processes of human communication.

The **parietal lobe** lies directly behind the frontal lobe and is separated from it by the **fissure of Rolando.** This lobe receives stimuli from the skin receptors, and it helps to process information regarding bodily position. Even if you can't see your feet right now, you have some idea of where they are thanks to receptors in the parietal lobe. Damage to the parietal lobe may produce numbness and may cause a person to perceive his or her own body as wildly distorted and to be unable to perceive spatial relationships in the environment.

By probing the brain with electrodes, it has been possible to determine exactly which area of the cerebrum is involved in the body's various sensory and motor activities. We can see the results of such mapping in the rather grotesque Figure 22.6. The pictures are distorted not out of any appreciation of the macabre, but to demonstrate that the area of the cerebrum devoted to each body part is dependent not on the size of the part, but on the importance it has come to have in the natural history of humans. Thus, we have the greatest number of sensory receptors in the face, hands, and genitals, but the greatest amount of control only in the face and hands (as you may have already discovered).

Also note that the sensory and motor areas are not randomly scattered through the cortex, but that proximity in brain areas reflects the proximity of the parts of the body they control. Thus the index finger control area lies near the thumb control area, and the elbow area lies closer to the finger area than does the shoulder area.

One important fact about the cerebral hemispheres is that although they are roughly equal in potential, the two hemispheres are not identical. Moreover, any chemical or structural differences between them are accentuated by learning, because the control for certain learned patterns takes place primarily in only one hemisphere. One example of the difference between the hemispheres is seen in handedness. (About 90 percent of people are right-handed but many other mammals also show handedness, the ratio being about 50:50 in nonhumans. It is interesting that less cerebrated animals such as rats and parrots also show handedness. Willy, my chinchilla, was a southpaw.) Other functions, such as speech, aspects of "I.Q.," and perception are also more likely to be located in one hemisphere than the other. Generally, it is believed that speech and language centers are located in the left hemisphere and that visual and spatial processing occurs in the left. However, such generalities are, for some reason, more valid for right-handers. For example, the primary speech center of right-handed people is located in the left hemisphere, but in lefties, on the other hand (no pun), damage to the right temporal lobe may result in poor performance on I.Q. tests, while perceptual-test performance remains relatively unaffected. In general, the brains of left-handers seems to have more diffuse control areas scattered through both hemispheres.

The Split Brain

If we learn a visual discrimination (such as correct and incorrect shapes) with one eye covered, we can make the same discrimination with the other eye. Anatomically, the reason is because fibers from the inner (medial) halves of each eye cross over the **optic chiasma** to the other side of the brain, as shown in Figure 22.7. Thus, the visual centers in both halves of the cerebrum receive information from both eyes. If the optic

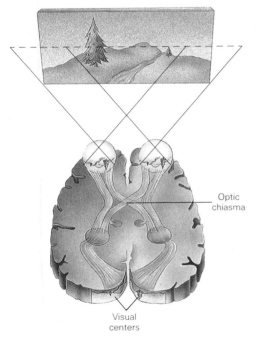

Optic
chiasma

Visual
centers

FIGURE 22.7 The optic chiasma is formed where tracts of visual nerves cross in the brain. The two visual centers of the brain each receive visual input from both eyes. The images from the inner areas of the two eyes cross at the optic chiasma to innervate opposite sides of the brain.

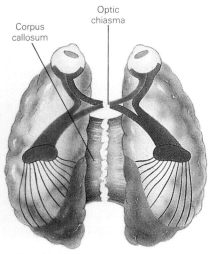

Corpus callosum

Optic chiasma

Left hemisphere   Right hemisphere

FIGURE 22.8 The major connection between the halves of the brain is the **corpus callosum.** If the corpus callosum in monkeys is cut, the animals will continue to behave normally, but experimentation reveals that the halves of the brain can function as separate and roughly equivalent units, although each has its own special qualities. If the optic chiasma is also cut, images seen by one eye can't be transferred to the opposite cerebral hemisphere. When the right eye is covered, the monkeys can learn tasks and make discriminations using only the left eye. But if the left eye is then covered, the animal can no longer perform the same tasks and has to be retrained.

chiasma is split, the right eye can still make the same discriminations as the left, and vice versa (Figure 22.8). However, the brain halves are joined by a flattened band called the **corpus callosum.** If the corpus callosum is also split, then things learned by one eye remain unknown to the other eye. In fact, *nothing* learned by one half of the brain is transferred to the other half. This means, in effect, that mammals have two brains that can act independently. It is even possible to train both halves of the brain separately. A split-brain monkey can be trained to approach an object if it is seen with one eye and to withdraw from the same object if it is seen with the other eye. If the monkey sees the object with both eyes, however, usually one side of the brain will take over and the monkey will respond without hesitation by either approaching or withdrawing.

It should be pointed out that a certain degree of compensation is possible between the two halves of the brain. For example, if one hand is injured, it is possible to learn to use the other hand. So, in spite of the natural handedness you were born with, retraining is possible as muscles controlling movement of the other hand are developed and neural events in the opposite side of the brain begin to permit the efficient use of that hand. It has also been demonstrated that if one hemisphere of the brain is injured (even severely), the other side may be able to take over some of its tasks.

THE PERIPHERAL NERVOUS SYSTEM

Pairs of thick, white nerves emerge from the brain and spinal cord and innervate every receptor and effector in the entire body. These nerves comprise the **peripheral nervous system.** There are twelve pairs of **cranial nerves,** which primarily innervate the senses, glands, and muscles of the head. These twelve pairs of nerves are found in virtually all reptiles, birds, and mammals (Figure 22.9). Fish and amphibia have only the first ten pairs.

The human spinal cord gives rise to thirty-one pairs of tough, **white spinal nerves.** Each nerve is formed from the union of a dorsal and ventral nerve root that emerge directly from the spinal cord. The **dorsal nerve root** (comprised of sensory neurons) is swollen by a huge ganglion at about the level where it enters the spinal column. The ganglion houses the cell bodies of all the sensory neurons entering the spinal column. The cell bodies of the neurons comprising the **ventral nerve root** (comprised of motor neurons) lie embedded in the spinal column, so these nerves are not disfigured by ganglia. The two great nerves fuse just outside the spinal column and travel together for a way before giving rise to increasingly smaller nerves—nerves that will ultimately branch into delicate neurons that reach every part of the body.

The peripheral nervous system consists of the somatic nervous system and the autonomic nervous system. The autonomic has only motor neurons, but the somatic has both motor and sensory neurons. The somatic nervous system carries the impulses that we are most conscious of, sensations and commands to our voluntary muscles. The autonomic nervous system is more concerned with our unconscious and involuntary internal workings.

FIGURE 22.9 Human cranial nerves. These twelve large nerves originate in the brain, and most have sensory and motor functions in the head and neck. An exception is the vagus nerve, which passes down into the trunk and serves the heart (slowing it) and part of the digestive tract. These nerves stem directly from the brain. Similar nerves rise from the spinal cord, further back, and service other parts of the body.

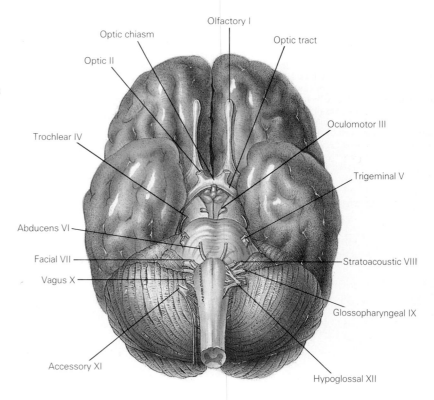

The Autonomic Nervous System

The autonomic nervous system is formed from a special set of peripheral nerves that serve the heart, lungs, digestive tract, and other internal organs. This system is largely under the control of the hypothalamus.

The autonomic nervous system can be divided into the **sympathetic** and **parasympathetic nervous systems,** which act antagonistically (Figure 22.10). Generally, the sympathetic system works to speed up certain body processes, with an increase in energy expenditure, and the parasympathetic system works to slow them down.

The sympathetic nervous system is activated in what has been called "fight-or-flight" reactions. It also functions in reproductive behavior, but perhaps we should avoid further alliteration. The fight-or-flight syndrome becomes apparent in certain emergency situations. For example, if a bear rushes into the room where you are quietly reading, your body will react "sympathetically." The pupils of your eyes will dilate, the better to see the bear; peripheral blood vessels will decrease in diameter, so blood loss will be minimized in case the bear swats you on your way out; heart and breathing rate will increase, bringing oxygen to your running muscles; and digestion will almost stop, since your visceral blood is needed elsewhere. So much blood may go to your muscles that your blood supply to the brain may decrease, causing you to faint, in which case you will hope the bear is just there as part of some effort to help curb forest fires. If upon awakening, you should discover the bear

FIGURE 22.10 The autonomic nervous system, showing the sympathetic and parasympathetic components. The central nervous system is shown at left and next to it is a column of nervous tissue, which comprises the peripheral nervous system through which autonomic impulses must pass. Impulses can also originate in the peripheral system, as is attested to by the often-active sex lives of people who have suffered injury to the spinal cord. The internal organs are innervated by both sympathetic and parasympathetic nerves, the two systems having opposite effects.

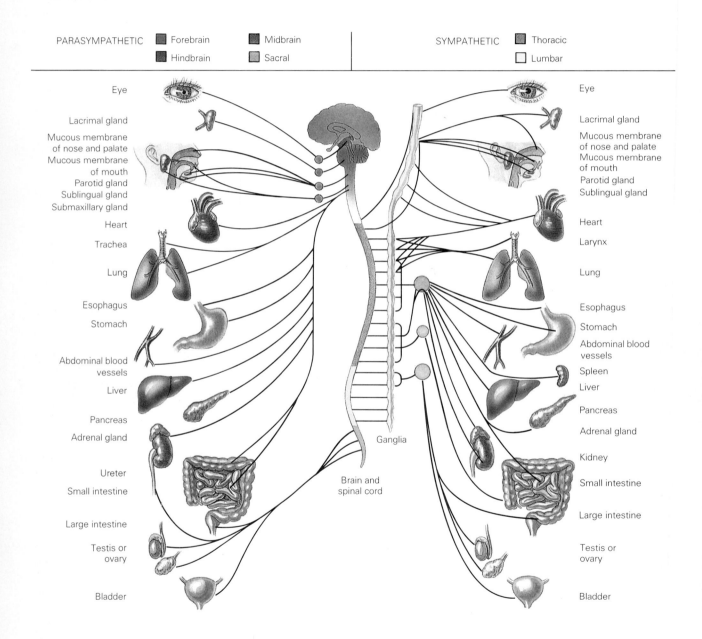

was only your roommate in a bear suit, your parasympathetic system will take over and reverse all these responses. You may then wish to activate the sympathetic system of your roommate.

Whereas the transmitter substance in the parasympathetic axon is **acetylcholine,** the sympathetic system's transmitter substance is **epinephrine** (adrenaline). You might wonder why epinephrine couldn't simply be released from the adrenal glands in an emergency situation, to travel in the blood and elicit these same changes throughout the body— why it is necessary to develop another, but similar, emergency system. Actually, the adrenal gland may secrete adrenaline into the bloodstream in such a stress situation, but the two emergency reactions (a *general* one, as the adrenal gland alerts the entire system, and a *specific* one, in which certain nerves activate only specific areas) illustrate the important difference between hormonal and neural regulation. The sympathetic system releases its adrenaline *directly* into the proper effector. Only small amounts are released, so although the response is immediate, it is of short duration. Greater amounts of adrenaline may be secreted by the adrenal gland, and while these take longer to reach the effector, their effect is more long-lasting. As was pointed out earlier, then, neural regulation is more immediate and short-term than hormonal regulation.

Autonomic Learning

The autonomic nervous system is usually described as "involuntary." It is assumed that we do not normally exercise conscious control over its functions, which include breathing rate, heartbeat, and urine formation. While it is true that the autonomic system can, and normally does, function in the absence of conscious control, there is evidence that some conscious control is possible. In other words, we may be able to learn to influence some of our autonomic reactions.

L. V. Dicara and N. E. Miller of Rockefeller University were able to teach rats to increase or decrease heart rate, blood pressure, intestinal contractions, blood vessel diameter, and even rate of urine formation. They did this by monitoring the animals' normal patterns and fluctuations in these parameters over a period of time. As natural variations occurred, the ones in the desired direction were rewarded by electrical stimulation of the "pleasure center" of the brain; variations in the other direction were punished by a slight but unpleasant shock. For example, if the experimenter wanted the rat to learn to slow its heart rate, when the heart slowed naturally, the animal would be rewarded; when it accelerated naturally, punishment would follow. Soon the heart beat more slowly. The researchers produced similar results in humans and have even taught some to control their blood pressure.

The possible applications of autonomic learning are quite varied and fascinating. For example, some of the amazing feats accomplished by some practitioners of yoga, such as when they drastically lower their metabolic rate, may be due to autonomic learning (Figure 22.11). In other cases, autonomic learning can aid in survival against the elements. It is known, for example, that mountain people of the Himalayas, such as the famed Sherpa guides, are able to withstand extreme cold. They show no effects from sleeping barefoot in the frigid mountains. It has

FIGURE 22.11 Yogis may work their wonders by using autonomic learning to alter the condition of their tissues. Western physiologists have not yet explored these possibilities as fully as they might.

FIGURE 22.12 Biofeedback techniques allow the subject to monitor his or her brainwaves and thus to increase the likelihood of generating waves of a specific type, such as those associated with relaxation. Here, an entire football team is learning to relax through biofeedback techniques. In the graphs we see the distinct differences produced in the brainwaves of a person who is (A) wakeful and (B) at rest.

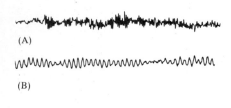

been suggested that such feats are possible because they have "learned" to withdraw body fluids from their extremities so that cell membranes cannot be ruptured by the formation of ice crystals. Some have suggested that autonomic learning has not received the attention it deserves in the Western world. However, a phenomenon known as *biofeedback* has received a good deal of attention, especially among harried professionals who see themselves caught up in the frenzy of the rat race. Here, people are taught to monitor their brainwave activity in order to consciously generate those waves that indicate restfulness and peace (Figure 22.12).

PENFIELD'S MAPPING

Wilder Penfield, of the Montreal Neurological Institute, was one of the group of researchers who fascinated the scientific world by stimulating the human brain with electrodes. Most of Penfield's work was done on epileptic patients in the course of surgical attempts to correct their condition. Epileptic seizures, which vary in intensity from a brief tingling or numbness to convulsions and unconsciousness, can result from an injury to the cerebral cortex. The attacks are caused by uncontrolled neural discharges from the injured area sporadically surging through the brain. Penfield's surgery was an attempt to identify and remove the damaged area without damaging other areas of the brain. The patients were conscious throughout the operation (the brain cannot feel pain) and could respond to questions when various parts of their brains were stimulated. By carefully positioning the electrodes and recording the patient's response, Penfield was able, in effect, to "map" the brain, to determine which parts are associated with specific functions. He found that such maps are, in a sense, similar to human faces. That is, the general structures are much the same from one person to the next, but the details are highly individual. Such mapping has resulted in the type of information described in Figure 22.6, and has yielded fascinating evidence regarding the nature of memory.

One of the most startling of Penfield's findings was that the human brain may "remember" virtually everything it has ever experienced, and that these memories can be elicited in very specific terms. We all have what are called "photographic memories" to an extent. Whatever stimuli we encounter indeed register with us—every aspect of them. For most of us, it is believed that these impressions actually last less than a second. The line of research begun by Penfield, however, suggests that perhaps much of the stimuli constantly barraging our subconscious minds are indelibly stamped there and, under certain conditions, can be recalled. Penfield found, for example, that stimulation of certain areas could cause patients to "relive" certain experiences. One patient *heard* an old familiar strain of music, as if it were being played in the operating room, when a certain point of the brain was stimulated. When the electrode was removed and reapplied to the same place, the music did not pick up where it had left off but started over again from the beginning.

Another patient saw himself sitting with relatives and friends in their house in South Africa. The event had actually occurred years before. Another patient "watched" a play she had seen years before, and another "heard" again Christmas carols being sung in her old church in Holland.

It was fascinating to learn that it is not only the important events that can be summoned from the past, but even everyday, humdrum occurrences. It seems that even extremely minute details of our lives are stored in the brain, literally millions of bits of information, which we may not be able to summon through normal attempts at remembering.

The incredibly detailed information that we store away has also been demonstrated by hypnosis. An elderly bricklayer under hypnosis described in detail the bumps on the bricks in a wall he had built when he was in his twenties. The researchers located the wall and found each bump and notch on one brick after another, just as he had said.

The events elicited by Penfield's probings are more than simple vivid recollections. The patients seemed actually to relive the experiences. Penfield himself has described the storage of such memories as being similar to recording by a tape recorder. The recorder is able to run only forward, and it has only one speed, normal. In other words, electrode stimulation cannot elicit events in any sequence other than the one in which they happened, and when elicited, these remembered events are reexperienced at the same rate they occurred. As long as the electrode is held in place, the experience of a former day goes forward. It cannot be held still, turned back, or crossed with other periods.

As Penfield and Lamar Roberts state in their book, *Speech and Brain Mechanisms:*

> The thread of time remains with us in the form of a succession of "abiding" facilitations. This thread travels through the ganglion cells and synaptic junctions. It runs through the waking hours of each man, from childhood to the grave. On the thread of time are strung, like pearls in unending succession, the "meaningful" patterns that can still recall the vanished content of a former awareness.

Such work has indeed provided surprises, but it has also given us new information on the vexing question of what memory really is. One hypothesis is that memory stems from RNA changes in neurons brought about by experience; it has also been suggested that memory involves the formation of new neural circuits developed in a learning situation. We now know that parts of the old brain called the hippocampus and amygdala are involved in processing memory, but we don't know just how. (These areas are also the seats of some emotions, such as rage and sexual desire.) Moreover, there is some evidence that there is more than one kind of memory, and that each kind involves different processes. In any case, the eventual explanation will have to account for the sometimes bizarre information provided by Penfield and his coworkers.

Now that we know something about the astonishing array of information stored in the brain, it is exciting to imagine its possible uses if we could develop a precisely controlled and less drastic means of retrieval. On long train rides (remember trains?), or in boring company, we could reread a favorite book. There would be little need for testing pure "memory" as is so often done in classroom examinations, since each exam would, in effect, be an open-book test. (In a hypnosis experiment that I performed, a high school girl was able to "see" her notes on the blackboard during an examination after having read them only once. She considered turning herself in for "cheating" but thought better of it.)

Other possibilities also come to mind. In cases of terminal disease, instead of facing death in apprehension and racked with pain, perhaps selective brain stimulation could enable a dying person to "relive" happier youthful days, to experience again long hikes with loved ones, robust optimism, and the companionship of old friends, now gone, as life dwindles away.

THE SPINAL CORD AND THE REFLEX ARC

If you pride yourself on being a "thinking animal," you may be a little disappointed to realize that your spinal cord can often receive information from the body's receptors, process it, and initiate the proper response before your brain even "knows" what has happened. Such a chain of events describes the **reflex arc** and illustrates something about the makeup of the spinal cord.

In the reflex arc shown in Figure 22.13, the impulse is generated in a special receptor called a **stretch receptor** that responds when it is elongated. The impulse is then transmitted to a sensory neuron, which enters the spinal column over the **dorsal nerve root.** The impulse is then transferred to an **association neuron,** which in turn transfers it to the proper motor neuron. The motor neuron leaves the spinal cord over the **ventral nerve root** and travels outward to the effector (usually a muscle group).

Note that impulses from one side of the body can cross the spinal cord so that effectors in the other side of the body are stimulated. In addition,

FIGURE 22.13 A simple reflex arc is shown in the familiar knee-jerk response, often suspected to be merely a physician's form of amusement. The tendon that attaches the large muscle in the upper leg to the upper shin is stretched by a light blow from a mallet. Receptors in the muscle and tendon sense this change and transmit impulses to the spinal cord, which synapses with the appropriate motor neurons, and association neurons that transmit the impulse to ascending neurons travelling to the brain. The motor neurons immediately signal the muscle to shorten and the lower leg snaps forward if the neuromuscular systems are in order. The brain is not involved, although it is made aware of the situation after the fact. The rapid shortening is a mechanism that restores the proper state of contraction to the muscle, even if a bit vigorously.

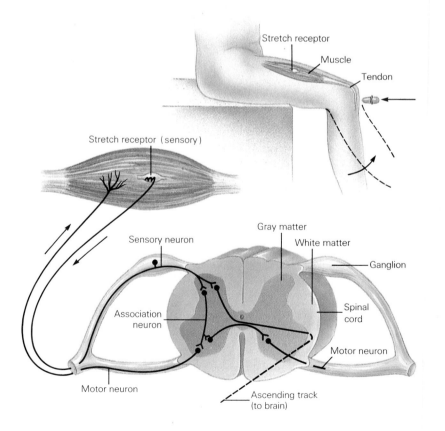

the incoming impulse can be transmitted to the dendrites of yet other neurons, **ascending neurons,** which go up the spinal cord to the brain. In this way, even if the response has already been accomplished reflexively, the brain is informed of the change. Direct neural routing from the spinal cord in a reflex arc saves time because the distance the impulse has to travel from receptor to effector is shorter. Furthermore, no time is spent in deliberating over the decision.

We have considered the brain and spinal cord in the simplest of terms, but even so, the complexity of this great central nervous system becomes apparent. As more of its mysteries are solved even more intriguing questions are revealed. It gives up its secrets slowly, but researchers continue to probe at the central nervous system with the greatest tool of all, their own.

Many of the brain's precise mechanisms remain a mystery. And while we often do not understand just *how* the brain is affected by various chemicals, we are often keenly aware of their results. So, let's briefly consider some of the more common means of chemically altering brain function and behavior.

MINDBENDERS

For some reason, a lot of people don't seem to like the minds they were born with. A great deal of our energy, it seems, is spent in finding ways to ''bend'' our minds and change our moods. And (in spite of popular beliefs) the search is not a new one. With all the universal problems facing our ancestors, they immediately set about learning to make booze. Also, we find hallucinogens were important in many of our earliest known cultures. The search for mindbenders continues today, however, in ways our ancestors could only imagine. (The ancient Greeks, for example, were not big on sniffing transmission fluid.) So some of the things we will mention here have been with us for ages, while others are so new that we really don't know how they act or what their long-term effects are.

We can begin by noting that ''drugs'' are **psychoactive agents.** That is, they can alter mood, memory, attention, control, judgment, time-and-space sense, emotion, and sensation. Fortunately, probably none do all of these at once. Most can be placed along a continuum between **stimulation** (an excitatory state) and **depression** (a state of reduced mental activity—see Figure 22.14). Here, we will review a few general principles about a few of the major groups. Not unexpectedly, we will find that all have potential for abuse.*

Abuse may arise because of our tendencies to overdo the drug or to overly rely on the drugs. Overuse can cause **dependence** or **addiction.** There are two major forms of drug dependence. **Physical dependence** occurs when the drug is necessary simply to maintain bodily comfort. It's disuse causes sometimes agonizing discomfort (called **withdrawal symptoms**). **Psychological dependence** exists where a drug is necessary for mental or emotional comfort. Withdrawal symptoms here can be as severe as they are for physical dependence.

---

*Even the *term* ''drug'' is abused. It technically refers to a narcotic (*narco,* sleep) or depressant, but largely due to its erroneous application by the federal government, it now refers to any psychoactive agent.

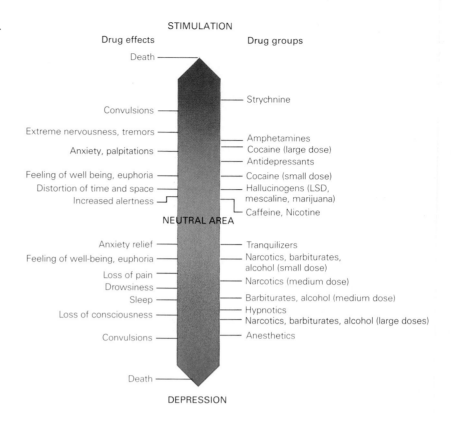

FIGURE 22.14 A continuum of drug action. The effects are shown at left, the various agents at right. The neutral area is drug-free.

STIMULATION

Drug effects   Drug groups

Death ———

——— Strychnine

Convulsions ———

Extreme nervousness, tremors ———

Anxiety, palpitations ———
——— Amphetamines
——— Cocaine (large dose)
——— Antidepressants

Feeling of well being, euphoria ———
——— Cocaine (small dose)
Distortion of time and space ———
——— Hallucinogens (LSD, mescaline, marijuana)
Increased alertness ———
——— Caffeine, Nicotine

NEUTRAL AREA

Anxiety relief ———
——— Tranquilizers
Feeling of well-being, euphoria ———
——— Narcotics, barbiturates, alcohol (small dose)
Loss of pain ———
——— Narcotics (medium dose)
Drowsiness ———
Sleep ———
——— Barbiturates, alcohol (medium dose)
——— Hypnotics
Loss of consciousness ———
——— Narcotics, barbiturates, alcohol (large doses)
Convulsions ———
——— Anesthetics

Death ———

DEPRESSION

Tobacco

Tobacco is the dried leaf of the plant *Nicotiana tabacum.* It is usually rolled, shredded, or flaked and then burned. The smoke is inhaled, allowing its products to cross the thin-walled alveoli of the lungs and to enter the bloodstream. Over 6,800 different chemicals are found in tobacco smoke, many of them carcinogens (cancer-causers). There is evidence that the major psychoactive product **nicotine** is carcinogenic. In large doses, nicotine may also cause cramps, vomiting, diarrhea, dizziness, confusion, and tremors. It can cause respiratory failure and death, the ultimate lesson. The poison is particularly dangerous to nonsmokers who have not developed a tolerance to nicotine. For them 75 milligrams of nicotine could threaten life (that is, the amount in about 3.5 packs of cigarettes).

The development of the smoking habit is curious because the first attempts can be ghastly. Perhaps, though, its continuance is even more curious because smoking has been clearly linked to lung cancer, emphysema, heart and circulatory ailments, and birth defects. (It also contributes to premature aging of the skin.)

Smoke can permanently paralyze the tiny cilia that sweep the breathing passages clean and can cause the lining of the respiratory tract to thicken irregularly. The body's attempt to rid itself of the smoking toxins may produce a deep, hacking cough in the person next to you at the lunch counter. Console yourself with the knowledge that these hackers are only trying to rid their bodies of nicotines, "tars," formaldehyde, hydrogen sulfide, resins, and who knows what. Just enjoy your meal.

Smoking may cause physiological dependence on the products in the smoke. Withdrawal may produce a variety of unpleasant reactions, and some people are apparently unable to stop, no matter what the results. The American Cancer Society estimates that about 90 percent of all lung cancer (which has a low cure-rate and which is now the number one cause of cancer deaths) is due to smoking and that if smoking ceased, the incidence of all cancers in the U.S. would fall about 25–30 percent. (You should be aware that so-called "smokeless tobacco," such as snuff or chewing plugs, has also been linked to cancer, especially of the mouth.)

Caffeine

Caffeine is a component of coffee, tea, chocolate, and many colas. Caffeine is a stimulant, affecting the central nervous system. It works by blocking chemicals that inhibit neural activity in the brain. Such drinks were not always so popular. In fact, at one time in the Near East, coffee drinkers were put to death—perhaps, some would say, a fate not worse than having to start the day without coffee.

Caffeine increases alertness and decreases fatigue and boredom. It also speeds the heart rate, increases blood pressure, increases urine formation, and dilates some blood vessels while contracting others. In small to moderate amounts (two to four cups), it may improve performance in boring or repetitive tasks, but it does not help in more complex intellectual tasks, such as reading or doing long division. However, it may help keep you awake so you can perform those tasks, since it inhibits sleep. In higher doses, it causes nervousness, irritability, and a "jangled" feeling. Very high doses can cause convulsions, but you would have to drink about 100 cups before you run a risk of dying. (By then, you probably would have already talked yourself to death.)

People who consume very large amounts of coffee, say ten to twenty cups a day, may develop **caffeinism.** The symptoms are insomnia, high blood pressure, increased body temperature, racing heart, and chills. (Caffeine is also known to encourage the development of breast cysts in women.)

One develops **tolerance** for caffeine so that increasingly higher doses are needed to produce the same effect. Withdrawal symptoms include headaches and irritability. Withdrawing coffee drinkers are generally not considered dangerous.

Marijuana and Hashish

Marijuana (also known as dope, pot, grass, reefer, killer weed, or Mary Jane) is a form of Indian hemp (*Cannabis sativa*) (Figure 22.15), and was cultivated in the United States during World War II to produce fibers for ropes after the supply of hemp from the Philippines was cut off. The wild progeny of those plants has driven crusading law enforcement officers up the wall, and the attempt to control the drug continues to absorb enormous amounts of public money.

The active ingredient in marijuana is a group of chemicals called tetrahydrocannabinols (THC). THC is highest in a preparation called **hashish** (or **hash**), which is the concentrated resinous exudate collected from

FIGURE 22.15 Marijuana plants were once a common weed. The species, however, has undergone tremendous artificial selection to produce plants high in THC. This is the leaf (right) that you are to report if found growing wild.

marijuana plants. Marijuana induces a mild euphoria, sometimes expressed as a happy or giggly mood. It may also produce mild hallucinations (if there is such a thing), forgetfulness and reduce mental agility. Its effect varies among individuals and may be partly dependent upon the setting in which it is used.

Some of the statements, official and otherwise, about marijuana have sometimes been ill-conceived, incorrect, and irrational (Figure 22.16). Because of the ludicrous nature of earlier official warnings about pot, there has been a tendency to reject all such warnings. Perhaps this is unfortunate because marijuana can, indeed, produce problems. For one thing, THC is fat-soluble and cannot be flushed out of the body by the kidneys. Instead it collects in the fatty tissues, such as the brain and reproductive organs. It may alter the production of some sex hormones. THC is slowly released as the fat is metabolized.

Marijuana smoke is a powerful lung irritant and with regular use can cause not only bronchitis, but apparently precancerous changes in lung cells. (We must keep in mind that heavy marijuana use is a recent trend and it was only after sixty years of heavy cigarette smoking that its devastating effects became apparent.)

Other studies show that marijuana temporarily reduces sperm production and may cause the production of abnormal sperm. In female monkeys, it can disrupt ovulation and cause abnormal cycles. THC can cross the placenta in some laboratory animals and is associated with a higher rate of miscarriages. (It should definitely be avoided by pregnant women.)

Marijuana is generally not considered addictive, and there seems to be little "tolerance," so doses need not be continuously increased. There may be psychological "dependence" on the drug because of what the user regards as its rewarding effects.

Cannabis sativa has been cultivated in the Near East for centuries. There, its THC is extracted to produce the resinous hashish. Hashish gets

FIGURE 22.16 An early governmental warning of the hazards of marijuana use. Such exaggerations have led to a general skepticism regarding such warnings.

FIGURE 22.17   Alcohol has become an integral part of many cultures, including our own. In low doses it is considered socially acceptable. In higher doses, it is not—particularly if it is associated with poverty.

its name because it was used by the hashshashin, a Moslem terrorist group whose notorious violence was thought to be a result of addiction to the drug. However, more careful studies of the hashshashin (the source of the word assassin), indicate that hashish was actually given as a reward for their murderous deeds—perhaps to produce the "visions of glory" promised by their leader, who, by the way, was a classmate of Omar Khayyam's.

Alcohol

Adolescent cynics have for years noted that while their parents were almost rabid in their opposition to other drugs, alcohol was often quite acceptable to them. The kids have a point. Alcohol is, indeed, a drug, even in the technical sense. And it is probably far more harmful than some of the drugs alcoholics fear.

No matter what the drink—beer, wine, or Singapore slings—the active ingredient is ethanol (CH_3CH_2OH). Ethanol, in spite of its reputation, is not a stimulant, it is a depressant. It may stimulate in a sense, however, because it can depress inhibitions and release the clever fellow within us all. It is also not an aphrodisiac, although it can in smaller amounts reduce anxieties. In larger doses, it interferes with the sexual act in a most frustrating manner. As Shakespeare observed, it "provokes the desire, but it takes away the performance."

People develop a tolerance for alcohol rather quickly. That is, as the body "handles it," more and more is required to produce the same effect. Essentially, the "handling" is mainly done by the liver, which can oxidize about an ounce of alcohol per hour. (Five to ten percent is excreted by the lungs and kidneys.) Finally, though, the liver may become so damaged that it can detoxify very little alcohol.

Alcohol is very addictive, and its disuse can produce severe withdrawal symptoms, the most drastic of which are **delirium tremens,** or **DTs.** Long-term use can damage the central and peripheral nervous systems, the liver, the stomach, and the intestines, but the effects vary greatly from one individual to the next (Figure 22.17). One alcoholic's mind may go, while his or her drinking companion may only die of cirrhosis of the liver. Interestingly, heavy "binge" drinking may be safer than daily drinking of smaller amounts, perhaps because the layoffs give the liver time to recover.

Opiates

Opiates fall within the realm of what are referred to as "hard drugs." They are technically narcotics in that they depress the nervous system, and they can relieve pain and produce sleep or a stupor. The group includes heroin, opium, and morphine. Users rapidly develop tolerance, and continued use often results in addiction. Consequently, the user not only needs the drug continually, but in increasingly higher doses. Because of the cost of opiates, addicts often resort to crime in order to finance their habit.

Injection of opiates produces a "rush," or sudden pleasurable sensation. (In beginners, it may also produce severe nausea.) The rush is followed by a great sense of well-being, accompanied by a marked

Egyptian hieroglyphics depict both priests and physicians administering to victims of hangovers. If you've had one, you may prefer the former to the latter. And after all this time no one knows how to remedy the problem. (But Puerto Ricans may recommend rubbing the underarm with lemon.)

A hangover is an apparently interminable but temporary chemical imbalance in the body, caused by alcohol acting as an anesthetic on the central nervous system. Usually the results involve a dilation of blood vessels in the brain; the movement of water, potassium, and other ions from the cells outward to the intercellular spaces; depletion of magnesium from the kidneys; and inflammation of the stomach lining. In addition, the sleep that follows is strangely de-void of the rapid eye movement (REM) that is characteristic of the most restful sleep.

The results are insatiable thirst, upset stomach, fatigue (many by-products of heavy physical exertion appear in the blood), grouchiness, and perhaps remorse (depending on who saw you).

The morning-after drink ("the hair of the dog that bit you") provides superficial relief while slowing down recovery, and coffee stimulates the already exhausted nervous system and prevents needed sleep.

Hangovers, however, have produced more philosophers than all the world's great books, the conventional wisdom being, "There ain't nothing, NOTHING, worth a hangover."

decrease in physical drive. An accompanying drowsiness produces the nodding you can see in almost any New York subway. Withdrawal symptoms following abrupt discontinuation of the drug are usually very violent. With heavy addiction, withdrawal may produce such an intense shock to the system that death results.

Cocaine

Cocaine (toot, coke, lady, girl) is another of those recently fashionable drugs that we don't know enough about. However, much of what we do know points to an insidious and potentially very dangerous substance. But, it was not always regarded so. In fact, it was once sold in the United States under its own name or used to lace other products, such as Coca Cola. It seems to stimulate neural activity in the brain by inhibiting the breakdown of neurotransmitters.

Cocaine is an alkaloid extract usually derived from two species of the coca plant, a South American shrub of the genus *Erythroxylon* (Figure 22.18). Historically, the leaves were chewed by pre-Columbian South American Indians, and the practice continues in that area to this day. The effect of chewing the leaves is a generally elevated intensity and an increase in apparent energy. Today, however, chemical extraction techniques can produce a white, crystalline powder that is usually "snorted" (inhaled through the nose), "shot" (injected into veins), or "based" (where the alkaloid is chemically freed and smoked—"free basing"). One particularly addictive form of based cocaine is called "crack." Snort-

FIGURE 22.18 Coca is one of the most lucrative crops in South America. Many pounds of leaves are necessary to produce a single gram of cocaine. Cocaine was once a legal stimulant and could be found in Coca Cola (once a cola with coca leaf extract), as well as in syrups.

ing produces the least drastic effects; basing produces the most powerful. In all cases there is a euphoria, mood elevation, and general stimulation. The result is talkativeness and a general intensity that may cause other people to tiptoe away. As it wears off, a depression sets in that compels the user to seek another "hit."

The long term effects of regular cocaine use can be severe. The user may focus on the drug while other facets of life are neglected. The drug was once considered physiologically nonaddictive, but new evidence indicates that it can be powerfully addictive. There is little tolerance, so increasingly larger doses are not required.

Cocaine also has an artificially elevated expense. Whereas it costs only a few cents to manufacture a gram of cocaine, the street price may be $60 to $150. Furthermore, there may be little of the drug left in the street powder by the time it reaches the user. As a rule, it is "stepped on" (cut) by each person through whose hands it passes.

Amphetamines

Amphetamines, often loosely referred to as "speed" or "uppers," include a number of commercial drugs such as Benzedrine, Dexedrine, and Methedrine. Their chemical properties are similar to those of epinephrine—that is, they cause great bursts of energy that can overcome feelings of fatigue. Students and truck drivers have been known to take low doses in order to stay awake during midnight cramming sessions and long hauls, respectively. Weight watchers also use them to decrease appetite. Because amphetamines are effective in improving performance on rigorous physical tasks, they are sometimes used by athletes. Low doses do not impair skills or judgment.

The greatest abuse of amphetamines is by injection of the drug. This produces an initial rush, followed by a feeling of vigor and euphoria that may last several hours. After this, however, come the dues. They appear in the form of aching, discontent, and irritability. To delay this letdown, the user may boost himself or herself with another injection. The high may thus last for days, during which time the user usually fails to eat or sleep. The end of this period may be followed by exhaustion, severe depression, paranoia, aggressiveness, extreme irritability, and emotional overreaction.

Users develop a tolerance for amphetamines, and cessation after prolonged use may produce withdrawal symptoms, although they are less severe than those associated with opiate withdrawal.

Barbiturates

Barbiturates, or "downers," are sold under a variety of trade names, including Nembutal, Seconal, Tuinal, and Amytal. They are all **sedative-hypnotics** that act on the cerebral cortex, midbrain, and brainstem areas. Their effect is to reduce anxiety and induce drowsiness and sleep. These results are accompanied by loss of muscular coordination and slurring of speech, similar reactions to those induced by alcohol.

Barbiturates are highly addictive and rapidly produce tolerance. Withdrawal symptoms may be as severe as with opiate or alcohol withdrawal. The heavy barbiturate user is likely to be confused, obnoxious, stubborn, and irritable. In contrast to the placid disinterest of the opiate user, barbiturate users may be particularly aggressive and violent.

Barbiturates and alcohol acting together are particularly dangerous. The combination may cause death by suppressing the breathing centers. In addition, because each drug causes mental confusion, accidental deaths in those who mix them are all too common.

Phencylidine

Phencylidine, also called PCP or angel dust, is one of the more dangerous drugs to make the rounds in recent years, finding its place among certain abysmally ignorant young people. Unfortunately, many people use this powerful animal tranquilizer without realizing it, since it is used to "lace" marijuana and even cocaine. Its users may feel euphoric and extremely relaxed, but the side effects of the drug include extreme violence, psychosis, confusion, and, at high doses, coma.

Methaqualone

Methaqualone (Quaalude, or lude) is a synthetic barbiturate. When it was introduced in 1963, it was believed to be nonaddictive and was prescribed freely. However, all this has changed; it is indeed addictive. It often causes the user to be relaxed, uninhibited, and receptive (as a result it has been touted as an aphrodisiac). When taken with alcohol, one may fall into a stupor ("lude out") and lose control of movement. This is accompanied by a feeling of "pins and needles" in the extremities and around the mouth.

Psychedelics

Psychedelics are comprised of a group of drugs that produce hallucinations and various other phenomena that very closely mimic certain mental disorders. These drugs include lysergic acid diethylamide (LSD), mescaline, peyote, psilocybin, and various commercial preparations such as Sernyl and Ditran.

Of these, LSD is probably the best known, although its use has apparently diminished since its heyday in the late 1960s. LSD is synthesized from lysergic acid produced by a fungus (ergot) that is parasitic on cereal grains such as rye. It usually produces responses in a particular sequence. The initial reactions may include weakness, dizziness, and nausea. These symptoms are followed by a distortion of time and space. The senses may become intensified and strangely intertwined—that is, sounds can be "seen" and colors "heard." Finally, there may be changes in mood, a feeling of separation of the self from the framework of time and space, and changes in the perception of the self. The sensations experienced under the influence of psychedelics are unlike anything encountered within the normal range of experiences. The descriptions of users therefore can only be puzzling to nonusers. Some users experience bad trips or "bummers," which have been known to produce long-term effects. Bad trips can be terrifying experiences and can occur in experienced users for no apparent reason.

Our nervous systems are the products of natural selection, and so they are specialized to help us deal successfully with the part of the world with which humans must interact. We see, though, that various aspects of our psyches can be satisfied, and perceived "needs" can be met, by a wide range of chemicals that our natural bodies are not prepared to deal with. Some of these chemicals have been around so long that they, and their dangers, have become familiar—even accepted. Others, though, are new, and we await data on the effects of their long-term use. And we are an inventive species, so we must assume that other "mindbenders" lie just over the horizon—new chemicals, new experiences, new behaviors, and new dangers.

SUMMARY

The evolution of nervous systems probably began with a netlike arrangement of nerves that was followed sequentially by paired longitudinal nerves with anterior ganglia, a single ventral nerve cord with a brain, and finally a dorsal nerve cord giving rise to branches. An enlarged "thinking center" was paralleled by parental care and, often, sociality. The human brain is divided into three parts: the hindbrain, midbrain, and forebrain. The hindbrain (composed of the medulla, cerebellum, and the pons) evolved first and is largely responsible for unconscious activity. The midbrain registers some stimuli, but its role is largely to connect the hindbrain and forebrain. The forebrain is composed of the thalamus, with its reticular system, the hypothalamus, and the cerebrum. Conscious thought is centered in the cerebrum, which is composed of two roughly (but not exactly) equivalent hemispheres connected by the corpus callosum. The hemispheres can be separated by severing the corpus callosum, resulting in the two halves of the brain behaving independently. Visual information from either eye can still reach both halves unless the optic chiasma is also severed.

The central nervous system is comprised of the brain and spinal cord. Where the dorsal and ventral spinal nerves join just outside the central nervous system, the peripheral nervous system is formed. The peripheral nervous system is composed of (1) the somatic nervous system, which is largely concerned with conscious functioning such as sensations and motor commands, and (2) the autonomic nervous system, which is concerned largely with subconscious internal workings. The autonomic nervous system is divided into two groups, the sympathetic (fight or flight) responses and the opposing group, the parasympathetic responses. Some kinds of learning are evidently possible in the autonomic system.

Brain mapping has indicated a general similarity in the arrangement of most brains and the possibility that every experience is recorded in our memory. The simplest neural loop is the reflex arc.

A variety of substances can influence mood and behavior. Some are addictive. Some cause tolerance.

Chapter 23

THE SENSES

What kind of world is this and how do we know? It seems obvious that whatever we know, we know through our senses. Immediately, however, this answer brings us into difficulties. This is partly because our senses monitor only certain aspects of the environment, so what about those things we can't detect? Aren't they, too, part of our world?

If you were to ask a variety of animals to describe what the earth is like, there would be little consensus. A fly might describe swirling eddies, delicious surfaces, and a kaleidoscopic world of shimmering mosaics. A dog might describe a gray, drab world dominated by surges of sounds, accented by thundering odors. A tapeworm might speak vaguely of a warm, wet world preceded by harsh light and withering dryness as it lay for months near death. Some insects see deep violet colors in flowers that we call yellow (Figure 23.1), but the red rose appears to them as black, lacking color at all. Each of these animals is sensitive to very limited aspects of their environment, and so are we. As with all species, we generally can sense those aspects of the environment that are important to our survival.

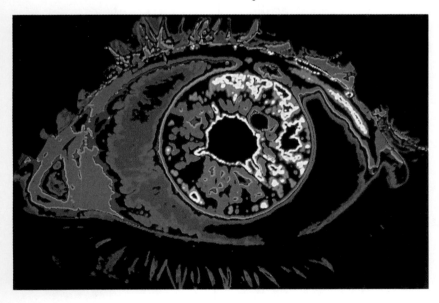

FIGURE 23.1 These appear to us as yellow flowers. However, color is partly a function of receptive abilities, and our receptors fail to register the deep violets that the insect sees and that are only apparent to us by use of a special filter.

The second problem in describing the senses is that sensory abilities may differ widely from one individual to the next, even within the same population. This area is rife with anecdotes and untested claims, but there are many substantiated cases of remarkable and unusual keenness of vision, hearing, touch, and so on. What are we to make of people who apparently can "feel" the color beneath their fingers? Of the ability of twins in different rooms to silently communicate? Of people who can read newspapers across the room or see the moons on other planets? Here, we're dangerously close to leaving the realm of acceptable scientific topics and, of course, above all things, we want to be acceptable. So let's just admit that there may be great differences in the sensory abilities of people and move on to consider some of the basic mechanisms of perception. We will review only a few basic types of **receptors,** those neural structures that are capable of responding to environmental stimuli, in a few representative groups of animals.

THERMORECEPTORS

Thermoreception is the ability to sense temperature. Such an ability is important because the delicate chemical processes of life can normally be conducted only within certain temperature ranges. A heat-sensitive animal is able to adjust itself in its environment so as to position itself within those ranges.

Not much is known about heat detection in invertebrates, and it is assumed that most of them lack thermoreceptors. Some, however, do have these sensors. For example, cockroaches have heat receptors on their legs, with which they can locate optimal places to live, like your house. In addition, sensitivity to temperature may be important in helping parasites locate their warm-blooded hosts.

The mechanisms of heat detection in vertebrates are better understood, but there are still wide gaps in our information. Some species, such as the pit vipers (Figure 23.2), have been intensively studied, and there is general agreement about the mechanisms involved. In other groups, however, there is strong disagreement. For example, there are strong differences of opinion regarding just how heat is detected in mammals. Some argue that specific receptors are involved, while others say these sensations are registered by simple free nerve endings. The latter explanation would imply that the same neurons register both heat

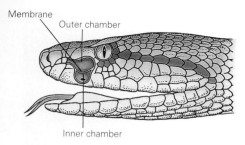

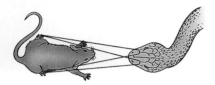

FIGURE 23.2 Heat sensors in the pit viper (rattlesnake, *Crotalus viridis*). Pit vipers detect their prey through a pair of heat-sensing devices located in the depressions near the eyes. Each pit consists of an outer chamber that ends in a thin membrane covering an inner chamber below. Extending over the membrane is a highly branched, heat-sensing nerve. The recessed structure of the pits permits the snake to zero in on its prey, just as though it were using its eyes. By moving the head back and forth, it determines the exact location of the prey by the intensity of neural impulses from the membrane. There is some evidence that the snakes also assess the size of the target by the heat generated.

and cold, and that differences in these sensations involve variations in the level of stimulation of the same receptors. Perhaps different temperatures simply trigger different rates of firing in the same neuron. It has also been suggested that neurons specialized to react to different temperatures lie at different levels beneath the surface of the skin, so that more extreme temperatures are required to trigger those receptors at deeper levels.

TACTILE RECEPTORS

Tactile receptors respond to touch. They fire whenever their shape is altered or distorted, and they trigger extremely sensitive, fast-firing neurons. In some cases, bristles, whiskers, or hairs extend from the tissues around these receptors so that objects are perceived by these feelers before they contact living tissue.

Many invertebrates have such sensory feelers, as we see in jumping spiders (Figure 23.3). Web-building spiders have hairy legs that react to vibrations set up by trapped insects. Tiny hairs on the abdomen of cockroaches are extremely sensitive to light air-currents—such as those produced by a descending human foot. Many invertebrates use touch in finding food and mates and in avoiding predators.

Vertebrates have two kinds of tactile receptors, those that register pressure and those that respond to touch. The pressure receptors are located deeper under the skin. They are essentially encapsulated nerve endings called **Pacinian corpuscles.** Near the surface of the skin are **Meissner's corpuscles,** which are believed to respond to light touch (Figure 23.4). In mammals, certain parts of the body are especially sensitive to touch. A sleeping dog can immediately be aroused if you touch the hairy area just under its tail (preferably with a stick). In humans, the most sensitive parts include the hairy areas and the genitals, as you may have noticed. In all primates, the areas most sensitive to touch include the lips, the area around the eyes, and uncalloused fingertips.

AUDITORY RECEPTORS

Audition, or hearing, involves the detection of sound, usually a distant stimulus. Most invertebrates lack specific receptors for detecting the vibrating columns of air that produce sound. However, many are sensitive to the vibration of the air, water, or soil in which they live.

Insects are an exception among the animals without backbones in that some of them can hear quite well. Some, such as grasshoppers and crickets, have **tympanal membranes** that respond to sound much as does the human eardrum. One of the best stories regarding the evolution of hearing in insects involves that of noctuid moths (Essay 23.1).

In general, the hearing structures of land vertebrates include an **auditory canal** and a **tympanic membrane** (ear drum) that vibrates from one to three moveable **middle ear** bones. The vibrations of the bones then stimulate receptors that carry impulses to the hearing centers of the brain.

The auditory apparatus of mammals is somewhat distinct among the vertebrates. For one thing, most mammals have an **external ear.** This is

FIGURE 23.3 Jumping spiders are extremely sensitive to touch because of the hairs that cover their bodies. When a hair is bent or moved, it activates a neuron that transmits an impulse.

FIGURE 23.4 Specialized touch receptors in the skin of humans. Meissner's corpuscles are located close to the surface and register light touch. These are most numerous in the finger-tips and around the lips. Pacinian corpuscles (below) are in deeper skin locations, and their complex end bulbs register pressure. Generalized sensory neurons (no distinct structures) surround the hair follicles of mammals and are stimulated by hair movement. (Try to move a single hair on your forearm without feeling it.)

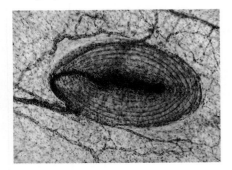

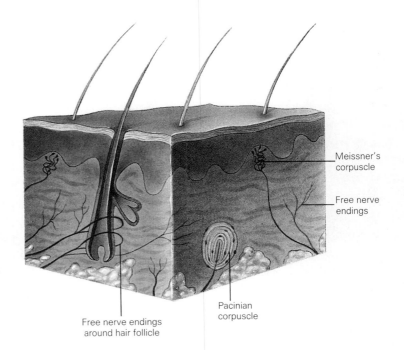

Meissner's corpuscle

Free nerve endings

Pacinian corpuscle

Free nerve endings around hair follicle

the part that can be moved, as we see when a dog focuses on some sound. (Humans have largely lost the ability to move their ears, but those who can manage it are in great demand at social events.) Whereas most other vertebrates have only one bone in the middle ear, mammals have three: the so-called hammer **(malleus),** anvil **(incus),** and stirrup **(stapes).** Sound vibrations move the hammer, which is pressed against the vibrating eardrum. The movement is transferred to the anvil which then vibrates the stirrup (Figure 23.5). The stirrup vibrates against the inner ear's **oval window,** which, in turn, sets up movements of the fluid inside the long, coiled **cochlea.** The cochlea is divided along its length by a **basilar membrane** from which arise modified **cilia** ("hairs") that push up against a more flexible **tectorial membrane** and connect directly to the auditory neurons that lead to the brain. (Some hearing impairment is directly related to the loss of these hairs, and they can be lost by exposure to loud noise, such as amplified rock music.) The basilar membrane itself is thicker and more flexible near the tympanum, becoming much thinner further away. The pitch of a sound is registered according to which hairs are stimulated. High pitches stimulate hairs in the thicker regions of the basilar membrane; low pitches register at the thinner tip of the coiled tube, the part lying in the innermost part of the snaillike coil. Loudness is apparently detected by the number of neurons stimulated and the frequency with which an impulse is generated in each of these neurons.

CHEMORECEPTORS

The ability to detect the presence of chemicals is called **chemoreception,** and it varies widely in sensitivity throughout the animal world. If you encounter a smelly dog, you should keep in mind that your opinion of his odor pales before his perception of you. We can assume that either dogs don't mind our scent or they're just being polite.

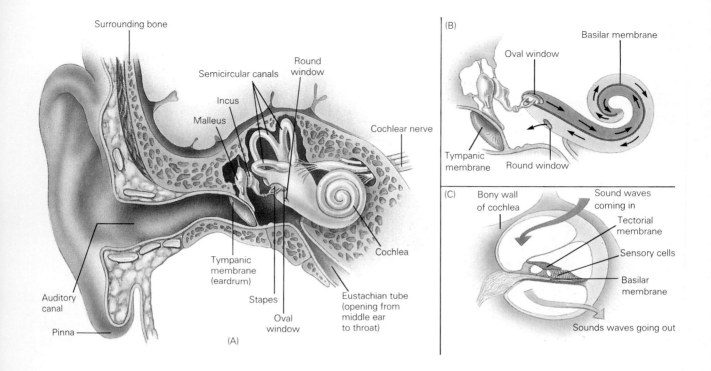

Surrounding bone
Semicircular canals
Round window
Incus
Malleus
Cochlear nerve
Tympanic membrane (eardrum)
Stapes
Oval window
Eustachian tube (opening from middle ear to throat)
Cochlea
Auditory canal
Pinna
(A)

(B)
Oval window
Basilar membrane
Tympanic membrane
Round window

(C)
Bony wall of cochlea
Sound waves coming in
Tectorial membrane
Sensory cells
Basilar membrane
Sounds waves going out

FIGURE 23.5 (A) When sound waves vibrate the human eardrum (tympanic membrane), they set in motion three tiny leverlike bones: the malleus, incus, and stapes. The stapes, attached to the oval window, sets fluids in motion within the snail-shaped cochlea. (B) The cochlea is actually a U-shaped tube, divided by the basilar membrane. (C) Sensory cells of the membrane are embedded in the gelatinous tectorial membrane. The two membranes and the sensory cells are called the organ of Corti. The sound impulses pass inward over one surface of the basilar membrane, turn a corner, and pass outward over the opposite surface of the membrane to be dissipated at the round window. Different regions of the basilar membrane are sensitive to different sound frequencies.

Chemoreception involves both olfaction (smell) and gustation (taste). The mechanisms are essentially similar, but olfaction usually involves distant stimuli and gustation registers those in which the source is in contact with the receptors.

The most remarkable chemoreceptive abilities are found among the insects. In this group, there are taste receptors on various parts of the body, including mouthparts, antennae, and forelegs (since some are known to eat what they walk on). Some can even taste with their egg-laying organs, an ability that enables them to lay eggs only on certain kinds of plants. Generally, however, most insect olfactory receptors lie at the ends of minute tubules that branch throughout the insect body. Molecules that diffuse into a tubule become dissolved in the insect's body fluids and can then cross the membranes of the tiny receptors. (One of the most amazing olfactory abilities is found in the Atlas moth, as is noted in Figure 23.6.)

Chemoreception in vertebrates usually involves moving chemicals into specialized sacs or tubes that are lined with receptors. The chemicals (as is almost always the case in cellular organisms) must first be dissolved before they can cross the membranes of the receptors, so these sacs and tubes are usually moist.

Among vertebrates, mammals have the best sense of smell, with the best smellers being the carnivores and rodents. However, some mammals, such as the toothed whales, have no sense of smell at all. The sense of smell is also rather poorly developed in us primates. Chimpanzees, for example, smell about like you do (no offense). In humans, the olfactory receptors in the nasal passage are connected directly to a slender, forward extension of the brain, called the **olfactory bulb** (Figure 23.7).

FIGURE 23.6 The oversized antennae of the atlas moth, *Artacus atlas*. The receptors in the antennae are extremely sensitive to a chemical attractant released by females. Even one molecule can generate an impulse.

The receptors involved in taste are neurologically very similar to those of smell. One distinction, however, is that among all land-dwelling vertebrates, taste receptors are confined to the mouth area. It is generally acknowledged that there are four basic tastes: sweet, sour, salty, and bitter. In humans, these are located in specific areas of the tongue. Salt can be tasted over the entire surface of the tongue, but sweet registers in the tip, sour on the sides, and bitter on the back. Research indicates that most cells can respond to three or four tastes, but those of each group are particularly sensitive to a single taste. Obviously, one might wonder, if a taste receptor can be activated by more than one kind of molecule, how does the brain respond properly to each taste? It turns out that each taste produces a distinctive pattern of neural firing in the receptors.

Chemoreception has been extremely important in the evolutionary history of humans. In essence, it gives us information about the environment before we are forced to interact with it. It lets us know what is good and desirable as well as those things that are to be avoided. For example, sweet is the taste of carbohydrates (ripe fruit tastes sweet). Unripe fruit, however, may be sour, and we have little tolerance for that taste. This means that we are not likely to eat unripe fruit but will probably wait until it has matured and its food value is higher. Bitter, we might mention, is the taste of a number of powerful poisons and can trigger a gag reflex at the back of the tongue (Figure 23.8).

We probably have a great deal to learn about the role of chemoreception in human life. For example, recent research has focused on the role of smell in sexual attractiveness. There is evidence that masking human odor with colognes might be counterproductive in increasing one's

FIGURE 23.7 The olfactory receptors in the human nose connect to the olfactory bulb of the brain. The receptors are able to distinguish a wider variety of stimuli than are the taste buds. The olfactory neurons are part of the nasal epithelium dispersed with other cells. Each neuron has numerous olfactory hairs that protrude from the epithelium. Each "hair" is a modified cilium.

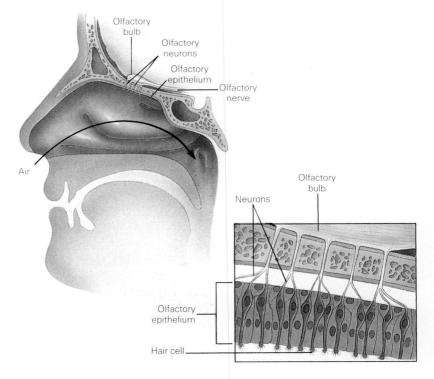

Noctuid moths have two **tympana,** one on either side of the thorax (the insect midsection), and each tympanum has only two receptors. Kenneth Roeder found that one, called the **A1 cell,** is sensitive to low-intensity (weak) sounds. The other, the **A2 cell,** responds only to loud sounds. Surprisingly, neither kind of receptor is very good at distinguishing frequencies (high versus low notes)—a sound of 20,000 Hertz (cycles per second) elicits the same neural action potential as one of 40,000 Hertz (a much higher sound).

As any sound becomes louder, however, the A1 cell fires more frequently and with shorter lag time between receiving the stimulus and firing. Also, the A1 cell shows a greater firing frequency in response to pulses of sound than to continuous sounds; it fires increasingly slower if subjected to a continuous sound. And it just so happens that the bats that prey on noctuid moths emit pulses of sound.

In a sense, the moth has beaten the bat at its own game. Its very sensitive A1 cell is able to detect bat sounds long before the bat is aware of the moth. The moth cannot only detect the distance of the bat, but it can tell whether the bat is coming nearer, since the sound of an approaching bat would grow louder.

In addition, the moth is able not only to detect the distance of the bat, but also its direction. The mechanism is simple. If the bat is on the left side, the left thoracic receptors of the moth will be exposed to the sounds, while the receptors on the right will be shielded. Therefore, the left receptor fires sooner and more frequently than the right if the bat is on the left. If the bat is directly behind, both neurons will fire simultaneously. Thus, the moth can determine the distance and direction of the bat. But what about its altitude?

If the bat is above the moth, the bat sounds will be deflected by the upward beat of the moth's wings. But if the bat is beneath the moth, the wing beats will have no effect on the pattern of neural firing. The moth, then, decodes the incoming data, probably in its thoracic ganglion (from which the auditory neurons emerge) so that it pretty well has the bat pinpointed.

What does it do with this information? If the bat is some distance away, the moth simply turns and flies in the opposite direction, thus decreasing the likelihood of ever being detected. The moth probably turns until the A1 cell firing from each ear is equalized. When the bat changes direction, so does the moth.

Bats fly faster than moths, though, and if a bat should draw to within 2.5 m (8 ft) of the moth, the moth's number is up—at least if it tries to outrun the bat. So it doesn't. If the bat and moth are on a collision course (that is, if the moth is about to be caught), the sounds of the onrushing bat will become very loud. At this point, the A2 fiber begins to fire—the signal of imminent danger. These messages are relayed to the moth's brain, which then apparently shuts off the thoracic ganglion that had been coordinating the antidetection behavior. Now the jig is up and the moth changes tactics. Its wings begin to beat in peculiar, irregular patterns or not at all. The insect itself probably has no way of knowing where it is going as it begins a series of unpredictable loops, rolls, and dives. But it is also very difficult for the bat to plot a course to intercept the moth. The erratic course may take the moth to the ground where it will be safe since the echoes of the earth will mask its own echoes.

The noctuid moth's evolutionary response to the hunting behavior of the bat serves as a beautiful example of the adaptive response of one organism to another. Also, it shows clearly that the sensory apparatus of any animal is not likely to respond to elements that are irrelevant to its well-being. It is not important for moths to be able to distinguish frequencies of sound, but it is important that they are sensitive to differences in sound volume. Anyone who tried to train a moth to respond to different sound frequencies could only conclude that moths are untrainable.

appeal. It has been suggested that the hair on some part of our bodies may be useful in trapping our odors and aiding in subtle communication with the opposite sex. (One might think that the communication had best be *very* subtle). In addition, preliminary experiments show that people can generally distinguish between males and females on the basis of both breath and body scent. The groundwork has been laid for what should be some fascinating lines of research.

The relationship of sound pulses from a hunting bat and auditory neural firing in the hunted moth. (A) When a hunting bat, emitting its high-pitched sounds, approaches a noctuid moth from the side, the receptors on that side fire slightly sooner and more rapidly than those on the shielded side. (B) When the bat is behind the moth, the moth's receptors on both sides fire with a similar rapid pattern. (C) When the bat is above the moth, the moth's auditory receptors fire when its wings are up, but not when its wings cover the receptors on the down stroke.

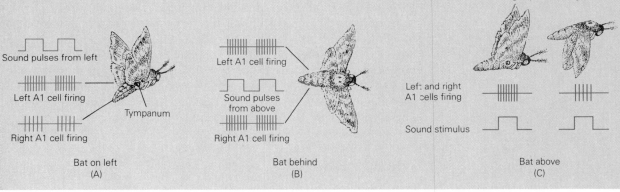

PROPRIOCEPTORS

The ability to distinguish how one's body is positioned is called **proprioception. Proprioceptors** also tell you where your body parts are relative to one another. (That seems a good thing to know.) This is an area that perhaps has not received enough research attention, but that may be because it normally works so efficiently.

FIGURE 23.8 Taste buds on the tongue are specialized to detect one of four flavors: bitter, salty, sweet, and sour. The photo, taken with a scanning electron microscope, reveals the flattened columns that contain the taste buds.

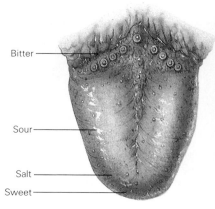
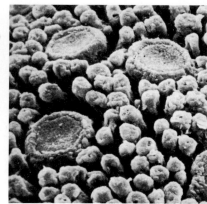

Bitter

Sour

Salt

Sweet

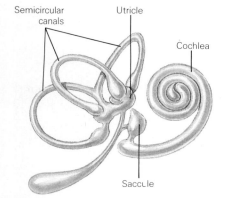

Semicircular canals

Utricle

Cochlea

Saccule

FIGURE 23.9 Semicircular canals. The middle ear is very sensitive to the body's position and movement. The semicircular canals lie at right angles to each other so that bodily movement in any direction shifts the fluid in at least one of them. It turns out, though, that humans are generally not particularly sensitive to accelerating motion, but very sensitive to vertical motion (and that's what usually makes you motion sick). The saccule and utricle are fluid-filled cavities in which grains (called otoliths) are embedded in a jellylike matrix. Movement of these grains send information to the brain regarding a change in direction or a change with respect to gravity.

Proprioception is common in many invertebrates and is achieved by a number of kinds of receptors. For example, some insects have sensors at the base of hairs, while crustaceans, such as lobsters, have receptors within the muscles.

Among vertebrates, many proprioceptors are found deep within the muscles. They are stimulated as the muscle stretches and places pressure on them. It is apparent that proprioception is particularly important among the more active or athletic vertebrates. Clumsy monkeys would tend to fall out of trees, and so we find that monkeys have, in fact, many proprioceptors.

Vertebrates have very precise equilibrium proprioceptors located in the inner ear. In most species, there are three fluid-filled loops, or **semicircular canals,** opening to two chambers (Figures 23.5 and 23.9). At the base of each canal is a bulge filled with jellylike fluid and lined with sensory hairs. Any change in the animal's position results in movement of the fluid, which then stimulates the hairs. Since the canals lie at right angles to each other, movement in any direction can be detected. The bulges are called the **utricle** and **saccule,** and they contain small granules that shift when the body moves, also stimulating the sensory hairs. If they are stimulated in certain ways, such as from rapid shifts in different directions, motion sickness may result.

VISUAL RECEPTORS

Humans are highly visual animals, and so we have a keen interest in sight. However, compared to some animals, especially birds, we don't see well at all. There is indeed a great deal of variation in visual ability among animals, and, not surprisingly, the ability is correlated with a species' simple *need* to see.

But how exactly is any visual receptor stimulated? And what stimulates it? In essence, it is sensitive to a particular part of the electromagnetic spectrum we call light. Light's wavelength ranges from about 430 nm (nanometers) to 750 nm (see Appendix B), but no animal can see more than part of this range. The shorter wavelengths, such as x-rays, beta rays, and gamma rays can't be detected by any animals, nor can the very long ones, such as radio waves. The detectable waves are absorbed by special **pigments** that then transform the wave energy into a neural stimulus.

In vertebrates, the light-sensitive part of the eye is the **retina.** It is composed of highly modified cilia of two kinds, **rods** (specialized to detect light) and **cones** (specialized for color) (Figure 23.10). The rods hold large amounts of a pigment called **visual purple.** When this pigment is activated by light, it "bleaches" (turns white) and the permeability of the rod changes. Depending on the kind of change, the rod either initiates an action-potential (fires) or it is inhibited from doing so. The interaction of many such rods sends a barrage of impulses to the brain where they are deciphered and integrated.

Whereas rods are sensitive to all wavelengths of visible light, cones respond only to specific wavelengths—that is, to specific colors. Humans and other primates have three kinds of cones that respond either to red, green, or blue. The multitide of colors we see depends on the interplay between these three.

The real question is, of course, do bulls see red? The answer is, not very well, if at all. (Cats may be able to barely detect red, but they cannot be depended upon to charge red capes.) In fact, real color vision is found only among some species of insects, fishes, reptiles, and birds, and among mammals. (In this last group, the primates are the color specialists.)

Since life is an opportunistic phenomenon, living things must have ways of detecting their opportunities (and minimizing their risks). This detection, we see, is the result of a dazzling interplay among a host of specialized receptors that comprise the senses. Next we will see just how animals use this information to get along in their special parts of the world.

FIGURE 23.10 The eye is actually a rather tough structure. The chamber in front of the lens contains a fluid called aqueous humour; the large chamber behind is filled with vitreous humour. The white of the eye is covered by the cornea and the colored part of the eye is the iris. The lens is focused by the muscles that support it. The sensory area, the retina, is composed of rods and cones that send impulses to the brain over the optic nerve. Where the nerve enters the retina there are no receptors (the "blind spot"). The most sensitive part of the retina is a tiny pit with a very high density of receptors, the **fovea.** When threading a needle, we usually turn our head in such a way that the image falls on the fovea. The longer structures in the photo are cones.

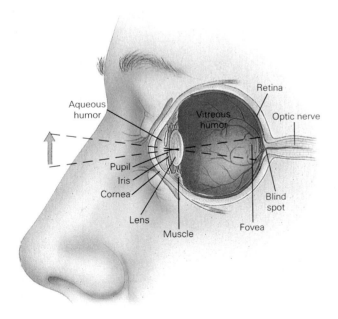

SUMMARY

Our senses are special groups of receptors that respond to very specific parts of the environment. We rely on our senses to tell us the nature of our surroundings—at least those parts of our surroundings that are important to us. Different kinds of species may sense different aspects of the environment. Thermoreceptors measure temperature. Most invertebrates lack them, and in the vertebrates there is disagreement about how heat is registered. Tactile receptors respond to touch and fire upon being distorted. They are common in invertebrates. Vertebrates can detect both touch and pressure. Auditory receptors detect sound, vibrating columns of air. Many invertebrates lack specific receptors, but insects detect sound by the vibration of special membranes. Land vertebrates, including humans, detect sound in this way, also. But here it is transferred over a series of three bones to the cochlea, where it vibrates fluids and moves hairlike structures that stimulate the auditory nerve. Chemoreceptors detect the presence of chemicals. They include both olfactory receptors (smell) and gustatory receptors (taste). The invertebrates boast the most sensitive chemoreceptors, especially the insects. Among vertebrates, the chemoreception is best developed in carnivores and rodents. Proprioceptors detect the body's position and are common in many invertebrates. In vertebrates, they are found in the muscles and are stimulated by the tension placed on the muscles. In the vertebrates, equilibrium is largely maintained by receptors in the semicircular canals. Visual receptors are stimulated by certain wavelengths of light as it falls on pigments that transform the lightwave into a visual stimulus. In vertebrates, the light sensitive part of the eye is the retina, which is composed of rods and cones, specialized to detect light and color, respectively.

Chapter 24

THE MECHANISMS OF BEHAVIOR

Those people who study animal behavior professionally must dread those times when their cover is blown at a dinner party. The unfortunate souls are sure to be seated next to someone with animal stories. The conversation will invariably be about some pet that did this or that, and nonsense is the *polite* word for it. The worst stories are about cats. The proud owners likes to talk about their ingenuity, what they are thinking and how they "miss" them while they're at the party. Those cats would rub the leg of a burglar if the burglar rattled the Friskies box.

The stories about dogs are sometimes not so outlandish. Dogs probably do miss their owners; it's probably easier to know what's crossing their minds; and some do tend to protect the house. Even serious scientists have investigated the stories of dogs that, after having been left behind, found their owners after travelling hundreds of miles to places the dogs had never been. Still, even some of the dog stories leave you in danger of falling face down into the soup.

Why should there be so much misunderstanding about animals? After all, we humans have lived in intimate association with other animals for thousands of years. Not only have animals been important in our cultures, but our evolutionary histories are inextricably linked as well. Yet, somehow, we seem to have learned very little about them. One reason may be that they often seem to be so much like us, and it is easy to assume that they are just somewhat eager but rather ineffective versions of humans. Not so, and our clichés about angry bees, fierce lions, wise elephants, proud eagles, cowardly hyenas, and loving cats may have impeded our attempts to truly understand animal behavior at any meaningful level.

Our goal here, then, will be twofold. First, we will try to take a close look at the structure of animal behavior. Then we will see how the various behaviors help the animals to fit more precisely into their corner of the world. Essentially, we will be asking what animals do and why they do it.

ETHOLOGY AND COMPARATIVE PSYCHOLOGY

We should begin by comparing the two major scientific attempts to unravel the basic questions of animal behavior. Keep in mind that the study of animal behavior certainly isn't new. Even the ancients had theories to account for such things as migration and territorial behavior. A comprehensive and unifying theory that covered cases and made sense, however, was not developed until the 1930s. The theory was encompassed in an explanation of the idea of **instinct.** The foundation of the modern concept of instinct is attributed primarily to the Austrian Konrad Lorenz and his coworkers. The Dutch zoologist Niko Tinbergen, at Oxford University, amended and clarified the scheme in 1951, and the result was an idea that, in spite of its imperfections, demanded serious consideration. And then the trouble started. Some people just couldn't accept instinct as a primal factor in animal behavior.

Lorenz and Tinbergen, who shared the 1973 Nobel prize in medicine and physiology with the German biologist Karl von Frisch, called the new field they had founded **ethology.** It was conceived and cultivated in Europe, and it addressed questions of innate behavioral patterns. Ethological studies, at least at first, were designed to learn about the evolution of behavior in the wild, and they were done primarily in the field, under conditions as natural as possible, and by observing tame or semi-tame animals.

Disagreement came from American **comparative psychologists,** whose research dealt primarily with learning processes in laboratory animals, typically the Norway rat. Many of the arguments continue to this day, but not often by the principal researchers. As each field has matured and as researchers have learned more about each other's work, the lines have become blurred. Ethologists have begun to come into the lab, and comparative psychologists are learning where birds go in winter. The final integration of the two approaches probably depends upon further data from another area of study, **neurophysiology,** dealing with how the nervous system works. At this point, however, let's consider the ethological interpretation of animal behavior.

THE DEVELOPMENT OF THE INSTINCT IDEA

It was once believed that human behavior results entirely from learning, but that other species respond to "instincts"—patterns that are indelibly stamped into their nervous systems at birth. But the term instinct was never precisely defined and, because of this, the entire concept fell into disrepute. Many behavioral scientists began to hesitate even to use the word. The problem was compounded by the fact that while ethologists were seeking to clarify the concept, the word instinct found its way into the general population. It was handled casually, twisted around, and tossed about. It was used to explain everything from a baby's grip to homemaking to swimming.

Many people seem to think that instinct is any **innate behavior**—something that appears naturally in an animal and is not due to learning. However, this is only part of the definition of an instinct. Instincts are in fact innate, but they are also characteristic of a species, just as bone structure is. Further, animals may experience "urges" to perform instinctive acts. Finally, the performance of such an act brings a measure of relief. Let's look at the various facets of instinct in more detail.

Appetitive and Consummatory Behavior

Instinctive behavior is usually divided into two parts, the **appetitive** and **consummatory** stages. Perhaps the clearest examples of appetitive behavior are seen in feeding patterns, from which the name is probably derived. Appetitive behavior is usually variable, or nonstereotyped, and involves searching—for example, for food. The specific objective of this appetitive behavior is the performance of consummatory behavior. Consummatory behavior is highly stereotyped and involves the performance of a fixed action pattern (discussed next). Swallowing is the consummatory pattern in feeding behavior and involves a complex and highly coordinated sequence of contractions of throat muscles (Figure 24.1). These contractions are the components of a fixed action pattern. The performance of consummatory behavior is followed by a period of rest or relief, during which the appetitive patterns cannot be so easily initiated again.

Fixed Action Patterns and Orientation

The first time a tern chick is given a small fish, it will manipulate it so that it goes down head first, the spines safely flattened. Dogs that have spent their entire lives indoors will attempt to bury a bone by making scratching motions on the carpet as if they were moving earth. They may also circle several times before lying down, although there is no grass to trample nor spiders to chase out. Behavioral patterns such as these that are innate and characteristic of a species are called fixed action patterns. It is such fixed action patterns—coupled with precise orienting movements—that produce instinctive behavior. These animals have had little or no opportunity to learn these behaviors, and all the members of their species perform them in about the same way. Thus, we can say that instinctive behavior (1) is specific to the species, (2) is genetically based, and (3) has appetitive and consummatory phases.

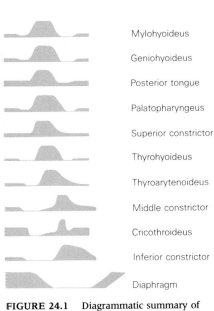

Mylohyoideus

Geniohyoideus

Posterior tongue

Palatopharyngeus

Superior constrictor

Thyrohyoideus

Thyroarytenoideus

Middle constrictor

Cricothroideus

Inferior constrictor

Diaphragm

FIGURE 24.1 Diagrammatic summary of the sequence, timing, and intensity of the muscular contractions involved in swallowing in the dog. A fixed pattern such as swallowing, then, may have many components. In order for the pattern to be performed correctly, each component must function in a precise way. Such precision and complexity builds a certain conservatism into the overall pattern since one part cannot change drastically without affecting the other parts.

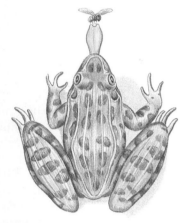

FIGURE 24.2 A leopard frog orienting to catch a fly. The tongue flick, a fixed action pattern, is not released until the central nervous system is stimulated by the sight of an insect in the proper position (close and in the midline). The orientation is obvious as the frog shifts its position, but once the tongue flick has started, no adjustment is possible; it will continue to its completion.

To see how the fixed and orienting patterns operate together in the performance of a consummatory act, consider the fly-catching movements of a frog (Figure 24.2). For the tongue flick (the fixed action pattern) to be effective, the frog must carefully orient itself according to the fly's position. Done correctly, the frog will catch a fly and be the envy of us all. However, the orientation must be precise, because once the fixed action pattern has begun, it cannot be altered. If the fly should move after the pattern starts, the frog will miss, but the sequence will be completed anyway. So we see the frog orient, shifting around so that the fly is directly ahead. Then we see the tongue flick. Thus we distinguish the orientation movements from the fixed action pattern.

Releasers

Instinctive behavior is released by specific signals from the environment. Stalking behavior in cats may be initiated by the sight of prey. In male ring doves, courtship bowing is normally released by the sight of an adult female. The environmental signals that evoke, or release, instinctive patterns are called **releasers.**

A releaser may be only a small part of any appropriate situation. For example, fighting behavior in territorial male European robins can be released not only by the sight of another male within their territories, but even by the sight of a tuft of red feathers at a certain height (Figure 24.3). Such a response usually "works" because red feathers at that height within the territory are normally on the breast of a competitor. The point is that the instinctive act can be triggered by only certain parts of the total environmental situation.

The strength of the stimulus necessary to release the innate pattern can change with time. Daniel Lehrman of Rutgers University found that a courting male dove isolated from females soon began to bow and coo to a stuffed model of a female—a model that it had ignored previously. Later, when the model was replaced by a rolled-up cloth, he began to court the cloth; and when this was removed, the sex-crazed dove directed his attention to a corner of the cage, where he could at least focus his gaze. It seems that the threshold for release of the behavior pattern became increasingly lower as time went by without the sight of a live female dove. It was almost as though an "energy" for performing courting behavior were building up within the male dove. Finally, the minimum stimulus necessary to elicit a response lowered to a point at which almost any stimulus would initiate his courting behavior. (This might be referred to as the "singles bar syndrome.")

Supernormal Releasers

As we saw in the case of the courting male dove, the specific signal that can act as a releaser can change in time so that weaker, or nonadaptive, releasers come to be accepted. Conversely, stronger-than-normal releasers may elicit responses more readily than normal ones. For example, an oyster-catcher may preferentially choose to brood an abnormally large egg with spots more distinct than normal, in preference to its own egg (Figure 24.4). Such abnormally strong signals are called **supernormal releasers.**

FIGURE 24.3 A male European robin in breeding condition will attack a tuft of red feathers placed in his territory. Since red feathers are usually on the breast of a competitor, it is to his reproductive advantage to behave aggressively at the very sight of them. The phenomenon illustrates that releasers of instinctive behavior need to meet only certain criteria—in other words, need to represent only a specific part of the total situation.

(A) (C)

(B) (D)

Innate Releasing Mechanisms

No one knows precisely how releasers work, or where they operate in the nervous system, but one theory is that there are certain neural centers, called **innate releasing mechanisms,** or **IRMs,** that, when stimulated by the perception of a releaser, trigger a chain of neuromuscular events. According to the theory, these movements produce instinctive behavior. Many IRMs may be involved in complex patterns, so that a number of releasers must be perceived in the proper sequence. There is also some evidence that in complex sequences, the performance of one instinctive act may serve as the releaser for the next act.

A Falcon on the Hunt: A Summary

Now let's consider a drama that might put some of this together for us. After daylight a peregrine falcon begins its hunt. At first its behavior is highly variable; it may fly high or low, or bank to the left or the right. It is driven by hunger and would be equally pleased by the sight of a flitting sparrow or a scampering mouse. Suddenly it sees a flock of teal flying below. Its behavior now becomes less random as it swoops toward them in what is the next stage of ''teal-hunting behavior.'' First the random search, and then the swoop.

Almost all falcons perform this dive in just about the same way, although there are some variations. This is a sham pass to scatter the flock, and the falcon is likely to fall through any part of it. The teal, at the sight of the falcon, have closed ranks for mutual protection. (After all, it is hard to pick a single individual out of a tightly packed group. A predator is likely to shift its attention from one individual to another, and in failing to concentrate on a single prey, it may come up empty-handed.)

FIGURE 24.4 An oyster-catcher *(Haematopus ostralegus)* shows a preference for a giant egg rather than a normal egg (foreground) or a herring gull's egg (left).

The sham pass works though, and the terrified teal scatter. The falcon immediately beats its way upward, and after picking out an isolated target below, it begins a second dive.

If we were to watch this same falcon perform such a dive on several occasions, we might notice that the procedure is much the same every time. And if we were to watch several different falcons, we might notice that they all perform this action in much the same way. The greatest amount of variation is during the earliest part of such dives, since, at this stage, what the falcon does will be dictated in part by what the teal does. The falcon's options become fewer, and its behavior becomes more and more stereotyped, as it descends. Finally, there is a point at which the falcon's action is no longer variable. It now initiates the last part of the attack, the part that is no longer influenced by anything the teal does. At this instant, the falcon's feet are clenched, and it may be traveling at over 150 miles an hour. Unless the teal makes a last-second change in direction, it will be knocked from the sky, and the falcon's hunt will have been successful.

The falcon's energy will then be spent in following the teal to the ground, where, with rather unvarying and stereotyped movements, it will pluck the teal, tear away its flesh with specific movements of its head, and swallow the meat. The sequence of muscle actions in swallowing is very precise and always the same. The swallowing action itself seems to bring a measure of relief, so that the falcon does not feel like hunting again for a while. Theoretically, it is the desire to perform swallowing movements that brought the falcon out to hunt in the first place. In other words, the falcon was searching for a situation that would enable it to swallow. This probably seems a bit bizarre, but it's easy to see how such a system would work—that is, provide food.

We can derive several points from this example. First, the state of the animal is significant. The falcon had not eaten for a time and was hungry. Hunger thus provided the impetus, or "motivation." In general, if an instinctive action pattern has not been performed for a time, there is an increasing likelihood of its appearance. Second, the earlier stages of this instinctive sequence were highly variable, but the behavior became more and more stereotyped until finally, the bird had the opportunity to orient a morsel of food and to perform the final, stereotyped, and satisfying act of swallowing. This stage would have been repeated until the bird lost the urge to perform it; that is, until the bird's hunger was satisfied.

LEARNING

The very word *learning* is revered among humans. We sacrifice and excuse a great deal at its altar. (In fact, it's what you're supposed to be doing at this very moment.) We even like to see it in other species ("I tell you, Chester, that dog's *smart*."). But what is learning? How does one learn? And what does "smart" mean?

The problem is that we haven't learned much about learning in spite of a great deal of research effort. Not only are we vague on how it occurs in humans, but each species seems to have its own learning propensities, so establishing general principles is exceedingly difficult.

This is one of the problems in attempting to compare the "I.Q." of various species. We tend to consider the one possessing abilities closest

FIGURE 24.5 Chimpanzees are highly social creatures that live in a complex, variable, and changing world. Intelligence is important under such circumstances.

to ours the more intelligent. So, in spite of perhaps being able to learn more tricks, a pig is not necessarily smarter than a horse. And what about the clearly unintelligent species? Tapeworms, for example, have never been considered particularly profound. But then, why should they be? They live in an environment that is soft, warm, moist, and filled with food. The matter of leaving offspring is also simplified; they merely lay thousands and thousands of eggs and leave the rest to chance—sheer blind luck.

In contrast, there are species for which intelligence or the ability to learn is critical. Chimpanzees live in complex and variable environments. They must be able to cope with a variety of conditions. It is no coincidence that they are long-lived, highly social, and mature slowly. These are the traits that are associated with high learning ability. They have time to learn and a social system that enables them to learn from each other (Figure 24.5).

There is increasing evidence that many species are able to learn more than has been assumed (Essay 24.1). Bees, for example, aren't usually thought to be very bright. But in experiments in which food is moved a certain distance further away from a beehive each day, bees seem to extrapolate. After the first few moves, they predict where it will be and fly directly to it. Some birds, it is reported, drop tiny white feathers onto the surface of a pond and then capture whatever little fish come to investigate. Squirrels are usually considered smarter than birds, but they have memory problems. They find very few of the nuts they bury in the summer. However, some birds that also bury nuts find almost every one, even months later. The point is that intelligence is both difficult to define and measure. As a general rule, as we stressed earlier, it seems that animals are able to learn those kinds of things that are important for them to learn. (Memory is certainly involved in learning and, by an odd twist, *forgetting* can also be important (Essay 24.2)).

Although there may be a number of fundamentally distinct ways animals learn, we will consider only three major types: habituation, classical conditioning, and operant conditioning.

Habituation

Habituation is, in a sense, learning *not* to respond to a stimulus. The first time an animal encounters a stimulus, it may respond vigorously. But if the stimulus is presented over and over, the response to it gradually lessens and may finally disappear altogether **(extinction).** Habituation is not necessarily permanent, however. If an animal habituated to some stimulus does not encounter it for a period of time, it may respond if the stimulus later reappears. As an example, bluejays become ill after eating monarch butterflies. After only one nauseating experience, the birds will avoid them. However, in the absence of further encounters, after awhile the birds seem to forget and must learn the lesson all over again.

Habituation may be quite important in the lives of many animals. For example, a bird must learn not to waste energy by taking flight at the sight of every skittering leaf. A coral-inhabiting fish may come to accept its neighbors, but will immediately attack a strange fish. The stranger might be seeking to displace the resident fish, and so the response is adaptive. It may also help to explain why animals respond to the sight of a furtive predator they see only rarely, while ignoring the harmless spe-

The octopus is a cephalopod, and like all cephalopods, is a vigorous predator. It can walk about on its "arms" or use jet propulsion by squirting water from its mantle cavity. It is soft-bodied and, thus, spends much of its time hidden safe in crevices of rocks. The octopus can quickly change its color or pattern to match its surroundings or blanch to a pale "white" (thus accenting its eyes) to escape predators. It may also project clouds of ink when threatened. The ink may temporarily blind an attacker and numb its smell receptors. If placed exposed into a tank, it will immediately build a house out of any material available. The octopus is perhaps the most intelligent invertebrate. It is also curious, active, and perpetually hungry, so it is an excellent animal to use in behavior tests.

The octopus has large eyes and it sees about as well as humans. It is primarily a visual hunter and at the sight of a crab, it turns dark, breaks out in blotches of color, and rushes toward its prey. If the prey is behind glass, the octopus will press against the glass in its excited efforts. If a tentacle should reach over the glass and contact the prey, the success doesn't seem to register with the octopus. It will continue to "pursue" the prey—almost as if the information from the tactile (touch) and visual receptors were not integrated.

The octopus can quickly learn to discriminate between horizontal and vertical visual images (possible because the neurons in its optic lobes are arranged at right angles rather than randomly as in most animals). But the octopus is unable to distinguish between cylinders A and B, which have both been cut away 30 percent. On the other hand, it can distinguish these from C, which has been cut away 20 percent. Apparently, discrimination is on the basis of what proportion of the object comes in contact with the suckers.

If an octopus learns to perform a pattern with one tentacle, there must be a time lag before that knowledge is transmitted to other tentacles.

---

cies they see more often. Habituation is often not given the attention it deserves in studies of learning, perhaps because it seems so simple. But it may well be one of the more important learning phenomena in nature.

Classical Conditioning

Classical conditioning was first described by the famed Russian biologist Ivan Pavlov (Figure 24.6). In classical conditioning, the response to a normal stimulus comes to be elicited by a substitute stimulus. In his experiments, Pavlov found that dogs would salivate at the sight of food. He then began to switch on a light five seconds before food was dropped onto the feeding tray. After doing this a few times, he presented the light without the food and found the dogs would continue to salivate. On the basis of numerous experiments, Pavlov found that the number of drops of saliva elicited by the light alone was in direct proportion to the number of previous trials in which the light had been followed by food.

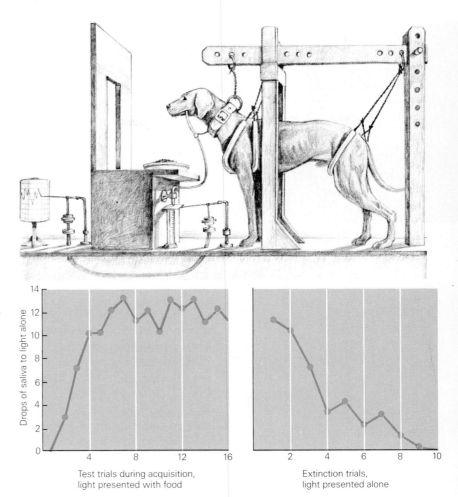

FIGURE 24.6 The Russian biologist Ivan Pavlov and the apparatus he devised to demonstrate classical conditioning. Upon presentation of a light, meat powder would be blown into the dog's mouth, causing it to salivate. Later it came to salivate at the sight of a light alone. The salivation, then, was conditional on the light. Note in the first graph that the dog salivated at maximal levels after only eight trials. When the experiment was reversed and food no longer followed the light, the dog stopped salivating after only nine trials.

This conditioning process also worked in reverse. When a dog that had come to associate light with food was shown the light over and over again without following it with food, salivation began to decrease in response to the light and finally stopped altogether.

Operant Conditioning

Operant conditioning differs from classical conditioning in several important ways. Whereas in classical conditioning the **reward** (or reinforcement, which is defined as something that increases the probability that an action will be repeated) such as food, follows the stimulus, in operant conditioning, the reward follows the behavioral response. Also, in classical conditioning the experimental animal has no control over the situation. In Pavlov's experiment, all the dog could do was wait for lights to go on and food to appear. There was nothing it could do one way or the other to make it happen. In operant conditioning, the animal's own behavior determines whether or not the reward appears.

In the 1930s, the noted psychologist B. F. Skinner demonstrated operant conditioning by employing a device now called a **Skinner box** (Figure 24.7). An animal placed inside a Skinner box must learn to press a small bar in order to receive a pellet of food from an automatic dis-

FIGURE 24.7 B. F. Skinner, one of the most important twentieth-century psychologists, building a "Skinner box," which is used to demonstrate operant conditioning.

penser. Skinner found that when an experimental animal (usually a hungry rat or hamster) was first placed in the box, it ordinarily began a random investigation of its surroundings. When it accidentally pressed the bar, a food pellet was delivered. The animal did not immediately show any signs of associating the two events, bar pressing and food, but in time it began to hang around near the bar. As more and more rewards appeared, its behavior became less random until finally it learned to press the bar. Eventually, it spent most of its time just sitting and pressing the bar. Skinner called learning through such a sequence operant conditioning. (The essential differences between classical and operant conditioning are shown in Figure 24.8.)

The relative importance of each type of learning to animals in the wild isn't known at this point. It is likely that most adaptive, or beneficial, behavior patterns arise in nature as interactions of several types of learning.

HOW INSTINCT AND LEARNING CAN INTERACT

Over the years, much of the argument in the field of behavior has been over one simple question: Is a certain behavior innate or learned? After years of bickering, the question is no longer regarded as valid. The supposition that a particular behavior stems entirely from one source or another neglects the myriad ways in which behavioral components interact. In a sense, it's like asking whether the area of a triangle is due to its height or its base. A better question is: How do innate and learned patterns *interact* in the development of a particular behavior?

As an example of such interaction, consider the development of flight in birds. Flight is usually considered a largely innate pattern. Obviously, a young bird must be able to manage it pretty well on the first attempt or it will crash to the ground as surely as would a launched mouse. It was once believed that the little fluttering hops of nestling songbirds were incipient flight movements, and that the birds were, in effect, learning to

FIGURE 24.8 In classical conditioning, a desirable commodity, such as food, comes to be associated with an irrelevant signal until the irrelevant signal alone can elicit an involuntary response normally associated with the commodity. Here, the animal learns passively. In operant conditioning, the animal can act when given a signal, but only one action is rewarded. This action then comes to predominate.

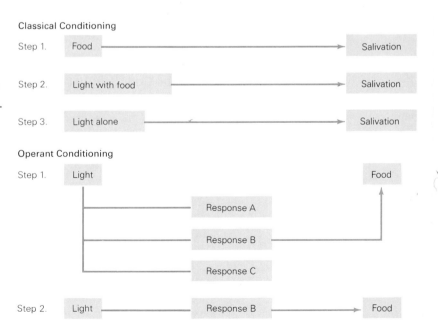

There are all sorts of strange abilites associated with memory in humans. **Idiot savants,** a special class of retardates, have very low I.Q.s, but some are able to accomplish incredible mathematical feats, such as multiplying two five-figure numbers in their heads. Others can immediately tell you the day of the week on which Christmas day fell in 1492—or any day in any year (although this is probably not a memory feat). A normally intelligent Russian man made his living giving stage performances as a **mnemonist.** He would sometimes memorize lines of fifty words. Once, in just a few moments he memorized the nonsense formula:

$$N \cdot \sqrt{d^2 \cdot \frac{85^3}{vx}} \cdot \sqrt[3]{\frac{276^2 \cdot 86x}{n^2v \cdot 264}} \cdot n^2 b$$

$$= sv \cdot \frac{1624}{32^2} \cdot r^2 s$$

Fifteen years later, upon request, he repeated the entire formula without a single mistake.

And what about those people with "photographic memories"? They exist, and they are called **eidetikers.** Proof of their abilities has been demonstrated with stereograms. These are apparently randomized dot patterns on two different cards; when they are superimposed by use of a stereoscope, a three-dimensional image appears. One person was asked to look at a 10,000 dot pattern with her right eye for one minute. After ten seconds, she viewed another "random" dot pattern. She then recalled the positions of the dots on the first card and conjured up the image of a T.

So, if such abilities are possible for our species, why haven't they been selected for so that by now we can all, more or less, perform such feats? On the surface, the advantages seem enormous. The reason we can't is, in part, because there are serious drawbacks to remembering everything. Both the Russian mnemonist and the eidetiker could look at a barren tree, "recall" its leaves, and, when they looked away, be confused over whether the tree was leafy or not. The mnemonist could watch the hands on a clock and not notice they had moved, remembering (or "seeing") only where they had

been. Also, what about all those insignificant events it is of no advantage to remember? The energetically expensive neural apparatus would be wasted retaining such information for recall (assuming such retention takes more energy than the storage of material we normally can't recall). And the mnemonist had trouble with discerning time lapse. He recalled everything so well that it seemed to him as if every event of his life had just occurred.

Obviously, the key here is that the reason we can't remember as well as the people in these examples is that we, as a species, have not found it necessary or useful. We generally don't need to recall every stone on the path to the place where we found food yesterday; we only need to remember the location of the path. In fact, remembering too much about our physical environment might mean that changes in the environment would not be adjusted to as quickly as if we were never quite sure what to expect. Perhaps it is best that we can't rely too strongly on our memories.

---

fly (practicing) before they left the nest. But then someone performed some experiments in which a group of nestlings were allowed to flutter and hop, while others were reared in boxes which prevented any such movement. Then, at the time the young birds would have normally begun to fly, both groups were released. The restricted birds flew just as well as the ones that had "practiced"!

It is apparent, then, that flight behavior is largely an innate, or unlearned, pattern. However, generally, young birds do not fly as well as adults (Figure 24.9). Even with flight, then, the innate pattern can be improved upon by learning through practice.

In other studies, the innate and learned components of a behavior have been more clearly differentiated. If you have ever watched a squirrel open a nut, you may have been impressed with its speed and efficiency. Red squirrels, for example, first cut a groove where the shell is thinnest, along the growth lines on the flat sides of the nut. Then they bite into the groove and break the shell open. In one study, baby squirrels were reared in an environment where they were never exposed to

solid bits of food that they could handle with the paws. Interestingly, when later given hazelnuts, they performed the correct gnawing actions and even inserted their incisors into the grooves correctly, demonstrating that these patterns are largely innate. The problem was that the young squirrels still had trouble actually opening the nuts. They gnawed all over the surface of the nut and started several grooves without completing any of them. After only a few such trials, however, they began to make their grooves at the thinnest part of the shell and easily opened the nuts. So most of the squirrels quickly became proficient at opening hazelnuts as their innate gnawing pattern was modified through learning, thereby producing the complex adaptive result.

White-crowned sparrows demonstrate a very specific kind of learning. In order to be able to sing the song of their species, they must hear the song at a particular time during a brief "sensitive period" early in life. If they are exposed to the song after this period, they will produce abnormal songs, lacking the finer details. At the period during which they must hear the song, they are not yet even able to sing.

White-crowned sparrows (Figure 24.10) are wide ranging, and the species forms subpopulations that tend to breed among themselves and sing their own local dialects of the basic song. If young birds from different populations are isolated at hatching and reared in soundproof chambers, they all begin to sing the same rather basic song. Thus, they are apparently born with the basic pattern; the local embellishments of each population are learned later.

Interestingly, isolated hatchling white-crowned sparrows can learn the recorded dialect of any subpopulation of the species while in their learning periods. But if they are exposed to the song of another species, they will not learn that song and will sing as if they had been reared in a soundproof cage. Finally, if a young white-crowned sparrow is deafened after it has heard the proper song but before it has had a chance to sing, it will sing only garbled passages. It apparently must hear its own song in order to match it to its inborn "template." The development of song in white-crowned sparrows, then, serves as an example of the interaction of learned and innate patterns in producing an adaptive response.

FIGURE 24.10 Studies of song development in white-crowned sparrows have led to a number of surprising findings, helping to shed light on how instinct and learning can interact.

ORIENTATION AND NAVIGATION

South Americans probably wonder where the robins go every spring. The robins, of course, are "our" birds; they simply vacation in the south each winter. Furthermore, they fly to very specific places in South America and will often come back to the same tree in your yard the following spring. The question is not why they would leave the cold of winter so much as how they find their way around. The question perplexed people for years, until in the 1950s a German scientist named Gustave Kramer provided some answers—and, in the process, raised new questions.

Kramer initiated important new kinds of research regarding how animals orient and navigate. **Orientation** is simply facing in the right direction; **navigation** involves finding one's way from point A to point B.

Early in his research, Kramer found that caged migratory birds became very restless at about the time they would normally have begun migration in the wild. Furthermore, he noticed that as they fluttered around in the cage, they often launched themselves in the direction of their normal migratory route. He then set up experiments with caged starlings and found that their orientation was, in fact, in the proper migratory direction—except when the sky was overcast. At these times, there was no clear direction to their restless movements. Kramer surmised, therefore, that they were orienting according to the position of the sun. To test this idea, he blocked their view of the sun and used mirrors to change its apparent position. He found that under these circumstances, the birds oriented with respect to the position of the new ''sun.'' They seemed to be using the sun as a compass. Of course, this was preposterous. How could a stupid bird navigate by the sun when we lose our way with road maps? Obviously, more testing was in order.

So, in another set of experiments, Kramer put identical food boxes around the cage, with food in only one of the boxes. The boxes were stationary, and the one containing food was always at the same point of the compass. However, its position with respect to the surroundings could be changed by revolving either the inner cage containing the birds or the outer walls, which served as the background. As long as the birds could see the sun through any window, no matter how their surroundings were altered, they went directly to the correct food box. Whether the box appeared in front of the right wall or the left wall, they showed no signs of confusion. On overcast days, however, the birds were disoriented and had trouble locating their food box (Figure 24.11).

In experimenting with artificial suns, Kramer made another interesting discovery. If the artificial sun remained stationary, the birds would shift their direction with respect to it at a rate of about 15 degrees per hour, the sun's rate of movement across the sky (see Essay 24.3). Apparently, the birds were assuming that the ''sun'' they saw was moving at that rate. When the real sun was visible, however, the birds maintained a constant direction as it moved across the sky. In other words, they were able to compensate for the sun's movement. This meant that some sort of biological clock was operating—and a very precise clock at that.

What about birds that migrate at night? Perhaps they navigate by the night sky. To test the idea, caged night-migrating birds were placed on the floor of a planetarium during their migratory period. A planetarium is essentially a theater with a domelike ceiling onto which a night sky can be projected for any night of the year. When the planetarium sky matched the sky outside, the birds fluttered in the direction of their normal migration. But when the dome was rotated, the birds changed their direction to match the artificial sky. The results clearly indicated that the birds were orienting according to the stars.

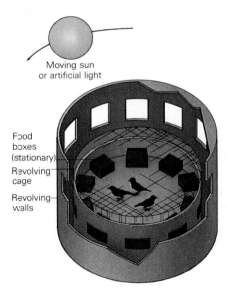

Moving sun
or artificial light

Food
boxes
(stationary)

Revolving
cage

Revolving
walls

FIGURE 24.11   Kramer's orientation cage. The birds can see only sky through the glass roof. The apparent direction of the sun can be shifted with mirrors.

It's the night before a big test. You are trying to study, but your pet hamster, active this night as usual, has a squeaky running wheel. The sound is so distracting that you decide to put your furry friend in the closet. A few nights later, the squeak coming from the closet reminds you that the hamster hasn't seen the sun, or anything else, for three days. Although it had no environmental time cues, such as a light-dark or temperature cycle, the hamster was still active at night, the customary time.

How did the hamster do it? How did it know when it was night? Most researchers assume many living things can measure the passage of time by an internal or biological clock.

Hamsters are not the only organisms with a biological clock. Indeed, the behavior and physiology of most organisms from protists to humans are rhythmic. In fact, rhythms are so common that rhythmicity should be considered a fundamental property of life.

The prevalence of biological rhythms is not surprising when we remember that life evolved in a cyclic environment. Behavior fluctuates in a repeating pattern so that any pattern ideally occurs at the appropriate time of day, in keeping, say, with the state of the tides, phase of the moon, or the season of the year. The rhythmicity on earth reflects the rhythmic movements of certain heavenly bodies, such as the earth, the moon, and the sun. The relative movements of the earth, moon, and sun cause regular changes in such things as light, temperature, geomagnetism, barometric pressure, humidity, and cosmic radiation. Because these environmental changes have been so regular and so predictable, evolution has been able to adjust behavior and physiology to match these cycles.

It has been argued that living things do not possess an internal clock, but they are merely responding to environmental stimuli. But the evidence does not support this. If you recorded the activity of your hamster while it was in the perpetual darkness of your closet, you would notice that, although bouts of activity regularly alternate with rest, the length of this activity cycle is slightly different than 24 hours. In other words, in constant conditions, daily rhythms are "about a day" in length, or circadian (*circa*, about; *diem*, day). If the hamster were responding to environmental cues, it would stay on a 24-hour cycle. Instead, environmental cues seem to keep its clock precisely "set."

Although biological rhythms continued without environmental cues, they are not completely independent of such cues. For example, light-dark cycles will set, or entrain, the rhythm so that its period length matches that of the environment.

There is accumulating evidence indicating that birds navigate by using a wide variety of environmental cues. Other areas under investigation include magnetism (Figure 24.12), landmarks, coastlines, sonar, and even smells. The studies are complicated by the fact that the data are sometimes contradictory and the mechanisms apparently *change* from time to time. Furthermore, one sensory ability may back up another.

COMMUNICATION

If you should come upon a large dog eating a bone, walk right up to it and reach out as if to take the bone away. You may notice several changes in its appearance. The hair on its back may rise, its lips may curl back to expose its teeth, and it may utter peculiar gutteral noises. The dog is communicating with you. To see what its message is, grab the bone.

It would be a remarkable understatement to say that communication is important in the animal world. Let's admit that and add that animals communicate in a variety of ways and that communication may have a number of advantages. We should also keep in mind that no matter what its immediate effects are, its ultimate function is to increase reproductive success. It can do this *directly*, such as in advertising for a mate or in precopulatory displays, or *indirectly*, as in warning offspring of danger.

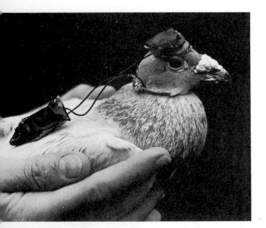

FIGURE 24.12   A pigeon with electrical coils on the head. Birds carrying coils with reversed polarity tended to home in opposite directions, indicating that pigeons can detect slight magnetic fields and that magnetism can influence their orientation.

Internal clocks have another advantage over clocks set by external cues. For example, some animals must be able to *anticipate* critical changes in their surroundings so that they have adequate time to prepare. A fiddler crab scurrying along the beach must return to its burrow before the tide returns or the waves will wash it away. Other animals use clocks to synchronize their behavior to an event that they cannot sense directly. This is the case for honeybees that travel to distant patches of flowers to gather nectar. Different types of flowers open their petals at different times of the day. The bees' clocks allow them to time their nectar gathering forays so that they arrive when the flowers are open. They might visit morning glories early in the day but want to visit four o'clocks in the afternoon.

Clocks can be used not only to determine the time of day, but how long it has been since some event.

For animals such as the birds and the bees, the second function of a clock, measuring the passage of time, is particularly important because it is essential for sun compass orientation (see text). A homing pigeon flying south would be required to keep its path of flight at a 45° angle to the right of the sun at 9 A.M., but would have to change that angle by about 15° an hour as the sun moved across the sky.

What makes the biological clock tick? We don't know. However, because rhythms exist in single cells and protists, we conclude that the clock must be intracellular. Cellular processes that may play a role in the timing process are protein synthesis on the cytoplasmic ribosomes, transport of ions across the plasma membrane or perhaps proton transport in the mitochondrion.

If a single cell can tell time, does every cell in a multicellular organism have its own "wristwatch"? Apparently so. Also, isolated tissues often remain rhythmic. For example, it has been found that if the heart of a hamster is removed and kept alive in a tissue culture, it will continue to beat more rapidly at night than during the day. Even a single heart cell will display a daily rhythm.

If there are many clocks in an animal, then they must be set to the same time or there would be internal chaos. Indeed, animals may have master clocks that synchronize the timepieces in individual cells. In mammals such as the rat, the master clock seems to be in a region of the brain called the suprachiasmatic nucleus. When neurosurgeons destroy this tiny group of cells, the rat's running activity, its drinking patterns, and several of its normal hormone rhythms disappear.

Judith Goodenough

Remember, the charge to all living things is "reproduce or your genes will be lost." Of course, the "charge" is really only an expression of simple arithmetic: the earth is now the home of the descendants of previous reproducers. The children of nonreproducers are not with us. (Put another way, if your parents didn't have any children, you probably won't either.) Communication, then, helps animals to carry out that "reproductive imperative."

Let's consider a few general methods of communication, and, in so doing, perhaps we can learn something about the ways in which animals influence the behavior (or the probability of behavior) of other animals. That, after all, is what communication is all about.

Visual Communication

Visual communication is particularly important among certain fish, lizards, birds, and insects—and some primates as well. Visual messages may be communicated by a variety of means, such as through color, posture or shape, movement, or timing. As we have seen, color serves as a component of the releaser of territorial fighting in European robins. In grackles, the female solicits the attention of the male by assuming a head up posture while fluttering her drooping wings. Some male butterflies

FIGURE 24.13 A male mandrill (left) has permanent markings that advertise his sex, but his signals of anger are temporary and dependent on his mood. It is probably often to his advantage to be perceived as a male, but less frequently advantageous to be perceived as an angry male. A male baboon (right) displaying his large canine teeth as a threat signal. These animals are powerful and dangerous.

are attracted to the female by her performance of a specific flight pattern. And as an example of communication by timing, fireflies are attracted to each other on the basis of their flash intervals, each species having its own frequency. (Some predatory fireflies flash at the frequency of females of another species and then eat whomever comes to call.)

Visual signals may be permanent or temporary. Permanent signals are seen in the elaborate coloration of the male pheasant or the striking facial markings of the male baboon. On appropriate occasions, they can emphasize their "machismo" by behavioral signals, such as the strutting of the pheasant and the glare of the baboon, or decrease it, for example by a baboon looking away. Visual signals may be more temporary, such as the reddened rump of female chimpanzees during estrus. A male baboon exposing his long canine teeth as a threat is producing an even more short-term signal (Figure 24.13).

Short-term visual signals have the advantage that they can be started or stopped immediately. If a displaying bird suddenly spots a hawk and "freezes," its position will not be given away by any lingering images in the environment. Also, visual messages reveal the exact location of the sender. A receiver can then respond in terms of its precise location, as well as its general presence and behavioral state (aggressive, romantic, or whatever).

Visual signals also have certain disadvantages. For example, if a sender can't be seen, it can't communicate. Visual signals are often useless at night or in dark places (except for light-producing species). Also, distance weakens visual signals, so the signal must become bolder and simpler. This means it can carry less information. The only way to intensify a visual signal is to increase its visibility, either by decreasing the distance or by increasing its size or contrast. It isn't possible to pump more energy into the signal itself, as it is with certain other forms of communication.

Sound Communication

Sound plays such an important part in communication in our own species that you may be surprised to learn that it is generally limited to

462   CHAPTER 24   THE MECHANISMS OF BEHAVIOR

arthropods and vertebrates. (Surprised, eh?) Insect sounds, such as those of the cricket and cicada, are usually produced by some sort of friction—for example, by rubbing the legs against each other or the legs against the wings. Arthropods rely largely on differences in cadence (beat or timing) rather than pitch or tone, as do most birds and mammals.

Vertebrates communicate by sound in a number of ways. Some fish produce sound by frictional devices in the head area or by manipulating the air bladder. Land vertebrates, on the other hand, usually produce sounds by forcing air through vibrating membranes in the respiratory tract, but they often communicate by sound in other ways. Rabbits thump the ground, gorillas pound their chests, and woodpeckers hammer drainpipes early on Sunday mornings.

Most of the lower vertebrates rely on signals other than sound, but some species of salamanders can squeak and whistle. One species barks, and another indelicately forces air out its cloaca, perhaps as an attempt to gross out its predators. Frogs and toads advertise territoriality and choose mates at least partly by sound. Most reptiles do not communicate by sound, but territorial bull alligators can still be heard roaring and "gulping" in southern lakes and swamps (if they have escaped becoming shoes). Darwin wrote about the roaring and bellowing of mating tortoises when he visited the Galapagos Islands in 1839.

Sound can vary in pitch (low or high, depending on the frequency of the sound waves), in volume, and in tonal quality. The last is demonstrated as two people hum the same notes, yet their voices remain distinguishable. It is possible to visually depict sounds by use of a **sound spectrogram.** This sort of translation makes it possible to more precisely analyze sounds. Figure 24.14 shows a sound spectrogram that indicates how a message can dictate the nature of a signal.

Sounds have a potentially high information load since they can be varied in frequency, volume, timing, and tonal quality. They are distinguishable at low levels, but they can also be made to carry over great distances by pumping more energy into them—that is, by becoming louder. In addition, sounds are transitory; they don't linger in the environment after they have been emitted. Thus an animal can stop its sounds should its situation change—for example, by the appearance of a

FIGURE 24.14 (A) Sound spectrograms of the mobbing calls of several species of British birds while attacking an owl. The sounds have qualities that make them easy to locate. Such calls are low-pitched *chuk* sounds. Different species of birds have developed similar calls through convergent evolution. That is, their calls serve much the same purposes and were developed under relatively similar conditions, thus the qualities of the sounds came to be somewhat alike.

(B) Sound spectrograms of five species of British birds when a hawk flies over. Such calls are high-pitched, drawn out *(tweeee)*, and difficult to locate. They, too, achieved their similarity through convergent evolution.

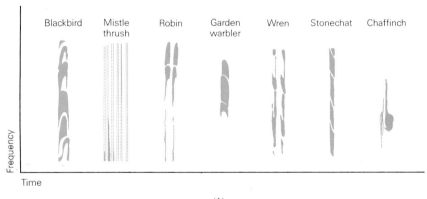

(A)

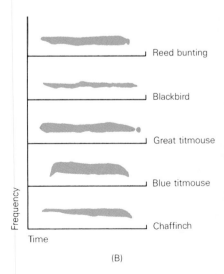

(B)

predator. A further advantage is that an animal doesn't ordinarily have to stop what it is doing to produce sounds. Furthermore, sounds, unlike visual images, can bypass many environmental objects.

The disadvantages of sound communication include its reduced effectiveness in noisy environments. (Some seabirds that live on pounding, wave-beaten shorelines rely primarily on visual signalling.) Also, sounds weaken with distance and are often hard to locate, especially underwater.

Chemical Communication

You have probably seen ants rushing along single file as they sack your cupboard. You may also have taken perturbingly slow walks with dogs that stop to urinate on every bush. In both cases, the animals are communicating by means of chemicals. The ants are following chemical trails laid down by their fellows, and the dog is advertising his presence to anyone who would possibly care.

Insects make extensive use of chemical signals, and certain of their signals can elicit very vigorous, if rigid and stereotyped, behavior. For example, ants produce an "alarm chemical" when they encounter some sort of threat. As the chemical permeates the area, other ants react by rushing about in a very agitated manner and attacking campers. Perhaps the best example of such communication is found among the moths. Female silk moths produce a "sex attractant" that can signal males from very long distances. The substance is called **bombykol,** and it is so potent that a male moth can be aroused by a single molecule or a glass rod dipped in a solution diluted to one part in a trillion.

The immediate and invariable response produced by certain chemical signals is similar to the response to hormones. However, hormones work *within* an individual, and these chemicals work *between* individuals. They are called **pheromones** (*pherin,* to carry; *horman,* to excite).

It was at first believed that pheromones existed only among the insects, but then it was found that pheromonal interactions also occur in mammals. If pregnant rats of certain species smell the urine of a strange male, they will abort their fetuses and become sexually receptive again. We are also aware that females of most mammal species signal sexual receptivity by some chemical means. There is evidence that humans may also communicate pheromonally. Some researchers maintain that scent can be a subtle sexual cue. Also, it has frequently been reported that women living together often begin to cycle in menstrual synchrony, apparently in response to pheromonal cues.

Chemical signals are extremely potent in very small amounts, and so they normally do not require great metabolic expense to produce sufficient quantities. Also, because they persist in the environment, the sender and receiver do not have to be precisely situated in time and space in order to communicate. In addition, chemicals can also move around many sorts of environmental obstacles.

Chemical communication, as does every other means of signalling, has certain disadvantages. For example, because chemicals are so precise in their configuration, their information load is limited. Moreover, because chemicals linger in the environment, they may advertise the presence of the signaler to predators as well as to the intended receiver.

We have indeed made great leaps, both conceptually and technically, in understanding the mechanisms of behavior. Yet this field remains one of the most perplexing, unresolved, and controversial in science. The problem is compounded because of the implications for our understanding of human behavior, perhaps the greatest challenge of all. We may someday find out how birds know where to go in winter, but we may never know what goes on in the mind of the investigator studying those birds.

SUMMARY

Ethology is an approach to animal behavior, developed in Europe, dealing primarily with instinctive actions of wild animals. Comparative psychology was first an American discipline that considered learning in laboratory animals. These fields are now overlapping. The instinct concept was developed by ethologists and includes the searching (appetitive) stage and the satisfying (consummatory) stage. Consummatory behavior has an orienting and fixed component. The orienting component is adjustable and responsive to whatever is happening in the environment. The fixed action pattern is stereotyped and, once initiated, continues to completion. Consummatory behavior is followed by a period of relief.

An instinctive act is species-specific, genetically based, and has appetitive and consummatory phases. An instinctive action is triggered by releasers in the environment. The releasers are believed to activate innate releasing mechanisms in the nervous system that send impulses to the proper effectors (muscle groups).

There are a number of kinds of learning. Habituation is learning not to respond to (usually harmless) stimuli. Classical conditioning involves the association of unrelated stimuli so that a normally irrelevant stimulus can initiate an autonomic response. Operant conditioning involves learning to perform a behavior in order to receive a reward. Normally learning and instinct interact at some level to produce an adaptive behavior.

Orientation and navigation are important in a variable world. Orientation involves facing the right direction; navigation involves finding one's way from point A to point B. Among birds, daytime migrators apparently often navigate by the sun and nighttime migrators by the night sky, but there is strong evidence that birds also navigate by other cues.

Communication is important in survival and reproduction among many kinds of animals. Visual communication can be accomplished through color, posture, shape, movement, or timing. It may be a permanent or temporary aspect of an animal's appearance. It can by started and stopped immediately, but is often easily blocked. Sound communication is most important to arthropods and vertebrates. It can vary in frequency, volume, timing, and tone. It can go around environmental obstacles, and energy can be added to the signal. Sounds are transitory—they do not linger—but they are less useful in noisy environments. Chemical communication does linger, and it can be used over very long distances and can go around obstacles. The information load is limited.

Chapter **25**

THE ADAPTIVENESS
OF BEHAVIOR

A striped skunk sniffs the air and squints into the cool mountain dusk,
then whirls and aims its rear at a lynx that leaps backward quickly and
then pads away, glancing over its shoulder. Some distance away, a burly
badger ceases from its digging and lopes, hairs abristle, toward a brows-
ing bear that has wandered too close. The bear retreats, and both return
to their foraging. Overhead, a flock of geese beats against a steely sky.

We may wonder how the lynx came to understand about the rear
ends of skunks (we can certainly imagine). And we may ask what triggered the charge of the badger and what motivates the geese. These
questions are difficult, but there are others, of another type, that may be
easier to answer. For example, we can ask how these behaviors are
useful, how they help the animals to succeed. This, then, will be our goal
here. We will ask questions about the adaptiveness of recognition,
aggression, cooperation, and altruism. The answers should provide us
not only with information about behavior, but they may also illustrate
some fundamental points about how natural selection influences animals.

RECOGNITION

Species Recognition

One of the more critical roles of communication is to enable an animal to identify members of its own species. After all, animals of two different species cannot normally produce healthy offspring. But sometimes species are very similar to each other, and in the absence of some precise means of identification, an animal might waste a lot of time and energy in trying to mate with a member of the wrong species or in driving away apparent competitors that are not a threat. Therefore, species identification must somehow be quickly established.

As we saw in Chapter 10, the golden-fronted and the red-bellied woodpecker may share the same woods in central Texas (Figure 25.1). The two species appear remarkably similar to the casual observer, the most conspicuous difference being a small yellow band along the base of the red cap of the golden-fronted woodpecker that is absent on the red-bellied. The habits of the two species are much alike, and their calls are very similar, at least to the human ear. Nevertheless, each of these birds is somehow able to recognize its own species. Cross mating has never been observed, nor have any hybrids. No one knows exactly what cue the birds use, but to them, apparently, the signals are clear.

Interestingly enough, whereas the two species ignore each other sexually, they have come to treat each other as competitors, and each will eject a member of the other species from a territory as quickly as they would a member of their own species. Apparently, the stimuli that release sexual and territorial behavior are quite different. Often the same signals that are employed in mating responses also are used to identify competitors (since the strongest competitors are likely to be individuals of the same species).

Population Recognition

In some cases, it is advantageous for an animal to recognize subgroups within the species, such as those of its own population. The advantage lies in enabling subgroups to adapt to different conditions across the

FIGURE 25.1   As we saw in our discussion of evolution (Chapter 10), golden-fronted (left) and red-bellied (right) woodpeckers are remarkably similar. However, they do not interbreed. They obviously have very specific ways of identifying their own species for reproductive purposes, while they also behave as if they recognize each other as strong competitors.

species' range. Consider a wide-ranging bird species that extends across different sorts of habitats. Some subpopulations might live in dark, moist lowlands and therefore be dark-colored, while other subpopulations live on dry mountaintops and are light-colored, making them less conspicuous against the light-colored dry earth. Should a lowland individual mate with a bird from the highland, their offspring might be conspicuous in both types of terrain, adapted to neither, and thus might be easily spotted by predators. In such a situation, there would be a tendency for members of each population to breed with their own type. This would not be a matter of prejudice or aesthetics, but just a result of the simple fact that such selective individuals would tend to produce more offspring. Of course, interbreeding often does occur where populations adjoin and intermediate types can be found there.

Individual Recognition

In some species, it is important to be able to recognize specific individuals within one's population. You may have noticed that the gulls at the beach all look alike—at least the adults do. This is because you aren't a gull (perhaps people all look alike to them). Experiments have shown, however, that gulls can recognize their own mates on sight and that they are able to filter out the calls of their mates from the raucous din of the clamoring gulls wheeling above. By recognizing one's mate, energy can be directed toward more efficient production of offspring through more precise cooperation with a reproductive partner (Figure 25.2).

In species in which the sexes come together only for copulation, the mate-recognition problem is reduced to sexual identification. Male sal-

FIGURE 25.2 Problems of individual recognition appear insurmountable to a human observer at a colony of penguins, such as this king penguin rookery on South Georgia Island. However, they can tell each other apart. After the female lays her single egg, she may wander off to fish for weeks or months, leaving the male to stand over the egg, living off his own body fat. When she returns to relieve him, she may approach many males before she recognizes the gaunt father of the egg.

FIGURE 25.3 A male salticid spider must approach the voracious female very carefully. She is larger and could easily overpower him. Of course a female that attacked her suitors would leave no genes in the next generation and that sort of behavior would die out. Still, he must approach her carefully. He advances a few careful steps at a time, stopping to wave specialized appendages in a very specific way. If he performs this correctly, he will cause changes in her nervous system that will inhibit her predatory behavior, at least long enough for him to climb astride her and rake a sperm-bearing arm across her genital opening until she is impregnated. He may then jolt her from her mesmerized state by a parting pinch as he scurries away. He may make it; he may not.

ticid spiders approach a female, *any* female, only for mating (Figure 25.3), and then only very carefully, waving special appendages as a display. After all, she is larger and stronger, but he must risk it simply because he is descended from generations that did. Mating is brief, and he quickly clears out. If she encounters him afterward, she will recognize him only as prey. So individual mate recognition is important only in species that establish pairs, and it increases in importance with the likelihood of a pair's interaction with other individuals.

Individual recognition is also important in the maintenance of hierarchies, or rank. Once animals in a group know their rank with respect to the others, there will be less wasteful and dangerous fighting over food or other commodities.

AGGRESSION

The old image of "Nature, red in tooth and claw" has recently been superseded by the popular notion that humans are the only animals that regularly kill members of their own species, while other animals get along with their own kind. Is this true? Let's see what is actually going on out there.

First, we should note that **aggression** is belligerent behavior that normally arises as a result of competition. An animal shows aggression mainly toward other individuals that tend to utilize the same resources. Thus, it is likely to be aggressive toward those most like itself. Those most like itself are, of course, of the same species and sex. And most aggressive interactions occur within species, between members of the same sex. Predatory behavior, by the way, is not aggressive. A cheetah is about as aggressive toward an antelope as you are toward a hamburger.

Fighting

The most blatant form of aggression is fighting. We can discount the old films of leopards and pythons battling to the death; such fights simply aren't likely to happen. What does a python have that a leopard would risk its life for, or vice versa? Fighting is much more likely between stronger competitors.

Animals of different species do sometimes fight, of course. For example, those golden-fronted and red-bellied woodpeckers that fail to find enough in common to breed, find enough similarities to exclude each other from their respective territories by fighting. And lions may attack and kill African cape dogs at the site of a kill. The lions don't eat the dogs; they just exclude them. In fact, interspecific aggression may be more common than we assume among animals (Figure 25.4).

The precise methods of fighting vary widely, but however animals fight, most species have means of avoiding injury to each other (Figure 25.5). Such avoidance has several benefits. First, no one is likely to get hurt. Although the opponent is permitted to continue its existence, the possibility of having to compete with him again is less risky than serious fighting.

Fighting between dangerous combatants is usually a stylized ritual and relatively harmless. For example, horned antelope may gore an

FIGURE 25.4 In some cases, fighting may occur between species that rarely interact at any level. Here, a raccoon approached a hare in the darkness and in the scuffle, the hare leaped over the raccoon and administered a severe drubbing with its powerful back legs to the raccoon's head. Actually, they may have known each other. There is evidence that animals of different species that share the same area may come to recognize each other individually and may react to each other on an individual basis.

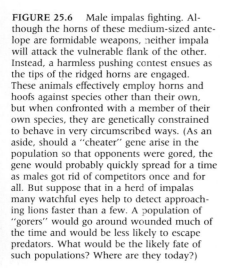

FIGURE 25.5 Male rattlesnakes fighting. The triangular heads show that their poison sacs are full. Each could kill the other, yet they do not bite. The fight is more a test of will and strength as the snakes press against each other, belly to belly. Finally, the weaker individual yields and his head is pushed to the ground by the stronger animal. He then retreats, and no one dies.

FIGURE 25.6 Male impalas fighting. Although the horns of these medium-sized antelope are formidable weapons, neither impala will attack the vulnerable flank of the other. Instead, a harmless pushing contest ensues as the tips of the ridged horns are engaged. These animals effectively employ horns and hoofs against species other than their own, but when confronted with a member of their own species, they are genetically constrained to behave in very circumscribed ways. (As an aside, should a "cheater" gene arise in the population so that opponents were gored, the gene would probably quickly spread for a time as males got rid of competitors once and for all. But suppose that in a herd of impalas many watchful eyes help to detect approaching lions faster than a few. A population of "gorers" would go around wounded much of the time and would be less likely to escape predators. What would be the likely fate of such populations? Where are they today?)

FIGURE 25.7 Hornless females of the Nilgai antelope have no inhibitions against attacking the flank of a competitor, but their butts are quite harmless. Keep in mind that "harmless" is used only in the immediate sense. The butt itself may not be dangerous, but if it establishes dominance so that the loser is deprived of commodities, the result can have far-reaching implications. An individual deprived of food, after all, is more likely to fall prey. If losing means that she is deprived of a mate, the result is genetic death, even if she should live a long life. Her genes will die with her. If losing one fight meant that the animal would have absolutely no chance of breeding, or contributing her genes to the next generation, they might be expected to fight to the death. A loss for this species, however, usually just means a temporary setback, so it behooves the loser to accept it gracefully and attempt to breed another time. Interestingly, horned males of the same species almost never attack in this way, nor do horned females of other species.

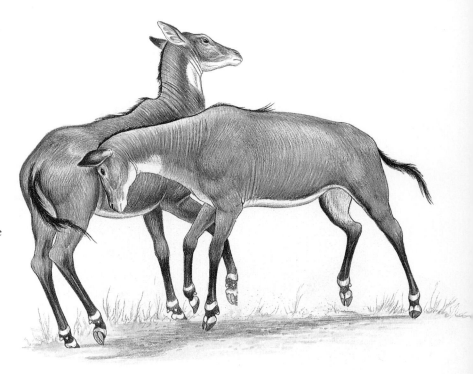

attacking lion, but when they fight each other, the horns are never directed toward the exposed flank of the opponent (Figure 25.6). Such stylized fighting enables the combatants to establish which is the stronger animal and, once dominance is established, the loser is usually permitted to retreat.

All-out fighting may occur between animals that cannot injure each other seriously, such as hornless female antelope (Figure 25.7). It may also occur between animals that are so fast that the loser can escape before serious injury, as is the case in house cats.

Why don't antelopes gore each other? An incurable romantic might assume that it is because they don't want to hurt each other. In all likelihood, what the two antelope "want" has little to do with it. The fact is, they *can't* hurt each other. When the system works, an antelope could no more gore an opponent than fly! Perhaps the sight of an opponent's exposed flank inhibits butting behavior. Conversely, the sight of an opponent head-on might release the stereotyped fighting behavior.

There are species, by the way, in which combatants do fight to the death. If a strange rat is placed in a cage with established rats, the group may chase and sniff at the newcomer carefully for a period, but eventually, they will attack it repeatedly until it is dead. Guinea pigs and mice may also fight to the death. The males of a pride of lions are likely to kill any strange male they find within their hunting area, and a pack of hyenas may kill any member of another pack that they can catch. Even roving bands of male chimpanzees may attack and kill a male from another band. The chimpanzee findings were quite surprising since it had long been assumed that chimpanzees were essentially peaceful animals. Nonetheless, fights to the death are rare in most species.

COOPERATION

Cooperation may seem to be at the opposite end of the behavioral spectrum from aggression—aggression is not nice, cooperation is nice. But we will see that both aggression and cooperation might be termed "enabling devices." After all, they both function in enabling individual animals to survive and reproduce. Furthermore, just as humans are not unique in their aggressive behavior, neither are they alone in cooperating with each other.

Cooperative behavior occurs both within and between species. As an example of **interspecific** (between species) cooperation, consider the relationship of the rhinoceros and the tickbird that may be found clinging to the rhino's thick hide. The little bird gets free food while the rhinoceros rids itself of ticks and harbors a wary little lookout. The highest levels of cooperation, however, are found among members of the same species.

As an example of **intraspecific** (within species) cooperation, consider the behavior of porpoises. Porpoises are air-breathing mammals, much vaunted in the popular press for their intelligence, and their behavior often seems to support the claim. Groups of porpoises will protectively circle a female in the process of giving birth, driving away any predatory sharks that might be attracted by the blood. They have also been known to carry a wounded comrade to the surface where it can breathe. Their behavior in such cases is highly flexible and is influenced by prevailing environmental conditions. Such flexibility indicates that their behavior is not a blind response to innate genetic influences.

Cooperation among mammals is probably most commonly found in their defensive and hunting behavior. For example, the adult musk oxen of the Arctic form a defensive circle around the young at the approach of danger, standing shoulder to shoulder with their massive horns directed outward (Figure 25.8). The defense is effective against all predators except humans, since the beasts try to maintain this stance while they are shot one by one. Wolves, African cape dogs, jackals, and hyenas

FIGURE 25.8 Musk oxen live above the Arctic circle and are preyed upon by wolves. If attacked, they immediately form a circle with the adults facing outward and the calves inside. The behavior is an effective defense since the oxen are not very large and, individually, could be brought down.

FIGURE 25.9 Weaver ants of the genus *Oecophylla*, working together to repair a damaged leaf nest. Note their high degree of coordination and cooperation. When the sides of the leaves are pulled together, they will be sewn tight as other ants pass silk-spinning larvae back and forth across the gap, pressing them against the margins of the leaves.

often hunt in packs and cooperate in bringing down their prey. In addition, the hunting animals may bring food to those mates or young that were unable to participate in the hunt.

It might seem that intelligent and purposeful animals, such as mammals, would be expected to show the highest levels of cooperative behavior. It is a bit surprising, therefore, that cooperation is most highly developed in the lowly insects (Figure 25.9). The behavior is generally considered to be genetically programmed, highly stereotyped, and usually not greatly influenced by learning.

In honey bees, the queen lays the eggs and all the other duties are performed by the workers, sterile females. Each worker has a specific job at any given time. For example, newly emerged workers prepare cells in the hive to receive eggs and food. Then, in a day or so, their "brood glands" develop, and they begin to feed larvae. Later they begin to accept nectar from field workers and to pack pollen loads into cells. At about this time, their wax glands develop, and they begin to build combs. Some of these "house bees" may become guards, patrolling the area around the hive. Eventually, each bee becomes a field worker, or forager. She flies afield and collects nectar, pollen, or water, according to the needs of the hive. These needs are indicated by the eagerness with which the field bees' different loads are accepted by the house bees.

If a large number of bees with a particular duty are removed from the hive, the normal sequence of duties can be altered. Young bees may shorten or omit certain duties and begin to fill in where they are needed. Other bees may revert to a previous job that is now required again.

The watchword in a beehive is *efficiency*. In the more "feminist" species, the drones (males) exist only as sex objects, reproductive partners. Once the queen has been inseminated, the rest of the drones are quickly killed off by the workers. They are of no further use. The females themselves live only to work. They tend the queen, rear the young, and maintain and defend the hive. When their wings are so torn and tattered that they can no longer fly, they either die or are killed by their sisters. But the hive goes on.

ALTRUISM

We don't wish to shatter anyone, but, well . . . most of the Lassie stories aren't really true. Consider what would happen to the genes of any dog that was given to rushing in front of speeding trains to save baby chickens. The reproductive advantages would be considerable to chickens, but dogs with those tendencies might be selected out of the population by the action of fast trains. In contrast, the genes of a dog that spent his energy, not in chivalrous deeds, but in seeking out estrous females, would be expected to increase in the population. So what kinds of dogs are likely to predominate in the next generation?

If an animal is going to engage in "unselfish" deeds, its best reproductive bet lies in those deeds that advance the genes of other members of its own species. And as we will see, it is here that we are most likely to see selfless behavior.

Altruism may be defined in a biological sense as an activity that benefits another organism at the individual's own expense. It seems to be common among animals, but is it? A more difficult question is, if it is at all common, how did it evolve?

It is easy to see how certain forms of altruism are maintained in any population. For example, pregnancy, in a sense, is altruistic. The prospective mother is swollen and slowed. Much of her energy goes to the maintenance of the developing fetus. At birth she is almost completely incapacitated and is in marked danger. Pregnancy is clearly detrimental to her. So why do females so willingly take the risk? It may help to understand the enigma if we remember that the population at any time is composed of the offspring of individuals who have made such a sacrifice. Thus, the females in the population may, to one degree or another, be genetically predisposed to make such a sacrifice themselves.

However, altruism on this basis doesn't explain why a bird may feed the young of *another* pair, or why an African hunting dog will regurgitate food to almost *any* puppy in the group. Why, also, would a bird that may have no offspring of its own give a warning cry at the approach of a hawk, alerting other birds at the risk of attracting the hawk's attention to itself?

To answer such questions we must look past the answers that first come to mind. It may seem cynical, but we must start with the premise that birds don't give a hoot about each other. A bird that issues a warning call isn't thinking, "I must save the others." At least, there is a simpler explanation for its behavior.

Keep in mind that the biologically "successful" individual is the one that maximizes its reproductive output. One way of accomplishing this is for the organism to behave in such a way as to leave as many individuals carrying *its own type of genes* as possible in the next generation. This would explain parental care. However—and this may not be so readily apparent—an individual can also leave its type of genes in the next generation by helping a *relative's* offspring to survive. An individual shares genes in common with a cousin, although fewer, of course, than with a son or a daughter. Hence, there is, theoretically, a point at which an individual could increase its reproductive output by saving its nieces and nephews (provided there were enough of them) rather than its own offspring. From the standpoint of reproductive output, the organism would be better off leaving a hundred nephews than one son.

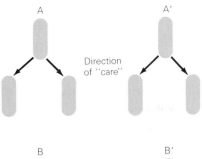

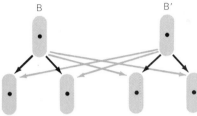

Direction of "care"

FIGURE 25.10 In this population, A and A' are nonaltruists. They behave in such a way as to maximize their own reproductive success, but do nothing to benefit the offspring of other individuals. In another segment of the population (B and B'), a gene for altruism has appeared that results in individuals benefiting the offspring of others in some way. It can be seen that, assuming the altruistic behavior is only minimally disadvantageous to the altruist, generations springing from B and B' are likely to increase in the population over those from A and A'. It should be apparent that the altruistic behavior is likely to be greatest where B and B' are most strongly related, so that B shares a maximum number of genes in common with the offspring of B' and vice versa. The idea is that B, for example, can increase its own reproductive success by caring for the offspring of a relative with whom it has some genes in common. After all, reproduction is simply a way of continuing one's own kind of genes.

FIGURE 25.11 An African vervet monkey giving an alarm call to warn others of the presence of a dangerous predator. Such risky behavior can be expected where the defender is likely to be strongly related to others in the group.

To illustrate, suppose a gene for altruism appears in a population (notice that this sets up the mechanism for the continuance of the behavior, but you can ask yourself if a behavioral tendency could also be passed along culturally in a verbal and social species such as our own). As you can see from Figure 25.10, altruistic behavior would most likely be maintained only in groups in which the individuals are related (that is, in which they have some kinds of genes in common). The behavior might be expected, then, in "nontransient" populations, those that don't move around much, and those in which there is little mixing with outside groups—in other words, where there is a high probability that proximity indicates kinship.

Keep in mind that *no conscious decision* on the part of the altruist is necessary. It simply works out that when conditions are right, those individuals that behave altruistically increase their kinds of genes in the population, including the "altruism gene." Nonrelatives would be benefited by the behavior of altruists, of course, but individuals near enough to receive the benefits of an act are likely, to some degree, to be relatives.

It has been determined mathematically that the probability of altruism increasing in a population depends on how closely the altruist and the recipient are related (Figure 25.11). In other words, the advantage to the recipient must increase as the kinship becomes more remote, or the behavior will disappear from the population. For instance, altruism toward siblings (brothers and sisters) that results in the death of the altruist will be selected for if the net genetic gain is *more* than twice the loss; for half-siblings, four times the loss; and so on. To put it another way, an altruistic animal would gain reproductively if it sacrificed its life for more than two siblings, but not for fewer, or for more than four half-siblings, but not for fewer, and so on. Therefore, we can deduce that in highly related groups, such as a small troop of baboons, a male might fight a leopard to the death in defense of the troop. By the same token, a bird will give a warning cry when the chance of attracting a predator to itself is not too great and when the average neighbor is not too distantly related.

This model helps to explain the extreme altruism shown by social insects, such as honey bees. Since workers are sterile, their only hope of propagating their own genotype is to maximize the egg-laying output of the queen. In some species, the queen is inseminated only once, so all the workers in a hive are sisters and have half their genes in common. (Why only half?) In such a system, then, almost *any* sacrifice is worth *any* net gain to the hive and to the queen.

The development of altruistic behavior is an admittedly somewhat esoteric topic. However, it seems important to consider even our most cherished and most despised behaviors in the context of their evolutionary history.

Reciprocal Altruism

In a brilliant essay, Robert Trivers expanded our understanding of the evolution of altruism by suggesting that in certain species, notably humans, altruism (outside of parental behavior) depends on the expectation of reciprocation: "I'll scratch your back if you'll scratch mine." **Reciprocal altruism** is an evolutionary strategy with some complex rules. Help (altruistic acts) is given to others—even offered to strangers—when the cost or risk is not too great to the giver. The expectation is that some kind of help will be reciprocated at another time. If the expected reciprocation is not forthcoming, the response is indignation and disapproval of the ingrate. The individual that fails to reciprocate is scorned, turned out of the social group, no longer aided. To get ahead, everyone has to play by the rules—or appear to.

Reciprocal altruism would be expected in those groups that are highly social (so the altruist is likely to encounter the beneficiary again) and relatively intelligent (so that individuals are recognized and their behavior can be remembered and repaid). Some evidence of intraspecific reciprocal altruism has been reported in troops of social mammals, such as hunting dogs and higher primates, but the evidence is not very strong for such behavior in any animals but humans. Trivers implies, in fact, that reciprocal altruism is the key to human evolution. The complexity of such behavior, entailing as it does memory of past actions, the calculation of risk, the foreseeing of the probable consequences of present actions, the possibility of advantageous cheating, and the need to be able to detect such cheating, all require a level of intelligence that is beyond most species. In the opinion of some anthropologists, it is exactly for the management of these elaborate social interactions that the human brain—and the conscious human mind—evolved.

SOCIOBIOLOGY AND SOCIETY

In the mid-1970s, an idea that had been around for decades was revived, reviewed, reanalyzed, refined, and reconsidered. Because it was a generally familiar concept and was stated very carefully in a professional format, the scientific community was surprised by the turmoil that followed. The turmoil has not yet subsided.

The idea is simply this: Social behavior can be influenced to some degree by natural selection, including the social behavior of humans. The idea had been around long enough that no one was really shocked by it, and its revival might have generated little discussion except for the reaction of a group of people who believed that the concept was based on a politically dangerous premise. They immediately set about to attack it, and their attack was so vigorous and well publicized that it drew a great deal of attention to an idea they didn't want to receive attention.

The defenders of the idea quickly rallied, and the arguments grew vitriolic as both sides firmed their position and stiffened their resolutions. It was an emotional argument, and some people apparently entered the fray with more opinions than information, and rather strange things were said. Furthermore, semantics became a large part of the problem because of sloganeering and loose definitions and because there was so little dialogue between the two groups.

So what was the problem? The opponents of the idea feared that a sociobiological explanation of human behavior would lead to a revival of the notion of **biological determination.** This is the idea that we are primarily the product of our evolution and that we therefore respond to the primitive calling of our heritage. As a result, our social behavior cannot be changed in any fundamental way and it is a waste of time to try. The opponents felt that the acceptance of sociobiology would mean resignation to the status quo. Furthermore, they said it supports such undesirable patterns as racism and sexism. They preferred to believe that our social patterns are molded by culture and learning and that we can change any undesirable trait through education, incentive, and social programs.

Sociobiologists strongly deny supporting biological determination and certainly denounce racism and sexism, but they believe that their approach holds real promise for building a better society. Perhaps it does seem a bit irreverent to some to suggest that human behavior, to any great degree, is programmed through evolution and that the forces of natural selection mold our own species as well as others. But the sociobiologists argue that if our behavior is, to *any* degree, genetically controlled or influenced, then we should know it. They note that we can't hope to find solutions if we don't understand the problem. And one of the reasons for our notable lack of success in improving society's condition is that we have ignored the role of our biological heritage. We must, they say, try to understand our social problems at every level and to use every tool we have to improve our lot. (We didn't develop the ability to fly by simply denying the existence of gravity.)

Sociobiology, in its refurbished form, is quickly maturing as data are now appearing from long term studies and as new researchers approach the problem from different angles. The next few years should be interesting as sociobiologists tighten their premises, precisely define their terms, and present us with new approaches to a weathered idea.

SUMMARY

The various types of behavior are adaptive in a number of ways. We consider recognition, aggression, cooperation, and altruism social behavior. Species recognition is important in mating and defense against competitors. Population recognition is important in the adaptation of subpopulations to specific habitats in a heterogeneous environment.

Individual recognition is important in mate recognition and hierarchies.

Aggression is usually a competitive interaction. Fighting among animals is rare but may occur between competitors, usually those that cannot do serious harm. Dangerous species usually solve conflicts stylistically in ritualistic shows or harmless tests of strength, but some species do fight to the death.

Cooperation is also an enabling device that helps individuals survive and reproduce. Among mammals, it is most common in defense and hunting behavior. Cooperative behavior is most developed in insects.

Altruism aids others at the altruist's expense. However, even the most apparently selfless acts may help the reproductive efforts of the altruist. Altruism can most be expected between related individuals because the beneficiary may carry the genes of the altruist. Usually, the greater the risk to the altruist, the greater the kinship between the altruist and the beneficiary. The less related the beneficiaries are to the altruist, the more of them must be saved if the altruistic act is to be performed.

Reciprocal altruism is a type of ''selfless'' act that is only likely to be performed if the favor is likely to be returned. It can only be expected in social and intelligent species, perhaps only among humans.

Sociobiology is the study of the effects of natural selection on social behavior. The approach is generally accepted for species other than humans, where it runs afoul of political theory.

Part Seven

LIFE'S BIG PICTURE

Chapter 26

BIOMES AND COMMUNITIES

If museums weren't generally large, modern artists like Jackson Pollock would be out of luck. A Pollock canvas viewed from close range is a confused, chaotic collection of drips, blobs, and smears. When the viewer can stand back so that the canvas is viewed as a whole, however, the various parts begin to fit and the organization appears.

So it is with life on this planet. Almost any aspect viewed up close seems to stand alone, a part of no grand scheme. But when one steps back, allowing the various parts to play together on the mind, the wholeness and completeness of it all becomes more apparent. The parts begin to be seen as one grand panorama.

Before the large picture could be seen, however, the smaller parts had to come into focus. Specialists have provided us with these narrow views, and those generalists called ecologists have begun to put them into place to form the larger picture. (**Ecology,** you recall, is the study of the interrelationship of the organism and the environment.) So let's take a look at the grand scheme of life and see how its components interact in the formation of it all.

HABITAT AND NICHE

It has been said that if the habitat is an organism's address, the niche is its profession. That's nice, but vague. Let's see if we can define the two terms a bit more precisely. The **habitat** is the place where an organism is found, and it can be described in several ways. For example, an animal may live in a desert habitat, or more specifically, in a briny desert pool. Furthermore, it may live in a certain part of that pool—its **microhabitat.** Wherever an organism lives, however, it interacts with its surroundings in myriad ways. The interaction involves the environment influencing the organism, which, in turn, influences the environment. The organism may also interact with the living things around it. The sum of all such interactions, along with the organism's own requirements, describe its **niche.** As you might imagine if you are beginning to gain some insight into the "scientific mind," there is some disagreement over precisely how these terms should be further defined, but let's not become involved.

It has become axiomatic that two species cannot occupy the same niche indefinitely. If two species were to find themselves in such a situation, it is generally predicted that one would be superior to the other under those conditions and would eventually replace the less fortunate species. Where species do coexist, then, we can assume that they are interacting with the environment in different ways (occupying different niches). This means that when you walk through the woods and see various species of small seed-eating birds in the same area, they are utilizing the habitat differently. This principle was nicely demonstrated by the ecologist Robert H. MacArthur, who showed that five species of American warblers that feed in spruce forests and seem to be occupying the same niche actually divide the trees into different feeding zones (Figure 26.1). Interestingly, the zones overlap more in times of food abundance.

Animals may also divide up a habitat in other ways. For example, they may utilize the same resources at different times of the day or at different times of the year, or they may utilize different commodities within the same part of the habitat.

Partitioning a habitat obviously reduces the level of competition between species. Generally, it is theorized, **competition** occurs where organisms utilize resources that are in short supply, or when they harm each other while seeking the same resource that is not in short supply. (Probably the latter cases are much more rare in nature.)

There is currently a vigorous disagreement among biologists regarding the evolutionary role of competition in nature. Part of the problem is that many of our ideas are based largely on intuition or supported by rather tenuous evidence. However, never deterred, we will assume that competition is important and now plunge ahead with a look at the earth's major habitats.

THE LAND ENVIRONMENT

It seems safe to say that no two organisms interact with their environments in identical ways. If they were asked, two roundworms living in a raccoon's intestine would probably describe their world differently, depending on their precise point of attachment and their individual per-

FIGURE 26.1 The feeding zones of five species of the North American warblers in spruce trees. The darker areas indicate where each species spends at least half its feeding time. By exploiting different parts of the tree, the species reduce their competition for food and thus they can occupy the same habitat. Studies such as this have been done on many species. It turns out that very similar species living in the same habitat tend to occupy different niches by subdividing the available resources. Since animals tend to behave so as to reduce competition, it has been suggested that competition must be an important factor in natural selection.

ceptive and reactive tendencies. Two wildebeest living on an African plain would not only disagree with the tapeworms about the nature of the world, but they would certainly have different experiences and might be expected to perceive their grassland environment quite differently. Even identical twins see the world from opposite sides of a baby carriage. Description, we see, is based on perception and interpretation, thus generalizations may not always be entirely valid. So now we will generalize.

We tend to think of our earth as a ponderous place that unfailingly provides its denizens with those things necessary for life. We often seem to forget, however, that life exists only in a thin film that veils the surface of this immense ball—a delicate shell wherein the wondrous forces of sunlight and water interact to permit life. This fragile film is responsive to a number of influences and, hence, is highly variable from one place to another. Furthermore, each place is likely to be unstable, so that it changes with time. Nonetheless, the different kinds of "places" in which life exists on the earth at present can be roughly categorized based on their present physical and biological properties. We should keep in mind, however, that these are merely arbitrary divisions of the great, complex, and intergrading areas of the earth.

Biomes

By using some imagination, we can divide the earth's land into several kinds of regions called **biomes** (Figure 26.2). Biomes are defined according to the plants they support. Of course, the makeup of the plant community is dependent on other factors, such as soil conditions, available

FIGURE 26.2 The world's biomes.

- Polar ice caps—white
- Tundra
- Taiga
- Coniferous forest
- Temperate deciduous forest
- Grasslands
- Chaparral
- Deserts
- Tropical rain forest
- Tropical deciduous forest
- Snow-capped mountains

FIGURE 26.3 Temperate deciduous forests are marked by distinct seasons. The winter, when limbs are bare, is quite different from the luxurious greenery of summer.

water, weather, day length, competition, and the nature and abundance of the resident plant eaters and their predators. Certain kinds of animals occupy each type of biome, since different species of animals are dependent on different sorts of plant communities for food, shelter, building materials, and hiding places. We will consider only the six largest types of biomes, but there are others, as we see in Figure 26.2.

Temperate deciduous forests (Figure 26.3) once covered most of the eastern United States and all of Central Europe. The dominant trees in these forests are hardwoods, such as oak, maple, beech, poplar, and hickory. The areas characterized by such plants are subject to harsh winters, times when the trees shed their leaves, and warm summers that mark periods of rapid growth and rejuvenation. Before the new leaves begin to shade the forest floor in the spring, a variety of herbaceous **annual plants** may appear, plants that live only one season—they bloom anew each spring only to die in the fall. Rainfall may average 30 to 50 inches or more each year in these forests.

People who live in temperate deciduous biomes often consider the seasonal changes as both moving and fascinating. They describe a certain joy that swells within them each spring and a secret pensiveness that overcomes them in the fall as the days darken and the forests become more silent. (Perhaps we are exceeding technical descriptions here, but these are my favorite places.)

Grasslands (Figure 26.4) occur in many parts of the world and are exemplified in the United States by the American prairie. Grasslands are characterized by grasses (not surprisingly), small bushes, shrubs, and, in some parts of the world, thickets of bamboo, which is a type of grass. The soil in such areas is usually porous. Trees may line streams and rivers. Rainfall averages between 10 and 40 inches each year, but may be erratic in its timing. Grasslands are found in both tropic and temperate zones. Interestingly, grasslands support more species of animals and a greater **biomass** (the total mass produced by all living things in the area) than any other kind of terrestrial habitat.

FIGURE 26.4 Grasslands support a surprising number of animals. Here, cape buffalo have left the shelter of the tall grass to move to water.

FIGURE 26.5 Deserts are fascinating places with great temperature variation because the air is unbuffered by water. Most of the species here have developed mechanisms for water conservation. Some, such as kangaroo rats, employ physiological mechanisms, while others, such as these pumas, tend to conserve water by behaving so as to minimize loss. For example, they may avoid direct sunlight except when hunting and may spend most of their time in higher, cooler areas.

Deserts (Figure 26.5) are characteristically hot in the daytime and cold at night. The 10 inches or less of rain that fall each year usually comes in sudden downpours, so that much of it runs off, sometimes causing flash floods and marked erosion; the rest quickly evaporates. Directly after a rainfall, the annual plants of the desert take advantage of the moisture and explode in an orgy of growth and seed production. Other desert plants meet the water-shortage problem in other ways, either by storing water in their tissues or by reaching far underground to tap water tables with their tap roots (of course). The most common animals of deserts are reptiles, arthropods, birds, and small mammals. They all must beat the heat and conserve water. One way they do this is by moving about mainly by night and reducing water loss through specialized excretory systems (see Chapter 17).

Tropical rain forests (Figure 26.6) are found mainly in the Amazon and Congo Basins and in Southeast Asia. The temperature in such biomes doesn't vary much throughout the year. Instead, the seasons are marked by the torrential rains that fall almost daily during the summer. In some areas, there may also be pronounced winter rainy seasons. These regions support many species of plants. Trees grow throughout the year and reach tremendous heights, with their branches forming a massive canopy overhead. The forest floor, which is quite open and easy to travel over, may be dark and steamy, often rather open and sparsely settled by smaller plants. Forests literally swarm with insects and birds. Animals may breed throughout the year as a result of the continual availability of food. Competition is generally considered to be very keen in such areas because of the abundance of species.

Where sunlight does manage to filter through the leafy canopy, **jungles** (Figure 26.7) are formed. These are the densely vegetated and tangled areas that movie companies love to hack their way through. Jungles often grow along river banks, and so river travellers who didn't want to get out and traipse around with the snakes once gained a mistaken impression of tropical rain forests and their descriptions colored the literature for years.

FIGURE 26.6 The interior of a tropical rain forest is often dark and quite open, permitting this Waorani hunter of Ecuador's Amazon basin to pursue monkeys and parrots with a blowgun and poisoned darts.

FIGURE 26.7 A Waorani hunter pursuing a capybara into the thick, jungle growth along the Cononaco River of Ecuador. Early river travellers believed the entire rain forest was this densely overgrown.

Tundra (Figure 26.8) is the northernmost land biome. It is covered throughout most of the year by ice and snow. This biome is most prevalent in the far north **(arctic tundra),** but it may also appear at high elevations in other parts of the world **(alpine tundra).** For example, in the United States, it may be seen in the high Rocky Mountains. Tundra appears in places where summer usually lasts two to four months, just long enough to thaw a few feet of the soil above the **permafrost,** or permanently frozen soil. Thaw brings soggy ground, and ponds and bogs appear in the depressions. The plant life consists mostly of lichens, herbs, mosses, and low-lying shrubs and grasses, as well as a few kinds of trees, such as dwarf willows and birches. Such plants obviously must be hardy, but their hardiness disguises their fragility. Once disturbed, these areas take very long periods to restore themselves. The wheel marks of wagons that passed over the arctic tundra a hundred years ago are clearly visible to this day.

FIGURE 26.8 After the snow melts in the brief arctic summer, the tundra often reveals a variety of plants, many with lovely, delicate hues that belie their rigorous existence.

FIGURE 26.9 The great evergreen forests of the taiga cover large areas of the northern hemisphere.

FIGURE 26.10 Where bogs and marshes form, the taiga is interrupted. These areas are often referred to as *muskegs*. They represent a perpetual tug-of-war between the aquatic and terrestrial environment, as plants continually invade the marshes, some failing, others becoming established.

Animal life on the tundra is surprisingly abundant for such a raw place. One finds not only ptarmigan, hare, ground squirrels, and lemming (along with their predators, such as owls, weasels, wolves, and foxes), but also large herbivores, including musk ox and caribou.

Taiga (tie-gah, Figure 26.9) is quite unmistakable; there is nothing else like it. It is confined almost exclusively to the Northern hemisphere and is identified by the great coniferous forests of pine, spruce, fir, and hemlock that extend across North America, Europe, and Asia. Some of these trees are the largest living things on earth (Essay 26.1).

Taiga is marked by long, cold, wet winters and short summer growing seasons. The forest is interrupted here and there by extensive bogs, or **muskegs,** which are the remains of large ponds (Figure 26.10). The forest floor is usually covered by a carpet of needles, with only a few shrubs and other plants able to push through. One may move silently on the muffling needles through the Canadian taiga observing a host of mammals, including porcupines, moose, bear, rodents, hares, and wolverines.

Ecosystems and Communities

Biomes are usually great complex expanses and are not always precisely identifiable. However, they are composed of smaller components that can be more precisely measured. For example, biomes can be theoretically broken down into ecosystems. Technically, an **ecosystem** is a group of interacting living things, together with all the environmental factors with which they interact. An ecosystem is considered an independent unit (although its independence would be hard to prove), and light is usually its only outside energy source. Conceptually, an ecosys-

It was 390 feet tall. Nothing on earth could match it. It had stood as a slender sapling in the cool coastal air, perhaps moving slightly in a light breeze, on the very day Caesar finally decided to move against Britain. But all that happened a long way from the area that would be called California. Great leaders were born as the tree grew. And they died as the tree became stronger and taller. Wars came and went, as well as plagues and famine. There were great celebrations and deep mournings here and there over the earth. The tree lived through it all.

As the ages passed, the tree continued to grow. No one marked the time when its crown reached above all the others, because it was only one of a vast forest of such trees. In time, however, it *was* noticed. Even before the start of the twentieth century, the straight, tall, and insect-resistant trees had caught the eyes of lumber companies. As the population of California grew, the trees began to be cut. Some citizens tried to establish a national park to save some of them, but the lumber industry blocked it in Congress. In what has been called "one of the greatest swindles of all time," they arranged to change nearly all the redwood lands from public to private ownership.

After World War II, California experienced a population surge of unprecedented dimensions. Factories and homes were being built feverishly and lumber was needed. Redwood was ideal for lawn furniture and tomato stakes, too. The conservationists, led by the Sierra Club, pushed again for parks to be set aside. By 1960, there were only two areas left that were relatively unscarred and of park caliber. While conservationists dickered among themselves over which area was the best, the lumber industry was busy in Sacramento and Wash-

ington. They effectively muddied the waters by cynically proposing a number of other sites. The easily confused public became confused. The lumber industry was confident. One official boasted that it takes five years to get a national park bill through Congress and in five years there wouldn't be anything worth fighting for. As the arguments continued, the trees kept falling.

Then, in the summer of 1965, the 390-foot giant was discov-

ered—by the lumber companies. With no announcement, they brought in their chain saws. They worked quickly and the great tree was felled, cut into twenty-foot lengths, and hauled away. In fact, every tree in the redwood stand at the junction of Bond and Redwood Creeks was brought down—right in the very heart of the Sierra Club's proposed park site. The crash of the great giant was drowned out by the cheerful ring of the cash register.

Table 26.1 Levels of Organization in Biomes

Level	Example	Description
Biome	Desert	Vast areas of a particular climate that support specific arrays of plants and animals.
Ecosystem	Sonoran desert, dominated by Sonoran cactus	Groups of organisms interacting together and with their environment.
Community	Sagebrush area in Sonoran desert	Identifiable groups of organisms within an ecosystem that generally interact independently of other such groups.
Population	Family of kangaroo rat	Groups of interbreeding organisms within communities. They are of the same species (Chapter 27).
Individual	Kangaroo rat	Individuals are . . . individuals. The sum of their interactions between themselves and their environment describes their niche.

tem might be a pond, a wooded area, or a field. But ecosystems are rather unwieldy entities themselves. So they, too, must be broken down into more manageable units. One such unit is called a community. A **community** is an identifiable, interacting group of organisms within an ecosystem. (Ecosystems include abiotic (or nonliving) factors; communities include only biotic (or living) factors.) For example, there are sage desert communities, and there are hardwood forest communities, each with their own defined and interacting populations. We will see in the next chapter that populations can be defined as interbreeding groups of organisms. And within a community, each population of a species has a particular role as individuals interact with the environment in a certain, theoretically circumscribed way (Table 26.1). The sum of those interactions is the niche. So there you are.

THE WATER ENVIRONMENT

The earth's water may be classified roughly as fresh water or salt water, although not all bodies of water fall neatly into one category or the other. For example, Lake Pontchartrain, near New Orleans, is **brackish,** or a mixture of salt and fresh water. So are **estuaries,** the places where rivers flow into seas. Fresh water has about 0.1 percent salt; seawater has about 3.5 percent salt; and, as we will see, each has its importance in the earth's drama.

Freshwater Bodies

Bodies of fresh water provide less stable habitats than does seawater. Their instability is often related to their smaller size. For example, some bodies of fresh water may evaporate to a fraction of their former size, thereby concentrating their solutes. By mixing with bottom material, they may then become muddied. Because of their relatively small volumes, they become polluted rather easily and their temperature fluctuates more. Smaller bodies of water are, in general, more susceptible to any outside influence.

Due to the peculiar vulnerability of freshwater bodies, human activity has, in many cases, drastically altered their character. The Great Lakes, for example, have been altered by human pollution so that they can no longer support great numbers of edible lake trout and whitefish. However, parasitic lamprey and the destructive alewife began to flourish. (Recently there has been some progress in cleaning up the Great Lakes.)

One way humans can drastically alter lakes is by speeding up a natural process called **eutrophication** ("true food"). This is a normal aging process of lakes, but the use of detergents and chemical fertilizers has increased the amounts of phosphate and nitrate that are washed into the water. These join other forms of human waste in our fresh waters and speed up the aging process of lakes and accelerate the production of organic matter in lakes. With increased plant growth, the lake begins to fill in, and as the plants die and sink to the bottom, the lake grows shallower. So much oxygen is used to break down the dead organic material that the water becomes oxygen-starved. In time, species that require a great deal of oxygen, such as trout, are replaced by species that don't, such as lampreys (Figure 26.11).

FIGURE 26.11 As a result of human activities the edible lake trout (top) and whitefish (bottom) are now almost nonexistent in the Great Lakes. On the other hand, the parasitic lamprey (attached to the trout) and the destructive alewife (center) have become common. One reason for this change is that the use of detergents and chemical fertilizers have increased the amount of phosphate and nitrate that is washed into the water. Bacteria can break these down but only with the use of great amounts of oxygen. Algae and other plants flourish under such conditions, but as they die, they return their nitrates and phosphates to the water and cause bacteria to use ever more oxygen. Finally, the lake cannot support animals that require a lot of oxygen, such as trout, but the new conditions are fine for alewives.

The Oceans

The oceans of the earth have fascinated humans since "men have gone down to the sea in ships." Not only are sailors attracted to the sea, but so are beach lovers, poets, and practically everyone else with a shred of romanticism about them. The oceans remain as mysterious as they are compelling. Nonetheless, we are free to make our attempts at describing them in some sort of organized way.

First, oceans not only cover about three-fourths of the earth's surface, but if the earth's surface were smoothed out, the planet would be covered entirely by water. The average depth of the oceans is about three miles, but there are places where the water is seven miles deep. So the ocean is deeper than the tallest mountains are high.

It has been suggested that the oceans will one day provide us with most of our food. However, we presently take only three to five percent of our food from the seas, and even if we were able to double that in the next decade, our food problem would not be solved.* The problem of increased reliance on food from the sea is complicated by the fact that the deep ocean waters, like the deserts, are not very productive. In fact, many parts of oceans are almost completely devoid of life.

Even the more desolate areas of the oceans are not simply huge, still bodies of water, broken only at the surface by waves. In fact, the deeper waters are continously in motion as great silent currents hold sway. These deep currents are primarily the result of a surface phenomenon called the **trade winds.** The "trades," once so important to commercial sailors, are prevailing winds—that is, winds that hold rather steadily from the same direction because of the difference in air temperature between the poles and the equator, coupled with the effects of the earth's rotation. The great ocean currents caused by these winds spin in wide circles, clockwise in the northern hemisphere, counterclockwise below the equator. One such ocean river is the famed Gulf Stream, which follows the Atlantic coast as far out as Bermuda and then swings eastward to northern Europe.

Where the currents are deflected upward by the mountains on the ocean floor, cold water from the ocean depths wells up to the warmer surface. This cold water carries with it ages of accumulated sediment from the floor, sediment that is rich in nutrients. The ancient debris then acts as a fertilizer and promotes the bloom of life where light dances through at the ocean's surface. The nutrients carried by upwellings are utilized by minute chlorophyll-bearing organisms called **phytoplankton,** which in turn serve as a food source for tiny, drifting animals called **zooplankton.** These minute organisms, barely visible to the naked eye, form the basis of the ocean's food pyramid (Figure 26.12). One cubic foot of seawater may contain over twelve million phytoplankton. These food producers, however, can live only near the ocean's surface, since the sunlight necessary for them to carry on photosynthesis cannot penetrate below about 600 feet.

Not only are phytoplankton important as primary food producers in the oceans, but they also manufacture much of the earth's oxygen. Thus, the importance of maintaining viable oceans is apparent. Some people

*(This is because with the world's population increasing at its present rate, doubling the intake from the sea would add only three to five percent more food to a population that had increased by over eighteen percent.)

FIGURE 26.12 The food pyramid of the ocean. The tiny phytoplankton at the base (not drawn to scale) capture the energy of the sun. These are eaten by animals larger than themselves, which are eaten in turn by larger animals. At the top are the largest carnivores of the sea. (There are no large herbivores in the open oceans as there are on land.) What is the position of man in the food chain of the sea? Note that each level is comprised of far fewer organisms than the one below it, for reasons we will discuss later.

FIGURE 26.13 A growing fleet of super-tankers is largely replacing the smaller oil tankers that once plied the world's oceans. Because of the enormous volume of oil they can carry, should their cargo be accidentally released, as we see here, the result can be an environmental disaster.

are increasingly concerned as we continue to use the world's oceans as a dump. The sight of oil companies' new supertankers, which carry dangerous, partly refined petroleum across the fragile oceans, may also cause frowns of consternation. Of course, the larger, more sophisticated tankers make fewer trips necessary, but an accident with one of these behemoths invites a clear disaster. Even among smaller, more maneuverable oil-bearing vessels, there were an estimated 5,000 collisions in one recent ten-year period (Figure 26.13).

People who have descended to the ocean's depths in pressurized bathyscaphs have reported that they can see light at depths as great as 2,000 feet. The light that reaches these depths is pale blue, since the reds and oranges of the light spectrum have been filtered out by the water above. Many of the fish that live there are reddish in color when they are viewed in the light at the surface. As a result, in their natural depths, they appear dark and shadowy, since there is no red light for their pigment to reflect and their red color absorbs what blue light there is.

It was once believed that nothing could live below 1,800 feet because of the tremendous pressure of the water at such depths. However, about the time Darwin wrote on the origin of species, a submarine cable broke in the Mediterranean and was hauled up for repairs from a depth of about 6,000 feet. To the astonishment of biologists, it was covered with all sorts of living things. So notions concerning the effects of pressure had to be revised. Now it is believed that living things can exist at any depth as long as they are able to develop pressure inside their bodies equal to the pressure outside. This pressure is so great that fish brought up alive from depths below 2,000 feet have literally exploded when they reached the surface.

The deepest reaches of the ocean must be peculiar places indeed. We are just beginning to learn something about them, and the new information is fascinating. It seems that the deepest areas, untouched by the currents above them, are calm places. So far, we know of only a few kinds of animals that live there, and because of the unusual demands placed on them, many are quite weird (Figure 26.14). Other life forms at

FIGURE 26.14 Viperfish, an example of a deep-sea fish. Such fish, while not noted for their personality, have enormous jaws and teeth. Thus, when they encounter another animal in their sparsely populated depths, there is a good chance that they can manage to eat it. There are luminescent spots along its sides. Such lighting is apparently useful in signaling between members of the species. This animal, fortunately, is only about a foot long.

The Water Environment

Most of the earth's surface is covered with water, both salty and fresh. It thus constitutes an important part of the biosphere, but we often neglect to consider its importance because in its naturally occurring form, it is essentially disruptive to much of our tissue, and our efforts to examine it are hindered in numerous ways. It puckers our skin, clogs our ears, is rarely at a comfortable temperature, and is usually hard to see through. So if we're not drinking or washing, we generally avoid it unless we are sports enthusiasts or specialists, thus we probably imagine we know more about its role in nature than we really do. Yet our lives depend on it and hence we are usually cheered by the sight of oxygen-laden mountain brooks or long, winding rivers, such as the Mississippi River, one which has escaped being "straightened" by the Army Corps of Engineers.

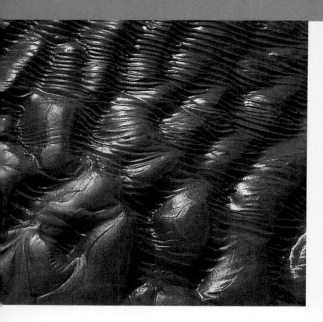

A mud flat (above left), a saltwater area that is covered at high tide and exposed at low tide. The soil is often hard and tightly packed. This is a very difficult place to adapt to because it is subjected to repeated drying and soaking. Predators stalk these areas in search of exposed prey. Such flats may occur at estuaries, the river currents producing the prominent rills seen here. The mud flat estuary harbors a variety of organisms if the river has not been poisoned upstream.

The rocky coast of Olympic National Park, Washington (above right). The incessant pounding of the water has carved out these striking monuments, which serve as attachment for a number of organisms, a particular type of animal dominating each vertical zone.

A cypress swamp in winter (above). Notice the water-swollen trunks and the "knees"—roots that have risen above the waterline into the oxygen-laden air. The water in such swamps moves very slowly and hence is low in oxygen.

Tules, or bulrushes, which thrive on over-flowed land in the American Southwest (left). Large tracts have been overgrown with these plants.

A glacial lake in Acadia National Park, Maine. Such lakes may be exceptionally deep and cold with a predictable turnover rate as their upper reaches are heated or cooled by the air temperature. One of the most beautiful of the world's mountain lakes is, or was, Tahoe, a name now synonymous with artifice and glitter as the struggle between developers and environmentalists continues.

The Channel Islands off the California coast. This is Anacapa Island. It is rather small and, as a rule, smaller islands harbor fewer kinds of plants and animals. Whereas Anacapa lacks forests, trees abound on Santa Cruz, her larger neighbor in the distance.

A salt marsh estuary in South Carolina (above). Salt marshes are exceedingly critical places in that the many tiny life forms living here initiate food chains that ultimately involve larger sea creatures. They are vulnerable in that they don't look important and so developers have gained easy access to them.

This sandy Hawaiian beach supports few life forms, probably because the shifting sands provide very poor attachment and shallow waters tend to surge strongly.

FIGURE 26.15 The ocean bottom is a still and quiet place that can support very fragile forms of life, such as these grass eels.

these depths include a few kinds of fungi, species that do not need light. Among the more interesting new findings on the ocean floor are the volcanic vents where bizarre and unexpected communities derive their energy from water being heated by the earth's core, as we saw in Chapter 5.

The ocean bottom itself is soft and composed of the chalky and glassy corpses of tiny plants and animals that have rained down steadily onto its surface throughout the ages. Animals that live on the bottom often have long appendages to keep them from sinking into this primal ooze. Because of the utter stillness of the deepest ocean bottom, very delicate and fragile creatures can live there, but we are probably aware of only a small part of the life in such places (Figure 26.15). Some animals, it seems, escape our attention by burrowing into the bottom. Divers have seen many strange holes and burrows that were made by no known creature.

COASTAL AREAS

The land-ocean interface has been called "the seashore." Those romanticists who like to write about it would compose new eulogies if they knew the full impact of the seashore on their lives. And who knows, perhaps coastal developers might even become a bit more cautious.

But why are these such important places? For one thing, life is much more abundant in coastal waters than in any other part of the ocean. The edges of the continents extend beyond the shore for anywhere from 10 to 150 miles as **continental shelves.** At these relatively shallow depths, usually less than 200 feet, sunlight easily penetrates and gives rise to a wide variety of plants. The plants, of course, provide many species of animals with food, shelter, and surfaces for attachment. Coastal areas are generally composed primarily of either rock, sand, or mud. And each type of area has its own special qualities.

Rocky coasts are perhaps the most dramatic and, to coastal navigators, the most awesome of the coastal areas, their jagged formations formed by years of pounding surf that often hides immense reefs below the surface. These coastal areas boast a remarkable array of plant and animal life that often is found in quite distinct zones. For example, the zone dominated by the periwinkle snail is clearly delineated from the barnacle zone, which is, in turn, strongly demarcated from the seaweed zone. The zones are not occupied exclusively by a single species; they are simply identified according to the most prevalent species occupying them.

Sandy seashores have relatively few forms of life, which is perhaps a good thing for timorous bathers. Perhaps you have waded out into the surf and were relieved that you didn't step on some mysterious and toothy bottom dweller. The reason they are so scarce is that wave action in such shallow areas causes a constant shifting of the sandy bottom, depriving the less mobile species of a fixed surface to which they can attach. Any animal that lives in such shifting areas must be able to get around quickly, like crabs, or be able to dig and withstand burial, like clams. (Unfortunately, stingrays can do both.) Animals that live in tidal areas and are not mobile probably encounter exceedingly severe conditions at low tide when they are subject to the parching sun and to terrestrial foragers such as shorebirds.

From an aesthetic standpoint, **mud flats** are probably the least appealing part of the seashore, and this perhaps explains the general lack of interest in protecting them. They aren't especially spectacular as scenery, and they are poor places to spread a towel. Also, they tend to smell peculiar (to say the least). Mud flats do not harbor as many life forms as rocky coasts, it is true, but they support much more life than do the other kinds of coastal areas.

Mud flats are submerged at high tide and exposed at low tide. Bottom dwellers, snails, insects, shrimp, crabs, fish, and birds abound in these areas. Each square yard of the mud flat can support thousands of such individuals. Many sea animals begin their lives in mud flats, only later moving out to take their place in the mysterious pageant of the open ocean. Hence, the life of the ocean itself is, in a very real sense, dependent on the preservation of the unsightly flats. Unfortunately, mud flats and saltwater marshes have attracted the eye of commercial developers who are aware of the public's lack of interest in them. Up and down the coasts, our bayous, marshes, and bays are being filled in for use as industrial sites, "waterfront" housing tracts, and high-rise apartment complexes and condominiums. San Francisco Bay alone is now only two-thirds of its original size, much of it having been replaced by long stretches of paved, chromed, and neon-lit areas designed to attract the tourist dollar.

Now that we have described some of the areas and habitats of the planet, let's consider some of the ways species interact. We can ask, how can one species influence the well-being of another? In what ways do species interact with their environment? Does the passage of any species from the earth affect us in any real way? Often there aren't specific answers to such questions, but we do already know enough to make some reasonably intelligent decisions about how we should, in our own best interest, interact with the life around us.

THE WEB OF LIFE

It should be clear that the processes of life are complex, interactive, and responsive. Because life is such an interactive process, it is important to keep in mind that if one part changes, the effect can ripple through other parts, sometimes in unexpected ways.

Let's consider an example of how interaction among living things can have far-reaching effects in maintaining the balance of life on the planet. We can begin with the premise that ecologically, simplicity means instability, and that in an unstable ecosystem, extinction rates can be expected to rise. Again we must deal in speculation to some degree because much of the essential data just isn't in.

Figure 26.16 shows a hypothetical and greatly simplified food web. In spite of the model's imperfections, however, it illustrates an important point. Suppose the only carnivorous (meat-eating) mammal in this food web were the fox. If some fox disease were to sweep the area, what would be the repercussions on the rest of the life there? We can see how the numbers of herbivores, such as rabbits, squirrels, mice, and some birds, might rise. These in turn might put new pressures on their plant food and, in so doing, destroy the habitat of many other species while depleting their own food supply. The resulting initial increase in the numbers of herbivores would provide more food for owls and hawks,

FIGURE 26.16 A hypothetical food web. Can you tell which organisms serve as food sources for the various species? This diagram is greatly simplified—for example, there are also omnivorous birds such as crows, and mice may sometimes eat bird food. The predatory fox may also have to share its food with lynx and wolves.

which could in turn be expected to increase their own numbers. The resulting abundance of such predators might then reduce the numbers of other animals, such as insect-eating birds and toads. What might then happen to the numbers of snakes? And the numbers of insects?

Now then, suppose the foxes had to share their food with lynx and wolves, as well as with hawks and owls. In this case, a rampant fox disease would have markedly less effect on the ecosystem. Of course, elimination of the foxes would result in some changes in the system, since the fox has a somewhat different niche than the wolf or the hawk. But perhaps the system would continue largely as before until the fox population was restored. The point here is that a simpler system—for example, one with a single "top" predator—is inherently easier to upset.

Let's draw on a more abstract example to show the effect of the extinction of a single species in an interdependent system. Figure 26.17 shows five species that depend on each other. Of course, any such dependency may be of a variety of types and levels. For example, fleas depend on dogs for living quarters and transportation, as well as for food. Starlings depend on old woodpecker holes for nests. A dog may put the tree to a different, more casual use, and the hawks that see starlings and woodpeckers as food have absolutely no use for dogs or their fleas.

Now, suppose that in this group of species that depend on each other in a variety of ways, system C somehow becomes extinct. Since A, E, and D depend on C, we might reasonably expect their numbers to be reduced. B, however, depends on A, E, and D. Even a slight reduction in the numbers of all three might have such an impact on B that it might follow C into oblivion. This would mean that the animals on which A, E, and D were dependent would be reduced by *half*. How might this affect their fate? If A is more susceptible to the loss than E or D, its numbers might then drop quickly. What effect might this have on the two remaining species?

Now, suppose that instead of a five-species system, we have a ten-species system, as in Figure 26.18. Here the system is much more complex, although it is still far simpler than any actual biotic system would be. It is easy to see, however, that the loss of any single component might be more easily absorbed in this system than in the simpler one. Thus, the more complex a system is, the more inherent stability it has.

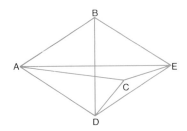

FIGURE 26.17 Interdependency of five theoretical species. This is a relatively simple system in which each actor has a major role. Because of their interdependency, the elimination of any could threaten the existence of all.

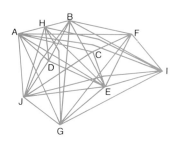

FIGURE 26.18 Interdependency of ten theoretical species. The system becomes vastly more complex by simply doubling the number of species. In such a case, the elimination of any one may not be crucial, since the complexity can act as a buffer. The point is, biological systems may become more unstable by becoming simplified and one way to simplify them is to reduce the numbers of their component species.

EXTINCTION AND US

Are there any lessons here for us? Should we care if over 150 known species of animals have disappeared from the earth in the last fifty years? Should we be concerned that there are literally thousands of species whose very existence is presently endangered—largely because of our activities? Extinction, after all, is the natural end of populations. Species are born, they mature, and they die. Some live a long time, perhaps millions of years; some die more quickly. We have hastened the extinction of many species we know about, and we have undoubtedly sealed the fate of others. In fact, there are undoubtedly many unobtrusive species that have lived among us during our time on earth, but that have

Indian farmers are incessantly waging war on rats. It has been calculated that the progeny of one pair of rats, in a year's time, can consume enough grain to feed five people. The killing is thereby justified. But who has calculated that five people can, in a year's time, consume enough grain to feed the entire year's progeny of a pair of rats? The question is patently ridiculous. But it hasn't been calculated only because rats can't calculate. Perhaps part of our problem lies here. We see ourselves as more "worthwhile" than other animals. Obviously, our worthiness is subjectively derived since it is we who are doing the measuring and it is to our own best interest to place a higher premium on ourselves and, to carry it further, on those most like us. The mechanism is self-protecting and perhaps justifiable. But can it lead to problems? Will it tend to make us, in very subtle ways, believe that we stand apart from the other species? Can such a belief, expressed or not, cause us to behave as if we are not subject to natural laws? That we are *better*, somehow *different*?

And by the same token, do we *lose* something by elevating ourselves? Do we feel uncomfortable being apart? Why is a deer's footprint in the snow "beautiful," but a man's bootprint "disfiguring"?

Why is a beaver dam natural, but a human dam unnatural? After all, we, too, evolved here. We are natural. We are a part of it all, and perhaps we should more comfortably take our place in the family of life.

disappeared as a result of our activities without our ever having known they existed.

It is hard to explain the rationale of many of us who are concerned about such matters. I have never seen a sei whale, yet I don't want them to become extinct. Moreover, I felt this way long before I understood anything about how they might be an important part of an ecosystem. Possibly such feelings merely reflect the cultural attitude that it is "nice" to wish other living things well; thus, the attitude is rewarded. I feel nice.

There are, of course, more rational reasons for mourning the extermination of any species. For one thing, the kind of attitude that encourages or sanctions the destruction of other species (Essay 26.2) constitutes an intrinsic threat to our own well-being. If such an attitude exists, we ourselves might fall victim to it. Living things (including us) might be expected to fare better where there is reverence for life. The extinction of other species could also threaten us indirectly by simplifying the system of which we are a part or by destroying parts of the ecosystem upon which we directly rely. For example, if we continue to poison the oceans because we are willing to believe only a few bottom dwellers are affected, we might eventually overstep some critical threshold and trigger the wholesale death of plankton, thus finding ourselves without a major source of the world's food and with our oxygen supplies dwindling.

This is all very rational, but none of this precipitated my gut response when I learned that some major whaling nations intended to resist the international ban on whaling. I simply found it very sad. For some reason, I wanted those whales to continue to share the planet. If you had the same reaction and you don't ever expect to see or eat a whale or use its oil, you might try to analyze the roots of your own response.

SUMMARY

Habitats are places where organisms are found. Niches describe the interaction between the organisms and the environment. It is generally assumed that two species cannot occupy the same niche indefinitely because one would eventually be competitively excluded. Competition occurs where organisms utilize resources that are in short supply, or when they harm each other while seeking the same resource that is not in short supply.

The land environment is divided into general, intergrading areas called biomes. The most prominent biomes are temperate deciduous forests, grasslands, deserts, tropical rain forests, tundra, and taiga. Each is defined by the plants found there. Associated with these are particular rainfall distributions and characteristic animal life.

The earth's water is generally divided into freshwater and salt water, although brackish areas are also important. Bodies of freshwater are generally the more unstable.

The coastal areas can be divided into somewhat distinct areas: the rocky coasts with their zones of life, the sandy seashores with their paucity of life, and the mud flats with their variety of life.

As life systems become simplified, such as by extinctions, they become more unstable. This is a biological rationale for the preservation of species.

Chapter 27

POPULATION DYNAMICS

It is a bit startling to hear population ecologists predict that within the next twenty years, about a million species of plants and animals will become extinct. Most of these extinctions will occur in the poorer, developing countries of the tropics. Here, expanding human populations encroach into the habitats of countless other species, species that may be so well adapted to their small areas that they can survive nowhere else. (It is hard to calculate the extinction rate because we simply don't know how many species there are. It is estimated that about three million species now exist in the tropics, but that only about a sixth of these are known.)

POPULATIONS, ETHICS, AND NECESSITY

It is sometimes argued that we shouldn't worry about *other* species while humans—people—need the land. However, there are at least a couple of ways that one might justify worrying about them. For one thing, these species may have something we can use, especially considering our emerging abilities in genetic engineering. Such species may serve as genetic reservoirs. Their demise may spell the end of unusual or intriguing combinations of genes that might have proven very useful later on. Second, it is argued that the other species of the earth may be our only companions in the entire universe and that they have a *right* to exist. But perhaps that argument is a bit esoteric for a Latin American farmer who has eight mouths to feed and one on the way, and who wants to clear more land. However, a recent incident illustrates just how valid some of the more esoteric arguments can be, even to the pragmatic farmer. On a rather ordinary hillside in Mexico, botanists discovered a new plant species. It was in the same genus as domestic corn and was named *Zea disploperinnis*. These few thousand plants were of great interest because they turned out to be perennial; thus, they do not require replanting each year. Furthermore, they are immune to a number of viruses that infect corn, and they are able to grow in wet areas where corn can't. This is all particularly interesting because there is some promise of being able to cross the wild strain with domestic corn. (It is of such things that agricultural revolutions are born.) Botanists had a devil of a time, however, in convincing the farmer who worked the land not to plow these plants under to make room for his usual crops (Figure 27.1).

This eleventh-hour discovery of a remarkable species can only suggest how many we have already unknowingly sent on their way to extinction. We are reminded of the fact that ninety-five percent of the genetic varieties of wheat native to Greece have become extinct in the last forty years. We may thus have lost forms with a genetic resistance to whatever disease might next strike the world's wheat crops.

FIGURE 27.1 Because of rapidly increasing human populations, land is at a premium throughout many areas of the world. In Latin America, native populations of plants are being destroyed at an alarming rate in order to make room for traditional farming efforts.

FIGURE 27.2 About the turn of the century, scientists reported the existence of a *Paleotragus* in the Congolese rain forest. The creature was believed to have been extinct for millions of years.

A rather bizarre story underscores the disregard of people in poorer countries for anything of no immediate value. It began with a 1959 report of an animal that some have referred to as an aquatic dinosaur of Loch Ness proportions. Fishermen on Lake Tele in Zaire reported that they found the beast and then killed it with spears because it was disturbing their fishing. It was then cut up and eaten. They complained about how long it took to prepare it because its neck and tail were so long. Dismayed scientists have offered a reward for any skeletal remains, but nothing has turned up. You may recall that the first coelocanths caught in modern times (Chapter 13) were left to rot on deck or were eaten. As a final reminder of how little we know of our fellow species, a giraffelike creature that lived twenty million years ago, called the **palaeotragus** (Figure 27.2) turned up alive and well in the Congolese rain forest about the turn of the century. (One wonders how many have been eaten since then.)

Extinctions, however, are admittedly only one aspect of population changes. Not all species are tail-spinning into the black hole of extinction. In fact, most species that we know about seem to be doing quite nicely. I know that in my house, there is no shortage of the huge Florida roaches they call palmetto bugs. One even inexplicably managed to find its way into a guest's shorts at a recent gathering here. (The bugs are still around, but I haven't seen much of my friend lately.) Bluejays and squirrels fight at the feeder, the possum visits nightly, and a big female raccoon (named Rosemary Cooney), always a bit testy, rambles around in blithe evidence that her species is doing fine.

Some species are not just holding their own, they're doing better than that. Kudzu, a large leafy vine, is choking off the trees in the Carolinas; local scientists tell us that fire ants are on the way; "killer bees" (a particularly aggressive strain of honeybee) are moving up from the south; our Florida waterways are becoming choked by water hyacinths (partly because the gentle manatee that feeds on them has become endangered, its numbers reduced in large by the slicing propellors of joy-riding boaters who insist on their rights to use the waterways as they like). Many species are indeed on the rise—their numbers growing, their densities increasing, and their ranges expanding.

We see, then, that while some populations are dwindling, others are holding their own, and yet others are soaring to new limits. So let's see if we can discover some of the underlying principles of population change. We might ask what are the signs of an impending change. The implications of the findings are enormous, as we will see. It is no secret that human population growth is quite simply out of control.

HOW POPULATIONS CHANGE

The subject of **population dynamics** (how populations change) is often a focal point of conversation, from scientific meetings to cocktail parties, perhaps because we *are* becoming increasingly aware of our own population problems. Some scientists insist on telling us that we are breeding ourselves to extinction. "Ourselves?" someone archly asks. "Extinction, you say? My! Care for some coffee?" Since some people are not particularly concerned, we might wonder if it is really time for concern, perhaps for drastic measures. Or is it time for that coffee? Just what *do*

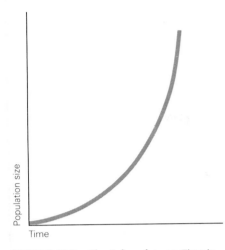

FIGURE 27.3 The J-shaped curve. Time is plotted against population size. In such a case, there is nothing operating to restrict growth, so the population is achieving its biotic (or reproductive) potential. The characteristic curve is produced because larger populations increase more rapidly than smaller ones.

the numbers mean? It is difficult to assess the data and understand the implications without some understanding of the basic principles of population dynamics.

Population Growth

The basic story of population growth can be told by two simple, sweeping lines on a graph; they are called the **J-shaped curve** and the **S-shaped curve.** Let's first consider the J-shaped curve.

If we were to place a few bacteria in a suitable medium, or a few sheep on an uninhabited but hospitable island, their numbers would increase in such a way as to produce a J-shaped curve (Figure 27.3). Since we are assuming that their new environment provides them with the necessities of life, their growth will be limited primarily by the rate at which they can reproduce, rather than by environmental factors.

Note that the population accelerates its *rate* of increase with time. This is because the number of reproducers keeps growing. The distinctive curve is produced because of this change in rate. If each individual in the original group leaves two offspring, the population will remain fairly small for several generations (one bacterium divides to leave two, these leave four, which leave eight—all low numbers). But after a few rounds of reproduction, the numbers begin to skyrocket. (What would be the result after only ten generations? Twenty generations?) The reproductive capacity of any population, when it is unrestricted, is called its **biotic potential,** and the J-shaped curve illustrates a population approaching its biotic potential.

The full biotic potential of any species is not likely to be reached because of the inherent limitations imposed by the environment. **Environmental resistance** includes those factors in the environment that act to reduce a population's increase. Environmental resistance increases as a population approaches the environment's **carrying capacity,** defined as the environment's ability to sustain a population over a long period of time. As a population approaches the carrying capacity of its environment, its numbers begin to level off, producing an S-shaped curve (Figure 27.4). Normally, populations oscillate around the environment's carrying capacity, rising above it or falling below, but staying somewhat stable (Figure 27.5). The relationship between these various factors is summarized in Figure 27.6.

FIGURE 27.4 The S-shaped curve generated when a population approaches the environment's carrying capacity.

FIGURE 27.5 Normally, populations increase until they reach the environment's carrying capacity. They may then oscillate at this level.

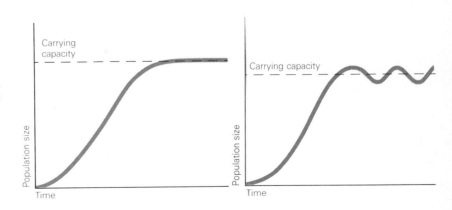

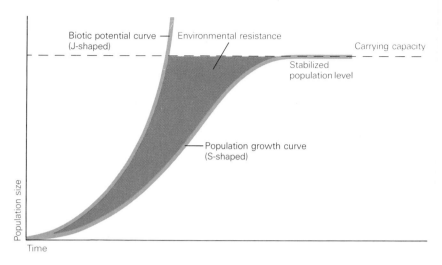

FIGURE 27.6 The theoretical relationships between biotic potential, environmental resistance, and carrying capacity. The biotic potential of a population is rarely, if ever, achieved due to the problems of living in a real world (the problems are collectively called environmental resistance). Some individuals don't reproduce and some fail to reproduce maximally. The characteristic growth curve is produced because a small population grows slowly, since a few individuals can generate only a small number of offspring. Once the breeding population reaches a critical size, however, it skyrockets (still well below its biotic potential) to be slowed only by the effect of its own numbers. Theoretically, it should stabilize around the carrying capacity of the environment.

In some cases, a population may rise so sharply that it drastically overshoots the carrying capacity of its environment. When the numbers are too large to be sustained by the environment, the resulting environmental overuse may cause the numbers to suddenly drop (a **crash,** Figure 27.7). Of course, the normal population cycles of many species involve surges and falloffs. For example, the numbers of annual plants, as well as many insects, fall each autumn, but their progeny appear each spring to reproduce madly in the few months they have available. Crashes, however, are not part of normal cycles. They may occur when the carrying capacity is overshot for some reason, such as when a population's predators are suddenly removed. In such a case, the prey may suddenly increase in numbers until it overexploits its environmental resources and dies back. Such a crash may be followed by a recovery to more moderate numbers.

Some crashes, however, are more permanent. For example, a bacterial population being grown in a petri dish on a laboratory shelf will rise sharply. However, it will eventually run out of food, even while it befouls its environment with its own wastes. Its numbers will then drop precipitously. Since the carrying capacity has also been lowered, the population cannot recover. As another example of permanent change in carrying capacity, oversized herds of elephants may eat all the young plants in their habitat that, if left alone, would have eventually provided more food and for longer periods, perhaps for generations afterward. Elephants, then, can lower the carrying capacity of their habitat, perhaps permanently. If they strip away so much foliage that they markedly increase erosion, the carrying capacity may become lower still, perhaps to the point that it is no longer suitable for this species at all. Fortunately, humans are a lot smarter than elephants.

Now we will take a look at just how the environment may influence population numbers. Essentially, we will find that populations can theoretically be regulated in two ways: (1) by the number of offspring produced and (2) by the mortality of the individuals in the population.

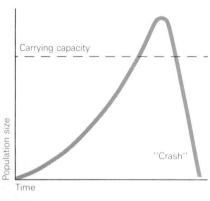

FIGURE 27.7 If a population should drastically overshoot the environment's carrying capacity, it may reduce the environment's ability to sustain it, and may "crash," as its numbers are drastically reduced.

CONTROLLING POPULATIONS THROUGH REPRODUCTION

The biotic potential of most species, even the slowest breeding ones, is surprisingly high. Charles Darwin estimated that a single pair of elephants could leave over 19,000,000 descendants in only 750 years. The fact that the entire world was not teeming with elephants indicated to him that some elephants were not reproducing and that the reproducers were somehow "selected" by the environment.

Whereas elephants are very slow breeders, other species are more prolific. For example, the startling reproductive potential of the housefly for a single year is shown in Table 27.1.

Among factors determining the rate at which an organism reproduces is energy. For example, simple energy demands would not allow an elephant to become pregnant several times a year. There is just no way that a female could find enough food to produce that many offspring. And what would happen to the helpless baby elephants already born as new brothers and sisters appeared on the scene?

If energy is a prime factor in determining reproductive output, it could exert its influence at two levels, **ultimately** and **proximately.** At the ultimate (evolutionary) level, natural selection could set the reproductive abilities of an organism according to the natural history of that line. Those that have traditionally had great resources and little demands might develop the physiological abilities to make quite grand attempts at reproduction. At the proximate (individual) level, reproductive attempts might be dictated by the immediate conditions. Thus, there are species that produce, say, ten offspring in a good year, but in hard times, their lack of food might cause them to attempt fewer.

Ecologists sometimes speak of the "reproductive strategy" of some species. This "strategy" is not some consciously derived tactic for leaving offspring, it simply refers to the means that have evolved for any species as the most effective way to leave offspring. The means, however, may differ from one species to the next depending on a number of factors in

---

Table 27.1 Projected populations of the housefly *Musca domestica* for one year, a period that encompasses about seven generations. Do all these offspring survive? The numbers are based on each female laying 120 eggs per generation, each fly surviving just one generation, and half of these being females.

Generation	Numbers If All Survive
1	120
2	7,200
3	432,000
4	25,920,000
5	1,555,200,000
6	93,312,000,000
7	5,598,720,000,000

Source: Adapted from E. V. Kormondy, *Concepts of Ecology* (New Jersey: Prentice-Hall, Inc., 1969), p. 63.

their natural histories. We can illustrate the point by considering two extremes, tapeworms and chimpanzees. Most species, though, fall somewhere between these and, therefore, have "mixed" strategies, as we will see in a final group, birds.

Tapeworms

The tapeworm living so comfortably in the idyllic environs of your intestine is lucky to be there (although its luck is the inverse of yours). It is lucky because the life of a tapeworm is fraught with risk and its success is largely dependent on chance. The odds are very much against its finding a host. In order for that tapeworm to accompany you on your dates and visit France with you in the summers, it had to successfully survive some rather unlikely events. First, it was passed, as an egg, from some previous human host. Then it lay around in the sun and rain until it was eaten by a pig. Not only must it survive the elements, but it must also be found by a pig, so already the odds are against it. Once inside the pig, it begins to grow and change—its appearance becoming wormlike. It then moves through the pig's intestinal wall and on through the bloodstream until it reaches the pig's muscle. Here it forms a protective capsule around itself and waits for a human to eat the pig's flesh without cooking it too well. If this should happen, the worm attaches to the intestine of its human host, grows into an adult, and begins to lay eggs.

There is obviously an overwhelming probability that not all these conditions will be met, and thus most eggs will not become adult tapeworms. So the tapeworm must have some means (a "strategy") of overcoming these odds. Through natural selection, the tapeworm has become capable of self-fertilization, a particular convenience when an individual is not likely to ever encounter another of its own species. It is also able to produce prodigious numbers of eggs. Its whole life is devoted to egg production. It exists only to lay eggs, eggs, EGGS! A few of these, with luck, will wend their way through the complex maze that is the tapeworm's life cycle. So tapeworms beat the odds against their reproducing by putting very little energy into any offspring, but laying so many eggs that some make it.

Chimpanzees

What about other animals? Are they *all* in the numbers game? The answer is no. The chimpanzee, for example, has developed a different evolutionary strategy. It reproduces much more slowly. In fact, during the time a female chimpanzee is sexually receptive, she may copulate with almost any male in the group, but once she has become pregnant, she may not become sexually receptive again for years. Flo, the aging but sexually active female studied by the famed British field researcher Jane Goodall, did not become sexually receptive for five years after delivering a baby. Of course, this meant she had plenty of time to attend to her offspring.

In the first months, Flo carried her baby everywhere and diligently guarded it against danger. Later, it was permitted brief forays on its own, but never far from her vigilant eye. The curious but wary baby would scurry back to her at any sign of real or imaginary danger. During its first

FIGURE 27.8 The reproductive strategy of intelligent and social creatures such as the chimpanzee involves carefully tending and training few offspring, giving them time to adapt to their complex world.

few years, the young chimpanzee gradually became able to care for itself and to associate less and less with its mother. Finally, Flo was free to mate again and to rear another baby. Among chimpanzees, we see, reproductive success does not involve producing large numbers of offspring. Instead, few offspring are produced and are given very careful attention until they are able to be independent. How did such a strategy evolve?

Chimpanzees have historically existed in a precarious and complex world. They have traditionally fallen prey to the big cats (whose numbers have now been drastically reduced by human hunters out to bag a member of some species "before they're all gone"). Furthermore, chimpanzees live in a seasonal world where various kinds of foods come and go. Their food includes seasonal fruits, leaves, shoots, baby baboons, small monkeys, and insects (including ants, which they may pick up by using tools made of broken twigs)—quite a varied menu with a different skill involved in acquiring each item. Also, chimpanzees must move around in order to exploit food sources as they become available. Each night they build a new nest of bent branches, twigs, and leaves. Through all of this, they are highly interactive, each animal recognized by the others and treated according to rank, sex, age, or whatever. Chimpanzees, we see, are constantly dealing with new, complex, changing, and sometimes threatening situations.

To cope with such a physical and social environment, the chimpanzee has developed a high intelligence and a system of extended parental care (Figure 27.8) that gives the young chimpanzee time to gain experience under the watchful eye of its mother as it learns about its complex world. Such extended parental care, however, involves a great deal of time and energy, and so chimpanzees are able to produce only a few offspring during their lifetimes. In this species, selection has resulted in the production of only a few infants that are then carefully attended to. So chimpanzees, like the tapeworm, seem to rear as many young as possible, but in accord with the demands of their particular kind of environment and history. So we see that the number of offspring attempted by members of any species is likely to be set by natural selection and that the "strategy" of natural selection will depend on a number of historical and environmental variables.

Birds

Bird studies have revealed reproductive strategies that provide a rather clear picture of how the environment can influence the number of offspring produced by any species. We will see that birds lay the number of eggs that will enable them to rear as many young as possible. In most cases, this is the number of young the parents can *feed* successfully.

Females of the common swift, *Apus apus*, normally lay one to three eggs. Apparently, the number has not yet stabilized through selection (stabilizing selection) or it is undergoing a shift due to some kind of new ecological pressures (directional selection). In years in which food is abundant, the number of young that reach feathering age was found to be 1.9 in those nests in which two eggs had been laid and 2.3 for nests with three eggs. When eggs were added to a number of nests so that the total was four, an average of only 1.4 young survived until feathering age. (Suppose the food supply were to remain relatively abundant and

FIGURE 27.9 The ground-nesting gulls, such as the herring gull (top), have been forced to assume a different reproductive strategy than the cliff-nesting kittiwake (bottom). The nest of the ground-nesting species is vulnerable to marauding predators, so they compensate by laying more eggs than they will be likely to rear. The ground nesters, for example, lay three eggs, whereas they can probably rear only two young, because at least one of the eggs is likely to fall to a predator. The kittiwake, on the other hand, nests on cliffs safe from predators. Since all their young are likely to survive, they lay only two eggs.

constant. At what number would the eggs laid by females of this population stabilize?)

The number of offspring attempted (eggs laid) by some species may also be influenced by the vulnerability of the young to predators. For example, some species of gulls nest on accessible beaches, and here the nest and young run a great risk of being discovered by prowling foxes or other nest marauders. Whereas these gulls can usually feed only two young successfully, there is a strong likelihood that at least one of the young will fall to some predator, so three eggs are laid. Since one is likely to be taken, the number will be adjusted to that which results in maximum reproductive success. On the other hand, the kittiwake, a species of gull that nests on cliff sides, normally lays only two eggs. A fox trying to get at these eggs is likely to break its neck. So two eggs are enough; less energy is expended by the parents, the chicks are safe from predators, and both young are likely to reach adulthood (Figure 27.9).

In some species of birds, the broods are small, even when the parents could successfully feed more young. One reason is that fewer eggs result in a safer nest. Conspicuous eggs, the activity of young, and the coming and going of parents might advertise the location of the nest and attract predators. Fewer eggs require less conspicuous activity around the nest.

In rather complex ways, time may also influence the number of eggs laid. For example, a given level of food can theoretically raise a small brood quickly or a large brood slowly. In areas with short nesting seasons, such as the polar regions, time is of the essence because food is abundant for only a very brief period. In such areas, there would be greater advantage in raising a small brood quickly than in portioning out the same amount of food among a larger brood that might not mature fast enough to escape the hastening winter.

The mechanism by which some birds regulate the number of eggs they lay may simply be one of energetics. After all, food provides the energy and substance to manufacture eggs. Thus, if food is scarce, the female is able to form fewer eggs. (A reduction in the number of eggs may actually mean that more young will be reared, since each one will receive a larger proportion of whatever food there is.) The following season, if the signs indicate an abundant food supply, she may initiate the formation of more eggs.

CONTROLLING POPULATIONS THROUGH MORTALITY

Just as the numbers that are born in a population are influenced by environmental factors, so are the numbers that survive. It is easy to see how population size can be regulated by death, but it is less obvious that death usually occurs through one of two fundamental means.

Abiotic and Density-independent Control

Death can be met by a number of means, as we all know. But they can basically be categorized according to whether they are nonliving (**abiotic**—*a*, without; *bios*, life) or living (**biotic**). First we will consider abiotic factors.

One of the greatest causes of death among many forms of life is the weather. We see this when normal seasonal changes bring on winter hardships that decimate populations. Such cycles can be hard on populations, but bad weather is especially dangerous when it is irregular or unusual. Almost yearly, we hear of the devastating effects of drought somewhere on the planet. Perhaps less spectacular, but just as deadly, is unseasonably cold weather. Many birds perish in some years because they begin their northward migration in the spring only to be caught by a late cold snap. Natural selection cannot effectively prepare organisms for novel, unusual, or unexpected conditions. Severe weather, then, is an example of abiotic control. A drought is not alive.

Population-depressing influences such as severe weather are **density-independent.** That is, such effects are independent of the population's density. In a severe drought, the parching sun is not influenced by how many corn plants are struggling in the field below. It kills them all. In an area saturated with DDT, insects die whether there are few or many. Their numbers mean nothing unless they somehow tend to protect or shelter each other.

The awesome impact of the human species on the planet's life-support system may soon bring the reality of density-independent population control into sharp focus. For example, many nations, including ours, routinely use the oceans as a dump for poisonous wastes that are not easily manageable on land. (Humans dump about ninety percent of their wastes into the earth's waters.) Is there some threshold, some point at which the sea's chain of life will no longer stand the insults? When will we cross that point of no return? We can't say, because we haven't a clue what that point is, but we keep dumping.

Even the soothing rain has been changed so that it has become a real threat in some areas (see Essay 27.1). If we continue to befoul our environment so callously and routinely, we may begin to more clearly understand the meaning of density-independent population control.

Biotic and Density-dependent Control

Biotic population control refers to living influences on population size. Biotic controls are widely varied and often more complex than abiotic control mechanisms. For example, the organism that causes bubonic plague can reduce populations, and so can a tiger. But biotic influences can operate in subtler ways than simply by killing outright. As an example of indirect influences, a territorial bird drives a competitor into an area where there is less food. When winter comes, the underfed competitor may be more likely to succumb to the rigors of the season. Thus, the territory holder has indirectly brought about the reduction of the population.

Biotic controls on populations are likely to be **density-dependent.** Here, the effect of the control mechanism depends on the density of the population. For example, with density-dependent control, an increase in population density (the numbers of individuals per unit area) increases the effects of mechanisms that reduce the density. Then, as the density falls, the pressures that had reduced the population begin to lessen, permitting the number to increase again. (You may recognize the familiar negative feedback system here.) Let's now consider a number of biotic, density-dependent effects on population, beginning with predation.

In the 1970s, it became clear that the rain was changing. In fact, in some areas the gentle raindrops were downright dangerous. The rain was becoming a dilute mixture of acids. It was first noticed in Scandinavia, then in the northeast United States and southeast Canada, then in Northern Europe and Japan.

Rainwater, of course, had always been slightly acidic because the water dissolved atmospheric carbon dioxide, forming carbonic acids. But now the rain was showing alarming concentrations of the more dangerous sulfuric acid and nitric acid. Where were they coming from? They were the result of accumulations of nitrous oxides and sulfuric oxides in the atmosphere. The nitrous oxides, it turned out, were from power plant and automobile emissions; the sulfuric oxides, mainly from power plants and smelters. Dissolved in the water of cloud formations, they formed nitric acid and sulfuric acid, then fell to earth to bathe our forests and cities and to fill our lakes with the corrosive mix.

The relative proportions of the two acids in our rain depends on where one lives. In the northeastern United States, the acidity is primarily due to sulfuric acid, in California, to nitric acid. So we do have a choice.

The rain has caused the reduction and even the elimination of fish in many of our lakes. The rain apparently doesn't kill the fish, it just keeps them from reproducing. So no young fish are found as the old ones gradually go the way of all flesh. In fact, about 700 lakes in southern Norway are now *entirely devoid* of fish, and our own northeastern lakes are following one by one. As our Adirondack lakes reach pH levels of 5 (not uncommon), 90 percent have no fish whatever.

They are also curriously devoid of frogs and salamanders.

Entire patches of forests worldwide are sickening and dying (see photos) as ecologists busily try to find out just what effects the rain is having. In fact, such studies have masked action by the polluting countries. The Reagan administration (undoubtedly under heavy attack by industrial lobbyists) refused steadfastly for years to take action. Instead, it initiated one "study" after another, finally admitting in 1985 that there was a problem and that it had to do with industrial pollution.

Interestingly, the solution is clear to everyone. We simply need to reduce the levels of our effluent from power plants, smelters, and automobiles. Most of the technology exists, but its implementation would be too expensive for the polluters to willingly bear. As they mull over their options, send people to Washington, and initiate new studies, the water passing through the pipes in our cities continues to dissolve the metals in those pipes and our drinking water becomes increasingly unusual.

Above, we see the devastating effects of acid rain on what had been dense woodland of Germany's black forest (below).

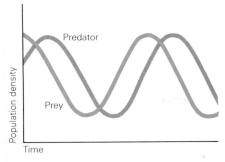

FIGURE 27.10 The classicial predator-prey fluctuation. Notice that as prey density falls, predator density falls also, but a little later in time. When predator density becomes low, the prey is allowed to recuperate and soon the predator density rises in response to its new food supply. This lag is given as the excuse for hunters to be able to resume shooting wolves. They argue that deer herds may never recover from the wolf predation. Look again at the curves. Does the argument make sense?

Predation

Interestingly, a predator normally doesn't eliminate its prey, but it may devastate its competitors. For example, when the vigorous and intelligent dingo, a wild relative of the dog, was introduced to Australia, it didn't kill off the primitive prey it found there. But its hunting prowess proved so superior to that of its competitors, the Tasmanian devil and Tasmanian wolf, that they disappeared from the Australian continent. In Tasmania, where the dingo was not introduced, the Tasmanian devil still survives. The question is why didn't the dingo wipe out its prey?

The usual explanation involves the cycling of such systems. For example, under undisturbed conditions, prey numbers steadily rise (Figure 27.10), thereby providing more food for predators. As a result, the predator numbers begin to rise. Of course, this doesn't happen immediately, because it takes time for the energy from food to be converted into reproductive efforts. As the predator numbers finally begin to rise, the pressure on the prey increases and their numbers begin to dwindle. As the prey are killed off, the predators find themselves with less food, and so their own numbers fall off. Because the fluctuation of predator numbers lags behind that of the prey, the prey may be well on the road to recovery before the predator population begins to rise again. In reality, the story is complicated a bit by such factors as prey-switching (when predators run out of one kind of prey and begin to find other, more available food) and the fact that prey populations usually recover more rapidly than do their predators.

Whereas the delicate and interacting density-dependent mechanisms may work quite well in the wild, our own species can provide us with remarkable examples of how the system can fail. Many of the great whales have been hunted to the brink of oblivion over the past few decades as modern whaling methods have reduced personal risk and increased profits. Furthermore, instead of relaxing the pressure on threatened species and allowing the whale populations to recover, the fleets have increased their efforts. Of course, as whales decrease in number, the price of whale products goes up and the whalers can be expected to respond even more feverishly to the merchants' pleading (Essay 27.2).

Parasitism

Parasitism exists when an organism, the **parasite,** exists on or in the body of another organism, the **host,** deriving benefit and doing harm, but usually without directly killing the host. The relationship between parasite and host is, in a sense, much more delicate than that between predator and prey, because it is to the parasite's advantage for the host to live. Also, the offspring of the parasite will have greater likelihood of success if the host itself lives to reproduce.

Whereas parasites are harmful, in certain cases, the parasite and host are so well adapted to each other that they actually exist quite nicely together. As a rule, the "newer" parasite-host relationships result in more damage to the host. In older associations, the two have had time to adapt to each other, "fine tuning" their relationship.

In some other cases, such adaptation never has a chance to take place because the initial effects of the parasite are so devastating, to the detriment of both. In 1904, the sac fungus *Endothica parasitica* was acciden-

Greenpeace is an organization dedicated to the preservation of the sea and its great mammals, notably whales, dolphins, and seals. Its ethic is nonviolent but its aggressiveness in protecting our oceans and the life in them is becoming legendary. In their roving ship, the *Rainbow Warrior*, Greenpeace volunteers have relentlessly hounded the profiteering ships of any nation harming the resources Greenpeace deems to be the property of the world community. Whales, they believe, belong to us all and have a right to exist no matter what the demand for shoe-horns, cosmetics, and machine oil. (In 1985, the *Rainbow Warrior* was sunk in a New Zealand harbor by French military saboteurs just before it was to sail into the South Pacific to protest French nuclear testing there, killing one member of the group.)

Greenpeace volunteers routinely place their lives in danger in many ways, such as by riding along the backs of whales in inflatable zodiacs, keeping themselves between the animal and the harpoons of ships giving chase. They have pulled alongside Dutch ships to stop the dumping of dangerous toxins into the sea. They have placed their zodiacs directly in the paths of ships disrupting delicate breeding grounds of the sea with soundings and have forced some to turn away or even abandon their efforts. They have confronted hostile sealers on northern ice floes to try to stop them from bludgeoning the baby seals on the birthing grounds, skinning them on the spot, and leaving the mother sniffing at the glistening red corpse of her baby as its skin is stacked aboard the ship on the way to warm the backs of very fashionable people who gather where the bartender knows their favorite drink. (The mother seal would be so proud to know that her dead baby had nearly impressed some bartender.) They have petitioned the International Whaling Commission to establish rules and enact bans.

But some conservationists feel that peaceful efforts are not enough. Such efforts did not, for example, stop a Japanese whaler called the *Sierra* (of all things). The *Sierra* sailed for twelve years, subscribing to no code but that of the merchant. It ruthlessly and efficiently killed every whale it encountered, including threatened species and nurslings. When accused or formally reprimanded, it simply changed flags and continued the killing.

In 1967, the ship was converted to a processor ship and killer all in one—the world's first factory-catcher. A long rear deck capable of holding several whales was added with a sloping stern slipway for hauling the whales aboard. Modifications below deck enabled it to move the tons of meat that they carved from each whale. New huge freezer compartments would accommodate the catch.

Renamed the *Run* and registered to the Atlantic Fishing Company in Nassau, Bahamas, the ship set out in 1968 to kill whales along the coast of southern Africa, an area totally off limits to IWC whaling. In the next three years, its crew harpooned 1,676 whales, most of them Bryde's whales, as well as humpbacks and the extremely rare southern right whale. For the sake of speed and efficiency, they only took about half the meat. When whales became scarce in one area, the ship would move on, often wiping out entire herds as it went.

---

tally introduced into North America from China. In China, the parasite was held in check by a variety of control mechanisms that do not exist in North America. Sadly, the majestic American chestnut tree proved defenseless to the fungus, and by the late 1940s, the great tree that had dominated our Appalachian forests was virtually extinct (Figure 27.11). Fortunately, researchers are still working to save the species and there has been recent talk of clear successes that may restore the tree to our forests.

We have seen that populations may be held in check by what might be called ecological factors, such as competition brought about by low food supply. The role of competition in the world of life has recently been hotly debated in academic circles. However, no one can deny that it exists. Competition, after all, results when two individuals utilize the same resource to the extent that when one acquires it, the other suffers in some way. The effects of competition on populations is clearly demonstrated in the behavioral phenomena called territoriality and hierarchies.

FIGURE 27.11   The American chestnut was nearly wiped out by an introduced fungus. Here, a new plant is grown by new techniques that may restore the tree to our forests.

Conservationists of a different stripe began to track the *Run* using everything from computers to bribery to keep it located. One group, aboard a vessel called the *Sea Shepherd,* apparently had enough of the wrist slapping. They wanted the killing stopped. Period. They encountered the *Run* one July day in 1979 off the coast of Portugal. With engines straining, they rammed the *Run.* It was rather unusual behavior for normally gentle conservationists, but although they failed to sink the killer ship, they had served a clear warning. But the point was not made absolutely clear until later. After the ramming, the *Run* had limped to port, unloaded its catch and then slowly made it back to Japan. Apparently the owners were not yet convinced. The boat was brought to Lisbon Harbor to be reoutfitted and resume its dreadful business. However, something strange happened. One dark night, powerful explosives managed to become attached to its hull and then somehow to become detonated. The *Run* found itself blown to smithereens and smiles found their way to faces around the world.

A Greenpeace volunteer carries a baby seal away from sealers as they club the pups for the fur trade on the birthing ground.

Territoriality

In **territorial** species, animals defend certain areas by chasing away competitors. In this way, the winner is likely to end up living in a better area than the loser. One area, for example, may hold more commodities, such as nest sites, hiding places, or food. If some areas hold more food than others, say, in times of food shortage those individuals with the best territories are more likely to survive. Territories may be important in other, perhaps related ways as well. For example, in some species of birds, only those males with the better territories are able to attract females. A female will not mate with a male with an inferior territory no matter what other qualities he has. If there aren't enough good territories to go around, then, not all individuals will be able to reproduce, and thus the population number is lowered.

Hierarchies

Hierarchies, or "pecking orders," also result in certain animals having freer access to commodities than do others. In hierarchies, each animal must be able to recognize others individually and to respond to each according to its rank. Hierarchies are established in various ways in different species. They may be established through fighting, ritualistic displays, or even play. Hierarchies are adaptive in that they reduce conflict within groups as a result of each animal yielding when confronted by a higher-ranking individual. In stressful times, such as periods of food shortage, then, the higher-ranking animals are more likely to survive than are subordinates because of their freer access to critical commodities.

In both territorial and hierarchical systems (which are not, by the way, mutually exclusive), harsh conditions are more likely to kill the lower-ranking animals, the "have-nots," thereby increasing the odds of survival for the "haves." This doesn't mean that the have-nots accept their role for the "good of the species." Quite the contrary. It is to their advantage to stay alive at all costs. However, a low-ranking individual may have better luck by searching for unclaimed commodities than by challenging a dominant, and probably more powerful, individual. The subordinate is likely to lose the conflict and the commodity as well, while risking being battered and weakened for its efforts.

Disease

We should have little trouble believing that disease can reduce the size of populations. (We will see in the next chapter how bubonic plague drastically reduced the human population, and we are aware that all sorts of less dramatic diseases claim thousands of lives daily.) Disease is density-dependent, as a general rule, for several reasons. The more closely a group is packed together, the easier many diseases are transmitted. Also, the more individuals there are in a population, the greater the number of potential reservoirs in which mutant strains of disease microorganisms can develop. If crowding results in some individuals being weakened, disease will have an even greater impact on the population.

In some cases, disease and predation may act together in reducing populations. For example, a two-week-old caribou fawn can outspring a full-grown timber wolf. However, caribou are subject to a hoof disease that quickly lames them. It is these lamed animals that wolves are likely to attack and cull from the herd. The "sanitizing" effect of predators is now well established. As evidence of such an effect, in those places where wolves have been poisoned to "protect" migrating caribou, the hoof disease has spread unchecked and, in only a few seasons, has wiped out entire herds.

THE ADVANTAGE OF DEATH

Have you ever wondered why you must die? Just the realization of that fact can be a real nuisance. So let's ask why death occurs at all. Isn't it logical that natural selection would have tended to extend life spans so that by now only disease or accident could claim us? Why are we slated to start falling apart, one system after the next, until we believe we simply can't go on? (And if we don't change our minds, they eventually bury us.) Why must we suffer the indignity of repairing one bodily breakdown after another, prisoner in a patchwork body until we finally expire, looking like the product of a quilting bee? Why can't all our systems hold? (Or all go at the same time so that one can at least be a handsome corpse?) If "life" is better than "death," then why is there not marked selection for longevity? Why are there delicate insects that don't even have mouths with which to feed themselves, but which must instead spend whatever energy their frail bodies possess in finding mates before their few precious hours of life are gone? Why do humans have so much trouble surviving past "threescore years and ten"? Why death?

People with all sorts of perspectives on life have ruminated on the "meaning" of death. Various notions have been proposed by philosophers, theologians, novelists, dramatists, poets, drunks and others of bad habit, and none of the explanations seem to coincide. Death to some is a fearful thing, while others welcome it as the gateway to a better life, but without modern conveniences. Some aboriginal societies deal with death casually and fearlessly by treating it as a kind of natural extension of life. Admittedly, the dullest dead person knows more about the meaning of death than all the living philosophers combined. But since it is impossible to prove us wrong, let's join the fray and consider death from a biological point of view. We might first account for its genetic advantages. (Note that last word. Let's see if it makes sense.)

The biological explanation of the advantage of death becomes apparent if we keep three things in mind. First, evolution at its basic level is a means whereby certain kinds of genes (not individuals) are perpetuated. Second, the earth changes, and third, it is a limited place.

To put these principles into the perspective of our argument we must adopt the rather irreverent notion that individuals are simply temporary caretakers of their genes. The evolutionarily more successful genes will be those found within individuals who propelled them into the next generation. None of today's *individuals* will be represented in the next generation, but some of today's *genes* will be. The point is that in the cool, unrepentant arithmetic of evolution, the individual is expendable.

Such "wastefulness" can be seen to be evolutionarily advantageous if we keep in mind the next two points—that is, the earth changes and it is limited. We have already noted that the variability of offspring is one way an individual might be able to leave its genes in a changing world. Put simply, an individual might not be malleable (changeable) enough to survive if conditions should change too drastically. But among that individual's variable offspring might be those that had just the traits necessary to continue on under the new conditions. However, since the earth is a finite place with limited resources, a parent that lingered on to compete with its offspring might damage the chances of success for those offspring. So, strange as it seems, an individual's best reproductive (or evolutionary) bet might be to reproduce its kinds of genes in various combinations and then to get out of the way so as not to interfere with them. Of course there are those who might say "To heck with kids. I want to live." No matter, that individual is here because countless earlier generations were programmed to make the sacrifice and thus that rebellious soul's time, too, will come.

By considering death as a mechanism, or tool, a way to perpetuate one's kinds of genes, we might be left with the idea that an organism is simply a vehicle that genes use in order to reproduce themselves. Of course, we would prefer to feel that our lives have a bit more meaning than that, that we are more than carrying cases for our gonads. Nonetheless, there is a line of logic that suggests that we are simply our genes' way of making more genes. Therefore, we primp, work, save, worship, rationalize, love, lust, avoid, hate, defend, retreat, and sunbathe. We have a variety of explanations for all the things we do. But perhaps what we are really doing is behaving in such a way that our genes end up in the next generation. If our primping helps us to reproduce our genes, primping it is. According to this notion, we are really acting at the behest of our genes, and we are given whatever latitude might most effectively enable them to be reproduced and passed on. Of course, we must ask, are we really just great lumbering robots, controlled by our genes, responding to these constant, wheedling messages that are delivered by other mindless intermediates called enzymes?

The question is definitely a bit tortured and overextended, but there is an elegant simplicity to its assumptions, once we can accept the irreverence of it all, even if just for the sake of entertaining an unfamiliar concept.

We've tossed around some rather weighty concepts fairly loosely here. Perhaps, however, we've managed to ask some rather fundamental questions. Perhaps, also, we have suggested new ways to investigate old problems. This, after all, is one of the roles of science in our increasingly complex world.

SUMMARY

Some populations are remaining rather stable while others are increasing and yet others becoming extinct. The study of how populations change is called population dynamics. The growth of populations under ideal conditions produces a J-shaped curve, which, with increasing environmental resistance, may level off around the carrying capacity, producing an S-shaped curve. Populations normally do not increase at their biotic potential because of environmental resistance. Should a population drastically overshoot the environment's carrying capacity, it may experience a rapid decline called a crash.

Populations are normally regulated through influences on reproduction and mortality. Species with little parental investment in any offspring can be expected to produce many offspring. Those with greater investment in each can be expected to produce fewer.

Control of populations through mortality (or through death) can be abiotic or biotic. Abiotic influences are generally density-independent. Biotic controls are generally density-dependent and may include predation, parasitism, territoriality, hierarchies, and disease.

Theoretically, death could have been selected for as a means of perpetuating hereditary lines because (1) the unit of evolution is the gene, not the individual, (2) the earth changes, and (3) it is limited. Thus, genes that exist in variable combinations are more likely to survive in a changing world, and once a suite of genes begins to succeed, the less well-adapted parents should ideally remove themselves through death so as not to compete with their offspring.

Chapter 28

HUMAN POPULATIONS

"Return with us now to the days of yesteryear." In fact, let's go back to between one and two million years ago to a far different planet than the one we call home today. Should we appear on that ancient earth, we probably wouldn't be able to recognize the place where our house sits today, and the array of plants and animals we would see would be strange indeed. One kind of animal, though, might have a faint air of familiarity about it. This animal we would know as a primate as it ambled in small bands over the countryside, stopping now and then to grab at some morsel of food. Who could have guessed that virtually every other kind of life, as well as the face of the earth itself, its air and waters, would yield before something as seemingly insignificant as the descendants of that hunched figure sitting in the shade, picking at the soft parts of an insect?

EARLY HUMAN POPULATIONS

We have no idea how many individuals comprised the human species in our earliest days, and we don't know much more about our numbers in more recent times. In fact, we know little about human populations before about 1650 B.C. However, we do have educated guesses. These are often based on what we know about modern groups that seem to be similar to those of earlier times. There are other lines of evidence as well. We know, for example, that agriculture was almost unknown before about 8000 B.C. (possibly a bit earlier in Malaysia and the Middle East). Before that time, most humans must have lived by hunting and gathering their food. Considering the inefficiency of such methods today, and assuming a land area of about 58 million square miles, we can infer that in 8000 B.C., the earth could have supported no more than about five million people.

It is important to realize that 90 percent of all the people who have ever lived on the earth have been hunter-gatherers, only 6 percent subsisted mainly by agriculture, and only a very small percentage have lived as industrial men and women. But this small industrial group has played an inordinate role in bringing us to our present situation. Another way to place this group, our group, in its proper perspective is to remind ourselves that about 99 percent of humankind's time on earth has been spent as hunter-gatherers. Obviously, then, our minds, attitudes, values, bodies, and behavior over the eons must have been geared for a different kind of life than we now have. This new kind of life has fallen upon us so suddenly that within the last 1 percent of the period that our species has been on the planet, only 1 percent of us continue to subsist as hunter-gatherers.

Hunter-gatherers have relatively little impact on the environment. They utilize a variety of plants and animals, and as one falls into short supply, they simply move on or switch to other resources, thus relieving the pressure on the scarcer species. Most of the food of hunter-gatherers, by the way, is taken by gathering. In modern groups, the men usually bring home the larger, swifter (or more dangerous) protein-laden meat, but the women, quietly gathering plant material, contribute the far greater part of the diet (Figure 28.1).

In time, the early hunter-gatherers became more efficient and their hunting began to pose more of a real threat to other species. Weapons and tools improved, and men began to hunt together in a more coordinated manner and to bring down larger game. Language would have had to become more sophisticated in order to maximize the results of group hunting ("Tomorrow, half of us will hunt here." And, in time, another concept: "Except Charlie; his wife won't let him go.") It was probably not a great step from someone saying, "You get his attention and I'll hit him with a stick," to the formation of rudimentary leadership and thence to political maneuvering. (Anthropologist Lionel Tiger has suggested that this is a reason for the overwhelming masculine domination of modern politics.)

We find early indirect signs of man's increasingly proficient hunting ability. For example, after the ice ages ended about 10,000 years ago, about 70 percent of the North American mammals became extinct. Their decline correlates suspiciously with the arrival of humans in those areas. In fact, many remains of such animals have been found with human

FIGURE 28.1 Among the most primitive of today's hunter-gatherers is the Waorani tribe of the Conanaco River of Ecuador. While the men hunt pigs, monkeys, and parrots and clear land with neolithic axes they find on the forest floor, the women plant the few crops that, with what they find growing wild, form the bulk of their simple diets. The jungle can sustain only small groups, and these must move on every few years to new hunting areas.

weapons nearby. Furthermore, as man grew more cunning, he apparently began to rely more on trickery and imagination. In one place near Soulutre, France, the fossilized remains of over 100,000 horses have been found where they were apparently stampeded over a cliff by primitive men. As people began to use fire, they could have made the habitat even more to their liking; some researchers believe that many of the world's great grasslands were created partly by the land being repeatedly burned over by early man. The point is that humans began very early to make sweeping changes in their environment, and they were apparently limited only by their primitive technology.

It is believed that these early human populations formed separate, but interbreeding, groups, possibly divided into small bands. From studies of present-day hunters and gatherers, it has been calculated that the average size of such groups would have been about 200 to 500 men, women, and children, of which about half were of breeding age. Population size was probably relatively stable and density-dependent. (Keeping in mind the food sources, can you tell why? See Chapter 27.) It is also generally believed that the number and size of the groups approached the carrying capacity of the environment.

THE ADVENT OF AGRICULTURE

About 8000 B.C., the earth's carrying capacity rose to a new level through the advent of agriculture. People began to roam less and to decrease their dependence on the uncertain presence of wild game. By learning to grow and store their food, they added a relatively stable element to their food supply. Stored food meant that the lean seasons no longer marked the deaths of so many children and old people. The result of the shift toward agriculture was a slow, steady increase in the size of human populations.

How could the shift to agriculture have raised the environment's carrying capacity? We have already noted that the ability to store foods, such as grain, meant a more dependable food supply and a greater likelihood of surviving the harsh seasons. But, in addition, agriculture meant a greater reliance on plant food, and so humans began to shift their diet to the "producer" level. As we will see in the next chapter, it is more efficient to eat plants than animals because energy is lost in converting plant food to animal food (Essay 29.1). All this is not to say that the agricultural revolution has been the only factor that has changed the carrying capacity of our environment. The advent of tool-making also had its marked effect, and we will learn shortly of the phenomenal impact of the recent scientific-industrial revolution.

As you can see from Fig. 28.2, the human population continued its slow rise through the beginning of the Christian era. It is believed that at the time of Christ, the earth's population was between 200 and 300 million and that it increased to about 500 million by about 1650, with only one period when it dropped significantly because of the impact of the bubonic plagues.

POPULATION CHANGES SINCE THE SEVENTEENTH CENTURY

Somewhere around the seventeenth century, the world's human population suddenly began to rise. There has been increases before, such as those associated with the advent of tool-making or the beginning of agriculture. But these were gentle rises; the population surge of about three hundred years ago was much more dramatic. Such an abrupt rise has heralded devastating population crashes in other species, and this increase might have had the same results if it had not been associated with an increase in the environment's carrying capacity. And what brought this about?

The sharp rise in the population of Europe between about 1650 and 1750 is attributed to innovations in agriculture and the beginning of the exploration and exploitation of the Western Hemisphere. The accompanying rise in Asian populations at the same time is harder to explain. India, for example, was in turmoil following the fall of the Mongol Empire. The simultaneous rise in the Chinese population might have been a result of a new political stability brought by the Manchu emperors, who also initiated new and better agricultural policies after the fall of the Ming Dynasty in 1644. Not much is known about Africa in this period, but its population is estimated to have been about 100 million in 1850—about the time that European medicine and technology were introduced. Between 1750 and 1850, the population increase in Asia

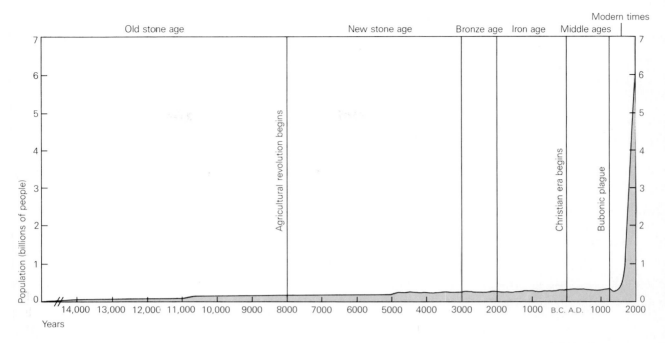

Population (billions of people)

7 · 6 · 5 · 4 · 3 · 2 · 1 · 0

Agricultural revolution begins

Christian era begins

Bubonic plague

14,000 13,000 12,000 11,000 10,000 9000 8000 7000 6000 5000 4000 3000 2000 1000 B.C. A.D. 1000 2000

Years

FIGURE 28.2 Human population growth during the last half million years. If the old stone age were drawn to scale, it would extend about eighteen feet to the left. Notice the very gradual rise as agriculture began about 8,000 years ago. Even 1,000 years ago, the earth's population was enormous by historical standards. But the events since the last bubonic plague and particularly in the last few hundred years could not have been predicted. Someone has rather whimsically calculated that at our current rate of increase, the end of the next thousand years will see human bodies completely filling the known universe and expanding outward in all directions at the speed of light! That probably won't happen. Where would they sit? Of course, we can depend on natural means to bring our population under control long before then. The very thought can be terrifying.

was only about 50 percent, far less than in Europe. European advances at this time included new agricultural techniques, better sanitation, and advances in medicine, such as the acceptance of bacteria as a source of infection and the development of smallpox vaccine.

Between 1850 and 1950, the populations of Asia, Africa, and Europe approximately doubled. The population of Latin America increased about five times over, and that of North America increased sixfold. But let's break this down a bit.

POPULATION CHANGES AFTER 1850

Between 1850 and 1900, the death rate in Europe continued to decline with advances in industry, agriculture, and medicine. Up until then, life in Europe was dismal almost beyond belief. The cities were dangerous, garbage-filled, and overrun with rats. Sewers ran freely through the streets, their delicate gurgling belying their burden. The stench was devastating. Life in the country was not much better. Outside the cities were scattered hamlets, little more than isolated slums inhabited by people of numbing ignorance. It is hard to imagine how they could be helped. Where to start? But, the changes in agriculture and health practices were having their effect, and people were indeed living longer. By the beginning of the twentieth century, the general situation improved even further, a result of even newer farming techniques that reduced the risk of crop failure. In addition, there were improvements in transportation that meant food could be more easily distributed and imported to locally stricken areas. Also, about this time the role of bacteria in infection became better known, and so more lives were saved. The death rate in Europe dropped from about 24 per 1,000 to 20 per 1,000 in just those fifty years.

By the beginning of the twentieth century, **demographers** (those who study human population changes) had identified a rather surprising trend. Their records showed that industrialization was generally fol-

In 1986, the Population Conference Bureau went beyond the usual predictions to the year 2000 and announced the results of a startling new study estimating the world population to the year 2100. By then, India is expected to pass China as the world's most populous nation (see figures). World population by then will top 10.4 billion, more than double the current 4.8 billion. This increase will take place even as growth rates are declining, because of the large number of young people, particularly in developing countries. In fact, the greatest increases will take place in some of the world's poorest regions—sub-Saharan Africa and Southeast Asia. Of the expected 3.4 billion people to be added to the earth by 2025 (how old will you be?), 3.1 billion (almost all) will be in Africa, Asia, and Latin America, among the poorest areas of the world. Furthermore, by 2025, about 83 percent of the world's populations will live in these areas, and 86 percent by the year 2100. Currently, 75 percent of the world's population lives there.

One of the greatest problems associated with such a distribution is the danger of a more polarized, two-tiered world, composed of the haves and have-nots. The have-nots can be expected to agitate for a larger share of the earth's resources. The more affluent countries will be islands in a sea of poverty and will be vulnerable, largely due to the export of their own technical advances in transportation, communication, and weaponry.

In the accompanying table are examples of population size in selected countries in 1986 and projected to 2100 (rounded off, in millions).

	1986	2100
India	785	1630
China	1050	1570
Nigeria	105	510
U.S.S.R.	280	375
Pakistan	100	315
United States	240	310
Bangladesh	105	300
Brazil	145	295
Mexico	80	195
Ethiopia	45	175
Vietnam	60	170
Kenya	20	115
South Africa	35	100
Uganda	15	80
France	55	60
England	57	58
Italy	56	57
Zimbabwe	9	40
Switzerland	7	6
Denmark	5	5
Libya	4	18
Haiti	6	14
Mauritania	7	8

lowed by a lower birth rate. The phenomenon, which has since received a great deal of attention, has never been entirely explained. However, it is suggested that in agricultural societies, children are a source of labor and a form of social security. They provide assistance before they leave home and they care for their aged parents later. In an industrial society, it is suggested, they are less likely to aid in production and are simply "more mouths to feed." Also, mobility, so important in industrial societies, is reduced by the presence of children.

Some researchers believed that simple development leading to industrialization was a sure form of population control. (A slogan of the 1974 World Population Conference in Budapest was "Development Is the Best Contraceptive.") But now it seems that the notion was perhaps a bit simplistic. Industrialization is often accompanied by a reduced population growth, but there are enough exceptions and stipulations that it is no longer held as a "truism."

About the time of World War II, many **developing** (poorer) **countries** saw a dramatic reduction in death rates, apparently due to imported public health programs, including new drugs, from the more developed

TABLE 28.1 World Population, Past and Projected

	1975	1985	1990	2000
World Total (in millions)	3,968	4,830	5,273	6,199
Africa	401	544	630	828
North and Central America	343	375	405	464
South America	219	278	313	392
Asia	2,256	2,923	3,307	4,775
Europe	473	492	501	520
USSR	255	280	295	312

Sources: Population Reference Bureau, U.N. Statistical Papers.

countries. (Among the most important imports was DDT, which was used to control malaria.) Such imported "death control" produced history's most sweeping increases in human populations. It is important to realize that the changes in death rate in these developing countries were not accompanied by the social and institutional changes that went hand-in-hand with the far slower reduction of death rates in the more developed countries (Table 28.1 and Essay 28.1).

THE HUMAN POPULATION TODAY

By the end of the 1970s, a curious and unexpected trend was emerging. The rate of **natural increase** (Table 28.2) was slowing. The key word here is *rate*. Rate changes can be illustrated by imagining a busload of people heading for a cliff at 30 miles per hour. As their rate slows, they may be moving at only 15 miles per hour as they go over the edge.

By the end of 1980, the world's population was estimated to be about 4½ billion. The percent annual growth had fallen slightly and the **doubling time** (the time in which a population doubles in size—Table 28.2 and Essay 28.2) was now about thirty-nine years. No one is sure how these changes came about, but several trends were apparent. Among them were the changing attitudes of women regarding their role in society, changes in the number of children that couples preferred, advances in birth control, and liberalized abortion laws in developed societies.

The lowered growth rates, however, were not worldwide. They occurred primarily in the **developed** (more industrialized and affluent) **countries.** That is, those that could best withstand increased populations. They included not only the United States, but Western Europe, the Soviet Union, and Japan. In the developing nations, the populations continued to swell. Their percent annual growth (excluding China) is now an ominous 2.4 percent, and their average doubling time only twenty-nine years.

The news from around the world varies from place to place. While the **crude birth rates** and **crude death rates** (Table 28.2) in the United States in 1980 were fifteen and seven per thousand, respectively, in Africa they were forty-six and nineteen, numbers that spelled much human suffering. And more misery is on the way with the African population expected to double in twenty-five years.

Table 28.2   Basic Population Arithmetic

Crude birth rate = number of births per year per 1,000 population

$$\text{determined by: } \frac{\text{total births}}{\text{midyear population}} \times 1{,}000$$

Crude death rate = number of deaths per year per 1,000 population

$$\text{determined by: } \frac{\text{total deaths}}{\text{midyear population}} \times 1{,}000$$

Rate of natural increase = crude BR − crude DR (or decrease)

$$\text{Percent annual growth} = \frac{\text{rate of natural increase}}{10}$$

$$\frac{\text{Doubling time}}{\text{(approximately)}} = \frac{70}{\text{percent annual growth}}$$

(Example: % annual growth in the world in 1983 was 1.8: $\frac{70}{1.8} = 38.89$ years)

FIGURE 28.3   The Chinese government has become keenly aware of the country's population problem. One-child families are encouraged by billboards such as this. Should this encouragement fail, prospective mothers are subjected to more drastic measures, such as being put under intense community pressure to undergo abortion.

At first glance, the news from Latin America seems more hopeful. The year 1980 saw a decrease in the crude birth rate in the region. But there's a small problem here: The birth rate is already four times the death rate and the doubling time is similar to that of Africa.

Assessing the problem in Asia is more difficult. The problem is that many areas are so primitive or so secretive that world demographers just don't have the basic information they need. We do know, however, that half the people on earth are Asian and half of these are Chinese. Because of China's highly publicized program to limit families to one child (Figure 28.3), the birth rate there has declined somewhat. Partly as a result of this trend, the doubling time in Asia has increased from thirty-one to thirty-five years. Similar efforts at controlling family size in India have been somewhat successful, but its population continues to swell dangerously out of control.

The small successes in controlling the Asian populations, unfortunately, will soon be swamped by another problem looming on the horizon. That is, almost half the Asian population is under the age of fifteen and have yet to enter the breeding population. The impact of this statement can be shown in the population age pyramids that we will see shortly.

THE FUTURE OF HUMAN POPULATIONS

Unfortunately, we can't be very confident about predicting future trends in human populations. The sad truth is that almost every major trend of the past few decades has caught even the most sophisticated demographers by surprise. Demography is, indeed, an inexact science at present.

ESSAY 28.2   DOUBLING TIMES

It is hard to imagine the growth of the human population on the earth. Perhaps it will help to consider "doubling times." Would you work for someone who paid you a penny your first day, two cents the second, four cents the third, and kept doubling your wages each day for one month? You *should*. Consider this: If this page is, say, 1,250th of an inch thick and you start doubling it, after only eight doublings it would be an inch thick, after twelve doublings it would be about sixteen inches, and after only twenty doublings, thicker than a football field is long. If we doubled the page forty-two times, the stack would reach to the moon, and after fifty times we would reach the sun, about 93,000,000 miles away.

Some countries have a much firmer grip on their population problems than do others, so evidenced by their doubling times. These are usually the older, more established nations. For example, Britain's population will double in 1,155 years and Sweden's in 1,386 years. The population of the United States is growing much faster and will double in about ninety-nine years, Canada's in eighty-eight years, and the Soviet Union's in eighty-two. But certain countries have far greater problems. For example, Brazil's population will double in twenty-five years, Egypt's in twen-

Doubling Times of the Human Population

Date	Estimated World Population	Time Required for Population to Double
8000 B.C.	5 million	1,500 years
A.D. 1650	500 million	200 years
A.D. 1850	1,000 million (1 billion)	80 years
A.D. 1930	2,000 million (2 billion)	45 years
A.D. 1985	4,800 million (4.8 billion)	38 years
A.D. 2010	8,000 million (8 billion)	?

ty-six, Kenya's in seventeen, and India's in thirty-six! China has instigated a monumental and rather oppressive program to encourage one-child families and still their population will double in about fifty-eight years. The next doubling, of course, will occur over a much briefer period. And the next, briefer yet. Historically, the time spans have been a bit longer. To go from five million people on earth in 8,000 B.C. (the present number in only three of New York City's five boroughs) to 500 million in A.D. 1650 took six or seven doublings over a period of 9,000 to 10,000 years. During that time, the human population doubled on an average of about every 1,500 years. A glance at the right-hand column of the table will show that, all other

things being equal, in only about thirty-eight years, we will need two cars, two schools, two roads, two houses, and two cities throughout the world for every one that presently exists. And that will only maintain our status quo as far as material goods are concerned. The problem is that not all things can remain equal over that time. Can India or Guatemala double its food? Can they distribute it equally? Can the people afford it? Can we double the population of our own eastern seaboard without doing severe damage to our urban populations? Can California stand another Los Angeles? Where will the water come from? If we double the number of oil wells, can we double the earth's oil?

However, increasingly sophisticated techniques do show some promise of improving our predictive abilities. One such tool is the increased reliance on fertility rates as indicators.

Fertility Rate

Two somewhat reliable indicators of future population trends are the general fertility rate and the total fertility rate (Figure 28.4). The **general fertility rate** is the number of live births per 1,000 women in the reproductive age bracket. (That is, 15–44 in the U.S., but considered to be 15–

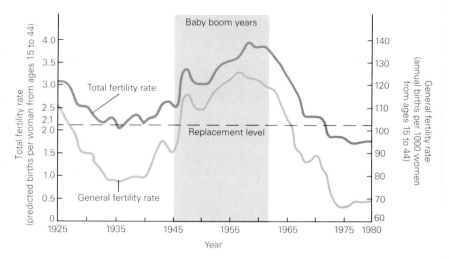

FIGURE 28.4 Fertility studies in the U.S. between 1925 and 1980. The two plots include the general fertility rate and total fertility rate. The depression years are seen as a valley, followed by a reproductive peak—the "baby boom" years—that came after World War II. The recent valley, following the boom years, is good news to those alarmed by population growth. But keep in mind that attitudes toward family size can change and that women born in the baby boom years are still in the reproductive period of life. The replacement level is the number of births that would simply replace the number of parents. Because of infant mortality, the replacement number is slightly over two.

49 in some other countries.) This figure, then, is based on historical fact. The **total fertility rate,** however, is a *prediction* of the average number of children that women will have in their reproductive years. This figure is a bit trickier to come by, of course.

The total fertility rate is based, we see, on a good measure of faith. But the faith is buttressed by data from economists, federal agencies, religious groups, and family planning agencies. Furthermore, it attempts to take into account such influences as attitude. For example, beginning in the 1960s, women in their reproductive phase in industrial societies began to reject reproduction as their sole function on the planet. They organized themselves politically; they entered the work force; they scrambled for rank and status; they challenged male enclaves; and all the while they placed less emphasis on children. Their changing attitudes were duly reflected by demographers and population predictions were revised downward. However, such attitudes can change quickly and we must stand ready for new shifts and swings of the pendulum of public consciousness.

Population Structure

Another powerful tool of demographers is the analysis of population structures. One way to show these is by use of **age structure pyramids.** For example, in Figure 28.5 we can see that populations can be described according to what percentage of their numbers are in various age groups. In Mexico, we see, the population is increasingly expanding in the younger groups. This is partly because of poor health care for older people and a generally high mortality rate throughout life, but it is mainly because of a strong social emphasis on larger families. This emphasis may be due to religious sanctions against certain forms of birth control and the fact that production of children is seen as a sign of virility in a male-dominated society. In the United States and Sweden, the age structures are more uniform, partly due to better health care and fewer children per family. Note that the pyramids reflect a difference in the age structures of males and females (Essay 28.3).

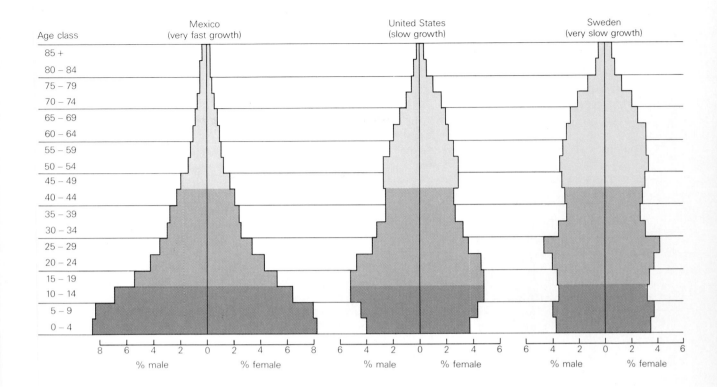

FIGURE 28.5 Age structure pyramids for Mexico, the United States, and Sweden reveal important differences between developing countries and those that are more economically advanced. Future growth rates can be predicted by studying the lowest levels of the pyramids, since these include females that have yet to reproduce. The middle region of a pyramid contains the reproducers and the labor force. These individuals have the burden of supporting both the lower and the upper levels, a burden whose size can be estimated by a look at the bases and tops of the pyramids.

There are important social implications in such pyramids. Note that the reproducers are also the workers in the population. In which kind of nation would programs such as Medicare, or other systems of care for the elderly, be more important? Who pays into Social Security programs and who draws from them? Who feeds the young? Would voting tendencies tend to differ in old and young populations? Would housing starts differ? Housing designs? Clothing designs? How about pressures on natural resources or the likelihood of belligerence toward neighboring countries? The point is that many facets of any society reflect the age structure of its population.

There is obviously a great deal of concern in the scientific community over the impact of rising human populations. But the reaction is not universal dismay. Some scientists believe we can absorb great increases in our numbers with only a few policy changes. Others profess an impressive faith in technology and suggest that the answer lies in increasing the carrying capacity of the earth through agricultural, technological, and medical "revolutions." It has recently been estimated that the human population could rise to eight to ten billion people—*temporarily*. From that point, the numbers would have to drop to more sustainable levels in order to avoid permanent ecological disasters.

We are an adaptable and innovative species and there is no doubt that we can successfully withstand all sorts of traumas, including great population increases. A dog can get along just fine underwater, too. But not for long. How long can we withstand the relentless pressures of increasing numbers of other people? Even if we can find ways to endure it for long periods of time, the question is, do we want to? How old will *you* be when the earth has six billion people? (This will occur before the year 2000.) At the current rate, how old would you guess your children will be when there are fourteen billion people on earth?

People are living longer than ever, but for some reason, women are living longer than men are. A baby boy born in 1984 can expect to live to be about seventy-one, a baby girl, about seventy-eight-and-a-half. This is indeed a wide gap, and no one really knows why it exists. (Some men have claimed it's because women aren't married to women and some women claim that anyone who can put up with a man has to be tough.)

The greater longevity of women has been known for centuries. It was, for example, described in the seventeenth century. However, the difference was smaller then—the gap is growing.

A number of reasons have been proposed to account for the differences. The gap is greatest in industrialized societies, so it has been suggested that women are less susceptible to the strains of "the rat race" that produce such debilitating conditions as heart disease and alcoholism. Sociologists also tell us that women are encouraged to be less adventurous than men (and that they are more careful drivers).

It was once suggested that working women are more likely to smoke and, as they entered the work force, the age-gap would begin to close. (Smoking is related to earlier deaths.) Now, however, we see more women smoking and they still tend to live longer although their lung cancer rate is climbing sharply. (One recent study indicated that if neither sex smokes, they have identical life spans.)

One puzzling aspect of the problem is that women do not appear to be as healthy as men. That is, they report far more illnesses. But when a man reports an illness, it is more likely to be serious.

Men may die earlier because their health is more strongly related to their emotions. For example, men tend to die sooner after losing a spouse than women do. Men even seem to be weakened by loss of a job. (Both of these are linked with a marked decrease in the effectiveness of the immune system.) Among men, death follows retirement with an alarming promptness.

Perhaps we are searching for the answers too close to the surface of the problem. Perhaps the answers lie deeper in our biological heritage. After all, the phenomenon is not isolated to humans. Females have the edge among virtually all mammalian species. Furthermore, the differences begin at the moment of conception; there are more male miscarriages. After birth, more baby boys than baby girls die.

Another biological explanation involves differences in the sex hormones. Estrogen, for example, may help protect against heart disease. Also, the female hormones apparently render the immune system more efficient.

SUMMARY

Historically, the earth has not supported great human populations, although we have little specific information before about 1650. Based on information on hunter-gatherers today, the human population is believed to have been no greater than five million in 8000 B.C. Ninety percent of the people who have ever lived were hunter-gatherers, probably in bands of 200 to 500 people. The population began to increase about 8000 B.C. with the advent of agriculture, which insured a more stable and efficient food supply. The earth's population increased to about 500 million people by about 1650, its growth interrupted only by the bubonic plagues. Around 1650, a dramatic surge in numbers began, largely due to agricultural innovations in Europe. The causes of the rise in other areas are not understood. Between 1850 and 1900, the death rate in Europe declined because of advances in industry, agriculture, and medicine. By World War II, death rates dropped in many developing (poorer) countries. By the end of the 1970s, the rate of natural increase was slowing, but the world's population continued to grow until in 1980 it was estimated to be about four-and-a-half billion. The doubling time by then was 39 years. The lowered rates were partly due to the changing attitudes of women, but they were not worldwide.

The greatest problem is in Latin America, with its great birth rate, and in Africa. Less is known about Asia, but half the world's population is Asian and half of these are Chinese. The Chinese have instigated a rigid code of one child per family.

Part of the problem of controlling populations lies in the age structure of the population. Developing countries have a disproportionate number of people of reproductive or prereproductive age. Scientists are divided over the question of the impact of the earth's swelling human population.

Chapter 29

RESOURCES, ENERGY, AND HUMAN LIFE

All the known field mice in the universe must live on whatever this planet can provide for them. They run little risk of depleting their planet's resources, however, because their needs are few: a warm hole, a little food, and a little social acceptance. It's a good thing for them that their requirements are so few, because they have only the resources that their planet can provide. Should they deplete or foul those resources, they would be in a bit of a fix, because *that's all there is*. The point is not too subtle, is it? We share the planet with the field mice. And our resources, too, are limited. But their gentle touch on the great, delicate globe is far from our sledgehammer impact.

FIGURE 29.1 The earth's resources are unequally distributed. The result is that various nations control the lion's share of certain commodities. For example, here we see the gold mining operations in South Africa. This country is blessed with approximately 55 percent of the world's supply of gold. The country is also blessed with unusual amounts of chromium and diamonds.

In this chapter, we will review the state of some of those resources necessary to life and try to see just how our impact is influenced by something very ephemeral and hard to pin down: our values. We will approach the question of whether our behavior and values could or should be based on our access to the earth's resources and energy. And how have we treated our heritage?

We will consider the earth's resources to be in one of two categories, renewable and nonrenewable. **Renewable resources** are those that do not exist in set amounts—they can be replenished or reused. If we cut a pine tree, we can grow another one (assuming that the top soil didn't wash away because we cut the first one). **Nonrenewable resources** are those that exist in set amounts on the planet and are not replenished. Once used, they are gone, or they are changed so that they are difficult to recover in their original form.

RENEWABLE RESOURCES: FOCUS ON FOOD AND WATER

Our heritage, or bequeathment, it could be said, is a generally bountiful earth. We heirs, then, should perhaps consider the earth a *home*—not an opportunity, nor a commodity, nor a bargaining chip, but a place that both blesses us and threatens us as a natural part of things. Strangely, it is the very bounty of a plentiful earth that has been at the root of some pressuring problems for our species. How could this situation have come about?

First, the earth's human population, we've seen, is expanding at an alarming rate, and each new face demands its coupons and catalogues. We are placing a severe and increasing demand on our earthly goods, bequeathed by ancestors who lived in a more sparsely populated and far different kind of world. Our numbers are increasing, but many of the earth's resources aren't. Part of the problem is as simple as that.

The second problem with our bequeathment is that the earth's resources are unequally divided (Figure 29.1). The Arabs got the oil, the South Africans got the chromium, Americans got the forests, and the Japanese got the little cars.

The unequal distribution of the earth's resources should provide us, one would think, with a marvelous opportunity to achieve world peace through the mutual dependence of trade. But instead of this unequal distribution being a great blessing, it seems as if it has set the stage for international confrontations. Perhaps we should apply our knowledge of the biology of our species to solving such problems. We must try to understand how such biological principles as competition, territoriality, carrying capacity, cooperation, altruism, and so on apply to our own group. After all, we are the product of the weeding of genes of countless generations by their environment. We have not been exempt from the blind and unprejudiced effects of natural selection.

So let's now review the state of our environment, our home, and let's see how the resources are holding out. We must always keep in mind that the pressure on our resources increases each day as our population grows.

FOOD

It may be hard for most Americans, who define hunger in terms of a late lunch, to understand true hunger. In fact, we seemingly have food to waste, as witnessed by our great attention to trivial detail regarding the earth's foods. We prefer our food this way or that; we hold some critical standard for our tomatoes. There are few experiences more numbing than spending an evening seated next to a food fetishist, particularly a vegetarian. Paradoxically, many vegetarians profess to be humanitarians also (the latter term *not* pertaining to diet). Many vegetarians I know decry the hunger on the planet, even as they pick and cull their way—with great expenditure of time, energy, and resources—through a variety of very specific green and red foodstuffs. Of course, they are denying their omnivorous heritage, and, in their mincing selections, they may discard enough to feed an entire family. They may also use a great deal of petroleum products in order to acquire their food and to steam it for hours in the wok. We shouldn't be too hard on vegetarians, however, in spite of their tendency toward self righteousness, because in this society we all waste food. Furthermore, they eat from lower on the food pyramid (Essay 29.1).

Interestingly, many culinary fetishists are abysmally ignorant about nutrition. Their preference for the "natural," for example, may lead them astray. Some actually believe that rose hips provide a better source of vitamin C than do the chemically synthesized pills. It's just not so. Nor is organically grown food better for you than food grown with chemical fertilizers (Figure 29.2). The point is that the dilettantes whose energy is now spent in cultivating their tastes may soon be doing the same to their soil. The food imbalance throughout the world is too great, carrying capacities shift too slowly, and the human population is growing too fast to allow us to continue business as usual. It is possible that in your lifetime (and maybe even sooner), social changes will bring our eating habits more in line with the realities of the worldwide condition.

The problem of food might be more easily solved if it weren't for the fact that people have a problem in utilizing the food that is available, especially the newer kinds of food. That is, we tend to be very conservative and traditional. It is very difficult to introduce a new food to a culture; no matter how nutritious the food is, people just won't eat it. For example, there are over 50,000 species of plants known to be edible. But only sixteen plants yield almost all the calories and three-fourths of the protein that humans receive from plants. The rest of the edible species are almost ignored, even in the face of hunger and starvation. The hungry, by the way, can be divided into two groups: the **undernourished,** those who don't receive enough food, and the **malnourished,** those who don't get the right kinds of food.

The Global Implications of Food Resources

A good diet is particularly important to children, pregnant women, and nursing mothers. Because the women essentially must eat for two, the risk of malnutrition is obvious. Children are especially vulnerable to two very dangerous nutritional diseases, marasmus and kwashiorkor. **Marasmus** is caused by a diet low in both calories and protein, and its results are a thin body with a bloated belly, wrinkled skin, and a startling

FIGURE 29.2 In the search for a better life, many people seek health through better diets. They may turn to self-styled "health food" stores that offer an array of herbs, grains, vitamins, minerals, and advice. Unfortunately, much of the dietary advice from such places may be misleading or simply wrong.

The relative energy efficiencies in using plant and animal food can be shown by use of a "food pyramid." We can begin to illustrate this principle with the premise that there is a certain amount of energy stored in the bodies of plants. However, when an animal, say a cow, eats the plant, only part of the plant energy is made available to the cow. If we, in turn, eat the cow, only part of the energy stored in the cow's flesh is made available to us. The transfer of energy, we see, is not 100 percent efficient. Thus, much of the food stored in living things is wasted. But the wastage varies depending on how far one's food supply is removed from the level of the primary producer, the plant.

In the food pyramid produced in the study described in the figure, we see that of the great amount of energy stored in the tissues of the plants less than half (8833/20810) can be utilized by animals, because plants have their own things to do; they aren't in the business of supporting animals. Thus, they turn much of the energy of sunlight, not into food reserves, but into structural materials or metabolic machinery that is of no use to the grazing animal. At each step in the pyramid, energy is lost to the level above. So, only a portion of the energy a herbivore gets from plants is stored in its own tissues and thus made available to the carnivores that eat it. Much of the energy stored by organisms at any food level is made indigestible, or is used in metabolic activities, or is lost as heat or in other ways. It is simply not made available to the next consumer level.

Another reason that animals do not recover all the energy in plant tissue is that animals are not 100 percent efficient at recovering the energy stored in the things they eat. Thus much of the energy that *is* available to them is not utilized. The result of all this is apparent in the figure. There we see how many producers are necessary to maintain even a few herbivores and even fewer carnivores. Obviously a massive amount of food energy must be produced by plants to maintain only a few secondary carnivores (carnivores that eat carnivores).

With this in mind, then, would you say it is more efficient for humans to eat the fish that ate the fish that ate the fish that ate the algae, *or* is it more efficient to eat the

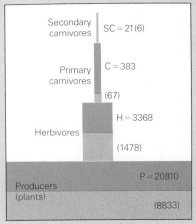

Kilo-calories/M^2/year

algae? Should we feed grain to beef cattle or should *we* eat the grain? The diet of North Americans places *four times* the demand on agricultural resources than do the poorer diets of the less fortunate, because of our interest in eating from higher levels of the food pyramid. With such wastage involved, can you see why steaks are so expensive (politics and market manipulation aside)? Nonetheless, in spite of costs, each American eats about 200 pounds of meat each year. (We seem to be shifting to fish and poultry with increasing rumors associating red meat consumption with cancer and other problems.)

appearance of old age. It is associated with drinking overdiluted formulas from unsterile bottles. **Kwashiorkor** results from protein deficiency and usually appears when a child is weaned—usually displaced at its mother's breast by a newborn sibling. Here, the entire body bloats, but especially the belly as it fills with fluid (Figure 29.3). In both cases, the children may suffer severe and permanent mental deficiencies because brains grow very rapidly in the early years and require high protein levels in the diet (See Essay 29.2). Both diseases could be prevented by extended nursing, but nursing mothers require a good diet that may not be available.

The worldwide distribution of food is very uneven. Not only are there poor, hungry countries, and rich, well-fed ones, but the same country may have rich and poor parts. Southern Brazil is rich and fertile, with affluent and well-fed citizens, while people in the less-fertile northern part of the country are more likely to be undernourished. The problem

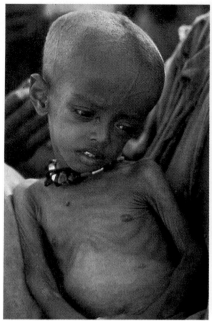

FIGURE 29.3 Children stricken with marasmus (left) and kwashiorkor (right). Marasmus is due to the mishandling of scarce food supplies, usually overdiluting babies' formulas, while kwashiorkor results from a lack of protein. Both conditions can permanently impair the victims, both physically and mentally.

also exists in the United States. As late as the mid-1980s, it was estimated that ten to fifteen percent of American citizens, mostly in impoverished areas such as northern ghettos and western deserts, were malnourished, while others, not far away, were sniffing at their plates and inquiring about the sauces.

The complexity of the food problem again becomes apparent when we realize that the global food problem involves more than simply finding means to feed people. Food also has its political implications. It is rapidly becoming a bargaining chip, as is the case with oil. For example, in 1972, the Soviet Union began a program to upgrade the diet of its citizens by supplying both more meat and grain products. Because large amounts of grain are needed to produce small amounts of meat, their grain requirements increased dramatically (at the same time that the world's weather took a sharp turn for the worse). The Soviets began negotiating with their old enemy the United States in an effort to buy our surplus grain. However, the United States refused to sell the grain to the Soviets because of tensions between the two countries. American farmers lost a lucrative market for their produce and perhaps an opportunity to bolster real peace through interdependency based on biological needs.

Some Global Realities

The developed (that is, the more affluent and industrial) countries use a disproportionate amount of the world's food, but their behavior cannot be called simple looting because they also grow most of the world's food. Furthermore, they tend to grow more food with each passing year. Interestingly, the developing countries, too, have been growing more food. In fact, from 1950 to 1980, the developing nations steadily increased their food production, but they actually became hungrier because their populations increased at more than twice the rate of the more affluent

What happens when someone starves to death? What does starvation do? The first thing is that the person loses weight. The living organism must metabolize, and if there are no nutrients entering the gut, the victim will burn his or her own tissues for fuel. The body literally digests itself.

On the average, people are considered to be starving when they have lost about thirty-three percent of their body weight. When they have lost forty percent, death is almost inevitable. As starvation progresses, various organs begin to lose their efficiency. The liver, kidneys, and endocrine system may cease to function properly. Lack of carbohydrates may affect the mind so that the starving person becomes listless and confused, often seemingly unable to understand his or her plight. Kwashiorkor and vitamin-deficiency diseases can set in. Soft bones may appear in children who lack vitamin D. Lack of thiamine may cause memory lapses. Lack of niacin causes skin inflammation, diarrhea, insanity, and finally death.

Children are often permanently affected since improved diet cannot straighten bones or build normal brain cells, but adults can often approach starvation and then largely recover, as we know from the rescue of concentration camp inmates. Interestingly, no matter how well such people recover, however, in most cases they die sooner than those who suffered no such trauma.

nations. In fact, the developed countries increased per capita food production nearly three times faster than the rest of the world. (The American population increased by a third while we doubled our food production.) The developing world (that is, the poorer countries), however, did not suffer as greatly as one might think because they had resources that they were able to trade for food. Of course, this meant that instead of the developing countries receiving the technology they needed so desperately, they were forced to trade for food, just to try to feed their booming populations.

It has been calculated that the differences between developed and developing countries in consumption of Calories* and proteins from plant products is only about thirteen percent. From plants alone, the poor receive an average of 2,016 Calories per person per day, near the minimum average requirement. However, the difference in consumption of the more expensive animal foods was quite striking; the richer countries consumed about five times as much as the poorer ones.

The most tragic story of recent years concerns northeast Africa, particularly Ethiopia, where a drought of several years has brought large-scale starvation and death, particularly affecting some six million people in the north (Essay 29.3). The Soviets responded, in 1984, by sending some thirty million dollars in arms to support the Marxist regime against insurgents. The rest of the world finally mobilized to send money and food, but there were complaints that the civil war and transportation problems left much of the intended aid stranded at ports (Figure 29.4).

Projections of Food Supply

By 1990, more than seventy-five percent of all people will live in developing countries, including China. People in these areas will produce ninety percent of the new mouths to feed. And worldwide, throughout

*Calorie is a measure of the energy in a unit of food. It is the amount of heat (from the chemical energy in food) necessary to raise the temperature of a kilogram of water by 1° C.

In late 1984, the plight of a hungry Africa was suddenly thrust upon Western consciousness. We were abruptly made aware that perhaps well over a million Africans had already died of starvation, thousands were dying daily, and many of the rest were on the move in search of food. For example, almost all of Mozambique's fourteen million people seemed to be deserting their homes and villages to move to "relief" centers where they hoped food could be found, but where many of them would meet their deaths.

The problem was not isolated to Mozambique, however. Much of Africa, by early 1985, was in the death-grip of famine and starvation. The area of greatest problem was a swath of land cutting across the northern desert area, with Chad and Ethiopia joining Mozambique as the countries in the greatest danger. A second swath cuts across Southern Africa. Some thirty countries in all are officially listed as hungry, and in the most desperate of these, whole populations are in danger of being wiped out. By early 1985, the famine had killed, or helped kill, some 200,000 people in Mozambique and 300,000 in Ethiopia (with another million Ethiopians in grave danger).

Africa has had famines before, of course. The last big famine, in fact, occurred in the mid 1970s (when 300,000 died). And, whereas some experts said such a thing could never happen again, the famine of the mid 1980s was the worst the continent has ever seen. In Mauritania in the last twenty years, more than three quarters of the grazing land has been lost to encroachment by the Sahara. The rainfall in 1983 was the lowest in seventy years, and in some areas, almost all the livestock has died.

How did all this happen and why was it so slowly brought to the West's collective consciousness? Of course, one reason it happened is because of a natural disaster: drought. But the problem was exacerbated, if not in part brought on, by human activities. The first warnings were given in 1982. They described an impending problem in Ethiopia, but the warnings were ignored, not only by the West, but by the Ethiopian government itself. Ethiopia was being torn by civil war and the government elected to spend its money on arms rather than food. Also, the government did not want to dampen a $200 million celebration of its Marxist takeover. In late 1984, a British Broadcasting Corporation film alerted the world to the condition that, by then, had spread to other black African countries. A massive relief effort was begun, but it was severly stymied by local corruption, mismanagement, and civil strife. Further, the countries suffering most from famine had been, not coincidentally, embroiled in civil war for years, their energies being dissipated in conflict.

As arriving food began to be distributed, in some camps doctors had to walk among the hungry masses making an "X" on the foreheads of the most healthy ones. This marked the ones most likely to be able to respond to help. The others couldn't be helped and no food could be wasted on them.

The rains returned in 1985, but hunger continues. The land had been overworked and overgrazed and its carrying capacity has been permanently lowered. Civil strife has not ended and a great deal of the national effort continues to be used to adjust people's political thinking. Problems of transportation and health dissemination are about as bad as ever, and there are some experts among us who expect the whole thing to happen again.

the 1980s an average of about eighty-six million people will be added to the planet each year.

Some researchers have suggested that in spite of such problems, a determined and careful worldwide agricultural program has the ability to produce sufficient food, mostly grain, to improve per capita consumption of food by a very limited amount. Of course, production of food is one thing and getting that food to hungry mouths is another. We must always consider the ability of the hungry nations to buy and distribute that food or the willingness of other nations to do it for them.

Most projections indicate that no trend is in sight that could in any way encourage the hungry people of the world. According to one prediction, by 1990 the developed world will comprise twenty-five percent of the earth's population, carry on eighty-five percent of its trade, and produce (and *consume*) about half the world's grain.

FIGURE 29.4 The great Ethiopian famine that began in the early 1980s was triggered by drought, but its basis is much more complicated than that. Changes in farming, herding, and demographics interacted in complex ways to promote the condition. Solving the problem also involves more than simply sending food. It will take a broad approach, apparently based on greater understanding than we have now.

It is important to keep in mind that the world produces enough Calories to feed its human inhabitants minimally. Thus, (ignoring protein, vitamin, and mineral needs), there is enough to go around—barely. The fact that people are starving means the problem is partly one of distribution and allocation. The rich have more than their share, the poor have less.

But there are problems other than production and distribution. For example, there is the problem of storage. At least ten percent of the world's available food is destroyed by pests, waste, and spoilage somewhere between the farm and the consumer. The wastage is greater in developing countries, the very ones that need the food the most.

In the best of worlds, it would seem that there would be a net flow of food from richer areas to poorer ones in an effort to alleviate the plight of the hungry. But is this the case on *our* world? It is true that the developed nations send tons of food to underdeveloped countries each year, much of it in the form of protein. (The gifts are well-publicized.) It is less well known, however, that the underdeveloped countries send tons of food to developed countries each year, much of it in the form of *higher quality* protein than they receive. Much of our shrimp, for example, comes from India, and protein-deficient Latin America sends us fish that we use to feed finicky cats with cute names. (As the food crisis worsens, we may have to place strong restrictions on the ownership of pets. Each day, the 500,000 owned dogs in New York City testify to good diets by depositing 90,000 gallons of urine and 150,000 pounds of feces on the city's streets, about half of which ends up on someone's shoes.) Much of the protein that we import from developed countries is also fed to our poultry and livestock as part of our effort to feed from higher levels of the food pyramid.

Farming the Earth's Jungles

It was once suggested that the food problem could be greatly alleviated simply by opening up the great jungles of the world to cultivation. Brazil even set up an agricultural colony in the Amazon Basin. The program

FIGURE 29.5 The soil in tropical rain forests is notoriously poor. However, it can be made productive by using the slash and burn agricultural technique employed by primitive tribes. This is a small clearing hacked out in the Ecuadorian jungle. Indigenous Indians will plant a few crops here, utilize whatever meager nutrients are available in the soil, then move on and let this plot be reclaimed by the jungle.

failed dismally because some very basic rules of agriculture ecology were ignored. The most fundamental of these was based on clear evidence that tropical soils are desperately poor in nutrients. This may be surprising, since tropical rain forests seem to be so lush and rich. There are great stores in tropical foliage, to be sure, but when the foliage falls to earth and decays, the nutrients are quickly washed from the soil by torrential rains or reabsorbed into new foliage.

The difficulty of farming such land is compounded by the problem of trying to clear an area in the face of almost overnight encroachment by rapid jungle growth. And then there is the problem of trying to till the jungle soil. In many areas, when such soil is exposed to sunlight and oxygen, it forms laterite, a hard, rocklike substance. The beautiful and enduring temples of Angkor Wat in Cambodia are made of laterite and sandstone.

This is not to say that tropical rain forests *cannot* be made to produce food for human populations, but the problem must be approached much more carefully than it has been thus far. We're going to have to make concentrated efforts to develop new farming techniques and to learn more about how indigenous peoples have been able to successfully cultivate such areas in the past (Figure 29.5).

Domestic Animals as Food

As you may recall, each time food is transferred to a higher level in the food pyramid, most of the energy it possessed in the preceding level is lost. It is therefore much more efficient to eat from low on the pyramid. However, the animal protein from high on the pyramid is advantageous for a number of reasons.

The most obvious is that since we are animals, the flesh of other animals provides us with the eight or ten essential amino acids in proportions that are relatively close to our needs. To get all the various proteins we need from plant tissue, we would have to consume large volumes of plant bulk from a wide variety of plant species.

The choice of which animals to raise for food is important. For example, goats are a good source of milk, but the goat is a destructive grazer, best suited for marginal areas that have no other use. In their overgrazing, goats have contributed to ecological problems by stripping the soil of its protective plant layer. Sheep also tend to overgraze; and they are not very popular as food. Also, because of their heavy wool coats, they are not well suited to tropical areas. Hogs, on the other hand, can be reared in a wide range of habitats and are extremely efficient in converting plant protein to animal protein. The trouble is that their dietary habits make them direct competitors with humans for the available plant food. In fact, like humans, hogs are omnivores; that is, they eat about anything—animals as well as plants. On the other hand, whereas they compete very directly with humans they can also subsist on human refuse.

Cattle, because of their complex digestive arrangement, are able to survive on plants that are not food sources for humans. Since they ordinarily don't wander from their grazing areas, they don't require constant attention. Moreover, they travel well, are large enough to be used as work animals, and provide leather as well as beef. In addition, cow milk is a food staple in many places, whereas people don't tend to get too excited about pig milk.

Perhaps the most efficient and least costly kind of animal to be raised for food is poultry. More flesh is produced per pound of plant tissue eaten than is produced by any of the other animals just mentioned. Chicken is relatively inexpensive because the less efficiently a pound of meat is produced, the more it will cost. (In our frantic efforts to impress each other, we like to feed our friends. But do we offer them beans? No, beans come from too low on the food pyramid. Their conversion efficiency makes them inexpensive and no one will think well of our provisions. However, they appreciate an offering from higher on the food chain. Maybe you should surprise them with something much higher, say, a nice vulture.)

Fishing

Perhaps fishing should be included in the discussion of nonrenewable resources (those that are not naturally replenished, such as oil or ore), because, in a very real sense, we do not replace the fish we take. Our fishing practices are more akin to mining. The fishing industries are diligently studying fish behavior, developing technical arsenals, and taking more fish as they continue their drive to meet the demands of our hungry numbers. Our biotic control, as predators, seems to be running amuck.

The management of the oceans is complicated by what biologist Garrett Hardin refers to as the "law of the commons." "Commons" have historically been areas that belonged to no one, but are available to all. In Victorian England, a common might be an area where everyone could graze sheep. The concept worked well if all the shepherds voluntarily limited their herds so that the commons would not be overgrazed. If someone added a few extra sheep, however, they would have a marketing advantage over the others while not doing too much harm to the grazing areas. But what were the others to do? Obviously, they would add sheep also, each hoping the others would not notice. Soon the grazing area would be lost to all.

The sea is a "common" today, but for the most part, nations are not behaving so as to ensure optimal productivity. For example, the decision to add another vessel to a nation's fishing fleet may be marginally productive to that country, but in the long run the action may contribute to a diminished catch because of overfishing. Yet if that nation chooses not to add another vessel, it may find itself on the short end as other nations continue to add vessels. Any nation caught in this game comes up short both ways. So how should such decisions be made? What will be the long-range effects of any alternative? What should be our priorities? Our logic? Can you think of some better way of apportioning the world's resources?

Solutions to the Hunger Problem

There are a number of largely untested, but fascinating ideas on ways to increase food supplies. For example, it is known that single-celled organisms high in protein can be grown on petroleum. If these organisms could be produced in large enough amounts and purified econom-

In the late 1960s, there was a great fanfare heralding the impending "green revolution." New kinds of grains had been developed that yielded plants heavy with seed. For example, one strain of dwarf rice, IR-8, produced over twice as many rice grains per plant as the conventional strains. The new miracle crops do, indeed, hold great promise for the world's populations. However, they have their problems.

First of all, the new strains require special handling; farmers must be carefully trained in new agricultural methods. However, there has traditionally been a reluctance of people steeped in their traditions to adopt new ways. They won't eat new foods and they won't change their farming practices. There is also a problem in underdeveloped countries of providing trained technicians to teach the farmers. In some cases, the technology itself is useless. How can controlled-temperature farming be implemented in areas that have no electricity?

Miracle crops also must be heavily fertilized and fertilizer, of course, costs money. Then there is the problem of the fertilizer runoff. As rains wash fertilizer from the fields into bodies of water, the nutrients may cause the water to eutrophy, or "age," prematurely. Nutrient-rich water may also be subject to great blooms of microscopic algae that can interfere with the balance of life and cause massive fish kills. The "red tides" of the oceans, when the waters turn strangely red and fish die from lack of oxygen, are the result of such blooms. An increase in agricultural production, then, may reduce the catch of fish in offshore waters.

Wheat, rice, and corn account for about three-fourths of the total yield of grain worldwide. Since grains are concentrated sources of energy and are easily stored, they probably represent a good focus of continued research effort. And, in fact, one new hybrid of wheat and rye, *triticale*, shows great promise (see the photo).

Ecologists have warned of the consequences of relying too heavily on artificial strains of grains. After all, they are rather homogenous, thus the low variation leaves little on which natural selection can operate. One of the attendant dangers is that wild populations of insects or disease organisms, replete with their wide genetic variation, might, through their own processes of natural selection, quickly accommodate themselves to any single strain and wipe it out in short order. Thus, in the long run, we might be better off paying the price of lower productivity for increased genetic variation in our crop species.

This point was dramatically illustrated by an unusual event in 1969. Specially developed high-yield corn plants were attacked by a new type of fungus called the southern corn blight. By 1970, the epidemic was so severe that the United States corn crop was reduced by fifteen to twenty percent (a loss calculated at more than a billion dollars). The tragedy jolted many agriculturalists into a new appreciation of how food supply is a function of biological realities. The lesson was a simple one: extreme uniformity of our food crops makes them particularly vulnerable to the appearance of new kinds of pathogens or predators.

Some global planners believe that we can indeed boost food pro-

ically, they could conceivably provide us with high-grade protein. Efforts are presently underway to develop new plants, through recombinant techniques, that can "fix" their own nitrogen from the air. Scientists are also busy trying to produce yet newer grains with higher quality proteins (that is, with a better balance of amino acids). A nutritious mixture of corn and cotton seed meal enriched with vitamins A and B, called incaparina, has been available in Central America for over fifteen years but unfortunately has not been accepted by the people there. As mentioned earlier, this is one of the great problems of food scientists—how to get people to eat new kinds of food. Historically, the hungriest people in the world are those that recognize the fewest items as food. In fact, even people in developed countries are warned to stock familiar foods for emergencies because even such "sophisticates" have been known to starve rather than try something new. (I would probably wait days before opening a can of broccoli.)

A variety of other ideas to increase food production are being considered, such as culturing algae, ranching various rodents or antelope, conduction on our planet by basically continuing the green revolution begun in the sixties. It was then that specially-bred food plants with exceptionally high yields were disseminated throughout the world. The greatest progress occurred in cereals, especially wheat and rice. Both were short, with upright leaves, and both were highly responive to fertilizers. In fact, without liberal doses of chemical fertilizers, the plants produced only a little better than traditional varieties. Because of these new crops, India's wheat production rose from about eleven million metric tons in 1964 to thirty-four million in 1979. The yields of other crops increased comparably, but from 1960 to 1983, the use of chemical fertilizers has increased eighteen-fold. Pesticide usage has also increased, and these chemicals must go somewhere, so a large part has run off into bodies of fresh waters on the continental shelf. Even in areas where miracle crops still hold promise, there is an acknowledged trade-off for immediate results.

Triticale (right) is a crossbreed of wheat (left) and rye (center).

verting weeds to cattle feed, extracting protein from plants or small fish, converting wood to cattle feed, and using sewage to grow edible slime. (That last one should be a big hit.)

It is interesting that of the thousands of species with which we share this planet, so few have been treated as food. In fact, it turns out that modern humans, with that vast ability and imagination, have discovered virtually *no* new crop, plant, or domesticable animal. Captain Bligh was directed to take the *Bounty* to bring back a "new" plant that had long been cultivated in the South Pacific. "New" kinds of fish meals are still made of fish. And the miracle grains are simply variations of plants that have been with us for thousands of years. More recently, a promising new "species," triticale, has been developed. Even this invention is a hybrid of wheat and rye (Essay 29.4).

One is tempted to ask whether our energies *should* be directed toward frantically trying to provide a minimal diet to teeming millions of people as their numbers grow increasingly faster. Should this be the role of our scientists and the lot of modern humans? Or should we begin to put strong pressures on the swell of our numbers? Should we stop asking how we can manage to keep more and more people barely alive and begin to ask how we can improve the quality of life for those already living? How can this be done? What would be the political, sociological, and moral results?

WATER AND THE COMING CRISIS

Obviously, the planet has a great deal of fresh water. We know where most of it is and we use it in a variety of ways—some personal, most on a grander scale. For example, most of the fresh water available in the United States is used in agriculture, with the next largest allocation going for steam generation, a trend that is expected to continue at least until the end of the century.

Our use of water in this country is prodigious and is still growing by leaps and bounds. In 1900, for example, Americans used 40 billion gallons of water each day (an average of 530 gallons per person). Now, however, we use over 400 billion gallons a day, an average of almost 2,000 gallons per person (Figure 29.6). On the other hand, a person in an underdeveloped country uses about twelve gallons per day.

Many ecologists predict severe shortages within a few years. The problem is pressing in several areas of the United States even now, including the western agricultural states, the northeast, Florida, and Louisiana. The problem is exacerbated because not only is water running short, but much of the remainder is becoming dangerously polluted and unusable.

Because of localized water problems, it has been suggested that we shift great waterways in the sparsely populated northern areas of the U.S. to the south and west where agriculturalists are running short. In fact, in North America, the amount of water available per capita is dwindling at an alarming rate. There are several reasons for the problem. First and foremost, of course, is our increasing population. Beneath the densely populated cities of the northeast, many critical underground water supplies are simply drying up. But, in addition, industries continue to dump dangerous chemicals into our waterways or let it seep into the water table from rusting barrels above, often rendering the water permanently unusable. In some coastal areas, the aquifers (underground supplies of fresh water) are being steadily encroached upon by salt water as the fresh water is pumped out. Now we are faced with reservoirs being refilled by acid rain, a plague of our own making.

In the west (the area that provides us with most of our vegetables, fruit, and grain), the major resources of irrigation are drying up. For example, the great Ogallala aquifer, an underground reservoir stretching from Texas to Nebraska, is being depleted twice as fast as the waters are being replaced by natural water cycles. The water table in the west is dropping by a disconcerting three feet each year. Above ground, the Colorado River, servicing seven states, is a salty trickle by the time it reaches Mexico.

The only practical measure seems to be conservation. But the ideas being considered are not of the Smokey the Bear variety. For instance,

FIGURE 29.6 This dry reservoir at Kensico, New York poignantly illustrates the earth's continuing problem with a dependable supply of fresh water. Perhaps such problems would be of less magnitude if we realized just how much water we actually use. For example, each American uses an average of almost 2,000 gallons of water each day. A bath takes thirty to forty gallons, and a five-minute shower, twenty-five gallons. Less obvious use includes an incredible forty gallons of water that goes into the production of a single egg, seventy-five gallons into one pound of flour, 150 gallons into a loaf of bread, 2,500 gallons into a pound of beef, 230 gallons into a gallon of bourbon, 280 gallons into a Sunday newspaper, and 100,000 gallons into a car.

we may be asked to switch our eating habits to foods that require less water to grow. We may be asked to largely give up meats because grain to feed the animals takes a great deal of water to grow. Also, we will eventually be asked to pay very dearly for water. (I, for one, don't want to pay more for anything, and I certainly don't want anyone making any more rules. But the interstate highway near my house just recently collapsed, forming a huge sinkhole, because the water underneath it had been pumped out. I don't like restrictions, but I don't like the earth caving in around me, either.)

NONRENEWABLE RESOURCES

The line between renewable and nonrenewable resources is a fine one. For example, water is considered cyclic in that it evaporates and falls again as rain and thus is "renewable." However, much water is tied up in various manufactured and natural products and is not easily retrievable. And we know that water may become so polluted that it is no longer usable. Thus, in a sense, it is at least partly nonrenewable.

In our consideration of these commodities, we must remember that they do not occur uniformly over the earth. Hence, their use may be manipulated by nationalistic policies. Consider the imbalances in metals. Asia is rich in tin, tungsten, and manganese. In fact, over half the world's tin is located in Indonesia, Malaysia, and Thailand. North America, on the other hand, is poor in these elements, but well-endowed with molybdenum. Gold, of course, occurs mostly in South Africa, so does platinum, but not much silver is found there. Over half the world's nickel is in New Caledonia and Cuba, and most mercury is found in Spain, Italy, and the Sino-Soviet areas. Because of the implications of such resource distribution, let's look at resource reserves both for the world and for the United States.

It is increasingly apparent that we can expect certain problems if we continue to take the earth's available resources on a "devil take the hindmost" basis. Each commodity is limited and we do not increase their supplies simply because we create a new deposit. Such thinking is short-sighted, but locating new sources does solve problems on a short-term basis. Problems arise because of our ignorance about the long-run. We simply don't know the extent of most of the earth's resources. Some will give out much sooner than expected, and others are probably much more abundant than we now believe. So what should be our policy? Do we gamble? What are the stakes?

Geologist Preston Cloud points out that we can make the best predictions for these substances that are found associated with specific rock layers. These include the fossil fuels—coal, natural gas, and oil—and, to a degree, iron and aluminum. The reserves, grades, locations, and recoverability of many other critical metals are much harder to estimate. Thus the problem of developing long-term policies regarding their use is magnified. Since we may not know how much of certain commodities are left, should we continue to drain each known deposit, confident that others will be found because they *must* be found?

ENERGY

Energy, you recall, is simply the capacity to do work. Just as plants must acquire energy from the sun and animals must derive energy from plants, humans must also utilize energy to be able to rearrange trees and oil and iron in order to build houses so that we may be shielded from disruptive elements in the environment and have a good place to read the newspaper. However, with each new house, or each new commodity of any sort, not only are resources required, but energy to rearrange the resources.

Of course, some of the demands placed on energy supplies are "legitimate" in that they relate to very basic needs such as food production or environmental shields. We may be reaching the point, however, where it will be necessary to distinguish such essential energy uses from nonessential demands (Figure 29.7). Do we really need elaborate neon-lit billboards, automatic can openers, six straight pins with each new shirt, oversized electronic amplifiers, electric toothbrushes, and powerful cars to carry one person to work? (What nonessential item most violates your own sensibilities? To which do you object the most?)

Now let's take a look at our energy resources and some of the demands being placed on them. Figure 29.8 can help us put the data in perspective by indicating the changes in energy consumption of humans over evolutionary time.

We should first make it clear that, at present, we are not running out of even the current forms of energy. The United States, for example, has vast deposits of coal. However, locating the coal is one thing; getting it out is another. There are heavy costs to our environment both in producing energy and in consuming it, and these costs must be carefully weighed in calculating the net benefit we receive. We will see again the

FIGURE 29.7 The "good life" for many people is based quite simply on more goods and services. As they climb that long ladder of success, however, they utilize disproportionate amounts of commodities and energy. Such artificial demands place a great burden on our reservoir of natural resources and create great amounts of waste that must be handled by the expenditure of yet more commodities and energy.

FIGURE 29.8 Estimated average daily per capita energy use at various stages of human cultural evolution.

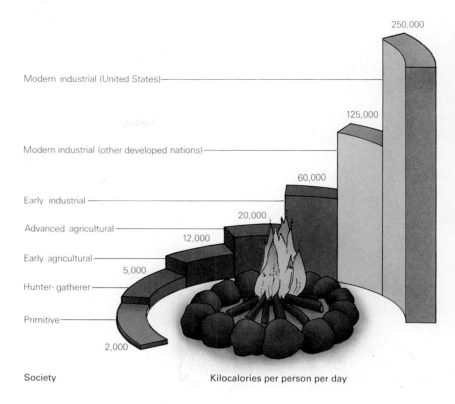

Modern industrial (United States) ——— 250,000

Modern industrial (other developed nations) ——— 125,000

Early industrial ——— 60,000

Advanced agricultural ——— 20,000

Early agricultural ——— 12,000

Hunter-gatherer ——— 5,000

Primitive ———

2,000

Society

Kilocalories per person per day

enormous impact of the industrialized countries, particularly the United States, on energy resources, because our attitudes and policies often have such wide-ranging effects worldwide.

Fossil Fuels

Fossil fuels are comprised of the partially decomposed remains of ancient plant life. As their bodies fell into the many shallow lakes that covered the earth in those days, great masses of once-living material accumulated and their great weight impeded the natural processes of decay. The partially decomposed masses produced liquified or gaseous material that held great stores of undissipated energy. Today, the fossil fuels (oil and gas), comprise an increasingly large part of the world's energy sources, especially among developed nations, such as the United States. However, human needs are placing inordinate demands on these reserves.

Whereas we have plenty of coal, as well as sunlight, wind, water, geothermal energy, waste energy (from farm and forest), and uranium, in the early 1980s, fully seventy percent of our energy came from gas and oil. It seems clear that some sort of change is in order (said the woman as her husband drove into the wall). We are not yet at the wall, but our options narrow as time passes.

Water, Wind, Earth, and Sun

Energy from Water

Moving water turned the wheels that ground the corn that fed our forefathers that gave them energy to have the descendants that became so abundant that they returned to the moving water in their search for a

FIGURE 29.9 Even the tides have been harnessed to provide people with inexpensive energy. Such sources have the great advantage of being renewable and nonpolluting, but the necessary alteration of the environment may produce serious problems with delicate ecosystems. This tidal power station is at Rance, France.

"new" form of energy. Obviously, moving water is not a new form of energy. But we are presently seeking, through a variety of means, to utilize the energy of these currents.

Today, we use rivers to turn great turbines that produce **hydroelectric energy** (electrical energy produced by the energy of moving water). Such water power is "clean" in that it doesn't pollute the atmosphere or create nuclear wastes, and it is efficient. However, it can create problems. For example, water falling over high dams can pick up nitrogen and kill fish downriver. Dams also trap sediment and thereby alter the natural courses of rivers and estuaries. Finally, many people just do not want to dam wild rivers for aesthetic reasons.

Harnessing the vast energy of moving tides is an alternative still largely on the drawing boards, except in England and France where several kinds of experimental systems utilizing tides are in operation (Figure 29.9). The problem with tidal energy is that the water has the greatest force where it moves through deep channels, and most of the natural channels are shallow estuaries where rivers meet oceans. Any disruption here could seriously upset the spawning grounds of some important links in the ocean's chain of life.

Energy from Wind

The energy of the wind has long been harnessed, as witnessed by the stilted monuments to an earlier effort, spindly structures standing braced against the winds that once served them. But the principle once utilized by windmills is being reconsidered. In some cases, not much change is in order. Even old-style windmills work well when the wind blows, and the technology to use the energy is already fairly well developed. Now, however, "wind farms" are being developed—huge tracts where a number of modern windmills produce energy that can be stored and sold commercially (Figure 29.10). Problems arise because there are relatively few places where winds are dependable. (The most promising places include the Pacific northwest, the Rocky Mountains, and the mountains

FIGURE 29.10 A wind farm. These unsightly and surprisingly noisy windmills produce electrical energy efficiently and without heavy pollution.

of North Carolina.) Wind power has a low environmental impact, but the new windmills are noisy and unsightly and interfere with radio and television transmission.

Energy from the Earth

It may have occurred to those who saw the awesome eruption of Mt. St. Helens that there is energy under the earth. In fact, one of the greatest explosions of all time was the 1883 eruption of an undersea volcano called Krakatoa between Java and Sumatra. It darkened the sky for months and changed sunsets around the world. This **geothermal energy** can be tapped, and it has the advantage of being available, to some extent, virtually everywhere. However, it may not produce enough energy for heavy use since the energy yield from this heat is low and it must be converted to electrical energy on the spot. Furthermore, we lack technological expertise for handling it, it is expensive, and it requires the use of a great deal of water. New Zealand is among the most successful developers of such energy.

Energy from the Sun

Solar energy is theoretically available wherever sunlight reaches the earth. The simple wisdom of using such an obvious energy source is becoming apparent, at long last, in the United States. Solar collectors are found on increasingly more rooftops throughout the country. These are generally used to heat water and, in fewer cases, space. The energy savings in individual homes by using the sun for heating can be substantial.

The energy of the sun can also be used to create electricity through **photovoltaics.** The problem with using solar energy to produce electricity is that large areas of land must be covered with collecting devices because sunlight is such a diffuse and dispersed form of energy.

Photovoltaics was actually conceived in 1839, but today it involves using sunlight to energize two layers of silicon. The sunlight disrupts the crystalline arrangement and unsettled electrons from one crystal move to the other, generating a slight electrical current. The lost electrons are replaced by attaching the donor crystal to some metal, such as a copper wire.

At present, electricity produced by photovoltaics is very expensive and there is such disagreement over its potential that we apparently still don't have a grasp on it. But since this sort of energy can be collected, localized, stored (in batterylike "fuel cells"), and then sold, we now see increasing research in the area.

Energy of the Atom

Three types of uranium occur on the earth, U238 (about 99.3 percent of all naturally occurring uranium), U235 (about 0.7 percent), and U234 (about 0.004 percent). U235 is the fissionable form (*fission,* break apart). Fission reactors cause U235 to break apart, or explode, by bombarding it with neutrons, making it very unstable. As a molecule of U235 explodes, it produces heat and three more neutrons, each of which may strike other uranium atoms, making them unstable, thus producing a chain reaction. The heat that is produced turns generators and produces elec-

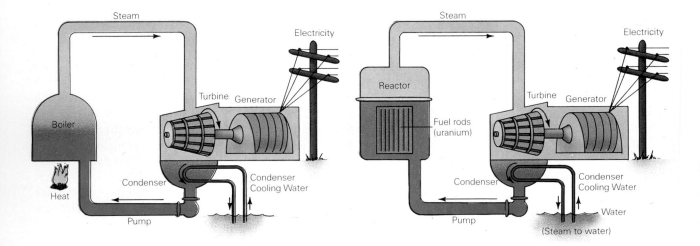

FIGURE 29.11 Compare the fossil fuel power plant at left with the nuclear power plant at right. Notice that both use steam to turn turbines that generate electricity. In the nuclear reactor, uranium bars comprise the fuel. As the decomposing uranium heats water, steam is produced, which turns turbines and operates electric generators. The fuel is highly reactive U235, which is not present in concentrations great enough to cause a nuclear explosion. As a molecule of U235 decomposes (or explodes), its escaping neutrons are allowed to strike other U235 molecules after having been slowed by some sort of dense material. The molecules that are struck then become unstable and explode, producing a controlled chain reaction.

U238, which is much more common, is very stable and hence cannot be used as a fuel. However, if it is struck by fast-moving neutrons, it is converted to highly reactive U239, or plutonium. Plutonium made in this way could provide our nuclear needs for another 60,000 years. But there are problems. The process takes place in breeder reactors, and these have yet to be perfected. Also it is a rather simple matter to make nuclear bombs from plutonium and it is estimated that if the program progresses according to schedule, within forty years there will be 100,000 shipments of nuclear material in the United States each year—a tempting target for terrorists. In mid-1977, the U.S. government admitted that it could not account for substantial amounts of nuclear fuel, but they added that they did not believe it had been stolen.

tricity (using the same principles as the generators that use fossil fuel—see Figure 29.11 for a comparison).

We obviously must view ourselves as custodians of our precious resources and not as exploiters, seeking ways to get at the earth's goods before someone else does. However, as our numbers grow, new demands must be met, and thus we must find new sources of commodities as we place greater pressures on the old ones. The race for goods, has produced, in many cases, a "devil take the hindmost" ethic. We can hope that we will develop a more careful approach to the use and distribution of the earth's resources. But any such change must begin soon because for now, the race is on.

SUMMARY

The earth has bountiful resources, but they are distributed unequally over its surface and as a result, international tensions can arise between those countries that hold certain commodities and those that do not. Renewable resources are those that can be replenished; they do not exist in fixed amounts. Nonrenewable resources are those that exist in fixed amounts, and that are depleted or changed by use. Food is a critical and renewable commodity, but relatively few items are recognized as food. Some areas are subjected to famine, although there is enough food (calories) worldwide to go around. Food is becoming increasingly important in international trade. Almost all areas are producing more food, but in developing countries, the growing population is outstripping the gains in food production. By 1990, more than three-fourths of the world's population will live in developing countries, but the rest will consume half the world's grain. An effort has been made to farm the world's tropical rain forests and jungles, but the soil is too poor. Certain domestic animals are far more efficient to raise as food than are others. Among the least efficient are goats and sheep. The most efficient include poultry, pigs, and cattle. Fishing practices currently do not allow fish populations to recover. Other solutions to the problem of hunger are proposed but few are truly new since people tend to resist novel foods.

The water problem in the United States is growing more severe for three reasons: our increasing numbers, our increasing usage per person, and our fouling of water supplies. Perhaps the best solution is conservation.

Energy sources are varied, and we rely on various sources to different degrees. Most of our energy needs are met by fossil fuels (gas and oil). Moving water has long provided energy, particularly as it falls over dams and turns electrical generators. The wind is also an old source that is currently being reconsidered. Geothermal energy comes from the heat within the earth, but it is still an emerging field. Solar energy is currently available for heating and may soon be able to be used to efficiently produce electricity. Atomic energy uses radioactive isotopes to make steam that turns turbines. We must begin to focus more on the development of renewable and nonpolluting energy sources.

Chapter **30**

BIOETHICS, TECHNOLOGY, AND ENVIRONMENT

In the blink of a geologic eye, it seems, the human species has produced for itself an unheard-of position of power and ability. We can move mountains, straighten rivers, cause extinctions, produce new species, and change the rain. We have newly developed abilities to do great damage or great good. With this ability, some say, comes a heavier responsibility than any species of any age has ever borne. The consideration of this new responsibility that has fallen upon us has produced a philosophical area that has come to be called **bioethics,** the ethics of life. Ethics itself implies inherent right and wrong, and so bioethics deals with the moral values of our behavior toward life on the planet.

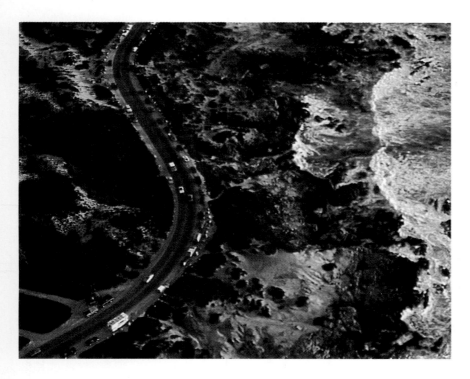

Bioethical questions are of many types, approaching many subjects dealing with life. But they all have one thing in common: They are difficult, if not impossible, to answer. Yet because we have become so powerful, so probing, and have touched so many areas of life with our technology, the questions must be asked and we must continue to seek answers. Perhaps just asking the questions has value.

The bioethical questions researchers must address are as varied as the researchers themselves, but let's mention a few that have been asked. Should we send food to countries whose population is spiralling out of control, thereby encouraging more childbearing? Should we build a dam to create electrical energy even if it means the extinction of yet another species? Should abortion be encouraged or even allowed? Should we spend public money to spare the lives of severely handicapped infants who may have to spend their lives in pain? The list could go on and on, but perhaps we have identified the problem.

Here, then, let's see where some of our "advances" have led us. We will delve into the untidy problem of pollution of the environment to identify some specific problems resulting from our traditional ethics. However, as we go, let's focus on how a shift in values, or point of view, might have spared us these problems and see how we might make changes now that would better guarantee our futures.

THE ETHICS OF DOORMATS

It has been said that we have seen more technological advances in the last ten years than in all our previous history. That may be true, but what does this information tell us? It could be cited with equal enthusiasm by an industrialist or an environmentalist. The former might happily note the increased efficiency of production while the latter might be concerned that our technological abilities have outstripped our wisdom, our ability to deal with our creations. We might certainly, in our quieter moments, wonder if Henry David Thoreau was right. Perhaps we have busied ourselves with trivialities and forgotten about the real business of living. Thoreau once refused a doormat for his cabin at Walden Pond because he figured he wouldn't have time to sweep it. He needed that time, he said, to sit or walk in the woods. Living. *Experiencing!*

Apparently, some people still enjoy the simple life, but even for them, things are becoming more complicated and they are being forced to deal with their own bioethical decisions or those of others. Every year, for example, people heave their backpacks to their shoulders and head for the mountains. Even there, though, they may run into scores of people on narrow trails who are also trying to reestablish their links with a more natural world. In fact, even the remote lakes of the high Sierra Nevada mountains are becoming seriously polluted by the wastes from so many backpackers in the area. Not far away, trees may be falling, their fibers that had once returned mountain rains to the cool air destined to become part of a gaily-colored toothpaste box. The box, of course, is designed to be discarded and to join those huge heaps of refuse whose disposal presents such problems of waste management. It's easy to forget that the sodden box had once been a tree.

What, then, is "quality of life"? Do we trade trees for toothpaste boxes? Is the quality of our lives to be measured by the waste we create? In a sense it is, since more affluent nations indeed create more waste.

FIGURE 30.1 We sometimes consider pollution as human-made. However, those steamy crevasses that belched sulfurous smoke when the earth was young were polluters, as are today's volcanoes that spew chemicals and ash into the air. Our task is to see that we minimize the human contribution to such natural pollution.

Let's now see how meeting our needs and our demands may exact a higher cost than we anticipated. As goods are provided, there are side-effects. One kind of side-effect may be the pollution of our environment.

ENVIRONMENTAL POLLUTION

A **pollutant** is any harmful substance that enters the environment. Pollutants not only take the form of chemicals or garbage, but also such things as noise and heat. Here, we will focus on the pollution produced by human activity, although we must be aware that nature can produce some of the same substances. For example, there is some natural oil seepage from undersea deposits, and it has been going on for thousands of years. Also, volcanoes (Figure 30.1) are among the greatest short-term air polluters on the planet.

Our present environment has been molded by such natural occurrences as fires, droughts, oil seepage, earthquakes, volcanoes, and even radiation throughout the earth's history. The life on earth has adapted to such events. The kind of pollution that concerns so many scientists today, however, is that which, in kind or degree, is new to the earth and its creatures.

There are four primary ways that these new pollutants can be dangerous or disruptive. First, they may alter critical aspects of our environment so rapidly that some forms of life will not have time to adapt to the changes before they are destroyed. Second, since thousands of known pollutants have never been tested for their risks, they may act subtly and slowly in our bodies, setting the stage for a tragedy down the line. Third, pollutants may have a **threshold effect.** That is, they may show no effect until they have accumulated to some critical level. Fourth, chemicals can act synergistically, interacting in such a way that their combined effects are far greater than they would have had individually. For example, both sulfuric acid and ammonium sulfate are rather dangerous pollutants alone, but together they form a far more dangerous combination.

Air Pollution

If you live in a large city, you might be particularly concerned about air pollution because you can *feel* it. It can burn your eyes, nose, and throat on the worst days. It may even make days darker in a literal sense. Air pollution, for example, frequently cuts by forty percent the amount of sunlight reaching Chicago. A huge stone obelisk from Egypt, a gift to New York City, has deteriorated more in the few years it has stood in Central Park than in the hundreds of years it stood in the desert.

The greatest source of air pollution in the United States is the automobile. However, fuel burned in stationary sources, such as in heating units for buildings and industrial furnaces, is also largely responsible for our foul air.

We should realize, also, that air pollution is not just the problem of big cities, nor is it restricted to industrialized countries. Meteorologists are now describing a thin veil of pollutants that hangs over the entire earth.

Table 30.1   Basic Types of Smog

Characteristic	Industrial Smog	Photochemical Smog
Typical city	London, Chicago	Mexico City, Los Angeles
Climate	Cool, humid air	Warm, dry air
Chief pollutants	Sulfur oxides, particulates	Ozone, aldehydes, nitrogen oxides, carbon monoxide
Main sources	Industrial and household burning of oil and coal	Motor vehicle gasoline combustion
Time of worst smog	Winter months (especially in the early morning)	Summer months (especially around noontime)

Industrial smog operates in the form in which it is released. Photochemical smog becomes more active when it is hit with sunlight.

(In recent years, they have repeatedly detected polluted air over the North Pole.) Generally, however, the foulest air is found in the industrialized countries, where it is particularly concentrated over large cities (Table 30.1 and Essay 30.1).

Carbon Monoxide

Carbon monoxide is the most prevalent air pollutant. It is biologically important because of its tendency to combine with the blood's hemoglobin in the place of oxygen. The effect is to cut down the blood's supply of oxygen to the tissues, thus causing the heart to have to work harder to oxygenate the body. The increased demands on the heart place a severe strain on people with heart or respiratory ailments. Looked at another way, carbon monoxide inhalation can have the same effect as loss of blood. In fact, spending eight hours in an atmosphere with eighty parts per million of carbon monoxide has the same effect as losing over a *pint* of blood. It may interest you to know that in a traffic jam, the air may contain nearly 400 parts per million of carbon monoxide. The people breathing this air may experience headache, loss of coordination, nausea, abdominal cramping, and even partial blindness. (Carbon monoxide is also a major component of cigarette smoke.)

Nitrogen Oxides

Nitrogen oxides are produced primarily from gasoline engines, power plants, and industry. It has been pointed out that nature produces ten times more oxides of nitrogen than humans do, but the problem is that human products are concentrated in urban areas. Although nitrogen dioxide irritates lungs and withers plants, the greatest danger arises when it combines with hydrocarbons in the presence of sunlight to produce **photochemical smog** (Figure 30.2). This smog is particularly reactive and is one reason pantyhose and automobile tires don't last long in the city.

Sulfur Oxides

Sulfur oxides, produced mainly by burning coal, are particularly severe during heavy smog. Sulfur oxides can combine with water and produce

FIGURE 30.2   Photochemical smog hanging over Mexico City. Many of the chemicals released from traffic and industrial sites are made more reactive by the energy of sunlight, and, thus, their corrosive effects are increased.

Under certain conditions, the polluted air of cities can be amplified by **temperature inversions.** Normally, air temperature decreases with altitude, but when a layer of warm air moves above the cooled air, it limits the normal upward flow of the air from below. As a result, polluted air nearer the ground is not allowed to escape into the upper atmosphere. This means the foul air is trapped and the unfortunate people below must breathe their own pollutants instead of releasing them into the air for other citizens to breathe. When temperature inversions occur over cities, the number of deaths from respiratory ailments often rises sharply.

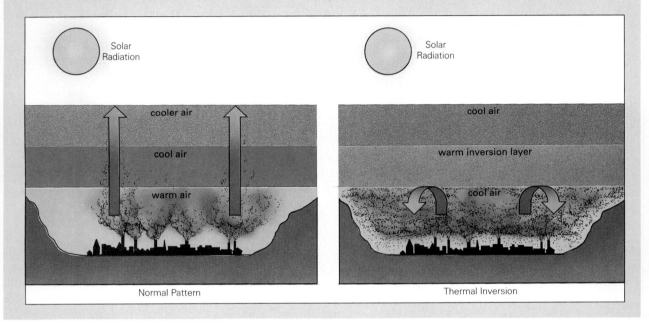

sulfuric acid, which damages not only lungs and plants, but can dissolve marble (Figure 30.3), iron, and steel. Sulfur oxides are very harsh and may bring on bouts of coughing, wheezing, and choking. The most serious damage is done when sulfur oxides interact synergistically with ammonia and metallic salts in the air. They are believed to figure importantly in the increased incidents of bronchitis, asthma, and emphysema, which are so frequently suffered by people in high-smog areas. (Emphysema is a progressive lung disease, generally considered fatal.) Globally, sulfur oxides may not present a great proglem because rain washes them from the air, but locally, they may be the most dangerous single pollutant in the atmosphere.

Hydrocarbons

Hydrocarbons, as you know, are a diverse and variable group of elements, some of which are implicated in the development of respiratory problems and some cancers. They may form from natural processes, such as plant decay, but most synthetic hydrocarbons come from the incomplete combustion of fossil fuels. Interestingly, they may help decrease the abundance of another pollutant. This is because the hydrocarbons emitted in automobile exhaust are rather easily trapped by our antismog devices. However, if those hydrocarbons were released into the

FIGURE 30.3 The smiles are gradually fading from many of the ancient carved figures in Venice. Sulfur dioxide in the air mixes with soot to form a film over the marble and to gradually change it to calcite and gypsum, which then crumble away.

air, they would react with nitrous oxides, which are not filtered out, and form relatively harmless compounds. Thus, partly because of improper antismog devices and partly because automobile manufacturers have increased the compression ratios of their engines, nitrous oxide pollution has soared in those areas with heavy automobile traffic (even though such engines get better mileage and therefore lower the amount of fuel used).

Particulate Matter

Certain industrial processes release dusts, minerals, plant products, or other *waste particles into the air that are large enough to be filtered.* Particles are also released from automobile clutches and brake pads and from tires as they are worn down. Some of these particles may contain chemically active groups of sulfates, nitrates, fluorides, or ammonia. Others are essentially metallic, often containing cobalt, iron, lead, or copper. Some can cause cancer and aggravate lung or heart conditions. Some also can interfere with plant photosynthesis.

Water Pollution

Among the greatest environmental problems facing us today is the contamination of our water supplies. Essentially, the problem is of two types: contamination from organic sources and contamination from inorganic sources. Let's first consider organic contamination, the sort that stems from living things.

Sewage

The fresh waters of the United States are generally considered, on an intuitive basis, to be the property of all, a kind of commons. Yet, those waters are routinely used for cooling, cleansing, and as dumps for indus-

ESSAY 30.2   SEWAGE TREATMENT

When our population was smaller, sewage could be safely emptied into moving rivers where bacteria and fungi (decomposers) could digest it. Those large organic molecules from excretion, fallen leaves, and other by-products of life were broken down into their elemental constituents. These constituents could then be used as building blocks by primary producers in the recycling of life's materials. When the numbers of waste producers along such rivers became too high, the natural processes of decomposition proved too slow, and sewage-treatment plants were devised. These plants are, in effect, places where the natural processes are speeded up within the confines of a restricted area.

Ideally, a modern two-stage sewage treatment plant collects the wastes discharged into our waters, gets rid of the waste, treats the water, and pumps it back into our supply. First, the largest pieces of waste are screened out. Then the water is pumped into settling tanks where finer material sinks to the bottom as sludge. The supernatant (fluid over the sludge) is pumped into tanks where air is bubbled through it to provide oxygen for bacteria and fungi, which break down the organic molecules. Sludge is again allowed to settle out and the supernatant is chlorinated to kill the bacteria. Then it is ready to drink again.

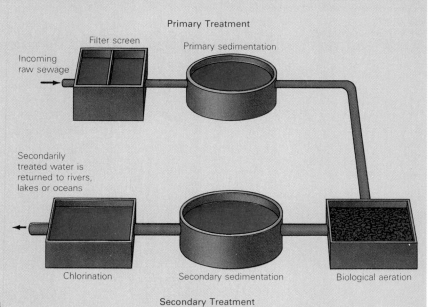

Unfortunately, many of our cities lack facilities for complete processing of sewage. This means that partly treated or untreated wastes are pumped into our waterways. But even those plants with adequately treated sewage present certain environmental problems. For example, the effluent from the best sewage plants is high in phosphorus, nitrogen, and certain other elements. These can act as fertilizers by enriching the water into which they are dumped. (Their effect is added to that of phosphate from detergents, which accounts for about half the phosphate overload.) The result can be cloudy, smelly water covered with algae scum. Many species of algae that respond so well to inorganic fertilizers are not used as food by zooplankton and thus do little to initiate a food chain. Also, it has been determined that where pollutants raise the overall "productivity" of water, in terms of mass, it decreases the diversity of species present. Certain cyanobacteria produce poisons that can kill cattle and cause rashes and vomiting in humans.

trial and municipal effluents. By law, when water is utilized for such purposes, it must return to the waterways in a pure, unpolluted form. However, we are aware that, in many cases, our waterways are treated as open sewers and our underground water tables as invisible rugs under which we sweep our poisonous wastes (Essay 30.2 and Table 30.2).

It has been said that the tap water of New York City contains many interesting ingredients—not the least of which is chow mein. A more precise, but no less fascinating statement is that in some places in the United States, the water one drinks has already passed through the bod-

Table 30.2 Major Water Pollutants

Pollutant	Sources	Effects	Control Methods
Oxygen-demanding wastes	Sewage; natural runoff; animal wastes; decaying plants; industrial wastes	Oxygen consuming bacteria depletes oxygen in water; fish and plants destroyed; foul odors; livestock sickened	Better treatment of waste water; reduce agricultural fertilizer runoff
Disease-causing wastes	Sewage; animal wastes	Outbreaks of diseases, including typhoid, infectious hepatitis, cholera, and dysentery	Treat waste water; reduce agricultural and ranching runoff
Acids	Mining (especially coal); industrial wastes	Kills plant life, some animals; increases solubility of other harmful minerals	Control surface mining and smoke stack effluent; treat wastewater

ies of seven or eight other good people. Under the best of conditions, that water will have been filtered and treated with chlorine to kill dangerous microorganisms. (Chlorine has been questioned as a water additive because certain chlorine compounds are known to produce mutations, and people who drink chlorinated water have a thirteen to ninety-three percent greater chance of developing colon and bladder cancers than those who drink untreated water. But the consensus is that, considering the risk of impure water, chlorination is by far the lesser of two evils.)

In many locales in the U.S., the water is simply not safe to drink, as evidenced by the continuing outbreaks of infectious hepatitis in the United States. Infectious hepatitis is believed to be caused by a virus carried in human waste, usually through a water supply that is contaminated by sewage. There is some disturbing evidence that this virus may be resistant to chlorine, especially in the presence of high levels of organic material. Despite our national pride in walk-in bathrooms, sewage treatment for many communities in the United States is inadequate, and waste that has been only partially treated is regularly discharged into waterways.

Chemical Effluents

Perhaps the most dangerous water pollutants are from inorganic sources, because many of them are so new that we know almost nothing of their long-term effects. However, we do know of the existence of over five million chemicals, about 45,000 of which are used commercially, with literally hundreds more being added to the list each year. We also know that well over 250 synthetic chemicals were isolated from the drinking water of only eighty American cities.

The dangers of releasing chemicals we know nothing about is underscored by evidence that some familiar compounds can be more dangerous than we thought. For example, we learned only around 1950 that nitrates, a common constituent of agricultural fertilizers and water supplies, could alter hemoglobin so as to impair its oxygen-carrying capacity. Even later, it was found that certain bacteria in the digestive tract can convert nitrates to highly dangerous and carcinogenic nitrites. Nitrate water pollution is especially dangerous in certain areas of California, Illinois, Wisconsin, and Missouri.

Heat Pollution

A rather subtle and often ignored form of water pollution is *heat.* Water absorbs heat, for example, when the water is used as a coolant or lubricant in manufacturing or energy production (including nuclear energy). The problem is that when water is returned to its source at this higher temperature, it may alter the forms of life in the water. Furthermore, a temperature rise of only a few degrees is sufficient to cause such changes. Some species may perish, while others begin to thrive under these new conditions, and the organisms affected may range from the producers to the highest level of consumers. Thus, delicate balances, which have been established as the result of long periods of evolution, can be seriously disrupted.

But why is the addition of only a little heat so critical? How does it exert its effect? Basically, heat pollution works in two ways: (1) by increasing the rate at which aquatic organisms consume oxygen, while, at the same time, lowering the oxygen-carrying capacity of the water (reducing the ability of the biological system to decompose waste), and (2) by increasing the rate of evaporation of the water, thus raising the concentration of pollutants left behind.

Pollution by Pesticides

Pesticides are biologically rather interesting substances. Most have no known counterpart in the natural world, and most didn't even exist thirty years ago. Today, however, a metabolic product of DDT, called DDE, may be the most common and widely distributed synthetic chemical on earth. It has been found in the tissues of living things from polar regions to the remotest parts of the oceans, forests, and mountains. Although the permissible level of DDT in cow's milk, set by the U.S. Food and Drug Administration, is 0.05 parts per million, it often occurs in human milk in concentrations as high as five parts per million and in human fat at levels of more than twelve parts per million (Figure 30.4).

Pesticides, of course, are products that kill pests. But what is a pest? Biologically, the term has no meaning. The Colorado potato beetle, for example, was never regarded as a pest until it made its way (as a result of human activity) to Europe, where it began to seriously interfere with potato production. Perhaps this episode best illustrates a definition of a pest: it is something that competes with, or otherwise disturbs, humans.

It seems that the greatest pesticidal efforts have been directed at insects and, clearly, much of it has been beneficial. The heavy application of DDT since World War II has caused sharp decreases in malaria and yellow fever in certain areas of the world. But DDT and other chlorinated hydrocarbons have continued to be spread indiscriminately any place in which insect pests are found. The result of course, is a kind of (is

FIGURE 30.4 We often eat from the top of the food chain and thereby encounter food that may, itself, have concentrated DDT in its tissues. DDT is soluble in fat and is often stored in human fatty tissue. Sometimes the concentration there may reach such levels that some researchers have suggested that it is unwise for an overweight person to attempt to lose weight and thus metabolize his or her own poisonous tissues.

Heron
3.50

Osprey
13.80

Gull
75.50

Tern
6.40

Marsh grass
2.30

Silversides
0.23

Marsh weed
0.08

Garfish
2.07

Algae
0.04

Shrimp
0.167

FIGURE 30.5 Once DDT enters the food chain, it becomes increasingly concentrated as it moves along from one link to the next. Whereas its concentration may be very low in such organisms as algae, fish that eat algae tend to store most of the DDT they eat in their tissues. Other fish that eat those fish, then, encounter an increased concentration of DDT in their food supply and, in turn, store most of it in their tissues. They eventually pass these boosted amounts along to their own predators. Each level, then, ends up with more DDT in its tissues than the species on which it feeds. The numbers in the figure indicate the parts per million of DDT in the tissues of various species that have been tested in the chain. Where would humans enter the system? The use of DDT has been curtailed in the United States, but it is still widely used in some other countries.

it artificial or natural?) selection. The problem is that some insects had a bit more resistance to these chemicals than did others. These resistant ones then reproduced and, in turn, the most resistant of their offspring continued the line. The result is that we now have insects that can almost bathe in these chemicals without harm.

There are also other risks involved in such wide use of insecticides. For example, most are unselective in their targets; they kill virtually *all* the insect species they contact. Many insects, of course, are beneficial and may form an important part of large ecosystems. Also, chemical insecticides move easily through the environment and can permeate far larger areas than intended. Another particularly serious problem with pesticides is that many of them persist in the environment for long periods. In other words, the chemicals are very stable and it is difficult for natural processes to break them down to their harmless components. Newer chemical pesticides are deadly in the short run, but quickly break down into harmless by-products.

The tendency of DDT to be magnified in food chains has been particularly disastrous for predatory birds that feed high on the food pyramid. This is because as one animal eats another in the food chain, the pesticide from each level is added to the next. Thus, species high on the food chain, the predators, tend to accumulate very high levels of these chemicals (Figure 30.5). Reproductive failures in peregrine falcons, the brown pelican, and the Bermuda petrel have been attributed to ingesting high levels of DDT. The problem is that the pesticide interferes with the birds' ability to metabolize calcium. As a result, they lay eggs with shells too thin to support the weight of a nesting parent.

Pollution by Radiation

Life on earth evolved under ionizing radiation in one form or another, and, in fact, we still live in a veritable sea of radiation. One source of radiation is the earth's crust, where radioactive substances undergo spontaneous nuclear disintegration, emitting both high-speed atomic particles and penetrating electromagnetic rays (similar to x-rays) (Essay 30.3). These types of radiation are more prevalent in some regions of the earth than in others. In addition, we are constantly bombarded by fast-moving atomic fragments as our atmosphere is struck by atomic particles from space. They are referred to as **cosmic rays.** Other radioactive substances, such as one called potassium 40, normally circulate through living systems. Such naturally occurring radiation is collectively referred to as **background radiation.** We do not know, at present, the degree to which background radiation is responsible for mutations resulting in either cancer or birth defects in the human population, but we do know that such ionizing radiation can produce these conditions under experimental conditions.

A much more serious problem is the radiation that we produce ourselves through medical radiation, nuclear testing, nuclear reactors, and nuclear accidents.

Nuclear Power Plants

In the United States, there is some disagreement (to say the least) over the risks and benefits of nuclear power. There can be no question that with our electrical power needs increasing rapidly, we cannot rely indefinitely on the earth's remaining fossil fuel supply. The question is, can we safely shift our reliance to nuclear fission power plants—considering the present state of our knowledge and technology? The American public has, in the past few years, developed a rather strong consensus regarding this question. Nuclear power has fallen into disfavor. The risks seem too great, the reward too small—at least for the immediate future.

It is not likely that an explosion of the type produced by atomic bombs can occur in the kinds of nuclear reactors being used today. However, we still don't know how close we came to a major tragedy at Three Mile Island (Fig. 30.6). A far greater tragedy occurred at a nuclear plant at Chernobyl, in the Soviet Union's Ukraine in 1986 (Essay 30.4). The risks associated with nuclear power, however, are not always so spectacular. Some are of a far subtler nature. For example, radioactivity could be released into the environment from activities related to mining and processing nuclear fuel, from the transportation and recycling of the fuel, and from storage of the radioactive wastes. We frequently hear of steam or gas leaks from the reactors themselves. And even the safest reactors

FIGURE 30.6 The worst nuclear accident in the United States occurred at Three Mile Island in the Susquehanna River, Pennsylvania. One of the cooling water pumps failed, immediately activating backup systems. However, valves on emergency water pumps had been accidentally left closed, and a faulty valve (scheduled for replacement) failed to work. Operators miscalculated the amount of water in the cooling system and prematurely shut off the water supply, uncovering the nuclear core for several hours. The housing of the fuel rods oxidized, producing an enormous hydrogen bubble that experts feared could explode. The bubble had not been anticipated. No one yet knows how close we came to a complete meltdown at Three Mile Island.

The nuclei of some atoms are unstable. Of the more than 320 isotopes that exist in nature, about sixty are unstable, or **radioactive.** In addition, humans have created about 200 more. Radioactive isotopes of any element are called **radioisotopes.** When a radioisotope decays (or explodes), certain particles (neutral, or positively or negatively charged) and rays are emitted from its nucleus. The resulting atom may also be radioactive, or it may be stable.

Radioisotopes decay at predictable rates. These rates are usually expressed as the element's **half-life.** Half-life refers to the time that must expire in order for half the atoms in any amount of the isotope to decay to another kind of atom. Half-lives may be very long. The half-life of natural uranium (U238) is 4,500,000,000 years! As a rule of thumb, the time it takes for a substance to become nonradioactive is considered to be twenty of its half-lives.

When radioisotopes decay, the products with shorter wavelengths may strike other atoms in the environment in such a way as to tear electrons away from them, leaving them with a positive charge. Thus, the process is called **ionizing radiation.** Genetically, the danger lies in the molecules of DNA being altered by ionizing radiation, thereby producing mutations. Such changes in DNA can result in a variety of damage to the body (including cancer), or if the change occurs in the DNA of a gamete, the result can produce abnormal offspring (see figure). It is generally assumed that

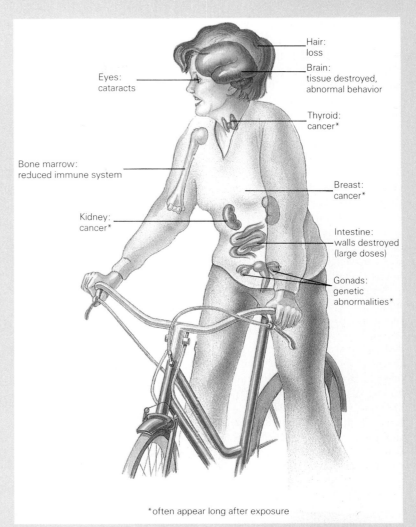

Eyes: cataracts

Bone marrow: reduced immune system

Kidney: cancer*

Hair: loss

Brain: tissue destroyed, abnormal behavior

Thyroid: cancer*

Breast: cancer*

Intestine: walls destroyed (large doses)

Gonads: genetic abnormalities*

*often appear long after exposure

there is *no* threshold below which radiation is genetically harmless; there are only "acceptable" risks. Also, the genetic effects of radiation are accumulative in the exposed individual and his or her offspring. A "snowball effect" operates so that an increased radiation level, which has little effect in the first few generations, if maintained, may be-

come magnified in following generations. The figure shows the effects of radiation on various parts of the body.

Radioactivity of any substance can only be reduced by allowing the radioisotopes to decay naturally. The process cannot be slowed or hastened.

normally leak small amounts of radiation into their immediate environment. (The problem with this is that there are no "safe" levels of radioactivity—only "acceptable" levels.)

We could probably greatly reduce the risks associated with nuclear power by simply exercising more care and common sense. There are numerous published accounts that attest to our carelessness, however.

The first warning came at 9:00 a.m. on Monday, April 28, 1986. Technicians at a nuclear plant 60 miles north of Stockholm began to see alarming blips crossing their computer screens. The blips meant one thing: high levels of radiation. They assumed a serious leak at their own plant and began a frantic search for the problem. They found nothing, but the radiation levels kept increasing both at the plant and in the surrounding countryside. They concluded the problem was not with their own facilities and immediately cast a suspicious eye to their powerful neighbor to the south, the Soviet Union.

For days, breezes had been blowing from the south and into Scandinavia, but when the Swedes demanded an explanation, the Soviet response was a stony silence. Later that day, they broke their silence and blandly said that nothing had happened. On the news that night, an expressionless newscaster read a four-sentence statement admitting that there had been a nuclear accident at the Chernobyl nuclear power (80 miles north of Kiev) damaging a reactor and that "measures were being taken."

The Soviets played down the incident, denying reports that thousands had been killed, but admitting two deaths in the first days. As the story unfolded, largely through the intense questioning of other countries and American satellite photographs, it became clear that the reactor had experienced a dreaded meltdown. A meltdown occurs when the cooling system fails and the radioactive core units overheat to the melting point. As the molten mass accumulates on the reactor floor, there is a risk of burning through the floor and into the ground below, contaminating the water table (a meltthrough). At Chernobyl, the problem was compounded by the type of encasement

around the fuel rods. It was graphite, and graphite can burn, producing temperatures over 5,000 degrees Fahrenheit. The Soviet technicians then flooded the area with water, but it was not only too late, it was the wrong move. The steam reacted with the uranium fuel, the graphite, and the zirconium sleeves that house the rods producing a flammable gas that blew up the reactor and released billowing clouds of radioactive gases into the atmosphere (see map).

In the days that followed, the Soviets grudgingly released bits of information telling of an increasing death toll, medical treatment (especially bone marrow transplants), and mass evacuations of large parts of what had been one of the Soviet Union's greatest agricultural areas.

We still do not know the full effects of the accident on Soviet citizens (or those in other parts of the world) because low doses of radiation may take years to exact their toll. Only two months before the disaster, the Soviets had calculated that a meltdown of this type could statistically be expected only every 10,000 years. The Chernobyl reactor had been operating three years.

For example, it has been revealed that the Diablo Canyon nuclear power plant in California was built on an earthquake fault line. Of course it was girded for that risk (inasmuch as one can gird for a large earthquake.) Incredibly, however, the blueprints were somehow reversed and the earthquake supports were put in backwards. Furthermore, the mistake was not noticed for four years. At the Comanche Peak plant in Texas, supports were constructed forty-five degrees out of line. At the Marble Hill plant in Indiana, the concrete surrounding the core was found to be full of air bubbles. At the WNP-2 plant in Washington state, the concrete contained air bubbles and pockets of water as well as shields that had been incorrectly welded. At the San Onofre plant in California, a 420-ton reactor vessel was installed backwards and the error was not detected for months. In 1981, the Nuclear Regulatory Commission inspected forty-three plants that were under construction (Figure 30.7) and rated seven "below average" and thirty-six "average." None were rated even "above average."

Completely apart from the possibility of accidents, there is the unsolved problem of what to do with the radioactive wastes generated in the course of normal nuclear plant reactions. The problem is a tough one since such wastes can only be rendered safe by the passage of time. The waste radioactivity is generated in the fuel system of the reactors because only a part of the fuel is fissionable and, for technical reasons, not all of the fissionable elements are spent. Much of the spent fuel materials removed from the reactor can be reused. However, some of the radioactive fuel in the spent elements cannot be recovered, and this material adds to the radioactive waste.

FIGURE 30.7 Sites of operating and proposed nuclear reactors in the United States. Only Washington state has reactors with graphite and without a containment building (similar to that at Chernobyl). Some sites have more than one reactor.

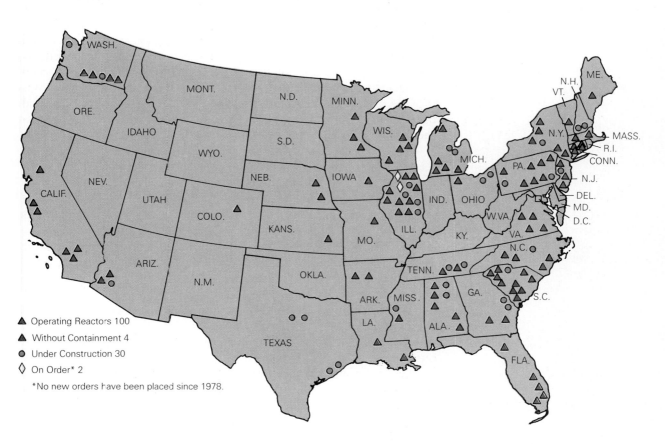

▲ Operating Reactors 100
▲ Without Containment 4
● Under Construction 30
◇ On Order* 2

*No new orders have been placed since 1978.

We have already generated over 10,000 tons of nuclear waste, with another 47,000 tons expected by 1995. Most of the waste is in the form of fuel rods which are, for now, stored in baths filled with a solution of neutron-absorbing boric acid. The problem is that these are only temporary repositories and, unless new space is found, existing plants must begin closing for lack of space. The rods can be reprocessed, but the technology could lead to the spread of fuel for nuclear weapons (see Essay 30.5).

The reprocessing also leaves "high-level liquid waste" that must be stored. The prevailing idea at the moment is to dry the liquid and mix it with molten glass that, when it hardens, can be stored in stainless steel containers. The problem then is what to do with the tanks. Suggestions range from burying it in the Antarctic to sending it into space.

HIDDEN DECISIONS

Perhaps in these pages you have come to learn more about the complexities of this great living system of which you are a part. Perhaps, also, you have come to appreciate just how inextricably linked are the various aspects of life. Because life's grand processes are complex and interrelated, it should be apparent now that life is easily touched. A move here, a decision there, and parts of our biosphere head off in some new direction. Sometimes we know about it, sometimes, however, we may not.

We are by now informed enough about life's processes to avoid unnecessary radiation, to choose certain foods, and to deplore extinctions. When the choice is ours, we choose life. But sometimes the choice is ours without our knowing it. Precisely because life is a sensitive, responding, and far-flung phenomenon, changing any part of it can have far-reaching and often unexpected effects. It is important to understand that we can change the nature of life on the planet and our place in its pageant without realizing we were involved in such things at all. It is the role of the educated person to uncover these hidden decisions and bring them to light so that we as a society begin to understand the cascading impact of our behavior.

As an example, we can influence the environment by our vote. Sometimes this is apparent, such as when we elect some staunch environmentalist (or a developmentally-minded industrialist). In other cases, though, our political choices may have less apparent, but just as important, biological effects. For example, if elected officials subsidize farmers with public funds, do they encourage people to stay on the land and out of cities? Is there a greater sense of community among people in large populations, or do such conditions encourage anonymity and reduce the sense of responsibility? What happens to an environment where people do not feel individually responsible?

The United States is proud of its system of public education. Every child in this country has the right, even the obligation, to be educated to some degree. The schools exist, and they are funded. Education is obviously good, so what lawmaker wants to oppose it? However, some say our system of public education has an unexpected side effect—that it subtly encourages people to have children by guaranteeing that the cost of their children's education will be paid for by public money. What are the biological effects of such encouragement?

The most sadly brutal and devastating effects of a nuclear exchange may not be due to the blast itself or the ionizing radiation that follows. The worst killer may have a different face. Imagine a darkened world blanketed by a constant winter. This is the scene that some scientists* now warn us about. They say that one effect of even a limited exchange of missiles would be **nuclear winter.**

According to their computer analysis, nuclear explosions at a few target cities would send clouds of smoke and ash high into the atmosphere, where it would hang suspended for long periods, blocking the sun's radiation and throwing the world into a dark and starving winter. Originally, they estimated that these conditions might be restricted to the northern hemisphere, lasting perhaps six months to a year. The early calculations suggested that debris would be lofted perhaps five miles high, but would eventually be washed from the sky by cleansing rains.

However, new data are emerging that paint an even more dismal picture. It seems that the great clouds

may be lifted as high as ten miles, from where they could spread their darkening quilt over the entire planet, bringing winter even to the tropics and stopping photosynthesis virtually everywhere. Further, such clouds would rise far above the cleansing effects of weather.

Some scientists disagree with all such projections, saying they encompass too many assumptions and that even newer data show the effects would be far less drastic than first estimated. It is also argued that rains would probably soon cleanse the skies and that the temperature changes would be only moderate. While the argument continues, a great number of people are hoping it will never be resolved.

*The best known is a group comprised of Richard Turco, O. Brian Toon, Thomas Ackerman, James Pollack, and Carl Sagan (TTAPS, they call themselves).

Even our system of taxation can have biological repercussions. In some years, single people get tax breaks, and in other years, the nod goes to married couples. If marriage is encouraged by government policies such as preferential tax schedules, and if (as is the case) earlier marriages tend to produce more children, then legislators again can subtly influence population growth and social patterns, often without realizing it.

Do tax deductions for home mortgages encourage building? Does home building affect forests? What are the effects of building booms on national parks and outdoor recreation? On land use policies? On sewage? On water supplies? On road building and its handmaidens? What are the social effects of governmental encouragement of home ownership as opposed to apartment rental?

The list of our hidden decisions is endless. The point of all this is that not only *can* we make biological decisions, we *do* make biological decisions. And we do it almost casually on a daily basis. Obviously, we must make such decisions. We have come too far to try to leave well enough alone. However, we must become aware of the far-reaching effects of our manipulations. In other words, our decisions must begin to make biological sense. They must become conducive not only to perpetuating life, but to increasing our quality of life. Perhaps the answer lies in educating our lawmakers in the biological implications of their decisions and policies. Or perhaps the answer lies in first educating ourselves and then electing lawmakers from among us who are well aware of what they're doing and who can confidently expect the support of an enlightened public.

THE FUTURE

We were once the future. We were among the generations our ancestors worked for, hoped for, and fretted about. They wanted to make the world safe for us, and they tried to prepare the way for our coming. The future, however, has a way of becoming the present. And now here we are, the much-loved progeny of people who never knew us. We are indeed reaping what they sowed, and it works both ways. In some ways, we benefit greatly from their concern and foresight. In other ways, we are paying the bill for their binge. However, perhaps they can be forgiven for their unwise choices because there is no way they could have imagined such a world as this.

Now it is our turn. And we, too, are blinded by the present light. We can't see ahead either. But we are to be forgiven less easily for our follies. We must bear greater responsibility than do our ancestors. The reason is quite simple: We know more than they did. We are aware of our present global situation. We know more about our world than any generation that preceded us. We know, in a general way, how bad it is and how good it can be. We have enough information to enable us to understand, more than any other generation, the implications of our decisions. We can see more clearly than ever where we're going and what any charted course will lead us to. Whereas we can't predict what the future holds, we can influence that future to a degree unimagined by our ancestors. Had they had this ability, the world might be a far different place today. But with this ability, born of sheer information, comes an added responsibility. We have but a few more years of this odd interlude in history in which we will be able to ignore our biological imperatives, to fail to recognize that we are a recent upstart among the venerable and ancient species, and that our future is not insured. It is important that as this period of borrowed time runs out, we become aware that our limitations are those of fragile cytoplasm, that we see the urgency of a steady-state global economy, an economy no longer based on a mortgaged future. Will it be said by our descendants that we foreclosed their future and demeaned their existence? Or will generations to come stand in admiration of our strength, dignity, and wisdom at a time when decisions were hard?

SUMMARY

Bioethics is the emerging study of ethical questions relating to biology; it is the ethics of life. A number of bioethical problems have been related to how and why we have altered our environment through environmental pollution (a pollutant being any harmful substance that enters the environment). Air pollution is largely caused by exhaust from automobiles and effluents from building heating and residential furnaces. Major pollutants include carbon monoxide, nitrous oxides, sulfur oxides, hydrocarbons, and particulate matter, each with specific sources and effects. Water pollution has two primary sources—organic and inorganic contaminants. Sewage is largely organic waste. Chemical effluents from agriculture and industry contribute largely to inorganic wastes. Heat contributes a more subtle form of pollution. Pesticides are particularly dangerous because they are too recently developed for us to ascertain long-term risks. Chlorinated hydrocarbons, such as DDT and its derivatives, are being replaced by shorter-lived and more specific chemicals. Problems of radiation are associated with nuclear energy. There have been numerous problems, and a few catastrophes, at nuclear power plants.

If we are to concern ourselves properly with the question of bioethics, we must realize the far-reaching implications of hidden decisions, choices available to us that may have profound biological impact that may go unrecognized.

Appendix A
CLASSIFICATION OF ORGANISMS

THE PROKARYOTES

This is one of several common ways of dividing this group.

Kingdom Monera

Single-celled organisms, tough cell wall, no membrane-bounded organelles or organized nucleus. Circular DNA, not joined with protein. Reproduction mostly by fisson; some by conjugation. Any flagella stiff and rotating.

Subkingdom Archaebacteria: Unique, proteinaceous cell walls. Cell membrane with branched fatty acids. Mostly anaerobic; includes methanogens, halophiles, and thermophiles.

Subkingdom Cyanobacteria: (formerly called blue-green algae) Photosynthetic. Colonial. Some species fix nitrogen. Membranous photosynthetic lamellae.

Subkingdom Eubacteria: ("true" bacteria) Cell walls of peptidoglycan, membranes with straight-chain fatty acids. Includes many pathogens, free-living reducers, phototrophs, and chemotrophs. Found in coccus, bacillus, and spirochaete forms. Occurs singly, in clusters, or in chains.

THE EUKARYOTES

All other organisms. Membrane-bounded cellular organelles, linear chromosomes join protein. Cell division by mitosis and meiosis, sexual reproduction common. Any flagella or cilia are microtubular. Single-celled, colonial, and multicellular.

Kingdom Protista

Includes photosynthetic, plantlike (algal) and heterotrophic, animallike (protozoan) forms. Primarily single-celled or colonial.

Phylum Pyrrophyta: (1100 species; the dinoflagellates) Single-celled, flagellated, chitinous cell walls, phototrophic.

Phylum Chrysophyta: (11,500 species; yellow-green and golden-brown algae) Single-celled or colonial, glass walls, phototrophic.

Phylum Protozoa:

> *Class Flagellata* (2500 species; flagellated protozoans) Single-celled, heterotrophic.

> *Class Sarcodina* (11,500 species; amoeboid protozoans) Single-celled, phagocytic heterotrophs; includes marine radiolarians and foraminiferans.

> *Class Sporozoa* (6000 species; nonmotile protozoans) Single-celled, parasitic spore formers.

> *Class Ciliata* (7200 species; ciliated protozoans) Single-celled, extremely complex and diverse heterotrophs.

Phylum Acrasiomycota: (26 species; cellular slime molds) Heterotrophic, individual amoeba that live alone or join to form multicellular plasmodia.

Phylum Myxomycota: (450 species; acellular slime molds) Heterotrophic, form multinucleate feeding plasmodia.

Kingdom Fungi

Multicellular heterotrophs, including reducers and parasites. Extracellular digestion. Mycelial organization that may or may not include walls between cells. Either sexual or asexual spores; primarily haploid with brief diploid stage.

Division Oomycota: (475 species; water molds) Flagellated fungi, many parasitic.

Division Zygomycota: (600 species; bread molds) Mycelium without cell end walls; simple sexual zygospores.

Division Ascomycota: (30,000 species; sac fungi) Extensive mycelium with cell end walls. Complex sexual dikaryotic asci, ascospores produced through meiosis. Many species symbiotic with cyanobacteria or algae, forming lichens.

Division Basidiomycota: (25,000 species; club fungi) Extensive mycelium with cell end walls. Large, complex, dikaryotic basidiocarp in which basidiospores are produced.

Division Deuteromycota: (also, *Fungi Imperfecti*; 25,000 species) Various fungi with no known sexual stage.

Kingdom Plantae

Primarily nonmotile, multicellular organisms with dense cell walls or cellulose. Specialized tissues and organs. Most phototrophic. Alternating generations.

Nonvascular Plants

Division Rhodophyta: (4000 species; red algae) Coastal seaweeds, floridean or carageenan as storage carbohydrates.

Division Phaeophyta: (1000 species; brown algae) Coastal seaweeds and kelps, fucoxanthin pigments, laminarin and mannitol storage carbohydrates. Often large, some with vascular tissue.

Division Chlorophyta: (7000 species; green algae) Single-celled, colonial, and multicellular, pigments mostly chlorophylls and carotenoids, usually aquatic.

Division Charophyta: (250 species; stoneworts) Freshwater alga with apical growth, calcareous cell walls.

Division Bryophyta: (16,000 species; mosses, liverworts, hornworts) Multicellular, nonvascular, terrestrial. Simple aerial spore, motile sperm, predominant gametophyte. Generally small with little supportive tissue.

Vascular Plants

Division Tracheophyta: (vascular plants) Vascular tissue in roots, stems, and/or leaves. Predominant sporophyte; some with distinct generation.

Seedless Vascular Plants

Class Psilophyta (4 species; whisk ferns) Simple, with few surviving species. Vascular stem, scalelike leaves. Aerial spore, motile sperm, distinct sporophyte and gametophyte.

Class Lycophyta (1000 species; club mosses) Vascular roots, stems, and leaves. Usually with aerial spore, motile sperm, and distinct sporophyte and gametophyte.

Class Sphenophyta (12 species; horsetails) One surviving genus, upright vascular stems, prominent nodes and tiny, scalelike nonphotosynthetic leaves; Aerial spore, motile sperm, and distinct sporophyte and gametophyte.

Class Pterophyta (Filicinae) (11,000 species; ferns) Most with aerial spore, motile sperm, predominant sporophyte, photosynthetic gametophyte, and vascular roots, stems, and leaves.

Vascular Plants with Seeds

Subdivision Spermatophytes: (seed plants) Gametophytes develop within sporophyte tissue, male and female spores produced, and microspores released within pollen grains. Embryo develops within seed that includes stored foods and protective seed coats.

Class Gymnospermae (naked seeds)

Cycads (100 species; the cycads) Palmlike leaves, exposed seeds, wind-dispersed pollen, flagellated sperm within pollen tube.

Ginkgos (1 species; ginkgo) Trees with fan-shaped leaves, exposed seeds, wind-dispersed pollen, motile sperm within pollen tube.

Conifers (500 species) Usually large trees with needlelike or scalelike leaves. Evergreens, exposed seeds borne upon cones. Nonmotile sperm.

Gnetophytes (71 species) Gymnosperms with angiosperm features: xylem vessels, pollen cones. Exposed seeds, nonmotile sperm.

Class Angiospermae (flowering plants) Flowers present, seeds enclosed by fruit, xylem vessels present. Nonmotile sperm.

> *Dicotyledons* (200,000 species; the dicots) Diverse, net-veined leaves, secondary growth common, floral parts in fours, fives, or multiples of these, two cotyledons.

> *Monocotyledons* (50,000 species; the monocots) Diverse, parallel-veined leaves, secondary growth rare, floral parts in threes or multiples of threes, one cotyledon.

Kingdom Animalia

Multicellular, heterotrophic eukaryotes. Specialized tissues, most with organ systems. Most highly responsive. Diploid except for gametes. Fertilization without intervening haploid life cycle. Small, flagellated sperm and large stationary egg typical.

Subkingdom Parazoa: Animals of flagellate origins, simple developmental progression.

Phylum Porifera: (5000 species; the sponges) Forms tissues. Nonmotile adults, filter feeding, skeletal elements of calcium carbonate, silicon dioxide, or spongin. Asexual reproduction by budding, sexual reproduction by fertilization of internalized egg.

Subkingdom Metazoa: Animals of ciliate origin; includes all animals except Porifera.

Radiate, Acoelomate Phyla
(radial symmetry, no coelom, diploblastic)

Phylum Coelenterata: (9000 species) Radial body of two cell layers, saclike gastrovascular cavity, tentacles and stinging cells. May alternate between medusa and polyp stages, or only one stage may be present. Three classes: Hydrozoa (hydroids), Scyphozoa (jellyfish), Anthozoa (corals and anemones).

Phylum Ctenophora: (90 species; comb jellies) Radial body of two cell layers. Tentacles with glue cells.

Bilateral, Acoelomate Phyla
(bilateral symmetry, no coelom, triploblastic)

Phylum Playthelminthes: (13,000 species; flatworms) Flattened body, branching gastrovascular cavity, dense bodies with many cell layers. Three classes: Turbellaria (free-living planarians), Trematoda (parasitic flukes), and Cestoda (tapeworms).

The Bilateral, Pseudocoelomate Phyla
(bilateral symmetry, pseudocoelom)

Phylum Nematoda: (12,000 species named—estimated half million unnamed; the roundworms and rotifers) Nematodes (roundworms) include free-living and parasitic species; slender body, pseudocoelom (not completely mesodermally lined), tube-within-a-tube body plan (complete gut).

Phylum Rotifera: ("wheel animals") Free-living, minute, with complex organ systems.

Phylum Nematomorpha: (230 species; horsehair worms)

Phylum Rynchocoela: (650 species; proboscis or ribbon worms)

The Bilateral, Coelomate, Protostome Phyla
(bilateral symmetry, true coelom, embryologically "mouth-first" animals)

Phylum Mollusca: (47,000 species) Controversial classification—segmentation and coelom may not exist. Diversification through modifications of head, foot, mantle, and radula. Includes seven classes: Aplacophora (solenogasters—wormlike, radula only clear characteristic), Monoplacophora (*Neopalina*, deep-sea form once believed to be extinct), Scaphopoda (tooth shells), Polyplacophora (chitons), Bivalvia (bilvalves: two shells—clams, etc.), Gastropoda (snails, slugs), and Cephalopoda (octopus, squid—rather intelligent, fast, predators, foot subdivided into tentacles). Covering mantle, large brain, keen eyesight.

Phylum Annelida: (9000 species; segmented worms) Body subdivided into repeating segments, true coelom, well-developed digestive system, closed circulatory system. Three classes: Oligochaeta (earthworms), Hirudinea (leeches), and Polychaeta (marine worms).

Phylum Priapulida: (9 species; proboscis worms)

Phylum Pogonophora: (100 species; beard worms)

Phylum Sipuncula: (300 species; peanut worms)

Phylum Tartigrada: (350 species; water bears)

Phylum Arthropoda: (800,000 to 1,000,000 species; "jointed-footed" animals) Paired, jointed appendages with chitinous exoskeleton, varied segmentation, wide distribution.

> *Subphylum Chelicerata:* Six pairs of appendages, four pairs being legs, with paired chelicerae (fangs). Three classes: Meristomata (horeshoe crabs), Arachnida (spiders, ticks, scorpions, mites, daddy-longlegs), and Pycnogonida (sea spiders).

> *Subphylum Mandibulata:* Most with three pairs of walking legs, mandibles, compound eyes, antennas, some with wings. Four classes: Crustacea (aquatic with crusty exoskeleton, gills), Chilopoda (centipedes), Diplopoda (millipedes), and Insecta (insects—commonly three pairs of legs, wings at some time, three-part body, specialized mouth parts).

Phylum Onycophora: (70 species; *Peripatus*) Possessing both annelid and arthropod characteristics.

Phylum Brachiopoda: (250 species; lampshells) Appear similar to bivalves, but shell mounted differently, lopophore (ring of ciliated tentacles) present.

Phylum Phoronida: (18 species) Lopophore present.

Phylum Ectoprocta: (4000 species; moss animals or bryozoans) Lopophore present.

The Bilateral, Coelomate, Deuterostome Phyla
(bilateral symmetry, true coelom, embryologically "mouth second")

Phylum Echinodermata: (6000 species) Spiny, skinned animals, five-part radial symmetry as adults, bilateral larvae, endoskeleton, water vascular system. Five classes: Crinoidia (sea lilies), Holothuroidia (sea cucumbers), Echinoidea (sea urchins, sand dollars), Asteroidea (sea stars, basket stars), and Ophiuroidea (serpent stars, brittle stars).

Phylum Hemichordata: (80 species; acorn worms) Gill slits show relatedness to chordates.

Phylum Chordata: (43,000 species) Gill slits, notochord, postanal tail, dorsal hollow nerve cord all present at some time.

> *Subphylum Urochordata* (1300 species; sea squirts) Chordate characteristics seen mainly in bilateral larva.

> *Subphylum Cephalochordata* (28 species; lancelet) Fishlike body, permanent notochord and gill slits, filter feeder.

> *Subphylum Vertebrata* (41,700 species; the vertebrates) Vertebral column of bone or cartilage, heads well developed, ventral heart, dorsal aorta, two pairs of limbs. Seven classes: the fishes: Agnatha (jawless fishes), Chondrichthyes (cartilagenous fishes: sharks, rays, chimera), and Osteichthyes (bony fishes). Also, Amphibia (frogs, toads, salamanders), Reptilia (reptiles), Aves (birds), and Mammalia (mammals).

Appendix B
METRIC MEASUREMENTS

	Symbol	Fundamental unit	Quantity	Numerical unit	American measurement
Time		second (sec)			second
	msec		millisecond	0.001 sec	
	μsec		microsecond	0.000001 sec	
Length		meter (m)			39.4 inches
	km		kilometer	1,000 m	0.62137 miles
	cm		centimeter	0.01 m	0.3937 inches
	mm		millimeter	0.001 m	
	μm		micrometer	0.000001 m	
	nm		nanometer	0.000000001 m	
	Å		angstrom	0.0000000001 m	
Volume (liquids)		liter(l)			1.06 quarts
	ml		milliliter	0.001 liter	
	μl		microliter	0.000001 liter	
Volume (solids)		cubic meter (m^3)			1.308 cubic yards
	cm^3		cubic centimeter	0.000001 m^3	0.061 cubic inches
	mm^3		cubic millimeter	0.000000001 m^3	
Mass		gram (g)			0.035 ounces
	kg		kilogram	1,000 g	2.2 pounds
	mg		milligram	0.001 g	
	μg		microgram	0.000001 g	

Appendix C
TEMPERATURE
CONVERSION CHART

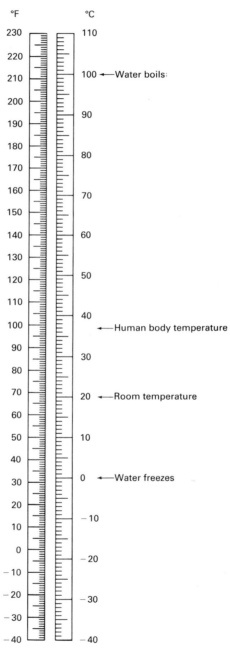

To convert fahrenheit to
centigrade, use this formula:
°C = ⁵⁄₉ (°F) − 32

To convert centigrade to
fahrenheit, use this formula:
°F = ⁹⁄₅ (°C) + 32

GLOSSARY

A

A1 cell: A sound receptor in nocturnal moths that responds only to weak pulses.

A2 cell: A sound receptor in nocturnal moths that responds only to loud pulses.

Abdomen: 1. In mammals, the body cavity between the diaphram and the pelvis. 2. In other vertebrates, the body cavity containing the stomach, intestines, liver, and reproductive organs. 3. In arthropods, the posterior section of the body.

Abductor: Any muscle that moves or draws away from the axis of the body or from one of its parts. Compare *adductor*.

Abiotic(ā'bī ot'ik): Characterized by the absence of life.

Abortion: The expulsion or removal of an embryo or fetus before it can survive on its own.

Acetic acid(ə sē'tik as'id): $C_2H_4O_2$, the ionized form of which joins with coenzyme A to form acetyl-CoA before it enters the Krebs cycle.

Acetylcholine(ə sē'təl kō'lēn): A chemical agent that transmits neural impulses across synapses from one neuron to another.

Acetylcholinesterase (ə sē'təl kō lēn es'tə rās): A membrane-bound enzyme that hydrolyzes acetylcholine in the course of synaptic nerve impulse transmission.

Acetyl-CoA (ə sē'təl kō ā): A key intermediate in metabolism, consisting of an acetyl group covalently bonded to coenzyme A.

Acid: A substance that releases hydrogen ions into a solution. It has a sour taste and unites with bases to form salts.

Acquired immune deficiency syndrome (AIDS): A viral disease that results in severely reduced immunity.

Acromegaly (ak'rə meg'ə lē): A chronic disease characterized by enlargement of the bones of the head, the soft parts of the feet and hands, and sometimes other structures, due to dysfunction of the pituitary gland.

Actin: A cytoplasmic protein and a constituent of muscle, known in both globular and fibrous forms. See also *myofilaments*.

Actinomycete (ak' tən ō mī sēt'): Any of several rod-shaped or filamentous, aerobic or anaerobic bacteria, certain species of which are pathogenic for humans and other animals.

Action potential: In a neuron, a traveling, depolarizing wave. A short-lived change in membrane potential that produces a neural impulse, or, in a muscle, contraction. Compare *resting potential*.

Active site: That part of an enzyme directly involved in specific enzymatic activity.

Active transport: The transport of a substance across the cell membrane, usually against the concentration gradient, that requires energy.

Addiction: The state of being given up or devoted to a habit or practice or to something that is habit-forming to such an extent that stopping may cause trauma.

Adductor: Any muscle that moves or draws toward the axis of the body or one of its parts. Compare *abductor*.

Adenine (ad'nēn): A purine, one of the nitrogenous bases found in both DNA and RNA, as well as in ATP and several coenzymes.

Adenosine diphosphate (ə den'ə sēn'dī'fos'fāt): ADP, a product formed by the hydrolysis of ATP, adenosine triphosphate, with the release of energy.

Adenosine triphosphate (ə den'ə sēn'trī'fos'fāt): ATP, a compound found in all cells, that serves as an energy source as it is broken down into ADP plus phosphate with the release of energy.

Adenyl cyclase (ad'ə nil sī'clās): An enzyme, usually incorporated into the cell membrane, that is capable of transforming ATP into cyclic AMP and pyrophosphate.

ADP: Adenosine diphosphate, a compound of adenine, ribose, and two phosphate groups.

Adrenal gland (ə drēn'əl gland): A vertebrate endocrine gland. The outer area produces steroid hormones, the inner produces epinephrine.

Adrenalin (ə dren'əl in): See *epinephrine*.

Afferent neuron: A nerve cell that carries impulses toward the brain.

Afterbirth: The placenta with its associated membranes, which is expelled from the uterus after childbirth.

Age structure pyramid: Also called *age profile* or *population pyramid*. A pyramidal graph of a population, divided into age groups. Each age group is represented by a horizontal bar with that of the youngest forming the base.

Aggression: Hostility, attack, or threat, especially unprovoked, usually against a competitor or potential competitor.

Agnatha (ag'nə thə): The class of vertebrates comprising the lampreys, hagfishes, and several extinct forms, having no jaws or paired appendages.

AIDS: See *Acquired immune deficiency syndrome*.

Albumen: The white of an egg.

Alcoholic fermentation: The anaerobic breakdown of glucose by yeast to form alcohol and carbon dioxide.

Aldosterone (al'dō sti rōn'): A steroid hormone produced by the adrenal cortex, involved in potassium reabsorption by the kidney.

Alga (pl. algae): Any photosynthetic member of the kingdom *Protista*.

Alimentary canal: A tubular passage functioning in the digestion and absorption of food in humans and most animals, beginning at the mouth and terminating at the anus.

Allantois (ə lan'tō is)**:** One of the extraembryonic membranes. In birds and reptiles, it serves as a repository for the embryo's nitrogenous wastes.

Allele (ə lēl')**:** One of two or more genes that produce a specific characteristic such as blood type, hair color, and so on.

Allopatric speciation (al'ə pat'rək spē'shē ā'shən)**:** The formation of new species from populations that are geographically separated. Compare *sympatric speciation.*

Alpha linkage: The type of bonding between glucose molecules of starch.

Alpine tundra: A biome found at elevations above the tree line, characterized by dense growth of mosses, lichens, dwarf herbs, and shrubs and covered·throughout most of the year by ice and snow.

Alternation of generations: The existence in the life cycle of an individual of a haploid(1N) gametophyte stage that alternates with a diploid(2N) sporophyte stage.

Altruism: Behavior that is directly beneficial to others at some cost or risk to the altruistic individual.

Alvarez hypothesis: A hypothesis that proposes that much of the massive extinction of life that accompanied the end of the Mesozoic era was produced by the aftereffects of a gigantic asteroid's collision with the earth.

Alveolus (al vē'ə ləs) **(pl. alveoli):** One of the tiny air sacs that occur in grapelike clusters in the vertebrate lung, in which carbon dioxide and oxygen are exchanged.

Amino acid (ə mē'nō as'id)**:** An organic molecule that contains at least one carboxyl group, COOH, and one amino group, NH_2.

Amino group (ə mē'nō group)**:** The univalent group NH_2, often ionized as NH_3+, which can form amines.

Amnion (am'nē ən)**:** The innermost extraembryonic membrane of reptiles, birds, and animals.

Amoeba (ə mē'bə)**:** 1. Any protozoan of the large genus *Amoeba,* characterized by lobose pseupods and the lack of permanent organelles or supporting structures. 2. Any ameboid protist such as the ameboid stage of a flagellate or sporozoan.

Amoebocyte (ə mē'bə sīt)**:** In many individuals, an ameboid cell that functions in reproduction, digestion, and so on.

Amphetamine (am'fet'ə mēn)**:** A drug that stimulates the central nervous system. Also called *speed* or *uppers.*

Amphibia (am fib'ē ə)**:** The class of cold-blooded vertebrates (frogs, salamanders, etc.) the larva of which are typically aquatic, breathing by gills, and the adult of which are terrestrial, breathing by lungs and through moist, glandular skin.

Amphioxus (am'fē ok'səs)**:** A lancelet of the genus *Brachiostoma,* having such vertebrate characteristics as a notochord and a dorsal cord of nerve tissue.

Anaphase (an'ə fāz)**:** The third phase of mitosis in which the chromatids separate and move to opposite sides of the cell.

Angina pectoris (an jī'nə pek'tə ris)**:** A syndrome characterized by constricting pain below the sternum, most easily brought about by exertion or excitement and caused by insufficient blood flow to the heart muscle, usually due to coronary artery disease, such as arteriosclerosis.

Angiosperm (an'jē ə spėrm)**:** A plant in which the seeds are enclosed in an ovary; a flowering plant.

Animal pole: That end of an animal zygote that has relatively little yolk and experiences more rapid cell division.

Annelid (an'ə lid)**:** Any segmented worm of the phylum *Annelida,* including earthworms, leeches, and various marine forms.

Annual plant: A plant that completes its life cycle in a single season.

Antagonist: One of a pair of skeletal muscles (or groups of muscles) the actions of which oppose one another.

Anterior lobe: The lobe of the pituitary gland that lies toward the front of the brain.

Anther: In a flower, the pollen-producing organ of the stamen.

Antheridium (an'thə rid'ē əm)**:** In plants, a male reproductive organ that produces and stores motile sperm.

Anthozoa (an'thə zō'ə)**:** The class of marine coelenterates that contains the corals sea anemones, sea pens, and so on.

Antibiotic: Any of a large number of substances, produced by various microorganisms and fungi, capable of inhibiting or killing bacteria and usually not harmful to higher organisms (e.g., *penicillin, streptomycin,* etc.).

Antibody: A protein molecule of the immune system that can recognize and bind to a foreign substance or invader (such as a bacterium or virus). See also *immunoglobulin.*

Anticodon (an'ti kō'don)**:** A region of a *t*RNA molecule consisting of three sequential nucleotides that will have a matching codon in *m*RNA.

Antidiuretic hormone (ADH) (an'tē dī'yōō ret'ik hormone)**:** A polypeptide hormone secreted by the posterior pituitary, the action of which is to increase the reabsorption of fluid from the kidney filtrate. Also called *vasopressin.*

Antigen recognition region: The separated part of a Y-shaped antibody that matches specific antigens.

Anus (ā'nəs)**:** The posterior opening of the digestive tract (gut) through which digestive wastes are eliminated.

Aorta (ā ôr'tə)**:** In vertebrates, the principal or largest artery. It carries oxygenated blood from the heart to the body.

Apgar test series: A series of tests given infants at birth to help discover any congenital problems.

Apical meristem (ā'pə kəl mer'ə stem)**:** The undifferentiated tissue at the stem and root tip that contributes cells for primary growth.

Apodeme (ap'ə dēm)**:** An ingrowth of the arthropod exoskeleton that serves as the point of attachment for a muscle.

Appendicular skeleton (a'pen dik'yə lər skeleton)**:** In vertebrates, bones of the pectoral and pelvic girdles and of the appendages. Compare *axial skeleton.*

Appetitive behavior (ap'i tī'tiv behavior)**:** A variable, nonstereotyped part of instinctive behavior involving searching (for food, water, a mate) for the opportunity to perform.

Arachnida (ə rak'ni də)**:** The class of arthropods that contains spiders, scorpions, mites, ticks, and others.

Archaebacteria (är'kə bak tir'ē ə)**:** One of the two prokaryote kingdoms (or phyla in some schemes); it contains primitive bacteria such as methanogens (methane producers), thermophiles (heat lovers), and acidophiles (acid lovers). See also *Eubacteria.*

Archegonium (är kə gō'nē əm)**:** The female (egg-producing) reproductive structure in the gametophytes of ferns and bryophytes.

Archenteron (är ken'tə ron')**:** In embryology, the primitive digestive cavity of the gastrula.

Arctic tundra: A biome characterized by level or gently undulating treeless plains of the far north, supporting dense growth of mosses and lichens, as well as dwarf herbs and shrubs. It is underlain by permafrost and seasonally covered by snow.

Arteriole (är tir'ē ōl)**:** A small artery, usually giving rise directly to capillaries.

Artery: A vessel carrying blood away from the heart, toward a capillary bed.

Arthropoda (är throp'ə də): The phylum containing segmented invertebrates with jointed legs, including insects, arachnids, crustaceans, and myriapods.

Artificial selection: The deliberate selection for breeding by humans of domesticated animals or plants on the basis of desired characteristics.

Artiodactyla (är'tē ō dak'tə lə): "Even toes"; the order comprised of hoofed, even-toed mammals such as pigs, hippopotamuses, camels, deer, giraffes, sheep, goats, antelope, and cattle.

Ascending colon: The first region of the large intestine.

Ascending neuron: A nerve cell that carries impulses up the spinal cord toward the brain.

Ascomycete (as'kə mī sēt'): A fungus of the class that includes the yeasts, mildews, truffles, and so on, characterized by bearing the sexual spores in a sac, the ascus.

Ascomycota (as'kə mī cot'ə): See *Ascomycete.*

Ascus (as'kəs): In ascomycetes, the sac in which meiosis occurs and in which four or eight ascospores are subsequently formed.

Association neuron: A nerve cell that connects an afferent neuron to an efferent neuron or to an ascending neuron.

Asteroidea (as'tə roi'dēə): The class of echinoderms that contains starfishes.

Atherosclerosis (ath'ər ō sklə rō'sis): A form of arteriosclerosis characterized by fatty deposits (plaques) in the inner lining of the arterial walls.

Atom: The smallest indivisible unit of an element still retaining the element's characteristics.

Atomic number: The number assigned to a particular element, determined by the number of protons in its atoms.

ATP (Adenosine triphosphate): A ubiquitous small molecule involved in many biological energy exchange reactions, consisting of the nitrogenous base adenine, the sugar ribose, and three phosphate residues.

Atrium (ā'trē əm): The smaller compartment of the heart that receives venus blood from the body or lungs.

Audition: The act, sense, or power of hearing.

Auditory canal: The open, bony canal from the outer ear to the eardrum.

Autosome (ô'tə sōm): Any chromosome other than a sex (X or Y) chromosome.

Autotroph (ô'tə trōf): Self-feeder, able to manufacture food such as through photosynthesis.

Auxin (ôk'sin): A class of natural or artificial substances that acts as the principle growth hormone in plants.

Aves (ā'vēz): The class containing the birds.

Axial skeleton: In vertebrates, the skull, vertebral column, and bones of the chest. Compare *appendicular skeleton.*

Axon: The extension of a neuron that conducts nerve impulses away from the cell body.

B

Bacillus (bə sil'əs): 1. An aerobic, rod-shaped, spore-producing bacterium of the genus *Bacillus.* 2. Any rod-shaped bacterium.

Backbone: The spinal or vertebral column.

Background radiation: Naturally occurring radiation that is present almost everywhere.

Ball-and-socket joint: A joint allowing maximal rotation and flexion, consisting of a ball-like termination on one part, held within a concave, spherical socket on the other (e.g., the hip joint).

Barbiturate (bär bich'ə rāt): Any of a group of barbituric acid derivatives used in medicine as sedatives and hypnotics; "downers."

Barr body: A dark-staining feature in the nuclei of the cells of female mammals, representing the condensed X chromosome.

Base: Any chemical that releases hydroxyl (OH-) ions in water or joins with free protons. Bases leave a bitter taste and form salts when they react with acids.

Basidiomycete (bə sid'ē ō mī sēt'): A fungus of the class that includes the smuts, rust. mushrooms, puffballs, and so on, characterized by bearing the spores on a basidium.

Basidiomycota (bə sid'ē ō mī cot'ə): See *Basidiomycete.*

Basidiospore (bə sid'ē ō spōr): An aerial spore produced by meiosis in Basidiomycetes.

Basidium (bə sid'ē əm): The meiotic cell of Basidiomycetes such as mushrooms; it produces basidiospores by budding.

Basilar membrane (bas'ə lər membrane): In the vertebrate ear, a membrane that conducts sound waves.

Beta linkage (bā'tə linkage): The linkage between the glucose molecules of cellulose.

Biceps: 1. A muscle on the front of the arm that acts to bend the elbow. 2. The hamstring muscle on the back of the thigh

that assists in bending the knee and extending the hip joint.

Biceps brachii (biceps brā'kē ī): The large muscle lying along the front (ventral) side of the humerus.

Bile: A bitter-tasting, highly pigmented, alkaline liquid secreted by the liver, containing bile salts and bile pigments, that functions in fat digestion.

Bimodal curve (bī mōd'əl curve): A "two-humped" graphic distribution.

Binary fission (bī'nə rē fission): A form of asexual reproduction; fission (splitting) into two organisms of approximately the same size; cell division in prokaryotes.

Binucleate cell (bī nyoo'klē āt' cell): A cell having two nuclei.

Bioethics (bī'ō eth'iks): The philosophical area that deals with the moral implications of decisions that affect life overall.

Biological determination: The influence of genetics, particularly on the social behavior of an organism.

Biomass (bī'ō mas'): The total weight of all the organisms in a prescribed area.

Biome (bī'ōm): A geographical area that is characterized by relatively similar plant and animal representatives; e.g., a tundra.

Biotic (bī ot'ik): Pertaining to life.

Biotic population control: Population control by living factors, including both intraspecific and interspecific influences.

Biotic potential: The maximum growth rate of a population when it is unrestricted by environmental resistance.

Birth control pill: An oral steroid contraceptive that inhibits ovulation, fertilization, or implantation causing temporary infertility in women.

Blastocoel (blas'tō sēl'): The cavity of a blastula, arising in the course of cleavage.

Blastopore (blas'tə pōr): In a gastrula, the opening of the archenteron produced by the involution of cells during gastrulation.

Blastula (blas'chə lə): An early embryonic stage in mammals consisting of a single layer of cells that forms a hollow ball enclosing a central cavity, the blastocoel.

Bombykol (bom'bi kol): The sex attractant produced by the female silkworm moth.

Bowman's capsule: A curved sac at the beginning of the kidney nephric unit that surrounds the glomerulus.

Brackish: Slightly salty.

Bronchiole (brän'kē ōl'): A small airway in the lung that is a branch of a bronchus and part of the respiratory tree.

Bronchus (brän′kəs) (pl. *bronchi*): Either of the two main branches of the trachea.

Bryophyta (brī′ō fīt′ə): The plant group that includes the mosses, liverworts, and hornworts.

Budding: Asexual reproduction seen in yeasts and other organisms where smaller cells bud or grow from a parent cell.

Bulk flow: The movement of water or another liquid brought about by pressure or gravity, generally from areas of greater water potential to areas of lesser water potential.

C

Caffeine: A white, crystalline, bitter alkaloid, usually derived from coffee or tea, which acts as a stimulant and diuretic.

Caffeinism: A condition found in people who consume very large amounts of coffee, characterized by insomnia, high blood pressure, increased body temperature, racing heart, and chills.

Calcitonin (kal si′tō nin): One of the two major hormones of the thyroid gland whose major action is to inhibit the release of calcium from the bone.

Caltrops (kal′trəps): A configuration in which four "spikes" extend from a common center so that each is the greatest possible distance from the others.

Calvin cycle: A pathway of carbon dioxide that occurs during the dark phase of photosynthesis as carbohydrate is produced.

Calyx (kā′liks) (pl. *calyces*): The outermost whorl of floral parts (the sepals), usually green and leaflike.

Cambium (kam′bē əm) (pl. *cambia*): Undifferentiated meristematic tissue in a plant, including *cork cambium, procambium,* and *vascular cambium.*

Canaliculus (kan′ə lik′yə ləs) (pl. *canaliculi*): A small canal or tubular passage, as in bone.

Capillary (kap′ə ler′ē): A small blood vessel (with walls one cell thick) in which exchanges between the blood and tissues occur. Capillaries are located between arterioles and venules.

Carbohydrate (kär′bō hī′crāt): An organic substance containing carbon, hydrogen, and oxygen in the ratio of CH_2O. Includes sugars, starches, cellulose, and so on.

Carbon monoxide: A colorless, odorless, poisonous gas, CO, that burns with a pale blue flame; produced when carbon burns in insufficient air.

Carbonic anhydrase (kär bon′ik an hī′drās): An enzyme that catalyses the reversible conversion of carbonic acid to carbon dioxide gas and water.

Carboxyl group (kär bok′sil group): The univalent group, =COOH, present in and characteristic of organic acids.

Cardiac muscle: Specialized involuntary muscle of the heart whose fibers are striated and branching.

Carnivora (kär niv′ərə): "Meat eaters"; the order comprising chiefly flesh-eating mammals, such as dogs, cats, bears, seals, weasels, and so on.

Carnivore (kär′nə vōr′): Any flesh-eating organism.

Carotene (kar′ə tēn): A red or orange hydrocarbon, found in most plants as an accessory photosynthetic pigment.

Carotenoid (kə rot′ə noid): Any of a group of red, yellow, and orange plant pigments chemically and functionally similar to carotene.

Carpal (kär′pəl): Any wrist bone.

Carpel (kär′pəl): In flowers, a simple pistil, or a single member of a compound pistil; one sector or chamber of a compound fruit.

Carrying capacity: A property of the environment defined as the size of a population that can be maintained indefinitely.

Catalase (kat′ə lās′): An enzyme that decomposes hydrogen peroxide to molecular water and oxygen, in the reaction $2H_2O_2 \rightarrow 2H_2O + O_2$.

Catalysis (kə tal′i sis): a chemical reaction initiated by catalysts.

Catalyst (kat′list): A substance, such as an enzyme, that increases the rate of a reaction without entering into the reaction.

Cecum (sē′kəm): In vertebrates, a blind pouch or diverticulum of the intestine at the juncture of the small and large intestines.

Cell body: The region of a neuron containing the cell nucleus and most of the cytoplasm and organelles.

Cell membrane: The semipermeable membrane that surrounds all cells and consists of a double layer of phospholipids, with proteins, glycoproteins, and glycolipids interspersed in a mosaic arrangement.

Cell theory: The universally accepted proposal that cells are the functional units of organization in living organisms and that all cells today come from preexisting cells.

Cellular respiration: The energy-yielding metabolism of foods in which oxygen is used. Also see *respiration* and *glycolysis.*

Cellulose (sel′yə lōs): An insoluble carbohydrate comprised of glucose units; forms cell walls in plants.

Cenozoic (sē′nə zō′ik): Noting or pertaining to the present geological era, beginning about 75,000,000 years ago and characterized by the appearance of mammals.

Central nervous system: In vertebrates, the brain and spinal cord to which sensory impulses are transmitted and from which motor impulses are sent.

Centriole (sen′trē ōl′): A paired organelle near the nucleus of animal cells that functions in cell division.

Centromere (sen′trə mir): A structure on the chromosome to which the spindle fiber attaches when chromatids separate in meiosis and mitosis.

Cephalic ganglia (sə fal′ik gang′glē ə): Neural aggregations at the anterior ends of some invertebrates.

Cephalochordate (sef′ə lə kôr′dāt): "Head cord animal"; a lancelet (Branchiostoma) of a chordate subphylum in which the notochord persists throughout life and extends through what would be the head if it had one.

Cerebellum (ser′ə bel′əm): The double walnut-shaped portion of the hindbrain that coordinates voluntary movement, posture, and balance.

Cerebral cortex (ser′ə brəl kor′teks): The outermost region of the cerebrum, the "gray matter," consisting of several dense layers of neural cell bodies and including numerous conscious centers, as well as regions specializing in voluntary movement and sensory reception.

Cerebrum (sə rē′brəm): The anterior portion of the vertebrate brain; the largest portion in humans, consisting of two *cerebral hemispheres* and controlling many localized functions, among them voluntary movement, perception, speech, memory, and thought.

Cervical cap (sûr′vi kəl cap): A birth control device consisting of a small cap that encloses the cervix.

Cestoda (se stō′də): The class composed of parasitic flatworms, like the tapeworm.

CF_1 particle: An ultramicroscopic structure on the outer surface of the hylakoid, the site of the chemiosmotic phosphorylation in photosynthesis.

Chelicerata (chəli′ sə ra′tə): Arthropods with no jaws, no antennae, four pairs of legs, and book lungs, including the spiders.

Chemiosmosis (kem′ē oz mō′sis): The process in mitochondria, chloroplasts, and aerobic bacteria in which an electron transport system utilizes the energy of photosynthesis or oxidation to pump hydrogen ions across a membrane. resulting that can be utilized to produce ATP.

Chemiosmotic phosphorylation (kem′ē oz mot′ik fos′fər ə lā′shən): The production of ATP using the energy of protons passing across a membrane and through F_1 and CF_1 particles.

Chemoreception (kēm′ ō ri sep′shən): The ability to detect the presence of chemicals in the environment.

Chemosynthesis (kem′ə sin′the sis): The synthesis of organic compounds with energy derived from inorganic chemical reactions.

Chilopoda (kī′lə pod′ə): The class comprising the centipedes.

Chiroptera (kī rop′tərə): "Hand wings"; the order comprising the bats.

Chlorophyll (klôr′ə fil): The light-absorbing pigment of plants and some bacteria that traps light energy for photosynthesis.

Chlorophyll A: A type of chlorophyll common to all plants.

Chlorophyll B: A type of chlorophyll common in certain land plants and in green algae.

Chlorophyta (klôr′ə fī′tə): The green algae.

Chloroplast (klôr′ə plast): A chlorophyll-containing organelle that functions in photosynthesis in plant cells.

Choanocyte (kō an′ə sīt′): Also *collar cell*; a type of cell in all sponges and certain protists in which a single flagellum is surrounded at its base by a screen of fused cilia that filter food from the water current created by the flagellum.

Chondrichthyes (kon drik′thē ēz′): The class containing the cartilaginous fishes.

Chordate (kôr′dāt): Belonging or pertaining to the phylum *Chordata*, which contains the true vertebrates and those animals having a notochord, like the lancelets.

Chorioallantoic membrane (kôr′ē ō al′ən tō′ik membrane): A highly vascular extraembryonic membrane of birds, reptiles, and some mammals, formed by the fusion of the chorion and the allantois.

Chorion (kôr′ē on): The outermost extraembryonic membrane of birds, reptiles, and mammals. It contributes to the formation of the placenta in placental mammals.

Chorionic villi (kôr′ē on′ik vil′ī): Highly branched and pouched growths of the placenta, where the fetal and maternal blood are separated by a membrane's thickness.

Chromatid (krō′mə tid): One of two duplicated strands in a chromosome.

Chromatin (krō′mə tin): Indistinct, granular chromosomal material seen at interphase; DNA.

Chromosome (krō′mə sōm): Threadlike, condensed chromatin (DNA), visible at cell division only, occurring in pairs with specific numbers for each species.

Chromosome mutation: A massive spontaneous change in DNA, generally breakage involving a whole chromosome that has not been repaired or has been repaired improperly.

Chrysophyta (kris′ə fī′tə): The yellow-green and golden-brown algae ("golden plants"), named for their yellow carotenoid pigments although they also possess chlorophyll A and B. This group includes the diatoms.

Chyle (kīl): A milky fluid containing emulsified fat and other products of digestion, formed from the chyme in the small intestine and conveyed by the lacteals and the thoracic duct to the veins.

Chyme (kīm): The semifluid mass into which food is converted by gastric secretion and which passes from the stomach into the small intestine.

Ciliophora (si′li ə fôr′ə): The ciliated protozoans.

Cilium (sil′ ē əm) (pl. *cilia*): A hairlike cellular organelle that beats rhythmically on the cell surface.

Citric acid: A 6-carbon compound formed in the Krebs cycle; the nonionized form of citrate.

Citric acid cycle: A cyclic series chemical transformation in the mitochondria by which pyruvate is degraded to carbon dioxide, NAD and FAD are reduced to $NADH_2$ and $FADH_2$, and ATP is generated. See also *Krebs cycle.*

Class: A major taxonomic subdivision of a phylum, consisting of subordinate groups known as orders.

Classical conditioning: Learning in which a condition stimulus presented together with an unconditional stimulus initiates a reaction to the condition stimulus.

Classical genetics: The study of genetics using the techniques essentially suggested by Mendel.

Clavicle (klav′ə kəl): One of a pair of bones of the pectoral girdle, articulating with the sternum and scapula. Also called *collar bone.*

Climax: An orgasm.

Cline (klīn): Small variations in genetic characters across the range of a population.

Clitellem (kli tel′əm): A thickened, ring-like glandular portion of the body wall of earthworms that secretes mucus to form a cocoon for eggs.

Clitoris (klit′ər əs): An external organ of the female genitalia considered to be homologous to the penis in the male.

Cloaca (klō ā′kə): The common cavity into which the intestinal, urinary, and reproductive canals open in all vertebrates except placental and marsupial mammals.

Clone: A genetically identical organism derived from a single individual by asexual reproduction.

Closed circulatory system: A circulatory system in which the blood elements remain in blood vessels and do not leave to percolate through tissue spaces. Compare *open circulatory system.*

Clotting: The coagulation of blood.

CoA: See *coenzyme A.*

Coacervate (kō as′ər vit): An organized kind of droplet that may have been the forerunner of primitive life.

Cocaine: A bitter, crystalline alkaloid, obtained from coca leaves and used medically as a local anesthetic.

Coccus (kok′əs): Any spherical bacterium (principally eubacteria).

Cochlea (kok′lē ə): In mammals, a spiral cavity of the inner ear containing fluid, vibrating membranes, and sound-sensitive neural receptors.

Codon (kō′don): A sequence of three nucleotide bases required to form one amino acid in a protein chain synthesized on the RNA messenger.

Coelacanth (sē′lə kanth): A fish thought to have been extinct since the Cretaceous period, but found in 1938 off the southern coast of Africa.

Coelenterate (si lent′ə rāt′): An animal characterized by only one digestive opening, such as the jelly fish.

Coelom (sē′ləm): A principal body cavity, or one of several such cavities, between the body wall and gut, entirely lined with mesodermal epithelium. Compare *pseudocoelom.*

Coenzyme A: A biocatalyst required by certain enzymes to produce their reactions.

Coitus interruptus: Coitus that is intentionally interrupted by withdrawal before ejaculation.

Collagen (kol′ə jən)**:** A common, tough, fibrous animal protein occurring principally in connective tissues.

Collar cell: One of certain flagellated cells in sponges that create a current and ingest food particles from the water.

Collenchyma (kə leng′kəmə)**:** In plants, a strengthening tissue; a modified parenchyma consisting of elongated cells with greatly thickened cellulose walls.

Colloid (kol′oid)**:** A solution in which the molecules are large and suspended rather than dissolved.

Colon: The large intestine from the cecum to the rectum; including the ascending, transverse, descending, and sigmoid regions, and the rectum.

Community: In ecology, an assemblage of interacting populations forming an identifiable group within a biome; e.g., a sage desert community or a beech-maple deciduous forest community.

Companion cell: In plants, a nucleated cell adjacent to a sieve tube member and believed to assist in its functions.

Comparative psychologist: One who studies behavior in animals from a comparative point of view, usually in a laboratory environment.

Competition: In ecology, the utilization by two or more individuals or species of the same limiting resource.

Complement: A group of blood proteins that interact with antibody-antigen complexes to destroy foreign cells.

Compound: A chemical substance composed of one kind of molecule.

Concentration gradient: 1. A slow, consistent decrease in the concentration of a substance along a line in space. 2. For any spatial difference in concentration, the direction away from the region of greater concentration.

Condensation: Of chromosomes, the coiling and supercoiling that transforms diffuse chromatin into compact, discrete bodies in mitosis.

Condom: A thin sheath of rubber or animal membrane worn over the penis during sexual intercourse to prevent conception or venereal infection. Also called *rubber* or *prophylactic*.

Cone: 1. A male or female reproductive structure of conifers, consisting of a cluster of scalelike modified leaves and either pollen or ovules or the seeds. 2. One of a class of cone-shaped photoreceptors in the retina that detect color, consisting of a highly modified cilium with specialized pigments for detecting red, green, or blue wavelengths.

Conidiophore (kō nid′ē ə fōr)**:** A specialized branch of the mycelium-bearing conidia.

Conidium (kō nid′ē əm)**:** An asexual spore borne on the tip of a fungal hypha.

Conifer (kō′nə fər)**:** An evergreen gymnospore of the order *Coniferales*, bearing ovules and pollen in cones; included are spruce, fir, pine, cedar, and juniper.

Conjugation: Sexual reproduction in which organisms (usually single-celled) fuse to exchange genetic material; in ciliates, a temporary cytoplasmic fusion in pairs, accompanied by meiosis and the exchange of haploid nuclei.

Consummatory behavior: A part of instinctive behavior that involves fixed, highly stereotyped (invariable) behavior that brings relief (for example, swallowing food).

Continental drift: The slow movement of the continents relative to one another on the earth's surface. See also *plate tectonics*.

Continental shelf: The part of a continent that is submerged in relatively shallow sea.

Contraception: Any process or method intended to prevent the sperm from reaching and fertilizing the egg or preventing ovulation or implantation. Also called *birth control*.

Contractile vacuole (kən trak′təl vak′yoo ōl)**:** An organelle of many freshwater protists that maintains the cell's osmotic equilibrium by bailing excess water through an active, ATP-powered process. Also called a *water vacuole*.

Control: A standard of comparison in scientific experiment; a replicate of the experiment in which a possibly crucial factor being studied is omitted.

Convergent evolution: A similarity in genetically different organisms resulting from their adaption to similar habitats.

Copulation (kop′yə lā′shən)**:** Sexual intercourse or coitus.

Coracoid process (kôr′ə koid process)**:** In man and other higher mammals, a reduced bony process of the scapula having no connection with the sternum.

Cork cambium (cork kam′bē əm)**:** The perpetually growing tissue of a plant or stem that forms the bark.

Corolla (kə rol′ə)**:** All the petals of a flower.

Corpus callosum (kôr′pəs kə lō′səm)**:** A connecting strip of nerve fibers that coordinates the activities of the cerebral hemispheres.

Corpus luteum (kôr′pəs lu′tē əm)**:** "Yellow body"; a temporary endocrine body that develops in the ovarian follicle after ovulation. It secretes estrogen and progesterone, which maintain the endometrium. It continues secretion if pregnancy occurs, and regresses quickly if it does not occur.

Cortex: 1. In plants, the portion of the stem between the epidermis and the vascular tissue. 2. In animals, the outer layer or rind of an organ, such as the *adrenal cortex*.

Cortisone (kôr′ti sōn)**:** A steroid hormone of the adrenal cortex active in carbohydrate and protein metabolism.

Cosmic ray: A radiation of extremely high penetrating power that originates in outer space and consists of high-energy atomic nuclei.

Cotyledon (kot′l ē′dən)**:** A food-storing structure in dicot seeds, sometimes emerging as first leaves; a food-digesting organ in most monocot seeds; first leaves in a gymnosperm embryo. Also called *seed leaf*.

Covalent bond (kō vā′lənt bond)**:** A bond formed between two atoms as they share electrons.

Cranial nerve: In humans, one of the twelve pairs of major nerves that emerge directly from the brain and pass through skull openings to the body's periphery.

Cranium (krā′nē əm)**:** 1. The skull. 2. The part of the skull enclosing the brain.

Crash: Population crash; a sudden die-off, particularly after the carrying capacity of the environment has been exceeded and resources have been depleted.

Cretinism (krēt′ə niz′əm)**:** A recessive genetic abnormality that results in the inability to produce thyroxine. Affected persons are extremely retarded physically and mentally unless thyroxine is administered from early infancy.

Crinoidea (krin oi′dē ə)**:** The class including the echinoderms with cup-shaped bodies to which are attached branched, radiating arms, such as the sea lilies and feather stars.

Crista (kris′tə)**:** A fold of the inner mitochondrial membrane, containing numerous electron transport systems and F_1 particles.

Crop: A food storage sac located below the esophagus in birds, earthworms, and other animals.

Cross bridges: Extensions of the myosin fibers of muscles that attach to actin fibers and pull the fibers past each other in a ratchet fashion as the muscle contracts.

Crossing over: The exchange of chromatid (DNA) segments by enzymatic breakage and reunion during meiotic prophase.

Crude birth rate: The number of live births per 1000 individuals in the population at midyear.

Crude death rate: The number of deaths per 1000 individuals in the population at midyear.

Crustacea (kru stā'shē ə): The class composed of aquatic arthropods typically having the body covered with a hard shell or crust, such as the lobsters, shrimps, crabs, barnacles, and so on.

Cuticle: A tough, often waterproof, non-living covering, usually secreted by epidermal cells.

Cyclic AMP (cAMP): The "second messenger," adenosine monophosphate. It is synthesized in response to certain hormones arriving at the cell membrane, stimulating further activity in the target cell.

Cytokinesis (sī'tō ki ne'səs): The division of cell cytoplasm following mitosis or meiosis.

Cytokinin (sī'tō kin in): A plant cell hormone, mitogen, and plant tissue culture growth factor that interacts with other plant hormones in the control of cell differentiation.

Cytology (sī tol'ə jē): The scientific study of cells.

Cytopyge (sī'to pij): An excretory structure of paramecia.

Cytosine (sī'to sēn): A pyramidine, one of the four nucleotide bases of DNA and RNA.

Cytostome (sī'to stōm): The mouth of a protozoan.

Cytotoxic killer cell (sī'to tocks'ik killer cell): A type of white blood cell that attacks specific antigens after receiving instructions from a macrophage.

D

D and C: See *dilation and curettage.*

Daughter cell: Either of the two cells created when one cell divides.

Deductive method: A logical process in which specific conclusions are drawn from generalities. Compare *inductive method.*

Dehydration: Chemically, an enzymatic reaction during which water is lost and a covalent bond forms between the reactants.

Deletion: In genetics, the loss of any segment of a chromosome or gene.

Delirium tremens (DTs) (di lēr'ē əm trē'mənz): A violent restlessness due to excessive indulgence in alcoholic beverages, characterized by trembling, terrifying visual hallucinations, and so on.

Demographer (di mog'rəfer): One involved in the science of statistics of populations.

Denaturation: The alteration of a protein so as to destroy its properties, through heating or chemical treatment.

Dendrite: A short, branched extension of a neuron that receives impulses and conducts them toward the cell body.

Density-dependent effects: Factors affecting populations the severity of which is dependent on the densities of the populations. Often an effect of competition.

Density-independent effects: Those factors affecting population size that are independent of population density, e.g., temperature, salinity, and meteorites.

Deoxyribonucleic acid (DNA) (de ok'sə rī'bō nü kle'ik acid): A nucleic acid containing deoxyribose, which is found chiefly in the nucleus of cells. DNA is the genetic message functioning in protein synthesis.

Dependence: The state of needing someone or something for aid, support, comfort, or the like.

Depolarization (dē pō'lər ə zā'shən): In neural transmission, a loss of polarization that is characteristic of the resting state; a shift in the neuron from a polarized condition (postively charged exterior and negatively charged interior) to a nonpolarized condition as a neural impulse is generated. See also *action potential.*

Depression: A condition of general emotional dejection, sadness, and withdrawal.

Depressor: A muscle that draws down some part of the body.

Descending colon: A part of the large intestine that descends to the rectum.

Desert: A region characterized by scanty rainfall, especially less than 25cm (10 in) annually.

Deuterostome (dü'tə rə stōm): A bilateral animal (for example, an echinoderm or chordate) whose anus arises from the first embryonic opening (blastopore) and whose mouth arises later as a second embryonic opening. Compare *protostome.*

Developed country: A country that is industrialized, with high per capita income and high literacy. Compare *developing country.*

Developing country: A nation that is characterized by low industrialization, low per capita income, and low literacy. Compare *developed country.*

Diaphragm (dī'ə fram): 1. A birth control device; a thin rubber cup with a springlike rim that covers the cervix. 2. The large, flattened muscle between the thoracic and abdominal cavities that aids in breathing.

Diatomaceous earth (dī'ə tə mā'shəs earth): Geological deposits consisting largely of the cell walls of diatoms; used in filtration and as an abrasive.

Dicot (dī'kot'): See *dicotyledon.*

Dicotyledon (dī kot'l ēd'n): A flowering plant (angiosperm) of the class *Dicotyledonae,* characterized by seeds with two cotyledons. Also called *dicot.*

Diencephalon (dī'en sef 'ə lon'): The posterior section of the forebrain.

Differentiation: In development, the poorly-understood process by which a cell or cell line becomes structurally or physiologically specialized.

Diffusion: The movement of particles from an area of greater concentration to lesser as a result of random movement.

Digestion: The hydrolic cleavage (adding of water), through enzyme action, of complete food molecules into the molecular subunits, permitting absorption to occur.

Digit: In terrestrial vertebrates, any of the terminal divisions of the limbs; a finger, thumb, or toe.

Dihybrid (dī hī'brid): The result of a genetic cross in which two different traits are involved.

Dilator (dī lā'tər): A muscle that dilates some cavity of the body.

Dilation and curettage (D and C) (dī lā'shən and kyoor'i täzh): A surgical operation in which the cervix is forcefully dilated and the uterine mucosa scraped with a curette.

Dimorphic (dī môr'fik): 1. Having two forms distinct in structure, coloration, and so on among animals of the same species. 2. Having two different forms of flowers, leaves, and so on on the same plant or on distinct plants of the same species.

Dinoflagellate (din'ə flaj'ə lāt): A flagellated, photosynthetic, marine protist of the group *Dinoflagellata.*

Diploid (dip'loid): Having a double set of genes and chromosomes, one set originating from each parent. Compare *haploid, polyploid.*

Diplopoda dip'lə pod'ə): The class composed of millipedes.

Directional selection: In evolution, that selection that favors one or the other of two alleles in a population so that one or the other accumulates.

Disaccharide (dī sak'ə rīd'): Any of a group of carbohydrates, as sucrose or lactose, that hydrolyze into simpler sugar or monosaccharides.

Disruptive selection: Selection that favors more than one trait, such as the extremes of a condition, resulting in the median condition being less common than the extremes.

Divergent evolution: Evolutionary change away from the ancestral type, with selection favoring newly arising conditions.

DNA: See *deoxyribonucleic acid.*

Dominance, principle of: The principle of certain genes being expressed over their allelic alternative.

Dominant trait: A phenotypic characteristic that is always expressed when a certain allele is present.

Dorsal nerve root: Of the two large nerves extending from the spinal column, the one nearer the back surface, carrying afferent impulses.

Double bond: A covalent bond in which two pairs of electrons are shared between two atoms.

Doubling time: The number of years required for a population to double in size.

DTs: See *delirium tremens.*

Duodenum (doo'ə dē'nəm): The first segment of the small intestine posterior to the stomach.

Duplication: Genetically, a doubling in the number of certain genes, resulting in a mutation.

E

Early anaphase (early an'ə fāz): The mitotic stage when sister chromatids have separated and just begun to move to opposite poles.

Echinoderm (i kī'nə dûrm): Any organism of the marine, coelomate, deuterostome phylum *Echinodermata;* i.e., starfishes, sea urchins, sea cucumbers, and so on.

Echinoidea (ek'ə noi'dē ə): The class composed of the sea urchins, sand dollars, and so on.

Ecologist (i kol'ə jist): A scientist who studies the relations between organisms and their environment.

Ecology (i kol'ə jē): The branch of biology dealing with the relations between organisms and their environment.

Ecosystem (ē'kō, sis'təm): In ecology, a unit of interaction among organisms and between organisms and their physical environment.

Ectoderm (ek'tə dûrm'): The outer cell layer in an animal embryo that gives rise to skin, nervous tissue, and so on.

Ectoplasm (ek'tə plaz'əm): The outer, clear, thin layer of the cytoplasm of a cell.

Ectotherm (ek'tə therm): Lacking the ability to thermoregulate physiologically. These animals are often called "cold-blooded," but their body temperature is usually that of the surroundings. Compare *endotherm.*

Effector: Any structure that elicits a response to neural stimulation (for example, a muscle or gland).

Efferent neuron: A neuron conducting impulses from the brain or spinal cord to an effector, such as a *motor neuron.*

Egg nucleus: The haploid plant cell that, at fertilization by the sperm nucleus, results in a zygote.

Eidetiker (ī det'i kər): A person with a "photographic" memory.

Ejaculation: The forceful expulsion of semen from the penis.

Electrical energy: A form of energy due to the flow of electrons.

Electromagnetic spectrum: The range of electromagnetic radiation, from low-energy, low-frequency radio waves to high-energy, high-frequency gamma rays.

Electron: A minute, negatively charged particle of an atom, in motion about the nucleus.

Electron acceptor: A substance that receives electrons through reduction.

Electron shell: A theoretical band around an atomic nucleus where electrons of a certain energy level are found.

Electron-transport chain: A series of reactions in which electrons are passed from one compound to the next in an energy-producing process.

Element: A substance consisting of atoms of one kind only, such as iron, gold, sulfur, and so on. There are 102 known elements.

Embryo: The developing stage of any organism.

Embryo sac: In flowering plants, the mature megagamtophyte after division into six haploid cells and one binucleate cell, enclosed in a common cell wall.

Endergonic: Any chemical reaction characterized by the absorption of energy from an outside source.

Endocrine gland: A discrete gland that secretes hormones directly into the blood. Also called a *ductless gland.*

Endocytosis (en' dō sī tō'sis): The process of taking food or solutes into the cell (into vacuoles) by engulfment (a form of active transport). See also *phagocytosis,* compare *exocytosis.*

Endoderm: The innermost cell layers of an embryo that differentiate into such organs as the digestive tract, liver, and pancreas.

Endometrium (en'dō mē'trē əm): In mammals, the tissue lining the cavity of the uterus.

Endoplasm: The granular inner portion of the cytoplasm of a cell, containing numerous crystals and mitochondria.

Endoplasmic reticulum (ER) endoplasmic re tik'yə ləm): The layers of membrane folded through the cytoplasm of a cell, forming complex inner surfaces, after being covered with ribosomes.

Endoskeleton: A mesodermally derived supporting skeleton inside the organism, surrounded by living tissue, as in vertebrates and echinoderms.

Endosperm: A nutritive tissue of seeds, formed around the embryo in the embryo sac.

Endosperm nuclei: A binucleate cell in the plant embryo sac which, when fertilized, produces the triploid endosperm.

Endospore: A resistant, thick-walled spore formed from a bacterial cell. See also *spore.*

Endotherm: An organism with the ability to metabolically thermoregulate; also called "warm-blooded." See also *ectotherm.*

Energy of activation: The energy boost required to initiate a chemical reaction.

Entropy (en'trə pē): The opposite of "free energy." The degree of randomness in a system; a measure of disorder, which means that as the available energy in a system decreases, the system becomes increasingly disorganized.

Enucleate (i nyoo'klē āt'): Having no nucleus.

Environmental resistance: The sum of environmental factors (e.g., limited resources, drought, disease) that restrict the growth of a population below its biotic potential (maximum possible population size).

Enzyme: An organic catalyst, usually a protein, capable of changing the rate of chemical reaction in cells.

Epidermis: 1. In plants, the outer protective cell layer in leaves and in the primary root and stem. 2. In animals, the outer epithelial layer of the skin.

Epinephrine (ep'ə nef'rən): A hormone with numerous effects produced by the adrenal medulla and by nerve synapses of the autonomic nervous system. Also called *adrenaline*.

Epithelium (ep'ə thē'lē əm): A tissue that consists of tightly adjoining cells that cover a surface or line a canal or cavity and that serves to enclose and protect.

Erythrocyte (i rith'rō sīt): A hemoglobin-filled, oxygen-carrying, circulating red blood cell.

Esophagus: The muscular tube connecting the pharynx to the stomach.

Estivate (es'tə vāt): To pass the summer in a state of dormancy or torpor induced by heat or dryness.

Estrogen: A female hormone; one of the hormones involved in the production of secondary sex characteristics and the menstrual cycle.

Estrus: The mating period of female mammals when they become sexually receptive.

Estuary (es'choo er'ē): The part of the mouth of a river in which the river's current meets the sea.

Ethology (e thol'ə jē): The scientific study of animal behavior, usually under natural conditions.

Eubacteria: "True bacteria"; The better-known prokaryote kingdom, containing many familiar human pathogens and important soil and water bacteria. See also *archaebacteria*.

Eukaryotic cell (yoo'kar'ē ō tik cell): A cell having among its traits a true nucleus and inner membrane structures.

Eutheria (yoo thēr'ē ə): The group composed of placental mammals.

Eutrophication (yoo trof'ə kā'shən): The aging process whereby a body of water supports increasing numbers of organisms, often causing lakes to become marshes and then terrestrial communities.

Evolution: The continuous changes occurring in populations, primarily as a result of adapting to the environmental changes.

Exergonic: Any chemical reaction that releases energy.

Exocytosis (ek'sō sī tō'sis): The process of expelling material from vacuoles through the cell membrane. Compare *endocytosis*.

Exoskeleton: An external skeleton or supportive covering, as in arthropods.

Extensor: A muscle that serves to extend or straighten a part of the body.

External: Pertaining to the outer part, or outside, of the body.

External ear: All parts of the ear external to the eardrum, comprising the external ear canal, the external auditory meatus, and the pinna.

External respiration: The process by which gases reach the exchange surfaces, as in breathing.

Extinction: Behaviorally, the loss of a conditioned response as a result of the absence of reinforcement.

F

F₁ particle: An ultramicroscopic mitochondrial organelle attached to the inner surface of the crista; the site of chemiosmotic phosphorylation.

Facilitated diffusion: The diffusion of molecules across a cell membrane assisted by a reversible association with carrier molecules; it differs from active transport in that no energy is expended and net movement follows the concentration gradient.

Fallopian plug (fə lō'pē ən plug): A birth control process in which silicone rubber plugs are inserted into the Fallopian tubes to prevent the eggs from reaching the uterus.

Family: In taxonomic classification, a grouping smaller than *order* but larger than *genus*.

Fascia (fāsh'ē ə): A heavy sheet of connective tissue covering or binding together muscles or other internal structures of the body, often connecting with ligaments or tendons.

Fatty acids: Constituents of fats and oils; long chains of carbon atoms with hydrogen attached and ending with carboxyl groups.

Feces (fē'sēz): Bodily waste discharged through the anus.

Feedback: The return of part of the output of a system back to the input, which regulates or affects the system's functioning. In *negative feedback,* the amount of output is sensed by the system and acts in an inhibiting manner, slowing activity and thus reducing output. In *positive feedback,* the output stimulates activity, further increasing the flow of output.

Fermentation: Anaerobic respiration with the usual end product being alcohol and carbon dioxide.

Fibroblast (fī'brə blast'): A cell in the connective tissue group that produces fibers and matrix substances such as collagen.

Fibrous connective tissue (fī'brəs connective tissue): A vertebrate connective tissue formed from cablelike fibers running in different directions.

Filament (fil'ə mənt): In flowers, the slender stalk of the stamen on which the anther is situated.

Filicinae (fil i'sin ē): The class of plants that contains the ferns.

First law of thermodynamics: The physical law that states that energy cannot be created or destroyed; later amended to allow for the interconversion of matter and energy.

Fissure of Rolando (fish'ər of rō lan'dō): The groove separating the frontal and parietal lobes of the cerebrum.

Fissure of Sylvius (fish'ər of sil'vē əs): The groove separating the frontal, temporal, and parietal lobes of the cerebrum.

Flagellum (flə jel'əm): A threadlike organelle that extends from the cell surface; it may beat with whiplike motions.

Flexing: The process by which the angle between two joined bones is increased, such as by bending the arm.

Fluid-mosaic model (fluid mō zā'ik model): A description of the plasma membrane as a phospholipid bilayer stabilized by specifically oriented proteins, with some proteins extending through to both surfaces and other proteins specific for the inner or outer surfaces.

Follicle stimulating hormone (fol'ə kəl stimulating hormone): A hormone produced by the anterior pituitary that stimulates the growth and maturation of eggs and sperm; also called *FSH.*

Food vacuole (food vak'yoo ōl): In animals and protists, an intracellular vacuole that arises by the phagocytosis of solid food materials and in which digestive processes occur.

Foot: In the moss sporophyte, the anchoring base.

Forebrain: 1. The anterior of the three primary divisions of the vertebrate embryonic brain. 2. The parts of the adult brain developed from the embryonic forebrain.

Foreplay: Sexual stimulation of another person, intended as a prelude to sexual intercourse.

Founder principle: The chance assortment of genes carried out of the original population by colonizers (or founders) who subsequently give rise to a large population.

Free energy: The energy that is available to do work.

Free radical: A highly reactive atom or molecule unit with an unpaired electron.

Frond: An often large, finely divided leaf, as seen in ferns and certain palms.

Frontal lobe: An anterior division of the cerebral hemisphere, believed to be a site of higher cognition.

Fucozanthin (fū kō zan'thin): A brown carotenoid pigment characteristic of brown algae.

Functional group: A group of atoms in an organic molecule that are free to participate in chemical reactions.

Fungus: An organism of the kingdom *Fungi*, including yeasts, mushrooms molds, mildews, part of the lichen symbiosis, rusts, smuts, sac fungi, puffballs, water molds, and sometimes the slime molds.

G

Gametangium (gam'i tan'jē əm): In algae and fungi, the structure (cell or organ) in which gametes are formed.

Gamete (gam'ēt): A male or female sex cell with half the chromosomal material. Fertilization is the union of gametes, producing the full amount of hereditary material.

Gametophyte (gə mē'tō fīt): In plants with alteration of generations, the haploid form in which gametes are produced. Compare *sporophyte*.

Gametophytic (gam'i tə fit'ik): Referring to the sexual generation of a plant in which the cells have half the normal chromosome numbers.

Gap junction: A dense structure that physically connects membranes of adjacent cells along with channels for cell-to-cell transport.

Gastric cavity: The digestive area, such as the saclike gut of coelenterates.

Gastrodermis (gas'trō dûr'mis): The inner cell layer of the body of a coelenterate.

Gastrovascular cavity: The cavity of coelenterates, ctenophores, and flatworms, which has only one opening to the outside and functions as a digestive cavity and crude circulatory system. Also called *coelenteron*.

Gastrula: An early metazoan embryo consisting of a hollow, two-layered cup with an inner cavity (archenteron) opening out through a blastopore.

Gastrulation (gas'troo lā'shən): The formation of the gastrula stage of an embryo. This is the stage in which the embryo acquires its three germ layers.

Gene: The portion or portions of DNA that produce a recognizable effect or trait; e.g., genes produce enzymes, color pigments, and so on.

Gene frequency: The fractional distribution of alleles for a specific trait in a population.

Gene linkage: The condition of genes being physically linked on the same chromosome.

Gene mutation: A mutation involving alterations in the genes, occurring during the duplication of DNA.

General fertility rate: The average number of live births per 1000 females in their reproductive years (in the U.S., women 15–44).

Genetic code: The biochemical specification (DNA) for producing a particular genetic trait of the entire species' characteristics.

Genetic drift: Changes in the composition of a gene pool caused by random factors rather than by selection.

Genetic engineering: Modern techniques of gene management, including gene cloning, gene splicing, amino acid, DNA and RNA nucleotide sequencing, and gene synthesis.

Genotype (jen'ə tīp): The assortment of genes that make up the genetic characteristics of an organism.

Genus (jēnəs): A major subdivision of a family, consisting of one or more species.

Geothermal energy: Energy obtained from the internal heat of the earth.

Germ layer: Any of the three layers of undifferentiated embryonic cells formed at gastrulation. See *ectoderm, endoderm,* and *mesoderm*.

Germinal spot: A very early stage of development of the bird embryo.

Gibberellin (jib'ə rel'lən): Any of a family of plant growth hormones that control cell elongation, bud development, differentiation, and other growth effects.

Gill: 1. In fungi, the thin, flattened structures on the underside of a mushroom (basidiocarp) that bear the spore-forming basidia. 2. In aquatic organisms, the surface over which external respiration occurs.

Gizzard: A muscular sac in the digestive system that mechanically changes food by mashing it against sand particles; found in birds, earthworms, and so on.

Glial cell (glē'əl cell): One of the numerous nonconducting supporting cells of the central nervous system that may play some role in information storage. Also called *neuroglia*.

Glomerulus (glə mer'ə ləs) (pl. glomeruli): A cluster of capillaries within the Bowman's capsule of the kidney.

Glycerol (glis'ə rôl): A three carbon alcohol molecule that combines with fatty acids to form fats and oils.

Glycogen (glī'kə jən): A complex chain of glucose units produced by animals.

Glycolysis (glī kol'ə sis): The first phase of respiration where glucose is broken down with a small energy yield.

Golgi body (gōl'jē body): A secretory structure in some cells consisting of a complex system of folded membrane.

Gonad (gō'nad): An ovary or testis.

Granum (grā'nəm): A green granule within chloroplasts, appearing as stacks of thylakoids; contains chlorophyll and carotenoids; the site of the light reaction.

Grassland: One of the natural biomes of the earth, characterized by perennial grasses, limited seasonal rainfall, and a great number of herbivorous mammals, birds, and insects.

Guanine (gwä'nēn'): One of the nitrogenous bases of RNA and DNA.

Gullet: The esophagus or food-receiving tube in an animal.

Gymnosperm (jim'nə spèrm): A nonflowering seed plant of the group *Gymnospermae*, producing seeds that lack fruit; included are conifers, cycads, gnetophytes, and ginkgos.

H

Habitat: The specific place where an organism lives; the native environment of a plant or animal.

Habituation (hə bich'ōo ā shən): A simple form of learning in which an organism's response to a stimulus diminishes.

Half life: In radioisotopes, the time it takes for half of the atoms in a sample to undergo spontaneous decay.

Haploid (hap'loid): Having a single set of genes and chromosomes. Compare *diploid* and *polyploid*.

Hashish (hash'ēsh): The flowering tops, leaves, and so on of Indian hemp, smoked, chewed, or drunk as a narcotic and intoxicant. Also called *hash*.

Haversian canal (hə vûr'shən canal): In bone, a small canal through which a blood vessel runs.

Heart: A hollow, muscular organ which by rhythmic contractions and relaxations keeps the blood in circulation throughout the body.

Heat energy: Energy produced by the random movement of molecules that increases with a rise in temperature.

Helper T-cell: A cell that activates elements of the immune system.

Heme (hēm): A deep-red, iron-containing pigment found in hemoglobin.

Hemichordate (hem'i kôr'dāt): An animal of the deuterostome phylum *Hemichordata*, such as an acorn worm.

Hemoglobin (hē'mə glō'bən): A protein; a respiratory pigment consisting of one or more polypeptide chains, each associated with a heme (iron-containing) group.

Hemophilia (hē'mə fil'ē ə): A sex-linked condition in which the blood does not clot normally. Also called *bleeder's disease*.

Hepatic portal vein (hi pat'ik portal vein): In all vertebrates, a large blood vessel that collects blood from the capillaries and venules of the esophagus, stomach, and intestine and carries it to the liver, where it once more divides into capillaries.

Herbivore (hür'bə vōr): Any organism that feeds on plants.

Hermaphrodite (hər maf'rə dīt): An individual animal or plant with both male and female reproductive parts.

Heterotroph (het'ər ə trōf): An organism that depends on an external source of organic substances for its food.

Heterozygous (het'ər ō zī'gəs): A mixed genotype; different alleles for a particular trait.

Hierarchy (hī'ə rär'kē): Any system of persons, animals, or things in which one is ranked above another; "pecking order."

High selection pressure: Strong selection against departure from the optimum of a phenotype.

Hindbrain: 1. The most posterior of three primary divisions of the embryonic vertebrate brain. 2. The parts of the adult brain derived from the embryonic hindbrain, including the cerebellum, pons, and medulla oblongata.

Hinge joint: A joint that moves in one plane, like a door hinge.

Hirudinea (hir'oo din'ē ə): The class composed of leeches.

Histamine (his'tə mēn): An amine compound released in allergic reactions that dilates blood vessels and reduces blood pressure, stimulates gastric secretions, and causes contraction of the uterus.

Holdfast: A rhizoid base of a seaweed, serving to anchor it to the ocean floor.

Holothuroidea (hol'ə thoo roi'dē ə): The class consisting of sea cucumbers.

Homeostasis (hō'mē ō stā'sis): The tendency of organisms to maintain a stable internal environment.

Homologue (hom'ə log): Either of the two members of each pair of chromosomes in a diploid cell.

Homozygous (hō mō zī'gəs): Having identical alleles for a particular trait.

Hormone (hôr'mōn): A chemical messenger transmitted in body fluids or sap from one part of the organism to another, producing a specific effect on target cells and regulating physiology, growth, differentiation, or behavior.

Host: A living organism that supports or sustains a parasite.

Hybrid: Offspring resulting from the cross of parents with different characteristics or parents of different species.

Hybrid swarm: Genetically diversified populations produced by introgressive hybridization; that is, the breeding of hybrids with parent stock, thus creating numerous partial hybrids.

Hydra: Any fresh-water polyp of the genus *Hydra*.

Hydrocarbon: Any organic compound that contains simply carbon and hydrogen.

Hydroelectric energy: Electric energy derived from falling water or any other hydraulic source.

Hydrogen bond: A weak attractive force between two molecules where one contains hydrogen.

Hydrolysis (hī drol'ə sis): The breakdown of a compound by the addition of water molecules.

Hydrophilic (hī'drə fil'ik): Water-loving, or having an attraction for water.

Hydrophobic (hī'drə fō'bik): Having little or no affinity for water; water-fearing, or repelling of water.

Hydroxyl group (hī drok'sil group): =OH, consisting of an oxygen and hydrogen covalently bonded to the remainer of the molecule; a constituent of alcohols, sugar, glycols, phenols, and other compounds.

Hydrozoa (hī'drə zō'ə): The class comprising the coelenterates such as solitary or colonial polyps and free-swimming medusae.

Hypha (hī'fə): One of the individual filaments that make up a fungal mycelium.

Hypothalamus (hī'pə thal'ə məs): The region that lies in front of the thalamus in the forebrain. It regulates the internal environment, pituitary secretions, and some of the basic drives.

Hypothesis (hī poth'ə sis): A proposition set forth as an explanation for a specified group of phenomena, either asserted merely as a provisional conjecture to guide investigation (e.g., working hypothesis) or accepted as highly probable in the light of established facts. See also *theory*.

I

Idiot savant (id'ē ət sa vänt'): A mentally defective person with a highly developed special talent, such as an ability to play music, to solve complex mathematical problems mentally at great speed, and so on.

Ileum (il'ē əm): A region of the small intestine.

Ilium (il'ē əm): The hip bone; one of the three pairs of fused bones forming the pelvis.

Immune system: In vertebrates, widely-dispersed tissues that respond to the presence of the antigens of invading microorganisms or foreign chemical substances.

Immunoglobulin (im yoo'nō glob'yoo lən): A protein antibody produced by T-cells in response to specific foreign substances; it consists of four subunits that are joined through disulfide linkages and have specific antigen binding sites.

Impulse: Neural impulse; a wave of excitement (transitory membrane depolarization) transmitted along a neuron and between neurons.

Incus: The middle one of a chain of three small bones in the middle ear of humans and other mammals. Also called *anvil*.

Inducer: In embryology, a substance that stimulates the differentiation of cells or the development of a particular structure.

Induction: The influence on cell differentiation or development that one embryonic tissue has on another.

Inductive method: A logical process in which a generalizing conclusion is proposed that contains more information than the observations or experience on which it is based. The truth of the conclusion is verifiable only in terms of future experience.

Inert (in ürt'): Chemically, showing a lack of bonding activity.

Inferior vena cava (inferior vē'nə kā'və): A large vein discharging blood from all parts of the body below the diaphragm into the right atrium of the heart.

Inhibitory neuron (in hib'i tōr'ē neuron): A nerve cell that functions by inhibiting the firing of the next nerve cell.

Inhibitory synapse (in hib'i tōr'ē sin'aps): The space between an inhibitory nerve cell and the next nerve cell.

Inhibitory transmitter substance (in hib′i tōr′ē transmitter substance): The neurotransmitter of an inhibitory neuron.

Innate behavior (i nāt′ behavior): Behavior that is inborn, not due to learning.

Innate releasing mechanism (IRM) (i nāt′ releasing mechanism): An ethological term that refers to a neural mechanism that produces a specific behavioral event when triggered by a particular stimulus from the environment.

Inner cell mass: In a mammalian blastocyst, the portion that is destined to become the embryo proper.

Insecta (in sek′tə): The class composed of insects.

Insectivora (in sek′tə vōr′ə): "Insect eaters"; the order composed of insect-eating animals such as moles and shrews.

Insertion: In anatomy, the distal attachment of a tendon or muscle (the attachment on the part to be moved). Compare *origin*.

Insoluble (in sol′yoo bəl): Incapable of being dissolved.

Instinct: Any inherent, unlearned behavioral pattern that is functional the first time it happens and can occur in animals reared in total isolation.

Intercourse: Sexual relations or a sexual coupling, especially coitus.

Interference phenomenon: A cellular substance that interferes with viral replication; interferon.

Interferon (in′tər fēr′on): See interference phenomenon.

Internal: Pertaining to the inside or inner part of something.

Internal respiration: The energy-producing chemical reactions within cells that utilize oxygen.

Interphase: The period between two mitotic events in the same cell.

Interspecific cooperation: Cooperation between members of different species.

Intraspecific cooperation: Cooperation between members of the same species.

Intrauterine device (IUD) (in′trə yoo′tər in device): A plastic device, sometimes containing copper, that is inserted into the uterus as a means of preventing conception.

Inversion: The transposition of a portion of a chromosome following breakage and repair, altering the relative order of gene loci.

Involution: An inward roll or curve; in gastrulation, the movement of cells toward the blastopore, over the blastopore lip, and into the archenteron.

Ion (ī′on): An atom or molecule that has gained or lost one or more electrons.

Ionic bond (ī on′ik bond): An attractive force between atoms that have different charges as a result of the transfer of electrons between them.

Ionizing radiation (ī′ə nīz′ing radiation): Energetic radiation that produces ions.

IRM: See *innate releasing mechanism.*

Isotope (ī′sə tōp): An atom with a lighter or heavier nucleus than the average; may exhibit radioactivity.

IUD: See *intrauterine device.*

J

Jejunum (ji jōō′nəm): The area of the small intestine that lies between the duodenum and the ileum.

Joint: A point where two bones articulate, often movable.

J-shaped curve: A plot of population growth where growth is rapid, approaching the environment's carrying capacity. Compare *S-shaped curve.*

Jungle: A wild land overgrown with dense vegetation, parts of tropical rain forests penetrated by sunlight.

K

Kinetic energy (ki net′ik energy): The energy of motion.

Kinetin (kin ē′tin): A plant hormone, one of the cytokinins.

Kingdom: Any of the five categories (monera, protista, fungi, plants, animals) in which organisms are usually placed.

Krebs cycle: The oxidative phase of respiration in aerobic reactions where fuel is broken down to CO_2 and H_2O. Also called *citric acid cycle.*

Kwashiorkor (kwash′ē ôr′kôr): A syndrome of severe protein deficiency in human infants and children, including failure to grow, deficiency of melanin pigment, edema, degeneration of the liver, anemia, and retardation.

L

Labia minora (lā′bē ə mī′nə rə): The smaller inner folds of skin that border the vagina.

Labor: 1. The efforts of childbirth. 2. The period during which these pains and efforts occur.

Lacteal (lak′tē′əl): A lymphatic vessel of the intestinal villi.

Lagomorpha (lag′ə môr′fə): "Rabbit shape"; the mammalian order including hares, rabbits, and pikas.

Lamella (lə mel′ə): A layer, or flattened sheet.

Lancelet: Any small, fishlike, chordate animal of the subphylum Cephalochordata, having a notochord in the slender, elongated body pointed at each end.

Large intestine: Also called *colon* or *bowel*; a division of the alimentary canal, primarily functioning in the resorption of water.

Late prophase: The mitotic stage when chromosomal material is highly condensed and ready to enter metaphase.

Lateral meristem (lat′ər əl mer′ə stem): Meristem located along the sides of a part, as a stem or root.

Latimeria (lat′ə mer′ē ə): A coelacanth; a primitive, lobe-finned fish once thought to be extinct.

Leaf: In vascular plants, a lateral outgrowth from the stem functioning primarily in photosynthesis, arising in regular succession from the apical meristem, consisting typically of a flattened blade jointed to the stem by a petiole.

Leucoplast (lü′kə plast′): A colorless body in plant cells that stores proteins, lipids, and starch.

Leukocyte (lü′kə sīt): A vertebrate white blood cell; it aids in resisting infections.

Levator: A muscle that raises a part of the body.

Ligament: A tough, flexible, but inelastic band of connective tissue that connects bones or supports an organ in place. Compare *tendon.*

Light-dependent reaction: That part of photosynthesis directly dependent on the capture of photons; specifically the photolysis of water, the thylakoid electron transport system, and the chemiosmotic synthesis of ATP and NADPH.

Light-independent reaction: That part of photosynthesis not immediately involved in chemiosmosis, specifically the fixation of CO_2 into carbohydrate from the NADPH and ATP produced by the light reaction (Calvin cycle).

Light microscope: The commonly used microscope that illuminates the viewed object with ordinary visible light.

Lignin: An amorphous substance that helps give wood its rigidity.

Lipid: A type of fatty organic compound.

Loop of Henle: The prominent U-shaped loop in the renal tubule of the mammalian kidney.

Low selection pressure: Weak selection against departure from the optimum of a phenotype.

Lumen: The cavity or channel of a hollow tubular organ or organelle.

Lung: In land vertebrates, one of a pair of compound, saclike organs that function in the exchange of gases between the atmosphere and the bloodstream; or any of several analogous organs in invertebrates.

Luteinizing hormone (LH) (loo′tē ə nīz′ing hormone): A pituitary hormone that causes ovulation and stimulates hormone production in the corpus luteum.

Lymph: The clear yellowish intercellular fluid or plasma in the lymphatic system.

Lymph capillary (limf kap′ə ler′ē): One of the many tiny, blind endings in the lymphatic system, including the lacteals of the small intestine.

Lymph duct: See *thoracic duct.*

Lymph node: A clump of cells in the lymphatic system that produces lymphocytes and acts as a filter for lymph.

Lymphatic system (lim fat′ik system): The system of lymphatic vessels and ducts and lymph nodes that serves to redistribute excess tissue fluids and to combat infections.

Lymphocyte: Any of several varieties of similar-looking leukocytes involved in the production of antibodies and in other aspects of the immune response; they are formed in the lymph nodes.

Lysosome (lī′sə sōm): A membrane-bounded organelle that forms or isolates disruptive enzymes.

M

Macronucleus (mak′rō nü′klē əs): The larger of two types of nuclei found in certain protozoans.

Macrophage (mak′rō fāj): A large, phagocyte; one of the leukocytes.

Malleus (mal′ē əs): The outermost of a chain of three small bones in the middle ear of humans and other mammals; also called *hammer.*

Malnourished: Not receiving the proper kinds of food for growth and development. Compare *undernourished.*

Mammalia (mə mā′lē ə): The class of vertebrates that feeds its young with milk from the female mammary glands, that has the body more or less covered with hair, and that, with the exception of the monotremes, brings forth living young rather than eggs.

Mandibulata (man di′byoo la′ta): The jawed arthropods.

Mantel: The shell-secreting organ in mollusks.

Marasmus (mə raz′məs): Gradual loss of flesh and strength, occurring chiefly in infants, caused by a diet low in both calories and protein.

Marijuana (mar′ə wä′nə): Dried leaves and flowers of the Indian hemp, used in cigarettes as a narcotic.

Marrow: *Bone marrow,* the soft tissue in the interior cavity of a bone; *blood marrow,* vascularized bone marrow in which white blood cells of all types are produced; *fatty marrow,* bone marrow consisting of adipose tissue.

Marsupial (mär sü′pē əl): A mammal of the subclass *Metatheria;* usually the female has a pouch (marsupium); included are the kangaroo, wombat, koala, Tasmanian devil, opossum, and wallaby.

Mastigophora (mas′tə gof′ə rə): The phylum consisting of the flagellates.

Matrix (mā′triks): In the mitochondrion, the enzyme-laden region within the highly convoluted inner membrane; the site of the citric acid or Krebs cycle.

Mechanical energy: The energy produced by one physical object's exerting pressure on another.

Mechanism: The theory that the processes of life are based on the same physical and chemical laws that apply to nonliving phenomena.

Medulla (mi dul′ə): 1. The inner portion of a gland or organ. Compare with *cortex.* 2. *Medulla oblongata:* a part of the brain stem developed from the posterior portion of the hindbrain and tapering into the spinal cord.

Medusa (mə dü′sə): The motile, free-swimming jellyfish form of coelenterate. Compare *polyp.*

Megagametophyte (meg′ə gə mēt′ə fīt): The female gametophyte produced from a megaspore in flowering plants; the *embryo sac* consisting of eight haploid nuclei or cells.

Megaspore mother cell: In the ovule, a large diploid cell that will give rise to the megaspore by meiosis and the degeneration of three of the four haploid nuclei.

Meiosis (mī ō′sis): Two successive nuclear divisions in which the chromosomes are shuffled and reorganized and the numbers of chromosomes per cell are halved.

Meissner's corpuscle (mīz′nərz corpuscle): In mammals, a small touch-responsive neural end organ.

Membrane: A thin, pliable sheet or layer of animal or vegetable tissue, serving to line an organ, connect parts, and so on.

Menstrual cycle: The cycle of hormonal and physiological events and changes involving growth of the uterine mucosa, ovulation, and the subsequent breakdown and discharge of the uterine mucosa in menstruation (menses). The cycle averages 28 days. Also called *ovarian cycle.*

Menstruation (men′stroo ā′shən): In nonpregnant females of the human species only, the periodic discharge of blood, secretions, and tissue debris resulting from the normal, temporary breakdown of the uterine mucosa in the absence of implantation following ovulation. Also called *menses.*

Meristem (mer′i stem′): Embryonic plant tissue; undifferentiated, growing, actively-dividing cells. See *apical meristem.*

Merostomata (mer′ō stō ma′ta): The class of arthropods that includes the horseshoe crabs.

Mesenchyme (mes′eng kīm): An aggregation of cells of mesodermal origin that are capable of developing into connective tissues, blood, and lymphatic and blood vessels.

Mesenteric artery (mes′ən ter′ik artery): Artery between the digestive tract and the liver.

Mesenteric vein (mes′ən ter′ik vein): The vein between the digestive tract and the liver.

Mesoderm (mes′ə dèrm): The middle layer of the three primary germ layers of the gastrula, giving rise in development to the skeletal, muscular, vascular, renal, and connective tissues, and to the inner layer of skin and the epithelium of the coelom (peritoneum).

Mesoglea (mes′ə glē′ə): The loose, gelatinous middle layer of the bodies of sponges and coelenterates, between the outer ectoderm and the inner endoderm.

Mesozoic (mez′ə zō′ik): Noting or pertaining to a geological era occurring between 75,000,000 and 220,000,000 years ago, characterized by the appearance of flowering plants and by the appearance and extinction of dinosaurs.

Messenger RNA (*m*RNA): RNA that carries the genetic code to the ribosomes where it is translated into protein production.

Metacarpal (met′ə kär′pəl): One of the usually five bones of the hand or forefoot, between the carpals and the digits.

Metaphase (met′ə fāz): The second phase in mitosis during which the chromosomes line up at the center of the cell along the equator of the spindle.

Metatarsal (met′ə tär′səl): One of the usually five bones of the foot or hind foot, between tarsals and toes, forming the instep in human and part of the rear leg in ungulates and birds.

Metatheria (met′ə thēr′ē ə): The group consisting of marsupial mammals.

Metazoa (met′ə zō′ə): All animals other than *Parazoa* or phylum *Porifera* (sponges).

Methaqualone (meth′ə kwä lōn): An animal tranquilizer that has often been misused by humans; also called *Quaalude*.

Microfilament: A submicroscopic filament in the cytoskeleton, involved in cell movement and shape.

Microhabitat: The immediate environment of a species.

Micronucleus: The smaller of two types of nuclei in certain protozoa.

Micropyle (mī′krə pīl′): In seed plants, a minute opening in the integument of an ovule through which the pollen tube enters.

Microspore mother cells: In the anther of a flowering plant, the diploid cells that will undergo meiosis to form the haploid microspores.

Microtrabecular lattice (mī′krō trə bek′yə lər lattice): In the cell cytoplasm, a weblike system of microtubules and microfilaments that forms a cytoskeletal framework upon which many organelles are suspended.

Microtubule: A cytoplasmic hollow tubule composed of spherical molecules of tubulin, found in the cytoskeleton, the spindle, centioles, basal bodies, cilia, and flagella.

Microvilli (mī′krō vil′ī)(sing. microvillus): Tiny, fingerlike out-pocketings of the cell membrane of various epithelial secretory or absorbing cells, such as those of kidney tubule epithelium and the intestinal epithelium.

Midbrain: 1. The middle of the three primary divisions of the vertebrate embryonic brain. 2. The parts of the adult brain derived from the embryonic midbrain.

Middle ear: The middle portion of the ear, consisting of the tympanic membrane and an air-filled chamber lined with mu-cous membrane, which contains the malleus, incus, and stapes.

Midlobe: The part of the pituitary gland connecting the anterior and posterior lobes.

Mitochondrion (mī′tə kon′drē ən) (pl. mitochondria): An organelle in the cytoplasm that is sometimes called the powerhouse of the cell because it functions in energy-producing processes.

Mitosis (mī tō′sis): The process by which each cell replicates its chromosomes and then divides into two new cells.

Mnemonist (nē mon′ist): A person with a phenomenal memory.

Molecule: The smallest unit of a compound, composed of two or more atoms.

Mollusk: Any invertebrate of the phylum *Mollusca*, typically having a hard shell that wholly or partly encloses the soft, unsegmented body, such as the chitons, snails, bivalves, squids, and so on.

Moneran (mə nir′ən): Referring to the bacteria or the prokaryotes.

Monocot (mon′ə kot′): See *monocotyledon*.

Monocotyledon (mon′ə kot′l ēd′n): A flowering plant of the class *Monocotyledonae*, characterized by seeds with only one cotyledon (for example, grasses, palms, orchids, lilies). Also called *monocot*.

Monosaccharide (mon′ō sak′ə rīd): A simple sugar, such as glucose or fructose.

Monotreme (mon′ə trēm): A platypus, echidna, or extinct egg-laying mammal of the order *Monotremata*, subclass *Prototheria*.

Morula (môr′yü lə): An early embryonic stage in which the cells are arranged in a solid ball.

mRNA: See *messenger RNA*.

Mud flat: A mud-covered, gently sloping tract of land, alternately covered or left bare by tidal waters.

Muscle: A contractile tissue in vertebrates including skeletal, smooth, and cardiac muscle.

Muscle fiber: In skeletal muscle, one of the multinucleate cells that take the form of a long, contractile cylinder.

Muskeg: A bog of northern North America, commonly having spagnum mosses, sedge, and sometimes stunted black spruce and tamarack trees.

Mutation: Any permanent genetic change.

Mycelium (mī sē′lē əm): The mass of interwoven hyphae that forms the vegetative body of a fungus.

Myelin sheath: A fatty sheath surrounding the axons of some vertebrate neurons.

Myocardial infarction (mī′ə kär dē əl in färk′shən): The necrosis of the muscular substance of the heart caused by blood deprivation. Also called *coronary thrombosis*.

Myofibril (mī′ə fī′brəl): The slender protein thread in the skeletal muscle that acts in contraction of the muscle.

Myofilament: The highly organized fibrous proteins of striated muscle, including the thin, movable *actin myofilaments* and the thicker, stationary *myosin myofilaments*.

Myosin (mī′ə sən): A protein involved in cell movement and structure, especially in muscle cells. See also *myofilament*.

N

Natural increase: The net gain determined by the crude birth rate minus the crude death rate.

Natural killer cells: Lymphocytes that attack invading antigens.

Natural selection: The differential survival and reproduction of certain individuals in a population. In the same process, other organisms less suited to the environment are eliminated.

Navigation: The process of finding one's way.

Negative feedback: See *feedback*.

Nematocyst (nem′ə tə sist): One of the minute stinging cells of the coelenterates, consisting of a hollow thread coiled within a capsule and an external hair trigger.

Nematode (nəm′ə tōd): Any unsegmented worm of the phylum or class *Nematoda*, having an elongated, cylindrical body, such as the roundworm.

Nephric unit (nef′rik unit): A division of the kidney including the glomerulus, a Bowman's capsule, the loop of Henle, and the collecting duct.

Nephridium (ne frid′ē əm) (pl. nephridia): An excretory organ found in annelids.

Nerve: A number of neurons following a common pathway, covered by a protective sheath and supporting tissue.

Neural crest: The ridge of a neural fold, migrating cells of which will give rise to spinal ganglia, the adrenal medulla, the autonomic nervous system, and pigment cells.

Neural folds: In early vertebrate embryology, a pair of longitudinal ridges that arise from the neural plate on either side of the neural groove and that fold over and give rise to the neural tube, which eventually becomes the spinal cord.

Neural groove: An ingrowth of the neural plate of a vertebrate embryo; it eventually forms the spinal cord in mammals.

Neural plate: A thick plate formed by rapid cell division over the notochord on the dorsal side of the embryo; it eventually forms the nervous system in mammals.

Neuron: A cell specialized for the transmission of nerve impulses. It consists of one or more branched *dendrites,* a nerve *cell body* in which the nucleus resides, and a terminally branched *axon.* Also called a *nerve cell.*

Neurophysiology (nyoor′ə fiz′ē ol′ə jē): The branch of physiology dealing with the nervous system.

Neurotransmitter: A short-lived, hormonelike chemical (such as acetylcholine) that, when released from an axonal knob into a synaptic cleft, crosses the space to stimulate the next neuron to transmit a nerve impulse. See also *synapse* and *synaptic cleft.*

Neutron: An atomic nuclear particle that has about the same mass as a proton, but is electrically neutral.

Niche: All aspects of the biological and physical environment that relate to the activities of an organism.

Nicotine (nik′ə tēn′): A colorless, oily, water-soluble, highly toxic, liquid alkaloid obtained from tobacco.

Nitrogen oxide: An oxide of nitrogen; a common air pollutant in some industrial areas.

Node: 1. In a plant stem, any point at which one or more leaves emerge. 2. In a growing stem tip, a region of potential leaf growth, containing meristematic tissue.

Nonrenewable resources: Resources that exist in set amounts on the planet and that cannot be replaced. Once used, they are gone or so changed that they are difficult or impossible to return to their original state.

Norepinephrine (nôr′ep′ə nef′rən): A compound that serves as a synaptic neurotransmitter and as an adrenal hormone. Also called *noradrenaline.*

Notochord (nō′tə kôrd): A turgid, flexible rod running along the back beneath the nerve cord and serving as a body axis; it exists in all chordates at some point in development, but is replaced in most vertebrates by the vertebral column.

Nuclear pore: A tiny hole in the nuclear membrane of cells.

Nuclear winter: The hypothetical condition that would prevail after a nuclear exchange, when the sunlight would be largely blocked from reaching the earth, thus lowering global temperatures.

Nucleolus (nü klē′ə ləs): An organelle that lies within the nucleus of the cell; it is high in RNA and protein and forms ribosomal RNA.

Nucleotide (nü klē′ə tīd): A molecule made up of phosphate, a 5-carbon sugar, and a purine or pyrimidine base.

Nucelus: An organelle with a cell bounded by a double membrane that contains most of the DNA of the cell. Also, in the atom, a positively charged mass composed of neutrons and protons.

O

Occipital lobe (ok sip′i tal lobe): One of the four major lobes of the brain. It is involved in the reception and processing of visual information.

Olfactory bulb (ol fak′tər ē bulb): An extension of the brain that receives neurons from the olfactory receptors in the nasal passage.

Oligochaeta (ol′ə gō kē′tə): The class composed of annelids that have locomotory setae sunk directly into the body wall, such as earthworms.

Omnivore (om′nə vōr): An organism that eats both animals and plants.

Open circulatory system: A circulatory system in which the vessels open into intercellular spaces, through which the blood percolates before returning to the heart. Compare *closed circulatory system.*

Operant conditioning: A form of learning in which an animal performs an act to receive rewards.

Ophiuroidea (o fi′yoo roi′dē ə): The brittle sea stars.

Opiate: A drug or medicine containing opium, medically used as a sedative or pain killer.

Optic chiasma (optic kī az′ mə): The area where the optic nerves meet and part of each crosses to serve the area of the brain opposite the nerve.

Oral groove: In ciliates, a ciliated fold in the body wall, leading into the cytostome or mouth.

Orbit, electron: The approximate and hypothetical path on which an electron moves around the nucleus of an atom.

Order: A taxonomic subdivision of a class in plants and animals that may consist of several families.

Organ: A distinct structure that consists of a number of tissues and carries out a specific function.

Organ of Corti: On the basilar membrane in the cochlea, an organ containing the neural receptors for hearing.

Organ system: A number of organs participating jointly in carrying out a basic function of life (i.e., respiration, excretion, reproduction, digestion).

Organelle: A specific structure within a cell, usually bounded by a membrane.

Orgasm: In humans, the climax of sexual excitement typically occurring toward the end of coitus, usually accompanied in men by ejaculation and in women by rhythmic contractions of the cervix.

Orientation (ōr′ē en tā′shən): The ability to determine direction in one's environment.

Origin: The nonmoving, skeletal base to which a muscle or tendon is attached. Compare *insertion.*

Osmosis (os mō′sis): The tendency of water to move across a membrane from an area of its lower concentration to one that has a higher concentration.

Osteichthyes (os′tē ik′thē ēz′): The class composed of bony fishes.

Oval window: In the cochlea, a membrane articulating with the stapes that moves in response to its vibrations, subsequently creating movement in the fluid within.

Ovary: 1. In flowering plants, the enlarged, rounded base of a pistil, consisting of a carpel or several united carpels, in which ovules mature and megasporogenesis occurs. 2. The female gonad in which the eggs are formed.

Ovule (ō′vyool): In seed plants, an oval body in the ovary that contains the female gametophyte and consists of the embryo sac surrounded by maternal tissue.

Oxaloacetate (ok sal′ō as′i tāt′): One of the intermediate molecules in the Krebs cycle.

Oxaloacetic acid (ok sal′ō ə sē′tik acid): The nonionized form of oxaloacetate (an intermediate in the Krebs cycle).

Oxidation: The loss of electrons in chemical reactions, or the combination of oxygen with another substance.

P

Pacinian corpuscle (pa sin′ē ən corpuscle): An oval pressure receptor containing the ends of sensory nerves, especially in the skin of hands and feet.

Palaeotragus (pā′lē ō trā′gəs): A giraffelike creature that lived twenty million years ago and still survives in the Congolese rain forest.

Paleozoic (pā'lē a zō'ik): Noting or pertaining to a geological era occurring between 220,000,000 and 800,000,000 years ago, characterized by the appearance of fish, insects, and reptiles.

Paramecium (par'ə mē'shē əm): Any ciliated freshwater protozoan having an oval body and a long, deep oral groove.

Parapatric (par'ə pa'trik): Living in separate but adjacent geographic regions; compare *allopatric* and *sympatric*.

Parapod (par'ə pod): Also *parapodium*; one of the short, unsegmented, paired leg-like or finlike locomotive organs borne on either side of each body segment in *Nereis* and certain other polycheate worms.

Parasite: An organism that lives in or on another organism and causes some harm to its host as it derives food from it.

Parasympathetic nervous system: One of the divisions of the autonomic nervous system.

Parathyroid gland: One of four small endocrine glands embedded in or adjacent to the thyroid gland and involved in the regulation of calcium ion levels in the blood.

Parathyroid hormone: Also *parathormone*; the internal secretion of the parathyroid glands, involved in maintaining normal calcium balance.

Parenchyma (pə reng'kə mə): A basic plant tissue type, consisting typically of thin-walled cells and commonly specializing in photosynthesis and storage.

Parent cell: The arbitrarily chosen cell used to initiate studies of cellular descendancy.

Parietal lobe (pə rī'ə təl lobe): One of the four major lobes of the cortex. The detection of body position and sensory input are the major functions.

Partial dominance: Where neither of a pair of alleles is totally dominant and the combined expression of the two alleles in the heterozygote produces an intermediate trait; e.g., in the blossoms of four o'clocks, red × white = pink.

Particle: Ecologically, any small unidentified object; often an invasive irritant or contaminant; chemically, any of the components of an atom.

Passive transport: The movement of a substance through the cell membrane without cellular energy (ATP) involved.

PCP: See *phencylidine*.

Pectoral girdle (pek'tər əl girdle): The bones or cartilage supporting and articulating with the vertebrate forelimb.

Pellicle (pel'i kəl): 1. The semirigid, proteinaceous integument of many protists.

Pelvic girdle: The bones or cartilage supporting and articulating with the vertebrate hind limbs; in humans, consisting of the fused bones of the pelvis.

Penicillium (pen'i sil'ē əm): The mold that produces penicillin.

Penis: The male sex organ.

Peptide (pep'tīd): A compound composed of two or more amino acids.

Peptide bond (pep'tīd bond): The dehydration linkage formed between the carboxyl group of one amino acid and the amino acid group of another.

Peptide chain (pep'tīd chain): Two or more amino acids linked by peptide bonds, most often seen as a partial digestive product of a protein or polypeptide.

Peripheral nervous system: One of the two divisions of the vertebrate nervous system; includes the nervous tissue outside the central nervous system.

Periphery (pə rif'ə rē): The external boundary of any surface or area.

Perissodactyla (per'ə sō dak'tə lə): "Odd toes"; vertebrates with odd numbers of parts to their hooves, such as horses.

Peristalsis (per'ə stal'sis): Successive waves of involuntary contractions passing along the walls of the esophagus, intestine, or other hollow muscularized tube, forcing the contents onward.

Permafrost: In arctic and high altitude tundra, the permanently frozen layer of soil and/or subsoil.

Permease (per'mē ās): An enzymelike carrier that functions in facilitated transport of a specific substrate across a plasma membrane.

Peroxisome (pə rok'si sōm): A cytoplasmic organelle involved in the detoxification of peroxides.

Pesticide: A chemical preparation for destroying insects such as flies, mosquitoes, and ants.

Petal: One of the usually white or brightly-colored leaflike elements of the corolla of a flower.

PGAL: Phosphoglyceraldehyde; an end product of photosynthesis.

Phaeophyta (fā o fīt'ə): The brown algae.

Phagocyte (fag'ə sīt): Any leukocyte that engulfs particles.

Phagocytosis (fag'ə sī tō'sis): The engulfment of solid materials into the cell and the subsequent pinching off of the cell membrane to form a digestive vacuole.

Pharynx (far'ingks): The organ below the mouth that is the common respiratory and digestive structure leading to the esophagus and larynx.

Phencylidine (PCP) (fēn sī li dēn'): A powerful animal tranquilizer that has a variety of undesirable effects on humans, accompanied by or preceded by euphoria.

Phenotype (fē'nə tīp'): The observable genetic characteristics of any organism.

Pheromone (fer'ə mōn'): A substance that is secreted by an organism that changes the behavior of another organism of the same species.

Phloem (flō'em): A complex vascular tissue of higher plants that consists of sieve tubes, companion cells, and phloem fibers, and functions in transport of sugars and other nutrients.

Phosphoglyceraldehyde (PGAL) (fos'fō glis'ə ral'de hīd'): An end product of photosynthesis.

Phospholipid (fos'fō lip'id): A fatlike substance of the cell membrane that contains phophorus, fatty acids, glycerol, and a nitrogenous base.

Photochemical smog: Smog that becomes more reactive when energized by sunlight.

Photon (fō'ton): A unit of light energy.

Photoreceptor: A receptor of light stimuli.

Photosynthesis (fō'tō sin'thə sis): The process in which plants convert carbon dioxide and water into carbohydrates through the use of solar energy.

Photosystem I: The second in the two photosystems in the electron pathway of photosynthesis in cyanobacteria and chloroplasts, and the one involving the reduction of NADP to $NADPH_2$; believed to be evolutionarily more ancient than photosystem II.

Photosystem II: The first of the two photosystems in the electron pathway of photosynthesis in cyanobacteria and all photosynthetic eukaryotes, and the one involving the photolysis of water.

Photovoltaic (fō'tō vol tā'ik): Providing a source of electric current from light or similar radiation.

Phycocyanin (fī'kō sī'ə nin): A blue protein pigment found in algae that contributes to the process of photosynthesis.

Phycoerythrin (fī'kō i rith'rin): A reddish protein pigment found in algae that contributes to the process of photosynthesis.

Phycomycete (fī'kō mī'sēt): A fungus from the class that contains downy mildews and so on.

Phycomycota (fī'kō mī kot'ə): See *phycomycete*.

Phylum (fī'ləm): The major primary subdivision of a kingdom, composed of classes.

Physical dependence: The state that exists when a drug is necessary simply to maintain body comfort.

Phytoplankton (fī'tō plangk'tən): In the aquatic environment (marine and fresh waters), minute photosynthesizing organisms such as diatoms; the base of the marine food chain.

Pigment: Any coloring matter or substance.

Pinocytosis (pin'ō sī to'sis): Taking dissolved molecular food materials, such as proteins, into the cell by adhering them to the plasma membrane and invaginating portions of the plasma membrane to form digestive vacuoles; a form of active transport.

Pistil: In flowering plants, the female reproductive structure, composed of one or more carpels and ovaries, and a style and a stigma. See also *carpel*.

Pistillate (pis'tə lit): Having pistils but no stamens.

Pituitary gland: A tiny endocrine gland at the base of the brain.

Placenta (plə sen'tə): Embryonic membranes in the uterus of pregnant mammals through which the embryo transports wastes and receives essential nutrients.

Planaria (plə nâr'ē ə): The genus of free-swimming flatworms.

Planula (plan'yə lə): The early, ciliated, free-swimming larva of a coelenterate.

Plasma: The fluid matrix of blood. It is 90 percent water and 10 percent various other substances, including plasma proteins, ions, and foods.

Plasma membrane: The external semipermeable limiting layer of the cytoplasm.

Plasmid: A small circle of bacterial DNA that exists outside the single large circular chromosome.

Plasmodesmata (plaz'mo dez ma' tə): The tiny cytoplasmic bridges that extend through cell walls and connect the cytoplasm of adjacent living plant cells.

Plasmodium (plaz mō'dē əm): 1. A motile, multinucleate mass of protoplasm produced by the fusion of uninucleate slime mold ameboid cells. 2. The malarial parasite.

Plastid: A small body in a plant cell.

Plate: Any of the great land and ocean floor masses comprising the surface of the earth.

Plate tectonics: The movement of great land and ocean floor masses (plates) on the surface of the earth relative to one another, occurring largely in the Cenozoic era; also called *continental drift*.

Platelets: Minute, fragile noncellular discs present in the vertebrate blood. Upon injury, they are ruptured, releasing factors that initiate blood clotting and wound healing.

Platyhelminth: Any worm of the phylum *Platyhelminthes*, having bilateral symmetry and a soft, solid, usually flattened body, such as the planarians, tapeworms, and trematodes.

Point mutation: A mutation involving a minor change in a DNA sequence, such as a base substitution, addition, or deletion.

Polar body: A small, functionless egg formed during meiosis.

Pollen: The male microspores of flowering plants consisting of reproductive cells in a protective case.

Pollen grain: The male gametophyte of a seed plant; contains a generative nucleus and a tube nucleus and is enclosed in a hardened, resistant case.

Pollen sac: One of two or four chambers in an anther in which pollen develops and is held.

Pollen tube: A tube that extends from a germinating pollen grain and grows down through the style to the embryo sac, into which it releases sperm nuclei.

Pollen tube cell: The cell directly behind the pollen tube as it grows through the style of a flower before actual fertilization by one of the nuclei that follows it.

Pollutant: Anything that makes another thing unclean, impure, or contaminated.

Polychaeta (pol'ē kē'tə): The class composed of annelids having unsegmented swimming appendages with many chaetae or bristles.

Polymorphic: Having, assuming, or passing through different forms or stages.

Polyp: The typical attached, nonswimming form of coelenterate. Compare *medusa*.

Polypeptide: A group of two or more amino acids; a protein fragment.

Polyploid: Having more than two complete sets of chromosomes per cell.

Polyploidy: Multiple sets of chromosomes, often occurring in plants.

Polysaccharide (pol'ē sak'ə rīd): A large molecule of sugar subunits.

Pons: Adjacent to the medulla; it contains nerve cells that relate the spinal motor nerves.

Population: An interbreeding group of organisms. Usually a subset of a species.

Population dynamics: The study of how populations change.

Portal: In circulation, a vessel between two capillary beds.

Positive feedback: See *feedback*.

Posterior lobe: The lobe of the pituitary gland nearer the dorsal surface.

Post-Mendelian genetics: The study of genetics after the classical phase suggested by Mendel.

Potential energy: Energy stored in chemical bonds, in nonrandom organization, in elastic bodies, in elevated weight or any other static form in which it can theoretically be transformed into another form or into work.

Pre-Cambrian (prē kam'brē ən): Noting or pertaining to the earliest geological era, ending 600,000,000 years ago, during which the earth's crust was formed and the first life appeared.

Prefrontal area: The area situated at the anterior of the frontal lobe.

Primary growth: The initial growth or elongation of a plant stem or root, resulting in an increase in length and the addition of leaves, buds, and branches. Compare *secondary growth*.

Primary immune response: The relatively slow response of the immune system upon its first contact with an invading organism or foreign protein. Compare *secondary immune response*.

Primary phloem (primary flō'em): Phloem developed from apical meristem; that is, the phloem of primary growth.

Primary structure: The amino acid sequence of proteins.

Primary xylem (primary zī'lem): In primary growth, xylem produced by procambium rather than vascular cambium.

Primata (prī mā'tə): "First"; the order comprising the primates, such as humans, the apes, monkeys, lemurs, tarsiers, and marmosets.

Proboscidea (prō'bə sīd'ē ə): "Front feeder"; the order that includes elephants.

Procambium (prō kam'bē əm): In plants, the primary tissue that gives rise to primary xylem and primary phloem.

Progesterone (prō jes'tə rōn): An ovarian hormone, produced by the corpus luteum; assists in the preparation of the uterus for implantation of a fertilized egg.

Proglottid (prō glot'id): Any segment of a mature tapeworm, containing male and female organs and being shed when full of mature fertilized eggs.

Prokaryotic cell (prō kar'ē ō'tik cell): A cell lacking a membrane-bounded nucleus and membrane-bounded organelles; it has a single circular chromosome of nearly naked DNA.

Pronator (prō nā'tor): A muscle that turns a surface downward.

Prophase: The first stage of mitosis or meiosis.

Proprioception (prō'prē ə sep' shən): The ability to sense and respond to stimuli from within the body, including the integration of the movement of body parts, balance, and stance.

Proprioceptor (prō'prē ə sep'tər): A sensory receptor that responds to changes in body position, muscle tension, or internal chemistry.

Prostaglandin (pros'tə gland'ən): A kind of fatty acid hormone; may play a role in fertilization.

Protein: A specific chain of amino acids; essential in producing cell structures and a variety of enzymes.

Protista (prə tis'tə): The group of organisms that includes all the unicellular animals and plants.

Proton: An atomic nuclear particle that has a mass similar to a neutron, but with a positive charge.

Protostome (prō'tə stōm): An animal in which the mouth derives from the first embryonic opening (the blastopore). Compare *deuterostome.*

Prototheria (prō'tə thēr'ē ə): The group that includes the monotremes like the platypuses.

Protozoan (prō'tə zō'ən): 1. Any of a large group of protists. 2. Any nonphotosynthetic protist.

Proximate level: The influences on an individual as a result of individual experience.

Pseudocoelom (sü'də sē'ləm): In nematodes and rotifers, the body cavity between the body wall and the intestine that is not entirely lined with mesodermal epithelium; the gut is entirely endodermal and thus not muscularized.

Pseudopod (sü'd pod): Literally, a false foot; a temporary projection from the cell used in ameboid cells for feeding and movement.

Psychedelic (sī kə del'ik): A substance that may produce sensory distortions, bizarre thought patterns, and abnormal behavior.

Psychoactive agent: Any of a number of chemicals that alters mood or perception.

Psychological dependence: The state in which a drug is necessary for mental or emotional comfort.

Pulmonary artery (pul'mə ner'ē artery): A branching artery that conveys deoxygenated blood from the right ventricle to the lungs.

Pulmonary vein (pul'mə ner'ē vein): In birds and mammals, the only vein that carries oxygenated blood, returning it from the lungs to the left atrium.

Punctuated equilibrium: A theory stating that evolution does not proceed in a gradual manner but rather in sudden bursts of activity, followed by very long time intervals during which there is little evolutionary activity.

Punnett square (pun'et square): A graphic device used to predict genetic ratios.

Purine (pyür'ēn): Nitrogenous base of DNA and RNA, containing two nitrogen rings—adenine and guanine.

Pus: A thick, yellowish-white accumulation of dead phagocytes, dead or living bacteria, and tissue debris in a fluid exudate.

Pyrimidine (pir'ə mid'ēn): Nitrogenous base of DNA and RNA, containing one nitrogen ring—thymine, cytosine, and uracil.

Pyrogen (pī'rə jen'): Any fever-producing substance.

Pyrrophyta (pir'ə fi'tə): "Fire-colored plants"; microscopic, photosynthetic algae with two tinsellike flagella.

Pyruvic acid (pī rü'vik acid): A water-soluble liquid, important in many metabolic and fermentation processes.

Q

Quaternary structure: The interaction of two or more polypeptides through disulfide linkages.

R

Radioisotope (rā'dē ō ī'sə tōp): An unstable isotope that spontaneously breaks down with the release of ionizing radiation; also called *radioactive isotope.*

Radula (raj'u lə): In all mollusks except bivalves, a toothed, chitinous band that slides backward and forward, scraping and tearing food and bringing it into the mouth.

Reactant: Any element, ion, or molecule participating in a chemical reaction.

Receptacle: The end of a floral stalk that forms the base on which the flower parts are borne.

Receptor: A sensory structure that detects a particular form of stimulus in the environment, such as light or sound.

Recessive traits: Traits due to alleles that will be masked in the presence of dominant alleles.

Reciprocal altruism: Benefitting another individual at some cost in the expectation (and probability) that the favor will be returned.

Recombinant DNA (rē kom'bə nənt DNA): The general term for laboratory manipulation of DNA; includes gene splicing and gene cloning.

Rectum: The terminal part of the intestine, used for the temporary storage of feces.

Red marrow: The regions within the ribs, sternum, vertebrae, and hip bones where red blood cells are produced.

Reflex arc: A simple nervous pathway that involves a sensory cell, a connecting neuron, and a muscle cell.

Releaser: A stimulus that acts as a cue, releasing a certain behavior in an animal.

Renal artery (rē'nəl artery): The artery that feeds blood to the glomerulus.

Renewable resources: Resources that do not exist in set amounts; they can be replenished or reused.

Replica: A copy or reproduction.

Repolarization: In the neuron, reestablishment of the resting potential or polarized state following an action potential.

Reproductive isolation: The state of a population or species in which successful mating outside the group is impossible because of anatomical, geographic, or behavioral mating barriers.

Reptilia (rep til'ē ə): "Crawlers"; the class composed of cold-blooded vertebrates such as the turtles, lizards, snakes, crocodilians, and the tuatara.

Respiration: 1. The physical and chemical process by which an organism supplies oxygen to its tissues and removes carbon dioxide. 2. Also *aerobic respiration,* any energy-yielding reaction in living matter involving oxygen.

Resting potential: The charge difference across the membrane of a neuron or muscle fiber while it is not transmitting an impulse. Compare *action potential*.

Reticular system (ri tik′yə lər system): A major neural tract in the brain stem containing neural pathways to other parts of the brain and to the reticular activating system (RAS) and arousal center.

Retina (ret′ə nə): The layer of light sensitive cells in the vertebrate eye.

Reward: Reinforcement; something positive given to strengthen the probability of a response to a given stimulus.

Rhizoid (rī′zoid): 1. A portion of a fungal mycelium that penetrates its food medium. 2. A rootlike structure that serves to anchor the gametophyte of a fern or bryophyte to the soil.

Rhizome (rī′zōm): An underground, horizontal plant stem that is often thickened by deposits of reserve food material, produces shoots above and roots below, and is distinguished from a true root in possessing buds, nodes, and usually scalelike leaves.

Rhodophyta (rō′dō fī′tə): The red algae.

Rhynia (rin′ē ə): One of the oldest known fossil plants, flourishing about 400,000,000 years ago.

Rhythm method: A method of contraception whereby copulation is avoided during periods when conception is likely; also called *natural birth control*.

Ribonucleic acid (RNA) (rī′bō nü klē′ik acid): A nucleic acid containing ribose sugar and the nitrogen base uracil; functions in protein synthesis.

Ribosomal RNA (rī′bə sō′məl RNA): The RNA that forms the matrix of ribosome structure; its actions are not clearly known, but it is believed to function in protein synthesis.

Ribosome (rī bə sōm): A grainy structure found in great number in cells; important in protein synthesis.

Ribulose diphosphate (rī′byoo lōs dī fos′fāt): A form of ribulose that acts as the carbon dioxide acceptor in the Calvin cycle of photosynthesis.

RNA: See *ribonucleic acid*.

Rocky coast: A seacoast area characterized by abundant rock formations, pounding surf, and numerous life forms that occur in distinct zones.

Rod: One of the numerous long, rod-shaped sensory bodies in the vertebrate retina; it contains many membrane layers bearing visual pigments and is responsive to faint light. Compare *cone*.

Rodentia (rō den′shē ə): "Gnawers"; the order of gnawing or nibbling mammals, including the mice, squirrels, beavers, and so on.

Rough ER: Endoplasmic reticulum studded with ribosomes, as opposed to smooth ER.

Roundworm: Any nematode that infests the intestine of humans and other mammals.

S

Saccule (sak′yool): 1. The smaller of two sacs in the membranous labyrinth of the ear; compare *utricle*. 2. A small sac anywhere in the body of a plant or animal.

Salting out: A method of abortion used in the second trimester in which the fetus is killed with an injection of a concentrated salt solution into the amniotic cavity. Also called *saline abortion*.

Sandy seashore: A coastal zone subject to a constant shifting of the sand bottom; has relatively few life forms.

Saprobe (sap′rōb): An organism that reduces dead plant and animal matter.

Sarcodina (sär′kə din′ə): The class of protozoans that move and capture food by forming pseudopods.

Sarcomere (sär′kə mir): The contractile unit of striated muscle bounded by Z-line partitions; consists of actin filaments bound to the Z-line partitions and myosin filaments regularly interspersed between them.

Saturated: Having accepted as many hydrogens as possible. *Saturated fat*, a triglyceride lacking carbon-carbon double bonds.

Scanning electron microscope: A device for visualizing microscopic objects by scanning them with a moving beam of electrons, recording impulses from scattered electrons, and displaying the image by means of the synchronized scan of an electron beam in a cathode ray (television) tube.

Scapula (skap′yə lə): Either of the two flat, triangular bones, forming a shoulder blade.

Schwann cell: One of the many cells that constitute the myelin sheath, wrapped around the axon of a myelinated neuron.

Sclerenchyma (skli reng′kə mə): A protective or supporting plant tissue composed of cells with greatly thickened, lignified, and often mineralized cell walls.

Scyphozoa (sī′fə zō′ə): The class consisting of marine jellyfishes.

Secondary growth: Growth in dicot plants that results from the activity of secondary meristem, producing chiefly an increase in the diameter of stem or root. Compare *primary growth*.

Secondary immune response: The more rapid production of antibodies and conquest of an invader during a second or subsequent infection. Compare *primary immune response*.

Secondary meristem: The tissue that causes growth in plant girth, the vascular and cork cambiums.

Secondary structure: The pattern of folding of adjacent residues of a macromolecule, usually in the form of helices or sheets.

Second law of thermodynamics: The statement in physics that, left alone, all systems proceed toward entrapy (disorganization).

Sedative-hypnotic: A drug that acts on the cerebral cortex to reduce anxiety and induce drowsiness and sleep.

Seed: The fertilized and ripened ovule of a seed plant, comprising an embryo, including one or two cotyledons, and usually a supply of food in a protective seed coat: capable of germinating under proper conditions and developing into a plant.

Seed coat: The outer protective covering of a seed, composed of one or more layers of tissue derived from the parent sporophyte.

Segregation, principle of: The separation of allelic genes in different gametes during meiosis, resulting in the separation of their characters in the offspring.

Semicircular canal: Any of the three curved tubular canals in the labyrinth of the ear, associated with the sense of equilibrium.

Seminal receptacle: A storage organ for sperm in certain invertebrate females. The sperm is released for fertilization after mating.

Senses: The faculties, such as sight, hearing, smell, taste, or touch, by which organisms perceive stimuli originating from outside or inside the body.

Sepal: The leaflike floral parts surrounding the base of the petals. See also *calyx*.

Sequence: In proteins, the linear arrangement of amino acids in the formation of large protein molecules; defines a protein's primary structure.

Seta (sē′tə) (pl. setae): 1. In the moss sporophyte, the stalklike growth that supports the sporangium. 2. The bristlelike, chitinous structure in the body wall of annelids, arthropods, and certain other invertebrates.

Sexual dimorphism: The differences—in size, color, anatomy, etc.—between the sexes.

Sickle cell anemia: A hereditary disease caused by a form of hemoglobin that is defective in its nature and causes sickle-shaped red blood cells.

Sieve tube: In phloem, a thin-walled tube consisting of an end-to-end series of enucleate, living cells joined by sieve plates; a channel through which sap flows.

Sigmoid: S-shaped. *Sigmoid colon,* an S-shaped part of the large intestine.

Simple sugar: A carbohydrate molecule containing a single sugar group. See also *monosaccharide.*

Single-gene effect: Trait due to the expression of a single gene; relatively rare.

Sinus: A cavity that forms part of an animal body; e.g., any of the several air-filled, mucous-membrane lined cavities of the skull.

Skeletal muscle: Muscle attached to the skeleton, under direct and conscious control, striated with multinucleate unbranched fibers. Also called *voluntary muscle, striated muscle.* Compare *cardiac muscle, smooth muscle.*

Skinner box: A device for investigating operant conditioning, named after B. F. Skinner, who invented it.

Sliding filament theory: The widely accepted explanation of skeletal muscle contractions in which actin myofilaments in the sacromere are actively drawn through myosin myofilaments, thus shortening the contractile unit.

Small intestine: In vertebrates, the region of the alimentary canal between the stomach and the cecum; the region in which most food absorption occurs.

Smooth ER: Endoplasmic reticulum that is not studded with ribosomes.

Smooth muscle: The muscle tissue of the glands, viscera, iris, piloerectors, and other involuntary structures; consists of masses of uninucleate, unstriated, spindle-shaped cells, usually occurring in thin sheets. Also called *involuntary muscle.*

Sodium/potassium ion exchange: A poorly understood molecular entity in the plasma membrane, capable of actively transporting sodium out of the cell and potassium in, at a cost of ATP energy. Also called *sodium pump.*

Solar energy: Energy obtained from the sun.

Solute (säl′yüt)**:** Any substance dissolved in a solvent; substance in solution.

Somatotropin (sō′mə tə trō′pin)**:** A growth hormone.

Somite (sō′mīt)**:** One of the paired mesadermal segments along the neural axis of an embryo.

Specialization: The process of narrowing abilities so that fewer roles are performed with greater efficiency; the process of becoming increasingly differentiated.

Speciation (spē′shē ā′shən)**:** An evolutionary process by which new species are formed, normally by the division of one species into two.

Species: The major subdivision of a genus. Individuals of a species are able to breed among themselves under natural conditions.

Sperm cell: The male gamete.

Sperm nucleus: 1. The nucleus of a sperm. 2. Either of the two nuclei that arise from the generative nucleus of a pollen tube and function in the double fertilization characteristic of seed plants.

Sphincter: A ring of muscles that controls the movement of materials through passages; e.g., sphincters at either end of the stomach.

Spicule (spik′yo͞ol)**:** 1. One of the many tiny calcareous or siliceous pointed bodies embedded in and serving to stiffen and support the tissues of various invertebrates, including sponges and sea cucumbers. 2. Any small, stiff, spikelike or needlelike body part.

Spinal cord: The complex band of neurons that runs through the spinal column of vertebrates to the brain.

Spinal nerve: Any of the many nerves that enter and leave the spinal cord, including both somatic and autonomic. Compare *cranial nerve.*

Spindle: A structure formed in mitosis and meiosis that consists of fine threads radiating from the centrosomes along which the chromosomes appear to move.

Spiracle (spī′rə kəl)**:** An external opening to the respiratory system of terrestrial arthropods.

Spirillum (spī ril′əm)**:** Any of several spirally twisted, aerobic bacteria of the genus *Spirillum,* certain species that are pathogenic for man.

Sponge: A birth control device; a small, absorbent polyurethane sponge that is saturated with a spermicide and inserted into the vagina before intercourse.

Sporangium (spə′ran jē′əm)**:** A hollow structure in which spores are formed.

Spore: An asexual reproductive cell that can develop into an adult.

Sporophyte: In plants having an alternation of generations, a diploid individual capable of producing haploid spores by meiosis; the prominent form of ferns and seed plants. Compare *gametophyte.*

Sporophytic (spōr′ə fit′ik)**:** Referring to the spore-producing, diploid generation in plants that alternates with the sexual or gametophyte generation.

Sporozoa: The class composed of parasitic protozoans.

S-shaped curve: A plot of population growth where growth is rapid at first but then slows when *environmental resistance* is met and levels off at some point near or below the *carrying capacity.* Compare *J-shaped curve.*

Stabilizing selection: Selection against both extremes of a continuous phenotype, favoring an intermediate optimum.

Stalk: A stem, shaft, or slender supporting part of the structure of a plant or animal.

Stamen (stā′mən)**:** The male reproductive structure of a flower, consisting of a pollen-bearing anther and the filament on which it is borne.

Staminate (stam′ə nit)**:** Having stamens but no pistils; an exclusively male flower.

Stapes (stā′pēz)**:** The outer, stirrup-shaped bone of a chain of three small bones in the middle ear of man and other mammals. Also called *stirrup.*

Sterilization: The destruction of the ability to reproduce by removing the sex organs or inhibiting their functions.

Sternum: A median ventral bone or cartilage in land vertebrates, connecting with the ribs, shoulder girdle, or both.

Steroid: A type of lipid consisting mainly of fatty acids attached to complex alcohols.

Stigma: In flowers, the top, slightly enlarged and often sticky end of the style, on which pollen grains adhere and germinate.

Stimulation: The state of being excited or invigorated.

Stoma: (pl. stomata): One of the minute pores in the epidermis of leaves, stems, and other plant organs; formed by the concave walls of two guard cells; allows the diffusion of gases into and out of the intercellular spaces.

Stomach: A saclike enlargement of the alimentary canal, as in humans and certain animals, forming an organ for storing, diluting, and digesting food.

Stretch receptor: A sensory receptor that is stimulated by stretching, as in a tendon, muscle, or bladder wall.

Striation (strī ā′shən)**:** A striped condition or appearance.

Strobilation (strob'ə lā'shən): Asexual reproduction by traverse division of the body into segments which break free as independent organisms, occurring in certain coelenterates and flatworms (tapeworms).

Stroma: Matrix of a chloroplast in which the grana are imbedded.

Style: The stalk of the pistil in a flower connecting the stigma with the ovary.

Substrate: A substance that is acted upon by an enzyme.

Sudden death: An unanticipated demise, possibly due to congenital defects.

Sugar: A general term for certain larger carbohydrates.

Sulfur oxide: An oxide of sulfur; a common air pollutant in some industrial areas.

Superior vena cava (superior vē'nə kā'və): The anterior large vein by which blood is returned to the right atrium of the heart of land vertebrates.

Supernormal releaser: An environmental stimulus with exaggerated features that produces an instinctive response; not normally encountered in nature.

Supinator (soo'pə nā'tər): A muscle that rotates the hand or forearm so that the palm is facing upward. Compare *pronator*.

Suppressor T-cell: A type of white blood cell that shuts down the immune response after the risk of infection has passed.

Suture line (soo'cher line): In anatomy, immovable joints formed by the articulation of skull bones.

Symbiosis hypothesis (sim'bē ō'sis hī poth'ə sis): The hypothesis that the eukaryotic cell evolved from the mutualistic union of various prokaryotic organisms, one of which gave rise basically to the cytoplasm, nucleus, and motile membranes, a second to mitochondria, a third to chloroplasts and other plastids, and a fourth to cilia, eukaryotic flagella, basal bodies, centrioles, the spindle, and all other microtubule structures.

Sympathetic nervous system: A subdivision of the autonomic nervous system that increases energy expenditure and prepares the body for emergency situations.

Sympatric (sim pat'rik): Occupying the same geographical area.

Sympatric speciation (sim pat'rik spē'shē ā'shən): Speciation in populations that are not geographically separated. Compare *allopatric speciation*.

Synapse (sin'aps): The junction between the axon of one neuron and the dendrite or cell body of another; crossed by neural impulses.

Synaptic cleft: The minute space between the synaptic knob of one neuron and the dendrite or cell body of another. Neurotransmitters are released into it when nerve impulses are transmitted between cells.

T

Tactile receptor: A sensory receptor responsive to light touch.

Taiga (tī'gə): A subarctic forest biome dominated by spruce and fir trees; it is found in Europe and North America and at high altitudes elsewhere.

Tapeworm: A parasitic flatworm of the class *Cestoda*.

Tarsal: One of the smaller bones of the ankle, between the talus and the metatarsals.

Taxonomist (tak son'ə mist): Person concerned with the identification, naming, and classification of organisms.

Tectorial membrane (tek tor'ē əl membrane): A membrane of the cochlea, overlying and contacting the hair cells of the organ of Corti.

Telophase: The final stage of mitosis or meiosis that includes total separation of chromosomes, cytoplasmic division, and the return of the interphase nucleus.

Temperate deciduous forest: A forest biome of the temperate zone, in which the dominant tree species and most other trees are deciduous and are bare in winter months.

Temperature inversion: A reversal in the normal temperature lapse rate, the temperature rising with increased elevations instead of falling.

Temporal lobe (tem'pər əl lobe): One of the four lobes in the human brain; contains auditory and visual centers.

Tendon: A tough, dense cord of fibrous connective tissue that is attached at one end to a muscle and at the other to that part of the skeleton that moves when the muscle contracts. Compare *ligament*.

Terminal bud: The dormant bud at the stem tip, representing the next season's potential growth.

Territorial behavior: Behavior associated with the defense of a territory, in most territorial species primarily by the male.

Tertiary structure (tėr'shē er'ē structure): The pattern of folding of a polypeptide upon itself, which is generally quite specific for each protein type.

Testosterone (tes täs'tə rōn): A male hormone, produced in the testes, important in the sex drive and producing secondary sex characteristics.

Tetraploid (tet'rə ploid): Having four complete sets of chromosomes in each cell.

Thalamus (thal'ə məs): The middle part of the diencephalon through which sensory impulses pass the reach the cerebral cortex.

Thallus (thal'əs): A plant body of a multicellular alga, that does not grow from an apical meristem, shows no differentiation into distinct tissues, and lacks stems, leaves, or roots.

Theory: A proposed explanation whose status is still conjectural and unproven, but highly likely and supported by evidence, in contrast to well-established propositions that are regarded as facts.

Thermoreception: The ability to sense temperature or changes in temperature.

Thoracic duct (thō ras'ik duct): The main trunk of the lymphatic system, passing along the spinal column in the thoracic cavity, and conveying a large amount of lymph and chyle into the venous circulation.

Thorax: 1. In animals, the part of the body anterior to the diaphragm and posterior to the neck, containing the lungs and the heart. 2. The middle of the three parts of an insect body, bearing the legs and wings.

Threshold value: The point at which a stimulus is intense enough to initiate a response.

Thylakoid (thī'lə koid): A saclike, membranous structure in the chloroplasts. Stacks of these form grana.

Thymine (thī'mēn'): A pyrimidine, one of the four nitrogenous bases of DNA.

Thyroid (thī'roid): A large endocrine gland in the lower neck region of all vertebrates, the secretion of which regulates the rates of metabolism and body growth.

Thyroxine (thī räk'sin): A thyroid hormone that functions in metabolism.

Tissue: A group of contiguous cells of similar origin, structure, and function. Compare *organ*.

Tobacco: Any plant of the genus *Nicotiana*, whose leaves are prepared for smoking or chewing or as snuff.

Tolerance: The power of enduring or resisting the action of a drug, poison, and so on. With drugs, increasingly greater amounts are needed to produce the same effects.

Total fertility rate: A projection of the average number of children women aged 14–44 will bear.

Trachea (trā′kē ə) (pl. tracheae): 1. In land vertebrates, the air passage between the lungs and the larynx, usually stiffened with rings of cartilage. 2. One of the air-conveying tubules in the respiratory system of an insect, millipede, or centipede.

Tracheal system (trā′kē əl system): The respiratory system of insects, composed of thin-walled air conducting tubules opening to spiracles and extending to finer, branched tracheoles, terminating in air sacs.

Tracheid) (trā′kē əd): A long, tubular xylem element that functions in support and water conductions. It is distinguished from xylem vessels by having tapered, closed ends and communicating with other tracheids through pits.

Tracheophyta (trā′kē ō fī′tə): The vascular plants.

Trade winds: The nearly constant easterly winds that dominate most of the tropics and subtropics throughout the world, blowing mainly from the northeast in the northern hemisphere and from the southeast in the southern hemisphere.

Transfer RNA (*t*RNA): Lightweight nucleic acid molecules that identify with specific amino acids during protein synthesis.

Transmitting electron microscope (TEM): A device for creating magnified images of small specimens by bombarding them with an electron beam and by subsequent magnetic focusing.

Transverse colon: See *colon.*

Transverse nerve: A laterally branching nerve, such as those connecting the longitudinal nerves in flatworms.

Tree fern: Any of various, mostly tropical ferns that reach the size of trees, sending up a straight, trunklike stem with fronds at the summit.

Trematoda (trem′ə tō′də): The class composed of parasitic flatworms having one or more external suckers.

Triceps brachii (trī′seps brā′kē ē): The muscle on the back of the arm, the action of which extends the elbow.

Trichina (tri kī′nə): A nematode, the adults of which live in the intestine and produce embryos that encyst in the muscle tissue, especially in pigs, rats, and humans.

Trichocysts (trik′ə sist′): In some ciliates, minute harpoonlike bodies below the pellicle that can be extruded.

Tridacna (tri dak′nə): A genus of giant clams found on reefs in the South Pacific, attaining a diameter of four feet or more.

***t*RNA:** See *transfer RNA.*

Trophoblast (trof′ə blast′): The thin wall side of a blastocyst that forms the chorion when implantation occurs in mammals.

Tropical rain forest: A tropical woodland biome that has an annual rainfall of at least 250 cm (98 in) and often much more; it is typically restricted to lowland areas and characterized by a mix of many species of tall, broad-leaved evergreen trees that form a continuous canopy, with vines and woody epiphytes, and by a dark, rather bare forest floor.

Tubal ligation: Female sterilization by cutting and tying the Fallopian tubes.

Tubal occlusion: A birth control method that consists of inserting silicone rubber plugs into the Fallopian tubes, which prevents the egg from reaching the uterus; see also *Fallopian plug.*

Tubule: Any slender, elongated channel in an anatomical structure.

Tubulin: A protein consisting of two dissimilar polypeptides making up the subunit of microtubules.

Tumescent (tōō mes′ənt): Swollen.

Tundra: A biome characterized by level or gently undulating treeless plains of the arctic and subarctic that support dense growths of mosses and lichens, as well as dwarf herbs and shrubs; it is underlain by permafrost and seasonally covered by snow.

Turbellaria (tûr′bə lâr′ē ə): A class of platyhelminths or flatworms, mostly aquatic, and having cilia on the body surface.

Tympanal membrane (tim pan′əl membrane): A thin, tense membrane of an organ of hearing in an insect.

Tympanum (pl. tympana): A resonating covering associated with hearing.

U

Ulna: The bone of the forearm on the side opposite the thumb, or the corresponding bone in the forelimb of other land vertebrates.

Ultimate level: Influences on an individual as a result of the evolutionary experience of the historical line.

Undernourished: Not receiving enough food.

Ungulate (ung′gyə lāt): Having hoofs.

Uracil (yü′rə sil): One of the nitrogenous bases of RNA.

Urea (yü′rē ə): A highly soluble nitrogenous compound that is the principle nitrogenous waste of the urine of animals.

Ureter (yü rēt′ər): The tube that conducts urine from the kidney to the bladder in higher animals.

Urethra (yü rē′thrə): The tube that conducts urine from the bladder to the outside of the body in higher animals.

Uric acid (yü′rik acid): A relatively insoluble purine; a principal nitrogenous excretion product of reptiles, birds, and insects. It is excreted in small quantities as a product of nucleic acid breakdown in mammals.

Urinary bladder (yü′rə ner′ē bladder): A membranous sac in which urine is retained until it is discharged from the body.

Urine (yü′rin): The liquid-to-semisolid matter excreted by the kidneys, in humans, being a yellowish, slightly acid, watery fluid.

Urochordata (yü′rə kôr da′tə): A subphylum of the chordates that includes the tunicates, salps, and larvaceae.

Utricle (yü′tri kəl): The chamber of the membranous labyrinth of the middle ear into which the semicircular canals open.

V

Vacuole: A space within a cell, bounded by a membrane.

Vacuum curettage (vacuum kyoor′i täzh): A means of abortion by which the embryonic mass is removed by suction.

Vagina (və jī′nə): The female copulatory organ and birth canal.

Variable: Experimental variable; the focus of an experiment to be tested and compared with a control.

Vascular cambium: The cylinder of meristematic tissue that in secondary growth produces xylem on its inner side and phloem on its outer side, thus contributing to growth in circumference.

Vascular tissue: Any tissue that contains vessels through which fluids are passed.

Vasectomy (vas ek′tə mē): Male sterilization by cutting and tying the seminal ducts.

Vegetal pole: The lower, more yolk-filled end of a zygote or early cleavage blastula, determining the ventral side in development.

Vein: A vessel that carries blood toward the heart.

Ventilation: Exposure to air in the lungs or gills in respiration; oxygenation.

Ventral nerve root: Any of the large motor nerves extending from the ventral area of the vertebrate spinal nerves.

Ventricle (ven'tri kəl): A cavity of a body part or organ; one of the large muscular chambers of the four-chambered heart.

Venule: A small vein.

Vertebra (vü'tə brə): Any of the bones or segments of the spinal column.

Vertebral column (vür'tə brəl column): The articulated series of vertebrae connected by ligaments and separated by intervertebral discs that in vertebrates forms the supporting axis of the body and of the tail in most forms.

Vertebrate (vür'tə brāt): An animal in the subphylum *Vertebrata*, phylum *Chordata*; an animal with a vertebral column or backbone.

Vessel: In botany, a conducting tube in a dicot formed in the xylem by the end-to-end fusion of a series of cells (vessel elements) followed by the loss of adjacent end walls and of cell cytoplasm. Compare *tracheid*.

Villus (vil'ə s) (pl. villi): A small protrusion of the intestinal wall, greatly increasing the absorbing surface of that organ.

Virus: An infectious, submicroscopic parasite that consists of an RNA or DNA core with a protein coat.

Visible light: Electromagnetic wavelengths longer than about 400nm and shorter than about 750nm, which can serve as visual stimuli to most photoreceptive organisms.

Visual purple: Rhodopsin; a bright-red photosensitive pigment found in the rods of the retina of certain fishes and most vertebrates.

Vitalism: The notion that life has unique mystical properties that are distinct from those ascribed by chemical and physical laws.

W

Water potential: The potential energy of water to move as a result of concentration, gravity, pressure, or solute content.

Water vascular system: A system of vessels in echinoderms that contains sea water and is used as a hydraulic system in the movement of tentacles and tube feet.

Wave: In physics, a progressive disturbance moving from point to point in a medium or space without progress or advance by the points themselves, as in the transmission of sound of light.

Wax: A dense, hard, lipid-soluble ester of a long-chain alcohol and a fatty acid.

Whorl: 1. A group of parts repeated in a circle. 2. Any of the four basic radially repeated groups of flower parts.

Withdrawal symptom: Any of a number of physical and psychological disturbances, such as sweating and depression, experienced by a narcotic addict deprived of a required drug dosage.

X

Xylem: One of the two complex tissues in the vascular system of plants; consists of the dead cell walls of vessels, tracheids, or both, often together with sclerenchyma and parenchyma cells; functions chiefly in water conduction and strengthening the plant. See also *tracheid, vessel*. Compare *phloem*.

Y

Yellow marrow: Yellow, fatty material within the central cavity of long bones.

Yolk: The nutrient portion of the egg.

Yolk plug: A pluglike mass of yolk-filled endoderm cells left protruding from the blastopore of an amphibian embryo after the cresentic blastopore enlarges to form a complete circle.

Z

Z line: In striated muscle, the partition between adjacent contractile units to which actin filaments are anchored.

Zooplankton (zō'ə plangk'tən): The non-photosynthetic animal life drifting at or near the surface of the open sea.

Zygospore (zī'gə spōr): A diploid fungal or algal spore formed by the union of two similar sexual cells; it has a thickened wall and serves as a resistant resting spore.

Selected Readings

Part 1

Albert, B. et al. 1983. *Molecular Biology of the Cell.* Garland Publishing Co., New York.

Amerine, M. A. 1964. Wine, *Scientific American* 211(2):46–56.

Asimov, I. 1968. *Photosynthesis.* Basic Books, New York.

Asimov, I. 1954. *The Chemicals of Life.* Abelard-Schuman, New York.

Baker, J. J. W. and Allen, G. E. 1968. *Hypothesis, Prediction and Implication in Biology.* Addison-Wesley, Reading, Mass.

Barry, J. M. and Barry, E. M. 1969. *An Introduction to the Structure of Biological Molecules.* Prentice-Hall, Englewood Cliffs, N.J.

Conant, J. B. 1951. *Science and Common Sense.* Yale Univ. Press, New Haven, Conn.

Darwin, C. R. 1962. *The Voyage of the Beagle.* Doubleday, Garden City, N.Y.

Dautry-Varsat, A. and Lodish, H. 1984. How Receptors Bring Proteins and Particles Into Cells, *Scientific American,* May.

Eiseley, L. 1956 Charles Darwin. *Scientific American* 194(2):62–72.

——————— 1958. *Darwin's Century.* Doubleday, Garden City, N.Y.

——————— 1960. *The Firmament of Time.* Atheneum, New York.

Farago, P. and Lagnado, J. 1972. *Life in Action.* Vintage, New York.

Folsome, C. E. 1979. *Life: Origin and Evolution.* Freeman, San Francisco.

Gingerich, O. 1982. The Galileo Affair, *Scientific American,* July.

Goldsby, R. A. 1967. *Cells and Energy.* MacMillan, New York.

Jastrow, R. 1967. *Red Giants and White Dwarfs.* Harper and Row, New York.

Jorgensen, J. J. ed. 1970. *Biology and Culture in Modern Perceptive.* Readings from Scientific American. Freeman, San Francisco.

Koestler, A. 1972. *The Roots of Coincidence.* Random House, New York.

Lederman, L. 1984. The Value of Fundamental Science, *Scientific American,* November.

Margaria, R. 1972. The Sources of Muscular Energy, *Scientific American* 226(3):84–91.

Mayr, E. 1970. *Populations, Species and Evolution.* Harvard University Press, Cambridge, Mass.

Miller, S. L. 1955. Production of Some Organic Compounds Under Possible Primitive Earth Conditions, *Journal of the American Chemical Society* 77:2351–2361.

Moorehead, A. 1969. *Darwin and the Beagle.* Harper and Row, New York.

Porter, E. 1971. *Galapogos.* Ballatine, New York.

Rothman, J. 1985. The Compartmental Organization of the Golgi Apparatus, *Scientific American,* September.

Smith, H. W. 1961. *From Fish to Philosopher.* Doubleday, Garden City, N.Y.

Part 2

Crick, F. H. C. 1966. The Genetic Code: III, *Scientific American* 215(4):55–62.

Crow, J. R. and Kimura, J. 1970. *An Introduction to Population Genetics Theory.* Harper and Row, New York.

Darnell, J. 1985. RNA, *Scientific American,* October.

Dobzhansky, T. 1963. Evolutionary and Population Genetics, *Science* 142:3596.

Eckhardt, R. B. 1972. Population Genetics and Human Origins, *Scientific American* 226(2):94–102.

Feder, J. and Tolbert, W. 1983. The Large-Scale Cultivation of Mammalian Cells, *Scientific American,* January.

Hayflich, L. 1980. The Cell Biology of Ageing, *Scientific American,* January.

Koller, P. C. 1971. *Chromosomes and Genes.* Norton, New York.

Lawn, R. and Vehar, G. 1986. The Molecular Genetics of Hemophilia, *Scientific American,* March.

Ledbetter, M. C. and K. Porter. 1970. *Introduction to the Fine Structure of Plant Cells.* Springer, New York.

Leeson, C. R. et al. 1985. *Textbook of Histology.* Saunders, Philadelphia.

Lewin, R. 1983. A Naturalist of the Genome, *Science* 222:402.

Lowey, A. G. and Siekevitz, P. 1969. *Cell Structure and Function.* Holt, Rinehart and Winston, New York.

McKusick, V. A. 1965. The Royal Hemophilia, *Scientific American* 213(2):88–95.

Menoskz, J. A. 1981. The Gene Machine, *Science 81* July/August.

Pestka, S. 1983. The Purification and Manufacture of Human Interferons, *Scientific American,* August.

Watson, J. D. 1968. *The Double Helix.* Atheneum, New York.

——————— . 1970. *Molecular Biology of the Gene.* Benjamin, New York.

Woese, C. R. 1967. *The Genetic Code, the Molecular Basis for Genetic Expression.* Harper and Row, New York.

Part 3

A Scientific American book. 1978. *Evolution.* W. H. Freeman, San Francisco.

Alexopoulos, C. J. 1962. *Introduction to Mycology.* John Wiley, New York.

Anderson, H. T. 1969. *Biology of Marine Mammals,* Academic Press, New York.

Bellairs, A. 1970. *Life of Reptiles.* Vol. II Universe Books, New York.

Bishop, J. A. and Laurence M. Cook. 1975. Moths, Melanism and Clean Air, *Scientific American*, January.

Budker, Paul. 1971. *The Life of Sharks.* Columbia University Press, New York.

Dobzhansky, T. 1970. *Genetics of the Evolutionary Process.* Columbia University Press, New York.

Dodson, E. O. 1974. Phylogeny, in *The New Encyclopedia Brittanica*, 15th ed. 14:376.

Esau, K. 1977. *Anatomy of Seed Plants*, 2d ed. Wiley, New York.

Gilbert, L. 1982. The Coevolution of a Butterfly and a Vine, *Scientific American*, August.

Haldane, J. B. S. 1932. *The Causes of Evolution.* Cornell University Press, New York (paperback).

Hutchinson, J. 1969. *Evolution and Phylogeny of Flowering Plants.* Academic Press, New York.

Jensen, W. A. 1973. Fertilization in Flowering Plants, *Bioscience* 23:21.

Kaplan, D. R. 1983. The Development of Palm Leaves, *Scientific American*, July.

King, J. L. and T. H. Jukes. 1969. Non-Darwinian Evolution, *Science* 164:788.

Lehner, R. and J. Lehner. 1962. *Folklore and Odysseys of Food and Medicinal Plants.* Tudor, New York.

Marshall, N. B. 1966. *The Life of Fishes.* The World Publishing Co., Cleveland.

Ptashne, M. et al. 1982. A Genetic Switch in a Bacterial Virus, *Scientific American*, November.

Rockstein, M. ed. 1973. *Physiology of Insecta*, Vol. 1 Academic Press, New York.

Romer, A. S. 1977. *The Vertebrate Body*, 5th ed. W. B. Saunders, Philadelphia.

Russell, D. A. 1982. The Mass Extinctions of the Late Mesozoic, *Scientific American*, January.

Schultz, A. 1969. *Life of Primates.* Weidenfeld and Nicolson.

Sporne, K. R. 1971. *The Mysterious Origin of Flowering Plants.* Carolina Biological Supply Company, Burlington, N.C.

Stanier, R. Y. and M. Douderoff. 1973. *The Microbial World*, 3rd ed. Prentice-Hall, Englewood Cliffs, N.J.

Stanley, S. 1984. Mass Extinctions in the Ocean, *Scientific American*, June.

Wallace, R. et al. 1986. *Biology The Science of Life.* Scott, Foresman, Glenview, IL.

Welty, J. C. 1975. *The Life of Birds*, 2d ed. W. B. Saunders, Philadelphia.

Young, J. Z. 1975. *The Life of Mammals*, 2d ed. Clarendon Press, Oxford.

Part 4

Asdell, S. A. 1964. *Patterns of Mammalian Reproduction.* Cornell University Press, Ithaca, N.Y.

Browder, L. 1980. *Developmental Biology*, Saunders, Philadelphia.

Bullough, W. S. 1961. *Vertebrate Reproductive Cycles.* Barnes & Noble, New York.

Corner, E. H. H. 1968. *The Life of Plants.* New American Library, New York.

Flickinger, R. A. ed. 1966. *Developmental Biology.* Brown, Dubuque, Iowa.

Frazer, J. F. D. 1959. *The Sexual Cycles of Vertebrates.* Hutchinson University Library, London.

Gehring, W. 1985. The Molecular Basis of Development, *Scientific American*, October.

Katchadowrian, H. and Lunde, D. T. 1972. *Fundamentals of Human Sexuality.* Holt, Rinehart and Winston, New York.

Landau, B. R. 1980. *Essential Human Anatomy and Physiology, 2nd ed.* Scott, Foresman, Glenview, IL.

Masters, W. and Johnson, V. 1966. *Human Sexual Response.* Little, Brown, Boston.

Michelmore, S. 1965. *Sexual Reproduction.* Natural History Press, Garden City, N.Y.

Odell, W. D. and Moyer, D. L. 1971. *Physiology of Reproduction.* Mosby, St. Louis.

Rugh, R., and Shettles, L. B. 1971. *From Conception to Birth: The Drama of Life's Beginnings.* Harper and Row, New York.

Shell, E. R. 1982. The Guinea Pig Town, *Science 82*, December.

Shodell, M. 1983. The Prostaglandin Connection, *Science 83*, March.

Wallace, R. A. 1980. *How They Do It.* Morrow, New York

Wilmoth, J. H. 1967. *Biology of Invertebrata.* Prentice-Hall, Englewood Cliffs, N. J.

Wilson, J., et al. 1981. The Hormonal Control of Sexual Development, *Science:* 211 (4488), 1278–1285.

Part 5

Beck, W. S. 1971. *Human Design: Molecular, Cellular and Systematic Physiology.* Harcourt Brace Jovanovich, New York.

Caravoli, E. and Penniston, J. 1985. The Calcium Signal, *Scientific American*, November.

Case, J. 1966. *Sensory Mechanisms.* MacMillan, New York.

Clegg, P. C. and Clegg, A. C. 1968. *Hormones, Cells and Organisms.* Stanford University Press, Stanford, Calif.

Crawshaw, L. et al. 1981. The Evolutionary Development of Vertebrate Thermoregulation, *Scientific American*, September–October.

Currey, J. 1970. *Animal Skeletons.* St. Martin's Press, New York.

Day, R. H. 1971. *Perception.* Brown, Dubuque, Iowa.

Greene, R. 1970. *Human Hormones.* McGraw-Hill, New York.

Hoyle, G. 1970. How is Muscle Turned On and Off? *Scientific American* 222(4):72–82.

Jaret, P. 1986. Our Immune System, The Wars Within, *National Geographic*, June: 169(6), 702–734.

Kolata, G. 1982. New Theory of Hormones Proposed, *Science* 215:1383.

Macey, R. I. 1968. *Human Physiology.* Prentice-Hall, Englewood Cliffs, N.J.

Morton, J. E. 1967. *Guts.* St. Martin's Press, New York.

Ramsay, J. A. 1968. *Physiological Approach to the Lower Animals.* Cambridge University Press, New York.

Schmidt-Nielsen, K. 1964. *Animal Physiology*, 2d ed. Prentice-Hall, Englewood Cliffs, N.J.

————. 1964. *Desert Animals.* Oxford University Press, New York.

————. 1972. *How Animals Work.* Cambridge University Press, New York.

Part 6

Barnard, C. J. 1983. *Animal Behavior.* John Wiley and Sons, New York.

Bright, M. 1984. *Animal Language.* Cornell Univ. Press, Ithaca, New York.

Gould, J. L. 1982. *Ecology: The Mechanism and Evolution of Behavior.* Norton, New York.

Heimer, L. (1971. Pathways in the Brain, *Scientific American* 225(1):48–60.

Jolly, A. 1972. *The Evolution of Primate Behavior.* MacMillan, New York.

Kimble, D. P. 1973. *Psychology as a Biological Science.* Goodyear, Santa Monica, Calif.

Klopfer, P. H. and Hailman, J. P. 1967. *An Introduction to Animal Behavior: Ethology's First Century.* Prentice-Hall, Englewood Cliffs, N.J.

_____. 1973. *Behavior Aspects of Ecology.* Prentice-Hall, Englewood Cliffs, N.J.

Marler, P., and Hamilton, W. 1966. *Mechanisms of Animal Behavior.* Wiley, New York.

Morris, D. 1967. *The Naked Ape.* Dell, New York.

Snider, R. S. 1958. The Cerebellum, *Scientific American* 199(2):84–90.

Springer, S. P. and G. Deutsch. 1985. *Left Brain, Right Brain.* Freeman, New York.

Tiger, L. 1969. *Men in Groups.* Random House, New York.

Tinbergen, N. 1951. *The Study of Instinct.* Oxford University Press, Oxford.

_____. 1953. *Social Behavior in Animals.* Methuen. London.

Wallace, R. A. 1978. *The Ecology and Evolution of Animal Behavior,* 2d ed. Goodyear, Santa Monica, Calif.

Wallace, R. A. 1979. *The Genesis Factor.* Morrow, New York.

Wallace, R. A. 1979. *Animal Behavior, Its Development, Ecology and Evolution.* Scott, Foresman, Glenview, IL.

Whalen, R. E. 1967. *Hormones and Behavior.* Van Hostrand-Reinhold. New York.

Wilson, E. O. 1984. *Biophilia.* Harvard Univ. Press, Cambridge.

Wilson, E. O. 1975. *Sociobiology,* Belknap, Cambridge, Mass.

Wood-Gush, D. G. M. 1983. *Elements of Ethology.* Chapman and Hall, London.

Part 7

Beddington, J. and R. May 1982. The Harvesting of Interesting Species in a Natural Ecosystem, *Scientific American,* November.

Bergerud, A. 1984. Prey-Switching in a Simple Ecosystem, *Scientific American,* December.

Calahan, D. 1972. Ethics and Population Limitation, *Science* 175:487–492.

Calhoun, J. R. 1962. Population Density and Social Pathology, *Scientific American* 206(2):139–148.

Carson, R. 1962. *Silent Spring.* Houghton Mifflin, Boston.

Cloud, P. 1983. The Biosphere, *Scientific American,* September.

Clutton-Brock, T. 1985. Reproductive Success in Red Deer, *Scientific American,* February.

Commoner, B. 1966. *Science and Survival.* Viking Press, New York.

_____. 1971. *The Closing Circle.* Knopf, New York.

Cornell, J., Mertz, D. and Murdoch, W. eds. 1970. *Readings in Ecology and Ecological Genetics.* Harper & Row, New York.

Edmond, J. and Von Damm, K. 1984. Hot Springs on the Ocean Floor, *Scientific American,* April.

Ehrlich, P. 1968. *The Population Bomb.* Ballantine, New York.

Ehrlich, P. and A. Ehrlich. 1979. What Happened to the Population Bomb? *Human Nature,* January.

Hazen, W. 1970, 1975. *Readings in Population and Community Ecology,* 1st and 2d eds. Saunders, Philadelphia

Hulett, H. 1970. Optimum World Population, *BioScience* 20:160–161.

Langer, W. 1972. *Checks on Population Growth:* 226(2):92–99.

Lappe, F. 1971. *Diet for a Small Planet.* Ballantine, New York.

McNaughton, S. J. and Wolf, L. L. 1973. *General Ecology.* Holt, Rinehart and Winston, New York.

Meadows, D. et al. 1972. *Limits to Growth.* Universe, New York.

Miller, G. T. 1975. *Living in the Environment: Concepts, Problems, and Alternatives.* Wadsworth, Belmont, Calif.

Moen, A. 1973. *Wildlife Ecology.* Freeman, San Francisco.

Odum, E. P. 1983. *Basic Ecology.* Saunders, Philadelphia.

Perry, D. 1984. The Canopy of the Tropical Rain Forest, *Scientific American,* November.

Pianka, E. 1974. *Evolutionary Ecology.* Harper & Row, New York.

Revelle, R. 1982. Carbon Dioxide and World Climate, *Scientific American,* July.

Ricklefs, R. 1973. *Ecology.* Chiron, Newton, Mass.

Tschirley, F. 1986. Dioxin, *Scientific American,* February.

Wagar, J. A. 1970. Growth versus the Quality of Life, *Science* 168:1179–1184.

Acknowledgments

Photo Credits

Unless otherwise acknowledged, all photos are the property of Scott, Foresman and Company.

Abbreviations: L left, C center, R right, T top, B bottom
PR, Photo Researchers, BPS Biological Photo Service

Page
Covers, IX, 12   Robert P. Carr
VIII–IX   Jane Burton/Bruce Coleman Inc.
X   © David Muench
XII T   Gerald G. Weland/Berg & Associates
XII C,B   Dwight R. Kuhn
XIV   Bill Longcore/Science Source, P.R.
XVI   NASA
XVII T   Michael Fox
XVII B   Milt & Joan Mann/Cameramann International

Chapter 1
page 1   Norman Myers/Bruce Coleman Inc.
page 2   Tui A. De Roy/Bruce Coleman Inc.
1.1 L   Charles Darwin. Watercolor by George Richmond, 1840. © Darwin Museum and The Royal College of Surgeons of England
1.1 R   History of Cambridge, 1815. Prints Division, New York Public Library
1.2   Des & Jen Bartlett/Bruce Coleman Inc.
1.3   © Darwin Museum and The Royal College of Surgeons of England
1.4 TC,BR   E. R. Degginger
1.4 TR   Stephen J. Krasemann/DRK Photo
1.4 BC   Bob & Miriam Francis
1.5   The Bettmann Archive
1.6 L   © Darwin Museum and The Royal College of Surgeons of England
1.6 R   Staircase of the Old British Museum, Montague House. George Scharf, the Elder. The Trustees of the British Museum

1.7   Sir Charles Lyell by L. Dickinson (replica), 1883. By courtesy of the National Portrait Gallery, London
1.8   Frans Lanting
Photoessay, p. 13   M. P. L. Fogden/Bruce Coleman Inc.
14 TL   Norman Owen Tomalin/Bruce Coleman Inc.
14 TR   Tui A. De Roy/Bruce Coleman Inc.
14 BL   K. W. Fink/Bruce Coleman Inc.
15 T   Tui A. De Roy/Bruce Coleman Inc.
15 B   Norman Owen Tomalin/Bruce Coleman Inc.
16 L   Mark I. Jones/Bruce Coleman Inc.
16 TR   Tui A. De Roy/Bruce Coleman Inc.
16 B   Elliott Varner Smith
17 TL   Tui A. De Roy/Bruce Coleman Inc.
17 TR   Norman Owen Tomalin/Bruce Coleman Inc.
17 CL,BL   George H. Harrison
18 TL   Keith Gunnar/Bruce Coleman Inc.
18 TC,TR   George H. Harrison
18 BR   Jessica Anne Ehlers/Bruce Coleman Inc.
19 TL   Jen & Des Bartlett/Bruce Coleman Inc.
19 R   Norman Owen Tomalin/Bruce Coleman Inc.
19 BL   Larry Keenan, Jr.
20 TL   Dotte Larsen/Bruce Coleman Inc.
20 TR   Leonard Lee Rue III/Bruce Coleman Inc.
20 BR   Tui A. De Roy/Bruce Coleman Inc.

1.9   Richard Howard (2)
1.11   Alfred Russel Wallace by Thomas Sims (original photograph), c1863–6. By courtesy of the National Portrait Gallery, London
1.12A   Charles Robert Darwin by John Collier, 1883. By courtesy of the National Portrait Gallery, London
1.12B   Mark Kauffman/LIFE Magazine © 1958 Time Inc.
1.13   Thomas Henry Huxley by John Collier, 1883. By courtesy of the National Portrait Gallery, London

Chapter 2
page 26   Dan McCoy/Rainbow
2.1   Sir Isaac Newton by Sir Godfrey Kneller, 1702. By courtesy of the National Portrait Gallery, London
2.2A   Solarfilma, Reykjavik, Iceland
2.2B   Peter Veit/Sygma
2.3 p. 32 TL   Stephen J. Krasemann/DRK Photo
32R   Jane Burton/Bruce Coleman Inc. (5)
32 BL   Frans Lanting
33 T   © Agence Nature/NHPA
33 B   Dr. Frank Carpenter
2.4   © James H. Pickerell 1986
2.5   Dupuy/Black Star

Chapter 3
page 36   NASA/Jet Propulsion Laboratory
3.1 L   Phyllis Greenberg/Nat'l Audubon Society Collection, P.R.
3.1 R   Raymond A. Mendez/Animals Animals
3.2   UPI/Bettmann Newsphotos
3.3   © California Institute of Technology

3.4 Herb Orth/LIFE Magazine © 1985 Time Inc. Painting by Chesley Bonestell
3.5 © Roger Ressmeyer/Starlight
3.6 Manfred Kage/Peter Arnold, Inc.
3.8 p. 48L Rod Planck
48 C M.I. Walker/Science Source, P.R.
48 R William E. Ferguson
49 L Robert & Linda Mitchell
49 C Norman Myers/Bruce Coleman Inc.
49 R Johnny Johnson/DRK Photo
3.9 © Archie Lieberman
Es. 3.2 p. 50 Copyright Smithsonian Books 1982, Smithsonian Institution. Photograph by Charles H. Phillips
51 L U.S.G.S.
51 R NASA/Jet Propulsion Laboratory

Chapter 4

page 52 Dan McCoy/Rainbow
4.1 TR,BR Dan McCoy/Rainbow
4.1 CR Peter Angelo Simon/Phototake
4.1 BL Frans Lanting
4.1 BC Tony Duffy/All-Sport Photographic Ltd.
4.4 National Air and Space Museum, Smithsonian Institution
4.7 B Bureau of Land Management, U.S. Dept. of the Interior
4.10 B Lynn M. Stone
Es. 4.2 AP/Wide World

Chapter 5

page 80 NASA
5.1 © David Muench
5.2 NASA
Es. 5.1 John Shaw
Es. 5.2 p. 90 B Dudley Foster/Woods Hole Oceanographic Institution
91 L Robert D. Ballard/Woods Hole Oceanographic Institution
91 TR Robert Hessler, SIO/Woods Hole Oceanographic Institution
91 BR Rod Catanach/Woods Hole Oceanographic Institution
5.9 John Zoiner/Peter Arnold, Inc.
Es. 5.3 Runk-Schoenberger/Grant Heilman Photography

Chapter 6

page 101 Science Photo Library/Science Source, P.R.
page 102 CNRI/SPL/Science Source, P.R.
6.1 From Micrographia by Robert Hooke, 1664
6.2 Ed Reschke (4)
6.4 Ed Reschke (3)
6.4 BR Biophoto Associates/Science Source, P.R.
Es. 6.1 University of Oregon photograph
6.8 Dr. Eva Frei and Prof. R. D. Preston
6.10 Dr. Keith Porter
6.11 Taylor, J. W., and M. S. Fuller. 1981. *Exp. Mycology* 5:42
6.12 Richard Chao
6.13 From S. E. Frederick and E. H. Newcomb. 1969. *J. Cell Biology* 43:343
6.14 Biophoto Associates/Science Source, P.R.
6.15 Biophoto Associates/Science Source, P.R.
6.16 Dr. Keith R. Porter
6.18 Dr. William E. Barstow
6.19 L. E. Roth, U. of Tennessee/BPS
6.20 Dr. William E. Barstow
6.21 Dr. Daniel Branton
6.23 L Kenneth W. Fink/Berg & Associates
6.23 R Stephen J. Krasemann/DRK Photo
6.28 David R. Frazier
6.30 John Walsh, SPL/Science Source, P.R.

Chapter 7

page 126 Biophoto Associates/Science Source, P.R.
7.1 Dr. Andrew Bajer for SF (10)
7.7 Watson and Crick in front of DNA model. Photographer: A. C. Barrington Brown. From J. D. Watson, 1968; *The Double Helix*. New York: Atheneum, p. 215. © 1968 by J. D. Watson
Es. 7.1 Dan McCoy/Rainbow (2)

Chapter 8

page 148 Dan McCoy/Rainbow
8.1 Wellcome Institute Library, London
8.2 E. R. Degginger
Es. 8.1 SCALA/Art Resource, NY.
8.6 T Keystone Press
8.6 B AP/Wide World
8.10 Dr. Murray L. Barr (2)
Es. 8.2 The Bettmann Archive
8.18 Courtesy *Collected Papers of G. H. Hardy*, The Clarendon Press, Oxford
Es. 8.3 Courtesy of Carmen Raventos-Suarez and Ronald L. Nagel, M.D.

Chapter 9

page 168 Dan McCoy/Rainbow
9.1 Dr. Tony Brain/Elsa Hemming, SPL/Science Source, P.R.
Es. 9.1 L Centers for Disease Control, Atlanta
Es. 9.1 R Valentine, R. C., and Pereira, H. G. 1965. *J. Molecular Biology* 13:13–20. © 1965 by Academic Press, Inc.
9.2 Charles C. Brinton, Jr. and Judith Carnahan
9.3 Tom Broker/Rainbow
Es. 9.2 Hank Morgan/Rainbow
Es. 9.3 L Dan McCoy/Rainbow
Es. 9.3 R Douglas Kirkland/Sygma

Chapter 10

page 179 Robert Frerck/Odyssey Productions, Chicago
page 180 Paläontologisches Museum, Museum für Naturkunde der Humboldt-Universität, Berlin, DDR
page 181 David L. Brill/ © National Geographic Society
10.1 T Robert P. Carr
10.1 B Stephen J. Krasemann/DRK Photo
10.2 Scott Fehr
10.3 L Jim Tallon
10.3 R Michael H. Francis
10.6 Milt & Joan Mann/Cameramann International
10.9 Kjell B. Sandved
10.10 Norman R. Lightfoot/Nat'l Audubon Society Collection, P.R.

A–2   ACKNOWLEDGMENTS

10.10   Joseph T. Collins/Nat'l Audubon Society Collection, P.R.

Es. 10.2   Stephen J. Krasemann/DRK Photo

10.11   Kim Taylor/Bruce Coleman Inc. (2)

Chapter 11

page 199   Manfred Kage/Peter Arnold, Inc.

11.3 L   Carl O. Wirsen/Woods Hole Oceanographic Institution

11.3 C   Centers for Disease Control, Atlanta

11.3 R   Dr. Tony Brain, SPL/Science Source, P.R.

11.4 L, C   Dr. Tony Brain, SPL/Science Source, P.R.

11.4 R   Reprinted with permission of the present publisher, Jones and Bartlett Publishers, Inc., from Shih and Kessel: *Living Images*, Science Books International, 1982, page 3

Es. 11.2   M. Abbey/Science Source, P.R.

11.5   Courtesy of D. L. Findley, P. L. Walne, and R. W. Holton, U. of Tennessee, Knoxville. From *J. Phycology* 6:182–188, 1970

11.6   B. Ben Bohlool

Es. 11.3   Nathan Benn/Woodfin Camp, Inc.

11.9   E. R. Degginger

11.10   Ed Reschke

11.11 L   L. E. Roth, U. of Tennessee/BPS

11.11 C, R   Biophoto Associates/Science Source, P.R.

11.12 L   Prof. Gordon Leedale, Biophoto Associates/Science Source, P.R.

11.12 R   Georg Gerster/Nat'l Audubon Society Collection, P.R.

11.13   Heather Angel/Biofotos

11.17   Martin W. Miller, U. of California, Davis

11.19   Larry West

11.20   Larry West

Es. 11.4 L, BR   Ed Cooper

Es. 11.4 TR   Rod Planck

Es. 11.4 C   Larry West

Chapter 12

page 221   Rod Planck

12.2   William E. Ferguson

12.3   Doug Wechsler

12.4 L   Nancy Sefton

12.4 R   Anne Hubbard/Nat'l Audubon Society Collection, P.R.

12.5   James Bell/Science Source, P.R.

12.6 L   William E. Ferguson

12.6 R   William E. Ferguson

12.7   Don & Pat Valenti

12.9 L   William E. Ferguson

12.9 R   Larry West

12.10   Ed Cooper

12.11   Ed Cooper

12.12   Ed Cooper

12.13 L   William E. Ferguson

12.13 R   Ed Cooper

12.14   Wardene Weisser/Berg & Associates

12.15   William E. Ferguson

12.16 TC   Gerald G. Weland/Berg & Associates

12.16 CL, CR   Larry West

12.16 BL, BR   Rod Planck

Chapter 13

page 233   F. S. Mitchell/Tom Stack & Associates

13.2   John C. Deitz/Photo Researchers

13.5   Charles Seaborn/Odyssey Productions, Chicago

13.6   Charles Seaborn/Odyssey Productions, Chicago

13.7 L   Nancy Sefton

13.7 R   Carleton Ray/Nat'l Audubon Society Collection, P.R.

13.11   Reprinted with permission of the present publisher, Jones and Bartlett Publishers, Inc., from Shih and Kessel: *Living Images*, Science Books International, 1982, page 84

13.12   Jack Kath

13.15   Robert Frerck/Odyssey Productions, Chicago

13.16   Dr. C. F. E. Roper. Photo: Chip Clark

13.18   David Hughes/Bruce Coleman Inc.

13.19   Charles Seaborn/Odyssey Productions, Chicago

13.22   Dwight R. Kuhn (2)

13.23 TL   Rod Planck

13.23   Dwight R. Kuhn (3)

13.25   Jeffrey L. Rotman

13.28 L   OSF/Animals Animals

13.28 R   Nancy Sefton

13.29   Heather Angel/Biofotos

13.31 L   Tom McHugh, courtesy Steinhart Aquarium/Nat'l Audubon Society Collection, P.R.

13.31 R   Dr. Giuseppe Mazza

13.32   Flip Nicklin/Nicklin & Associates

13.34   Peter Scoones/Planet Earth Pictures—Seaphot

13.35   Peter Scoones/Planet Earth Pictures—Seaphot

13.36   George H. Harrison

13.37 A, D   Rod Planck

13.37 B   John Gerlack/DRK Photo

13.37 C   Don and Pat Valenti/DRK Photo

13.38 TL   Stephen J. Krasemann/DRK Photo

13.38 TC, BC   M. P. Kahl/DRK Photo

13.38 TR, BL   Rod Planck

13.38 BR   Wardene Weisser/Berg & Associates

13.39   Dr. Carl Welty

13.40   Eric Crichton/Bruce Coleman Inc.

13.41 L   Malcolm S. Kirk/Peter Arnold, Inc.

13.41 C   Stephen J. Krasemann/DRK Photo

13.41 R   M. P. Kahl/DRK Photo

13.42   Dwight R. Kuhn

13.43   E. R. Degginger

13.44   F. S. Mitchell/Tom Stack & Associates

13.46   © Merlin D. Tuttle/Bat Conservation International

13.47   Stephen J. Krasemann/DRK Photo

13.48   Gerald Corsi/Tom Stack & Associates

Chapter 14

page 261   Stephen J. Krasemann/DRK Photo

page 262   Marty Cooper/Peter Arnold, Inc.

page 263   Brian Milne/First Light

14.1   Dwight R. Kuhn

14.2   Macmillan Science Co., Inc.

14.4   G. R. Roberts

14.7 L   Ben Goldstein/Valenti Photography

14.7 C   John Shaw

14.7 R   W. K. Fletcher/Nat'l Audubon Society Collection, P.R.

Es. 14.1 Sean Morris, OSF/Animals Animals
14.9 Alice J. Belling, Ph.D.
14.10 J. N. A. Lott, McMaster Univ./BPS
14.15 J. N. A. Lott, McMaster Univ./BPS (2)
14.17 Grant Heilman Photography
14.18 Dr. Sylvan H. Wittwer

Chapter 15
page 280 Craig Blacklock
15.2 Tom Branch/Science Source, P.R.
15.3 R. Andrew Odum/Peter Arnold, Inc.
Es. 15.1 T Leonard Lee Rue III/DRK Photo
Es. 15.1 C Don & Pat Valenti/DRK Photo
Es. 15.1 B Stephen J. Krasemann/DRK Photo (2)

Chapter 16
page 300 Petit Format/Nestle/Science Source, P.R. (2)
Es. 16.1 Bill Gage
Es. 16.2 Manfred Kage/Peter Arnold, Inc.
16.3 © Lennart Nilsson (3)
Photoessay Petit Format/Nestle/Science
p. 315 Source, P.R.
316–322 C. Bevilacqua/CEDRI
323 T Hal Stoelzle/Mickey Pfleger
323 B Howard Dratch/The Image Works

Chapter 17
page 327 Co Rentmeester/LIFE Magazine, © 1965 Time Inc.
page 328 John Shaw
page 329 Ben Goldstein/Valenti Photography
17.2 Jeff Foott/DRK Photo
17.3 Rod Planck
17.5 Thomas Eisner, Cornell University (2)
Es. 17.1 Frans Lanting

Chapter 18
page 342 Heidi Hoeffer
18.2 Cosmos Blank/Nat'l Audubon Society Collection, P.R.
18.12 Ed Reschke (3)
18.16 Charles Seaborn/Odyssey Productions, Chicago (2)

Chapter 19
page 359 M. P. L. Fogden/Bruce Coleman Inc.
19.3 Nancy Sefton
Es. 19.1 Nilsson, Lennart. 1974. *Behold Man.* Boston: Little, Brown and Co.
Es. 19.2 Lou Lainey/© *Discover* Magazine 3/84, Time Inc.

Chapter 20
page 380 Bill Longcore/Science Source, P.R.
20.1 Bruce Wetzel and Harry Schaefer, NCI
20.2 James D. Hirsch
Es. 20.1 Centers for Disease Control, Atlanta

Chapter 21
page 392 David Parker/Science Source, P.R.
21.3 T BBC Hulton Picture Library/The Bettmann Archive
21.3 B Wide World
21.5 Dr. Cedric S. Raine

Chapter 22
page 407 CNRI
page 408 Nilsson, Lennart. 1974. *Behold Man.* Boston: Little, Brown
page 409 Roe Di Bona
22.3 A. Glauberman/Science Source, P.R.
22.11 Jonathan Wright/Bruce Coleman Inc.
22.12 Dan McCoy/Rainbow
22.15 L, R Frank Oberle/Photographic Resources, Inc.
22.16 Historical Pictures Service
22.17 Charles Gatewood/The Image Works
22.18 Kjell B. Sandved/Nat'l Audubon Society Collection, P.R.

Chapter 23
page 436 Howard Sochurek/Woodfin Camp & Associates
23.1 Thomas Eisner, Cornell University (2)
23.3 Lynn M. Stone
23.4 Ed Reschke
23.6 M. P. L. Fogden/Bruce Coleman Inc.
Es. 23.1 Jane Burton and Kim Taylor/Bruce Coleman Ltd.

23.8 Omikron/Science Source, P.R.
23.10 Omikron/Science Source, P.R.

Chapter 24
page 447 Glenn D. Prestwich
24.3 BBC Natural History Unit. From *The Discovery of Animal Behavior* by John Sparks. 1982. A Collins Publishers/BBC Co-production
24.5 Warren & Genny Garst/Tom Stack & Associates
24.6 The Bettmann Archive
24.7 Joe McNally/Wheeler Pictures
24.9 Stephen J. Krasemann/DRK Photo (2)
24.10 Tom Bledsoe/DRK Photo
24.12 Charles Walcott
24.13 L George H. Harrison
24.13 R M. P. Kahl

Chapter 25
page 466 Stephen J. Krasemann/DRK Photo
25.1 L Robert P. Carr
25.1 R Stephen J. Krasemann/DRK Photo
25.2 M. P. Kahl
25.4 Lynwood Chace/Photo Researchers
25.5 Gordon Wiltsie/Bruce Coleman Inc.
25.6 Peter Davey/Bruce Coleman Inc.
25.8 Stephen J. Krasemann/DRK Photo
25.9 Kjell B. Sandved/Nat'l Audubon Society Collection, P.R.
25.11 Peter Davey/Bruce Coleman Inc.

Chapter 26
page 479 NASA
page 480 Robert P. Carr
page 481 David Muench
26.3 Frank Oberle/Photographic Resources, Inc.
26.4 © Michael Fox
26.5 Robert P. Carr
26.6 Robert A. Wallace
26.7 Robert A. Wallace
26.8 Johnny Johnson/DRK Photo
26.9 Frans Lanting
26.10 Lynn M. Stone

Es. 26.1	Robert Frerck/Odyssey Productions, Chicago
26.13	Martin Rogers/Woodfin Camp, Inc.
26.14	J. M. Bassot, H. Chaumeton/Nature
Photoessay p. 495	Baron Wolman
496 L	Gary Braasch
496 R	Robert Frerck/Odyssey Productions, Chicago
497 T	Lynn M. Stone
497 B	Gary Braasch
498 L	Everett C. Johnson
498 R	Frans Lanting
499 T	Lynn M. Stone
499 B	Frans Lanting
26.15	Al Giddings/Ocean Images, Inc.
Es. 26.2	Marc & Evelyne Bernheim/Woodfin Camp & Associates

Chapter 27

page 506	© Michael Fox
27.1	Robert Frerck/Odyssey Productions, Chicago
27.2	L. L. T. Rhodes/Animals Animals
27.8	M. Austerman/Animals Animals
27.9 A	Lynn M. Stone
27.9 B	C. C. Lockwood/DRK Photo
Es. 27.1	Stern/Black Star (2)
27.11	Carson Baldwin/Animals Animals, Earth Scenes
Es. 27.2	Sygma

Chapter 28

page 523	Baron Wolman
28.1	James A. Yost
28.3	Owen Franken/Sygma

Chapter 29

page 536	Baron Wolman
29.1	Milt & Joan Mann/Cameramann International
29.2	Alan Carey/The Image Works
29.3 A	Anthony Suau/Black Star
29.3 B	Joseph P. Shapiro/U.S. News & World Report
29.4	Anthony Suau/Black Star
29.5	Robert A. Wallace
Es. 29.4	U.S. Dept. of Agriculture
29.6	Nachtwey/Black Star
29.7	Fred Ward/Black Star
29.9	Milt & Joan Mann/Cameramann International
29.10	Milt & Joan Mann/Cameramann International

Chapter 30

page 556	Baron Wolman
30.1	James Mason/Black Star
30.2	Milt & Joan Mann/Cameramann International
30.3	Osvaldo Bohn Photo Studios © Discover Magazine 2/86, Time Inc. (2)
30.4	Mickey Pfleger
30.6	Robin Moyer/Black Star
Es. 30.5	M. P. Kahl

Illustration Credits

Scott Thorn Barrows, AMI, Clinical Assistant Professor, University of Illinois Medical Center: Figures 4.21, 5.3B, 5.4, 5.5, 5.6, 5.7, 5.12, 6.6, 6.7, 6.10A, 6.17, 6.19B, 6.20B, 6.24, 6.25, 6.29, 7.8, 7.9, 7.10, 19.8, 19.9.

Lewis E. Calver: Essay 20.1B, C, D.

©Teri J. McDermott, M.A., Clinical Assistant Professor, University of Illinois Medical Center: Figures 15.4, 15.5, 16.7, 17.8, 18.4, 18.5, 18.6, 18.7, 18.9, 18.10, 18.11, 18.13, 18.14, 18.15, 18.18, 19.5, 19.6, 19.10, 19.11, 19.12, 19.13, 21.2.

Sandra McMahon, Biomedical Illustration: Figures 21.4, 21.6, 21.7, 21.10, 22.4, 22.5, 22.6, 22.8, 22.9, 22.10, 22.13, 23.4B, 23.5, 23.7, 23.9, Essay 30.3.

Precision Graphics/Karen Shannon: Figures 3.5B, 3.7, 4.2, 4.3, 4.5, 4.6, 4.7A, 4.8, 4.9, 4.10A, 4.11, 4.12, 4.13, 4.14, 4.15, 4.16, 4.17, 4.18, 4.19, 4.20, 4.22, Essay 5.1B, 5.8, 5.10, 5.11, 6.3, 6.5, 6.9, 6.16, 6.22, 6.27, 7.1, 7.2, 7.3, 7.4, 7.5, 7.6, 8.3, 8.4, 8.5, Essay 8.2B, 8.9, 8.11, 8.13, 8.14, 9.3A, 9.4, 9.5, 9.6, 10.4, 10.5, Essay 10.1, 10.7, 10.12, 11.2, 12.1, 13.3, 14.9A, 14.10, 14.11, 14.12, 14.13, 14.16, 15.6, 15.7, 15.8, 15.9, 18.8, 20.3, 20.4, 20.5, 21.5B, 22.14, Essay 23.1B, 24.6C, 24.8, 24.11, 24.14, 25.10, 26.2, 27.3, 27.4, 27.5, 27.6, 27.7, 27.10, 28.2, 28.4, 28.5, 29.8, 29.11, Essay 30.1, Essay 30.2, Essay 30.4, 30.7.

Lewis Sadler: Essay 5.2A.

Sarah Forbes Woodward: Figures 6.31, 11.1, 11.5B, 11.7, 11.8, 11.9B, 11.10, 11.14, 11.15, 11.16, 11.17B, 11.18, 12.8, 13.4, 13.8, 13.9, 13.10, 13.14, 13,17, 13.20, 13.21, 13.24, 13.25B, 13.26, 13.29A, 13.30, 13.33, 13.39B, 14.2B, 14.3, 14.5, 14.6, 14.8, 15.2A, 16.1, 16.2, 16.4, 16.5, 16.6, 17.6, 17.7, 18.1, 18.17, 19.1, 19.2, 19.4, 19.7, 21.1, 22.1, 22.2, 23.2, 24.2, 24.4, Essay 24.1, 25.7, 26.1, 26.11, 26.12, 26.16, 30.5.

Lisa Zucker: 1.10, 13.1, 13.13, 13.27, 17.4.

INDEX

buffer, 67
Buffon, George-Louis Leclerc De, 8
bulbourethral gland, 287
bulk flow, 119, 122–23
butane, **64**

C

caffeine, 429
caffeinism, 429
calcitonin, 395, 397
calcium
    atomic number, 58
    bone, 345
    regulation by parathyroid hormone,
        395–96, 397
California redwood, 489–90
calorie, 541
caltrop, 63
Calvin cycle, 88, **89**
calyx, of flower, 271
cambium, 275, **276**
camel, 338
canaliculus, 345
cancer
    risk of birth control pills, 295
Cannabis sativa, 429–30, **430**
capillary, 365
    lymph, 372
Capsella (Shepherd's purse), 274
capybara, 258
carbohydrate, 70–72
    structure, 71–72
carbon
    atomic number, 58
    covalent bonding, 62
    functional groups, 64–65
carbon dioxide (CO_2)
    gas exchange, 360–63, 365–66
carbon monoxide, 559
carbonic acid
    gas exchange, 365
carbonic anhydrase, 365–66
carboxyl group, 65
cardiac muscle, 351, **351**
cardiopulmonary resuscitation (CPR), 373
Carnivora, 258, **258**
carnivore, 48
carotene, 84
carotenoid, 114
carpal, **347,** 349
carpel, of flower, 271
carrying capacity, of environment, 509–10,
    509, 510
cartilage, 345
    cartilaginous fish. 252
Castle, W. E., 165
cat, 417
catalase, 113
catalysis, 68
catalyst, 68

cell
    amoeboid, 364, 375
    animal, **104,** 119, **120**
    bone, **104,** 345
    cartilage, **345**
    contractile, 106, **106**
    defined, 104
    fat, **104**
    plant, **104, 119, 120**
    secretory, 106, **106**
    sensory, 106, **106**
    skin, **147**
    specialization, 106
    sperm, **106**
    theories of development, 44
    wall, 110–11, **111**
cell division, 301, **302, 303**
    chick, 301, **302, 303**
    frog, 301, **302, 303**
    human, 301, **302, 303**
cell membrane, 107–9, **109**
    fluid-mosaic model, 108, **109**
    permeability, 108–9, 123, **124**
    semipermeable, 122, **122**
    structure, 107–8
cell theory, 104
cell-mediated immunity, 384
cellular respiration
    chemiosmotic phosphorylation, 99–100
    glycolysis, 92–95, **93, 98**
    Kreb's cycle, 86, 94, 95, **96,** 98
cellulose, **69,** 72
    in cell wall, 110–11
Cenozoic era, 185, 195
centipede, 244–45, **245**
central nervous system, 398, 411, **422**
    brain, 411
    spinal cord, 411
centriole, 116, **116**
centromere, 127, 130, **130**
centrum
    of vertebra, **348**
cephalic ganglia, of flatworm, 410, **410**
Cephalochordata, 248, 250
cephalosporin, 207
cephalothorax, **244**
cerebellum, 411, **412, 414,** 415
cerebral cortex, 415, 417
    frontal map, **418**
    motor area, **418**
    sensory area, **418**
cerebral hemispheres, 416–17, 419
cerebrum, 411, **412,** 415, 416–17
cervical cap, 293, **293**
cervix, **287,** 291, 293, **293**
cesarian section, 325
Cestoda, 238–39, **239**
CF_1 particle, 84, **85,** 86–87, **87, 89**
Chalazae, **309**
Chelicerata, 244
chemical communication, 464–65
chemical energy, 82
chemical messengers. *See* hormone

chemical reaction, 65
chemiosmosis, 88
chemiosmotic phosphorylation, 99–100
chemoreceptor, 439–42
    insect, 440
chemosynthesis, 90
chemosynthetic bacteria, **204**
Chernobyl, 566, 568
chewing, 375
chick
    embryo, 309–10, **309, 310**
Chilopoda, 245
chimpanzee, 453, **453**
    reproductive strategy, 512–13, **513**
Chiroptera, 258–59
chiton
    respiratory system, 360, **360,** 361
chlorine
    atomic number, 58
chlorophyll, 84–86, **85, 89,** 205
chlorophyll a, 86, **224,** 225
chlorophyll b, 86, **224,** 225
chlorophyte, 224–25
chloroplast, 84, **85,** 86, **89,** 114, **120**
    symbiosis hypothesis, 115
CHNOPS, 54
choanocyte, 235, **344**
cholesterol, 73, **74**
Chondrichthyes, 252
Chordata, 248
chordate, 248–60
    common features, 248
    evolutionary tree, **249**
chorioallantoic membrane, 310, **310**
chorion, **302**
    chick embryo, 310, **310**
    human embryo, **316**
chorionic villi, 312
chromatid, 127, 130, **130**
chromatin, 119, 127, **128,** 132
chromatography, 77
chromosome
    crossing over, 132, **132,** 160, **160**
    DNA replication, 138–39, **139**
    meiosis, 130–33, **136, 137**
    mitosis, 127, **128, 129,** 130, **130, 136,**
        137
    numbers in different species table, 131
    pairing in female fruit fly, 131
chromosome mapping, 162, **162**
chromosome mutations, 162, **163**
    deletion, 163, **163**
    duplication, 163, **163**
    inversion, 163, **163**
Chrysophyte, 212
cilia, 117–18, 208, **208**
Ciliophore, 208, **208**
circulatory system, 363–71
    annelid, 242–43
    closed, 364–65
    coelenterate, 364
    flatworm, 364
    human embryo, 313
    open, 364–65

density-dependent control on population, 515

density-independent control on population, 515

deoxygenated blood, 366, **367**

deoxyribonucleic acid. *See* DNA

dependence, drug, 427

depolarization, 402, 404

depressor muscle, 353

desert, **484,** 486, **486,** 490

deuterostome, 239

development, animal, 301–11
   chick, 309–10, **309, 310**
   human, 311–26, **312**

diaphragm
   birth control, 293, **293**
   contraction of, in breathing. **363**

diatom, 212, **212**

diatomaceous earth, 212

Dicara, L. V., 423

dicotyledon, 274, **275**

differentiation, 274

diffusion, 119, 121
   facilitated, 119, 121
   rate, 121

digestion, 373

digestive enzyme, 69, **69,** 72, 373, 375–78, **376**

digestive system, 373–79
   amoeba, 373, **374**
   earthworm, 373, **374, 375**
   flatworm, 238, 373, **374**
   human, 373, 375–79, **376**
   hydra, 373, **374**
   salamander, **374,** 375
   vertebrate, **374,** 375

digit, 349, **350**

dihybrid, 153, **153**

dilation and curettage (D&C), 297

dilator, 353

dimorphism, 190

dinoflagellate, 212, **213**

dinosaur, 196–97

Diplopoda, 245

disease
   nutritional, 538–39
   and population control, 520

display behavior, 469, **469**

divergent evolution, 184, **184**

DNA, 133–45
   base pairing, 138, **139**
   compared to RNA, 140–41
   diagrammed, **139**
   double helix, 133, 138–39, **138, 139**
   gene cloning methods, 176–77
   history of discovery, 133, 138
   replication, 138–39, **139**
   rungs of ladder design, 138, **139**

dog
   Pavlov's, 454–55, **455**

dominance, principle of, 150–51

dominant trait, 151, 156

dorsal nerve root, 420, 426

double bond, 61

double helix of DNA, 133, 138–39, **138, 139**

doubling time, of human population, 529–31

Drosophila melanogaster (fruit fly), 155–59, **157, 158, 159,** 162

drug abuse, 427

drug dependence, 427

drug use, 427–35

drumstick, 158, **158**

duodenum, 377

E

E. coli. *See* Escherichia coli

ear, human, 438–39, **440**
   embryo, **320**

eardrum, 438–39, **440**

earth
   theories of formation, 40–43

earth calendar, 196
   earthworm, 242–43, **243**
   closed circulatory system, 242, **243,** 365, **365**
   digestion, 373, 375
   excretory system, 336, **336, 337**
   nervous system, 243, **243, 410,** 411
   reproduction, 243, **243**

echinoderm, 246–47, **247**

Echinoidea, 246

ecology
   biomes, 484–88, 490
   competition, 482
   defined, 481
   density-dependent control, 515
   ecosystems, 488, 490
   pollution, 558–66
   population control, 514–20
   populations, animal, 506–22
   populations, human, 523–33

ecosystem, 488, 490

ectoderm, **302, 303,** 304
   derivatives, **302, 303,** 304
   lens induction, 308, **308**
   neurulation, 307–8

ectoplasm, 210, **210**

ectothermy, 254–55, 331–32

effector, 401

efferent neuron, 400

egg
   conception, 290–91
   frog, 301, **302, 303**
   human, 301, **302, 303**
   meiosis in, 130–33
   menstrual period, 289
   number of human, 313
   types of animal, 301, **302, 303**

egg nucleus, 269

eidetikers, 457

ejaculation, 286–87, **286**

Eldredge, Niles, 194–95

electrical energy, 82

electrical potential, 401

electromagnetic spectrum, 97

electron, 54
   excitability of, 56, **56**
   orbits, 55–56
   shell, 55–56

electron acceptor, 85–86, **85, 89**

electron microscope, 107, 108

electron transport system, 87, **87, 89**

electron-transport chain, 96, **98,** 99

electronegativity
   of hydrogen bond, 61

element
   defined, 54
   inert, 57
   letter symbol of, 54
   number of, 54
   table of elements found in living matter, 58

elephant, 259, **260**

elephantiasis, 239

embryo
   of angiosperm, 274, **274**
   chick, 309–10, **309, 310**
   human, 311–14, **316, 317, 318, 319, 320**
   sex of human, 314

embryo sac
   of flower, 271

embryology, animal, 301–11
   bird, 309–11
   cell division, 301, **302, 303,** 304
   gastrulation, **302, 303,** 304
   human, 311–25
   sequencing, 306–7

endergonic reaction, 70

endocrine system, 393–97, **397**
   adrenal, 396, 398
   human, 395–97
   insect, 393, **393**
   parathyroid, 395, 397
   plant, 395
   sex hormones, 393
   table, of human hormones, 397
   thyroid, 395

endocytosis, 124, **125**

endoderm
   derivatives, 304

endometrium, **286,** 289, **289**

endoplasm, 210, **210**

endoplasmic reticulum (ER), 115, **115, 116**

endoskeleton, 343, 345

endosperm, 272, **273**

endosperm nucleus, 271

endospore, 205

endothermy, 255, 332, 366, 367

energy
   chemical, 82
   currency, 83
   electrical, 82
   forms, 81–82
   free, 82
   geothermal, 551
   heat, 82
   kinetic, 82

renewable, 537–49
   water, 548–49
respiration, 360
   aerobic, 94
   anaerobic respiration, 94
   cellular, 92
   external, 360
   internal, 360
respiratory system, 360–71
   chiton, 360, **360,** 361
   earthworm, 365, **365**
   fish, 360–61, **360, 361**
   grasshopper, **247**
   insect, 361–62, **362**
   mammal, 362–63, **363, 364**
   mollusk, 365
   shark, 361
   spider, **244**
   vertebrae, 365
resting potential, 401,
reticular system 415–16, **416**
retina, 445, **445**
retrovirus, 388
reward, 455
rhizoid, 214, 215, **226,** 227
   moss, 225, **226**
rhizome, 227
Rhizopus nigricans, 214, **215, 216**
Rhodophyta, 223–24
Rhynia, 226, **226**
rhythm method, 292,
rhythmicity, 460–61
rib cage, **347, 348**
rib, 348, **348, 349**
   floating, 348
ribonucleic acid. *See* RNA
ribosomal RNA, 143
ribosome, 115, **115**
ribulose diphosphate (RuDP), 88, **89**
right-handedness, 419
RNA, 140–45
   messenger, 140, 141–45, **144**
   ribosomal, 143
   transfer, 143, **144**
Roberts, Lamar, 425
robin, European male, 450, **451**
rockweed, 224, **224**
rod, retinal, 445, **445**
Rodentia, 258
Roeder, Kenneth, 442
root, 274–75, **275**
rough ER, 115
roundworm, 239, **239,** 241, 357, **357**
   germ layers, 357, **357**
   support system, 357, **357**
rust, 218

S
S-shaped curve of population growth, 509, **509**
sac fungus *(Endothia parasitica),* 517–18, **518**
saccule, 444, **444**

salamander, **374,** 375
salivary gland, 378
Sally lightfoot crab, **16**
salt. *See* sodium chloride
sand dollar, 246
Sanger, Frederick, 77, **77**
saprobe, **204**
saprophyte, 213, **214**
Sarcodine, 209–10, **210**
Sargasso Sea, 224
Sargassum, 224
saturated fat, 73, **74**
scanning electron microscope (SEM), 109
scapula, **347,** 349
Schleiden, Matthias Jakob, 104
Schwann cell, 398, **399, 400**
Schwann, Theodor, 104
scientific knowledge, history of, 6–9
schlerenchyma, 276, **277**
scolex, 238, **239**
scrotum, **286,** 296
Scyphozoa, 236
sea anemone, 235–36, **235, 236**
sea cucumber, 246
sea lion, Galapagos, **18**
sea squirt, 248, **250**
sea urchin, 246
secondary sex characteristics, 396, 397
sedative-hypnotic drug, 433
seed, **268,** 269
seed coat, 273, **273**
segmentation, 242–44
segregation, principle of, 150–52
self-fertilization
   tapeworm, 512
self-pollination, Mendelian genetics, 151–52
semen, 287, 290
semi-circular canal, 444, **444**
seminal receptacle, 243
seminal vesicle, **286,** 287
seminiferous tubule, **286**
sense organs, 410–11, 436–45
   chemoreceptors, 439–42, **441**
   hearing, 438–39, **440**
   olfaction, 440, **441**
   proprioception, **440,** 443–44, **444**
   tactile, 438, **438, 439**
   temperature, 437–38, **438**
   vision, 444–45, **445**
sensory areas of brain, **418,** 419
sensory hair
   inner ear, 439
sensory neuron, 400
sensory receptor, 419
sepal, of flower, 271
sequoia, giant, 228
serine, structural formula, **75**
serotonin, **381**
setae, 242, **243**
sewage treatment, 562
sex
   of embryo, 314

sex attractant
   silkworm moth, 464
sex chromosome, 157
sex determination, 157–58
sex hormone, 393, 396, 397
sex-linkage, 158–59
sexual dimorphism, 192
sexual reproduction, 282–83
   bread mold, 214–15, **215, 216**
   disadvantages and advantages, 264–65
   frog, 282, **283**
   human, 285–91
   shark, 252, **252**
   compared to bony fish, 360–61, **360, 361**
   respiratory system, 361, **362**
shell
   electron, 55–56
   of mollusk, 241
sickle-cell anemia, 166–67
silkworm moth, 440, **441,** 464
simple sugar. *See* monosaccharide
single-gene effect, 189
sinus, 365
skeletal muscle, 351, **351**
skeletal systems
   clam, 343, 344
   frog, **348**
   human, 314, **320, 321,** 346, **347,** 348–49
   lobster, 344
   sponge, 344, **344**
   types of, 344–45
   vertebrate, 345
Skinner box, 455–56, **456**
Skinner, B. F., 455–56
skull, **347,** 348
sliding filament theory, 355, **356**
small intestine, 377–78
smell, 440
smog, 559
smoking, 428–29
smooth ER, 115
smooth muscle, 349, 351, **351**
smut, fungi, 218
snail
   reproductive behavior, 285–86
snake, 254, **254,** 437, **438**
snowshoe hare, 118, **119**
sociobiology, 476–77
sodium
   atomic number, 58, **59**
sodium chloride, 58, 65
   ionic bonding, 59, **60**
sodium/potassium ion exchange pump, 401
solar energy, 553
solute, 122, 123
somatic nervous system, 420
somatotropin, 395
somite, **302,** 308
sorus
   of fern, **267**
sound communication, 462–64
sound spectrogram, 463, **463**

A Biological Lexicon

Being a list of Greek and Latin prefixes, suffixes, and word roots commonly used in biological terms, alphabetized by the most common combining form in English; with examples illustrating each usage.

a-, an- (Gk. *an-*, not, without, lacking): anaerobic, abiotic, anemia

ad- (L. *ad-*, toward, to): adhesion, adrenaline

allo- (Gk. *allos*, other): allele, allopatric, allotetraploid

ammo- (from ammonium salts produced from camel dung near the Libyan shrine to the god Ammos): ammonia, amino group, amino acid

amphi- (Gk. *amphi-*, two, both, both sides of): amphibian, Amphineura

andro- (Gk. *andros*, an old man): androecium, androgen

anti- (Gk. *anti-*, against, opposite, opposed to): antibiotic, antibody, antigen, antidiuretic hormone

archeo- (Gk. *archaios*, beginning): archegonium, archenteron

arthro- (Gk. *arthron*, a joint): arthropod, arthritis

auto- (Gk. *auto-*, self, same): autoimmune, autotroph, autosome

auxo- (Gk. *aux*, to grow or increase): auxin, auxillary

bi-, bin- (L. *bis*, twice; *bini*, two-by-two): binary fission, binocular vision, binomial

bio- (Gk. *bios*, life): biology, biome, biosphere

blasto-, -blast (Gk. *blastos*, sprout; now pertaining to the embryo): blastopore, blastula, trophoblast, osteoblast

broncho- (Gk. *bronchos*, windpipe): bronchus, bronchiole, bronchitis

carcino- (Gk. *karkin*, a crab, cancer): carcinogen, carcinoma

cardio- (Gk. *kardia*, heart): cardiac, electrocardiogram

cephalo- (Gk. *kephale*, head): cephalochordate, cephalopod, cephalothorax

chloro- (Gk. *chloros*, green): chlorophyll, chloroplast, chlorine

chole- (Gk. *chole*, bile; Gk. *cholecyst*, gall bladder): cholesterol, cholera, acetylcholine

chromo- (Gk. *chroma*, color): chromosome, chromatin

coelo-, -coel (Gk. *koilos*, hollow, cavity): coelacanth, coelenterate, coelom

cuti- (L. *cutis*, skin): cutaneous, cuticle, cutin

cyclo-, -cycle (Gk. *kyklos*, circle, ring, cycle): cyclostome, pericycle

cyto-, -cyte (Gk. *kytos*, vessel or container; now cell): cytoplasm, cytology, erythrocyte, leucocyte, cytosine

derm- (Gk. *derma*, skin): dermis, epidermis, ectoderm, endoderm, mesoderm

di- (Gk. *dis*, twice): dicotyledon, dioxide, diatom

diplo- (Gk. *diploos*, two-fold): diploid, diploblastic

eco- (Gk. *oikos*, house, home): ecology, androecium, ecosphere, ecosystem

ecto- (Gk. *ektos*, outside): ectoderm, ectoparasite

endo- (Gk. *endon*, within): endocrine, endoderm, endometrium, endoskeleton, endoplasmic

epi- (Gk. *epi*, on, upon, over): epicotyl, epidemic, epidermis, epididymis, epithelium

eu- (Gk. *eus*, good; *eu*, well; now true): eubacterium, eukaryote

extra- (L., outside of, beyond): extracellular, extraembryonic

-fer (L. *ferre*, to bear): fertile, fertilization, conifer

gam-, gameto- (Gk. *gamos*, marriage; now usually in reference to gametes (sex cells)): gamete, gametogenesis, isogamete

gastro- (Gk. *gaster*, stomach): gastric, gastrula, gastrin, gastrovascular cavity

gen- (Gk. *gen*, born, produced by; Gk. *genos*, race, kind; L. *genus*, *generare*, to beget): gene, polygenic, genotype, glycogen, florigen, estrogen

gluco-, glyco- (Gk. *glykys*, sweet; now pertaining to sugar): glucose, glycogen, glycolysis

gyn-, gyno-, gyneco- (Gk. *gyne*, woman): gynecology, gynoecium

hemo-, hemato- (Gk. *haima*, blood): hematology, heme, hemoglobin, hemophilia, hemorrhage

hepato- (Gk. *hepar*, *hepat-*, liver): hepatitis, hepatic

hetero- (Gk. *heteros*, other, different): heterogeneous, heterozygote

histo- (Gk. *histos*, web of a loom, tissue; now pertaining to biological tissues): histology, histamine, antihistamine

homo-, homeo- (Gk. *homos*, same; Gk. *homios*, similar): homeostasis, homogeneous, homologue, homozygote